# Student Solutions Manual
# for Chemistry

# Student Solutions Manual

# for

# CHEMISTRY

## SECOND EDITION

**Prepared by**

E. A. Robinson
University of Toronto

Judith Poë
University of Toronto

Ronald J. Gillespie
McMaster University

ALLYN AND BACON
Boston
London
Sydney
Toronto

A Division of Simon & Schuster
160 Gould Street
Needham Heights, Massachusetts 02194



Printed in the United States of America

94 93 92 91 10 9 8 7 6 5 4 3 2

CONTENTS

# CHAPTER 1

You should be able to answer questions 1-4 without consulting the text.

1. (a) Hydrogen, H (b) Oxygen, O (c) Carbon, C (d) Nitrogen, N (e) Chlorine, Cl (f) Magnesium, Mg (g) Sodium, Na (h) Potassium, K (i) Sulfur, S (j) Phosphorus, P

2. (a) Iron, Fe (b) Nickel, Ni (c) Mercury, Hg (d) Silicon, Si (e) Chromium, Cr (f) Helium, He (g) Barium, Ba (h) Lead, Pb (i) Uranium, U (j) Calcium, Ca

3. (a) H, hydrogen (b) He, helium (c) Ne, neon (d) F, fluorine (e) Mg, magnesium (f) Al, aluminum (g) P, phosphorus (h) S, sulfur (i) K, potassium (j) Na, sodium

4. (a) Ca, calcium (b) Br, bromine (c) Fe, iron (d) Mn, manganese (e) Cu, copper (f) Ag, silver (g) Au, gold (h) Zn, zinc (i) As, arsenic (j) Pt, platinum

5. Iron - Ir has already been used for iridium; Io would be suitable.
   Copper - Co has already been used for cobalt; Cp would be suitable.
   Lead - Le would be suitable, or Ld; La has already been used for lanthanum.
   Tin - Ti has already been used for titanium; Tn would be suitable.

6. You should recognize iron (b) and sulfur (g) as elements, and water (a), sugar (d), and magnesium oxide (j) as compounds. The remainder are mixtures. See if you can find out the typical constituents of each mixture.

7. The pure substances are nitrogen (an element), iron (an element), carbon dioxide (a compound of carbon and oxygen), carbon (an element found in several forms, such as diamond and graphite), the polymer nylon (a compound), oxygen (an element), sodium chloride (a compound), and distilled water (a compound). The remainder are mixtures. Salad dressing, wet sand, iodized table salt, vegetable soup, smog, cottage cheese, milk,automobile tires, and concrete are heterogeneous mixtures. Gasoline, filtered seawater, and (filtered) black coffee are homogeneous mixtures.

8. Soda water is a homogeneous solution of carbon dioxide in water.
   Efficiently filtered coffee is a complex homogeneous solution.
   Wood is a heterogeneous mixture of fibers, other organic substances, and water.
   (Clean) snow is pure solid water; a homogeneous substance.
   Soil is a heterogeneous mixture of silicates, aluminosilicates, other minerals, decayed organic matter, and microorganisms.
   "Dry-Ice" is the commercial name for solid carbon dioxide, a compound, and hence a homogeneous substance.
   Vinegar is essentially a dilute solution in water of the compound acetic (ethanoic) acid, and is thus a homogeneous mixture.
   (Clean) air is a homogeneous mixture of gases, mainly nitrogen and oxygen.

9. (a) Tin, Sn, and (e), uranium, U, are elements.
(g) Sodium nitrate, $NaNO_3$, and (j), hydrogen peroxide, $H_2O_2$, are compounds.
(b) Lemon juice (essentially citric acid, water and solids), (c) beer (mainly a dilute solution of ethanol, $C_2H_6O$, in water), (d) natural gas (a mixture of gaseous hydrocarbons - mainly methane, $CH_4$, ethane, $C_2H_6$, and propane, $C_3H_8$), (f), 3% aqueous hydrogen peroxide (a solution of the compound $H_2O_2$ in water), (h) a solution of iodine, $I_2$, in tetrachloromethane, $CCl_4$, and (i) air (a mixture of gases - mainly nitrogen, $N_2$, and oxygen, $O_2$) are mixtures.

10. Some examples of physical properties are given below. You may find others by consulting standard tables of data, such as those given in "The Handbook of Chemistry and Physics":

Water Colorless liquid, m.p. $0^\circ C$, b.p. $100^\circ C$, density 1.00 g $cm^{-3}$
Sugar Colorless solid, melts to a sticky liquid, soluble in water
Mercury Silver liquid metal, good conductor of electricity, density 13.6 g $cm^{-3}$, insoluble in water, m.p. $-39^\circ C$, b.p. $357^\circ C$
Copper Reddish solid metal, good electrical conductor, insoluble in water, m.p. $1083^\circ C$, b.p. $2600^\circ C$, density 9.0 g $cm^{-3}$
Oxygen, ($O_2$) Colorless gas, slightly soluble in water, m.p. $-219^\circ C$, b.p. $-183^\circ C$
Bromine, ($Br_2$) Reddish-brown liquid, m.p. $-7^\circ C$, b.p. $58^\circ C$, moderately soluble in water, very soluble in organic solvents such as $CCl_4(l)$, density 3.1 g $cm^{-3}$
Magnesium oxide, MgO Colorless solid, insoluble in water, density 3.60 g $cm^{-3}$, m.p. $2800^\circ C$, b.p. $3600^\circ C$, electrical conductor when molten
Sodium chloride, NaCl Colorless solid, soluble in water, insoluble in organic solvents, density 2.2 g $cm^{-3}$, m.p. $801^\circ C$, b.p. $1465^\circ C$, electrical conductor in molten state and in solution in water

11. Some typical chemical properties are given below. Many other examples of the chemical reactions of these substances are given in the textbook:

Water, $H_2O$ - Reacts with many metals to give hydrogen gas
Copper, Cu - Reacts with nitric acid to give a blue solution of copper nitrate, $Cu(NO_3)_2$, and oxides of nitrogen (NO and $NO_2$), depending on the concentration of the acid and temperature
Iron, Fe - Reacts with dilute sulfuric acid to give a pale green solution of iron(II) sulfate, $FeSO_4$, with the evolution of hydrogen
Magnesium, Mg - Burns brilliantly in air or oxygen to give colorless magnesium oxide, MgO(s)
Hydrogen, $H_2$ - Burns with a pale blue flame in air of oxygen, to give water
Hydrogen peroxide, $H_2O_2$ - Decomposes readily to give water, $H_2O$, and oxygen, $O_2(g)$

12. (a) Water is a nonconductor of electricity, while solutions of salts, such as sodium chloride, NaCl, are good electrical conductors. If each is used in turn to complete the circuit of an electric lamp attached to a battery, only the salt solution will cause the lamp to light.

(b) Chalk, calcium carbonate, $CaCO_3(s)$, is insoluble in water, while sodium chloride, NaCl(s), is quite soluble. They are easily distinguished by testing their solubilities in water.

13. The molecular formula gives the total numbers of each atom in a molecule; the empirical formula is the simplest formula that gives the ratios of the different atoms in a molecule as whole numbers. Thus:

| | molecular formula | empirical formula |
|---|---|---|
| (a) | $As_4$ | As |
| (b) | $C_3H_6$ | $CH_2$ |
| (c) | $P_4O_{10}$ | $P_2O_5$ |
| (d) | $XeF_4$ | $XeF_4$ |

14. The empirical formula is the simplest formula that gives the ratios of the different atoms in one molecule as whole numbers. Thus:

(a) $NO_2$ (b) $N_2O_5$ (c) P (d) HgCl (e) $AlCl_3$ (f) $C_3H_8$

(g) $C_2H_5$ (h) $CH_2O$ (i) $CH_2$

15. The empirical formula is the simplest formula that gives the ratios of the different atoms in one molecule as whole numbers. Thus:

(a) HO (b) $H_2O$ (c) $Li_2CO_3$ (d) $CH_2O$ (e) S (f) $C_3H_7$ (g) $BH_3$ (h) O

16. The formulas of these common substances are to be found in various places in Chapter 1, together with their names: (see e.g. pages 8,13,14).

(a) $H_2O$, water (b) $H_2O_2$, hydrogen peroxide (c) $CH_4$, methane

(d) $C_2H_6$, ethane (e) $CO_2$, carbon dioxide (f) CO, carbon monoxide

(g) $C_{12}H_{22}O_{11}$, sucrose (a sugar)

17. The formulas of these common substances are to be found in various places in Chapter 1, together with their names: (see e.g. pages 8,13,32-36).

(a) Mg, magnesium (b) MgO, magnesium oxide (c) NaCl, sodium chloride

(d) $P_4$, phosphorus* (e) $S_8$, sulfur** (f) $HNO_3$, nitric acid

(g) $H_2SO_4$, sulfuric acid (h) $KNO_3$, potassium nitrate

(i) $Na_2SO_4$, sodium sulfate (j) $BaSO_4$, barium sulfate

*specifically white phosphorus, because there are other forms

**specifically orthorhombic or monoclinic sulfur, because there are other forms

18. (a) Hydrogen, $H_2$ (b) Oxygen, $O_2$ (c) Nitrogen, $N_2$ (d) Water, $H_2O$

(e) Hydrogen peroxide, $H_2O_2$ (f) Nitric acid, $HNO_3$

(g) Sulfuric acid, $H_2SO_4$ (h) Potassium nitrate, $KNO_3$

19. (a) Magnesium oxide - MgO (b) Carbon monoxide - CO (c) Carbon dioxide - $CO_2$
(d) Bromine - $Br_2$ (e) Phosphorus - $P_4$ (f) Sulfur - $S_8$ (g) Ammonia - $NH_3$
(h) Methane - $CH_4$ (i) Ethane - $C_2H_6$ (j) Sodium chloride - NaCl

20. A diatomic molecule consists of two atoms, a triatomic molecule consists of three atoms, and a polyatomic molecule consists of more than three atoms. Thus:

| | |
|---|---|
| Diatomic molecules: | $F_2$, HCl |
| Triatomic molecules: | $NO_2$, $N_2O$, $O_3$ |
| Polyatomic molecules: | $C_2H_6$, $P_4$, $H_2O_2$, $P_4O_{10}$, $CH_4$ |

21.

By Pythagoras's theorem

$$r_{HH}^2 = r_{SH}^2 + r_{SH}^2$$

$$= 2(134\ \text{pm})^2$$

Taking the square root of each side gives

$$r_{HH} = \underline{190\ \text{pm}}.$$

22. Selecting a suitable scale, in each case, we have:

(a)

Scale: 2.00 cm = 100 pm

(b)

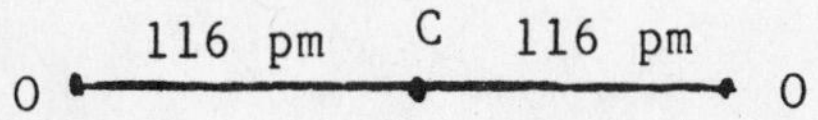


Scale: 2.00 cm = 100 pm

Linear means that all three atoms are on a straight line (colinear) The C atom is equidistant from each O atom, since both CO distances (bond lengths) are 116 pm. The OCO angle is 180°.

(c)

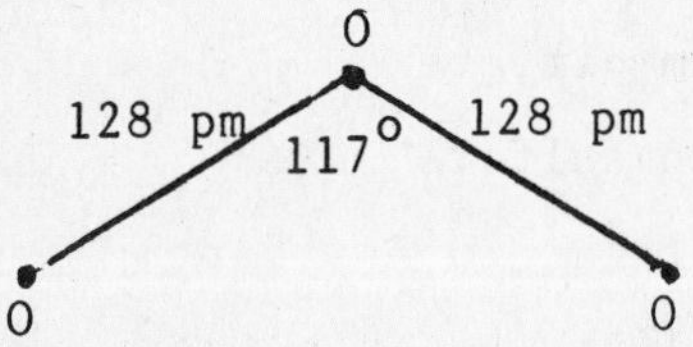


When three atoms are not colinear, the molecule is described as angular

Scale: 2.00 cm = 100 pm

(d)

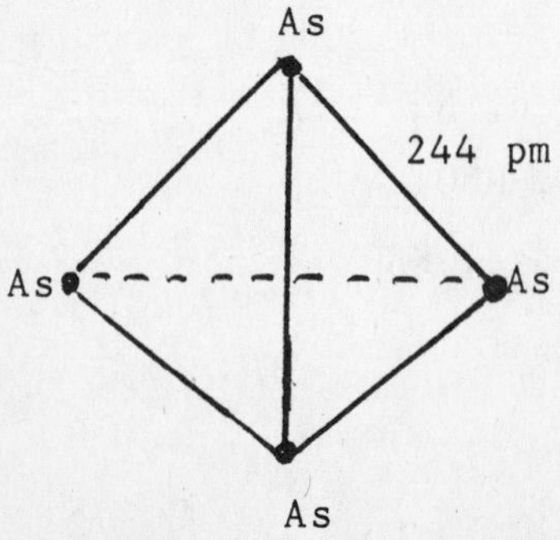


It is difficult to accurately represent a three-dimensional shape in two-dimensions All six As-As distances are 244 pm; each of the four faces of the tetrahedron is an equilateral triangle, i.e., each of the twelve AsAsAs angles is 60°.

Scale: 1.00 cm = 100 pm

23. $H_2O$ is angular since the three atoms are not in a straight line.

$XeF_2$ is linear since the three atoms are in a straight line.

$PH_3$ is a triangular pyramid (trigonal pyramid) since the P atom is above the center of an equilateral triangle formed by the three H atoms.

$BCl_3$ is triangular planar (trigonal planar) since the B atom is at the center of an equilateral triangle formed by the three Cl atoms.

24. The tetrahedron ACHF is constructed inside the cube ABCDEFG by drawing two face diagonals of the cube, such as AC and FH that are at right angles to each other, and joining up the points ACHF. The center of the cube, O, is the center of the tetrahedron.

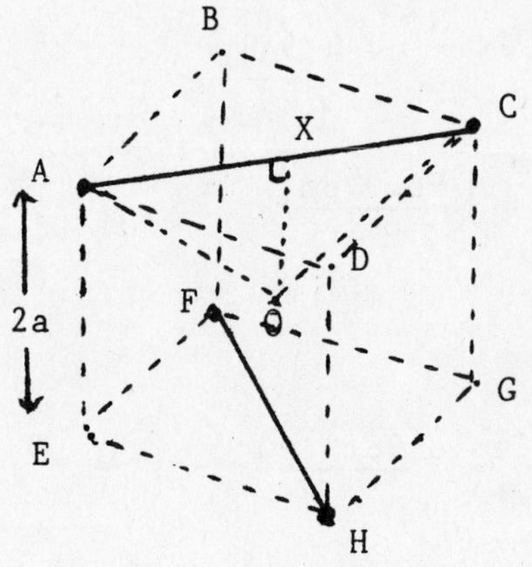


The angle in question is, for example, < AOC, and we can construct the triangle AOC in which AC is a face diagonal of the cube and, since O is at the center of the cube, the distance XO is one-half of the length of the edge of the cube, 2a, = a, and since X is at the center point of the diagonal, AX is one-half the length of the face diagonal.

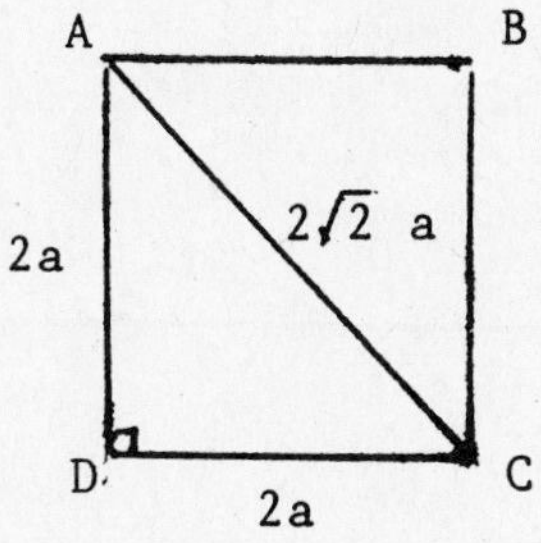


From Pythagoras's theorem

$$AC^2 = AD^2 + DC^2 = (2a)^2 + (2a)^2 = 8a^2$$

$$AC = \sqrt{8a^2} = \underline{2\sqrt{2}\ a}$$

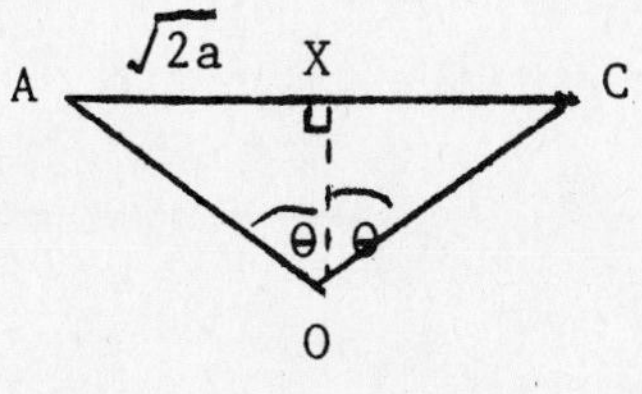


$$AX = \frac{AC}{2} \quad \text{and } OX = \frac{AE}{2}$$

$AX = \sqrt{2}a$ and $OX = a$,

$2\theta$ is the required angle, and

$$\tan\theta = AX/OX = \sqrt{2}a/a = 1.414$$

$$\theta = 54^\circ\ 44'$$

$$\underline{2\theta = 109^\circ\ 28'}$$

25. The geometry is shown below. The problem is to calculate the distance OH.

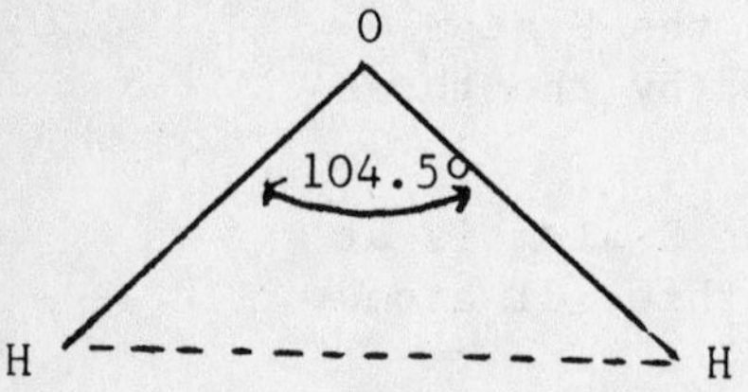


Draw a line from O to the midpoint of HH, to obtain the right-angled triangle OXH, where

$$< \theta = \frac{104.5^o}{2} = 52.3^o$$

and

$$XH = \frac{HH}{2} = \frac{153.3 \text{ pm}}{2} = 76.65 \text{ pm}$$

Whence,

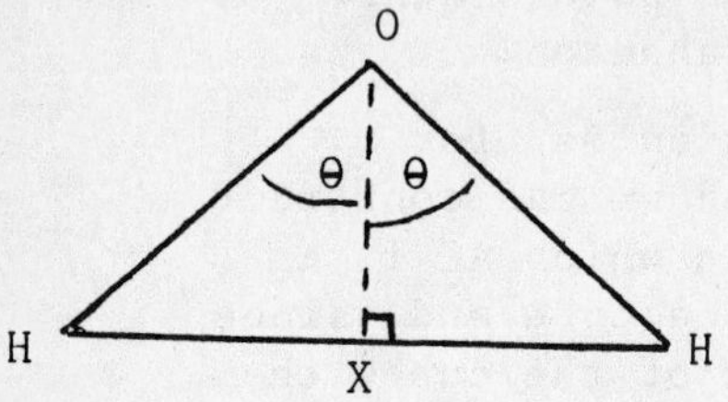


$$\sin \theta = \frac{XH}{OH} = \frac{76.65 \text{ pm}}{OH} = \sin 52.3^o$$

$$= \underline{0.7912}$$

$$\text{i.e., } OH = \frac{76.65 \text{ pm}}{0.7912} = \underline{96.9 \text{ pm}}$$

The O-H distance in the angular water molecule is 96.9 pm

26. For the tetrahedron (drawn below), we have:

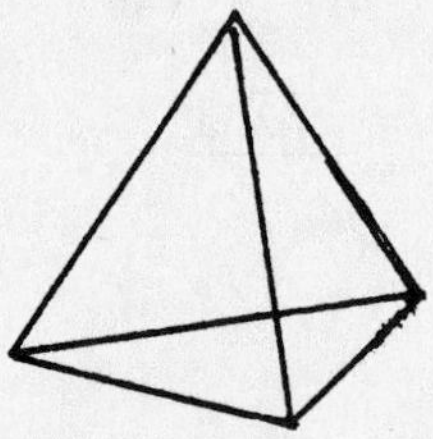

(a) Four equilateral triangular faces, with

(b) all the angles exactly equal to $60^o$, and

(c) there are twelve such angles.

27. The exact numbers are those for which there is absolutely no uncertainty:

(a) There are 12 eggs in one dozen; 12 is an exact number
(b) There are 24 hours in one day; by definition an exact number
(c) The newspaper gave the attendance at a football game as 64 000; it is unlikely that the number was precisely 64 000 - not an exact number
(d) Peter weighs 165 lb; clearly Peter does not weigh 165 lb exactly - not an exact number
(e) There are 1760 yards in a mile; 1760 is an exact number by definition
(f) There are 268 students in the chemistry class; 268 is an exact number that is obtained by counting the students.

28. In each case, we move the decimal point until there is one non-zero digit which precedes it. If moved n places to the left, we multiply the resultant number by $10^n$; if moved n places to the right, we multiply the resultant number by $10^{-n}$. Thus:

(a) 12 000 becomes $1.200\ 0 \times 10^4$ significant figures = 5

(b) 1 740 312.29 becomes $1.740\ 312\ 29 \times 10^6$ significant figures = 9

(c) 0.004 04 becomes $4.04 \times 10^{-3}$ significant figures = 3

(d) -0.049 00 becomes $-4.900 \times 10^{-2}$ significant figures = 4

Note that the number of significant figures remains the same as in the original number (for numbers less than 1 it is the number of numbers after the zeros in the decimal expression.

29. In each case the process of Problem 28 is reversed. The number of significant figures is the same as in the number that appears before the exponent $10^n$. Thus:

(a) $3.0 \times 10^2$ becomes 300; 2 significant figures

(b) $1.162 \times 10^5$ becomes 116 200; 4 significant figures

(c) $4.8 \times 10^{-3}$ becomes 0.004 8; 2 significant figures

(d) $-6.440 \times 10^{-2}$ becomes -0.064 40; 4 significant figures

30. The answer is $1.695\ 188\ 295\ 44 \times 10^{29}$

31. (a) becomes $(4 \times 10^{-5})(1 \times 10^{-3})(5 \times 10^3) = 20 \times 10^{-5} = \underline{2 \times 10^{-4}}$

Note that $10^n \times 10^m = 10^{n+m}$

(b) becomes $(1.2 \times 10^{-4})/(6 \times 10^{-2}) = 0.2 \times 10^{-2} = \underline{2 \times 10^{-3}}$

Note that $10^n/10^m = 10^{n-m}$

(c) becomes $(2 \times 10^2)^3(9 \times 10^{-4})^{1/2} = (8 \times 10^6)(3 \times 10^{-2}) = 24 \times 10^4$
$= \underline{2.4 \times 10^5}$

Note that $(10^n)^m = 10^{n \times m}$

(d) becomes $\dfrac{(2 \times 10^{-2})(6 \times 10^{2})(5 \times 10^{1})}{(3 \times 10^{-3})(5 \times 10^{2})} = \underline{4 \times 10^{2}}$

(e) becomes $(9 \times 10^{-6})^{1/2}(2 \times 10^{1})^{3} = (3 \times 10^{-3})(8 \times 10^{3}) = \underline{24}$

32. (a) $\dfrac{(6 \times 10^{-6})(3 \times 10^{14})}{(2 \times 10^{3})(1 \times 10^{6})} = \dfrac{18 \times 10^{8}}{2 \times 10^{9}} = \underline{9 \times 10^{-1}}$

(b) Express each number in terms of $10^{23}$ and then add them:
$(6.022 \times 10^{23}) + (7.7 \times 10^{21}) = (6.022 + 0.077) \times 10^{23} = \underline{6.099 \times 10^{23}}$

(c) $\dfrac{(5000)(0.06)}{0.000\ 3} = \dfrac{(5 \times 10^{3})(6 \times 10^{-2})}{3 \times 10^{-4}} = \underline{1 \times 10^{6}}$

(d) $\dfrac{(3.6 \times 10^{-5})^{1/2}(0.000\ 12)}{3000} = \dfrac{(36 \times 10^{-6})^{1/2}(1.2 \times 10^{-4})}{3 \times 10^{3}}$
$= \dfrac{(6 \times 10^{-3})(1.2 \times 10^{-4})}{3 \times 10^{3}} = \underline{2.4 \times 10^{-10}}$

(e) Either convert to decimals and add, or convert to a common power of 10 and add:

| | |
|---|---|
| 119.2 | $1.192 \times 10^{2}$ |
| 204.12 | $2.041\ 2 \times 10^{2}$ |
| 0.000 373 4 | $0.000\ 003\ 734 \times 10^{2}$ |
| 323.3 | $3.233 \times 10^{2}$ |
| $= \underline{3.233 \times 10^{2}}$ | |

The number of significant figures to which the answer can be quoted is the smallest number of significant figures that appears in any of the numbers (in this case 4)

Thus, the answer is $\underline{3.233 \times 10^{2}}$

33. (a) Basic SI unit of length is the meter, m

(b) Basic unit of mass is the kilogram, kg

(c) Basic SI unit of volume is the meter cubed, $m^{3}$

(d) Basic SI unit of density is kilogram per cubic meter, $kg\ m^{-3}$

34. (a) kilo = $10^{3}$ (b) milli = $10^{-3}$ (c) centi = $10^{-2}$ (d) pico = $10^{-12}$
(e) micro = $10^{-6}$

35. (a) $10^{3}$ = kilo, k (b) $10^{-1}$ = deci, d (c) $10^{-2}$ = centi, c
(d) $10^{-6}$ = micro, $\mu$ (e) $10^{-12}$ = pico, p

36. This is an exercise in using unit factors, with which you should become very familiar.

(a) 1 mg = $10^{-3}$ g, and 1 kg = $10^{3}$ g, so that we may write

$$\frac{1\ mg}{10^{-3}\ g} = 1 \text{ or } \frac{10^{-3}\ g}{1\ mg} = 1 \quad \text{and} \quad \frac{1\ kg}{10^{3}\ g} = 1 \text{ or } \frac{10^{3}\ g}{1\ kg} = 1$$

We wish to convert mg to kg, so we write:

$$5.84 \times 10^{-3}\ \text{mg} = (5.84 \times 10^{-3}\ \text{mg})\left(\frac{10^{-3}\ \text{g}}{1\ \text{mg}}\right)\left(\frac{1\ \text{kg}}{10^{3}\ \text{g}}\right) = \underline{5.84 \times 10^{-9}\ \text{kg}}$$

Note that it is not too difficult to decide which way up the unit factor has to be used, if you remember that the units must cancel to give the required final units.

(b) $1\ \text{Mg} = 10^{6}\ \text{g}$, and $1\ \text{kg} = 10^{3}\ \text{g}$.

$$54.34\ \text{Mg} = (5.434 \times 10^{1}\ \text{Mg})\left(\frac{10^{6}\ \text{g}}{1\ \text{Mg}}\right)\left(\frac{1\ \text{kg}}{10^{3}\ \text{g}}\right) = \underline{5.434 \times 10^{4}\ \text{kg}}$$

(c) $1\ \text{kg} = 10^{3}\ \text{g}$

$$0.345\ \text{g} = (3.45 \times 10^{-1}\ \text{g})\left(\frac{1\ \text{kg}}{10^{3}\ \text{g}}\right) = \underline{3.45 \times 10^{-4}\ \text{kg}}$$

(d) $1\ \text{kg} = 10^{3}\ \text{g}$

$$1.673 \times 10^{-21}\ \text{g} = (1.673 \times 10^{-21}\ \text{g})\left(\frac{1\ \text{kg}}{10^{3}\ \text{g}}\right) = \underline{1.673 \times 10^{-24}\ \text{kg}}$$

37. (a) $1\ \text{inch} = (1\ \text{inch})\left(\frac{1\ \text{ft}}{12\ \text{inch}}\right)$ ; unit factor is $\underline{1\ \text{ft}/12\ \text{inch}}$

(b) $1\ \text{ft} = (1\ \text{ft})\left(\frac{12\ \text{inch}}{1\ \text{ft}}\right)$; unit factor is $\underline{12\ \text{inch}/1\text{ft}}$

(c) $1\ \text{mile} = (1\ \text{mile})\left(\frac{1\ \text{km}}{0.6214\ \text{mile}}\right)$; unit factor is $\underline{1\ \text{km}/0.6214\ \text{mile}}$

(d) $1\ \text{km} = (1\ \text{km})\left(\frac{0.6214\ \text{mile}}{1\ \text{km}}\right)$ ; unit factor is $\underline{0.6214\ \text{mile}/1\ \text{km}}$

(e) Remember that $1\ \text{L} = 1\ \text{dm}^3$, $1\ \text{m} = 10\ \text{dm}$, and $1\ \text{L} = 10^{3}\ \text{mL}$

$$1\ \text{m}^3 = (1\ \text{m})^3\left(\frac{10\ \text{dm}}{1\ \text{m}}\right)^3\left(\frac{1\ \text{L}}{1\ \text{dm}^3}\right)\left(\frac{10^{3}\ \text{mL}}{1\ \text{L}}\right) = 10^{6}\ \text{mL}$$

unit factor is $\underline{10^{6}\ \text{mL}/\text{m}^3}$

(f) $$(1\ \text{mile})^2 = (1\ \text{mile})^2\left(\frac{1\ \text{km}}{0.6214\ \text{mile}}\right)^2 = 2.590\ \text{km}^2$$

unit factor is $\underline{2.590\ \text{km}^2/1\ \text{mile}^2}$

(g) $$1\ \text{cm}^2 = (1\ \text{cm})^2\left(\frac{1\ \text{m}}{10^{2}\ \text{cm}}\right)^2 = 10^{-4}\ \text{m}^2$$

unit factor is $\underline{10^{-4}\ \text{m}^2/1\ \text{cm}^2}$

38. (a) $1\ \text{g} = (1\ \text{g})\left(\frac{1\ \text{kg}}{10^{3}\text{g}}\right)$ unit factor is $\underline{1\ \text{kg}/10^{3}\text{g}}$

(b) $1\ \text{mg} = (1\ \text{mg})\left(\frac{1\ \text{g}}{10^{3}\ \text{mg}}\right)$ unit factor is $\underline{1\ \text{g}/10^{3}\ \text{mg}}$

(c) $1\ \text{cm} = (1\ \text{cm})\left(\frac{1\ \text{m}}{10^{2}\text{cm}}\right)$ unit factor is $\underline{1\ \text{m}/10^{2}\ \text{cm}}$

(d) $1\ \text{dm}^3 = (1\ \text{dm})^3\left(\frac{10\ \text{cm}}{1\ \text{dm}}\right)^3$ unit factor is $\underline{10^{3}\ \text{cm}^3/1\ \text{dm}^3}$

(e) $1\ \text{g cm}^{-3} = (1\ \text{g cm}^{-3})\left(\frac{1\ \text{kg}}{10^{3}\ \text{g}}\right)\left(\frac{1\ \text{m}}{10^{2}\ \text{cm}}\right)^{-3}$ unit factor is $\underline{\frac{10^{3}\ \text{kg m}^{-3}}{1\ \text{g cm}^{-3}}}$

39. (a) $$24\ 000 \text{ mile} = (2.4\text{x}10^4 \text{ mile})\left(\frac{1 \text{ km}}{0.6214 \text{ mile}}\right)\left(\frac{10^3 \text{ m}}{1 \text{ km}}\right) = \underline{3.9\text{x}10^7 \text{ m}}$$

(b) $$150 \text{ lb} = (1.5\text{x}10^2 \text{ lb})\left(\frac{1 \text{ kg}}{2.205 \text{ lb}}\right) = \underline{68 \text{ kg}}$$

(c) $$14 \text{ lb in}^{-2} = \left(\frac{14 \text{ lb}}{1 \text{ in}^2}\right)\left(\frac{1 \text{ kg}}{2.205 \text{ lb}}\right)\left(\frac{1 \text{ in}}{2.540 \text{ cm}}\right)^2\left(\frac{10^2 \text{ cm}}{1 \text{ m}}\right)^2 = \underline{9.8\text{x}10^3 \text{ kg m}^{-2}}$$

(d) $$60 \text{ mile h}^{-1} = \left(\frac{60 \text{ mile}}{1 \text{ h}}\right)\left(\frac{1 \text{ km}}{0.6214 \text{ mile}}\right)\left(\frac{1 \text{ h}}{60 \text{ min}}\right)\left(\frac{1 \text{ min}}{60 \text{ s}}\right)$$

$$= \underline{2.7 \text{ x } 10^{-2} \text{ km s}^{-1}}$$

40. (a) $$0.0254 \text{ cm} = (2.54\text{x}10^{-2} \text{ cm})\left(\frac{1 \text{ m}}{10^2 \text{ cm}}\right) = \underline{2.54\text{x}10^{-4} \text{ m}} \text{ (3 sig fig )}$$

(b) $$0.30\text{x}10^5 \text{ g} = (3.0\text{x}10^5 \text{ g})\left(\frac{1 \text{ kg}}{10^3 \text{ g}}\right) = \underline{3.0 \text{ x } 10^2 \text{ kg}} \text{ (2 sig fig )}$$

(c) $$365.0\text{x}10^{-5} \text{ mm} = (3.650\text{x}10^{-3} \text{ mm})\left(\frac{1 \text{ m}}{10^3 \text{ mm}}\right) = \underline{3.650\text{x}10^{-6} \text{ m}} \text{ (4 sig fig)}$$

(d) $$0.065\text{x}10^{-2} \text{ pm} = (6.5\text{x}10^{-4} \text{ pm})\left(\frac{1 \text{ m}}{10^{12} \text{ pm}}\right) = \underline{6.5\text{x}10^{-16} \text{ m}} \text{ (2 sig fig)}$$

(e) $$637.1 \text{ mL} = (6.371\text{x}10^2 \text{ mL})\left(\frac{1 \text{ cm}^3}{1 \text{ mL}}\right)\left(\frac{1 \text{ m}}{10^2 \text{ cm}}\right)^3 = \underline{6.371\text{x}10^{-4} \text{ m}^3} \text{ (4 sig fig)}$$

(f) $$13.43\text{x}10^4 \text{ L} = (1.343\text{x}10^5 \text{ L})\left(\frac{1 \text{ dm}^3}{1 \text{ L}}\right)\left(\frac{1 \text{ m}}{10 \text{ dm}}\right)^3 = 1.343\text{x}10^2 \text{ m}^3 \text{ (4 sig fig)}$$

(g) $$0.52\text{x}10^{-2} \text{ kg L}^{-1} = (5.2\text{x}10^{-3} \text{ kg L}^{-1})\left(\frac{1 \text{ L}}{1 \text{ dm}^3}\right)\left(\frac{10 \text{ dm}}{1 \text{ m}}\right)^3$$

$$= \underline{5.2 \text{ kg m}^{-3}} \text{ (2 sig fig)}$$

(h) $$63.43 \text{ x } 10^3 \text{ dm h}^{-1} = (6.343\text{x}10^4 \text{ dm h}^{-1})\left(\frac{1 \text{ m}}{10 \text{ dm}}\right)\left(\frac{1 \text{ h}}{60 \text{ min}}\right)\left(\frac{1 \text{ min}}{60 \text{ s}}\right)$$

$$= \underline{1.762 \text{ m s}^{-1}} \text{ (4 sig fig)}$$

41. (a) $$(0.1998\text{x}10) \text{ Gm s}^{-1} = (1.998 \text{ Gm s}^{-1})\left(\frac{10^9 \text{ m}}{1 \text{ Gm}}\right) = \underline{1.998\text{x}10^9 \text{ m s}^{-1}}$$

(b) $$(6.022\text{x}10^{23})(1.673\text{x}10^{-24} \text{ g})\left(\frac{1 \text{ kg}}{10^3 \text{ g}}\right) = \underline{1.007\text{x}10^{-3} \text{ kg}}$$

(c) $$32\ 150 \text{ Å min}^{-1} = (3.215\text{x}10^4 \text{ Å min}^{-1})\left(\frac{10^{-10} \text{ m}}{1 \text{ Å}}\right)\left(\frac{1 \text{ min}}{60 \text{ s}}\right)$$

$$= \underline{5.358 \text{ x } 10^{-8} \text{ m s}^{-1}}$$

(d) $$22.41 \text{ mL} = (2.241\text{x}10 \text{ mL})\left(\frac{1 \text{ cm}^3}{1 \text{ mL}}\right)\left(\frac{1 \text{ m}}{10^2 \text{ cm}}\right)^3 = \underline{2.241\text{x}10^{-5} \text{ m}^3}$$

42. $$20 \text{ gallon} = (20 \text{ gallon})\left(\frac{4 \text{ quart}}{1 \text{ gallon}}\right)\left(\frac{1 \text{ L}}{1.06 \text{ quart}}\right) = \underline{75 \text{ L}}$$

43. (a) $2.998 \times 10^{7}$ km = $(2.998 \times 10^{7}\ \text{km})(\frac{10^{3}\ \text{m}}{1\ \text{km}})$ = $\underline{2.998 \times 10^{10}\ \text{m}}$

(b) 143 pm = $(143\ \text{pm})(\frac{1\ \text{m}}{10^{-12}\ \text{m}})$ = $143 \times 10^{-12}$ m = $\underline{1.43 \times 10^{-10}\ \text{m}}$

(c) 0.001 nm = $(1 \times 10^{-3}\ \text{nm})(\frac{1\ \text{m}}{10^{9}\ \text{nm}})$ = $\underline{1 \times 10^{-12}\ \text{m}}$

(d) 1.54 Å = $(1.54\ \text{Å})(\frac{10^{-10}\ \text{m}}{1\ \text{Å}})$ = $\underline{1.54 \times 10^{-10}\ \text{m}}$

44. $20.0\ \text{km L}^{-1} = (2.00 \times 10\ \text{km L}^{-1})(\frac{0.6214\ \text{mile}}{1\ \text{km}})(\frac{1\ \text{L}}{1.06\ \text{quart}})(\frac{4\ \text{quart}}{1\ \text{gallon}})$

$= \underline{46.9\ \text{mile gallon}^{-1}}$

45. $100\ \text{km h}^{-1} = (100\ \text{km h}^{-1})(\frac{0.6214\ \text{mile}}{1\ \text{km}}) = \underline{62.1\ \text{mile h}^{-1}}$

46. Circumference of earth rotates once in 24 h

Distance traveled in 24 h = $(1039\ \text{mile h}^{-1})(24\ \text{h})$ = $\underline{2.5 \times 10^{4}\ \text{mile}}$

$2.5 \times 10^{4}\ \text{mile} = (2.5 \times 10^{4}\ \text{mile})(\frac{1\ \text{km}}{0.6214\ \text{mile}})(\frac{10^{3}\ \text{m}}{1\ \text{km}}) = \underline{4.0 \times 10^{7}\ \text{m}}$

47. Average speed of runner = $\frac{1500\ \text{m}}{3\ \text{min}\ 50\ \text{s}} = (\frac{1500\ \text{m}}{230\ \text{s}})(\frac{1\ \text{km}}{10^{3}\ \text{m}})(\frac{0.6214\ \text{mile}}{1\ \text{km}})$

$= 4.052 \times 10^{-3}\ \text{mile s}^{-1}$

Time to run 1 mile at the same speed is given by $\frac{1\ \text{mile}}{4.052 \times 10^{-3}\ \text{mile s}^{-1}}$

= $\underline{247\ \text{s}}$ or $\underline{4\ \text{min}\ 7\ \text{s}}$

48. Price of gasoline = 40.2 ¢ Can per liter; \$1.00 US = \$1.20 Can.

$40.2\ ¢\ \text{Can L}^{-1} = (40.2\ ¢\ \text{Can L}^{-1})(\frac{\$1\ \text{Can}}{100\ ¢\ \text{Can}})(\frac{\$1\ \text{US}}{\$1.20\ \text{Can}})(\frac{1\ \text{L}}{1.06\ \text{quart}})(\frac{4\ \text{q}}{1\ \text{gal}})$

$= \underline{\$1.26\ \text{US gallon}^{-1}}$

49. Gasoline consumption = $25.4\ \text{miles gallon}^{-1}$

$= (25.4\ \text{miles gallon}^{-1})(\frac{1\ \text{km}}{0.6214\ \text{mile}})(\frac{1\ \text{gallon}}{4\ \text{quart}})(\frac{1.06\ \text{quart}}{1\ \text{L}})$

$= \underline{10.83\ \text{km L}^{-1}}$

Gasoline consumption in L per 100 km = $\frac{100\ \text{km}}{10.83\ \text{km L}^{-1}} = \underline{9.2\ \text{L}}$

50. We convert the data given to metric units, preferably SI units.

$27.32\ \text{days} = (27.32\ \text{days})(\frac{24\ \text{h}}{1\ \text{day}})(\frac{60\ \text{min}}{1\ \text{h}})(\frac{60\ \text{s}}{1\ \text{h}}) = \underline{2.360 \times 10^{6}\ \text{s}}$

$238\ 850\ \text{mile} = (2.388\ 50 \times 10^{5}\ \text{mile})(\frac{1\ \text{km}}{0.6214\ \text{mile}}) = \underline{3.844 \times 10^{5}\ \text{km}}$

$1081 \text{ mile} = (1.081\text{x}10^3 \text{ mile})(\frac{1 \text{ km}}{0.6214 \text{ mile}}) = \underline{1.740\text{x}10^3 \text{ km}}$

$8.1\text{x}10^{19} \text{ ton} = (8.1\text{x}10^{19} \text{ ton})(\frac{2000 \text{ lb}}{1 \text{ ton}})(\frac{453.6 \text{ g}}{1 \text{ lb}})(\frac{1 \text{ kg}}{10^3 \text{ g}}) = \underline{7.3\text{x}10^{22} \text{ kg}}$

Thus, "The moon revolves around the earth with a period of $\underline{2.360\text{x}10^6 \text{ s}}$ at a distance of $\underline{3.844\text{x}10^5 \text{ km}}$. The moon has a radius of $\underline{1.734\text{x}10^3 \text{ km}}$, and its mass is estimated to be $\underline{7.3\text{x}10^{22} \text{ kg}}$"

51. Convert the data given to metric units, preferably SI units.

$11.86 \text{ yr} = (11.86 \text{ yr})(\frac{365 \text{ day}}{1 \text{ yr}})(\frac{24 \text{ h}}{1 \text{ day}})(\frac{3600 \text{ s}}{1 \text{ h}}) \quad \underline{3.74\text{x}10^8 \text{ s}}$

$4.84\text{x}10^8 \text{ mile} = (4.84\text{x}10^8 \text{ mile})(\frac{1 \text{ km}}{0.6214 \text{ mile}})(\frac{10^3 \text{ m}}{1 \text{ km}}) = \underline{7.79\text{x}10^{11} \text{ m}}$

$1.330 \text{ g cm}^{-3} = (\frac{1.330 \text{ g}}{1 \text{ cm}^3})(\frac{1 \text{ kg}}{10^3 \text{ g}})(\frac{10^2 \text{ cm}}{1 \text{ m}})^3 = \underline{1.330\text{x}10^3 \text{ kg m}^{-3}}$

Thus, "Jupiter revolves around the sun with a period of $\underline{3.74\text{x}10^8 \text{ s}}$, at a distance of approximately $\underline{7.79\text{x}10^{11} \text{ m}}$. Jupiter has an average density of $1.330\text{x}10^3 \text{ kg m}^{-3}$."

52. $4.0 \text{ light yr} = (4.0 \text{ yr})(\frac{365 \text{ day}}{1 \text{ yr}})(\frac{24 \text{ hr}}{1 \text{ day}})(\frac{3600 \text{ s}}{1 \text{ hr}})(\frac{3.00\text{x}10^8 \text{ m}}{1 \text{ s}})(\frac{1 \text{ km}}{10^3 \text{ m}})$

$= \underline{3.78\text{x}10^{13} \text{ km}}$

$4.0 \text{ light yr} = (3.78\text{x}10^{13} \text{ km})(\frac{0.6214 \text{ mile}}{1 \text{ km}}) = \underline{2.35\text{x}10^{13} \text{ mile}}$

53.

| | Canada | U.S.A. | Australia | U.K. |
|---|---|---|---|---|
| Population | $2.3845\text{x}10^7$ | $2.2009\text{x}10^8$ | $1.451\text{x}10^7$ | $5.5819\text{x}10^7$ |
| Area ($\text{km}^2$) | $9.976139\text{x}10^6$ | $9.519617\text{x}10^6$ | $7.686849\text{x}10^6$ | $2.244013\text{x}10^5$ |
| Density ($\text{km}^{-2}$) | 2.3902 | 23.120 | 1.888 | 248.75 |

54. $1 \text{ g cm}^{-3} = (\frac{1 \text{ g}}{1 \text{ cm}^3})(\frac{1 \text{ kg}}{10^3 \text{ g}})(\frac{10^2 \text{ cm}}{1 \text{ m}})^3 = \underline{10^3 \text{ kg m}^{-3}}$ ; $\underline{\text{Factor} = 10^3}$

55. (a) $75.0 \text{ g CHCl}_3 = (75.0 \text{ g})(\frac{1 \text{ cm}^3}{1.49 \text{ g}}) = \underline{50.3 \text{ cm}^3}$

(b) $125 \text{ cm}^3 \text{ CHCl}_3 = (125 \text{ cm}^3)(\frac{1.49 \text{ g}}{1 \text{ cm}^3}) = \underline{186 \text{ g}}$

Note that in each case we use the density as the appropriate unit factor which cancels the units appropriately.

56. (a) density = $\frac{129.5 \text{ g}}{165.0 \text{ mL}}$ = $\underline{0.7848 \text{ g mL}^{-1}}$

(b) Mass of ethanol = $(350 \text{ mL})(\frac{0.7848 \text{ g}}{1 \text{ mL}})$ = $\underline{275 \text{ g}}$

57. Density of NaCl = $2.16 \text{ g cm}^{-3}$

a) Volume of crystal = $(1.34 \text{ mg})(\frac{1 \text{ g}}{10^3 \text{ mg}})(\frac{1 \text{ cm}^3}{2.16 \text{ g}})$ = $\underline{6.20\text{x}10^{-4} \text{ cm}^3}$

Density of mercury = $13.6 \text{ g cm}^{-3}$

b) Volume of mercury = $(21.34 \text{ g})(\frac{1 \text{ cm}^3}{13.6 \text{ g}})$ = $\underline{1.57 \text{ cm}^3}$

Density of benzene = $0.880 \text{ g cm}^{-3}$

c) Volume of benzene = $(1.00 \text{ kg})(\frac{10^3 \text{ g}}{1 \text{ kg}})(\frac{1 \text{ cm}^3}{0.880 \text{ g}})$ = $\underline{1.14\text{x}10^3 \text{ cm}^3}$

Density of water = $1.00 \text{ g cm}^{-3}$

d) Volume = $(1.00 \text{ quart})(\frac{1 \text{ L}}{1.06 \text{ quart}})(\frac{10^3 \text{ mL}}{1 \text{ L}})(\frac{1 \text{ cm}^3}{1 \text{ mL}})$ = $\underline{943 \text{ cm}^3}$

58. (a) Mass = $(1.00 \text{ L})(\frac{10^3 \text{ mL}}{1 \text{ L}})(\frac{1 \text{ cm}^3}{1 \text{ mL}})(\frac{1.84 \text{ g}}{1 \text{ cm}^3})(\frac{1 \text{ kg}}{10^3 \text{ g}})$ = $\underline{1.84 \text{ kg}}$

(b) Mass = $(25.00 \text{ mL})(\frac{0.785 \text{ g}}{1 \text{ cm}^3})(\frac{1 \text{ cm}^3}{1 \text{ mL}})(\frac{1 \text{ kg}}{10^3 \text{ g}})$ = $\underline{1.96\text{x}10^{-2} \text{ kg}}$

(c) Mass = $(1 \text{ quart})(\frac{1 \text{ L}}{1.06 \text{ quart}})(\frac{10^3 \text{ mL}}{1 \text{ L}})(\frac{1 \text{ cm}^3}{1 \text{ mL}})(\frac{1.000 \text{ g}}{1 \text{ cm}^3})(\frac{1 \text{ kg}}{10^3 \text{ g}})$

= $\underline{0.943 \text{ kg}}$

(d) Mass = $(1.00 \text{ inch})^3(\frac{2.54 \text{ cm}}{1 \text{ inch}})^3(\frac{2.16 \text{ g}}{1 \text{ cm}^3})(\frac{1 \text{ kg}}{10^3 \text{ g}})$ = $\underline{3.54\text{x}10^{-2} \text{ kg}}$

59. We need to calculate mass per unit volume, so first calculate the volume of the silver bar,

Volume = $(40 \text{ cm})(25 \text{ cm})(15 \text{ cm}) = 1.5\text{x}10^4 \text{ cm}^3$

Density = $\frac{(157.5 \text{ kg})(\frac{10^3 \text{ g}}{1 \text{ kg}})}{1.5 \text{ x } 10^4 \text{ cm}^3}$ = $\underline{11 \text{ g cm}^{-3}}$

60. Convert lb to g, and then $\text{gallon}^{-1}$ to $\text{cm}^{-3}$,

$6.56 \text{ lb gal}^{-1} = (\frac{6.56 \text{ lb}}{1 \text{ gal}})(\frac{453.6 \text{ g}}{1 \text{ lb}})(\frac{1 \text{ gal}}{4 \text{ quart}})(\frac{1.06 \text{ quart}}{1 \text{ L}})(\frac{1 \text{ L}}{10^3 \text{ mL}})(\frac{1 \text{ mL}}{1 \text{ cm}^3})$

= $\underline{0.789 \text{ g cm}^{-3}}$

61. 10 000 gal = $(1\text{x}10^4\ \text{gal})(\frac{4\ \text{qt}}{1\ \text{gal}})(\frac{1\ \text{L}}{1.06\ \text{qt}})(\frac{10^3\ \text{cm}^3}{1\ \text{L}})(0.785\ \text{g cm}^{-3})(\frac{1\ \text{kg}}{10^3\ \text{g}})$

$= \underline{2.96 \times 10^4\ \text{kg}}$

62. First calculate the volume of mercury in $\text{cm}^3$.

For a cylinder of radius r and height h, the volume is given by $\underline{V = \pi r^2 h}$

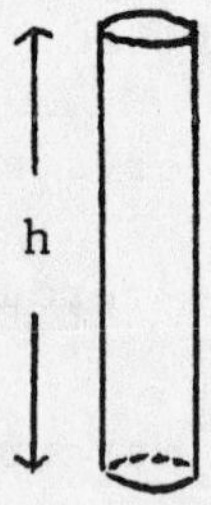


$V = \pi(4.0\ \text{mm})^2(\frac{1\ \text{cm}}{10\ \text{mm}})^2\ (78.3\ \text{cm}) = 39.4\ \text{cm}^3$

Mass of mercury = $(39.4\ \text{cm}^3)(13.954\ \text{g cm}^{-3})$

$= \underline{5.5 \times 10^2\ \text{g}}$

63. First calculate the volume of air in the room, in $\text{m}^3$, and convert this volume to liters, remembering that 1 L = 1 $\text{dm}^3$.

Vol of air = $(8.5\ \text{m})(13.5\ \text{m})(2.8\ \text{m})(\frac{10\ \text{dm}}{1\ \text{m}})^3(\frac{1\ \text{L}}{1\ \text{dm}^3}) = \underline{3.2\text{x}10^5\ \text{L}}$

Mass of air = $(3.2\text{x}10^5\ \text{L})(1.19\ \text{g L}^{-1})(\frac{1\ \text{kg}}{10^3\ \text{g}}) = \underline{3.8\text{x}10^2\ \text{kg}}$

64. Assuming that the earth is a perfect sphere, its volume is given by $\frac{4}{3}\pi r^3$, where r is the radius.

Volume of earth = $\frac{4}{3}\pi(6.34\text{x}10^3\ \text{km})^3 = \underline{1.067\text{x}10^{12}\ \text{km}^3}$

Density = $\frac{\text{mass}}{\text{volume}} = \frac{(6.00 \times 10^{24}\ \text{kg})}{(1.067\text{x}10^{12}\ \text{km}^3)}(\frac{10^3\ \text{g}}{1\ \text{kg}})(\frac{1\ \text{km}}{10^3\ \text{m}})^3(\frac{1\ \text{m}}{10^2\ \text{cm}})$

$= \underline{5.62\ \text{g cm}^{-3}}$

(The actual density of iron is 7.9 g $\text{cm}^{-3}$)

65. Volume of aluminum cube = $(1.04\ \text{m})^3 = 1.125\ \text{m}^3$

Volume of 1 Al atom = $(\frac{1.125\ \text{m}^3}{6 \times 10^{28}\ \text{atom}}) = \underline{1.9 \times 10^{-29}\ \text{m}^3\ \text{atom}^{-1}}$

Volume of sphere = $\frac{4}{3}\pi r^3 = 1.9 \times 10^{-29}\ \text{m}^3$

$r^3 = \frac{3}{4\pi}(1.9 \times 10^{-29}\ \text{m}^3) = 4.53 \times 10^{-30}\ \text{m}^3$

$r = (1.7 \times 10^{-10}\ \text{m})(\frac{10^{12}\ \text{pm}}{1\ \text{m}}) = \underline{170\ \text{pm}}$

66. (a) Mass of cube = volume x density

$$= (5.34 \times 10^{-2}\ m)^3(2.7\ g\ cm^{-3})\left(\frac{10^2\ cm}{1\ m}\right)^3 = \underline{411\ g}$$

(b) The volume of the sphere is $\frac{4}{3}\pi r^3$, and is required in $cm^3$.

$$\text{Volume} = \frac{4}{3}\pi\left(\frac{5.42\ mm}{2}\right)^3\left(\frac{1\ cm}{10\ mm}\right)^3 = \underline{8.33 \times 10^{-2}\ cm^3}$$

$$\text{Mass of sphere} = (8.33 \times 10^{-2}\ cm^3)(13.60\ g\ cm^{-3}) = \underline{1.13\ g}$$

67. The solvent is the major component and the solute is the minor component of a solution. (For a solution of a solid in a liquid, the solid is always considered to be the solute and the liquid the solvent).

68. To find the density $= \frac{\text{mass}}{\text{volume}}$, determine the total mass of the solution (the volume is given).

Mass of solution = mass of $NH_4Cl$ + mass of water

$$= (31.4 + 73.8)\ g = 105.2\ g$$

$$\text{Density} = \frac{\text{mass}}{\text{volume}} = \frac{105.2\ g}{82.4\ mL} = \underline{1.28\ g\ mL^{-1}}$$

69. 5% acetic acid means 5 g of acetic acid in 100 g of aqueous solution.

$$\text{Mass acetic acid in 1 kg sol'n} = \left(\frac{5\ g\ \text{acetic acid}}{100\ g\ \text{soln}}\right)\left(\frac{10^3\ g\ \text{soln}}{1\ kg\ \text{soln}}\right)$$

$$= \underline{50\ g\ \text{acetic acid}\ kg^{-1}}$$

70. Mass % is the mass of a component in 100 g of all components.

(a) Mass of ethanol = 5.34 g

Total mass of solution = (5.34 + 121.51) g = 126.9 g

$$\text{Mass \% ethanol} = \frac{5.34\ g\ \text{ethanol}}{126.9\ g\ \text{solution}} \times 100 = \underline{4.21\ \text{mass \%}}$$

<u>or</u>

$$\text{mass \%} = \frac{\text{mass of the component}}{\text{total mass of all components}} \times 100\%, \ \underline{\text{in one step}}$$

$$\text{mass \% ethanol} = \frac{5.34\ g}{(5.34 + 121.51)\ g} \times 100\% = \underline{4.21\ \text{mass \%}}$$

(b) The total mass of NaCl and water in the solution is 250.0 g

$$\text{Mass \% NaCl} = \frac{18.12\ g\ \text{NaCl}}{250.0\ g\ \text{solution}} \times 100 = \underline{7.248\ \text{mass \% NaCl}}$$

(c) Mass % HCl = $\frac{30.1 \text{ g HCl}}{(30.1 + 3.42 + 250.0) \text{ g}}$ x 100 = $\underline{10.6 \text{ mass \% HCl}}$

Mass % NaCl = $\frac{3.42 \text{ g NaCl}}{(30.1 + 3.42 + 250.0) \text{ g}}$ x 100 = $\underline{1.21 \text{ mass \% NaCl}}$

(d) Here we are given the volume of the solution and its density. To find the mass of the solution, multiply volume by density.

Mass of solution = (250 mL)(1.35 g $mL^{-1}$) = $\underline{337.5 \text{ g}}$

Mass % $H_3PO_4$ = $\frac{178 \text{ g}}{337.5 \text{ g}}$ x 100 = $\underline{52.7 \text{ mass \% } H_3PO_4}$

(e) Assume that the density of water is 1.00 g $mL^{-1}$, then 100.0 mL of water has a mass of 100.0 g.

Mass % $H_2O$ = $\frac{100.0 \text{ g}}{(100.0 + 21.35) \text{ g}}$ x 100 = $\underline{82.41 \text{ mass \% } H_2O}$

71. There are 95 g of $H_2SO_4$ in 100 g of concentrated solution. We need to calculate the mass of $H_2SO_4$ in 500 mL of concentrated acid. Thus, we need first to convert the 500 mL to a mass in g, using the density:

Mass of $H_2SO_4$(conc) = (500 mL)(1.84 g $mL^{-1}$) = 920 g

Mass of $H_2SO_4$ in 920 g solution = (920 g)$\left(\frac{95 \text{ g acid}}{100 \text{ g}}\right)$ = $\underline{874 \text{ g}}$

72. 100 g of solution contains 50 g NaOH. Thus

20 g NaOH are contained in (20 g NaOH)$\left(\frac{100 \text{ g soln}}{50 \text{ g NaOH}}\right)$ = 40 g solution.

Volume of 40 g of solution = $\left(\frac{40 \text{ g}}{1.53 \text{ g cm}^{-3}}\right)$ = $\underline{26 \text{ cm}^3}$ (26 mL).

Note that here we have used $\frac{100 \text{ g solution}}{50 \text{ g NaOH}}$ = 1, as the unit factor

73. First we convert 1.00 g $O_2$ to the mass of seawater that contains this mass of $O_2$, using the unit factor (6.22 g $O_2$)/$10^6$ g seawater = 1.

Grams of seawater = (1.00 g $O_2$)$\left(\frac{10^6 \text{ g seawater}}{6.22 \text{ g } O_2}\right)$ = $\underline{1.61\text{x}10^5 \text{ g seawater}}$ $\underline{\text{or}}$ $\underline{161 \text{ kg}}$

$1.61\text{x}10^5$ g seawater = $\left(\frac{1.61 \times 10^5 \text{ g}}{1.03 \text{ g cm}^{-3}}\right)$ = $\underline{1.56\text{x}10^5 \text{ cm}^3 \text{ seawater}}$ $\underline{(156 \text{ L})}$

74. 1 pg = $10^{-12}$ g, and 1 Tg = $10^{12}$ g

$1.6\text{x}10^{12}$ Tg seawater contains ($1.6\text{x}10^{12}$ Tg)$\left(\frac{10^{12} \text{ g}}{1 \text{ Tg}}\right)$ = $\underline{1.6\text{x}10^{24} \text{ g seawater}}$

1.0 g seawater contains (4.0 pg Au)$\left(\frac{10^{-12} \text{ g}}{1 \text{ pg}}\right)$ = $\underline{4.0\text{x}10^{-12} \text{ g Au}}$

The oceans contain $(1.6\text{x}10^{24}\text{ g seawater})\left(\frac{4.0 \times 10^{-12}\text{ g Au}}{1.00\text{ g seawater}}\right)\left(\frac{1\text{ kg}}{10^3\text{ g}}\right)$

$= \underline{6.4 \times 10^9\text{ kg Au}}$ (or $\underline{6.4\text{ Tg}}$)

75. Concentration of $Br^-$ ion = 0.06 ppm = 0.06 g $Br^-$ in $10^6$ g seawater.

Mass of seawater containing 1.00 kg $Br^-$ ion (1.00 kg $Br_2$)

$= (1.00\text{ kg }Br^-)\left(\frac{10^3\text{ g}}{1\text{ kg}}\right)\left(\frac{10^6\text{ g seawater}}{6 \times 10^{-2}\text{ g }Br^-}\right)\left(\frac{1\text{ kg}}{10^3\text{ g}}\right)$

$= \underline{2 \times 10^7\text{ kg seawater}}$

76. Macroscopic means what we can observe with our unaided senses; microscopic refers to explanations at the atomic and molecular level.

(a) A sample of a solid has a definite shape, and volume, and is not easily deformed or compressed.

(b) A sample of a liquid has a definite volume but the liquid easily deforms (flows), so that liquids fit the shape of the vessel in which they are contained; they are fluids. Like solids, they are not easily compressed, and the density of a liquid is generally slightly less than that of the corresponding solid.

(c) A sample of a gas has no definite shape or volume unless it is contained in a closed vessel, which it completely fills. A gas is readily compressed and has a density which is much smaller than that of the corresponding liquid. Unlike a solid or a liquid, where the density changes only gradually with change in temperature, the density of a gas in a closed vessel depends markedly on the temperature.

77. See important terms.

78. (a) Sugar and Water can be separated by distilling off the water to leave a residue of sugar. Care would have to be taken not too heat too strongly because this could decompose the sugar. The possibility of decomposing the sugar could be minimized by distilling off the water at a temperature well below 100°C, by reducing the pressure.

(b) Water and gasoline are immiscible, so that their mixture forms two distinct layers if left to settle. Gasoline is less dense than water and forms a liquid layer on top of the water. Thus, the gasoline can be simply decanted. In the laboratory, a separatory funnel with a tap is usually used for this purpose. If very pure water was required, the water layer could then be distilled, and the gasoline layer (a mixture itself) could be dried to remove traces of water, using a suitable solid drying agent, such as calcium chloride.

(c) Iron filings and sawdust could be separated by making use of the special property of iron, that it is attracted by a magnet.

(d) Sugar and powdered glass can be separated using the property that sugar is soluble in water while glass is insoluble in water. Thus, when water is added and the mixture warmed, a solution of sugar containing solid powdered glass results. The glass may be separated by filtration, and dried. The sugar could be recovered from the filtrate as described in (a) above.

(e) Food coloring in water: A number of methods based on chromatography could be used, depending on the amount of food coloring. Paper chromatography would separate small amounts. For larger amounts, the solution could be passed down a column containing aluminum oxide. Very small amounts could be removed using activated charcoal, which has the property of absorbing "impurities" in solutions.

79. (a) Zinc powder and sufur: Sulfur is soluble in liquid carbon disulfide, while zinc is insoluble. Thus, separation could be made by warming the mixture with carbon disulfide and filtering off the zinc. The sulfur could be recovered by evaporating off the solvent and recrystallization from the same solvent.

(b) $CCl_4$ and $CH_2Cl_2$: Since these liquids have a relatively large difference in their boiling points, the preferred method would be fractional distillation.

(c) NaCl and $KNO_3$ from solution: Below about $20^{\circ}C$, potassium nitrate is much less soluble in water than is sodium chloride. Depending on the concentrations, separation could be achieved by fractional crystallization, although it would probably be necessary to concentrate the solution first by boiling off some of the water, or allowing the sample to evaporate. Further recrystallization from water would be necessary to obtain pure samples of each salt.

(d) Natural gas, which is mainly methane, $CH_4$, with some ethane, $C_2H_6$, and propane, $C_3H_8$, could be separated into its components by gas chromatography.

(e) Iodine and potassium iodide: Iodine, $I_2(s)$, is very soluble in organic solvents such as tetrachloromethane, $CCl_4$, in which the salt potassium iodide, KI(s) is insoluble. Thus, the separation could be achieved by dissolving the iodine in $CCl_4(l)$ and filtering off the potassium iodide.

80. For a saturated solution, the solubility of $K_2SO_4$ is 12 g in 100 g $H_2O$.

Thus, amount of $K_2SO_4$ that would dissolve in 30 g of water at $25^{\circ}C$

$$= (30 \text{ g } H_2O)\left(\frac{12 \text{ g salt}}{100 \text{ g } H_2O}\right) = \underline{3.6 \text{ g } K_2SO_4}$$

$$\text{Mass \% } K_2SO_4 = \frac{3.6 \text{ g}}{(30 \text{ g } H_2O + 3.6 \text{ g salt})} \times 100 = \underline{11 \text{ mass \%}}$$

81. Solubility of NaCl = 36.0 g in 100 g $H_2O$ at 25°C.

Maximum amount of NaCl that can be dissolved in 150 g $H_2O$ at 25°C

$= (150\ g\ H_2O)(\frac{36.0\ g\ NaCl}{100\ g\ H_2O}) = 54.0\ g\ NaCl$

Thus, maximum mass of NaCl left undissolved at 25°C

$= (60.0-54.0)\ g = \underline{6.0\ g}$

and the filtrate is a saturated solution of NaCl containing 36.0 g NaCl per 100 g $H_2O$.*

$$\text{Mass \% NaCl} = \frac{36.0\ g\ NaCl}{(36.0\ g\ NaCl + 100\ g\ H_2O)} \times 100 = \underline{26.5\ \text{mass \% NaCl}}$$

*or 54.0 g NaCl in 150 g $H_2O$ gives exactly the same result.

# CHAPTER 2

1. The text gives the following information:

| | electron | proton | neutron |
|---|---|---|---|
| mass (kg) | $9.1096 \times 10^{-31}$ | $1.6726 \times 10^{-27}$ | - |
| mass (u) | 0.000 548 | 1.007 28 | 1.008 66 |
| charge (C) | $1.6020 \times 10^{-19}$ | $1.6020 \times 10^{-19}$ | 0 |

No value is given for the mass of the neutron in kg; this is obtained from the kg equivalent of the proton mass, as follows

$$\text{neutron mass} = (1.008\ 66\ \text{u})\left(\frac{1.672\ 6 \times 10^{-27}\ \text{kg}}{1.007\ 28\ \text{u}}\right) = \underline{1.674\ 9\times10^{-27}\ \text{kg}}$$

Relatively, compared to the mass of the electron

$$\text{mass of proton} = \frac{1.672\ 6 \times 10^{-27}\ \text{kg}}{9.109\ 6 \times 10^{-31}\ \text{kg}} = \underline{1.8360 \times 10^{3}}$$

$$\text{mass of neutron} = \frac{1.674\ 9 \times 10^{-27}\ \text{kg}}{9.109\ 6 \times 10^{-31}\ \text{kg}} = \underline{1.8386 \times 10^{3}}$$

i.e., Ratio of masses of electron : proton : neutron

= 1 : 1836.0 : 1838.6

Relatively, the charges on the proton and the electron are the same, but of opposite sign.

2. The atomic number, Z, of an atom comes from its position in the periodic table. Z is the charge on the nucleus, equal to the number of protons inside the nucleus of the atom. In the periodic table, the elements are arranged in order of increasing Z.

| Element | Symbol | Atomic No. (Z) | Element | Symbol | Atomic No. (Z) |
|---|---|---|---|---|---|
| Hydrogen | H | 1 | Phosphorus | P | 15 |
| Oxygen | O | 8 | Chlorine | Cl | 17 |
| Fluorine | F | 9 | Calcium | Ca | 20 |
| Neon | Ne | 10 | Zinc | Zn | 30 |
| Magnesium | Mg | 12 | | | |

3. The number of protons in a nucleus gives the charge on the nucleus, which is equal to the atomic number of the respective element

| Atomic number | 5 | 9 | 32 | 54 | 92 |
|---|---|---|---|---|---|
| Element | B | F | Ge | Xe | U |

4. In the periodic table the elements are arranged in order of increasing atomic number, Z, where Z is the number of nuclear protons, which in a neutral atom is the same as the number of electrons outside of the nucleus.

| Element | Symbol | Z | Element | Symbol | Z |
|---|---|---|---|---|---|
| Boron | B | 5 | Oxygen | O | 8 |
| Nitrogen | N | 7 | Sulfur | S | 16 |
| Hydrogen | H | 1 | Potassium | K | 19 |
| Neon | Ne | 10 | Iron | Fe | 26 |
| Chlorine | Cl | 17 | | | |

5. The number of protons gives the position of the element in the periodic table, that is, its atomic number, Z. The number of electrons is also Z for a neutral atom. In general an atom contains Z protons and N neutrons in its nucleus, and the symbol is written as

$$^{Z+N}_{Z}X$$

where Z is the atomic number, and Z+N, the total number of protons and neutrons (nucleons) is called the mass number. Thus:

| | Number of Protons (Z) | Number of Neutrons (N) | Number of Nucleons (Z + N) | Number of Electrons (Z) | Element | Symbol |
|---|---|---|---|---|---|---|
| (a) | 1 | 1 | (2) | (1) | Hydrogen | $^{2}_{1}H$ |
| (b) | 8 | 9 | (17) | (8) | Oxygen | $^{17}_{8}O$ |
| (c) | 6 | 8 | (14) | (6) | Carbon | $^{14}_{6}C$ |
| (d) | (9) | (10) | 19 | 9 | Fluorine | $^{19}_{9}F$ |
| (e) | (16) | (16) | 32 | 16 | Sulfur | $^{32}_{16}S$ |
| (f) | (12) | 12 | (24) | 12 | Magnesium | $^{24}_{12}Mg$ |
| (g) | (35) | 45 | (80) | 35 | Bromine | $^{80}_{35}Br$ |

6. The atomic number, Z, identifies the position of the element in the periodic table and is placed as a subscript before the symbol for the element. The mass number (nucleon number) is placed as a superscript before the elemental symbol.

(a) $^{40}_{19}K$ (b) $^{30}_{14}Si$ (c) $^{40}_{18}Ar$ (d) $^{15}_{7}N$ (e) $^{32}_{16}S$ (f) $^{23}_{11}Na$ (g) $^{27}_{13}Al$

7. The position of the element in the periodic table comes from locating its symbol. The position gives the atomic number, Z, the number of protons in the nucleus (equal to the number of electrons outside the nucleus in the neutral atom). The superscript before the symbol gives the total number of nucleons, (protons and neutrons), M, so that the number of neutrons is given by M-Z.

| Symbol | ${}^{4}_{2}He$ | ${}^{27}_{13}Al$ | ${}^{14}_{6}C$ | ${}^{31}_{15}P$ | ${}^{37}_{17}Cl$ | ${}^{85}_{37}Rb$ | ${}^{108}_{47}Ag$ | ${}^{131}_{54}Xe$ |
|---|---|---|---|---|---|---|---|---|
| Protons (Z) | 2 | 13 | 6 | 15 | 17 | 37 | 47 | 54 |
| Neutrons (M-Z) | 2 | 14 | 8 | 16 | 20 | 48 | 61 | 77 |
| Electrons (Z) | 2 | 13 | 6 | 15 | 17 | 37 | 47 | 54 |

8. The subscript gives the number of protons (equal to the number of electrons) and the superscript gives the number of nucleons (equal to the number of protons plus the number of neutrons). Thus,

| | Protons | Neutrons | Electrons |
|---|---|---|---|
| ${}^{2}_{1}H$ | 1 | 1 | 1 |
| ${}^{19}_{9}F$ | 9 | 10 | 9 |
| ${}^{40}_{20}Ca$ | 20 | 20 | 20 |
| ${}^{112}_{48}Cd$ | 48 | 64 | 48 |
| ${}^{117}_{50}Sn$ | 50 | 67 | 50 |
| ${}^{131}_{54}Xe$ | 54 | 77 | 54 |

9.

| atomic symbol | ${}^{9}_{4}Be$ | ${}^{15}_{7}N$ | ${}^{18}_{8}O$ | ${}^{12}_{6}C$ | ${}^{23}_{11}Na$ |
|---|---|---|---|---|---|
| mass number (M) | 9 | 15 | 18 | 12 | 23 |
| atomic no. (Z) | 4 | 7 | 8 | 6 | 11 |
| protons (Z) | 4 | 7 | 8 | 6 | 11 |
| electrons (Z) | 4 | 7 | 8 | 6 | 11 |
| neutrons (M-Z) | 5 | 8 | 10 | 6 | 12 |

10.

| | $^{24}_{12}Mg$ | $^{106}_{47}Ag$ | $^{137}_{56}Ba$ |
|---|---|---|---|
| Atomic symbol | $^{24}_{12}Mg$ | $^{106}_{47}Ag$ | $^{137}_{56}Ba$ |
| Mass number (M) | 24 | 106 | 137 |
| Atomic No. (Z) | 12 | 47 | 56 |
| Protons (Z) | 12 | 47 | 56 |
| Electrons (Z) | 12 | 47 | 56 |
| Neutrons (M-Z) | 12 | 59 | 81 |

11. The formula of water is $H_2O$.
Thus, the two H atoms could be $^1H$, $^1H$; $^2H$, $^2H$, or $^1H$, $^2H$, and the one O atom could be $^{16}O$, $^{17}O$, or $^{18}O$, giving the possible molecules:

$^1H^1H^{16}O$ $^2H^2H^{16}O$ $^1H^2H^{16}O$

$^1H^1H^{17}O$ $^2H^2H^{17}O$ $^1H^2H^{17}O$

$^1H^1H^{18}O$ $^2H^2H^{18}O$ $^1H^2H^{18}O$

Thus, the total number of isotopically different molecules is nine.

12. If we first express the percentage abundances as fractions of 1, then the average mass = the sum of (isotopic fraction x isotopic mass), for all of the isotopes. The isotopic fraction for a particular isotope is simply its % abundance divided by 100:

| isotope | mass | fractional abundance | abundance x mass |
|---|---|---|---|
| $^{10}B$ | 10.012 94 u | 0.197 7 | 1.979 u |
| $^{11}B$ | 11.009 31 u | 0.802 3 | 8.832 u |
| | | Average mass | 10.81 u |

13. (a) $^{24}_{12}Mg$ The subscript gives the atomic number, and the superscript gives the mass number; the symbol tells us that this is the magnesium-24 isotope.

(b) The calculation of the average atomic mass is similar to that shown in Problem 12.

| isotope | mass | fractional abundance | abundance x mass |
|---|---|---|---|
| $^{24}Mg$ | 23.993 u | 0.7860 | 18.8585 u |
| $^{25}Mg$ | 24.994 u | 0.1011 | 2.5269 u |
| $^{26}Mg$ | 25.991 u | 0.1129 | 2.9344 u |
| | | Average mass = | 24.31 u |

14. Most elements consist of a mixture of atoms that have the same nuclear charge but different masses, that are called isotopes. The mass number of an isotope is the total number of protons and neutrons in its nucleus. The atomic number is the number of protons in the nucleus. These protons account for a nuclear charge of +Z, which is balanced by Z electrons surrounding the nucleus of a neutral atom with a total charge of -Z. It is the Z electrons that give an atom its characteristic chemical properties. For all practical purposes, isotopes are chemically identical, although they may have slightly different physical properties. Thus, in most cases an element is a mixture of isotopes with the same atomic number, Z.

Isotopes have masses in atomic units, u, that are close to whole numbers, because the proton and neutron have similar masses close to 1 u. Thus, the copper isotope of mass 62.929 8 u has a mass number of 63, and that with mass 64.929 8 u has a mass number of 65. Copper is element 29 in the periodic table, so its atomic number is 29. We write the symbol for an isotope with its mass number placed as a superscript before the symbol, and the atomic number placed as a subscript before the symbol. i.e.,

$$^{63}_{29}\text{Cu} \quad \text{and} \quad ^{65}_{29}\text{Cu}$$

The average atomic mass of an element is given by the sum of the products of the fractional abundance of each isotope multiplied by its atomic mass. Thus, if the fractional abundance of copper-63 is x, and that of copper-65 is (1-x), we can write

$$x(62.929\ 8\ \text{u}) + (1-x)(64.927\ 8\ \text{u}) = 63.55\ \text{u}$$

and solving this equation for x gives 1.998x = 1.378, and x = 0.6897

Thus, the fractional abundance of $^{63}$Cu is 0.6897, and that of $^{65}$Cu is (1-0.6897) = 0.3103. These are often multiplied by 100 to express them as % abundances. i.e.,

% abundance $^{63}$Cu = 68.97% ; % abundance $^{65}$Cu = 31.03%

15. The information needed to understand this problem is given above.

From the atomic masses of 68.926 u and 70.926 u, and the atomic number of 31, we can write for the symbols of the two isotopes

$$^{69}_{31}\text{Ga} \quad \text{and} \quad ^{71}_{31}\text{Ga}$$

where the subscripts are the atomic number (number of nuclear protons), and the superscript is the mass (nucleon) number (protons + neutrons),

$^{69}_{31}$Ga - 31 protons and (69-31) = 38 neutrons

$^{71}_{31}$Ga - 31 protons and (71-31) = 40 neutrons

As explained in Problem 14, if x is the abundance of $^{69}$Ga, the abundance of $^{71}$Ga is (1-x), and we can write:

$$x(68.926\ u) + (1-x)(70.926\ u) = 69.72\ u$$

$$\underline{x = 0.6030}, \text{ whence } \underline{1-x = 0.3970}$$

Fractional abundances: $\underline{^{69}Ga = 0.6030}$ $\underline{^{71}Ga = 0.3970}$

% abundances: $\underline{^{69}Ga = 60.30\%}$ $\underline{^{71}Ga = 39.70\%}$

16. See Problem 14 for method:

If the fractional abundance of $^{235}U$ is x, then that of $^{236}U$ is 1-x, and

$$x(235.044\ u) + (1-x)(238.051\ u) = 238.03\ u$$

$$\underline{x = 0.0070}$$

Thus, the fractional abundance of $^{235}U$ in natural uranium is 0.0070, and the % abundance is 0.70 %

17. See problem 14 for method:

The atomic mass of $^{35}Cl$ will be close to 35 u, and the atomic mass of $^{37}Cl$ will be close to 37 u. If the fractional abundance of $^{35}Cl$ is x, that of $^{37}Cl$ is 1-x, we may write, approximately:

$$x(35)\ u + (1-x)37\ u = 35.45\ u$$

$$\underline{x = 0.78}\ ;\ \underline{(1-x) = 0.22}$$

i.e., Fractional abundances are $\underline{^{35}Cl = 0.78}$; $\underline{^{37}Cl = 0.22}$

and the % abundances are $\underline{^{35}Cl = 78\%}$ ; $\underline{^{37}Cl = 22\%}$

The average mass of a $Cl_2$ molecule is twice that of a Cl atom

Average mass of $Cl_2$ molecule = 2(35.45 u) = 70.90 u

18. A molecular mass is given by the sum of the masses of the constituent atoms.

mass of BrCl = mass of Br + mass of Cl

Since, there are BrCl molecules with masses 114, 116, and 118 u, and two isotopes of chlorine, $^{35}Cl$ with mass 35 u, and $^{37}Cl$ with mass 37 u, there are six possibilities for the mass of the Br atom, corresponding to subtracting each Cl mass (35 and 37) from each of the BrCl masses (114,116, and 118)

| | | | |
|---|---|---|---|
| (1) | (114-35) u = 79 u | (2) | (114-37) u = 77 u |
| (3) | (116-35) u = 81 u | (4) | (116-37) u = 79 u |
| (5) | (118-35) u = 83 u | (6) | (118-37) u = 81 u |

However, of the four possibilities for the mass of the Br atom that the above calculation gives (77 u, 79 u, 81 u, and 83 u), only 79 u and 81 u are admissible, since in BrCl each Br isotope can form molecules containing each of the Cl isotopes, only the Br masses that appear twice are admissible.

Thus: (a) The isotopes of bromine are $^{79}Br$ and $^{81}Br$.

(b) The isotopically different BrCl molecules are:

$$\underline{^{79}Br^{35}Cl}\ ;\ \underline{^{79}Br^{37}Cl}\ ;\ \underline{^{81}Br^{35}Cl}\ ;\ \underline{^{81}Br^{37}Cl}$$

19. Electron charge = $-1.602\ 19 \times 10^{-19}$ C. For 1 mol electrons we use the identity 1 mol = $6.022\ 05 \times 10^{23}$ entities, thus:

$$\text{Charge on 1 mol electrons} = -(1.602\ 19\text{x}10^{-19}\ \text{C})\left(\frac{6.022\ 05 \times 10^{23}\ \text{el}}{1\ \text{mol el}}\right)$$
$$= \underline{-9.648\ 47 \times 10^{4}\ \text{C mol}^{-1}}$$

Since the magnitude of the charge on the proton is the same as that on the electron (but of opposite charge), charge on 1 mol protons is

$$\underline{+9.648\ 47 \times 10^{4}\ \text{C mol}^{-1}}$$

20. The atomic mass unit, u, is one-twelth of the mass of a $^{12}$C atom, and 1 mol $^{12}$C atoms has a mass of exactly 12 g. Thus, 1 mol of atomic mass units is exactly 1 g, so that in grams

$$1\ \text{u} = \left(\frac{1\ \text{g}}{1\ \text{mol u}}\right)\left(\frac{1\ \text{mol u}}{6.022\ 05\text{x}10^{23}\ \text{u}}\right) = \underline{1.660\ 56 \times 10^{-24}\ \text{g u}^{-1}}$$

Thus, for element X,

$$\text{Mass of X} = (3.155\text{x}10^{-23}\ \text{g atom}^{-1})\left(\frac{1\ \text{mol u}}{1.660\ 56\text{x}10^{-24}\ \text{g}}\right) = \underline{19.00\ \text{u atom}^{-1}}$$

Reference to the periodic table gives X as fluorine, which has only one naturally occurring isotope, $^{19}$F.

21. (a) Molar mass = [14(12.01)+18(1.008)+2(14.01)+5(16.00)] = $\underline{294.3\ \text{g mol}^{-1}}$

(b) $$\text{Mol aspartamine} = (6.22\ \text{g asp})\left(\frac{1\ \text{mol asp}}{294.3\ \text{g asp}}\right) = \underline{2.11\text{x}10^{-2}\ \text{mol asp.}}$$

(c) $$\text{Mass asp.} = (0.245\ \text{mol asp})\left(\frac{294.3\ \text{g asp}}{1\ \text{mol asp}}\right) = \underline{72.1\ \text{g asp.}}$$

(d) $$\text{Molecules} = (4.28\ \text{mg asp})\left(\frac{10^{-3}\ \text{g}}{1\ \text{mg}}\right)\left(\frac{1\ \text{mol asp}}{294.3\ \text{g asp}}\right)\left(\frac{6.022\text{x}10^{23}\ \text{molecules}}{1\ \text{mol asp}}\right)$$
$$= \underline{8.76\text{x}10^{18}\ \text{molecules}}$$

22. (i) $6.022\text{x}10^{23}\ \text{u} = 1\ \text{g}$

(ii) $$1\ \text{kg} = (1\ \text{kg})\left(\frac{6.022 \times 10^{23}\ \text{u}}{1\ \text{g}}\right)\left(\frac{10^{3}\ \text{g}}{1\ \text{kg}}\right) = \underline{6.022\text{x}10^{26}\ \text{u}}$$

(iii) $$1\ \text{lb} = (1\ \text{lb})\left(\frac{453.6\ \text{g}}{1\ \text{lb}}\right)\left(\frac{6.022\text{x}10^{23}\ \text{u}}{1\ \text{g}}\right) = \underline{2.732\text{x}10^{26}\ \text{u}}$$

Thus, the appropriate unit factors are:

(i) $\dfrac{1\ \text{g}}{6.022\text{x}10^{23}\ \text{u}}$ (ii) $\dfrac{1\ \text{kg}}{6.022\text{x}10^{26}\ \text{u}}$ (iii) $\dfrac{1\ \text{lb}}{2.732\text{x}10^{26}\ \text{u}}$

23. (a) Molar mass $O_2(g) = 2(16.00) = 32.00\ g\ mol^{-1}$

$7.1x10^{-3}\ mol\ O_2(g) = (7.1x10^{-3}\ mol)(\frac{32.00\ g}{1\ mol}) = \underline{0.23\ g}$

(b) Molar mass of Fe = $55.85\ g\ mol^{-1}$

$5.43\ mol\ Fe = (5.43\ mol)(\frac{55.85\ g}{1\ mol}) = \underline{303\ g}$

(c) Molar mass of $P_4 = 4(30.97) = 123.9\ g\ mol^{-1}$

$3.14x10^{-2}\ mol\ P_4 = (3.14x10^{-2}\ mol)(\frac{123.9\ g}{1\ mol}) = \underline{3.89\ g}$

(d) Molar mass of $S_8 = 8(32.06) = 256.5\ g\ mol^{-1}$

$9.6x10^{-4}\ mol\ S_8 = (9.6x10^{-4}\ mol)(\frac{256.5\ g}{1\ mol}) = \underline{0.25\ g}$

(e) Molar mass of $CuSO_4.5H_2O = [63.55+32.06+9(16.00)+10(1.008)]$
$= 249.7\ g\ mol^{-1}$

$1.263x10^{-2}\ mol\ CuSO_4.5H_2O = (1.263x10^{-2}\ mol)(\frac{249.7\ g}{1\ mol}) = \underline{3.154\ g}$

(f) Molar mass of $Mg_3(PO_4)_2 = [3(24.31)+2(30.97)+8(16.00)]$
$= 262.9\ g\ mol^{-1}$

$0.452\ mol\ Mg_3(PO_4)_2 = (0.452\ mol)(\frac{262.9\ g}{1\ mol}) = \underline{119\ g}$

24. Convert to moles of Hg atoms and then to Hg atoms.

$$\text{Hg atoms} = (1\ g\ Hg)(\frac{1\ mol\ Hg}{200.6\ g\ Hg})(\frac{6.022x10^{23}\ Hg\ atoms}{1\ mol\ Hg})$$
$$= \underline{3.002x10^{21}\ Hg\ atoms}$$

25. (a) Molar mass of $SO_2 = [(32.06)+2(16.00)] = 64.06\ g\ mol^{-1}$

$0.028\ g\ SO_2 = (0.028\ g\ SO_2)(\frac{1\ mol\ SO_2}{64.06\ g\ SO_2}) = \underline{4.4x10^{-4}\ mol\ SO_2}$

(b) $3\ mol\ SO_2 = (3\ mol\ SO_2)(\frac{64.06\ g\ SO_2}{1\ mol\ SO_2}) = \underline{192.2\ g\ SO_2}$

26.

| | molar mass ($g\ mol^{-1}$) | number of moles |
|---|---|---|
| (a) | 18.02 | $(10\ g)(\frac{1\ mol}{18.02\ g}) = \underline{0.5549\ mol\ water}$ |
| (b) | 64.06 | $(10\ g)(\frac{1\ mol}{64.06\ g}) = \underline{0.1561\ mol\ SO_2}$ |
| (c) | 60.05 | $(10\ g)(\frac{1\ mol}{60.05\ g}) = \underline{0.1665\ mol\ acetic\ acid}$ |
| (d) | 98.08 | $(10\ g)(\frac{1\ mol}{98.08\ g}) = \underline{0.1020\ mol\ H_2SO_4}$ |

27. (a) 180.2 u

(b) Moles of aspirin $= (500\ \text{mg})\left(\frac{1\ \text{g}}{10^3\ \text{mg}}\right)\left(\frac{1\ \text{mol aspirin}}{180.2\ \text{g aspirin}}\right)$

$= \underline{2.77\text{x}10^{-3}\ \text{mol}}$

Molecules of aspirin $= (2.77\text{x}10^{-3}\ \text{mol})\left(\frac{6.022\text{x}10^{23}\ \text{molecules}}{1\ \text{mol}}\right)$

$= \underline{1.67\ \text{x}\ 10^{21}\ \text{molecules}}$

28. Molar mass $= (2.653\text{x}10^{-22}\ \text{g molecule}^{-1})\left(\frac{6.022\text{x}10^{23}\ \text{molecules}}{1\ \text{mol}}\right)$

$= \underline{159.8\ \text{g mol}^{-1}}$

29. $2.00\ \text{g}\ O_2 = (2.00\ \text{g}\ O_2)\left(\frac{1\ \text{mol}\ O_2}{32.00\ \text{g}\ O_2}\right)\left(\frac{6.022\text{x}10^{23}\ O_2\ \text{molecules}}{1\ \text{mol}\ O_2}\right)$

$= \underline{3.76\text{x}10^{22}\ O_2\ \text{molecules}}$

Mol of O atoms $= (2.00\ \text{g})\left(\frac{1\ \text{mol O}}{16.00\ \text{g O}}\right) = \underline{0.125\ \text{mol}}$

30. (a) Average mass of 1 Mg atom $= \left(\frac{24.30\ \text{g}}{1\ \text{mol}}\right)\left(\frac{1\ \text{mol}}{6.022\text{x}10^{23}\ \text{atoms}}\right)$

$= \underline{4.037\text{x}10^{-23}\ \text{g atom}^{-1}}$

(b) Vol of 1 mol Mg atoms $= \left(\frac{24.31\ \text{g}}{1\ \text{mol}}\right)\left(\frac{1\ \text{cm}^3}{1.738\ \text{g}}\right) = \underline{13.99\ \text{cm}^3\ \text{mol}^{-1}}$

(c) Avge vol per Mg atom $= \left(\frac{13.99\ \text{cm}^3}{1\ \text{mol}}\right)\left(\frac{1\ \text{mol}}{6.022\text{x}10^{23}\ \text{atoms}}\right)$

$= \underline{2.323\text{x}10^{-23}\ \text{cm}^3\ \text{atom}^{-1}}$

(d) Volume of spherical atom of radius r is $\frac{4}{3}\pi r^3$

$r = \left[\frac{3}{4\pi}(2.323\text{x}10^{-23}\ \text{cm}^3)\right]^{1/3}\left(\frac{1\ \text{m}}{10^2\ \text{cm}}\right)\left(\frac{10^{12}\ \text{pm}}{1\ \text{m}}\right) = \underline{177.0\ \text{pm}}$

31. (a) Number of cells $= (4\text{x}10^9\ \text{bodies})\left(\frac{6\text{x}10^{13}\ \text{cells}}{1\ \text{body}}\right)\left(\frac{1\ \text{mol cells}}{6.022\text{x}10^{23}\ \text{cells}}\right)$

$= \underline{0.4\ \text{mol cells}}$

(b) Molecules of $H_2O$ body$^{-1}$ $= \left(\frac{80\ \text{g}}{100\ \text{g}}\right)(65\ \text{kg})\left(\frac{10^3\ \text{g}}{1\ \text{kg}}\right)\left(\frac{1\ \text{mol}\ H_2O}{18.02\ \text{g}\ H_2O}\right)$

$\text{x}\ \left(\frac{6.022\text{x}10^{23}\ \text{molecules}}{1\ \text{mol}}\right)$

$= \underline{1.7\ \text{x}10^{27}\ \text{molecules}\ H_2O}$

32. First calculate the volume of the garage in liters; then calculate the fatal mass of CO in this volume, in grams, and convert this mass to mol CO, and then to molecules of CO. The requisite unit factors are

$1\ \text{L} = 1\ \text{dm}^3$; $2.38\text{x}10^{-4}\ \text{g CO} = 1\ \text{L}$ ; $1\ \text{mol CO} = 28.01\ \text{g CO}$

Number of CO molecules =

$$(150\ m^3)(\frac{10\ dm^3}{1\ m})(\frac{1\ L}{1\ dm^3})(\frac{2.38x10^{-4}\ g\ CO}{1\ L})(\frac{1\ mol\ CO}{28.01\ g\ CO})(\frac{6.022x10^{23}}{1\ mol})$$

$= \underline{7.68x10^{23}\ \text{molecules}}$

33. $H_2O$, 18.02 g $mol^{-1}$; $H_2O_2$, 34.02 g $mol^{-1}$; NaCl, 58.44 g $mol^{-1}$; $MgBr_2$, 184.1 g $mol^{-1}$; CO, 28.01 g $mol^{-1}$; $CO_2$, 44.01 g $mol^{-1}$; $CH_4$, 16.04 g $mol^{-1}$; $C_2H_6$, 30.08 g $mol^{-1}$; $NH_3$, 17.03 g $mol^{-1}$; HCl, 36.46 g $mol^{-1}$

34. $NH_4Cl$, 53.49 g $mol^{-1}$ ; $Ca_3(PO_4)_2$, 310.2 g $mol^{-1}$; KI, 166.0 g $mol^{-1}$; $BaCl_2.2H_2O$, 244.2 g $mol^{-1}$; $PCl_5$, 208.2 g $mol^{-1}$; $C_4H_{10}$, 58.12 g $mol^{-1}$.

35. (a) 60.05 u, 30.03 u; (b) 46.03 u, 46.03 u; (c) 342.3 u, 342.3 u; (d) 58.12 u, 29.06 u; (e) 27.67 u, 13.84 u; (f) 34.02 u, 17.01 u;

36. Molecular mass of benzene = $\underline{78.11\ u}$

$$1\ cm^3\ C_6H_6 = (1cm^3)(\frac{0.880\ g}{1\ cm^3})(\frac{1\ mol}{78.11\ g}) = \underline{1.127x10^{-2}\ mol}$$

$$C_6H_6\ \text{molecules} = (1.127x10^{-2}\ mol)(\frac{6.022x10^{23}\ \text{molecules}}{1\ mol})$$

$$= \underline{6.79x10^{21}\ \text{molecules}}$$

Each $C_6H_6$ molecule splits into 12 atoms,

$$\text{Mol atoms} = (1.127x10^{-2}\ \text{mol molec})(\frac{12\ \text{mol atoms}}{1\ \text{mol molecules}})$$

$$= \underline{0.135\ \text{mol atoms}}$$

37. In calculating mass % elemental composition of substances, first calculate the molar mass, then,

$$\text{Mass \% X} = (\frac{\text{mass of X in 1 mol}}{\text{molar mass}}) \times 100$$

(a) For the hydrogen and oxygen in $H_2O$,

$$\text{Mass \% H} = \left(\frac{\text{mass of 2 mol H}}{\text{molar mass } H_2O}\right) \times 100 = \frac{2(1.008 \text{ g})}{18.02 \text{ g}} = \underline{11.19\% \text{ H}}$$

$$\text{Similarly, mass \% O} = \frac{1(16.00 \text{ g})}{18.02 \text{ g}} \times 100 = \underline{88.79\% \text{ O}}$$

(b) $\underline{NaCl}$ 39.34% Na, 60.66% Cl; (c) $\underline{C_2H_6}$ 79.89% C, 20.11% H

(d) $\underline{MgBr_2}$ 13.20% Mg, 86.80% Br; (e) $\underline{CO_2}$ 27.29% C, 72.71% O

38. See Problem 37 for the method:
(a) $\underline{NH_3}$ 82.25% N, 17.75% H; (b) 100.0% Cl;
(c) $\underline{NaOH}$ 57.48% Na, 40.00% O, 2.52% H;
(d) $\underline{C_2H_6O}$ 52.14% C, 13.13% H, 34.73% O;
(e) $\underline{C_6H_5NO_2}$ 58.53% C, 4.09% H, 11.38% N, 25.99% O

39. (a) $\underline{C_2H_4}$ 85.63% C, 14.37% H; (b) $\underline{C_2H_2}$ 92.26% C, 7.74% H;
(c) $\underline{MgCl_2.6H_2O}$ 11.96% Mg, 34.87% Cl, 5.95% H, 47.22% O;
(d) $\underline{FeSO_4.7H_2O}$ 20.09% Fe, 11.53% S, 5.08% H, 63.31% O;
(e) $\underline{[Cu(NH_3)_4]SO_4.2H_2O}$ 24.09% Cu, 21.24% N, 12.15% S, 36.39% O, 6.11% H

40. Convert the mass of $HNO_3$ to mol $HNO_3$, and since 1 mol $HNO_3$ contains 3 mol of O, use the unit factor

$$1 \text{ mol } HNO_3 = 3 \text{ mol O}$$

to convert mol $HNO_3$ to mol O. The mass of O is then found by multiplying mol O by the molar mass of O, as follows:

$$\text{Mass of O} = (3.40 \text{ g } HNO_3)\left(\frac{1 \text{ mol } HNO_3}{63.02 \text{ g } HNO_3}\right)\left(\frac{3 \text{ mol O}}{1 \text{ mol } HNO_3}\right)\left(\frac{16.00 \text{ g O}}{1 \text{ mol O}}\right)$$

$$= \underline{2.59 \text{ g O}}$$

41. See Problem 37 for the method:

$\underline{\% \text{ N} = 21.21\%}$

$$100 \text{ g N atoms} = (100 \text{ g N})\left(\frac{100 \text{ g } (NH_4)_2SO_4}{21.21 \text{ g N}}\right) = \underline{471 \text{ g } (NH_4)_2SO_4}$$

42. Mass % S = $\underline{28.30\%}$

$$28.4 \text{ g } Sb_2S_3 \text{ contains } (28.4 \text{ g } Sb_2S_3)\left(\frac{28.30 \text{ g S}}{100 \text{ g } Sb_2S_3}\right) = \underline{8.04 \text{ g S}}$$

Mass containing 64.4 g S = $(64.4 \text{ g S})(\frac{100 \text{ g } Sb_2S_3}{28.30 \text{ g S}})$ = $\underline{228 \text{ g } Sb_2S_3}$

43. The given composition, 7.00% C and 93.00% Br, is conveniently expressed as

100 g compound contains 7.00 g C and 93.00 g Br

In calculating the empirical formula, the masses of C and Br are converted to moles of C and moles of Br by dividing the masses by the respective molar masses:

$$\text{mol C} = \frac{7.00 \text{ g C}}{12.01 \text{ g mol}^{-1}} = \underline{0.583 \text{ mol C}}$$

$$\text{mol Br} = \frac{93.00 \text{ g Br}}{79.90 \text{ g mol}^{-1}} = \underline{1.164 \text{ mol Br}}$$

Thus, the ratio of mol C : mol Br, which is the same as the ratio of atoms, is 0.583 : 1.164, which we can rationalize by dividing each by the smaller number:

$$\text{mol C : mol Br = atoms C : atoms Br} = \frac{0.583}{0.583} : \frac{1.164}{0.583} = 1.00 : 2.00$$

Empirical formula is $CBr_2$ (empirical formula mass = 171.8 u)

44. The method is discussed in detail in Problem 43.

| | | C | H | |
|---|---|---|---|---|
| Mass % | | 94.33 | 5.67 | |
| 100 g compound contains | | 94.33 | 5.67 | g |
| | = | $\frac{94.33}{12.01}$ | $\frac{5.67}{1.008}$ | mol |
| | = | 7.854 | 5.625 | mol |
| Ratio mol C : mol H | = | $\frac{7.834}{5.625}$ | $\frac{5.625}{5.625}$ | |
| Ratio of mol = ratio of atoms | = | 1.40 : | 1.00 | |
| | = | 7.00 : | 5.00 | |

Empirical formula of anthracene is $C_7H_5$

Empirical formula mass = [7(12.01) + 5(1.008)] = 89.11 u

45. The empirical formula is $SF_4$

46. The empirical formula is $C_3OF_6$; empirical formula mass = 166.0 u

47.

| | C | H | N | O |
|---|---|---|---|---|
| Mass % | 49.5 | 5.2 | 28.8 | 16.6 |
| grams in 100 g compound | 49.5 | 5.2 | 28.8 | 16.6 |
| moles in 100 g compound | $\frac{49.5}{12.01}$ | $\frac{5.2}{1.008}$ | $\frac{28.8}{14.01}$ | $\frac{16.6}{16.00}$ |
| = | 4.12 | 5.16 | 2.06 | 1.04 |
| ratio of moles (atoms) = | $\frac{4.12}{1.04}$ : | $\frac{5.16}{1.04}$ : | $\frac{2.06}{1.04}$ : | $\frac{1.04}{1.04}$ |
| = | 3.96 : | 4.96 : | 1.98 : | 1.00 |

Which within the expected precision of the analytical data gives $\underline{C_4H_5N_2O}$ as the empirical formula; <u>empirical formula mass = 97.1 u</u>

48. In this type of problem, the aim is to obtain the ratio of the moles of atoms in the compound, which is the same as the ratio of the numbers of atoms.

The mass of the sample gives the total mass of the atoms

The mass of $CO_2$ obtained on combustion contains all of the carbon in the the original sample; the mass of $CO_2$ can be converted to moles of $CO_2$, and hence the moles of carbon in the original sample can be obtained

$$\text{moles of C} = (4.40\text{ g } CO_2)\left(\frac{1\text{ mol } CO_2}{44.01\text{ g } CO_2}\right)\left(\frac{1\text{ mol C}}{1\text{ mol } CO_2}\right) = \underline{1.00 \times 10^{-1}\text{ mol C}}$$

Similarly, the mass of $H_2O$ obtained on combustion contains all of the H in the original sample; the mass of $H_2O$ can be converted to moles of $H_2O$, and hence the moles of hydrogen in the original sample can be obtained.

$$\text{moles of H} = (2.70\text{ g } H_2O)\left(\frac{1\text{ mol } H_2O}{18.02\text{ g } H_2O}\right)\left(\frac{2\text{ mol H}}{1\text{ mol } H_2O}\right) = \underline{3.00 \times 10^{-1}\text{ mol H}}$$

<u>Note</u> that the unit conversion factor here is 1 mol $H_2O$ = 2 mol H, because 1 mol $H_2O$ contains 2 mol H.

We have obtained the number of moles of C and the number of moles of H in the original sample, now we need to calculate the moles of O. Since the compound contains only C, H, and O, if we convert moles of C, and moles H, to grams, we can subtract these masses from the original mass of the sample, to give grams of O, which we can then convert to moles of O

$$\text{grams of C} = (1.00 \times 10^{-1}\text{ mol})(12.01\text{ g mol}^{-1}) = \underline{1.201\text{ g C}}$$

$$\text{grams of H} = (3.00 \times 10^{-2}\text{ mol})(1.008\text{ g mol}^{-1}) = \underline{0.302\text{ g H}}$$

Thus, mass of O in sample = (3.10-1.20-0.30) = <u>1.60 g O</u>

$$\text{moles of O} = (1.60\text{ g O})\left(\frac{1\text{ mol O}}{16.00\text{ g O}}\right) = \underline{1.00 \times 10^{-1}\text{ mol O}}$$

i.e., mol C : mol H : mol O = 0.100 : 0.300 : 0.100 = 1 : 3 : 1 = atom ratio

The <u>empirical formula</u> is $\underline{CH_3O}$ (empirical formula mass = 31.03 u)

Since the empirical formula mass is one-half the molecular mass

<u>Molecular formula</u> = $\underline{C_2H_6O_2}$

49. The method is similar to that for Problem 48. First calculate moles,

$$\text{Mol C} = (5.19\text{ g }CO_2)\left(\frac{1\text{ mol }CO_2}{44.01\text{ g }CO_2}\right)\left(\frac{1\text{ mol C}}{1\text{ mol }CO_2}\right) = \underline{0.118\text{ mol C}}$$

$$\text{Mol H} = (2.83\text{ g }H_2O)\left(\frac{1\text{ mol }H_2O}{18.02\text{ g }H_2O}\right)\left(\frac{2\text{ mol H}}{1\text{ mol }H_2O}\right) = \underline{0.314\text{ mol H}}$$

Then convert mol of C and H to grams of C and H,

$$\text{Mass of C in sample} = (0.118\text{ mol})\left(\frac{12.01\text{ g C}}{1\text{ mol C}}\right) = \underline{1.42\text{ g C}}$$

$$\text{Mass of H in sample} = (0.314\text{ mol})\left(\frac{1.008\text{ g H}}{1\text{ mol H}}\right) = \underline{0.317\text{ g H}}$$

and since the compound contains C, H, and O, only, by difference

$$\text{Mass of O} = (3.62-1.42-0.32) = 1.88\text{ g} = (1.88\text{ g O})\left(\frac{1\text{ mol O}}{16.00\text{ g O}}\right)$$

$$= \underline{0.118\text{ mol O}}$$

$$\text{Ratio of moles C:H:O} = 0.118 : 0.314 : 0.118 = \text{ratio of atoms}$$

$$= \frac{0.118}{0.118} : \frac{0.314}{0.118} : \frac{0.118}{0.118} = 1.00 : 2.66 : 1.00$$

i.e., 3.0 : 8.0 : 3.0

<u>Empirical formula</u> = $\underline{C_3H_8O_3}$

50. The method is similar to that for Problem 48.

$$\text{Mol C} = (26.4\text{ g }CO_2)\left(\frac{1\text{ mol }CO_2}{44.01\text{ g }CO_2}\right)\left(\frac{1\text{ mol C}}{1\text{ mol }CO_2}\right) = \underline{0.600\text{ mol C}}$$

$$\text{Mol H} = (6.30\text{ g }H_2O)\left(\frac{1\text{ mol }H_2O}{18.02\text{ g }H_2O}\right)\left(\frac{2\text{ mol H}}{1\text{ mol }H_2O}\right) = \underline{0.700\text{ mol H}}$$

$$\text{Mol N} = (4.60\text{ g }NO_2)\left(\frac{1\text{ mol }NO_2}{46.01\text{ g }NO_2}\right)\left(\frac{1\text{ mol N}}{1\text{ mol }NO_2}\right) = \underline{0.100\text{ mol N}}$$

$$\text{Ratio of mol} = \text{ratio of atoms, C : H : N} = 0.600 : 0.700 : 0.100$$

$$= \underline{6 : 7 : 1}$$

<u>Empirical formula</u> is $\underline{C_6H_7N}$

51. $$\text{Mol C} = (3.14\text{ g }CO_2)\left(\frac{1\text{ mol }CO_2}{44.01\text{ g }CO_2}\right)\left(\frac{1\text{ mol C}}{1\text{ mol }CO_2}\right) = \underline{0.0713\text{ mol C}}$$

$$\text{Mol H} = (1.29\text{ g }H_2O)\left(\frac{1\text{ mol }H_2O}{18.02\text{ g }H_2O}\right)\left(\frac{2\text{ mol H}}{1\text{ mol }H_2O}\right) = \underline{0.143\text{ mol H}}$$

Ratio of mol (atoms), C : H = 0.0713 : 0.143 = <u>1.00 : 2.01</u>

<u>Empirical Formula</u> = $\underline{CH_2}$

52. This is similar to Problem 48.

Assume that all of the carbon in the sample reacts to give $CCl_4$, and all of the hydrogen reacts to give HCl (molar masses: $CCl_4$, 153.8 g $mol^{-1}$; HCl, 36.46 g $mol^{-1}$)

$$\text{Mol C} = (30.8 \text{ g } CCl_4)\left(\frac{1 \text{ mol } CCl_4}{153.8 \text{ g } CCl_4}\right)\left(\frac{1 \text{ mol C}}{1 \text{ mol } CCl_4}\right) = \underline{0.200 \text{ mol C}}$$

$$\text{Mol H} = (21.9 \text{ g HCl})\left(\frac{1 \text{ mol HCl}}{36.46 \text{ g HCl}}\right)\left(\frac{1 \text{ mol H}}{1 \text{ mol HCl}}\right) = \underline{0.600 \text{ mol H}}$$

So that 6.20 g of the compound contains

$(0.200 \text{ mol C})(12.01 \text{ g mol}^{-1}) = \underline{2.40 \text{ g C}}$, and

$(0.600 \text{ mol H})(1.008 \text{ g mol}^{-1}) = \underline{0.605 \text{ g H}}$

Thus, the mass of sulfur in the sample (the only other component) is

$$(6.20-2.40-0.60) = \underline{3.20 \text{ g S}} = (3.20 \text{ g S})\left(\frac{1 \text{ mol S}}{32.06 \text{ g S}}\right) = \underline{0.100 \text{ mol S}}$$

Ratio of moles (atoms) C : H : S = 0.200 : 0.600 : 0.100 = $\underline{2:6:1}$

Empirical formula is $\underline{C_2H_6S}$

53.

| Oxide | Mass % V | Mass % O | Mol V in 100 g | Mol O in 100 g | Mol V:Mol O |
|---|---|---|---|---|---|
| A | 76.10 | 23.90 | $\frac{76.10}{50.94} = 1.494$ | $\frac{23.90}{16.00} = 1.493$ | 1.0 : 1.0 |
| B | 67.98 | 32.02 | $\frac{67.96}{50.94} = 1.335$ | $\frac{32.02}{16.00} = 2.001$ | 1.0 : 1.5 |
| C | 61.42 | 38.58 | $\frac{61.42}{50.94} = 1.206$ | $\frac{38.58}{16.00} = 2.411$ | 1.0 : 2.0 |
| D | 56.02 | 43.98 | $\frac{56.02}{50.94} = 1.100$ | $\frac{43.98}{16.00} = 2.749$ | 1.0 : 2.5 |

Empirical formulas: A, VO ; B, $V_2O_3$; C, $VO_2$ ; D, $V_2O_5$

54. The oxide contains 3.04 g N, and (9.99-3.04) = 6.95 g O

$$\text{Mol (atom) ratio, N : O} = (3.04 \text{ g N})\left(\frac{1 \text{ mol N}}{14.01 \text{ g N}}\right) : (6.95 \text{ g O})\left(\frac{1 \text{ mol O}}{16.00 \text{ g O}}\right)$$

$$= 1.00 : 2.00$$

Empirical formula = $NO_2$ (empirical formula mass = 46.01 u)

Molecular formula = $N_2O_4$ (molecular mass = 92.02 u)

55.

| | | C | H | N | |
|---|---|---|---|---|---|
| 100 g nicotine contains | | 74.0 | 8.7 | 17.3 | g |
| | = | $\frac{74.0}{12.01}$ | $\frac{8.7}{1.008}$ | $\frac{17.3}{14.01}$ | mol |
| Mol (atom) ratio | = | 6.16 : | 8.6 : | 1.23 | |

Mol (atom) ratio $= \frac{6.16}{1.23} : \frac{8.6}{1.23} : \frac{1.23}{1.23} = \underline{5 : 7 : 1}$

Empirical formula of nicotine = $C_5H_7N$ (formula mass = 81.12 u)

Molecular formula = $C_{10}H_{14}N_2$ (molar mass 162.2 g $mol^{-1}$)

56. Both samples contain the same amount of copper, = 2.542 g Cu

i.e., 3.182 g black copper oxide contains 2.542 g Cu and 0.640 g O,

2.862 g red copper oxide contains 2.542 g Cu and 0.320 g O

| | Red oxide | | Black oxide | |
|---|---|---|---|---|
| | Cu | O | Cu | O |
| grams | 2.542 | 0.320 | 2.542 | 0.640 |
| mol | $\frac{2.542}{63.55}$ | $\frac{0.320}{16.00}$ | $\frac{2.542}{63.55}$ | $\frac{0.640}{16.00}$ |
| = | 0.040 | 0.020 | 0.040 | 0.040 |
| Mol (atom) ratio, Cu:O | 2.0:1.0 | | 1.0:1.0 | |
| The empirical formulas are: | $Cu_2O$ | | CuO | |

57. On heating, n moles of $Li_2SO_4.xH_2O$ gives n mol $Li_2SO_4$ + xn mol $H_2O$

| | $Li_2SO_4.xH_2O(s) \longrightarrow$ | $Li_2SO_4(s)$ | + $xH_2O(l)$ |
|---|---|---|---|
| initially | 3.25 g | 0 | 0 |
| after heating | 0 | 2.80 g | 0.45 g |

and molar masses are: $Li_2SO_4$, 109.9 g $mol^{-1}$ and $H_2O$, 18.02 g $mol^{-1}$

i.e., mol $Li_2SO_4$ = $(2.80\text{ g } Li_2SO_4)(\frac{1\text{ mol } Li_2SO_4}{109.9\text{ g } Li_2SO_4}) = \underline{2.55\times10^{-2}\text{ mol}}$

mol $H_2O$ = $(0.45\text{ g } H_2O)(\frac{1\text{ mol } H_2O}{18.02\text{ g } H_2O}) = \underline{2.5\times10^{-2}\text{ mol}}$

and since the ratio mol $Li_2SO_4$ : mol $H_2O$ = 1 : 1, then x = 1

The empirical formula of hydrated lithium sulfate is $Li_2SO_4.H_2O$

58. This problem is similar to Problem 57.

Let the formula of hydrated copper sulfate be $CuSO_4.xH_2O$,

| | $CuSO_4.xH_2O(s) \longrightarrow$ | $CuSO_4(s)$ | + $H_2O(l)$ |
|---|---|---|---|
| initially | 0.2800 g | 0 | 0 |
| after heating | 0 | 0.1789 g | 0.1011 g |

and molar masses are: $CuSO_4$, 159.6 g $mol^{-1}$ and $H_2O$, 18.02 g $mol^{-1}$

$$\text{mol CuSO}_4 = (0.1789\text{ g CuSO}_4)\left(\frac{1\text{ mol CuSO}_4}{159.6\text{ g CuSO}_4}\right) = \underline{1.121\times10^{-3}\text{ mol}}$$

$$\text{mol H}_2\text{O} = (0.1011\text{ g H}_2\text{O})\left(\frac{1\text{ mol H}_2\text{O}}{18.02\text{ g H}_2\text{O}}\right) = \underline{5.610\times10^{-3}\text{ mol}}$$

and the mol ratio $CuSO_4 : H_2O = 1.121 : 5.610 = \underline{1.00 : 5.00}$

The empirical formula of hydrated copper sulfate is $\underline{CuSO_4.5H_2O}$

59. For the empirical formula $C_{12}H_mCl_{10-m}$, the empirical formula mass is $[12(12.01) + m(1.008) + (10-m)(35.45)] = (498.6 - 34.44m)$ u

i.e., $\text{mass \% Cl} = \dfrac{[35.45(10-m)]\text{ u}}{(498.6-34.44m)\text{ u}} \times 100\% = 58.9\%$

Solving for m gives $\underline{m = 4.01}$.

Thus, the empirical formula of the PCB is $\underline{C_{12}H_4Cl_6}$.

60. (a) Unbalanced as written, $2SO_2 + H_2O + O_2 \longrightarrow 2H_2SO_4$

Only the S atoms are balanced, however, if we balance the H atoms by adding one $H_2O$ molecule to the left-hand side, both sides are balanced:

$$2SO_2 + 2H_2O + O_2 \longrightarrow 2H_2SO_4 \quad \underline{\text{balanced}}$$

(b) Unbalanced as written, $CH_3OH + 2O_2 \longrightarrow CO_2 + 2H_2O$

C atoms and H atoms are balanced, and we could balance the O atoms by reducing the number of O atoms on the LHS to 4, to give

$$CH_3OH + \tfrac{3}{2}O_2 \longrightarrow CO_2 + 2H_2O$$

Since we generally do not write fractional numbers of molecules, this can now be multiplied throughout by 2, to give

$$2CH_3OH + 3O_2 \longrightarrow 2CO_2 + 4H_2O \quad \underline{\text{balanced}}$$

(c) Unbalanced as written, $H_2O_2 \longrightarrow H_2O + O_2$

H atoms are balanced, but not the O atoms, which could be balanced by writing,

$$H_2O_2 \longrightarrow H_2O + \tfrac{1}{2}O_2$$

Now multiply throughout by 2 to obtain integral coefficients

$$2H_2O_2 \longrightarrow 2H_2O + 2O_2 \quad \underline{\text{balanced}}$$

(d) This equation is already balanced.

(e) Unbalanced as written, $Zn + HCl \longrightarrow ZnCl_2 + H_2$

Zn atoms are balanced. Balance Cl atoms by adding one HCl to the LHS, which also balances the H atoms,

$$Zn + 2HCl \longrightarrow ZnCl_2 + H_2 \quad \underline{\text{balanced}}$$

61. (a) Al atoms are balanced, add 2HCl to LHS to balance Cl atoms on RHS, to give

$$Al + 3HCl \longrightarrow AlCl_3 + \frac{3}{2}H_2$$

and then multiply throughout by 2

$$2Al + 6HCl \longrightarrow 2AlCl_3 + 3H_2 \quad \underline{\text{balanced}}$$

(b) Balance C atoms first and then H atoms, to give

$$C_5H_{12} + O_2 \longrightarrow 5CO + 6H_2O$$

then balance O atoms

$$C_5H_{12} + \frac{11}{2}O_2 \longrightarrow 5CO + 6H_2O$$

and finally multiply throughout by 2,

$$2C_5H_{12} + 11O_2 \longrightarrow 10CO + 12H_2O \quad \underline{\text{balanced}}$$

(c) Balance C atoms first, then O atoms, and finally H atoms,

$$C_3H_8 + H_2O \longrightarrow 3CO + H_2$$

$$C_3H_8 + 3H_2O \longrightarrow 3CO + 7H_2 \quad \underline{\text{balanced}}$$

(d) Balance Na atoms and then Cl atoms,

$$Na_2CO_3 + HCl \longrightarrow 2NaCl + H_2O + CO_2$$

$$Na_2CO_3 + 2HCl \longrightarrow 2NaCl + H_2O + CO_2 \quad \underline{\text{balanced}}$$

(e) Balance Al and O atoms on LHS, and then multiply throughout by 2,

$$2Al + \frac{3}{2}O_2 \longrightarrow Al_2O_3$$

$$4Al + 3O_2 \longrightarrow 2Al_2O_3 \quad \underline{\text{balanced}}$$

(f) Balance $SO_4$ on RHS by adding $2H_2SO_4$ to LHS,

$$Al_2O_3 + 3H_2SO_4 \longrightarrow Al_2(SO_4)_3 + H_2O$$

Then balance RHS by adding $2H_2O$ to balance H atoms, which also balances O atoms,

$$Al_2O_3 + 3H_2SO_4 \longrightarrow Al_2(SO_4)_3 + 3H_2O \quad \underline{\text{balanced}}$$

62. (a) S atoms are balanced. Balance O atoms and then multiply throughout by 2,

$$S + \frac{3}{2}O_2 \longrightarrow SO_3$$

$$2S + 3O_2 \longrightarrow 2SO_3 \quad \underline{\text{balanced}}$$

(b) Balance C atoms and then O atoms, and then multiply by 2,

$$C_2H_2 + O_2 \longrightarrow 2CO + H_2O$$

$$C_2H_2 + \frac{3}{2}O_2 \longrightarrow 2CO + H_2O$$

$$2C_2H_2 + 3O_2 \longrightarrow 4CO + 2H_2O \quad \underline{\text{balanced}}$$

(c) Balance Na atoms,

$$Na_2CO_3 + Ca(OH)_2 \longrightarrow 2NaOH + CaCO_3 \quad \underline{\text{balanced}}$$

(d) Na and S atoms are balanced and the four O atoms on the LHS must appear in $H_2O$ on the RHS. Add $3H_2O$ to the RHS, and then balance the H atoms,

$$Na_2SO_4 + H_2 \longrightarrow Na_2S + 4H_2O$$

$$Na_2SO_4 + 4H_2 \longrightarrow Na_2S + 4H_2O \quad \underline{\text{balanced}}$$

(e) Only the O atoms are unbalanced. Balance them by writing

$$Cu_2S + \frac{3}{2}O_2 \longrightarrow Cu_2O + SO_2$$

and then multiply throughout by 2,

$$2Cu_2S + 3O_2 \longrightarrow 2Cu_2O + 2SO_2 \quad \underline{\text{balanced}}$$

(f) The only source of O atoms is $Cu_2O$. Add one $Cu_2O$ to balance O atoms, and then balance the Cu atoms,

$$2Cu_2O + Cu_2S \longrightarrow Cu + SO_2$$

$$2Cu_2O + Cu_2S \longrightarrow 6Cu + SO_2 \quad \underline{\text{balanced}}$$

63. (a) Na and S atoms are balanced. Balance the O atoms first, and then the C atoms,

$$Na_2SO_4(s) + C(s) \longrightarrow Na_2S(s) + 2CO_2(g)$$

$$Na_2SO_4(s) + 2C(s) \longrightarrow Na_2S(s) + 2CO_2(g) \quad \underline{\text{balanced}}$$

(b) This equation is already balanced.

(c) Balance Cl atoms and then O atoms,

$$PCl_3(l) + H_2O(l) \longrightarrow H_3PO_3(aq) + 3HCl(aq)$$

$$PCl_3(l) + 3H_2O(l) \longrightarrow H_3PO_3(aq) + 3HCl(aq) \quad \underline{\text{balanced}}$$

(d) Balance H atoms and then N atoms,

$$NO_2(g) + H_2O(l) \longrightarrow 2HNO_3(aq) + NO(g)$$

$$3NO_2(g) + H_2O(l) \longrightarrow 2HNO_3(aq) + NO(g) \quad \underline{\text{balanced}}$$

(e) Balance P atoms and then H atoms,

$$P_4O_{10}(s) + H_2O(l) \longrightarrow 4H_3PO_4(aq)$$

$$P_4O_{10}(s) + 6H_2O(l) \longrightarrow 4H_3PO_4(aq) \quad \underline{\text{balanced}}$$

(f) Balance the Na atoms,

$$Na_2O(s) + H_2O(l) \longrightarrow 2NaOH(aq) \quad \underline{\text{balanced}}$$

64. (a) $S + O_2 \longrightarrow SO_3$, unbalanced; $2S + 3O_2 \longrightarrow 2SO_3$ balanced

(b) $CH_4 + O_2 \longrightarrow CO + H_2O$, unbalanced;

$2CH_4 + 3O_2 \longrightarrow 2CO + 4H_2O$ balanced

(c) $Mg + H_2O \longrightarrow MgO + H_2$ balanced

(d) $C_4H_{10} + 4H_2O \longrightarrow 4CO + 9H_2$ balanced

65. (a) $P_4(s) + 5O_2(g) \longrightarrow P_4O_{10}(s)$ balanced

(b) $Na(s) + H_2O(l) \longrightarrow NaOH(aq) + H_2(g)$ unbalanced

$2Na(s) + 2H_2O(l) \longrightarrow 2NaOH(aq) + H_2(g)$ balanced

(c) $NH_4NO_3(s) \longrightarrow N_2O(g) + H_2O(g)$ unbalanced

$NH_4NO_3(s) \longrightarrow N_2O(g) + 2H_2O(g)$ balanced

(d) $Pb(NO_3)_2(s) \longrightarrow PbO(s) + NO_2(g) + O_2(g)$ unbalanced

$Pb(NO_3)_2(s) \longrightarrow PbO(s) + 2NO_2(g) + \frac{1}{2}O_2(g)$

$2Pb(NO_3)_2 \longrightarrow 2PbO(s) + 4NO_2(g) + O_2(g)$ balanced

66. First write the balanced equation, $2Mg(s) + O_2(g) \longrightarrow 2MgO(s)$.

From the given mass of MgO, calculate the number of moles of MgO, then the moles og Mg, and finally the mass of Mg.

$$\text{Mol MgO} = (1.000 \text{ g MgO})\left(\frac{1 \text{ mol MgO}}{40.31 \text{ g MgO}}\right); \text{ Mol Mg} = (\text{mol MgO})\left(\frac{2 \text{ mol Mg}}{2 \text{ mol MgO}}\right)$$

$$\text{i.e., Mass Mg} = (1.000 \text{ g MgO})\left(\frac{1 \text{ mol MgO}}{40.31 \text{ g MgO}}\right)\left(\frac{2 \text{ mol Mg}}{2 \text{ mol MgO}}\right)\left(\frac{24.31 \text{ g Mg}}{1 \text{ mol Mg}}\right)$$

$$= \underline{0.603 \text{ g Mg}}$$

Normally this type of calculation is done in one step, as is shown in the final calculation above.

67. 100 g cesium chloride contains 78.94 g Cs and 21.06 g Cl

$$\text{mol Cs : mol Cl} = \frac{78.94}{132.9} : \frac{21.06}{35.45}$$

$$= 0.594 : 0.594$$

$$\text{ratio of moles (atoms)} = 1.00 : 1.00$$

Empirical formula of cesium chloride is CsCl, and the balanced equation for the reaction of Cs(s) with $Cl_2(g)$ is

$$2Cs(s) + Cl_2(g) \longrightarrow 2CsCl(s)$$

$$\text{Mass Cs} = (4.34 \text{ g CsCl})\left(\frac{1 \text{ mol Cs}}{168.4 \text{ g CsCl}}\right)\left(\frac{2 \text{ mol Cs}}{2 \text{ mol CsCl}}\right)\left(\frac{132.9 \text{ g Cs}}{1 \text{ mol Cs}}\right)$$

$$= \underline{3.43 \text{ g Cs}}$$

68. First write the balanced equation for the reaction

$$Xe(g) + 2F_2(g) \longrightarrow XeF_4(s)$$

$$\text{Mass of } F_2 = (2.50 \text{ g Xe})\left(\frac{1 \text{ mol Xe}}{131.3 \text{ g Xe}}\right)\left(\frac{2 \text{ mol } F_2}{1 \text{ mol Xe}}\right)\left(\frac{38.00 \text{ g } F_2}{1 \text{ mol } F_2}\right)$$

$$= \underline{1.45 \text{ g } F_2}$$

$$\text{Mass of } XeF_4 = (2.50 \text{ g Xe})\left(\frac{1 \text{ mol Xe}}{131.3 \text{ g Xe}}\right)\left(\frac{1 \text{ mol } XeF_4}{1 \text{ mol Xe}}\right)\left(\frac{207.3 \text{ g } XeF_4}{1 \text{ mol } XeF_4}\right)$$

$= \underline{3.95 \text{ g } XeF_4}$ (or $1.45 \text{ g } F_2 + 2.50 \text{ g Xe} \longrightarrow 3.95 \text{ g } XeF_4$)

69. The balanced equation is $NH_3(g) + HCl(g) \longrightarrow NH_4Cl(s)$

$$\text{Mass HCl} = (0.200 \text{ g } NH_3)\left(\frac{1 \text{ mol } NH_3}{17.03 \text{ g } NH_3}\right)\left(\frac{1 \text{ mol HCl}}{1 \text{ mol } NH_3}\right)\left(\frac{36.46 \text{ g HCl}}{1 \text{ mol HCl}}\right)$$

$= \underline{0.428 \text{ g HCl}}$

70. The balanced equation is $2P(s) + 3Cl_2(g) \longrightarrow 2PCl_3(l)$

$$\text{Mass of P} = (100 \text{ g } PCl_3)\left(\frac{1 \text{ mol } PCl_3}{137.3 \text{ g } PCl_3}\right)\left(\frac{2 \text{ mol P}}{2 \text{ mol } PCl_3}\right)\left(\frac{30.97 \text{ g P}}{1 \text{ mol P}}\right)$$

$= \underline{22.6 \text{ g}}$

71. The balanced equation is $2C_6H_6 + 15O_2 \longrightarrow 12CO_2 + 6H_2O$

$$\text{Mass of } CO_2 = (0.434 \text{ g } C_6H_6)\left(\frac{1 \text{ mol } C_6H_6}{78.11 \text{ g } C_6H_6}\right)\left(\frac{12 \text{ mol } CO_2}{2 \text{ mol } C_6H_6}\right)\left(\frac{44.01 \text{ g } CO_2}{1 \text{ mol } CO_2}\right)$$

$$\text{Mass of } H_2O = (0.434 \text{ g } C_6H_6)\left(\frac{1 \text{ mol } C_6H_6}{78.11 \text{ g } C_6H_6}\right)\left(\frac{6 \text{ mol } H_2O}{2 \text{ mol } C_6H_6}\right)\left(\frac{18.02 \text{ g } H_2O}{1 \text{ mol } H_2O}\right)$$

Mass of $CO_2$ = $\underline{1.47 \text{ g}}$ ; mass of $H_2O$ = $\underline{0.300 \text{ g}}$

72. The unbalanced equation is $Ca(OH)_2 + H_3PO_4 \longrightarrow Ca_3(PO_4)_2 + H_2O$

Balancing Ca atoms and then P atoms gives

$3Ca(OH)_2 + 2H_3PO_4 \longrightarrow Ca_3(PO_4)_2 + 6H_2O$ balanced

$$\text{Mass of } Ca(OH)_2 = (30.0 \text{ g } H_3PO_4)\left(\frac{1 \text{ mol } H_3PO_4}{97.99 \text{ g } H_3PO_4}\right)\left(\frac{3 \text{ mol } Ca(OH)_2}{1 \text{ mol } H_3PO_4}\right)$$

$$\times \left(\frac{74.10 \text{ g } Ca(OH)_2}{1 \text{ mol } Ca(OH)_2}\right) = \underline{34.0 \text{ g}}$$

$$\text{Mass of } Ca_3(PO_4)_2 = (30.0 \text{ g } H_3PO_4)\left(\frac{1 \text{ mol } H_3PO_4}{97.99 \text{ g } H_3PO_4}\right)\left(\frac{1 \text{ mol } Ca_3(PO_4)_2}{2 \text{ mol } H_3PO_4}\right)$$

$$\times \left(\frac{310.2 \text{ g } Ca_3(PO_4)_2}{1 \text{ mol } Ca_3(PO_4)_2}\right) = \underline{47.5 \text{ g}}$$

73. The unbalanced equation is $SiO_2 + Cl_2 + C \longrightarrow SiCl_4 + CO$

Balance O atoms first, and then C atoms and finally Cl atoms, to give

$SiO_2 + 2Cl_2 + 2C \longrightarrow SiCl_4 + 2CO$ balanced

$$\text{Mass of } SiCl_4 = (15.0 \text{ g } SiO_2)\left(\frac{1 \text{ mol } SiO_2}{60.09 \text{ g } SiO_2}\right)\left(\frac{1 \text{ mol } SiCl_4}{1 \text{ mol } SiO_2}\right)\left(\frac{169.9 \text{ g } SiCl_4}{1 \text{ mol } SiCl_4}\right)$$

$= \underline{42.4 \text{ g}}$

74. The formula of the unknown oxide must contain at least one Ba atom. Write it as $BaO_x$, where x is not necessarily integral, and we can write the balanced equation

$$BaO_x(s) \longrightarrow BaO(s) + \frac{x-1}{2}\, O_2(g)$$

From the given mass of BaO, calculate the moles of BaO,

$$\text{mol BaO} = (5.00 \text{ g BaO})\left(\frac{1 \text{ mol BaO}}{153.3 \text{ g BaO}}\right) = 3.26\text{x}10^{-2} \text{ mol BaO}$$

The difference in mass of BaO and the unknown oxide (0.53 g) is the mass of $O_2(g)$ evolved in the reaction. Thus,

$$\text{mol } O_2 = (0.53 \text{ g } O_2)\left(\frac{1 \text{ mol } O_2}{32.00 \text{ g } O_2}\right) = 1.66 \text{ x } 10^{-2} \text{ mol } O_2$$

$$= (3.26\text{x}10^{-2} \text{ mol BaO})\left(\frac{(x-1) \text{ mol } O_2}{2 \text{ mol BaO}}\right)$$

$$\text{i.e., } x-1 = 2\left(\frac{1.66 \text{ x } 10^{-2}}{3.26 \text{ x } 10^{-2}}\right) = 1.02 \quad \underline{\text{or}} \quad \underline{x = 2}$$

Empirical formula of unknown oxide = $\underline{BaO_2}$ (barium peroxide)

75. $2SO_2(g) + O_2(g) + 2H_2O(1) \longrightarrow 2H_2SO_4(1)$

$$\text{Mass } O_2 = (0.320 \text{ g } SO_2)\left(\frac{1 \text{ mol } SO_2}{64.06 \text{ g } SO_2}\right)\left(\frac{1 \text{ mol } O_2}{2 \text{ mol } SO_2}\right)\left(\frac{32.00 \text{ g } O_2}{1 \text{ mol } O_2}\right)$$

$$= \underline{0.080 \text{ g } O_2}$$

$$\text{Mass } H_2O = (0.320 \text{ g } SO_2)\left(\frac{1 \text{ mol } SO_2}{64.06 \text{ g } SO_2}\right)\left(\frac{2 \text{ mol } H_2O}{2 \text{ mol } SO_2}\right)\left(\frac{18.02 \text{ g } H_2O}{1 \text{ mol } H_2O}\right)$$

$$= \underline{0.090 \text{ g } H_2O}$$

$$\text{Mass } H_2SO_4 = (0.320 \text{ g } SO_2)\left(\frac{1 \text{ mol } SO_2}{64.06 \text{ g } SO_2}\right)\left(\frac{2 \text{ mol } H_2SO_4}{2 \text{ mol } SO_2}\right)\left(\frac{98.08 \text{ g } H_2SO_4}{1 \text{ mol } H_2SO_4}\right)$$

$$= \underline{0.490 \text{ g}}$$

<u>or</u>, since mass is conserved,

mass $H_2SO_4$ = mass $SO_2$ + mass $O_2$ + mass $H_2O$ = $\underline{0.490 \text{ g}}$

76. $CaCO_3(s) \longrightarrow CaO(s) + CO_2(g)$, and first calculate the mass of $CaCO_3(s)$ in the impure limestone.

$$\text{Mass of } CaCO_3 = \left(\frac{74.2 \text{ g}}{100 \text{ g}}\right)(1.000 \text{ kg})\left(\frac{10^3 \text{ g}}{1 \text{ kg}}\right) = \underline{742 \text{ g}}$$

$$\text{Mass of } CO_2 = (742 \text{ g } CaCO_3)\left(\frac{1 \text{ mol } CaCO_3}{100.1 \text{ g } CaCO_3}\right)\left(\frac{1 \text{ mol } CO_2}{1 \text{ mol } CaCO_3}\right)\left(\frac{44.01 \text{ g } CO_2}{1 \text{ mol } CO_2}\right)$$

$$= \underline{326 \text{ g}}$$

77. (a) First balance the S atoms, and then the O atoms,

$$As_4S_6(s) + 9O_2(g) \longrightarrow As_4O_6(s) + 6SO_2(g)$$

$$As_4S_4(s) + 7O_2(g) \longrightarrow As_4O_6(s) + 4SO_2(g)$$

(b) Suppose that initially we have x mol $As_4S_6$ and y mol $As_4S_4$, then since 1 mol $As_4S_6$ gives 1 mol $As_4O_6$, and 1 mol $As_4S_4$ also gives 1 mol $As_4O_6$, we can write,

$$(x \text{ mol } As_4S_6)\left(\frac{492.0 \text{ g } As_4S_6}{1 \text{ mol } As_4S_6}\right) + (y \text{ mol } As_4S_4)\left(\frac{427.9 \text{ g } As_4S_4}{1 \text{ mol } As_4S_4}\right)$$

$= 1.000$ g (the initial mass of the two sulfides)

i.e., $492.0x + 427.9y = 1.000$ ............. (1)

and, since the mass of $As_4O_6(s)$ after combustion is 0.905 g,

$395.7(x+y) = 0.905$ ............. (2)

Solving equations (1) and (2) for y gives

$64.1y = 0.125$, or $\underline{y = 1.95 \times 10^{-3} \text{ mol}}$

Thus, mass of $As_4S_4 = (1.95 \times 10^{-3} \text{ mol } As_4S_4)\left(\frac{427.9 \text{ g } As_4S_4}{1 \text{ mol } As_4S_4}\right)$

$= \underline{0.834 \text{ g}}$

Mass % $As_4S_4 = \left(\frac{0.834 \text{ g}}{1.000 \text{ g}}\right) \times 100 = \underline{83.4 \text{ mass } \%}$

78. Balancing the O atoms and then the Mg atoms gives,

$$6Mg(s) + B_2O_3(s) \longrightarrow 3MgO(s) + Mg_3B_2(s) \quad \underline{\text{balanced}}$$

For the reaction with HCl(aq), the unbalanced equation is

$$Mg_3B_2(s) + HCl(aq) \longrightarrow B_4H_{10}(g) + MgCl_2(aq) + H_2(g)$$

Balancing B atoms, Mg atoms, Cl atoms, and H atoms, in turn gives

$$2Mg_3B_2(s) + 12HCl(aq) \longrightarrow B_4H_{10}(g) + 6MgCl_2(aq) + H_2(g)$$

$\underline{\text{balanced}}$

From the initial reaction,

Mol $Mg_3B_2 = (10.00 \text{ g } B_2O_3)\left(\frac{1 \text{ mol } B_2O_3}{69.62 \text{ g } B_2O_3}\right)\left(\frac{1 \text{ mol } Mg_3B_2}{1 \text{ mol } B_2O_3}\right) = \underline{0.1436 \text{ mol}}$

From the second reaction,

Mass of $B_4H_{10} = (0.1436 \text{ mol } Mg_3B_2)\left(\frac{1 \text{ mol } B_4H_{10}}{2 \text{ mol } Mg_3B_2}\right)\left(\frac{53.32 \text{ g } B_4H_{10}}{1 \text{ mol } B_4H_{10}}\right)$

$= \underline{3.828 \text{ g}}$

79. Let X stand for H or D, then the balanced equations are

$$2X_2O(l) \xrightarrow{\text{electrolysis}} 2X_2(g) + O_2(g)$$

$$X_2(g) + Cl_2(g) \longrightarrow 2XCl(g)$$

$$XCl(aq) + AgNO_3(aq) \longrightarrow AgCl(s) + HNO_3(aq)$$

First calculate moles of AgCl(s) formed from 25 mL of XCl(aq),

$$0.3800 \text{ g AgCl} = (0.3800 \text{ g AgCl})\left(\frac{1 \text{ mol AgCl}}{143.4 \text{ g mol}^{-1}}\right) = \underline{2.650\text{x}10^{-3} \text{ mol}}$$

Then, moles XCl in 1 L XCl(aq) is given by

$$(2.650\text{x}10^{-3} \text{ mol AgCl})\left(\frac{1 \text{ mol XCl}}{1 \text{ mol AgCl}}\right)\left(\frac{1 \text{ L}}{25 \text{ mL}}\right)\left(\frac{10^3 \text{ mL}}{1 \text{ L}}\right) = \underline{0.1060 \text{ mol XCl}}$$

and moles of $X_2O$ in the original sample is given by

$$(0.1060 \text{ mol XCl})\left(\frac{1 \text{ mol } X_2}{2 \text{ mol XCl}}\right)\left(\frac{1 \text{ mol } X_2O}{1 \text{ mol } X_2}\right) = \underline{0.0530 \text{ mol } X_2O}$$

Let moles of $H_2O$ in sample = x, and moles of $D_2O$ = y, then

$$x + y = 0.0530 \quad \ldots\ldots\ldots\ldots\ldots\ldots \quad (1)$$

and for the original mass of 1.00 g, we can write

$$(x \text{ mol } H_2O)\left(\frac{18.02 \text{ g } H_2O}{1 \text{ mol } H_2O}\right) + (y \text{ mol } D_2O)\left(\frac{20.03 \text{ g } D_2O}{1 \text{ mol } D_2O}\right) = 1.00 \text{ g}$$

or $18.02x + 20.03y = 1.00 \quad \ldots\ldots\ldots\ldots\ldots\ldots \quad (2)$

Solving equations (1) and (2) for x and y gives $\underline{x = 0.0306 \text{ mol}}$

$\underline{y = 0.0224 \text{ mol}}$

i.e., mass of $D_2O$ = $(0.022 \text{ mol})(20.03 \text{ g mol}^{-1}) = \underline{0.449 \text{ g}}$

mass of $H_2O$ = $(0.031 \text{ mol})(18.02 \text{ g mol}^{-1}) = \underline{0.551 \text{ g}}$

Mass % $D_2O$ = $\left(\frac{0.449\text{g}}{1.00 \text{ g}}\right) \times 100 = \underline{44.9\%}$

80. $M(s) + H_2SO_4(aq) \longrightarrow MSO_4(aq) + H_2(g)$

Mol $H_2$ = $(0.248 \text{ g } H_2)\left(\frac{1 \text{ mol } H_2}{2.016 \text{ g } H_2}\right) = 0.141 \text{ mol } H_2$

and from the balanced equation, total moles of metal is also 0.141 mol.

Let moles of Mg = x, then moles of Zn = (0.141-x), and the total mass of metal is

$$(x \text{ mol Mg})\left(\frac{24.31 \text{ g Mg}}{1 \text{ mol Mg}}\right) + [(0.141\text{-}x) \text{ mol Zn}]\left(\frac{65.38 \text{ g Zn}}{1 \text{ mol Zn}}\right) = 5.00 \text{ g}$$

From which x = $\underline{0.103 \text{ mol}}$.

Mass of Mg = $(0.103 \text{ mol Mg})\left(\frac{24.31 \text{ g Mg}}{1 \text{ mol Mg}}\right) = \underline{2.50 \text{ g}}$; mass of Zn = $\underline{2.50 \text{ g}}$

and the composition of the mixture is $\underline{50.0 \text{ mass\% Mg}}$ and $\underline{50 \text{ mass\% Zn}}$

$$[MgSO_4] = \left(\frac{0.103 \text{ mol } MgSO_4}{100 \text{ mL solution}}\right)\left(\frac{10^3 \text{ mL}}{1 \text{ L}}\right) = \underline{1.03 \text{ mol L}^{-1}} \quad (1.03 \text{ M}).$$

Mol of $ZnSO_4$ in 100 mL solution = (0.141-0.103) = <u>0.038 mol</u>

$$[ZnSO_4] = \left(\frac{0.038 \text{ mol } ZnSO_4}{100 \text{ mL solution}}\right)\left(\frac{10^3 \text{ mL}}{1 \text{ L}}\right) = \underline{0.38 \text{ mol L}^{-1}} \quad (0.38 \text{ M})$$

81. Write the balanced equation and then calculate the moles of $O_2$ that react with 5.6 mol $SO_2$, and, if necessary the moles of $SO_2$ that react with 4.8 mol $O_2$.

| | $2SO_2$ | + $O_2$ | + $2H_2O$ | $\longrightarrow$ $2H_2SO_4$ | |
|---|---|---|---|---|---|
| initially | 5.6 | 4.8 | excess | 0 | mol |

$$5.6 \text{ mol } SO_2 \text{ reacts with } (5.6 \text{ mol } SO_2)\left(\frac{1 \text{ mol } O_2}{2 \text{ mol } SO_2}\right) = \underline{2.8 \text{ mol } O_2}$$

so that it is apparent that $O_2$ is present in excess, and the <u>limiting reactant</u> is <u>sulfur dioxide</u>

$$\text{Mol } H_2SO_4 = (5.6 \text{ mol } SO_2)\left(\frac{2 \text{ mol } H_2SO_4}{2 \text{ mol } SO_2}\right) = \underline{5.6 \text{ mol } H_2SO_4}$$

82. First calculate the moles of $PCl_3$ and the moles of $Cl_2$,

$$\text{Mol } PCl_3 = (5.15 \text{ g } PCl_3)\left(\frac{1 \text{ mol } PCl_3}{137.3 \text{ g } PCl_3}\right) = \underline{3.75\text{x}10^{-2} \text{ mol}}$$

$$\text{Mol } Cl_2 = (3.15 \text{ g } Cl_2)\left(\frac{1 \text{ mol } Cl_2}{70.90 \text{ g } Cl_2}\right) = \underline{4.44 \text{ x } 10^{-2} \text{ mol}}$$

Then write the balanced equation for the reaction,

| | $PCl_3$ | + $Cl_2$ | $\longrightarrow$ $PCl_5$ | |
|---|---|---|---|---|
| initially | $3.75\text{x}10^{-2}$ | $4.44\text{x}10^{-2}$ | 0 | mol |

From the balanced equation, 1 mol $PCl_3$ reacts with 1 mol $Cl_2$, so that $PCl_3$ is the <u>limiting reagent</u>

$$\text{Mass of } PCl_5 \text{ formed} = (3.75\text{x}10^{-2} \text{ mol } PCl_3)\left(\frac{1 \text{ mol } PCl_5}{1 \text{ mol } PCl_3}\right)\left(\frac{208.2 \text{ g } PCl_5}{1 \text{ mol } PCl_5}\right)$$
$$= \underline{7.81 \text{ g}}$$

83. $$\text{Mol } K_2SO_4 = (6.00 \text{ g } K_2SO_4)\left(\frac{1 \text{ mol } K_2SO_4}{174.3 \text{ g } K_2SO_4}\right) = \underline{3.44\text{x}10^{-2} \text{ mol}}$$

$$\text{Mol } Ba(NO_3)_2 = (8.00 \text{ g } Ba(NO_3)_2)\left(\frac{1 \text{ mol } Ba(NO_3)_2}{261.3 \text{ g } Ba(NO_3)_2}\right) = \underline{3.06\text{x}10^{-2} \text{ mol}}$$

and the balanced equation is,

| | $K_2SO_4(aq)$ | + $Ba(NO_3)_2(aq)$ | $\longrightarrow$ $BaSO_4(s)$ | + $2KNO_3(aq)$ | |
|---|---|---|---|---|---|
| initially | $3.44\text{x } 10^{-2}$ | $3.06\text{x}10^{-2}$ | 0 | 0 | mol |

Since 1 mol $K_2SO_4(aq)$ reacts with 1 mol $Ba(NO_3)_2(aq)$, the <u>limiting reagent</u> is $Ba(NO_3)_2(aq)$.

$$\text{Mass of } BaSO_4(s) = (3.06\text{x}10^{-2} \text{ mol } Ba(NO_3)_2)\left(\frac{1 \text{ mol } BaSO_4}{1 \text{ mol } Ba(NO_3)_2}\right)\left(\frac{233.4 \text{ g}}{1 \text{ mol}}\right)$$
$$= \underline{7.14 \text{ g } BaSO_4(s)}$$

84. Balance the equation to give

$$2Al(s) + 6HCl(g) \longrightarrow 2AlCl_3(s) + 3H_2(g)$$

and then calculate moles of reactants.

$$\text{mol Al} = (2.70 \text{ g Al})\left(\frac{1 \text{ mol Al}}{26.98 \text{ g Al}}\right) = \underline{0.100 \text{ mol}}$$

$$\text{mol HCl} = (4.00 \text{ g HCl})\left(\frac{1 \text{ mol HCl}}{36.45 \text{ g HCl}}\right) = \underline{0.110 \text{ mol}}$$

From the balanced equation, 1 mol Al reacts with 3 mol HCl, so that in this example HCl is the limiting reactant and Al is in excess.

$$\text{Mass of Al reacted} = (0.110 \text{ mol HCl})\left(\frac{2 \text{ mol Al}}{6 \text{ mol HCl}}\right)\left(\frac{26.98 \text{ g Al}}{1 \text{ mol Al}}\right) = \underline{0.99 \text{ g}}$$

$$\text{Mass of } AlCl_3 \text{ formed} = (0.110 \text{ mol HCl})\left(\frac{2 \text{ mol } AlCl_3}{6 \text{ mol HCl}}\right)\left(\frac{133.3 \text{ g } AlCl_3}{1 \text{ mol } AlCl_3}\right)$$

$$= \underline{4.89 \text{ g}}$$

$$\text{Excess Al} = (2.70-0.99) \text{ g} = \underline{1.71 \text{ g}}$$

85. $$Zn(s) + CuCl_2(aq) \longrightarrow ZnCl_2(aq) + Cu(s)$$

$$\text{Initial mol Zn} = (2.00 \text{ g Zn})\left(\frac{65.38 \text{ g Zn}}{1 \text{ mol Zn}}\right) = \underline{0.0306 \text{ mol}}$$

$$\text{Initial mol } CuCl_2 = (2.00 \text{ g } CuCl_2)\left(\frac{134.5 \text{ g } CuCl_2}{1 \text{ mol } CuCl_2}\right) = \underline{0.0149 \text{ mol}}$$

From the balanced equation, 1 mol Zn reacts with 1 mol $CuCl_2$. Thus, the limiting reactant is $CuCl_2$, and

$$\text{Mass of Cu formed} = (0.0149 \text{ mol } CuCl_2)\left(\frac{1 \text{ mol Cu}}{1 \text{ mol } CuCl_2}\right)\left(\frac{63.55 \text{ g Cu}}{1 \text{ mol Cu}}\right)$$

$$= \underline{0.947 \text{ g}}$$

86. $$C(s) + 2S(s) \longrightarrow CS_2(l)$$

$$\text{Theoretical yield of } CS_2 = (2.530 \text{ g C})\left(\frac{1 \text{ mol C}}{12.01 \text{ g C}}\right)\left(\frac{1 \text{ mol } CS_2}{1 \text{ mol C}}\right)\left(\frac{76.13 \text{ g}}{1 \text{ mol } CS_2}\right)$$

$$= \underline{16.04 \text{ g } CS_2}$$

$$\% \text{ Yield} = \left(\frac{12.50 \text{ g } CS_2}{16.04 \text{ g } CS_2}\right) \times 100 = \underline{77.93\%}$$

87. $$2KNO_3(s) \xrightarrow{\text{heat}} 2KNO_2(s) + O_2(g)$$

$$\text{Initial mol } KNO_3 = (2.500 \text{ g } KNO_3)\left(\frac{1 \text{ mol } KNO_3}{101.1 \text{ g } KNO_3}\right) = \underline{0.02473 \text{ mol}}$$

Thus, the theoretical yield of $O_2(g)$ if the reaction goes to completion is

$$(0.02473 \text{ mol } KNO_3)\left(\frac{1 \text{ mol } O_2}{2 \text{ mol } KNO_3}\right)\left(\frac{32.00 \text{ g } O_2}{1 \text{ mol } O_2}\right) = \underline{0.3957 \text{ g}}$$

The actual yield of $O_2(g)$ is the same as the change of mass between reactant and solid product

$$= (2.500 - 2.210) \text{ g} = \underline{0.290 \text{ g}}$$

$$\% \text{ Yield of } O_2 = \left(\frac{0.290 \text{ g } O_2}{0.396 \text{ g } O_2}\right) \times 100 = \underline{73.2\%}$$

88. $$Mg(s) + 2HCl(aq) \longrightarrow MgCl_2(aq) + H_2(g) \xrightarrow{H_2O} MgCl_2{\cdot}6H_2O(s)$$

$$\text{Theor. yield } MgCl_2{\cdot}6H_2O = (1.000 \text{ g Mg})\left(\frac{1 \text{ mol Mg}}{24.31 \text{ g Mg}}\right)\left(\frac{1 \text{ mol } MgCl_2{\cdot}6H_2O}{1 \text{ mol Mg}}\right)$$

$$\times \left(\frac{203.3 \text{ g } MgCl_2{\cdot}6H_2O}{1 \text{ mol } MgCl_2{\cdot}6H_2O}\right) = \underline{8.363 \text{ g}}$$

$$\% \text{ Yield of } MgCl_2{\cdot}6H_2O = \left(\frac{5.62 \text{ g}}{8.363 \text{ g}}\right) \times 100 = \underline{67.2\%}$$

89. $$BaCl_2(aq) + H_2SO_4(aq) \longrightarrow BaSO_4(s) + 2HCl(aq)$$

$$\text{Initial mol } BaCl_2(aq) = (27.42 \text{ mL})\left(\frac{1 \text{ L}}{10^3 \text{ mL}}\right)\left(\frac{0.112 \text{ mol}}{1 \text{ L}}\right) = \underline{3.07\text{x}10^{-3} \text{ mol}}$$

$$\text{Initial mol } H_2SO_4(aq) = (32.30 \text{ mL})\left(\frac{1 \text{ L}}{10^3 \text{ mL}}\right)\left(\frac{0.096 \text{ mol}}{1 \text{ L}}\right) = \underline{3.10 \times 10^{-3} \text{ mol}}$$

From the balanced equation, 1 mol $BaCl_2(aq)$ reacts with 1 mol $H_2SO_4(aq)$, thus, the limiting reactant is $BaCl_2(aq)$.

$$\text{Max. mass } BaSO_4(s) = (3.07\text{x}10^{-3} \text{ mol } BaCl_2)\left(\frac{1 \text{ mol } BaSO_4}{1 \text{ mol } BaCl_2}\right)\left(\frac{233.4 \text{ g } BaSO_4}{1 \text{ mol } BaSO_4}\right)$$

$$= \underline{0.717 \text{ g } BaSO_4}$$

90. (a) $$1.00 \text{ L } 0.0100 \text{ M NaOH(aq)} = (1.00 \text{ L})\left(\frac{0.0100 \text{ mol}}{1 \text{ L}}\right) = \underline{0.0100 \text{ mol NaOH}}$$

(b) $$250 \text{ mL } 0.0100 \text{ M NaOH(aq)} = (250 \text{ mL})\left(\frac{1 \text{ L}}{10^3 \text{ mL}}\right)\left(\frac{0.0100 \text{ mol}}{1 \text{ L}}\right)$$

$$= \underline{2.50\text{x}10^{-3} \text{ mol NaOH}}$$

(c) $$25.15 \text{ mL } 0.0100 \text{ M NaOH(aq)} = (25.15 \text{ mL})\left(\frac{1 \text{ L}}{10^3 \text{ mL}}\right)\left(\frac{0.0100 \text{ mol}}{1 \text{ L}}\right)$$

$$= \underline{2.52\text{x}10^{-4} \text{ mol NaOH}}$$

(d) $$25.15 \text{ mL } 0.0134 \text{ M NaOH(aq)} = (25.15 \text{ mL})\left(\frac{1 \text{ L}}{10^3 \text{ mL}}\right)\left(\frac{0.0134 \text{ mol}}{1 \text{ L}}\right)$$

$$= \underline{3.37\text{x}10^{-4} \text{ mol NaOH}}$$

91. (a) 4.00 L 0.100 M $H_2SO_4$ = $(4.00\ L)(\frac{0.100\ mol}{1\ L})$ = $\underline{0.400\ mol}$ $H_2SO_4$

(b) 125 mL 0.100 M $H_2SO_4$ = $(125\ mL)(\frac{1\ L}{10^3\ mL})(\frac{0.100\ mol}{1\ L})$ = $\underline{0.0125\ mol}$

(c) 31.46 mL 0.100 M $H_2SO_4$ = $(31.46\ mL)(\frac{1\ L}{10^3\ mL})(\frac{0.100\ mol}{1\ L})$
= $\underline{3.15\times10^{-3}\ mol}$

(d) 31.46 mL 0.151 M $H_2SO_4$ = $(31.46\ mL)(\frac{1\ L}{10^3\ mL})(\frac{0.151\ mol}{1\ L})$
= $\underline{4.75\times10^{-3}\ mol}$

92. Molarity is mol $L^{-1}$, thus, moles of solute and the volume of the solution are needed to calculate molarity. Here the volume is given as 2.00 L, and we need to calculate the number of moles of $KMnO_4$ in 12.00 g $KMnO_4$ (molar mass 158.0 g $mol^{-1}$).

$$\text{mol } KMnO_4 = (12.00\ g\ KMnO_4)(\frac{1\ mol\ KMnO_4}{158.0\ g\ KMnO_4}) = \underline{7.595\times10^{-2}\ mol}$$

$$\text{Molarity of } KMnO_4 = \frac{mol\ KMnO_4}{\text{volume of solution}} = \frac{7.595\times10^{-2}\ mol}{2.00\ L} = \underline{0.0380\ M}$$

93. (a) 0.250 L of 0.100 M glucose contains $(0.250\ L)(\frac{0.100\ mol}{1\ L})$ glucose

$$= (0.250\ L)(\frac{0.100\ mol}{1\ L})(\frac{180.2\ g}{1\ mol}) = \underline{4.50\ g\ glucose}$$

(b) Volume of solution containing 0.0010 mol glucose

$$= (0.0010\ mol)(\frac{1\ L}{0.100\ mol}) = \underline{0.010\ L}\ (\underline{or}\ 10.0\ mL)$$

94. This problem is best approached by first calculating the number of moles of $H_2SO_4$ in 500 mL of 0.175 M $H_2SO_4$, converting this to grams of $H_2SO_4$, and then, using the density, to mL $H_2SO_4$.

$$\text{mass } H_2SO_4 = (500\ mL)(\frac{1\ L}{10^3\ mL})(\frac{0.175\ mol}{1\ L})(\frac{98.08\ g\ H_2SO_4}{1\ mol\ H_2SO_4}) = \underline{8.582\ g}$$

From the data on $H_2SO_4$, 1.000 mL $H_2SO_4$ has a mass of 1.842 g and

contains $(1.842\ g)(\frac{98\ g\ H_2SO_4}{100\ g})$ = 1.805 g $H_2SO_4$ $mL^{-1}$

Thus, for 8.582 g $H_2SO_4$, we need

$$(8.582\ g\ H_2SO_4)(\frac{1\ mL}{1.805\ g\ H_2SO_4}) = \underline{4.75\ mL}$$

95. Proceed as in Problem 94 to calculate mol $H_3PO_4$ and g of $H_3PO_4$ in the solution.

$$\text{mass } H_3PO_4 = (2.50\ L)(\frac{1.50\ mol}{1\ L})(\frac{97.99\ g\ H_3PO_4}{1\ mol\ H_3PO_4}) = 367.5\ g\ H_3PO_4$$

From the data on $H_3PO_4(aq)$, 1.000 mL $H_3PO_4$ has a mass of 1.659 g.

$$\text{grams 85\% } H_3PO_4 = (367.5\text{ g})(\frac{100\text{ g}}{85\text{ g}})(\frac{1\text{ mL}}{1.659\text{ g}}) = \underline{260\text{ mL}}$$

96. (a) To prepare a more dilute solution, dilute the more concentrated solution. Suppose the number of mL of the more concentrated solution that has to be diluted is V mL, then

$$(V\text{ mL})(\frac{0.100\text{ mol}}{1\text{ L}}) = (6.3\text{ L})(\frac{10^3\text{ mL}}{1\text{ L}})(\frac{0.003\text{ mol}}{1\text{ L}})\ ;\ \underline{V = 189\text{ mL}}$$

Thus, to prepare the 0.003 M $Ba(OH)_2(aq)$ solution, take 189 mL 0.100 M $Ba(OH)_2(aq)$ and dilute with distilled water to a total volume of 6.30 L.

(b) First calculate the molarity of the more concentrated solution, then proceed as in part (a).

Molarity of 35 mass % $Cr_2(SO_4)_3(aq)$, assuming a volume of 1 L, is

$$(\frac{35\text{ g sulfate}}{100\text{ g sol'n}})(\frac{1\text{ mol sulfate}}{392.2\text{ g sulfate}})(\frac{1.412\text{ g sol'n}}{1\text{ cm}^3\text{ sol'n}})(\frac{1\text{ cm}^3}{1\text{ mL}})(\frac{1000\text{ mL}}{1\text{ L}})$$

$$= \underline{1.260\text{ mol } Cr_2(SO_4)_3\text{ L}^{-1}}$$

$$(V\text{ mL})(\frac{1.260\text{ mol}}{1\text{ L}}) = (750\text{ mL})(\frac{0.025\text{ mol}}{1\text{ L}})$$

$$\underline{V = 14.9\text{ mL}} \text{ of 35 mass \% } Cr_2(SO_4)_3(aq)$$

Thus 14.9 mL of the 35 mass % solution would have to be diluted with distilled water to a total volume of 750 mL.

97. 250 mL 0.100 M $HNO_3$ contains

$$(250\text{ mL})(\frac{1\text{ L}}{10^3\text{ mL}})(\frac{0.100\text{ mol}}{1\text{ L}})(\frac{63.02\text{ g } HNO_3}{1\text{ mol } HNO_3}) = 1.576\text{ g } HNO_3$$

and 1 mL 69 mass % $HNO_3$ contains $(1\text{ mL})(\frac{1.41\text{ g}}{1\text{ mL}})(\frac{69\text{ g } HNO_3}{100\text{ g}})$

$$= 0.973\text{ g } HNO_3\text{ mL}^{-1}$$

$$\text{mL 69 mass\% } HNO_3 = (1.576\text{ g } HNO_3)(\frac{1\text{ mL}}{0.973\text{ g } HNO_3}) = \underline{1.62\text{ mL}}$$

98.
$$AgNO_3(aq) + HCl(aq) \longrightarrow AgCl(s) + HNO_3(aq)$$

$$\text{initial mol } AgNO_3 = (25.00\text{ mL})(\frac{1\text{ L}}{10^3\text{ mL}})(\frac{0.068\text{ mol}}{1\text{ L}}) = \underline{1.70\times10^{-3}\text{ mol}}$$

$$\text{mass of } AgCl(s) = (1.70\times10^{-3}\text{ mol } AgNO_3)(\frac{1\text{ mol AgCl}}{1\text{ mol } AgNO_3})(\frac{143.4\text{ g AgCl}}{1\text{ mol AgCl}})$$

$$= \underline{0.24\text{ g AgCl(s)}}$$

99. $$Na_2CO_3(aq) + 2HCl(aq) \longrightarrow 2NaCl(aq) + H_2O(l) + CO_2(g)$$

Initial mol $Na_2CO_3$ = $(22.6\ g)(\frac{1\ mol}{106.0\ g})$ = $\underline{0.213\ mol}$

Mol HCl required = $(0.213\ mol\ Na_2CO_3)(\frac{2\ mol\ HCl}{1\ mol\ Na_2CO_3})$ = $\underline{0.426\ mol}$

Vol of HCl = $(0.426\ mol)(\frac{1\ L}{0.250\ mol})$ = $\underline{1.70\ L}$

100. $$Mg(s) + 2HCl(aq) \longrightarrow MgCl_2(aq) + H_2(g)$$

For complete reaction,

$$\text{mol HCl} = (0.1240\ g\ Mg)(\frac{1\ mol\ Mg}{24.31\ g\ Mg})(\frac{2\ mol\ HCl}{1\ mol\ Mg}) = 1.020\times10^{-2}\ mol$$

$$\text{volume of HCl} = (1.020\times10^{-2}\ mol\ HCl)(\frac{1\ L}{0.0120\ mol})(\frac{10^3\ mL}{1\ L})$$
$$= \underline{850\ mL}$$

101. Assuming one Zn in the product, write the balanced equation in the form

$$Zn(s) + xHCl(aq) \longrightarrow ZnCl_x(aq) + \frac{x}{2} H_2(g)$$

and for the combustion of hydrogen,

$$\frac{x}{2} H_2(g) + \frac{x}{4} O_2(g) \longrightarrow \frac{x}{2} H_2O(l)$$

(a) $$\text{mol Zn reacted} = (0.1573\ g)(\frac{1\ mol}{65.38\ g}) = 2.406\times10^{-3}\ mol\ Zn$$

$$\text{mol } H_2 \text{ produced} = (0.0434\ g\ H_2O)(\frac{1\ mol\ H_2O}{18.02\ g\ H_2O})(\frac{1\ mol\ H_2}{1\ mol\ H_2O})$$
$$= 2.40\times10^{-3}\ mol\ H_2$$

i.e., mol Zn reacted = mol $H_2$ produced, which gives x = 2 in the balanced equation, or

$Zn(s) + 2HCl(aq) \longrightarrow ZnCl_2(aq) + H_2(g)$ <u>balanced equation</u>

(b) $$\text{Initial mol HCl} = (25.00\ mL)(\frac{1\ L}{10^3\ mL})(\frac{0.300\ mol}{1\ L}) = 7.50\times10^{-3}\ mol$$

Mol HCl that reacted with $2.406\times10^{-3}$ mol Zn

$$= (2.406\times10^{-3}\ mol\ Zn)(\frac{2\ mol\ HCl}{1\ mol\ Zn}) = 4.812\times10^{-3}\ mol$$

Excess mol HCl in final solution = $(7.50-4.81)\times10^{-3}$ mol
$= 2.69\times10^{-3}$ mol

Concentration of unreacted HCl = $(\frac{2.69\times10^{-3}\ mol}{25\ mL})(\frac{10^3\ mL}{1\ L})$
$= \underline{0.108\ M}$

$$[ZnCl_2,aq] = (\frac{2.406\times10^{-3}\ mol\ Zn}{25.00\ mL})(\frac{1\ mol\ ZnCl_2}{1\ mol\ Zn})(\frac{10^3\ mL}{1\ L}) = \underline{0.096\ M}$$

102. (a) $UF_5 + H_2O \longrightarrow UO_2F_2 + UF_4 + HF$ unbalanced

Note that O atoms occur only in $UO_2F_2$ and $H_2O$, so balance O atoms first,

$$UF_5 + 2H_2O \longrightarrow UO_2F_2 + UF_4 + HF$$

and then H atoms,

$$UF_5 + 2H_2O \longrightarrow UO_2F_2 + UF_4 + 4HF$$

and finally U atoms,

$$2UF_5 + 2H_2O \longrightarrow UO_2F_2 + UF_4 + 4HF \quad \underline{\text{balanced}}$$

(b) Maximum mass $UF_4 = (10.00 \text{ g } UF_5)(\frac{1 \text{ mol } UF_5}{333.0 \text{ g } UF_5})(\frac{1 \text{ mol } UF_4}{2 \text{ mol } UF_5})$

$$\times (\frac{314.0 \text{ g } UF_4}{1 \text{ mol } UF_4}) = \underline{4.71 \text{ g}}$$

103. (a) $2\,^2_1H \longrightarrow \,^3_2He + \,^1_on$

The masses are: $^2_1H = 2.014\ 10$ u, $^3_2He = 3.016\ 03$ u, $^1_on = 1.008\ 66$ u

i.e., mass loss $= 2(2.014\ 10)-(3.016\ 03+1.008\ 66)$ u

$= \underline{3.51\text{x}10^{-3} \text{ u}}$

Mass loss per gram $^2_1H = (3.51\text{x}10^{-3} \text{ u})[\frac{1.00 \text{ g}}{2(2.014\ 10 \text{ u})}] = \underline{8.71\text{x}10^{-4} \text{ g}}$

(b) The energy equivalent of mass is given by $E = mc^2$, where E is in joules, J, m is in kg, and c is the velocity of light, ( $c = 2.998\text{x}10^8 \text{ m s}^{-1}$),

$$E = (8.71\text{x}10^4 \text{ g})(\frac{1 \text{ kg}}{10^3 \text{ g}})(2.998\text{x}10^8 \text{ m s}^{-1})^2(\frac{1 \text{ J}}{1 \text{ kg m}^2 \text{ s}^{-2}})$$

$= \underline{7.83\text{x}10^{10} \text{ J}}$ or $\underline{7.83\text{x}10^7 \text{ kJ}}$

(Note that $1 \text{ J} = 1 \text{ kg m}^2 \text{ s}^{-2}$)

(c) $2H_2O(l) \longrightarrow 2H_2(g) + O_2(g)$

1 mol $H_2O$ contains $(1 \text{ mol } H_2O)(\frac{1 \text{ mol } H_2}{1 \text{ mol } H_2O})(\frac{2.016 \text{ g } H_2}{1 \text{ mol } H_2})(\frac{0.015 \text{ g } D_2}{100 \text{ g } H_2})$

$= \underline{3.02\text{x}10^{-4} \text{ g } D_2}$

For 1.00 g $D_2$, we need

$$(1.00 \text{ g } D_2)(\frac{1 \text{ mol } H_2O}{3.02\text{x}10^{-4} \text{ g } D_2})(\frac{18.016 \text{ g } H_2O}{1 \text{ mol } H_2O})(\frac{1 \text{ kg}}{10^3 \text{ g}}) = \underline{59.7 \text{ kg}}$$

Minimum mass of water = 59.7 kg

# CHAPTER 3

1. (a) The three most abundant gases in the atmosphere are nitrogen, $N_2$, oxygen, $O_2$, and argon, Ar.

   (b) $N_2$ and Ar are inert, but $O_2$ is chemically very reactive. Among the minor constituents, carbon dioxide, $CO_2$, methane, $CH_4$, and hydrogen, $H_2$, are reactive gases, as well as $H_2O(g)$.

   (c) Carbon dioxide plays an important role in the photosynthesis cycle. It also has an important role in trapping heat and prevents much of the heat from the sun from reradiating into space.

2. (a) The most abundant element in the earth's crust is oxygen, both on the basis of mass percentage and atom percentage. The next most abundant element is silicon.

   (b) Oxygen and silicon are both nonmetals; aluminum is the most abundant metal.(c) Hydrogen is the most abundant element.

3. (a) Any of the noble gases, e.g., Ar and He; (b) $O_2$ and $N_2$; (c) $O_3$ ; (d) the noble gases He, Ne, or Ar; (e) Gold, Au; (f) Hydrogen in the water gas reaction

$$H_2O(g) + C(s) \longrightarrow H_2(g) + CO(g)$$

4. (a) The four most abundant elements by mass are oxygen, silicon, aluminum, and iron; by atom percentage, the four most abundant elements are oxygen, silicon, hydrogen, and aluminum.

   (b) Oxygen, silicon, and hydrogen are <u>nonmetals</u>; aluminum and iron are <u>metals</u>.

5. (a) <u>Oxidation</u> is a process by which an element or a compound combines with oxygen.

   Examples include those in Table 3.3, such as

$$2Mg(s) + O_2(g) \longrightarrow 2MgO(s)$$

$$S(s) + O_2(g) \longrightarrow SO_2(g)$$

$$C(s) + O_2(g) \longrightarrow CO_2(g)$$

   <u>Reduction</u> is a process by which oxygen is removed partially or completely from a compound. Examples include the following reductions of metal oxides.

$$Fe_2O_3(s) + 3CO(g) \longrightarrow 2Fe(s) + 3CO_2(g)$$

$$2CuO(s) + C(s) \longrightarrow 2Cu(s) + CO_2(g)$$

$$CuO(s) + H_2(g) \longrightarrow Cu(s) + H_2O(l)$$

   Note that an oxidation is always accompanied by a reduction, and vice-versa.

6. (a) $S(s) + O_2(g) \longrightarrow SO_2(g)$ ; the product is sulfur dioxide.

(b) Magnesium reacts readily with the $O_2(g)$ in air, and to some extent with the $N_2(g)$ in air:

$2Mg(s) + O_2(g) \longrightarrow 2MgO(s)$ - magnesium oxide

$3Mg(s) + N_2(g) \longrightarrow Mg_3N_2(s)$ - magnesium nitride

(c) $CH_4(g) + 2O_2(g) \longrightarrow CO_2(g) + 2H_2O(g)$ ; the products are carbon dioxide and water.

7. (a) Since $Fe_2O_3$ is reduced to Fe, the CO must be oxidized to $CO_2$:

$$Fe_2O_3(s) + 3CO(g) \longrightarrow 2Fe(s) + 3CO_2(g)$$

(b) Since $Fe_2O_3$ is reduced to Fe, the $H_2$ must be oxidized to $H_2O$:

$$Fe_2O_3(s) + 3H_2(g) \longrightarrow 2Fe(s) + 3H_2O(g)$$

(c) Since CuO is reduced to Cu, CO must be oxidized to $CO_2$:

$$CuO(s) + CO(g) \longrightarrow Cu(s) + CO_2(g)$$

(d) Mg must be oxidized to MgO, and water must be reduced to $H_2$:

$$Mg(s) + H_2O(g) \longrightarrow MgO(s) + H_2(g)$$

8. In each case steam, $H_2O(g)$, must be reduced to $H_2(g)$:

(a) $2Fe(s) + 3H_2O(g) \longrightarrow Fe_2O_3(s) + 3H_2(g)$

(b) $Mg(s) + H_2O(g) \longrightarrow MgO(s) + H_2(g)$

(c) $CH_4(g) + H_2O(g) \longrightarrow CO(g) + 3H_2(g)$

9. $N_2(g)$ is inert at ordinary temperatures but takes part in some nitrogen fixation reactions at high temperature:

(a) $3H_2(g) + N_2(g) \longrightarrow 2NH_3(g)$ - the product is ammonia gas

(b) $N_2(g) + O_2(g) \longrightarrow 2NO(g)$ - the product is nitrogen monoxide gas

This reaction takes place commonly in lightning discharges in thunderstorms and is responsible for fixing large amounts of nitrogen.

(c) Nitrogen reacts at high temperature with some metals, including Mg to give metal nitrides:

$3Mg(s) + N_2(g) \longrightarrow Mg_3N_2(s)$ - the product is magnesium nitride

10. A catalyst is a substance that increases the rate of a reaction without changing the nature of the products. Examples are the following:

(a) Manganese dioxide, $MnO_2(s)$, acting as a catalyst in the decomposition of hydrogen peroxide, $H_2O_2$, to water and oxygen gas.

$$2H_2O_2 \xrightarrow{MnO_2} 2H_2O + O_2$$

(b) In the production of hydrogen from water gas ($CO(g)$ plus $H_2(g)$), carbon monoxide is then oxidized to carbon dioxide using a catalyst at 500°C, according to the reaction:

$$\underbrace{[CO(g) + H_2(g)]}_{\text{water gas}} + H_2O(g) \xrightarrow[500^\circ C]{\text{catalyst}} 2H_2(g) + CO_2(g)$$

The $CO_2$ can then be removed by bubbling the gases through water; $CO_2$ is soluble in water while neither $H_2$ nor CO is soluble in water.

11. Neither $H_2(g)$ nor $CO(g)$ are soluble in water and there is no simple way of separating them. However, if the CO is oxidized to $CO_2$ using $H_2O(g)$ (steam) as the oxidizing agent, and a catalyst, at 500°C, the $CO_2(g)$ is easily separated from the $H_2(g)$ utilizing its solubility in water (see Problem 10).

12. (a) Hydrogen: A glowing splint or lighted match brought to the mouth of the test tube causes a mild explosion when the hydrogen reacts with the oxygen explosively to form water:

$$2H_2(g) + O_2(g) \longrightarrow 2H_2O(g)$$

(b) Oxygen: A glowing splint inserted into the mouth of the test tube will reignite and burst into flame when it comes into contact with pure $O_2$. The reaction is essentially:

$$C(s) + O_2(g) \longrightarrow CO_2(g)$$

13. (a) The unbalanced equation is:

$$N_2(g) + O_2(g) + H_2O(l) \longrightarrow HNO_3(aq)$$

Adding another $HNO_3$ molecule to the RHS balances the N atoms and the H atoms, and the O atoms are balanced by writing

$$N_2(g) + 5/2\ O_2(g) + H_2O(l) \longrightarrow 2HNO_3(aq)$$

and then the equation can be doubled throughout, to give

$$2N_2(g) + 5O_2(g) + 2H_2O(l) \longrightarrow 4HNO_3(aq)$$

(b) First we calculate moles of $N_2$ and moles of $O_2$ in 1.00 g air:

$$\text{Moles of } N_2 = \frac{0.76 \text{ g } N_2}{28.02 \text{ g mol}^{-1}} = 2.71 \times 10^{-2} \text{ mol } N_2$$

$$\text{Moles of } O_2 = \frac{0.24 \text{ g } O_2}{32.00 \text{ g mol}^{-1}} = 7.50 \times 10^{-3} \text{ mol } O_2$$

Moles $O_2$ that react completely with $2.71 \times 10^{-2}$ mol $N_2$

$$= (2.71 \times 10^{-2} \text{ mol } N_2)\left(\frac{5 \text{ mol } O_2}{2 \text{ mol } N_2}\right) = 6.78 \times 10^{-2} \text{ mol } O_2$$

so that $N_2$ is clearly in excess; $O_2$ is the limiting reactant

$$\text{Maximum moles of } HNO_3 = (7.50 \times 10^{-3} \text{ mol } O_2)\left(\frac{4 \text{ mol } HNO_3}{5 \text{ mol } O_2}\right)$$

$$= \underline{6.0 \times 10^{-3} \text{ mol}}$$

$$\text{Maximum mass } HNO_3 = (6.0\times10^{-3}\text{ mol})\left(\frac{63.02\text{ g }HNO_3}{1\text{ mol }HNO_3}\right) = \underline{0.38\text{ g}}$$

14. Hydrogen may be obtained from water by any reaction that reduces $H_2O$ to $H_2$.

(a) Using carbon (coke) as the reducing agent gives <u>water gas</u>,

$$C(s) + H_2O(g) \longrightarrow CO(g) + H_2(g)$$

(b) Reaction with a hydrocarbon such as methane gives <u>synthesis gas</u>

$$CH_4(g) + H_2O(g) \longrightarrow CO(g) + 3H_2(g)$$

(c) A number of metals may be used as the reducing agent, e.g.,

$$2Fe(s) + 3H_2O(g) \longrightarrow Fe_2O_3(s) + 3H_2(g)$$

Hydrogen may also be obtained from the electrolysis of water, in which electrons reduce the hydrogen to $H_2(g)$ and the oxygen is oxidized to $O_2(g)$,

$$2H_2O(l) \xrightarrow{\text{electric current}} 2H_2(g) + O_2(g)$$

15. (a) HCl(g) in the reaction

$$HCl(g) + NH_3(g) \longrightarrow NH_4Cl(s)$$

(b) A reactive metal such as Fe, Mg, or Al, e.g.,

$$2Fe(s) + 3H_2O(g) \longrightarrow Fe_2O_3(s) + 3H_2(g)$$

(c) Ammonia, $NH_3(g)$, is used to produce ammonium salt fertilizers,

e.g., $2NH_3(g) + H_2SO_4(aq) \longrightarrow (NH_4)_2SO_4(s)$

(d) Hydrogen burns with a light-blue flame in air,

$$2H_2(g) + O_2(g) \longrightarrow 2H_2O(g)$$

(e) Hydrogen forms an explosive mixture with oxygen,

$$2H_2(g) + O_2(g) \longrightarrow 2H_2O(g)$$

16. Ozone is formed in the upper atmosphere by the reaction of oxygen atoms formed by the photochemical decomposition of $O_2(g)$ molecules with molecular oxygen.

$$O_2(g) \xrightarrow{\text{sunlight}} O(g) + O(g)$$

$$O_2(g) + O(g) \longrightarrow O_3(g)$$

The ozone layer is extremely important to the protection of life on earth from the harmful effects of the ultraviolet radiation from the sun. The ozone layer prevents most of the UV light from the sun from reaching the surface of the earth by absorbing it. UV light reverses the reaction by which $O_3$ molecules are formed, and dissociates them into $O_2$ molecules and O atoms.

17. (a) Carbon reacts with the $O_2(g)$ in air,

$$2C(s) + O_2(g) \longrightarrow 2CO(g)$$

(b) $$2Ca(s) + O_2(g) \longrightarrow 2CaO(s)$$

(c) Potassium chlorate is $KClO_3(s)$ and $MnO_2(s)$ catalyzes the reaction and does not appear in the balanced equation,

$$2KClO_3(s) \longrightarrow 2KCl(s) + 3O_2(g)$$

(d) Complete combustion of a carbon compound in excess oxygen gives $CO_2(g)$ and $H_2O(g)$ as products,

$$C_3H_8(g) + 5O_2(g) \longrightarrow 3CO_2(g) + 4H_2O(g)$$

(e) $$C_2H_6O(l) + 3O_2(g) \longrightarrow 2CO_2(g) + 3H_2O(g)$$

18. (a) $$2CuO(s) + C(s) \longrightarrow 2Cu(s) + CO_2(g)$$

(b) $$2PbO_2(s) \longrightarrow 2PbO(s) + O_2(g)$$

(c) $$3Mg(s) + N_2(g) \longrightarrow Mg_3N_2(s)$$

(d) $$Ca(s) + H_2(g) \longrightarrow CaH_2(s)$$

(e) $$CaH_2(s) + 2H_2O(l) \longrightarrow Ca(OH)_2(s) + 2H_2(g)$$

19.

| Element | Type | State | Element | Type | State |
|---|---|---|---|---|---|
| (a) Magnesium | Metal | Solid | (f) Phosphorus | Nonmetal | Solid |
| (b) Nitrogen | Nonmetal | Gas | (g) Copper | Metal | Solid |
| (c) Oxygen | Nonmetal | Gas | (h) Hydrogen | Nonmetal | Gas |
| (d) Ozone | Nonmetal | Gas | (i) Bromine | Nonmetal | Liquid |
| (e) Sulfur | Nonmetal | Solid | | | |

20. For a given mass of gas at constant temperature apply Boyle's law,

$P_1V_1 = P_2V_2$ ; $(2.3 \text{ atm})(28 \text{ L}) = (1.0 \text{ atm})V_2$

$V_2$ = <u>64 L</u>; the air in the tire occupies <u>64 L</u> at 1.0 atmosphere.

21. Apply Boyle's law, $P_1V_1 = P_2V_2$; $(1.0 \text{ atm})(0.50 \text{ L}) = P_2(0.20 \text{ L})$

$P_2$ = 2.5 atm; the pressure in the cylinder is <u>2.5 atm</u>.

22. Apply Boyle's law, $P_1V_1 = P_2V_2$; $(1.00 \text{ atm})(150 \text{ L}) = (0.75 \text{ atm})V_2$

$V_2$ = 200 L; the volume of the balloon at a height of 2500 m = <u>200 L</u>.

23. Apply Boyle's law, $P_1V_1 = P_2V_2$, with $P_1$ = 0.98 atm, $V_1$ = 2.00 L, $V_2$ = (2.00+5.00) L = 7.00 L, and $P_2$ is unknown.

$(0.98 \text{ atm})(2.00 \text{ L}) = P_2(7.00 \text{ L})$ ; $P_2$ = <u>0.28 atm</u>

The pressure in each bulb is the same = <u>0.28 atm</u>.

24. Apply Charles's law, V/T = constant, remembering that the temperature T is the absolute temperature, i.e., in kelvins,

$$T \text{ K} = t^{\circ}\text{C} + 273.1^{\circ}\text{C}$$

In this problem $T_1 = (100 + 273) = 373$ K

and we can write Charles's Law in the form,

$$\frac{V_1}{T_1} = \frac{V_2}{T_2}$$

i.e., $$\frac{400 \text{ mL}}{373 \text{ K}} = \frac{200 \text{ mL}}{T_2}$$

$$T_2 = \frac{(200 \text{ mL})(373 \text{ K})}{400 \text{ mL}} = \underline{187 \text{ K}} \ \underline{\text{or}} \ \underline{-86^\circ\text{C}}$$

25. According to Charles's Law, $\frac{V_1}{T_1} = \frac{V_2}{T_2}$ with $T_1$ and $T_2$ in kelvins:

$T_1 = (25 + 273)$ K $$\frac{30.0 \text{ L}}{298 \text{ K}} = \frac{1.00 \text{ L}}{T_2}$$

$$T_2 = \frac{(1.00 \text{ L})(298 \text{ K})}{30.0 \text{ L}} = \underline{10 \text{ K}} \ \underline{\text{or}} \ \underline{-263^\circ\text{C}}$$

26. According to Charles's Law $\frac{V_1}{T_1} = \frac{V_2}{T_2}$ with $T_1$ and $T_2$ in kelvins:

$T_1 = (25.0 + 273.1) \text{ K} = \underline{298.1 \text{ K}}$ ; $T_2 = (-196 + 273.1) \text{ K} = \underline{77.1 \text{ K}}$

$$\frac{1.60 \text{ L}}{298.1 \text{ K}} = \frac{V_2}{77.1 \text{ K}}$$

$$V_2 = \frac{(1.60 \text{ L})(77.1 \text{ K})}{298.1 \text{ K}} = \underline{0.414 \text{ L}}$$

27. According to the combined gas law, $\frac{PV}{T}$ = const (mass of gas constant).

<u>Initially</u>, $P_1 = 1.02$ atm; $T_1 = (20 + 273) = 293$ K, and the volume is $V_1$.

<u>Finally</u>, $P_2$ is unknown, $T_2 = (200 + 273) = 473$ K, and $V_2 = V_1(\frac{110}{100})$

Thus, $$\frac{P_1V_1}{T_1} = \frac{P_2V_2}{T_2} \ ; \ \frac{(1.02 \text{ atm})(V_1)}{293 \text{ K}} = \frac{(P_2)(1.10\ V_1)}{473 \text{ K}}$$

$$P_2 = \frac{(1.02 \text{ atm})(473 \text{ K})}{1.10(293 \text{ K})} = \underline{1.50 \text{ atm}}$$

28. This problem also requires the combined gas law; STP is $0^\circ$C and a pressure of 1 atm., i.e., 273 K and 1 atm, (or 760 mm Hg):

$$\frac{P_1V_1}{T_1} = \frac{P_2V_2}{T_2} \ ; \ \frac{(850 \text{ mm Hg})(10.0 \text{ L})}{298 \text{ K}} = \frac{(760 \text{ mm Hg})(V_2)}{273 \text{ K}}$$

$$V_2 = \frac{(850 \text{ mm Hg})(10.0 \text{ L})(273 \text{ K})}{(298 \text{ K})(760 \text{ mm Hg})} = \underline{10.2 \text{ L}}$$

29. This problem also requires the combined gas law,

In Problem 22, the volume was 150 L at 1.00 atm; $T = 29^\circ$C.

i.e., initially, $V_1$ = 150 L; $P_1$ = 1.00 atm, $T_1$ = (29 + 273) = 302 K.

finally, $V_2$ is unknown; $P_2$ = 0.75 atm, $T_2$ =(-10 + 273) = 263 K.

Thus;

$$\frac{P_1V_1}{T_1} = \frac{P_2V_2}{T_2} \; ; \; \frac{(1.00 \text{ atm})(150 \text{ L})}{302 \text{ K}} = \frac{(0.75 \text{ atm})(V_2)}{263 \text{ K}}$$

$$V_2 = \frac{(1.00 \text{ atm})(150 \text{ L})(263 \text{ K})}{(0.75 \text{ atm})(302 \text{ K})} = \underline{174 \text{ L}}$$

30. This problem also requires the combined gas law.

Initially, $V_1$ = 0.840 L; $P_1$ = 0.450 atm; $T_1$ = (37 + 273) = 310 K

Finally, $V_2$ = 0.150 L; $P_2$ is unknown ; $T_2$ = (-13 + 273) = 260 K

$$\frac{P_1V_1}{T_1} = \frac{P_2V_2}{T_2} \; ; \; \frac{(0.450 \text{ atm})(0.840 \text{ L})}{310 \text{ K}} = \frac{P_2(0.150 \text{ L})}{260 \text{ K}} \; ; \; P_2 = \underline{2.11 \text{ atm}}$$

31. This problem also requires the combined gas law.

Initially, $V_1$ is unknown; $P_1$ is unknown; $T_1$ = (25 +273) = 298 K

Finally, $V_2 = V_1$; $P_2$ = 100 atm; $T_2$ = (300 + 273) = 573 K

$$\frac{P_1V_1}{T_1} = \frac{P_2V_2}{T_2} \; ; \; \frac{P_1V_1}{298 \text{ K}} = \frac{(100 \text{ atm})V_1}{573 \text{ K}} \; ; \; P_1 = \underline{52.0 \text{ atm}}$$

The cylinder can be filled to a maximum pressure of 52.0 atm at 25°C

32. This problem also requires the combined gas law.

Initially, $V_1$ = 0.500 L; $P_1$ = 730 mm Hg; $T_1$ = (25 + 273) = 298 K

Finally, $V_2$ is unknown; $P_2 = (730 \text{ mm Hg})(\frac{110}{100})$ = 803 mm Hg, $T_2$ = 293 K

$$\frac{P_1V_1}{T_1} = \frac{P_2V_2}{T_2} \; ; \; \frac{(730 \text{ mm Hg})(0.500 \text{ L})}{298 \text{ K}} = \frac{V_2(803 \text{ mm Hg})}{293 \text{ K}} \; ; \; V_2 = \underline{0.447 \text{ L}}$$

33. This problem also requires the combined gas law.

(a) Initially, $V_1$ is unknown; $P_1$ = 2,50 atm, $T_1$ = (20 + 273) = 293 K

Finally, $V_2 = V_1$, $P_2$ is unknown, $T_2$ = (30 + 273) = 303 K

$$\frac{P_1V_1}{T_1} = \frac{P_2V_2}{T_2} \; ; \; \frac{(2.50 \text{ atm})(V_1)}{293 \text{ K}} = \frac{P_2V_1}{303 \text{ K}} \; ; \; P_2 = \underline{2.59 \text{ atm}}$$

(b) The initial conditions are the same as in Part (a).

Finally, $V_2 = V_1$, $P_2$ is unknown, $T_2$ = (-10 + 273) = 263 K

$$\frac{P_1V_1}{T_1} = \frac{P_2V_2}{T_2} \; ; \; \frac{(2.50 \text{ atm})(V_1)}{293 \text{ K}} = \frac{P_2V_1}{263 \text{ K}} \; ; \; P_2 = \underline{2.24 \text{ atm}}$$

34. First convert the mass of $NO_2(g)$ to moles, and then calculate the pressure using the ideal gas equation, PV = nRT.

Remember that in using the ideal gas equation, the units of P, V, and T must be consistent with those for the value of the gas constant. When

$$R = 0.0821 \text{ atm L mol}^{-1} \text{ K}^{-1},$$

<u>P</u> must be in <u>atmospheres</u>, <u>V</u> in <u>liters</u>, and <u>T</u> in <u>kelvins</u>.

Here, V = 5.00 L, & = (30+273) K = 303 K, and the amount of gas, n, is given in moles.

Convert mass of $NO_2(g)$ to mol $NO_2$,

$$n = (5.29 \text{ g } NO_2)\left(\frac{1 \text{ mol } NO2}{46.01 \text{ g } NO_2}\right) = \underline{0.115 \text{ mol}}$$

and to calculate the pressure, PV = nRT, <u>or</u> $P = \frac{nRT}{V}$

$$P = \frac{(0.115 \text{ mol})(0.0821 \text{ atm L mol}^{-1} \text{ K}^{-1})(303 \text{ K})}{5.00 \text{ L}} = \underline{0.572 \text{ atm}}$$

35. This is a straightforward application of the ideal gas law. Note that under most conditions all gases obey the ideal gas law to a good approximation and the nature of the gas is irrelevant.

$$PV = nRT \ ; \ n = \frac{PV}{RT} = \frac{(1.00 \text{ atm})(4.00 \text{ L})}{(0.0821 \text{ atm L mol}^{-1} \text{ K}^{-1})(298 \text{ K})} = \underline{0.163 \text{ mol}}$$

36. Again, the nature of the gas is irrelevant. Remember that 20°C is 293 K, and convert pressure in mm Hg to pressure in atm.

$$PV = nRT \ ; \ V = \frac{nRT}{P} = \frac{(0.200 \text{ mol})(0.0821 \text{ atm L mol}^{-1} \text{ K}^{-1})(293 \text{ K})}{(740 \text{ mm Hg})\left(\frac{1 \text{ atm}}{760 \text{ mm Hg}}\right)}$$

$$= \underline{4.94 \text{ L}}$$

37. From the balanced equation, calculate moles of $CO_2(g)$, and use the ideal gas law to convert this to the volume of $CO_2(g)$.

$$\text{Mol } CO_2 = (7.20 \text{ g glucose})\left(\frac{1 \text{ mol glucose}}{180.2 \text{ g glucose}}\right)\left(\frac{6 \text{ mol } CO2}{1 \text{ mol glucose}}\right) = \underline{0.240 \text{ mol}}$$

$$V = \frac{nRT}{P} = \frac{(0.240 \text{ mol})(0.0821 \text{ atm L mol}^{-1} \text{ K}^{-1})(310 \text{ K})}{1.00 \text{ atm}} = \underline{6.11 \text{ L } CO_2}$$

38. Write the balanced equation for the reaction,

$$Mg(s) + H_2SO_4(aq) \longrightarrow MgSO_4(aq) + H_2(g)$$

and use the ideal gas law to calculate mol of $H_2(g)$ formed.

$$PV = nRT \ ; \ n = \frac{PV}{RT} = \frac{(1.00 \text{ atm})(174.1 \text{ mL})(1 \text{ L}/10^3 \text{ mL})}{(0.0821 \text{ atm L mol}^{-1} \text{ K}^{-1})(301 \text{ K})} = \underline{7.045\text{x}10^{-3} \text{ mol}}$$

$$\text{Mol Mg} = (7.045\text{x}10^{-3} \text{ mol } H_2)\left(\frac{1 \text{ mol Mg}}{1 \text{ mol } H_2}\right)\left(\frac{24.31 \text{ g Mg}}{1 \text{ mol Mg}}\right) = \underline{0.1713 \text{ g Mg}}$$

39. From the ideal gas law, PV = nRT, calculate mol $CO_2$ required to fill the 1.00 L vessel at 300 K and a pressure of 500 mm Hg.

$$PV = nRT \; ; \; n = \frac{PV}{RT} = \frac{(500 \text{ mm Hg})(\frac{1 \text{ atm}}{760 \text{ mm Hg}})(1.00 \text{ L})}{(0.0821 \text{ atm L mol}^{-1} \text{ K}^{-1})(300 \text{ K})} = 2.671\text{x}10^{-2} \text{ mol}$$

Now convert mol $CO_2$ to grams of $CO_2$,

$$\text{mass of } CO_2 = (2.671\text{x}10^{-2} \text{ mol } CO_2)(\frac{44.01 \text{ g } CO_2}{1 \text{ mol } CO_2}) = \underline{1.18 \text{ g } CO_2(s)}$$

40. Use the ideal gas equation to calculate the moles of HCl, and then convert mol HCl to mass of HCl.

$$PV = nRT \; ; \; n = \frac{PV}{RT} = \frac{(0.240 \text{ atm})(250 \text{ mL})(\frac{1 \text{ L}}{10^3 \text{ mL}})}{(0.0821 \text{ atm L mol}^{-1} \text{ K}^{-1})(310 \text{ K})} = \underline{2.357\text{x}10^{-3} \text{ mol}}$$

$$\text{mass of HCl} = (2.357\text{x}10^{-3} \text{ mol HCl})(\frac{36.46 \text{ g HCl}}{1 \text{ mol HCl}}) = \underline{0.0859 \text{ g HCl}}$$

41. Use the ideal gas equation to calculate mol $O_2(g)$ in a volume of 10 000 L at 0.20 atm and 25°C (298 K).

$$\text{mol } O_2 = \frac{PV}{RT} = \frac{(0.20 \text{ atm})(10\ 000 \text{ L})}{(0.0821 \text{ atm L mol}^{-1} \text{ K}^{-1})(298 \text{ K})} = \underline{82 \text{ mol}}$$

(a) From the balanced equation

$$2BaO_2(s) \longrightarrow 2BaO(s) + O_2(g)$$

$$\text{Mass of } BaO_2 = (82 \text{ mol } O_2)(\frac{2 \text{ mol } BaO_2}{1 \text{ mol } O_2})(\frac{169.3 \text{ g } BaO_2}{1 \text{ mol } BaO_2})(\frac{1 \text{ kg}}{10^3 \text{ g}})$$

$$= \underline{28 \text{ kg } BaO_2}$$

(b) 10 000 L of $O_2(g)$ at 25°C and 0.20 atm provides

$$(10\ 000 \text{ L})(\frac{293 \text{ K}}{298 \text{ K}}) = 9\ 832 \text{ L of } O_2(g) \text{ at } 20^\circ\text{C},$$

since, according to Charles's law, $V \propto T$, for a fixed mass of gas at constant pressure.

At a rate of consumption of 1.00 L $\text{min}^{-1}$, 9832 L of $O_2(g)$ would last for

$$(9832 \text{ L})(\frac{1 \text{ min}}{1 \text{ L}})(\frac{1 \text{ h}}{60 \text{ min}})(\frac{1 \text{ day}}{24 \text{ h}}) = \underline{6.83 \text{ days}}$$

42. This problem requires calculation of the volume of 1 mole of gas at 800°C (1073 K) and a pressure of 75 atm.

$$V = \frac{nRT}{P} = \frac{(1.00 \text{ mol})(0.0821 \text{ atm L mol}^{-1} \text{ K}^{-1})(1073 \text{ K})}{75 \text{ atm}} = \underline{1.2 \text{ L}}$$

The molar volume of an ideal gas on Venus would be approximately $\underline{1.2 \text{ L mol}^{-1}}$.

43. If the moles of gas at STP is $n_1$, then

$$n_1 = \frac{PV}{RT} = \frac{(1.00\ \text{atm})(1.00\ \text{L})}{R(273\ \text{K})} \quad \text{at STP} \quad \ldots\ldots (1)$$

and if the moles of gas is $n_2$ at 200°C (473 K), 1.25 atm pressure, and a volume of 1.00 L,

$$n_2 = \frac{PV}{RT} = \frac{(1.25\ \text{atm})(1.00\ \text{L})}{R(473\ \text{K})} \quad \ldots\ldots (2)$$

$$\text{i.e.,} \quad \frac{n_2}{n_1} = \frac{(1.25\ \text{atm})(1.00\ \text{L})}{R(473\ \text{K})} \cdot \frac{R(273\ \text{K})}{(1.00\ \text{atm})(1.00\ \text{L})} = 0.721$$

If the mass of gas in $n_2$ mol is x grams, then, since the mass of any substance is proportional to the number of moles,

$$\frac{n_2}{n_1} = \frac{x\ \text{g}}{1.89\ \text{g}} = 0.721\ ;\ x = \underline{1.36\ \text{g}}$$

44. First calculate the volume of the water bed in liters, then, using the ideal gas law convert this volume to mol He, and finally convert mol He to the mass of He.

$$\text{Volume} = (2.00\ \text{m})(1.50\ \text{m})(0.20\ \text{m})\left(\frac{10\ \text{dm}}{1\ \text{m}}\right)^3\left(\frac{1\ \text{L}}{1\ \text{dm}^3}\right) = \underline{600\ \text{L}}$$

$$\text{mol He} = n = \frac{PV}{RT}\ ;\ \text{g of He} = (n\ \text{mol He})\left(\frac{4.003\ \text{g He}}{1\ \text{mol He}}\right)$$

$$\text{g of He} = \left[\frac{(1.03\ \text{atm})(600\ \text{L})}{(0.0821\ \text{atm L mol}^{-1}\ \text{K}^{-1})(296\ \text{K})}\right]\left(\frac{4.003\ \text{g He}}{1\ \text{mol He}}\right) = \underline{102\ \text{g}}$$

45. This is a rather common type of problem, so it is useful first to establish the relationship between the densities of a gas under two different sets of conditions. Since density is mass per liter, we are comparing equal volumes of gas and if we calculate moles of gas per liter under each set of conditions

$$P_1V_1 = n_1RT_1 \quad \text{and} \quad P_2V_2 = n_2RT_2$$

and dividing the first expression by the second, with $V_1 = V_2$,

$$\frac{n_1}{n_2} = \frac{P_1T_2}{P_2T_1}$$; and since moles of gas are proportional to mass,

if the density under the first set of conditions is $d_1$, and $d_2$ under the second set of conditions, then

$$\frac{d_1}{d_2} = \frac{n_1}{n_2} = \frac{P_1T_2}{P_2T_1}$$

In other words, density is proportional to P and inversely proportional to T.

For $d_1 = 1.62\ \text{g L}^{-1}$ at STP (273 K and 1.00 atm), $P_1 = 1.00$ atm, and $T_1 = 273$ K, so that for $P_2 = 0.950$ atm and $T_2 = 575$ K,

$$\frac{1.62 \text{ g L}^{-1}}{d_2 \text{ g L}^{-1}} = \frac{(1.000 \text{ atm})(575 \text{ K})}{(0.950 \text{ atm})(273 \text{ K})} \quad ; \quad d_2 = \underline{0.731 \text{ g L}^{-1}}$$

46. The density of the gas mixture depends on its composition; from the data given, first calculate the total number of moles of gas in the mixture,

$$n = \frac{PV}{RT} = \frac{(0.980 \text{ atm})(1.00 \text{ L})}{(0.0821 \text{ atm L mol}^{-1} \text{ K}^{-1})(298 \text{ K})} = 0.0400 \text{ mol}$$

From the density, the mass of a mixture of 0.040 mol of $O_2$ and $N_2O$ is 1.482 g, and assuming x mol $O_2$ and 0.040-x mol $N_2O$,

$$\text{mass of gas} = (x \text{ mol } O_2)\left(\frac{32.00 \text{ g O}}{1 \text{ mol } O_2}\right) + [(0.040-x) \text{ mol } N_2O]\left(\frac{44.02 \text{ g N2O}}{1 \text{ mol } N_2O}\right)$$

$$= 1.482 \text{ g}$$

Whence, $32.00x + 1.761 - 44.02x = 1.482$ ; $\underline{x = 0.0232 \text{ mol}}$

$$\text{Mass of } N_2O = (0.0168 \text{ mol } N_2O)\left(\frac{44.02 \text{ g N O}}{1 \text{ mol } N_2O}\right) = \underline{0.739 \text{ g}}$$

$$\text{Mass \% } N_2O = \left(\frac{0.739 \text{ g}}{1.482 \text{ g}}\right) \times 100 = \underline{49.9 \text{ mass \%}}$$

47. The units of gas density are g $L^{-1}$, so calculate the mass of 1.00 L of $CF_2Cl_2$ at 1.00 atm and 20°C. First we calculate the number of moles,

$$n = \frac{PV}{RT} = \frac{(1.00 \text{ L})(1.00 \text{ atm})}{(0.0821 \text{ atm L mol}^{-1} \text{ K}^{-1})(293 \text{ K})} = 0.0416 \text{ mol}$$

$$\text{Mass of 0.0416 mol } CF_2Cl_2 = (0.0416 \text{ mol})\left(\frac{120.9 \text{ g}}{1 \text{ mol}}\right) = \underline{5.03 \text{ g}}$$

$$\text{Density} = \underline{5.03 \text{ g L}^{-1}}$$

48. Calculate the moles of noble gas, using the ideal gas equation.

$$n = \frac{PV}{RT} = \frac{(0.48 \text{ atm})(0.26 \text{ L})}{(0.0821 \text{ atm L mol}^{-1} \text{ K}^{-1})(300 \text{ K})} = \underline{5.07\text{x}10^{-3} \text{ mol}}$$

$$\text{Molar mass of gas} = \left(\frac{0.20 \text{ g}}{5.07\text{x}10^{-3} \text{ mol}}\right) = \underline{39 \text{ g mol}^{-1}}$$

The calculated molar mass is close to that of argon (39.95 g $mol^{-1}$)

49. Under the same conditions of temperature and pressure, 1 L of any (ideal) gas contains the same number of moles of gas. Thus, if the density of ozone is 1.50 times that of $O_2$, the molar mass of ozone must be 1.50 times that of $O_2$.

Molar mass of ozone = 1.50(32.00 g $mol^{-1}$) = $\underline{48.0 \text{ g mol}^{-1}}$

Since ozone contains only O atoms, its molecular formula must be $O_3$.

50. Use the ideal gas equation to calculate moles of gas in 1.00 L.

$$n = \frac{PV}{RT} = \frac{(740 \text{ mm Hg})(\frac{1 \text{ atm}}{760 \text{ mm Hg}})(1.00 \text{ L})}{(0.0821 \text{ atm L mol}^{-1} \text{ K}^{-1})(293 \text{ K})} = \underline{0.0405 \text{ mol}}$$

$$\text{Molar mass} = (\frac{1.134 \text{ g}}{0.405 \text{ mol}}) = \underline{28.0 \text{ g mol}^{-1}}$$

Empirical formula mass of $CH_2$ = 14.03 u, and the molecular mass of 28.0 u is close to 2(empirical formula mass).

<u>molecular formula is $C_2H_4$</u>

51. This problem is similar to problem 50. The data gives us the mass of 1 L of gas at 23.8°C and a pressure of 432 mm Hg.

$$n = \frac{PV}{RT} = \frac{(432 \text{ mm Hg})(\frac{1 \text{ atm}}{760 \text{ mm Hg}})(1.00 \text{ L})}{(0.0821 \text{ atm L mol}^{-1} \text{ K}^{-1})(296.8 \text{ K})} = \underline{0.0233 \text{ mol}}$$

$$\text{Molar mass} = (\frac{3.23 \text{ g}}{0.0233 \text{ mol}}) = \underline{139 \text{ g mol}^{-1}}$$

The chlorofluorocarbon contains one C atom and must have a molecular formula $CCl_xF_{4-x}$, and by trial and error x is clearly <u>3</u>.

The <u>molecular formula</u> is <u>$CCl_3F$</u> (molar mass 137.4 g $mol^{-1}$)

52. First calculate moles of gas in 1 L.

(a)
$$n = \frac{PV}{RT} = \frac{(740 \text{ mm Hg})(\frac{1 \text{ atm}}{760 \text{ mm Hg}})(1 \text{ L})}{(0.0821 \text{ atm L mol}^{-1} \text{ K}^{-1})(291 \text{ K})} = \underline{4.08\text{x}10^{-2} \text{ mol}}$$

$$\text{Molar mass} = (\frac{1.275 \text{ g}}{4.08\text{x}10^{-2} \text{ mol}}) = \underline{31.3 \text{ g mol}^{-1}}$$

(b) Molecules in 0.010 mL = $(0.010 \text{ mL})(\frac{1 \text{ L}}{10^3 \text{ mL}})(4.08\text{x}10^{-2}\text{mol})$

$$\text{x } (\frac{6.022\text{x}10^{23} \text{ molecules}}{1 \text{ mol}})$$

$$= \underline{2.5\text{x}10^{17} \text{ molecules}}$$

53. Calculate the moles of gas in 1 L.

$$n = \frac{PV}{RT} = \frac{(740 \text{ mm Hg})(\frac{1 \text{ atm}}{760 \text{ mm Hg}})(1 \text{ L})}{(0.0821 \text{ atm L mol}^{-1} \text{ K}^{-1})(293 \text{ K})} = \underline{4.05\text{x}10^{-2} \text{ mol}}$$

$$\text{Molar mass} = (\frac{1.402 \text{ g}}{4.05 \text{ x } 10^{-2} \text{ mol}}) = \underline{34.6 \text{ g mol}^{-1}}$$

54. From the information given, 0.750 g of volatile liquid occupies a volume of 350 mL at $100^{o}$C and a pressure of 0.980 atm. We can use the ideal gas law to calculate moles of gas:

$$PV = nRT \; ; \quad n = \frac{PV}{RT} = \frac{(0.980\ \text{atm})(\frac{1\ \text{L}}{1000\ \text{mL}})(350\ \text{mL})}{(0.0821\ \text{atm L mol}^{-1}\ \text{K}^{-1})(373\ \text{K})} = \underline{1.12 \times 10^{-2}\ \text{mol}}$$

$$\text{Molar mass} = \frac{0.750\ \text{g}}{(1.12\text{x}10^{-2}\ \text{mol})} = \underline{67.0\ \text{g mol}^{-1}}$$

55. (a) First calculate moles of gaseous compound:

$$PV = nRT \; ; \quad n = \frac{PV}{RT} = \frac{(1.00\ \text{atm})(440\ \text{mL})(\frac{1\ \text{L}}{1000\ \text{mL}})}{(0.0821\ \text{atm L mol}^{-1}\ \text{K}^{-1})(373\ \text{K})} = \underline{1.44 \times 10^{-2}\ \text{mol}}$$

$$\text{Molar mass} = \frac{1.673\ \text{g}}{1.44 \times 10^{-2}\ \text{mol}} = \underline{116\ \text{g mol}^{-1}}$$

(b) The empirical formula can be calculated from the mass % composition:

| | | C | H | O |
|---|---|---|---|---|
| Mass % | | 62.04 | 10.41 | 27.55 |
| Mass in 100 g of compound | | 62.04 | 10.41 | 27.55 |
| Moles in 100 g of compound | | $\frac{62.04}{12.01}$ | $\frac{10.41}{1.008}$ | $\frac{27.55}{16.00}$ |
| | = | 5.166 | 10.33 | 1.722 |
| Ratio of moles (atoms) | = | 3.00 : | 6.00 : | 1.00 |

The empirical formula is $C_3H_6O$ (empirical formula mass 58.08 u)

Thus, from part (a). Molar mass = 116 g $\text{mol}^{-1}$

Molecular formula = 2(Empirical formula)

Molecular formula is $\underline{C_6H_{12}O_2}$

56. From the information given, mass of ozone in the 1.500 L bulb, at 298 K, and a pressure of 287 torr, was (7.319 - 6.208) g = 1.111 g.

$$\text{Moles of ozone} = \frac{PV}{RT} = \frac{(287\ \text{torr})(\frac{1\ \text{atm}}{760\ \text{torr}})(1.500\ \text{L})}{(0.0821\ \text{atm L mol}^{-1}\ \text{K}^{-1})(298\ \text{K})} = \underline{2.32 \times 10^{-2}\ \text{mol}}$$

$$\text{Molar mass of ozone} = \frac{1.111\ \text{g}}{2.32 \times 10^{-2}\ \text{mol}} = \underline{47.9\ \text{g mol}^{-1}}$$

which corresponds to the molecular formula $O_3$ for ozone.

57.

$$\text{Moles of boron compound} = \frac{PV}{RT} = \frac{(730\ \text{torr})(\frac{1\ \text{atm}}{760\ \text{torr}})(2.22\ \text{L})}{(0.0821\ \text{atm L mol}^{-1}\ \text{K}^{-1})(298\ \text{K})} = \underline{0.0872\ \text{mol}}$$

$$\text{Molar mass} = \frac{2.41\ \text{g}}{0.0872\ \text{mol}} = \underline{27.6\ \text{g mol}^{-1}}$$

The compound contains only boron and hydrogen.

Presumably it must contain at least 2 B atoms per molecule, otherwise the number of H atoms would be unreasonably large, which suggests that

<u>Molecular formula $B_2H_6$</u> (molar mass 27.67 g $mol^{-1}$)

This is the known molecular formula of <u>diborane</u>

58. In any problem dealing with partial pressures, remember that the partial pressure of each gas is determined by the number of moles of that gas; the total pressure is determined by the total moles of all the gases.

(a) First calculate the moles of each gas:

$$\text{Moles of He} = \frac{0.200 \text{ g He}}{4.003 \text{ g mol}^{-1}} = 0.0500 \text{ mol He}$$

$$\text{Moles of } H_2 = \frac{0.200 \text{ g } H_2}{2.016 \text{ g mol}^{-1}} = 0.0992 \text{ mol } H_2$$

$$PV = nRT \; ; \; P = \frac{nRT}{V}$$

$$P_{He} = \frac{(0.0500 \text{ mol})(0.0821 \text{ atm L mol}^{-1} \text{ K}^{-1})(300 \text{ K})}{(225 \text{ mL})(\frac{1 \text{ L}}{1000 \text{ mL}})} = \underline{5.47 \text{ atm}}$$

$$P_{H_2} = \frac{(0.0992 \text{ mol})(0.0821 \text{ atm L mol}^{-1} \text{ K}^{-1})(300 \text{ K})}{(225 \text{ mL})(\frac{1 \text{ L}}{1000 \text{ mL}})} = \underline{10.9 \text{ atm}}$$

(b) The total pressure is the sum of the partial pressures (Dalton's Law):

$$P = P_{He} + P_{H_2} = (5.47 + 10.9) \text{ atm} = \underline{16.4 \text{ atm}}$$

59. This problem is similar to Problem 58.

$$\text{Moles of CO} = \frac{2.34 \text{ g } CO_2}{28.01 \text{ g mol}^{-1}} = 0.0835 \text{ mol CO}$$

$$\text{Moles of } CO_2 = \frac{1.56 \text{ g } CO_2}{44.01 \text{ g mol}^{-1}} = 0.0354 \text{ mol } CO_2$$

$$P_{CO} = \frac{(0.0835 \text{ mol})(0.0821 \text{ atm L mol}^{-1} \text{ K}^{-1})(303 \text{ K})}{1.00 \text{ L}} = \underline{2.08 \text{ atm}}$$

$$P_{CO_2} = \frac{(0.0354 \text{ mol})(0.0821 \text{ atm L mol}^{-1} \text{ K}^{-1})(303 \text{ K})}{1.00 \text{ L}} = \underline{0.881 \text{ atm}}$$

$$P = P_{CO} + P_{CO_2} = (2.08 + 0.881) \text{ atm} = \underline{2.96 \text{ atm}}$$

<u>Partial pressures</u>: CO, <u>2.08 atm</u>; $CO_2$, <u>0.881 atm</u>

<u>Total pressure</u>: <u>2.96 atm</u>

60. Each gas can be treated independently; since the temperature remains constant, Boyle's law applies.

For each gas $P_1V_1 = P_2V_2$; $(1.00 \text{ atm})(1.00 \text{ L}) = P_2(2.00 \text{ L})$

i.e., $P_2 = \underline{0.50 \text{ atm}}$

In the 2.00 L container, each gas exerts a pressure of 0.50 atm,

$$p_{O_2} + p_{N_2} + p_{H_2} = (0.50+0.50+0.50) \text{ atm} = \underline{1.50 \text{ atm}}$$

$$\underline{\text{Partial pressures: } p_{O_2} = p_{N_2} = p_{H_2} = 0.50 \text{ atm}}$$

61. 0.94 ppm of NO(g) by volume means 0.94 L NO(g) in $10^6$ L air. At constant temperature and pressure, the volume of gas is proportional to the number of moles of gas, and the partial pressure is also proportional to the volume of gas.

(a) $$\frac{p_{NO}}{p_{air}} = \frac{V_{NO}}{V_{air}} = \frac{0.94 \text{ L}}{10^6 \text{ L}} = \frac{p_{NO}}{750 \text{ mm Hg}}\ ;\quad p_{NO} = \underline{7.1\text{x}10^{-4} \text{ mm Hg}}$$

(b) From the partial pressure of NO, calculate moles of NO, and hence molecules of NO, in, for example, 1 L of air,

$$\text{mol NO in 1 L} = \frac{PV}{RT} = \frac{(7.1\text{x}10^{-4} \text{ mm})(\frac{1 \text{ atm}}{760 \text{ mm}})(1 \text{ L})}{(0.0821 \text{ atm L mol}^{-1} \text{ K}^{-1})(303 \text{ K})}$$

$$= \underline{3.755\text{x}10^{-8} \text{ mol}}$$

$$\text{molecules per m}^3 = (\frac{3.755\text{x}10^{-8} \text{ mol}}{1 \text{ L}})(\frac{1 \text{ L}}{1 \text{ dm}^3})(\frac{10 \text{ dm}}{1 \text{ m}})^3(\frac{6.022\text{x}10^{23}}{1 \text{ mol}})$$

$$= \underline{2.3\text{x}10^{19} \text{ molecules m}^{-3}}$$

62. The partial pressure of $O_2$ inhaled by the lungs is (159-115) mm Hg, = 44 mm Hg, at 37°C (310 K). Use this information and the volume of 1 L to calculate mol $O_2$ used on inhalation, and hence the number of $O_2$ molecules per liter used by the lungs.

$$O_2 \text{ molecules} = [\frac{(44 \text{ mm})(\frac{1 \text{ atm}}{760 \text{ mm}})(1 \text{ L})}{(0.0821 \text{ atm L mol}^{-1} \text{ K}^{-1})(310 \text{ K})}](\frac{6.022\text{x}10^{23}}{1 \text{ mol}})$$

$$= \underline{1.4\text{x}10^{21} \text{ molecules L}^{-1}}$$

63. Write the balanced equation for the reaction,

$$3CuO(s) + 2NH_3(g) \longrightarrow 3Cu(s) + N_2(g) + 3H_2O(g)$$

Then calculate initial mol CuO(s), and hence mol $N_2$(g) and mol $H_2O$(g) after complete reaction.

$$\text{mol CuO(s)} = (5.00 \text{ g CuO})(\frac{1 \text{ mol CuO}}{79.55 \text{ g CuO}}) = \underline{0.0628 \text{ mol}}$$

$$\text{mol } N_2\text{(g)} = (0.0628 \text{ mol CuO})(\frac{1 \text{ mol } N_2}{3 \text{ mol CuO}}) = \underline{0.0209 \text{ mol}}$$

$$\text{mol } H_2O(g) = (0.0628 \text{ mol CuO})\left(\frac{3 \text{ mol } H_2O}{3 \text{ mol CuO}}\right) = \underline{0.0628 \text{ mol}}$$

This assumes that all the CuO reacts, i.e., that CuO is the limiting reactant, so check to make sure that $NH_3(g)$ is initially in excess.

$$\text{initial mol } NH_3 = \frac{PV}{RT} = \frac{(1.00 \text{ atm})(80.0 \text{ L})}{(0.0821 \text{ atm L mol}^{-1} \text{ K}^{-1})(453 \text{ K})} = \underline{2.151 \text{ mol}}$$

Mol of $NH_3$ that react with 0.0628 mol CuO

$$= (0.0628 \text{ mol CuO})\left(\frac{2 \text{ mol } NH_3}{3 \text{ mol CuO}}\right) = \underline{0.0419 \text{ mol}}$$

Thus, $NH_3$ is initially <u>in excess</u>, as assumed, and after reaction,

$$\text{Final mol } NH_3 = (2.151 - 0.0419) = \underline{2.109 \text{ mol}}$$

Thus, <u>after reaction</u> is complete the 80.0 L vessel at 453 K contains

0.0209 mol $N_2(g)$ ; 0.0628 mol $H_2O(g)$ & 2.11 mol $NH_3(g)$.

To calculate the partial pressures, and the final pressure, use $P = \frac{nRT}{V}$, and since the factor $\frac{RT}{V}$ is the same throughout, calculate this first.

$$\frac{RT}{V} = \frac{(0.0821 \text{ atm L mol}^{-1} \text{ K}^{-1})(453 \text{ K})}{80.0 \text{ L}} = \underline{0.465 \text{ atm mol}^{-1}}$$

Then,

$$p_{N_2} = (0.465 \text{ atm mol}^{-1})(0.0209 \text{ mol}) = \underline{9.72 \times 10^{-3} \text{ atm}}$$

$$p_{H_2O} = (0.465 \text{ atm mol}^{-1})(0.0628 \text{ mol}) = \underline{2.92 \times 10^{-2} \text{ atm}}$$

$$p_{NH_3} = (0.465 \text{ atm mol}^{-1})(2.11 \text{ mol}) = \underline{0.981 \text{ atm}}$$

and the <u>total pressure</u> is the sum of the partial pressures

$$= \underline{1.02 \text{ atm}}.$$

64. 1.00 L of dry air, for example, at a given T and P, contains a certain number of moles of gases. If we add water to make it moist, then we are adding a certain number of moles of $H_2O$. But the pressure must remain constant, and this can only be achieved if the molecules in air (mainly $O_2$ and $N_2$) are replaced by an <u>equal</u> number of $H_2O$ molecules. Since the molar mass of $H_2O$ is less than the molar masses of both $N_2$ and $O_2$, the mass of 1.00 L of moist air must be <u>less</u> than that of dry air. Thus, <u>density of moist air must be less than that of dry air</u> (g $L^{-1}$).

65. The average speed of molecules depends on their masses. The average kinetic energies of gases depend only on the temperature. Thus, for two gases we can write,

$$\frac{1}{2} m_1 v_1^2 = \frac{1}{2} m_2 v_2^2 \;;\quad \frac{m_1}{m_2} = \frac{v_2^2}{v_1^2} \quad \text{or} \quad \frac{v_2}{v_1} = \sqrt{\frac{m_1}{m_2}}$$

and since the mass of a molecule is proportional to its molar mass, it follows that,

$$\frac{v_2}{v_1} = \sqrt{\frac{M_1}{M_2}} \quad \text{or} \quad v_2 = v_1\sqrt{\frac{M_1}{M_2}}$$

Use this formula to calculate the average speeds of molecules at 25°C from the average speed of oxygen molecules at 25°C:

(a) $v_{H_2} = (450 \text{ m s}^{-1})(\frac{32.00 \text{ g mol}^{-1}}{2.016 \text{ g mol}^{-1}})^{1/2} = \underline{1793 \text{ m s}^{-1}}$

(b) $v_{Cl_2} = (450 \text{ m s}^{-1})(\frac{32.00 \text{ g mol}^{-1}}{70.90 \text{ g mol}^{-1}})^{1/2} = \underline{302 \text{ m s}^{-1}}$

(c) $v_{CO} = (450 \text{ m s}^{-1})(\frac{32.00 \text{ g mol}^{-1}}{28.01 \text{ g mol}^{-1}})^{1/2} = \underline{481 \text{ m s}^{-1}}$

(d) $v_{H_2O} = (450 \text{ m s}^{-1})(\frac{32.00 \text{ g mol}^{-1}}{18.02 \text{ g mol}^{-1}})^{1/2} = \underline{600 \text{ m s}^{-1}}$

(e) $v_{CO_2} = (450 \text{ m s}^{-1})(\frac{32.00 \text{ g mol}^{-1}}{44.01 \text{ g mol}^{-1}})^{1/2} = \underline{384 \text{ m s}^{-1}}$

(f) $v_{H_2S} = (450 \text{ m s}^{-1})(\frac{32.00 \text{ g mol}^{-1}}{34.08 \text{ g mol}^{-1}})^{1/2} = \underline{436 \text{ m s}^{-1}}$

66\. Graham's law states that the rate of effusion (or diffusion) of a gas is proportional to the square root of its molar mass (see Problem 65).

Thus, the faster gas to effuse will be that with the lower molar mass.

Molar mass of $N_2$ = 28.02 g mol$^{-1}$; molar mass of $O_2$ = 32.00 g mol$^{-1}$

Thus, $\underline{N_2 \text{ effuses at a faster rate than } O_2}$.

$$\frac{r_1}{r_2} = \frac{M_2}{M_1}\ ;\quad \frac{r_1}{r_2} = (\frac{32.00 \text{ g mol}^{-1}}{28.02 \text{ g mol}^{-1}})^{1/2} = \underline{1.069}$$

Rate of effusion of $N_2$ : rate of effusion of $O_2$ = $\underline{1.069}$

67\. The atomic mass of deuterium is 2.014 g mol$^{-1}$ (Chapter 2).

$$\frac{r_{H_2}}{r_{D_2}} = (\frac{4.028 \text{ g mol}^{-1}}{2.016 \text{ g mol}^{-1}})^{1/2} = \underline{1.414}$$

68\. The molar masses are $^{235}UF_6$ = 349 g mol$^{-1}$; $^{238}UF_6$ = 352 g mol$^{-1}$.

$$\frac{r_{235}}{r_{238}} = (\frac{352 \text{ g mol}^{-1}}{349 \text{ g mol}^{-1}})^{1/2} = \underline{1.004}$$

69\. This problem is similar to Problem 65.

$$\frac{v_{N_2}}{v_{He}} = \frac{v_{N_2}}{0.707 \text{ mile s}^{-1}} = (\frac{4.003 \text{ g mol}^{-1}}{28.02 \text{ g mol}^{-1}})^{1/2} = 0.3780$$

$$v_{N_2} = \underline{0.267 \text{ mile s}^{-1}}$$

70. The distance each gas travels down the tube will be proportional to its rate of effusion. If the distances from each end of the tube are $d_{NH_3}$ and $d_{HBr}$, we have,

$$d_{NH_3} + d_{HBr} = 1.00 \text{ m} \quad \ldots\ldots (1)$$

and, $$\frac{r_{NH_3}}{r_{HBr}} = \frac{d_{NH_3}}{d_{HBr}} = \left(\frac{80.91 \text{ g mol}^{-1}}{17.03 \text{ g mol}^{-1}}\right)^{1/2} = \underline{2.180}$$

i.e., $$d_{NH_3} = 2.180\, d_{HBr} \quad \ldots\ldots (2)$$

From equations (1) and (2), $d_{NH_3} + d_{HBr} = 3.180 d_{HBr} = 1.00$ m

whence, $d_{HBr} = \underline{0.314 \text{ m}}$, and $d_{NH_3} = \underline{0.686 \text{ m}}$

A white cloud of $NH_4Br(s)$ will form at 31.4 cm from the end of the tube that HBr(g) enters, and at 68.6 cm from the end that $NH_3(g)$ enters.

71. The effusion rate of the unknown diatomic gas is $0.324r_{He}$,

$$\frac{r}{r_{He}} = 0.324 = \left(\frac{4.003 \text{ g mol}^{-1}}{M \text{ g mol}^{-1}}\right)^{1/2} \quad ; \quad \underline{M = 38.1 \text{ g mol}^{-1}}$$

The gas is a diatomic gas; the only possibility is $F_2(g)$, molar mass 38.00 g $mol^{-1}$.

<u>The diatomic gas is fluorine</u>.

72. The rate of effusion of each gas is inversely proportional to the <u>time</u> that it takes to effuse, thus

$$\frac{r_{O_2}}{r} = \frac{9.36 \text{ s}}{10.0 \text{ s}} = \left(\frac{M \text{ g mol}^{-1}}{32.00 \text{ g mol}^{-1}}\right)^{1/2} \quad ; \quad \underline{M = 28.0 \text{ g mol}^{-1}}$$

Possible gases with molar masses close to 28.0 g $mol^{-1}$ are: $N_2$ (28.02 g $mol^{-1}$), CO (molar mass 28.01 g $mol^{-1}$), $C_2H_4$, ethene, (molar mass 28.05 g $mol^{-1}$).

Nitrogen could be distinguished from carbon monoxide and ethene by its inertness. It cannot be burned in air or oxygen, while CO burns with a blue flame and $C_2H_4$ burns with a yellow flame,

$$2CO(g) + O_2(g) \longrightarrow 2CO_2(g)$$

$$C_2H_4(g) + 3O_2(g) \longrightarrow 2CO_2(g) + 2H_2O(g)$$

If the gas was ethene, the water from its combustion could be condensed on a cold surface, which would distinguish it from carbon monoxide.

73. Write the balanced equation in the form

$$B_xH_y + \frac{3x+y}{2} O_2 \longrightarrow \frac{x}{2} B_2O_3 + \frac{y}{2} H_2O$$

From the given mass of water produced, calculate mol $H_2O$, and hence mol of H and mass of H in the 0.492 g sample.

$$\text{mol H} = (0.617\text{ g }H_2O)\left(\frac{1\text{ mol H O}}{18.02\text{ g }H_2O}\right)\left(\frac{2\text{ mol H}}{1\text{ mol }H_2O}\right) = \underline{0.0685\text{ mol H}}$$

$$\text{mass H} = (0.0685\text{ mol H})\left(\frac{1.008\text{ g H}}{1\text{ mol H}}\right) = \underline{0.0690\text{ g}}$$

By difference, the mass of B in the sample is (0.492-0.069) g

$$= \underline{0.423\text{ g B}}$$

$$\text{and mol B} = (0.423\text{ g B})\left(\frac{1\text{ mol B}}{10.81\text{ g B}}\right) = \underline{0.0391\text{ mol B}}$$

The mol ratio B : H is the same as the atom ratio, i.e.,

$$\text{ratio B : H} = 0.0391 : 0.0685 = \frac{0.0391}{0.0391} : \frac{0.0685}{0.0391} = \underline{1.00 : 1.75}$$

$$\text{ratio B : H} = 4.00 : 7.00$$

and the <u>empirical formula</u> is $B_4H_7$ (empirical formula mass = <u>50.30 u</u>)

From the gaseous data, calculate the moles of hydride, and then the molar mass:

$$\text{mol hydride} = \frac{PV}{RT} = \frac{(110\text{ mm})\left(\frac{1\text{ atm}}{760\text{ mm}}\right)(250\text{ mL})\left(\frac{1\text{ L}}{10^3\text{ mL}}\right)}{(0.0821\text{ atm L mol}^{-1}\text{ K}^{-1})(300\text{ K})} = \underline{1.47\text{x}10^{-3}\text{ mol}}$$

$$\text{molar mass of hydride} = \left(\frac{0.147\text{ g}}{1.47\text{x}10^{-3}\text{ mol}}\right) = \underline{100\text{ g mol}^{-1}}$$

i.e., molecular mass = 2(empirical formula mass)

<u>molecular formula $B_8H_{14}$</u> (molar mass 100.6 g $mol^{-1}$)

74\. The balanced equation is

$$C_2H_5OH(l) + 3O_2(g) \longrightarrow 2CO_2(g) + 3H_2O(l)$$

$$\text{mol }O_2\text{ required} = (252\text{ g ethanol})\left(\frac{1\text{ mol ethanol}}{46.07\text{ g ethanol}}\right)\left(\frac{3\text{ mol }O_2}{1\text{ mol ethanol}}\right)$$

$$= \underline{16.41\text{ mol}}$$

now use the ideal gas equation to calculate the volume,

$$V = \frac{nRT}{P} = \frac{(16.41\text{ mol})(0.0821\text{ atm L mol}^{-1}\text{ K}^{-1})(298\text{ K})}{(770\text{ torr})\left(\frac{1\text{ atm}}{760\text{ torr}}\right)} = \underline{396\text{ L}}$$

$$\text{Volume of air required} = \left(\frac{100\text{ L air}}{21.0\text{ L }O_2}\right)(396\text{ L }O_2) = \underline{1.89\text{x}10^3\text{ L air}}$$

75\. $$4NH_3(g) + 5O_2(g) \longrightarrow 4NO(g) + 6H_2O(g)$$

Equal volumes of gases at the same temperature and pressure contain the same number of molecules (moles) (Avogadro's law). Thus, at 500°C and 750 mm Hg pressure,

$$\text{Vol of }O_2\text{ required} = (500\text{ L NO})\left(\frac{5\text{ mol }O_2}{4\text{ mol NO}}\right) = \underline{625\text{ L }O_2}$$

Now use the combined gas law to calculate the volume of $O_2$ at $25°C$ and 0.896 atm from the volume of 625 L at $500°C$ and 740 mm.

$$\frac{P_1V_1}{T_1} = \frac{P_2V_2}{T_2} \quad ; \quad \frac{(740 \text{ mm Hg})(\frac{1 \text{ atm}}{760 \text{ mm Hg}})(625 \text{ L})}{773 \text{ K}} = \frac{(0.896 \text{ atm})V_2}{298 \text{ K}}$$

$$V_2 = \underline{262 \text{ L of } O_2}$$

76. This is a straightforward application of Avogadro's law. First write the balanced equation for each reaction, and use the rule that under the same conditions of T and P, equal volumes of gases contain equal moles of molecules.

(a)
$$CH_4(g) + 2O_2(g) \longrightarrow CO_2(g) + 2H_2O(g)$$

| | | | | | |
|---|---|---|---|---|---|
| initially | 1.00 | 2.00 | 0 | 0 | L |
| finally | 0 | 0 | 1.00 | 2.00 | L |

$$\underline{\text{Volume of } O_2 \text{ required} = 2.00 \text{ L}}$$

(b)
$$2C_6H_{14}(g) + 19O_2(g) \longrightarrow 12CO_2(g) + 7H_2O(g)$$

| | | | | | |
|---|---|---|---|---|---|
| initially | 1.00 | 9.50 | 0 | 0 | L |
| finally | 0 | 0 | 6.00 | 3.50 | L |

$$\underline{\text{Volume of } O_2 \text{ required} = 9.50 \text{ L}}$$

77. First calculate mol $CaCO_3(s)$, and then mol $CO_2(g)$ produced, using the balanced equation.

$$CaCO_3(s) \longrightarrow CaO(s) + CO_2(g)$$

$$\text{mol } CO_2 = (1.00 \text{ kg } CaCO_3)(\frac{10^3 \text{ g}}{1 \text{ kg}})(\frac{1 \text{ mol } CaCO_3}{100.1 \text{ g } CaCO_3})(\frac{1 \text{ mol } CO_2}{1 \text{ mol } CaCO_3}) = \underline{9.99 \text{ mol}}$$

Thus, at STP,

$$\text{Volume of } CO_2 = \frac{nRT}{P} = \frac{(9.99 \text{ mol})(0.0821 \text{ atm L mol}^{-1} \text{ K}^{-1})(273 \text{ K})}{1.00 \text{ atm}}$$

$$= \underline{224 \text{ L}}$$

(or use the rule that 1 mol of gas at STP occupies a volume of 22.4 L)

78.
$$2Mg(s) + O_2(g) \longrightarrow 2MgO(s)$$

$$\text{mol } O_2 = (5.00 \text{ g } Mg)(\frac{1 \text{ mol } Mg}{24.31 \text{ g } Mg})(\frac{1 \text{ mol } O_2}{2 \text{ mol } Mg}) = \underline{0.103 \text{ mol}}$$

$$\text{volume of } O_2 = \frac{nRT}{P} = \frac{(0.103 \text{ mol})(0.0821 \text{ atm L mol}^{-1} \text{ K}^{-1})(298 \text{ K})}{(740 \text{ mm})(\frac{1 \text{ atm}}{760 \text{ mm}})}$$

$$= \underline{2.59 \text{ L}}$$

79. (a)
$$CaH_2(s) + 2H_2O(1) \longrightarrow Ca(OH)_2(s) + 2H_2(g)$$

(b) Convert 10.0 L $H_2(g)$ at STP to mol $H_2$; then calculate mol $CaH_2$ and finally the mass of $CaH_2(s)$.

$$\text{mol } H_2 = \frac{PV}{RT} = \frac{(1.00 \text{ atm})(10.0 \text{ L})}{(0.0821 \text{ atm L mol}^{-1} \text{ K}^{-1})(273 \text{ K})} = \underline{0.446 \text{ mol}}$$

$$\text{mass of } CaH_2(s) = (0.446 \text{ mol } H_2)\left(\frac{1 \text{ mol } CaH_2}{2 \text{ mol } H_2}\right)\left(\frac{42.10 \text{ g } CaH_2}{1 \text{ mol } CaH_2}\right)$$

$$= \underline{9.39 \text{ g}}$$

80. First write the balanced equation for the reaction.

$$2KClO_3(s) \longrightarrow 2KCl(s) + 3O_2(g)$$

$$\text{mol } O_2 = \frac{PV}{RT} = \frac{(1.00 \text{ atm})(5.00 \text{ L})}{(0.0821 \text{ atm L mol}^{-1} \text{ K}^{-1})(273 \text{ K})} = \underline{0.223 \text{ mol}}$$

$$\text{mass } KClO_3(s) = (0.223 \text{ mol } O_2)\left(\frac{2 \text{ mol } KClO_3}{3 \text{ mol } O_2}\right)\left(\frac{122.6 \text{ g } KClO_3}{1 \text{ mol } KClO_3}\right)$$

$$= \underline{18.2 \text{ g}}$$

81. To simplify the calculations, first calculate the volume of 1 mol $H_2$ at STP.

$$V_{O_2} = \frac{nRT}{P} = \frac{(1 \text{ mol})(0.0821 \text{ atm L mol}^{-1} \text{ K}^{-1})(273 \text{ K})}{1 \text{ atm}}$$

$$= \underline{22.41 \text{ L}}$$

(a) The balanced equation is

$$ZnO(s) + H_2(g) \longrightarrow Zn(s) + H_2O(g)$$

$$\text{mol } ZnO(s) = (1.00 \text{ kg } ZnO)\left(\frac{10^3 \text{ g}}{1 \text{ kg}}\right)\left(\frac{1 \text{ mol } ZnO}{81.38 \text{ g } ZnO}\right) = \underline{12.3 \text{ mol}}$$

Now calculate the volume of $H_2$ at STP required to react with this amount of $ZnO(s)$.

$$\text{Volume of } H_2 \text{ at STP} = (12.3 \text{ mol ZnO})\left(\frac{1 \text{ mol } H_2}{1 \text{ mol ZnO}}\right)\left(\frac{22.41 \text{ L}}{1 \text{ mol } H_2}\right) = \underline{276 \text{ L}}$$

(b)

$$2H_2(g) + O_2(g) \longrightarrow 2H_2O(l)$$

$$\text{Moles of } H_2O = (1.00 \text{ kg } H_2O)\left(\frac{1 \text{ mol } H_2O}{18.02 \text{ g mol}^{-1}}\right)\left(\frac{1000 \text{ g}}{1 \text{ kg}}\right) = \underline{55.5 \text{ mol}}$$

$$\text{Volume of } H_2 = (55.5 \text{ mol } H_2O)\left(\frac{2 \text{ mol } H_2}{2 \text{ mol } H_2O}\right)\left(\frac{22.41 \text{ L}}{1 \text{ mol } H_2}\right) = \underline{1.24 \times 10^3 \text{ L}}$$

(c)

$$2Li(s) + H_2(g) \longrightarrow 2LiH(s)$$

$$\text{Moles of LiH} = (1.00 \text{ kg LiH})\left(\frac{1000 \text{ g}}{1 \text{ kg}}\right)\left(\frac{1 \text{ mol LiH}}{7.949 \text{ g mol}^{-1}}\right) = \underline{126 \text{ mol}}$$

$$\text{Volume of } H_2 = (126 \text{ mol LiH})\left(\frac{1 \text{ mol } H_2}{2 \text{ mol LiH}}\right)\left(\frac{22.41 \text{ L}}{1 \text{ mol } H_2}\right) = \underline{1.41 \times 10^3 \text{ L}}$$

82.

$$2H_2O_2 \longrightarrow 2H_2O + O_2$$

First calculate the moles of oxygen produced:

$$\text{Moles } O_2 = \frac{PV}{RT} = \frac{(1.00 \text{ atm})(100 \text{ L})}{(0.0821 \text{ atm L mol}^{-1} \text{ K}^{-1})(298 \text{ K})} = \underline{4.09 \text{ mol}}$$

$$\text{Mass of } H_2O_2 = (4.09 \text{ mol } O_2)\left(\frac{2 \text{ mol } H_2O_2}{1 \text{ mol } O_2}\right)\left(\frac{34.02 \text{ g } H_2O_2}{1 \text{ mol } H_2O_2}\right) = \underline{278 \text{ g}}$$

83.

$$CaCO_3(s) \longrightarrow CaO(s) + CO_2(g)$$

Theoretical yield of $CO_2$ from 1.000 g $CaCO_3(s)$

$$= (1.000 \text{ g } CaCO_3)\left(\frac{1 \text{ mol } CaCO_3}{100.1 \text{ g } CaCO_3}\right)\left(\frac{1 \text{ mol } CO_2}{1 \text{ mol } CaCO_3}\right) = 9.99 \times 10^{-3} \text{ mol}$$

$$\text{Actual yield of } CO_2 = \frac{PV}{RT} = \frac{(755 \text{ torr})\left(\frac{1 \text{ atm}}{760 \text{ torr}}\right)(215 \text{ mL})\left(\frac{1 \text{ L}}{1000 \text{ mL}}\right)}{(0.0821 \text{ atm L mol}^{-1} \text{ K}^{-1})(298 \text{ K})}$$

$$= \underline{8.73 \times 10^{-3} \text{ mol}}$$

$$\text{Mass \% } CaCO_3 = \left(\frac{8.73 \times 10^{-3} \text{ mol}}{9.99 \times 10^{-3} \text{ mol}}\right)(100\%) = \underline{87.4\%}$$

84.

| | $2CO(g)$ | + $O_2(g)$ | $\longrightarrow 2CO_2(g)$ | |
|---|---|---|---|---|
| initially | 2.064 | 1.032 | 0 | moles |
| finally | 0 | 0 | 2.064 | moles |

(a) The pressure is proportional to the total moles of gases:

$$\frac{P_{final}}{P_{initial}} = \frac{P_{final}}{2.30 \text{ atm}} = \frac{2.064 \text{ mol}}{(2.064 + 1.032) \text{ mol}} = \frac{2}{3}\ ;\ P_{final} = \underline{1.53 \text{ atm}}$$

(b) From the ideal gas law, PV = nRT, for a fixed mass and volume, P is proportional to T:

$$\frac{P_1}{T_1} = \frac{P_2}{T_2} \quad ; \quad \frac{1.53 \text{ atm}}{298 \text{ K}} = \frac{P_2}{573 \text{ K}} \quad ; \quad P_2 = \underline{2.94 \text{ atm}}$$

85. From the data given

$$0.55 \text{ L Cyclopropane(g)} \rightarrow 1.65 \text{ L } CO_2(g) + 1.65 \text{ L } H_2O(g)$$

i.e., 1 vol → 3 vol + 3 vol

and, since at constant T and P, equal volumes of gases contain equal numbers of moles (molecules)

$$1 \text{ Cyclopropane} \rightarrow 3CO_2(g) + 3H_2O(g)$$

so that, 1 molecule of Cyclopropane must contain 3 C atoms and 6 H atoms

i.e., molecular formula = $C_3H_6$

and the balanced equation for the reaction with $O_2(g)$ is:

$$2C_3H_6(g) + 9O_2(g) \rightarrow 6CO_2(g) + 6H_2O(g)$$

86. First calculate the moles of gas in, for example exactly 1 L of ideal gas at 1 atm and 27°C.

$$n = \frac{PV}{RT} = \frac{(1 \text{ atm})(1 \text{ L})}{(0.0821 \text{ atm L mol}^{-1} \text{ K}^{-1})(300 \text{ K})} = 0.0406 \text{ mol}$$

and the number of molecules in 1 L is given by

$$(0.0406 \text{ mol})(6.022 \times 10^{23} \text{ molecules mol}^{-1}) = \underline{2.44 \times 10^{22} \text{ molecules}}$$

So that the volume occupied by 1 molecule is

$$V_1 = \frac{(1 \text{ L})\left(\frac{1 \text{ dm}^3}{1 \text{ L}}\right)\left(\frac{1 \text{ m}}{10 \text{ dm}}\right)^3\left(\frac{10^{12} \text{ pm}}{1 \text{ m}}\right)^3}{(2.44 \times 10^{22} \text{ molecules})} = \underline{4.10 \times 10^{10} \text{ pm}^3.}$$

For an actual molecule of radius 100 pm, the volume is given by

$$V_2 = \frac{4}{3}\pi r^3 = \frac{4}{3}\pi(100 \text{ pm})^3 = \underline{4.19 \times 10^6 \text{ pm}}$$

Taking the ratio of the effective volume to the actual volume, we have

$$\frac{V_1}{V_2} = \frac{(4.10 \times 10^{10} \text{ pm}^3)}{(4.19 \times 10^6 \text{ pm}^3)} \simeq \underline{10^4}$$

The available volume per molecule is about $10^4$ times the actual molecular volume.

87. For convenience, compare the volume of 1 mole of gaseous water at 100°C and 1 atm pressure with the volume of 1 mole of liquid water at 100°C.

Gaseous water $V = \frac{nRT}{P} = \frac{(1\ \text{mol})(0.0821\ \text{atm L mol}^{-1}\ \text{K}^{-1})(373\ \text{K})}{1\ \text{atm}}$

$= \underline{30.62\ \text{L}}$

Liquid water $V = (1\ \text{mol}\ H_2O)(\frac{18.02\ \text{g}\ H_2O}{1\ \text{mol}\ H_2O})(\frac{1\ \text{cm}^3}{0.96\ \text{g}}) = \underline{18.8\ \text{cm}^3}$

Expansion factor $= (\frac{30.62\ \text{L}}{18.8\ \text{cm}^3})(\frac{10^3\ \text{mL}}{1\ \text{L}})(\frac{1\ \text{cm}^3}{1\ \text{mL}}) = \underline{1630}$

88. Mass of Xe in bulb = (67.8883-67.6259) g = $\underline{0.2624\ \text{g}}$

Density $= \frac{\text{mass}}{\text{volume}} = (\frac{0.2624\ \text{g}}{50.00\ \text{mL}})(\frac{10^3\ \text{mL}}{1\ \text{L}}) = \underline{5.248\ \text{g L}^{-1}}$

For the mixture of xenon and oxygen,

Mass of 65 volume % Xe $= (\frac{5.248\ \text{g}}{1\ \text{L}})(50.00\ \text{mL})(\frac{1\ \text{L}}{10^3\ \text{mL}})(\frac{65}{100}) = \underline{0.1706\ \text{g}}$

Mass of 35 volume % $O_2 = (\frac{1.308\ \text{g}}{1\ \text{L}})(50.00\ \text{mL})(\frac{1\ \text{L}}{10^3\ \text{mL}})(\frac{35}{100}) = \underline{0.0229\ \text{g}}$

Mass of bulb = (67.6259+0.1706+0.0229) g = $\underline{67.8194\ \text{g}}$

89. First calculate the empirical formula of the hydrocarbon.

| | | C | H |
|---|---|---|---|
| grams in 100 g | = | 82.7 | 17.3 |
| mol in 100 g | = | $\frac{82.7}{12.01}$ | $\frac{17.3}{1.008}$ |
| ratio of moles (atoms) | = | 6.89 : 17.2 = 1.00 : 2.50 | |
| | = | 2.00 : | 5.00 |

empirical formula = $C_2H_5$ (empirical formula mass = 29.06 u)

Mol of gas in 1 L $= \frac{PV}{RT} = \frac{(750\ \text{mm})(\frac{1\ \text{atm}}{760\ \text{mm}})(1.00\ \text{L})}{(0.0821\ \text{atm L mol}^{-1}\ \text{K}^{-1})(303\ \text{K})} = \underline{0.0397\ \text{mol}}$

Molar mass $= (\frac{2.3080\ \text{g}}{0.0397\ \text{mol}}) = \underline{58.1\ \text{g mol}^{-1}}$

Molecular mass = 2(empirical formula mass)

Molecular formula = $\underline{C_4H_{10}}$ (molar mass = 58.12 g mol$^{-1}$)

The balanced equation for the combustion in oxygen is

$$2C_4H_{10}(g) + 13O_2(g) \longrightarrow 8CO_2(g) + 10H_2O(g)$$

Mass of $CO_2$ $= (10.00\ \text{g}\ C_4H_{10})(\frac{1\ \text{mol}}{58.12\ \text{g}})(\frac{8\ \text{mol}\ CO_2}{2\ \text{mol}\ C_4H_{10}})(\frac{44.01\ \text{g}\ CO_2}{1\ \text{mol}\ CO_2})$

$= \underline{30.29\ \text{g}}$

90. According to the kinetic molecular theory of gases, the gas molecules are in a constant state of motion and exert no force on each other, or on the walls of a containing vessel, unless they collide with each other. When collisions occur, they are elastic, so that the total energy of all the molecules remains constant. The average kinetic energy of the molecules of a gas is proportional to the absolute temperature.

(a) Pressure is the force per unit area exerted by the gas molecules and results from their collisions with the walls.

(b) The number of collisions with the walls, and thus the pressure, depends on the number of collisions per second, which in turn depends on the density of the gas molecules (number per unit volume, or concentration). Decreasing the volume increases the density of gas molecules proportionally, and leads to more collisions with the walls per second, which increases the pressure.

(c) Gases readily mix together because the actual molecules are very small compared to the total volume that they occupy; most of the volume is empty space and the molecules are free to move randomly in this space. By virtue of their high velocities, gases readily mix to give a homogeneous mixture. The pressure depends only on the total density of molecules and not on the type of molecule. Thus, if $n_1$ mol of one gas exerts a pressure of $p_1$, and $n_2$ mol of another gas exerts a pressure of $p_2$, $n_1+n_2$ mol of molecules exert a pressure $P = p_1+p_2$. In all cases the pressure is proportional to the total number of molecules, which is proportional to to total number of moles of molecules.

(d) The molecules of a gas move at different speeds, but with an average velocity at a given temperature, which determines their average kinetic energy (K.E. = $1/2\ mv^2$). As a gas is cooled, the average kinetic energy of its molecules decreases. But this average kinetic energy cannot decrease below zero, where all of the molecules are at rest, and the temperature at which this occurs is $-273.15^{\circ}C$ (0 K).

(e) Real gases differ from an ideal gas in two respects: (1) the volume occupied by the molecules of a real gas is not zero as assumed in the kinetic molecular theory. In other words, the measured volume is slightly greater than the actual volume in which the gas molecules are free to move, and (2) the molecules of a real gas attract each other slightly because of the weak attractive intermolecular forces between them. Thus, real gas molecules travel at average speeds slightly less than assumed for an ideal gas, so that the experimentally measured pressures are slightly less than the ideal pressures. Such effects are most pronounced at high pressures and low temperatures, where the molecules are relatively close to each other and their molecular volume is significant compared to the volume they occupy.

# CHAPTER 4

1.

| | Element | Group | Period | Type |
|---|---|---|---|---|
| (a) | He | 8 | 1 | Nonmetal |
| (b) | P | 5 | 3 | Nonmetal |
| (c) | K | 1 | 4 | Metal |
| (d) | Ca | 2 | 4 | Metal |
| (e) | Te | 6 | 5 | Nonmetal |
| (f) | Br | 7 | 4 | Nonmetal |
| (g) | Al | 3 | 3 | Metal |
| (h) | Sn | 4 | 5 | Metal |

2.

| Z | Element | Group | Period | Type |
|---|---|---|---|---|
| 2 | He | 8 | 1 | Nonmetal |
| 7 | N | 5 | 2 | Nonmetal |
| 9 | F | 7 | 2 | Nonmetal |
| 11 | Na | 1 | 3 | Metal |
| 16 | S | 6 | 3 | Nonmetal |
| 19 | K | 1 | 4 | Metal |

3. 1) Ar, Z = 18 (atomic mass 39.95 u); K, Z = 19, (atomic mass 39.10 u)

2) Co, Z = 27 (atomic mass 58.93 u); Ni, Z = 28 (atomic mass 58.70 u)

3) Te, Z = 52 (atomic mass 127.6 u); I, Z = 53 (atomic mass 126.9)
The position of an element in the periodic table is determined by its atomic number, Z, (the number of protons in the nuclei of its atoms, which is the same as the number of electrons surrounding the nucleus in the neutral atoms), rather than by the atomic mass (which is the weighted average of the masses of its isotopes). Note that there is also an apparent reversal at Th, Z = 90 (atomic mass 232.0 u) and Pa, Z = 91, (atomic mass approximately 231 u).

4. All of the given properties are those characteristic of a metal. In the periodic table the element with Z = 22 is titanium, Ti, a member of the series of transition metals in Period 4.

5.

| Element | Type | Group | Type |
|---|---|---|---|
| Se, selenium | main group | 6 | Nonmetal |
| P, phosphorus | main group | 5 | Nonmetal |
| Mn, manganese | transition | - | - |
| Kr, krypton | main group | 8 | Nonmetal |
| W, tungsten | transition | - | - |
| Al, aluminum | main group | 3 | Metal |
| Pb, lead | main group | 4 | Metal |

6.

| Element | Type | Group | Type |
|---|---|---|---|
| Ar, argon | main group | 8 | Nonmetal |
| Rb, rubidium | main group | 1 | Metal |
| V, vanadium | transition | - | - |
| Br, bromine | main group | 7 | Nonmetal |
| Ba, barium | main group | 2 | Metal |
| Fe, iron | transition | - | - |
| Au, gold | transition | - | - |

7.

| Symbol | Element | Group | No. of Valence Electrons* | Common Valence** |
|---|---|---|---|---|
| Li | Lithium | 1 | 1 | 1 |
| Mg | Magnesium | 2 | 2 | 2 |
| S | Sulfur | 6 | 6 | 2 |
| P | Phosphorus | 5 | 5 | 3 |
| Br | Bromine | 7 | 7 | 1 |
| Ne | Neon | 8 | 8 | 0 |
| As | Arsenic | 5 | 5 | 3 |
| Se | Selenium | 6 | 6 | 2 |
| Cl | Chlorine | 7 | 7 | 1 |
| Ba | Barium | 2 | 2 | 2 |

The metals are Li, Mg and Ba; the remainder of the elements are nonmetals.

*Number of valence electrons = group number

**Groups 1 to 4, common valence = number of valence electrons
Groups 5 to 8, common valence = 8 - (number of valence electrons)

8.

| Symbol | Element | Group | No. of Valence Electrons* | Common Valence** |
|---|---|---|---|---|
| Ca | Calcium | 2 | 2 | 2 |
| Al | Aluminum | 3 | 3 | 3 |
| F | Fluorine | 7 | 7 | 1 |
| Cs | Cesium | 1 | 1 | 1 |
| N | Nitrogen | 5 | 5 | 3 |
| I | Iodine | 7 | 7 | 1 |
| K | Potassium | 1 | 1 | 1 |
| Ar | Argon | 8 | 8 | 0 |
| O | Oxygen | 6 | 6 | 2 |
| Si | Silicon | 4 | 4 | 4 |

The metals are Ca, Al, Cs, and K; the remainder are nonmetals

*Number of valence electrons = group number

**Groups 1 to 4, common valence = number of valence electrons
Groups 5 to 8, common valence = 8 - (number of valence electrons)

9. (a) $Mg(s) + H_2O(g) \longrightarrow MgO(s) + H_2(g)$

(b) $S(s) + H_2(g) \longrightarrow H_2S(g)$ (c) $2Na(s) + I_2(s) \longrightarrow 2NaI(s)$

(d) $2K(s) + 2H_2O(l) \longrightarrow 2KOH(aq) + H_2(g)$

(e) $H_2(g) + Cl_2(g) \longrightarrow 2HCl(g)$ ; (f) NO REACTION

10. Group 3   B   Al   Ga   In   Tl

The group valence is 3, and the valence of Cl (group 7) is 1,

Hence: $BCl_3$, $AlCl_3$, $GaCl_3$, $InCl_3$, and $TlCl_3$

11. Francium is in group 1 (valence 1); the hydride is FrH.
Tin is in group 4 (valence 4); the hydride is $SnH_4$.
Radon is in group 8 (valence 0); no compound is expected.
Astatine is in group 7 (valence 1); the hydride is AtH.

Cl in group 7 has a valence of 1, the expected chlorides are FrCl, $SnCl_4$, and AtCl; Rn will form no chloride.

12. Astatine is a rare element of group 7 that comes in the periodic table below iodine. The trend in properties observed in descending the group is expected to continue from iodine to astatine.

(a) Astatine is expected to be a solid, continuing the trend fluorine and chlorine, gases; bromine liquid, and iodine solid.

(b) As an element of group 7, the common valence is expected to be 1. The formula of the potassium salt is KAt.

(c) All of the other halogens are found as diatomic molecules; Gaseous astatine consists of $At_2$ molecules.

(d) Fluorine is a pale yellow gas, chlorine is greenish-yellow, bromine is brown, iodine is violet, and astatine would be expected to be an even deeper color than iodine.

13.

| Element | Group | Valence | Type | Hydride |
|---|---|---|---|---|
| Carbon | 4 | 4 | Nonmetal | $CH_4$ |
| Calcium | 2 | 2 | Metal | $CaH_2$ |
| Helium | 8 | 0 | Nonmetal | None |
| Boron | 3 | 3 | Nonmetal | $BH_3$ |
| Chlorine | 7 | 1 | Nonmetal | HCl |
| Lithium | 1 | 1 | Metal | LiH |
| Oxygen | 6 | 2 | Nonmetal | $H_2O$ |
| Fluorine | 7 | 1 | Nonmetal | HF |
| Phosphorus | 5 | 3 | Nonmetal | $PH_3$ |
| Magnesium | 2 | 2 | Metal | $MgH_2$ |

14. (a) $Mg(s) + Br_2(l) \longrightarrow MgBr_2(s)$

(b) $2Ca(s) + O_2(g) \longrightarrow 2CaO(s)$

(c) $2Na(s) + I_2(s) \longrightarrow 2NaI(s)$

(d) $3Mg(s) + N_2(g) \longrightarrow Mg_3N_2(s)$ (at high temperature)

(e) $2K(s) + 2H_2O(l) \longrightarrow 2KOH(aq) + H_2(g)$

15. (a) $2Li(s) + S(s) \longrightarrow Li_2S(s)$

(b) $Ca(s) + 2H_2O(l) \longrightarrow Ca(OH)_2(aq) + H_2(g)$

(c) $Ne(g) + H_2O(l) \longrightarrow$ no reaction (Ne is inert)

(d) $Sr(s) + Br_2(l) \longrightarrow SrBr_2(s)$

(e) $Mg(s) + H_2(g) \longrightarrow MgH_2(s)$ (high temperature reaction)

16. (a) Although we should also consider diagonal relationships between elements in the periodic table, the greatest similarities are expected between elements in the same group, especially when the elements are adjacent to each other.

| Element | Na | Mg | C | Cl | Ca | Si | K | F |
|---|---|---|---|---|---|---|---|---|
| Group | 1 | 2 | 4 | 7 | 2 | 4 | 1 | 7 |
| Period | 3 | 3 | 2 | 3 | 4 | 3 | 4 | 2 |

Thus we have group 1, Na and K ; group 2, Mg and Ca;
group 4, C and Si ; group 7, Cl and F ;

and in their respective groups the elements are in adjacent periods so that each member of each pair will have a close similarity in chemical and physical properties.

(b) $2Na(s) + H_2(g) \longrightarrow 2NaH(s)$ ; $2K(s) + H_2(g) \longrightarrow 2KH(s)$

$4Na(s) + O_2(g) \longrightarrow 2Na_2O(s)$; $4K(s) + O_2(g) \longrightarrow 2K_2O(s)$

$Mg(s) + H_2(g) \longrightarrow MgH_2(s)$ ; $Ca(s) + H_2(g) \longrightarrow CaH_2(s)$

$2Mg(s) + O_2(g) \longrightarrow 2MgO(s)$ ; $2Ca(s) + O_2(g) \longrightarrow 2CaO(s)$

$C(s) + 2H_2(g) \longrightarrow CH_4(g)$ ; $Si(s) + 2H_2(g) \longrightarrow SiH_4(g)$

$C(s) + O_2(g) \longrightarrow CO_2(g)$ ; $Si(s) + O_2(g) \longrightarrow SiO_2(s)$

$Cl_2(g) + H_2(g) \longrightarrow 2HCl(g)$ ; $F_2(g) + H_2(g) \longrightarrow 2HF(g)$

$2Cl_2(g) + O_2(g) \longrightarrow 2Cl_2O(g)$ ; $2F_2(g) + O_2(g) \longrightarrow 2F_2O(g)$

17.

| | Element | Mg | K | Br | P | Si | Al | S | Pb |
|---|---|---|---|---|---|---|---|---|---|
| (a) | Group | 2 | 1 | 7 | 5 | 4 | 3 | 6 | 4 |
| | Valence | 2 | 1 | 1 | 3 | 4 | 3 | 2 | 4 |
| (b) | Fluoride | $MgF_2$ | KF | BrF | $PF_3$ | $SiF_4$ | $AlF_3$ | $SF_2$ | $PbF_4$ |
| | Sulfide | MgS | $K_2S$ | $SBr_2$ | $P_2S_3$ | $SiS_2$ | $Al_2S_3$ | S | $PbS_2$ |

18. Hydrogen is placed at the top of group 1 because it has one valence electron, as do the alkali metals of group 1. However it differs from the alkali metals in that it is a nonmetal and exists as gaseous $H_2$ molecules. In these respects it resembles $F_2$ in group 7, which like hydrogen has a valence of 1 , is a nonmetal and exists as gaseous diatomic molecules, thus the top of group 7 also seems a logical place in the periodic table for hydrogen. Hydrogen atoms can also gain an electron to give $:H^-$, hydride, ions, and in this respect also resemble the halogens, which can form $X^-$, halide, ions. (It also seems

that group 7 is a logical place for H when one considers the next element, helium. Although He has two valence shell electrons, like the alkaline earth metals of group 2, its chemical inertness places it in group 8, rather than in group 2. In fact, like He, H is unique among the elements because its valence shell can accommodate a maximum of only two electrons.

19. Fluorine (group 7, valence 1) and oxygen (group 6, valence 2) form a compound of empirical formula $F_2O$.

Aluminum (group 3, valence 3) and sulfur (group 6, valence 2) form a compound of empirical formula $Al_2S_3$.

Boron (group 3, valence 3) and chlorine (group 7, valence 1) form a compound of empirical formula $BCl_3$.

Carbon (group 4, valence 4) and sulfur (group 6, valence 2) form a compound of empirical formula $CS_2$.

Magnesium (group 2, valence 2) and nitrogen (group 5, valence 3) form a compound of empirical formula $Mg_3N_2$.

20. This type of problem requires knowing the maximum number of electrons that each electron shell can accommodate. For the first three shells, it is 2.8.8, which takes us as far as the noble gas Ar.

(a) Phosphorus, Z = 15 = 2.8.5
Carbon, Z = 6 = 2.4
Potassium, Z = 19 = 2.8.8.1

(b) The core charge of an atom is equal to its atomic number, Z, minus the number of electrons in inner filled shells.

Phosphorus, core charge = 15-10 = 5
Carbon, core charge = 6-2 = 4
Potassium, core charge = 19-18 = 1

More exactly, we should give these as +5,+4, and +1

(More simply, the core charge of an atom is the same as its <u>group</u> number:

P, group 5, core charge +5
C, group 4, core charge +4
K, group 1, core charge +1)

21. The <u>valence shell</u> of an atom is the outermost shell of electrons. The <u>number of electrons in the valence shell</u> of a neutral atom is the same as its group number (which is the same as its core charge). Thus,

(a) Boron, group 3, valence electrons = 3
(b) Any halogen, group 7, valence electrons = 7
(c) Neon, group 8, valence electrons = 8
(d) Magnesium, group 2, valence electrons = 2
(e) Helium is an exception to the general rule. Although it is in group 8, like the other noble gases it has a filled valence shell, but in this case it is the n = 1 shell, which accommodates a maximum of 2 electrons. Valence electrons = 2

22. First locate the element in question with respect to its group and period in the periodic table. The group number = number of valence electrons, and the period gives the total number of electron shells, of which the outer is the valence shell. Thus:

(a) A neutral Cl atom: Cl is in group 7 and period 3; the outermost shell contains 7 electrons; of the 3 shells the first two are filled, and the outermost (valence) shell contains 7 electrons. Thus, we have the electron arrangement 2.8.7, for a total of 17 electrons.

(b) A negatively charged Cl atom must have one more electron than a neutral Cl atom. The shell arrangement must be 2.8.8, = 18 electrons in total.

(c) A (neutral) Si atom: Si is in group 4 and period 3; the outermost shell contains 4 electrons; of the 3 shells the first two are filled, and the outermost (valence) shell contains 4 electrons. Thus, we have the arrangement 2.8.4, for a total of 14 electrons.

(d) A positively charged Ne atom must have one electron less than a neutral Ne atom. Ne is in group 8 and period 2, with the configuration 2.8. The electron configuration of $Ne^+$ must be 2.7, for a total of 9 electrons.

We usually write these configurations as:

(a) Cl 2.8.7 (b) $Cl^-$ 2.8.8 (c) Si 2.8.4 (d) $Ne^+$ 2.7

23. (a) The valence shell of an atom is the outermost electron shell.
(b) For all of the atoms in a particular group of the periodic table, the number of valence electrons is the same as the group number.
(c) We locate each element in the periodic table with respect to group and period. The period gives the number of electron shells and the group number gives the number of electrons in the outermost (valence) shell, for the neutral atom. For negatively charged species, we add a number of electrons equal to the negative charge to the number of valence electrons of the neutral atom; for a positively charged species we subtract a number of electrons equal to the postive charge from the number of valence electrons of the neutral atom. Thus:

N, group 5, period 2, configuration 2.5, valence electrons = 5

$N^{3-}$, configuration 2.8, valence electrons = 8

Be, group 2, period 2, configuration 2.2, valence electrons = 2

$Be^{2+}$, configuration 2.0, valence electrons = 0

O, group 6, period 2, configuration 2.6, valence electrons = 6

$O^{2-}$, configuration 2.8, valence electrons = 8

Na, group 1, period 3, configuration 2.8.1, valence electrons = 1

$Na^+$, configuration 2.8, valence electrons = 8

Cl, group 7, period 3, configuration 2.8.7, valence electrons = 7

$Cl^-$, configuration 2.8.8, valence electrons = 8

(d) Isoelectronic species have the same electron configuration. Thus, of the above species:

$N^{3-}$ $O^{2-}$ and $Na^+$ are isoelectronic

24. Core charge is equal to the atomic number Z minus the number of electrons in inner shells. Note that for an ion the outer shell is often a filled shell. Deduce the shell configuration first and then subtract the number of inner shell electrons from Z.

C, Z = 6, shell configuration 2.4, core charge +6-2 = +4
Mg , Z = 12, shell configuration 2.8.2, core charge +12-10 = +2
$Mg^{2+}$, Z = 12, shell configuration 2.8, core charge +12-2 = +10
Si, Z = 14, shell configuration 2.8.4, core charge = +14-10 = +4
O, Z = 8, shell configuration 2.6, core charge +8-2 = +6
$O^{2-}$, Z = 8, shell configuration 2.8, core charge = +8-2 = +6
$S^{2-}$, Z = +16, shell configuration 2.8.8, core charge = +16-10 = +6
Br, Z = +35, shell configuration 2.8.18.7, core charge = +35-28 = +7

25.

| | Element | Group | Core Charge | | Element | Group | Core Charge |
|---|---|---|---|---|---|---|---|
| (a) | F | 7 | +7 | (f) | Mg | 2 | +2 |
| (b) | N | 5 | +5 | (g) | Cs | 1 | +1 |
| (c) | S | 6 | +6 | (h) | Si | 4 | +4 |
| (d) | Li | 1 | +1 | (i) | I | 7 | +7 |
| (e) | O | 6 | +6 | | | | |

26. The ionization energy is the energy required to remove an electron from an atom in the gas phase. In going from left to right across any period, the electron that is removed is a valence shell electron and for all the elements in the period its distance from the nucleus is approximately the same, but it becomes progressively more difficult to remove because it is attracted by a core charge that increases progressively across the period (from +1, group 1, to +8, group 8). In any group, the core charge remains constant but the valence shell is progressively further from the nucleus as the number of filled inner shells increases, and thus the attraction for an outer electron decreases, making it progressively easier to remove an electron.

27. Use the generalizations given in Problem 26. Consider first the core charge of each atom, and then the distance of the valence shell from the nucleus:

F, group 7, core charge +7, period 2
Ne, group 8, core charge +8, period 2
Na, group 1, core charge +1, period 3

Thus, of the three atoms, Na has the smallest core charge and its valence shell is relatively the farthest from the nucleus. Ne and F both have n = 2 valence shells but the core charge of Ne is greater than that of F. Thus in increasing order we have for the ionization energies

$$\underline{Na < F < Ne}$$

28. This problem is similar to Problem 27. Placing the atoms in order of their increasing periods gives the increasing order of the distance of the valence shells from the nuclei. Then, when necessary, consider the relative core charges for atoms in the same period.

| Atom | Period | Group | Core Charge |
|---|---|---|---|
| Ba | 6 | 2 | +2 |
| Cs | 6 | 1 | +1 |
| F | 2 | 7 | +7 |
| S | 3 | 6 | +6 |
| As | 4 | 5 | +5 |

Period number increases in the order F < S < As < Ba = Cs, which gives the order of decreasing ionization energies, except for Ba and Cs, where in terms of core charge Ba > Cs.

In order of increasing ionization energy,

Cs < Ba < As < S < F

29. The ionization energy of sodium is 0.50 MJ $mol^{-1}$, thus for 1.00 g of sodium atoms

$$\text{Energy} = (1.00\text{ g Na})\left(\frac{1\text{ mol Na}}{22.99\text{ g Na}}\right)\left(\frac{0.50\text{ MJ}}{1\text{ mol Na}}\right) = \underline{0.022\text{ MJ}}\ \ (22\text{ kJ})$$

30. The valence shell of He is the smallest possible, n = 1, so that the only other atom to consider is H. But the core charge of He is +2, while that of H is +1, so the ionization energy of He must be the largest for any atom. In going from period 1 to any other period, the ionization energy must decrease, independent of core charge, because of the increasing distance of the valence shell from the nucleus.

31. The total number of valence electrons is the same as the group number so that the total number of electrons in the Lewis dot symbol is given by the group number. The symbol shows the valence electrons arranged singly if there are from 1 to 4 electrons, and as electron pairs and single electrons if there are more than 4 valence electrons.

32. The group number gives the total number of electrons. Arrange these singly or as pairs according to the total number.

| Element: | K | Ca | B | Sn | Sb | Te | Br | Xe | As | Ge |
|---|---|---|---|---|---|---|---|---|---|---|
| Group | 1 | 2 | 3 | 4 | 5 | 6 | 7 | 8 | 5 | 4 |
| Symbol | K· | ·Ca· | ·B· | ·Sn· | :Sb· | :Te· | :Br· | :Xe: | :As· | ·Ge· |

33. (a) Each dot represents a valence shell electron; the total number of valence electrons gives the group number: A, group 1, D, group 4, E, group 2, and G, group 7.

(b) Cations are formed from metal atoms by the loss of electrons to give a noble gas structure with no electrons left in the valence shell of the metal atom. Commonly 1 to 3 electrons may be lost in this way. Thus, A and E are metals that are expected to form $A^+$ and $E^{2+}$ ions, respectively.

Anions are formed from nonmetal atoms by the gain of electrons, to give ions with a noble gas structure in which the valence shell of the nonmetal atom is complete; commonly 1 to 3 electrons may be added in this way. Thus, G, is a nonmetal that would be expected to form the $G^-$ ion.

D with 4 valence electrons is not expected to form either $D^{4+}$ or $D^{4-}$.

34. The number of valence electrons is given by the group number; cations are formed by the loss of all of the valence electrons, and anions are formed by adding sufficient electrons to completely fill the valence shell.

(a)

| atom | Li | H | Al | Na | Ca | Mg | Rb | Sc |
|---|---|---|---|---|---|---|---|---|
| group | 1 | 1 | 3 | 1 | 2 | 2 | 1 | * |
| electrons lost | 1 | 1 | 3 | 1 | 2 | 2 | 1 | 3 |
| ion | $Li^+$ | $H^+$ | $Al^{3+}$ | $Na^+$ | $Ca^{2+}$ | $Mg^{2+}$ | $Rb^+$ | $Sc^{3+}$ |

*Sc (scandium) is the first transition metal in period 4 and has 3 valence electrons (1 more than Ca).

(b)

| atom | O | H | Cl | N | P | S | F | I |
|---|---|---|---|---|---|---|---|---|
| group | 6 | 1 | 7 | 5 | 5 | 6 | 7 | 7 |
| electrons gained | 2 | 1* | 1 | 3 | 3 | 2 | 1 | 1 |
| ion | $O^{2-}$ | $H^-$ | $Cl^-$ | $N^{3-}$ | $P^{3-}$ | $S^{2-}$ | $F^-$ | $I^-$ |

*In all cases except H, the valence shell is complete with a total of 8 electrons. H, in group 1 is unique in that its valence shell is complete with a total of 2 electrons.

35. The metal atoms (Mg, Rb, Al, and Li) lose electrons to form cations; the nonmetal atoms (Br and S) gain electrons to form anions.

| | Mg | Rb | Br | S | Al | Li |
|---|---|---|---|---|---|---|
| group | 2 | 1 | 7 | 6 | 3 | 1 |
| electrons lost | 2 | 1 | - | - | 3 | 1 |
| electrons gained | - | - | 1 | 2 | - | - |
| ion | $Mg^{2+}$ | $Rb^+$ | $Br^-$ | $S^{2-}$ | $Al^{3+}$ | $Li^+$ |

36. In each case the metal atoms lose all of their valence electrons and the nonmetal atoms gain sufficient electrons to complete their octets:

(a) $Li\cdot + \cdot\ddot{\underset{\cdot\cdot}{Cl}}: \longrightarrow Li^{+}\,[:\ddot{\underset{\cdot\cdot}{Cl}}:^{-}]$

(b) $2Na\cdot + \cdot\dot{\underset{\cdot\cdot}{O}}: \longrightarrow [Na^{+}]_2[:\ddot{\underset{\cdot\cdot}{O}}:^{2-}]$

(c) $\cdot\dot{Al}\cdot + 3\ \cdot\ddot{\underset{\cdot\cdot}{F}}: \longrightarrow Al^{3+}[:\ddot{\underset{\cdot\cdot}{F}}:^{-}]_3$

(d) $\cdot Ca\cdot + \cdot\dot{\underset{\cdot\cdot}{S}}: \longrightarrow Ca^{2+}[:\ddot{\underset{\cdot\cdot}{S}}:^{2-}]$

(e) $\cdot Mg\cdot + 2\ \cdot\ddot{\underset{\cdot\cdot}{Br}}: \longrightarrow Mg^{2+}[:\ddot{\underset{\cdot\cdot}{Br}}:^{-}]_2$

37. In each case, we combine sufficient cations and anions to give a neutral compound. In general for a $M^{n+}$ cation and an $A^{m-}$ anion, where the compound is $M_xA_y$, we have:

$$xn = ym$$

e.g., for $Fe^{3+}$ and $O^{2-}$, $3x = 2y$, or $\frac{y}{x} = \frac{3}{2}$, and the formula is $Fe_2O_3$

(a) $(NH_4)_2S$ (b) $Fe_2O_3$ (c) $Cu_2O$ (d) $AlCl_3$

38. First we deduce the charges on the respective ions using the method used in Problem 34, then use the method outlined in Problem 37 to deduce the formula of the compound, and write the Lewis structure:

(a) Li is a metal from group 1 and S is a nonmetal from group 6. Li can lose its one electron to form $Li^+$, and sulfur can complete its octet by gaining two electrons to form $S^{2-}$. Hence, we have:

$2Li^{+} + S^{2-} \longrightarrow Li_2S$ Lewis structure: $[Li^{+}]_2[:\ddot{\underset{\cdot\cdot}{S}}:^{2-}]$

(b) Be in group 2 loses two electrons to give $Be^{2+}$, and O in group 6 gains two electrons to give $O^{2-}$:

$Be^{2+} + O^{2-} \longrightarrow BeO$ Lewis structure: $Be^{2+}\ :\ddot{\underset{\cdot\cdot}{O}}:^{2-}$

(c) Mg is in group 2 and loses two electrons to give $Mg^{2+}$; Br in group 7 gains one electron to give $Br^-$:

$Mg^{2+} + 2Br^{-} \longrightarrow MgBr_2$ Lewis structure: $Mg^{2+}[:\ddot{\underset{\cdot\cdot}{Br}}:^{-}]_2$

(d) Na is the metal and H is the nonmetal; both are in group 1. Na loses an electron to give $Na^+$; H gains an electron to form $H^-$ :

$Na^{+} + H^{-} \longrightarrow NaH$ Lewis structure: $Na^{+}\ :H^{-}$

(e) Al in group 3 loses 3 electrons to give $Al^{3+}$; I in group 7 gains an electron to give $I^-$ :

$Al^{3+} + 3I^{-} \longrightarrow AlI_3$ Lewis structure: $Al^{3+}[:\ddot{\underset{\cdot\cdot}{I}}:^{-}]_3$

39. This problem is similar to Problem 38.

(a) Ca, metal, group 2, gives $Ca^{2+}$; I, nonmetal, group 7 gives $I^-$ :

$Ca^{2+} + 2I^{-} \longrightarrow CaI_2$ Lewis structure: $Ca^{2+}[:\ddot{\underset{\cdot\cdot}{I}}:^{-}]_2$

(b) Ca, metal, group 2, gives $Ca^{2+}$; O, nonmetal, group 6, gives $O^{2-}$ :

$Ca^{2+} + O^{2-} \longrightarrow CaO$   Lewis structure: $Ca^{2+} :\ddot{\underset{..}{O}}:^{2-}$

(c) Al, metal, group 3, gives $Al^{3+}$; S, nonmetal, group 6, gives $S^{2-}$ :

$2Al^{3+} + 3S^{2-} \longrightarrow Al_2S_3$   Lewis structure: $[Al^{3+}]_2[:\ddot{\underset{..}{S}}:^{2-}]_3$

(d) Ca, metal, group 2, gives $Ca^{2+}$; Br, nonmetal, group 7, gives $Br^-$ :

$Ca^{2+} + 2Br^- \longrightarrow CaBr_2$   Lewis structure: $Ca^{2+}[:\ddot{\underset{..}{Br}}:^-]_2$

(e) Rb, metal, group 1, gives $Rb^+$; Se, nonmetal, group 6, gives $Se^{2-}$ :

$2Rb^+ + Se^{2-} \longrightarrow Rb_2Se$   Lewis structure: $[Rb^+]_2 :\ddot{\underset{..}{Se}}:^{2-}$

(f) Ba, metal, group 2, gives $Ba^{2+}$; O, nonmetal,group 6, gives $O^{2-}$ :

$Ba^{2+} + O^{2-}$ —— $BaO$   Lewis structure: $Ba^{2+} :\ddot{\underset{..}{O}}:^{2-}$

40. This is essentially the same kind of problem as Problems 38 and 39 : barium iodide is a compound of barium and iodine, aluminum chloride is a compound of aluminum and chlorine, etc.,

(a) Ba, metal, group 2, gives $Ba^{2+}$; I, nonmetal group 7, gives $I^-$ :

$Ba + I_2 \rightarrow BaI_2$

(b) Al, metal, group 3, gives $Al^{3+}$; Cl, nonmetal, group 7, gives $Cl^-$ :

$2Al + 3Cl_2 \rightarrow 2AlCl_3$

(c) Ca, metal, group 2, gives $Ca^{2+}$; O, nonmetal, group 6, gives $O^{2-}$ :

$2Ca + O_2 \longrightarrow 2CaO$

(d) Na, metal, group 1, gives $Na^+$; S, nonmetal, group 6, gives $S^{2-}$ :

$2Na + S \longrightarrow Na_2S$

(e) Al, metal, group 3, gives $Al^{3+}$; O, nonmetal, group 6, gives $O^{2-}$ :

$4Al + 3O_2 \longrightarrow 2Al_2O_3$

41. The strategy here is to first deduce the possible ions that these elements can form:

H, group 1, can form $H^+$ and $:H^-$ ; Mg, group 2, can form $Mg^{2+}$

O, group 6, can form $O^{2-}$ ; Al, group 3, can form $Al^{3+}$

F, group 7,can form $F^-$ ; Ca, group 2, can form $Ca^{2+}$

Thus, the binary ionic compounds in question are those that can be formed between each of the cations $Mg^{2+}$, $Al^{3+}$, and $Ca^{2+}$, and each of the anions $H^-$, $O^{2-}$, and $F^-$. ($H^+$, the proton is excluded because it forms no ionic compounds but only covalent species). Taking each cation in turn:

$Mg^{2+} + 2H^- \longrightarrow MgH_2$   Lewis structure: $Mg^{2+}[:H^-]_2$

$Mg^{2+} + O^{2-} \longrightarrow MgO$   Lewis structure: $Mg^{2+} :\ddot{\underset{..}{O}}:^{2-}$

$Mg^{2+} + 2F^- \longrightarrow MgF_2$   Lewis structure: $Mg^{2+}[:\ddot{\underset{..}{F}}:^-]_2$

$Al^{3+} + 3H^- \longrightarrow AlH_3$ Lewis structure: $Al^{3+}[:H^-]_3$

$2Al^{3+} + 3O^{2-} \longrightarrow Al_2O_3$ Lewis structure $[Al^{3+}]_2[:\ddot{O}:^{2-}]_3$

$Al^{3+} + 3F^- \longrightarrow AlF_3$ Lewis structure $Al^{3+}[:\ddot{F}:^-]_3$

$Ca^{2+} + 2H^- \longrightarrow CaH_2$ Lewis structure $Ca^{2+}[:H^-]_2$

$Ca^{2+} + O^{2-} \longrightarrow CaO$ Lewis structure $Ca^{2+}:\ddot{O}:^{2-}$

$Ca^{2+} + 2F^- \longrightarrow CaF_2$ Lewis structure $Ca^{2+}[:\ddot{F}:^-]_2$

42.* (a) Ca, ·Ca· ; (b) $Ca^{2+}$, $Ca^{2+}$; (c) Ne, :N̈e: ; (d) $O^{2-}$, $:\ddot{O}:^{2-}$

(e) $S^{2-}$, $:\ddot{S}:^{2-}$ ; (f) $Cl^-$, $:\ddot{Cl}:^-$ ; (g) Ar, :Är: ; (h) $N^{3-}$, $:\ddot{N}:^{3-}$

43. All of the molecules are composed of nonmetal atoms, which must be joined by <u>covalent</u> (electron pair) bonds.

(a) H· + ·H ⟶ H-H (b) H· + ·C̈l: ⟶ H-C̈l: (c) H· + ·Ï: ⟶ H-Ï:

(d) 3H· + ·Ṗ· ⟶ H-P̈-H (with H below P)

(e) 4:F̈· + ·Si· ⟶ :F̈-Si-F̈: (with :F̈: above and below Si) (f) 2:F̈· + ·Ö: ⟶ :F̈-Ö-F̈:

(g) :C̈l· + ·C̈l: ⟶ :C̈l-C̈l:

44. As in problem 43, all the molecules consist of nonmetal atoms and the atoms must be joined by <u>covalent</u> bonds.

(a) 2H· + ·Ċ· + ·Ö: ⟶ H-C=Ö: (with H below C) (each H atom forms a single C-H bond, leaving two unpaired electrons to form a CO double bond)

(b) :Ṗ· + ·Ṗ: ⟶ :P≡P: (three unpaired electrons on each P atom combine to form a triple bond)

(c) Here it is useful to accommodate the negative charge on $CN^-$ by first adding an electron to carbon, then we can write

:Ċ·⁻ + ·Ṅ: ⟶ ⁻:C≡N:

(d) 2 H· + ·N̈· + ·N̈·+ 2·H ⟶ H—N̈—N̈—H (with H above each N)

(e) H· + ·Ö· + ·Ö· + ·H ⟶ H—Ö—Ö—H

45. The bonds between identical atoms, as in $S_6$, must be <u>covalent</u> bonds.

*isoelectronic species: (b),(e), (f), and (g), <u>and</u> (c), (d), and (h).

Each ·S: atom can form two electron pair bonds to other sulfur atoms. Thus, in the $S_6$ ring, all the bonds are single electron pair bonds; in addition to forming two covalent bonds, each S atom has two unshared electron pairs.

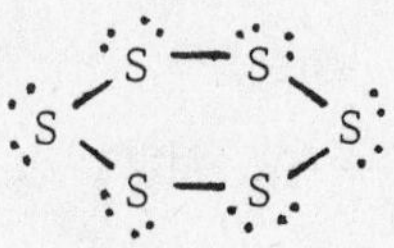

46. Antimony is in group 5, bromine is in group 7, tellurium is in group 6, and tin is in group 4. Thus the valences are Sb, 3; Br, 1; Te, 2, and Sn, 4, respectively. Since the valence of H is 1, the predicted empirical formulas are:

$$SbH_3 \quad HBr \quad H_2Te \quad \text{and} \quad SnH_4$$

The Lewis structures may be obtained as follows:

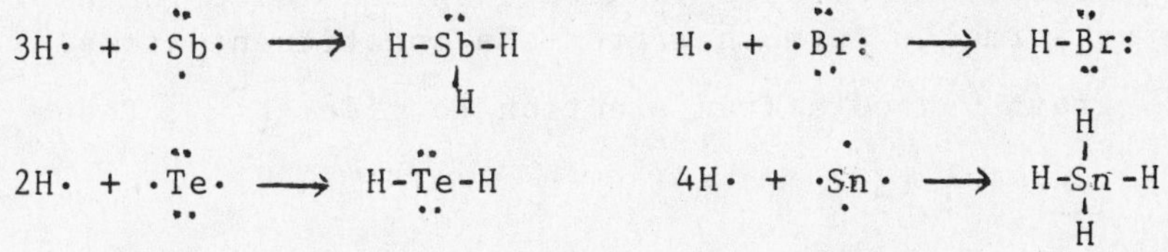

We first draw the Lewis structures, and then assign formal charges. The formal charge on each atom in a molecule is calculated by summing up the atom's share of the electrons. To do this

(i) unshared (lone) pairs of electrons are assigned entirely to the atom of which they constitute part of the valence shell

(ii) each shared (bonding) pair of electrons is shared equally between the atoms it bonds together.

Then the total number of electrons assigned to any atom is compared to the core charge of the atom.

(a) $NH_4^+$ We can deduce the Lewis structure as follows:

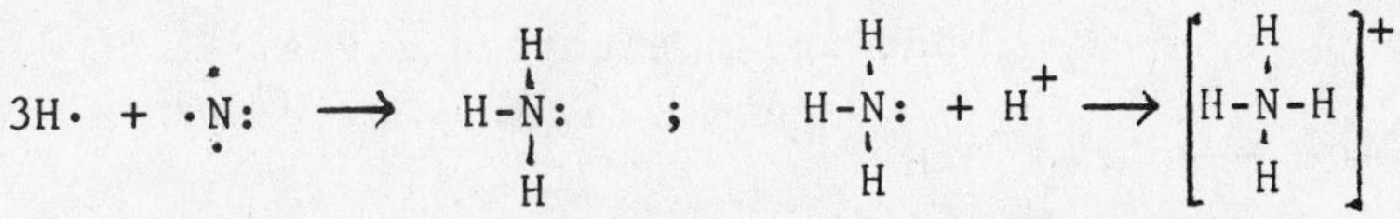

Each of the H atoms in $NH_4^+$ has a core charge of +1, and is assigned one-half of one bonding pair of electrons, or a charge of -1. i.e., the formal charge on each H atom is +1-1 = 0
The N atom is assigned one-half of four bonding pairs, or a charge of -4. The core charge on N (group 5) is +5. i.e., the formal charge on N is +5-4 = +1. Thus, we write:

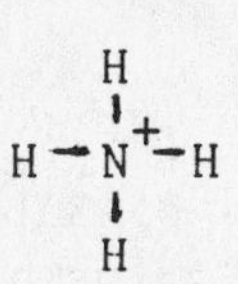 for the complete Lewis structure

Writing Lewis structures for charged species sometimes presents students with some difficulty. A useful approach is to first write the Lewis symbol for the central atom, and immediately allow for any charges on the ion it is part of by <u>subtracting</u> one electron for each + charge, or <u>adding</u> one electron for each - charge. For example for $NH_4^+$, we write:

$$\cdot \overset{\displaystyle \cdot}{\underset{\displaystyle \cdot}{N}} : \quad \xrightarrow{-e^-} \quad \cdot \overset{\displaystyle \cdot}{\underset{\displaystyle \cdot}{N}}^+$$

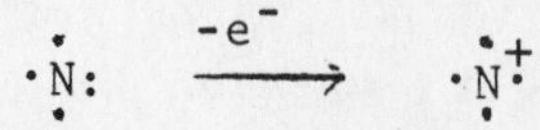

Then we can form the four N-H bonds by sharing each of the four unpaired electrons with a H atom:

$$4H\cdot \; + \; \cdot \overset{\displaystyle \cdot}{\underset{\displaystyle \cdot}{N}}^+ \; \longrightarrow \; H-\underset{\displaystyle H}{\overset{\displaystyle H}{N^+}}-H$$

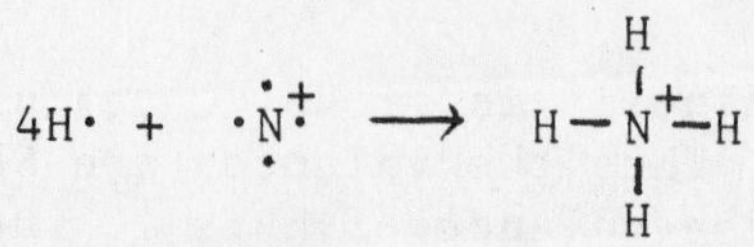

Both ways of deducing the Lewis structure of the $NH_4^+$ ion are equally valid; remember that once the molecule is formed it has no notion of which atoms formally supplied the electrons in its Lewis structure. The four N-H bonds of the ammonium ion are identical electron pair bonds, however we formally form them from the constituent atoms.

(b) <u>BH$_4^-$</u> Start with $\cdot \overset{\displaystyle \cdot}{B} \cdot$ (group 3) and add an electron to give $\cdot \overset{\displaystyle \cdot}{B} \overset{\displaystyle +}{:}$ , then:

$$4H\cdot \; + \; \cdot \overset{\displaystyle \cdot}{B} \overset{-}{:} \; \longrightarrow \; H-\underset{\displaystyle H}{\overset{\displaystyle H}{B}}-H$$

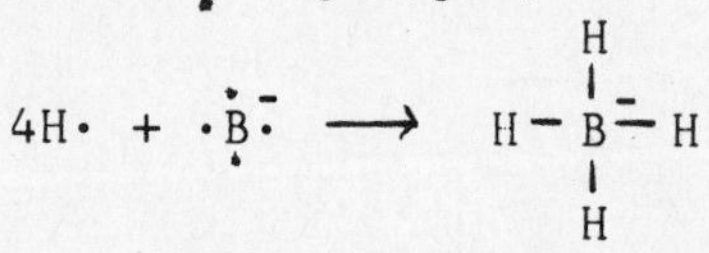

We can check that the formal charge is indeed on B, by calculating it formally. The core charge of boron (group 3) is +3, and boron has a half-share of four bonding pairs = -4:

$$\text{Formal charge on boron} = +3-4 = \underline{-1}$$

(c) <u>CH$_3^+$</u> Carbon (group 4) is $\cdot \overset{\displaystyle \cdot}{C} \cdot$; removing one electron gives $\cdot \overset{\displaystyle \cdot}{C} \overset{+}{:}$

$$3H\cdot \; + \; \cdot \overset{\displaystyle \cdot}{C} \overset{+}{:} \; \rightarrow \; H-\underset{\displaystyle H}{C^+}-H$$

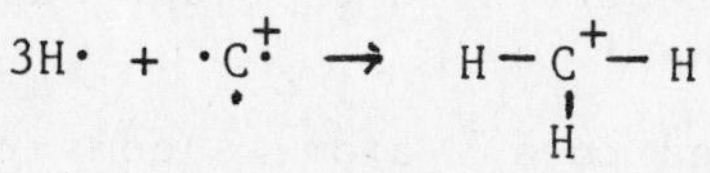

C has a half-share of 3 bonding pairs, -3, and a core charge of +4.
$$\text{Formal charge on C} = +4-3 = \underline{+1}$$

(d) <u>CH$_3^-$</u> Carbon (group 4) is $\cdot \overset{\displaystyle \cdot}{C} \cdot$ ; adding an electron gives $\cdot \overset{\displaystyle \cdot\cdot}{C} \overset{-}{:}$

$$3H\cdot \; + \; \cdot \overset{\displaystyle \cdot\cdot}{C} \overset{-}{:} \; \longrightarrow \; H-\underset{\displaystyle H}{\overset{\displaystyle \cdot\cdot}{C}}^--H$$

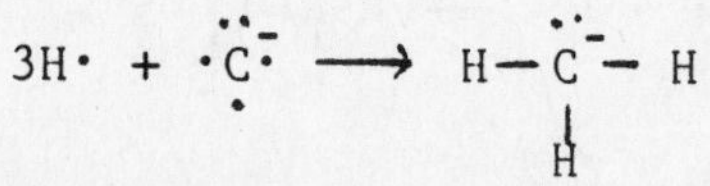

C has one lone pair and a half-share of 3 bonding pairs, -2-3 = -5; the core charge of C is +4 (group 4).
$$\text{Formal charge on C} = -5+4 = \underline{-1}$$

(e) <u>SCl$_3^+$</u> Sulfur (group 6) is $\cdot \overset{\displaystyle \cdot\cdot}{S} :$ ; removing an electron gives $\cdot \overset{\displaystyle \cdot\cdot}{S} \overset{+}{:}$

$$3:\overset{\displaystyle \cdot\cdot}{\underset{\displaystyle \cdot\cdot}{Cl}}\cdot \; + \; \cdot \overset{\displaystyle \cdot\cdot}{S} \overset{+}{:} \; - \; :\overset{\displaystyle \cdot\cdot}{\underset{\displaystyle \cdot\cdot}{Cl}}-\underset{\displaystyle \overset{\displaystyle |}{:\overset{\displaystyle \cdot\cdot}{\underset{\displaystyle \cdot\cdot}{Cl}}:}}{\overset{\displaystyle \cdot\cdot}{S}}^+-\overset{\displaystyle \cdot\cdot}{\underset{\displaystyle \cdot\cdot}{Cl}}:$$

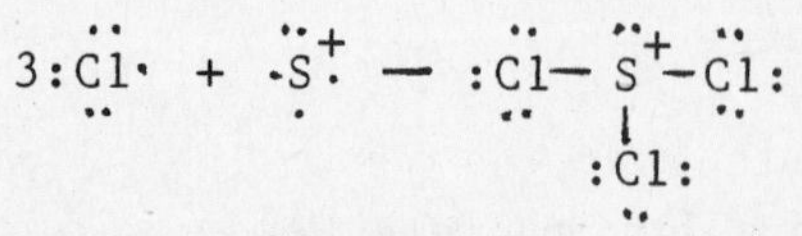

S has one lone pair and a half-share of 3 bonding pairs, -2-3 = -5; the core charge of S is +6 (group 6); formal charge = -5+6 = $\underline{+1}$

48. Follow the method given in Problem 47.

(a) N (group 5) is $\cdot \overset{\displaystyle \cdot}{\underset{\displaystyle \cdot}{N}} :$, and adding an electron gives $\cdot \overset{\displaystyle \cdot \cdot}{N} : ^-$

$$2H \cdot \; + \; \cdot \overset{\displaystyle \cdot \cdot}{N} : ^- \longrightarrow \; H - \overset{\displaystyle \cdot \cdot}{N} - H$$

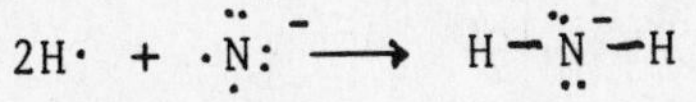

<u>Formal  charge on N</u> is +5+(-4-3) = <u>-1</u>

(b) O (group 6) is $\cdot \overset{\displaystyle \cdot}{O} :$, and removing an electron gives $\cdot \overset{\displaystyle \cdot \cdot}{\underset{\displaystyle \cdot}{O}} \cdot \, ^+$

$$3H \cdot \; + \; \cdot \overset{\displaystyle \cdot \cdot}{\underset{\displaystyle \cdot}{O}} \cdot \, ^+ \longrightarrow \; H - \overset{\displaystyle \cdot \cdot}{\underset{\displaystyle |}{O}} {}^{+} - H$$
$$H$$

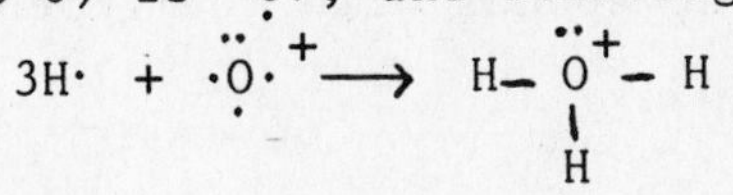

<u>Formal charge on O</u> is +6+(-2-3) = <u>+1</u>

(c) F (group 7) is $\cdot \overset{\displaystyle \cdot \cdot}{\underset{\displaystyle \cdot \cdot}{F}} :$, and removing an electron gives $\cdot \overset{\displaystyle \cdot \cdot}{\underset{\displaystyle \cdot \cdot}{F}} \cdot \, ^+$

$$2H \cdot \; + \; \cdot \overset{\displaystyle \cdot \cdot}{\underset{\displaystyle \cdot \cdot}{F}} \cdot \, ^+ \longrightarrow \; H - \overset{\displaystyle \cdot \cdot}{\underset{\displaystyle \cdot \cdot}{F}} {}^{+} - H$$

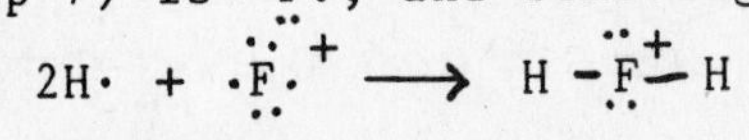

<u>Formal charge on F</u> is +7+(-4-2) = <u>+1</u>

(d) P(group 5) is $\cdot \overset{\displaystyle \cdot \cdot}{P} \cdot$ and removing an electron gives $\cdot \overset{\displaystyle \cdot}{\underset{\displaystyle \cdot}{P}} \cdot \, ^+$

$$4H \cdot \; + \; \cdot \overset{\displaystyle \cdot}{\underset{\displaystyle \cdot}{P}} \cdot \, ^+ \longrightarrow \begin{array}{c} H \\ | \\ H - \overset{}{\underset{|}{P}} {}^{+} - H \\ | \\ H \end{array}$$

<u>Formal charge on P</u> is +5-4 = <u>+1</u>

(e) B (group 3) is $\cdot \overset{\displaystyle \cdot}{B} \cdot$ and adding an electron gives $\cdot \overset{\displaystyle \cdot}{\underset{\displaystyle \cdot}{B}} \cdot \, ^-$

$$4 : \overset{\displaystyle \cdot}{\underset{\displaystyle \cdot \cdot}{F}} \cdot \; + \; \cdot \overset{\displaystyle \cdot}{\underset{\displaystyle \cdot}{B}} \cdot \, ^- \longrightarrow \begin{array}{c} : \overset{\displaystyle \cdot \cdot}{F} : \\ | \\ : \overset{\displaystyle \cdot \cdot}{F} - \overset{}{\underset{|}{B}} - \overset{\displaystyle \cdot \cdot}{F} : \\ | \\ : \overset{\displaystyle \cdot \cdot}{F} : \end{array}$$

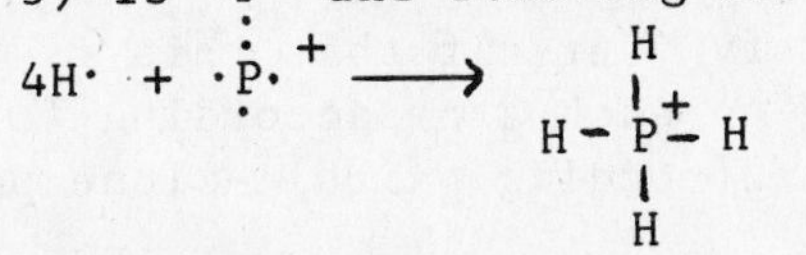

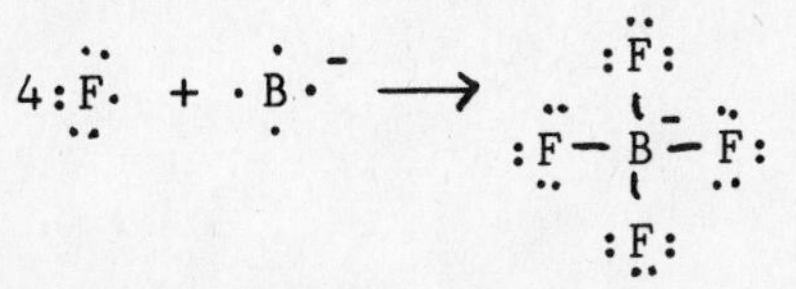

<u>Formal charge on B</u> is +3-4 = <u>-1</u>

(f) P (group 5) is $\cdot \overset{\displaystyle \cdot \cdot}{P} \cdot$ and removing an electron gives $\cdot \overset{\displaystyle \cdot}{\underset{\displaystyle \cdot}{P}} \cdot \, ^+$

$$4 : \overset{\displaystyle \cdot \cdot}{\underset{\displaystyle \cdot \cdot}{Cl}} \cdot \; + \; \cdot \overset{\displaystyle \cdot}{\underset{\displaystyle \cdot}{P}} \cdot \, ^+ \longrightarrow \begin{array}{c} : \overset{\displaystyle \cdot \cdot}{Cl} : \\ | \\ : \overset{\displaystyle \cdot \cdot}{Cl} - \overset{}{\underset{|}{P}} {}^{+} - \overset{\displaystyle \cdot \cdot}{Cl} : \\ | \\ : \overset{\displaystyle \cdot \cdot}{Cl} : \end{array}$$

<u>Formal charge on P</u> is +5-5 = <u>+1</u>

Note that in each case we have calculated only the formal charge on the central atom, which is in each case the unique atom, which normally carries the formal charge in such species. It we calculate the formal charges on all the other atoms they are shown to be O.

49. The octet rule applies strictly to the elements of the <u>second</u> period, Li to Ne, where eight electrons is the maximum number of electrons that can be accommodated in their valence shells, but even then not all the species formed have complete octets, as is the case, for example, for Be in compounds such as $BeH_2$ and $BeCl_2$, or boron in $BH_3$ or $BCl_3$. In general, we can say that for the elements C, N, O, and F, of the second period, the Lewis Octet Rule strictly applies.

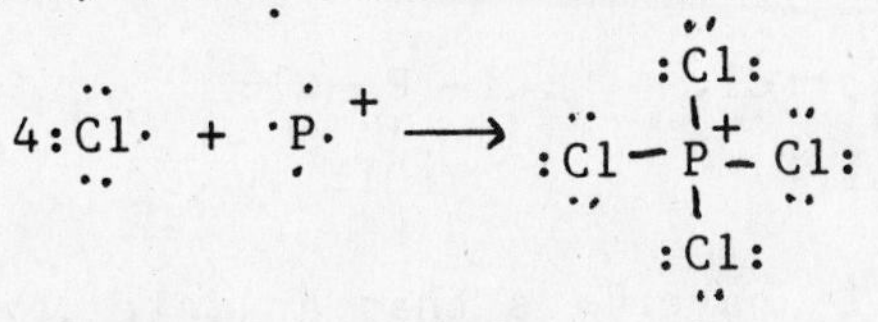

50. Elements:    Na   Mg   Al   Si   P    S    Cl   (Ar)

Hydrides:    NaH  $MgH_2$  $AlH_3$  $SiH_4$  $PH_3$  $H_2S$  HCl   none

Oxides    $Na_2O$  MgO  $Al_2O_3$  $SiO_2$  $P_2O_3$  SO  $Cl_2O$  none

Lewis structures:

Hydrogen and oxygen are both nonmetals; their compounds with the group 3 _metals_ will be ionic and their compounds with the group 3 _nonmetals_ will be covalent.

$Na^+$ :H⁻    $Mg^{2+}$[:H⁻]$_2$    $Al^{3+}$[:H⁻]$_3$    H-Si-H    H-P̈-H    H-S̈-H    H-C̈l:
                                                                  |         |
                                                                  H         H

[$Na^+$]$_2$:Ö:²⁻    $Mg^{2+}$:Ö:²⁻    [$Al^{3+}$]$_2$[:Ö:²⁻]$_3$    :Ö=Si=Ö:    :Ö=P-Ö-P=Ö:

:S̈=Ö:    :C̈l-Ö-C̈l:

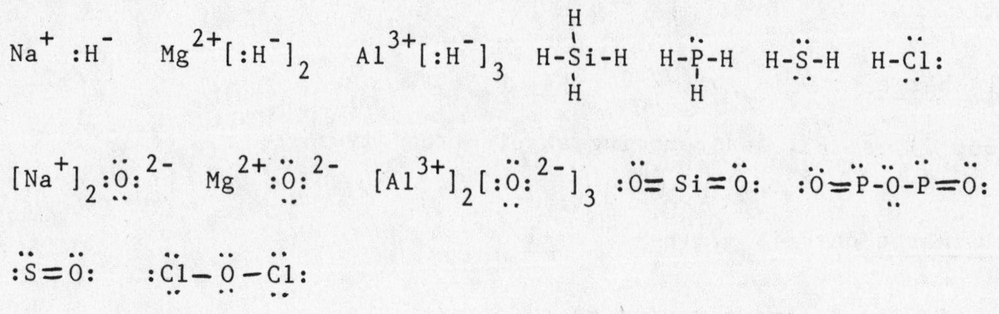

51. The formal charges are as shown (satisfy yourself that this is the case by calculating the formal charge on each atom according to the rule: formal charge = core charge - 1/2(bonding pairs) - lone pairs.

H-Ö⁺-H    H-Ö:⁻    :F̈:    :F̈-S̈-F̈:    :S̈=C=S̈:    :F̈-Ö-F̈:
   |                 |
   H              :F̈-N⁺-F̈:
                     |
                  :F̈:         :F̈:
                               |
                            :F̈-B-F̈:
                               |
                            :F̈:

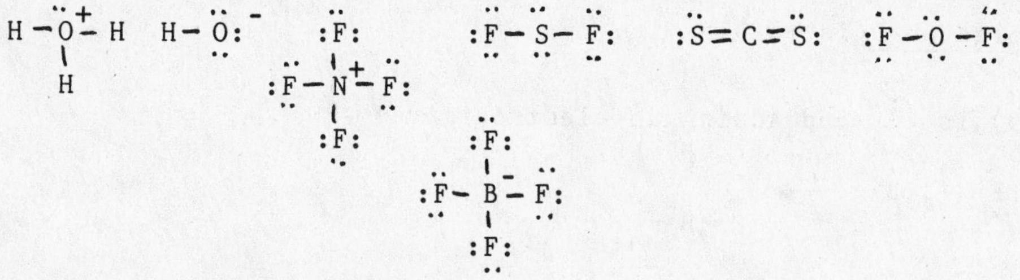

52. Write the Lewis structure for each and count up the electrons on each central atom.

:C̈l-Ö-C̈l:    :C̈l-Be-C̈l:    :C̈l-Al-C̈l:    :C̈l-P̈-C̈l:
                                     |              |
                                  :C̈l:           :C̈l:

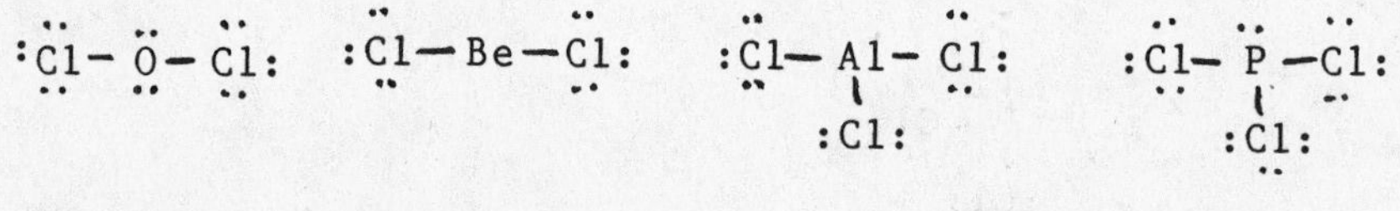

The central atoms in $BeCl_2$ and $AlCl_3$ have less than an octet, and they are exceptions to the octet rule.

53. Across any _period_ from left to right, the core charge increases and the valence shell electrons are attracted progressively more strongly to the nuclei, which results in decreasing covalent radii from left to right. From the top to the bottom of any _group_, the core charge remains constant (equal to the group number) but the number of filled inner electron shells increases progressively, so that the valence shell is ever more distant from the nuclei, and the covalent radii increase in going down any group. He, Ne, and Ar form no known compounds; their covalent radii cannot be measured.

54. The length of a bond is to a good approximation given by the sum of the covalent radii of the atoms forming the bond. The covalent radius of C is 77 pm, and that of Br is given by one-half the bond length in the Br-Br molecule. Thus,

$$\text{C-Br bond length in } CBr_4 = r_C^{cov} + r_{Br}^{cov} = 77 \text{ pm} + \frac{227 \text{ pm}}{2} = \underline{192 \text{ pm}}$$

55.

| Molecule | Bond Length (pm) | Covalent radius (pm) |
|---|---|---|
| $Cl_2$ | 198 | $r_{Cl} = \frac{198}{2} = 99$ |
| $CCl_4$ | 176 | $r_C = (176-99) = 77$ |
| $CBr_4$ | 194 | $r_{Br} = (194-77) = 117$ |
| $CI_4$ | 215 | $r_I = (215-77) = 138$ |

$\underline{Br_2}$, predicted Br-Br bond length = $2r_{Br} = \underline{234 \text{ pm}}$

$\underline{BrCl}$, predicted Br-Cl bond length = $r_{Br} + r_{Cl} = (117+99) = \underline{216 \text{ pm}}$

$\underline{I_2}$, predicted bond length = $2r_I = 2(138 \text{ pm}) = \underline{276 \text{ pm}}$

56. This problem is similar to Problem 27 where we considered trends in ionization energies. Covalent radii decrease from left to right across any period and increase in descending any group (see Problem 53). Thus, by comparing the group number (core charges) of the atoms and the period to which each belongs, the relative magnitudes of their covalent radii can be estimated.

(a) F and Cl are both in group 7 (core charge +7), but Cl is in period 3, below F in period 2. Thus, $r_{Cl} > r_F$.

(b) Boron and carbon are both in period 2 but the +4 core charge of C (group 4) is greater than the core charge of +3 of B (group 3). Thus, $r_B > r_C$

(c) Both C and Si are in group 4 (core charge +4) but Si, in period 3, is below C in period 2. Thus, $r_{Si} > r_C$.

(d) P (group 5, core charge +5) is in period 3, and Al (group 3, core charge +3) is also in period 3. Thus, $r_{Al} > r_P$.

(e) Si (group 4, core charge +4) is in period 3, while O (group 6, core charge +6) is in period 2. Thus, $r_{Si} > r_O$.

57. In each case the covalent radius is one-half of the single bond length for identical bonded atoms, and the bond lengths in other species are given by the sum of the covalent radii of the atoms forming the bonds.

(a) C-C (diamond) = 154 pm; $r_C = \frac{154 \text{ pm}}{2} = \underline{77 \text{ pm}}$

P-P ($P_4$) = 220 pm; $r_P = \frac{220 \text{ pm}}{2} = \underline{110 \text{ pm}}$

S-S ($S_8$) = 208 pm; $r_S = \frac{208\ pm}{2} = \underline{104\ pm}$

Cl-Cl ($Cl_2$) = 198 pm; $r_{Cl} = \frac{198\ pm}{2} = \underline{99\ pm}$

(b) $\underline{PCl_3}$, $r_{PCl} = r_P + r_{Cl}$ = (110+99) pm = $\underline{209\ pm}$

$\underline{CCl_4}$, $r_{CCl} = r_C + r_{Cl}$ = (77+99) pm = $\underline{176\ pm}$

$\underline{SCl_2}$, $r_{SCl} = r_S + r_{Cl}$ = (104+99) pm = $\underline{203\ pm}$

$\underline{P(CH_3)_3}$, $r_{PC} = r_P + r_C$ = (110+77) pm = $\underline{187\ pm}$

58. (a) $\underline{\text{linear } AX_2}$ (b) $\underline{\text{equilateral triangular } AX_3}$

X — A — X

<XAX = 180°

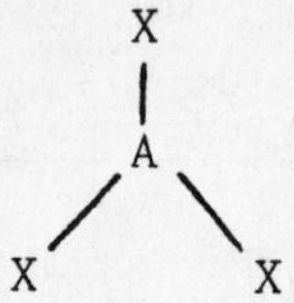


<XAX = 120°

(c) $\underline{\text{tetrahedral } AX_4}$

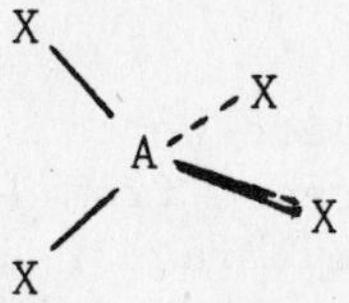


<XAX = 109° 28' (109.5°)

59.

| | Molecular Type | Geometry | Shape |
|---|---|---|---|
| (a) | $AX_3E$ | 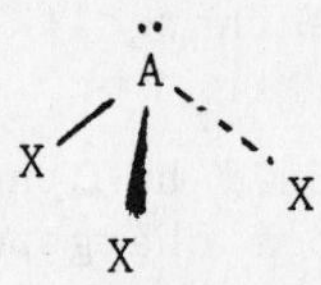 | triangular (trigonal) pyramid |
| (b) | $AX_2E_2$ |  | angular |
| (c) | $AXE_3$ | 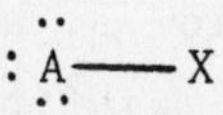 | (linear) |
| (d) | $AX_2E$ |  | angular |
| (e) | $AXE_2$ | 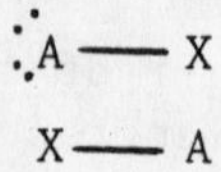 | (linear) |
| (f) | $AX_2$ | X— A — X | linear |

60. First draw the Lewis structures and then, by counting up the bonds to the central atom A, and the number of lone pairs on A, classify each molecule according to the $AX_nE_m$ nomenclature.

```
 ..     ..           ..  ..  ..
:Cl - B - Cl:       :Cl - P - Cl:
 ..   |   ..         ..   |   ..
     :Cl:                :Cl:
      ..                  ..
```

$BCl_3$ is an $AX_3$ type molecule and will be triangular (trigonal) planar, with the Cl atoms at the corners of an equilateral triangle of which the B atom is the center, so that each of the ClBCl bond angles must be 120°. In contrast to the B atom in $BCl_3$, the P atom in $PCl_3$ forms three bonds and has a lone pair of electrons in the phosphorus valence shell. Thus, $PCl_3$ is an $AX_3E$ type arrangement, based to a good approximation on a tetrahedron with the P atom at its center. Three of the corners of the tetrahedron are occupied by bonding pairs and the other by the lone pair of electrons. Thus the shape delineated by the P atom and the three Cl atoms is a triangular (or trigonal) pyramid with the chlorine atoms occupying an equilateral triangular base and the P atom occupying the apex of the pyramid. Ideally, the ClPCl angles should be the tetrahedral angle of 109.5°. The smaller values of 100° observed are accounted for in terms of the unshared electron pair on P taking up more space than each of the three bond pairs, and therefore pushing the bond pairs together to reduce the bond angle to less than 109.5°.

61. Write the Lewis structures and express the electron pairs in the valence shell of the central A atoms in terms of the $AX_nE_m$ nomenclature, where <u>n</u> is the number of X ligands and <u>m</u> is the number of lone (unshared) electron pairs

| Species | Lewis structure | Molecular Type | Shape |
|---|---|---|---|
| $H_2O$ | H-Ö-H (with lone pairs on O) | $AX_2E_2$ | angular |
| $H_3O^+$ | H-Ö-H (+), with H below O | $AX_3E$ | triangular pyramid |
| $PCl_3$ | :Cl-P-Cl: (:Cl: below P) | $AX_3E$ | triangular pyramid |
| $BCl_3$ | :Cl-B-Cl: (:Cl: below B) | $AX_3$ | triangular planar |
| $SiH_4$ | H-Si-H (H above and below Si) | $AX_4$ | tetrahedral |

62. This problem is similar to Problem 61.

| Species | Lewis structure | Molecular Type | Shape |
|---|---|---|---|
| $BH_4^-$ | H-B⁻-H (H above and below B) | $AX_4$ | tetrahedral |
| $H_2S$ | H-S-H (with lone pairs on S) | $AX_2E_2$ | angular |

| Species | Lewis Structure | Molecular Type | Shape |
|---|---|---|---|
| $NH_4^+$ | H above, H-N-N (+), H below | $AX_4$ | tetrahedral |
| $BeH_2$ | H-Be-H | $AX_2$ | linear |
| $BeH_4^{2-}$ | H above, H-Be-H (2-), H below | $AX_4$ | tetrahedral |

63. See Problem 61 for the approach.

| Species | Lewis Structure | Molecular Type | Shape |
|---|---|---|---|
| $BO_3^{3-}$ | :O: (−) above; $^-$:Ö-B-Ö:$^-$ | $AX_3$ | triangular planar |
| $O_3$ | $^-$:Ö-Ö$^+$=Ö: | $AX_2E$ | angular |
| $SnBr_2$ | :Br-Sn-Br: | $AX_2E$ | angular |
| $PbCl_4$ | :Cl: above; :Cl-Pb-Cl:; :Cl: below | $AX_4$ | tetrahedral |

64. Each of $BF_4^-$, $CF_4$, and $NF_4^+$ has $AX_4$ tetrahedral geometry; they have the same number of electrons and constitute an isoelectronic series.

65. The molecules in question are: $BeCl_2$, $BCl_3$, $CCl_4$, $NCl_3$, $OCl_2$ and ClF.

| :Cl-Be-Cl: | :Cl-B-Cl: / :Cl: | :Cl: / :Cl-C-Cl: / :Cl: | :Cl-N-Cl: / :Cl: | :Cl-O-Cl: | :Cl-F: |
|---|---|---|---|---|---|
| $AX_2$ | $AX_3$ | $AX_4$ | $AX_3E$ | $AX_3E_2$ | $AXE_3$ |
| linear | trigonal planar | tetrahedral | trigonal pyramid | angular | (linear) |

For diagrams - see text

66. The tetrahedron is constructed inside a cube by connecting together the corners of the top face diagonal and the corners of the face diagonal of the bottom of the cube that is perpendicular to the top face diagonal. Connecting together all four points gives the six sides of the tetrahedron. The intersection of any two of the body diagonals of the cube gives its center, which is also the center of the tetrahedron. Connection of each of the four corners of the tetrahedron to the center point gives the direction of the bonds in an $AX_4$ molecule. Note that the plane defined by one $AX_2$ arrangement is perpendicular to the plane defined by the other $AX_2$ group. In other words, any two of the AX bonds are in a plane perpendicular to the

to the plane formed by the other two AX bonds. Thus, if we draw one $AX_2$ angular arrangement in the plane of the paper, the other two AX bonds have an angular arrangement in the plane perpendicular to the paper - one of the AX bonds points out of the plane of the paper, and the other AX bond points out of the back of the paper.

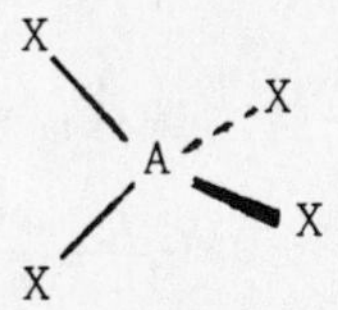


For the calculation of $<XAX = 109^{\circ}28'$, see Problem 1.24

67. From its position in the periodic table, scandium would also be expected to be a metal. Since Ca has two valence electrons and Ti has 4, Sc is expected to have three valence electrons. Although it is the first member of the first transition metal series, it should resemble aluminum in group 3. It compounds should be mainly ionic compounds of the $Sc^{3+}$ ion. The melting point, boiling point, and density should be intermediate between the respective values for Ca and Ti. Reference to Chapter 21, Table 21.3, shows this indeed to be the case: m.p. 1812 K, (b.p. 2730 K), and density 3.0 g $cm^{-3}$.

Indeed, these values are approximately the averages of the respective values for Ca and Ti.

68. a) In calculating the <u>core charge</u> it is assumed that the valence shell electrons are attracted by the nucleus of the atom (charge +Z), and repelled by all of the electrons in all of the inner filled shells of electrons. Numerically, Z minus the number of electrons in filled shells equals the number of valence electrons of the neutral atom. Thus, the core charge is +G, where G is the <u>group number</u>.

b) Core charge equals the number of the group to which the element belongs.

c)

| atom | H | C | N | F | Ne | Al | S | Cl | Ar | Ca |
|---|---|---|---|---|---|---|---|---|---|---|
| group | 1 | 4 | 5 | 7 | 8 | 3 | 6 | 7 | 8 | 2 |
| $Z_{core}$ | +1 | +4 | +5 | +7 | +8 | +3 | +6 | +7 | +8 | +2 |

(d) For elements in the same period, the ionization energy will increase as $Z_{core}$ increases. For elements in the same group (the same $Z_{core}$) the ionization energy decreases as the period increases, because the greater the number of inner filled shells of electrons, the farther away from the nucleus is the valence shell.

<u>Ne, Ar</u> Both have core charges of +8, but Ne is in period 2 while Ar is in period 3. <u>Ne > Ar</u>.

<u>Na, Cl</u> Na and Cl are both in period 3, but Na has a core charge of +1 while Cl has a core charge of +7. <u>Cl > Na</u>

Be, Mg Both are in group 2, but Be is in period 2 while Mg is in period 3. Be > Mg

F, Cl Both are in group 7, but F is in period 2 while Cl is in period 3. F > Cl

N, P Both are in group 5, but N is in period 2 while P is in period 3. N > P

69. Indium oxide contains 82.7 mass % In and 17.3 mass % O.

Thus 100.0 g indium oxide contains $(17.3 \text{ g O})(\frac{1 \text{ mol O}}{16.00 \text{ g O}}) = \underline{1.08 \text{ mol}}$

If the molecular formula is InO, 100.0 g oxide contains 1.08 mol In

Thus, molar mass In $= \frac{82.7 \text{ g In}}{1.08 \text{ mol In}} = \underline{76.6 \text{ g mol}^{-1}} = \underline{\text{atomic mass}}$

An atomic mass of this magnitude places In between As and Se, which suggests that In should be a nonmetal in presumably either group 5 or group 6. The empirical formula InO would be suitable for In in group 6, but InO would not resemble ZnO, which is a metal oxide.

On the basis of the empirical formula $In_2O_3$

100.0 g oxide contains $(1.08 \text{ mol O})(\frac{2 \text{ mol In}}{3 \text{ mol O}}) = \underline{0.720 \text{ mol In}}$

On this basis, atomic mass of In $= (\frac{82.7 \text{ g In}}{0.720 \text{ mol In}}) = \underline{115 \text{ g mol}^{-1}}$

An atomic mass of this magnitude places indium in group 3, between the transition metal Cd and Sn, another metal, in group 4, which is consistent with its valence of 3 in the oxide $In_2O_3$.

70. The given properties suggest that A is a reactive metal and the empirical formula $ACl_2$ for its chloride places it in group 2 (valence 2) so that the balanced equation for the reaction of A with HCl(g) is

$$A(s) + 2HCl(g) \longrightarrow ACl_2(s) + H_2(g)$$

and the balanced equation for the reaction of A with water must be

$$A(s) + 2H_2O(l) \longrightarrow A(OH)_2(aq) + H_2(g)$$

From the data for the $H_2(g)$ evolved, calculate mol $H_2$,

$$\text{mol } H_2 = n = \frac{PV}{RT} = \frac{(740 \text{ mm})(\frac{1 \text{ atm}}{760 \text{ mm}})(144.1 \text{ mL})(\frac{1 \text{ L}}{10^3 \text{ mL}})}{(0.0821 \text{ atm L mol}^{-1} \text{ K}^{-1})(298 \text{ K})} = \underline{5.73\text{x}10^{-3} \text{ mol}}$$

Thus, moles of A $= (5.73\text{x}10^{-3} \text{ mol } H_2)(\frac{1 \text{ mol A}}{1 \text{ mol } H_2}) = \underline{5.73\text{x}10^{-3} \text{ mol}}$

and molar mass of A $= \frac{0.230 \text{ g A}}{5.73\text{x}10^{-3} \text{ mol}} = \underline{40.1 \text{ g mol}^{-1}}$

Reference to the periodic table identifies A as Calcium, Ca.

This is confirmed by the data from the reaction where AgCl(s) is precipitated. The reactions involved must be

$$CaCl_2(s) \longrightarrow Ca^{2+}(aq) + 2Cl^-(aq)$$
$$Ag^+(aq) + Cl^-(aq) \longrightarrow AgCl(s)$$

From the mass of dry AgCl(s),

$$\text{mol AgCl} = (0.3760 \text{ g AgCl})\left(\frac{1 \text{ mol AgCl}}{143.4 \text{ g AgCl}}\right) = \underline{2.622\times10^{-3} \text{ mol}}$$

which must be the same as the number of moles of $Cl^-$ in the $CaCl_2$ sample.

$$\text{mol } CaCl_2 = (2.622\times10^{-3} \text{ mol } Cl^-)\left(\frac{1 \text{ mol } CaCl_2}{2 \text{ mol } Cl^-}\right) = \underline{1.311\times10^{-3} \text{ mol}}$$

$$\text{Molar mass of } CaCl_2 = \left(\frac{0.1456 \text{ g } CaCl_2}{1.311\times10^{-3} \text{ mol } CaCl_2}\right) = \underline{111.1 \text{ g mol}^{-1}}$$

which favorably compares with the calculated molar mass of $111.0 \text{ g mol}^{-1}$.

A is Calcium; (atomic mass $40.1 \text{ g mol}^{-1}$)

71. $:\ddot{O}:^{2-}$, $:\ddot{F}:^-$, $Na^+$, and $Mg^{2+}$ all have the same number of electrons, and the electron configuration of Ne, 2.8., and thus constitute an isoelectronic series.

Here the outer shell of electrons is the n = 2 shell of 8 electrons in each case, and we calculate the core charges by subtracting from the atomic numbers the two electrons in the n = 1 shell, to give

| Ion | Z | Electrons | Core Charge |
|---|---|---|---|
| $O^{2-}$ | 8 | 2.8 | +6 |
| $F^-$ | 9 | 2.8 | +7 |
| $Na^+$ | 11 | 2.8 | +9 |
| $Mg^{2+}$ | 12 | 2.8 | +10 |

Thus, the ions would be expected to decrease in size with increase in the magnitude of the core charge, i.e., from $O^{2-}$ to $Mg^{2+}$, as observed.

72. Choose your own examples from the text.

(a) The bond angle YAX is the angle formed by the three atoms YAX; the angle between the bonds AX and AY.

(b) The bond length AX is the distance between the nucleus of atom A and the nucleus of atom X forming an AX bond.

(c) The molecular geometry of a $AX_nE_m$ molecule is the geometric shape defined by the A atom and the n X atoms (ligands).

(d) A bonding electron pair is a pair of electrons that participates in the formation of a chemical bond.

(e) A nonbonding (unshared or lone) pair of electrons is a pair of electrons that does not take part in the formation of a chemical bond.

(f) A double bond is a bond formed by the sharing of two electron pairs.

(g) A triple bond is a bond formed by the sharing of three electron pairs.

73. Starting with $\cdot\dot{\underset{\cdot}{C}}\cdot$ we can form two single bonds to each of two H atoms, to give $H_2\underset{\cdot}{C}:$, and combination of two $H_2C:$ fragments gives $H_2C{=}CH_2$ with a double bond between the C atoms.

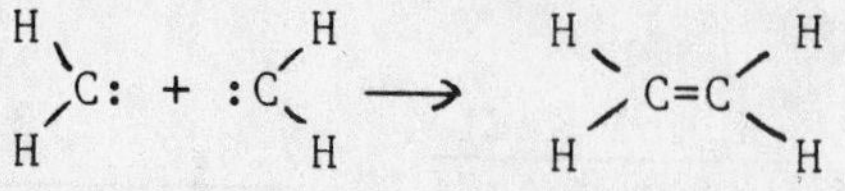

(ethene)

Starting with $\cdot\dot{\underset{\cdot}{C}}\cdot$ we can form a single bond to one H atom, to give $H\dot{\underset{\cdot}{C}}\cdot$, and combination of two $H\dot{\underset{\cdot}{C}}\cdot$ fragments gives $HC\equiv CH$ with a triple bond between the two C atoms.

$$H-\dot{\underset{\cdot}{C}}\cdot + \cdot\dot{\underset{\cdot}{C}}-H \longrightarrow H-C\equiv C-H$$ (ethyne)

If we add an additional electron to $\cdot\dot{\underset{\cdot}{C}}\cdot$ to give $:\dot{\underset{\cdot}{C}}\cdot^{-}$, we can combine this with an $:\dot{\underset{\cdot}{N}}\cdot$ atom to give $CN^-$, with a triple bond between the two atoms.

$$:\underset{\cdot}{C}\cdot^{-} + \cdot\dot{\underset{\cdot}{N}}: \longrightarrow {}^{-}:C\equiv N:$$

Combination of two $:\dot{\underset{\cdot}{N}}\cdot$ atoms gives $N_2$ with a triple bond.

$$:\dot{\underset{\cdot}{N}}\cdot + \cdot\dot{\underset{\cdot}{N}}: \longrightarrow :N\equiv N:$$

$NO^+$ is isoelectronic with $N_2$ and must have the same Lewis structure. Alternatively, if we combine $:\dot{\underset{\cdot}{N}}\cdot$ with $:\dot{\underset{\cdot}{O}}\cdot^{+}$, we have

$$:\dot{\underset{\cdot}{N}}\cdot + \cdot\dot{\underset{\cdot}{O}}:^{+} \longrightarrow :N\equiv O:^{+}$$

P(group 5), $:\dot{\underset{\cdot}{P}}\cdot$ , (like $:\dot{\underset{\cdot}{N}}\cdot$ ) can combine with itself to give a $P_2$ molecule with a triple bond.

$$:\dot{\underset{\cdot}{P}}\cdot + \cdot\dot{\underset{\cdot}{P}}: \longrightarrow :P\equiv P:$$

## CHAPTER 5

1. The most important sources of bromine are, in Israel, the water of the Dead Sea, and in the U.S.A., brine deposits found in Arkansas. These contain up to 5000 ppm of bromide ion, $Br^-(aq)$. Chlorine and steam are passed through the brine, which oxidizes bromide ion to bromine,

$$Cl_2(g) + 2Br^-(aq) \longrightarrow Br_2(l) + 2Cl^-(aq)$$

The bromine that is formed is carried from the solution, together with excess $Cl_2(g)$, in the steam. It is condensed from this mixture and purified by distillation (b.p. 59°C).

2. (a) $MnO_2(s) + 4HCl(aq) \longrightarrow MnCl_2(aq) + 2H_2O(l) + Cl_2(g)$

   (b) The direct reaction of $Cl_2(g)$ with $H_2(g)$ is very slow under ordinary conditions but occurs explosively when a mixture of $H_2(g)$ and $Cl_2(g)$ is exposed to strong sunlight,

   $$H_2(g) + Cl_2(g) \longrightarrow 2HCl(g)$$

   (c) HCl is a polar covalent molecule, $^{\delta+}H-Cl^{\delta-}$, which readily dissolves in water because of its attraction to polar water molecules. It then transfers its proton to a lone pair of electrons on the oxygen of $H_2O$, to give quantitatively $H_3O^+(aq)$ and $Cl^-(aq)$.

   $$^{\delta-}:\ddot{C}l-H^{\delta+} + 2\ ^{\delta-}:\ddot{O}-H^{\delta+} \longrightarrow :\ddot{C}l:^- + H-\overset{+}{\underset{H}{\ddot{O}}}-H$$

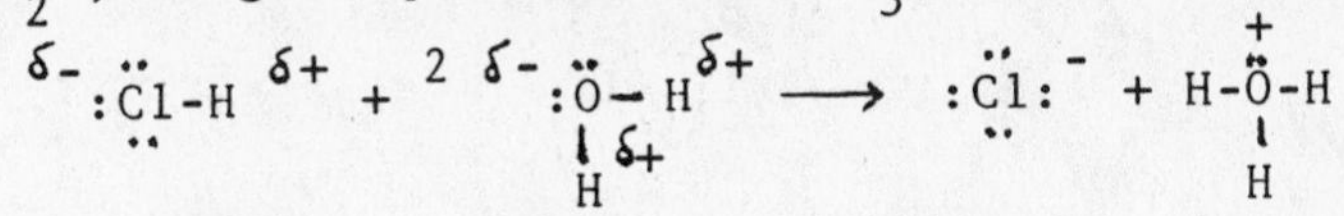

3. Sea salt consists mainly of sodium chloride but also contains small amounts of other salts, such as the chlorides, bromides, iodides, and sulfates of potassium, magnesium, and calcium. The metal cations that these contain, as well as anions such as $Cl^-$ and $I^-$, are important in the biochemistry of the body and, thus, essential to our good health, as well as adding flavor to our food. Table salt is essentially pure sodium chloride with a small amount of added potassium iodide, which provides sufficient iodide ion to promote the formation of the growth-regulating hormone thyroxine, which is produced by the thyroid gland. Lack of iodine in the diet leads to the condition called goiter in which the thyroid gland becomes enlarged.

4. $P_4(s) + 6Cl_2(g) \longrightarrow 4PCl_3(l)$

   $S_8(s) + 8Cl_2(g) \longrightarrow 8SCl_2(l)$; $S_8(s) + 4Cl_2(g) \longrightarrow 4S_2Cl_2(l)$

   $C(s) + 2F_2(g) \longrightarrow CF_4(g)$

   $As_4(s) + 6Br_2(l) \longrightarrow 4AsBr_3(l)$

5.  :$\ddot{C}l$-P-$\ddot{C}l$:     :$\ddot{C}l$-$\ddot{S}$-$\ddot{C}l$:     :$\ddot{C}l$-$\ddot{S}$-$\ddot{S}$-$\ddot{C}l$:          :$\ddot{F}$:          :$\ddot{B}r$-As-$\ddot{B}r$:

       :$\ddot{C}l$:                                                    :$\ddot{F}$-C-$\ddot{F}$:          :$\ddot{B}r$:

                                                                        :$\ddot{F}$:

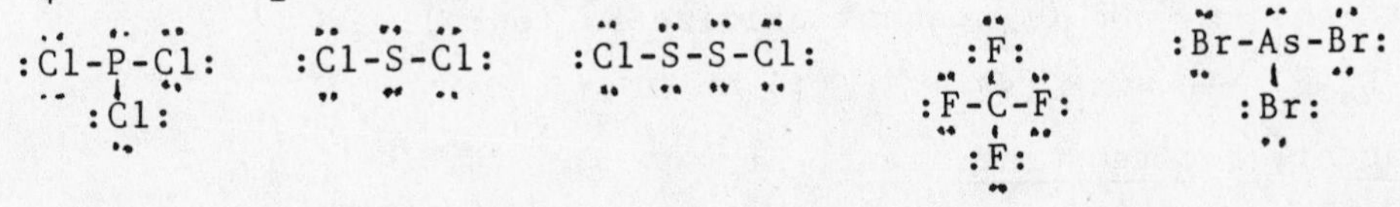

6. The most important mineral in bones and teeth is hydroxyapatite, $Ca_5(PO_4)_3OH$. Addition of a small amount of fluoride ion to our diet replaces $OH^-$ ion in hydroxyapatite by $F^-$ ion, to give fluorapatite, $Ca_5(PO_4)_3F$, which forms a hard more acid resistant layer on our teeth.

7. (a) $Ba(s) + Cl_2(g) \longrightarrow BaCl_2(s)$

   (b) $2Al(s) + 3Br_2(l) \longrightarrow 2AlBr_3(s)$

   (c) $2K(s) + I_2(s) \longrightarrow 2KI(s)$

   (d) $P_4(s) + 6Cl_2(g) \longrightarrow 4PCl_3(l)$, <u>and</u> $P_4(s) + 10Cl_2(g) \longrightarrow 4PCl_5(s)$

   (e) $P_4(s) + 6I_2(s) \longrightarrow 4PI_3(s)$

8.

| | Formula | Name | Lewis Structure | |
|---|---|---|---|---|
| (a) | $BaCl_2$ | Barium chloride | $Ba^{2+}[:\overset{\cdot\cdot}{\underset{\cdot\cdot}{Cl}}:^-]_2$ | (ionic) |
| (b) | $AlBr_3$ | Aluminum bromide | $:\overset{\cdot\cdot}{Br}-Al-\overset{\cdot\cdot}{Br}:$ $\overset{\cdot\cdot}{\underset{\cdot\cdot}{Br}}:$ | (covalent) |
| (c) | $KI$ | Potassium iodide | $K^+[:\overset{\cdot\cdot}{\underset{\cdot\cdot}{I}}:^-]$ | (ionic) |
| (d) | $PCl_3$ | Phosphorus trichloride | $:\overset{\cdot\cdot}{\underset{\cdot\cdot}{Cl}}-P-\overset{\cdot\cdot}{\underset{\cdot\cdot}{Cl}}:$ $:\overset{\cdot\cdot}{\underset{\cdot\cdot}{Cl}}:$ | (covalent) |
| | $PCl_5$ | Phosphorus pentachloride | $:\overset{\cdot\cdot}{\underset{\cdot\cdot}{Cl}}:$ $:\overset{\cdot\cdot}{\underset{\cdot\cdot}{Cl}}-P\overset{Cl:}{\underset{Cl:}{<}}$ $:\overset{\cdot\cdot}{\underset{\cdot\cdot}{Cl}}:$ | (covalent) |
| (e) | $PI_3$ | Phosphorus triiodide | $:\overset{\cdot\cdot}{I}-P-\overset{\cdot\cdot}{I}:$ $:\overset{\cdot\cdot}{\underset{\cdot\cdot}{I}}:$ | (covalent) |

9. (a) $Cl_2(g) + 2NaOH(aq) \longrightarrow NaOCl(aq) + NaCl(aq) + H_2O(l)$

   (b) Use the $Cl_2(g)$ data to calculate mol $Cl_2(g)$.

   $$\text{mol } Cl_2(g) = \frac{PV}{RT} = \frac{(1.00 \text{ atm})(1.00 \text{ L})}{(0.0821 \text{ atm L mol}^{-1} \text{ K}^{-1})(298 \text{ K})} = \underline{4.09 \times 10^{-2} \text{ mol}}$$

   $$\text{mass of NaOCl} = (4.09 \times 10^{-2} \text{ mol } Cl_2)(\frac{1 \text{ mol NaOCl}}{1 \text{ mol } Cl_2})(\frac{74.44 \text{ g NaOCl}}{1 \text{ mol NaOCl}})$$

   $$= \underline{3.04 \text{ g}}$$

10. (a) $Br_2(l) + 2NaCl(aq) \longrightarrow$ NR ($Br_2$ cannot oxidize $Cl^-$ to $Cl_2$)

    (b) $2NaCl(s) + F_2(g) \longrightarrow 2NaF(s) + Cl_2(g)$
        <u>Oxidizing agent</u>, $F_2$; <u>reducing agent</u>, $Cl^-$

    (c) $I_2(s) + 2NaF(aq) \longrightarrow$ NR ($I_2$ cannot oxidize $F^-$ to $F_2$)

    (d) $CaBr_2(s) + F_2(g) \longrightarrow CaF_2(s) + Br_2(l)$
        <u>Oxidizing agent</u>, $F_2$; <u>reducing agent</u>, $Br^-$

    (e) $2KF(s) + Br_2(l) \longrightarrow$ NR ($Br_2$ cannot oxidize $F^-$ to $F_2$)

    (f) $MgI_2(aq) + Cl_2(g) \longrightarrow MgCl_2(aq) + I_2(s)$
        <u>Oxidizing agent</u>, $Cl_2$; <u>reducing agent</u>, $I^-$

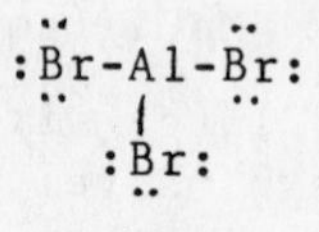

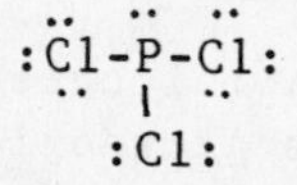

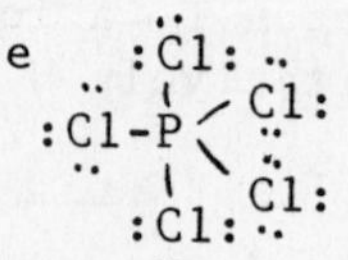

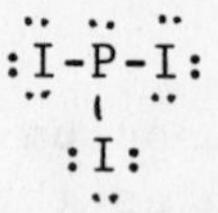

(g) $2LiI(aq) + Br_2(l) \longrightarrow 2LiBr(aq) + I_2(s)$

Oxidizing agent, $Br_2(l)$; reducing agent, $I^-(aq)$

(h) $I_2(s) + MgCl_2(aq) \longrightarrow$ NR ($I_2$ cannot oxidize $Cl^-$ to $Cl_2$)

11. Electronegativity is a measure of the ability of an atom in a molecule to attract the electrons of a covalent bond to itself. It is determined by the core charge and the distance of the valence shell electrons from the nucleus. It increases with increasing core charge and decreases with increasing distance of the valence shell from the nucleus. Thus, electronegativity increases from left to right in any period, and decreases from top to bottom down any group.

12. As the core charge increases in going from left to right in any period, the electronegativity increases; in going down any group, the core charge remains constant but the distance of the valence shell electrons from the nucleus increases as the number of inner filled shells increases, and the electronegativity decreases.

13. This problem is solved by remembering that the electronegativity is a function of the core charge (given by the group number) and the distance of the valence shell electrons from the nucleus (which depends on the period to which an element belongs).

(a) Fluorine (group 7, period 2) has a core charge of +7, while chlorine (group 7, period 3) also has a core charge of +7. Since the valence shell of a period 3 element is further from the nucleus than the valence shell of a period 2 element, F must be more electronegative than Cl. (In fact F is the most electronegative of all the elements). F > Cl.

(b) F, group 7, core charge +7, and O, group 6, core charge +6, are both in period 2. F > O.

(c) P, group 5, core charge +5, and S,group 6, core charge +6, are both in period 3. S > P.

(d) C and Si are both in group 4, core charge +4, but C is in period 2 and Si is in period 3. C > Si.

(e) O, group 6, core charge +6, is in period 2, while P, group 5, core charge +5, is in period 3. Since electronegativity is approximately proportional to the core charge but inversely proportional to the distance of the valence shell from the nucleus, O > P.

(f) Br and Se are both in period 4. Se, group 6, has a core charge of +6, while Br, group 7, has a core charge of +7. Br > Se.

(g) Al, group 3, has a core charge of +3 and is in period 3; P, group 5, has a core charge of +5, and is also in period 3. P > Al.

14. Al, group 3, core charge +3, is in period 3, S, group 6, core charge +6, is also in period 3; F, group 7, core charge +7, is in period 2, and Sr, group 2, core charge +2, is in period 5. Thus, F > S > Al >Sr.

15.

| Compound | Bond Type | Explanation |
|---|---|---|
| $Cl_2$ | nonpolar covalent* | bond between identical atoms |
| $PCl_3$, ClF | polar covalent | bonds between nonmetal atoms of different electronegativity |
| LiCl, $MgCl_2$ | ionic | bonds between a metal and a nonmetal atom of very different electronegativities |

*pure covalent

16.

| Compound | Bond Type | Explanation |
|---|---|---|
| $Li_2O$, MgO | ionic | bonds between a metal and a nonmetal atom of very different electronegativities |
| $O_2$ | nonpolar covalent | bond between identical atoms |
| $SO_2$, $Cl_2O$, NO | polar covalent | bonds between nonmetal atoms of different electronegativities |

17.

| Compound | Bond Type | Explanation |
|---|---|---|
| $I_2$, $H_2$ | nonpolar covalent | bonds between identical atoms |
| HBr, ClF | polar covalent | bonds between nonmetal atoms of different electronegativities |
| LiH | ionic | bond between a metal and a nonmetal of different electronegativities |

18. Sodium chloride is a solid ionic compound because when there is a large number of positive and negative ions formed together, each positive ion attracts as many negative ions as possible to surround it, and vice-versa. In NaCl, each $Na^+$ ion is surrounded by an octahedral arrangement of six $Cl^-$ ions, and each $Cl^-$ ion is surrounded by an octahedral arrangement of six $Na^+$ ions and form the sodium chloride lattice. The empirical formula is NaCl but a NaCl crystal is composed of equal and very large numbers of $Na^+$ and $Cl^-$ ions that attract each other strongly by electrostatic forces and are arranged in an infinite lattice. It takes a large amount of energy to pull the lattice apart, so that only in NaCl vapor formed when NaCl is boiled are $Na^+Cl^-$ ion pairs, and other small aggregates such as $(Na^+)_2(Cl^-)_2$ found.

19. The successive ionization energies of an atom rapidly increase since it becomes progressively more difficult to remove electrons from positively charged species. (Not only is it difficult to form highly charged cations, but when such an ion is formed it is relatively small and its high formal charge strongly attracts unshared electron pairs of nearby anions into its valence shell, to give a polar rather than an ionic bond. Thus, for example, even $Al^{3+}(Cl^-)_3$ shows very few of the properties of a typical ionic substance. It is soluble in nonpolar solvents and sublimes at the relatively low temperature of 180°C to give polar covalent $AlCl_3$ molecules in the gas phase. Similarly, none of, for example, the chlorides of the group 4 elements is ionic; $CCl_4$, $SiCl_4$, $GeCl_4$, $SnCl_4$, and even $PbCl_4$, are all polar covalent substances.)

20. Lithium fluoride is composed of $Li^+$ and $F^-$ ions as a consequence of the large difference in the electronegativities of Li and F. The ions attract each other strongly but in no particular preferred directions, so that the crystal is composed of $Li^+$ ions surrounded by $F^-$ ions, and vice-versa. The whole aggregate is one giant molecule and LiF has a high melting point and a very high boiling point. In contrast, two identical F atoms form an $F_2$ molecule by sharing a pair of electrons equally in a nonpolar covalent bond. These $F_2$ molecules attract each other only by weak intermolecular forces, and $F_2$ is a gas at room temperature.

21. (a) $K^+$ and $Ca^{2+}$ are isoelectronic, but $Ca^{2+}$ has a larger core charge than $K^+$. They are in the same period. Thus $K^+$ is a larger ion than $Ca^{2+}$. $K^+ > Ca^{2+}$.

(b) $S^{2-} > Cl^-$; they are isoelectronic and both are in period 3, but the core charge of Cl (+7) is greater than that of S (+6).

(c) $Cl^-$ and $K^+$ are isoelectronic; considering the outer filled shell in each case as the valence shell (n = 3), the core charge of $Cl^-$ is +7 and that of $K^+$ is +9. $Cl^- > K^+$.

(d) $Na^+ > Li^+$, because $Na^+$ has one more filled shell of electrons than $Li^+$.

(e) $I^- > Br^-$, for the same reason as in (d).

22. (a) Cl and $Cl^-$ have the same core charge (+7) but $Cl^-$ is larger than Cl because it has one more electron and there is an increased repulsion between the electrons in its more crowded valence shell.

(b) $Na > Na^+$ because the additional electron in Na is in a valence shell that is further from the nucleus than the outer shell electrons of $Na^+$.

(c) $Na > Mg^{2+}$ because the outer valence shell of Na is further from its nucleus than the outer electron shell of $Mg^{2+}$. Otherwise, since we saw in part (b) that $Na > Na^+$, and $Mg^{2+} < Na^+$ because the core charges are +9 and +10, respectively, then $Na > Mg^{2+}$.

(d) $Cl^- > K^+$, since these two ions are isoelectronic but the core charge of $K^+$ is +9 and that of $Cl^-$ is +7.

23. The bonds are expected to be polar covalent because all of the atoms involved are nonmetals. The electronegativity of N is 3.1, and those of F, Cl, Br, and I are 4.1, 2.8, 2.7, and 2.2, respectively. Thus, F > N, but in all the other cases the electronegativity of the halogen is less than that of nitrogen. Thus, the expected polarities of the bonds in the trihalides, $NX_3$, are:

$$^{\delta+}N\text{-}F^{\delta-} \qquad ^{\delta-}N\text{-}Cl^{\delta+} \qquad ^{\delta-}N\text{-}Br^{\delta+} \qquad ^{\delta-}N\text{-}I^{\delta+}$$

with an increasing polarity from N-Cl to N-I.

24. The elements of the second period are Li, Be, B, C, N, O, F, and Ne, but neon forms no compounds.

(a) Considering the relative electronegativities, we have

Li (1.0) < H (2.1); Be (1.5) < H (2.1); B (2.0) < H(2.1);
C (2.5) > H (2.1); N (3.1) > H (2.1); O (3.5) > H (2.1), and
F (4.1) > H (2.1).

LiH contains a metal and a nonmetal; ionic $Li^+ :H^-$

BeH$_2$ Although Be is a metal, the electronegativity difference is relatively small; the Be-H bond is polar covalent, $^{\delta+}$Be-H$^{\delta-}$

The remaining elements are nonmetals and the bonds with the nometal H are polar covalent:

| | | | | | |
|---|---|---|---|---|---|
| $BH_3$ | $^{\delta+}$B-H$^{\delta-}$ | $CH_4$ | $^{\delta-}$C-H$^{\delta+}$ (slighly polar) | | |
| $NH_3$ | $^{\delta-}$N-H$^{\delta+}$ | $H_2O$ | $^{\delta-}$O-H$^{\delta+}$ | HF | $^{\delta-}$F-H$^{\delta+}$ |

and the polarity increases from C-H to F-H as the electronegativity difference increases.

(b) Since F has a higher electronegativity than any other element, LiF will be ionic, $Li^+F^-$, as we might also expect $BeF_2$ to be, $Be^{2+}(F^-)_2$. The remainder of the elements of the second period form covalent fluorides, $BF_3$, $CF_4$, $NF_3$, and $OF_2$, with fluorine, all with polar covalent bonds, in the sense $^{\delta+}$X-F$^{\delta-}$, with decreasing polarity from B-F to O-F. In $F_2$, the F-F bond is nonpolar covalent.

25. The empirical formulas are readily written by inspection, bearing in mind that the compounds are overall neutral, so that the charges on the positive ions have to balance those on the negative ions in the empirical formulas:

(a) $(NH_4)_3PO_4$ (b) $Fe_2O_3$ (c) $Cu_2O$ (d) $Al_2(SO_4)_3$

26. Each salt contains a positive ion (cation) and a negative ion (anion). The charge on a positive metal ion is easily deduced from its position in the periodic table: Group 3, group 2, and group 1 metals give $M^{3+}$, $M^{2+}$ and $M^+$ cations, respectively, by the loss of 3, 2, and 1 electrons, respectively. The only nonmetal cation encountered so far is $NH_4^+$. A negative ion is formed by removing one or more protons from an acid.

(a) Barium iodide must contain $Ba^{2+}$ (group 2) and $I^-$ derived from HI, so the empirical formula is $BaI_2$, since the salt is neutral overall.

(b) Aluminium chloride must contain $Al^{3+}$ (group 3) and $Cl^-$ derived from HCl, so the empirical formula is $AlCl_3$

(c) Ammonium perchlorate must contain the ammonium ion, $NH_4^+$, and the $ClO_4^-$ ion derived from perchloric acid, $HClO_4$. Thus, the empirical formula is $NH_4ClO_4$

(d) Calcium nitrate is $Ca(NO_3)_2$ formed from $Ca^{2+}$ (group 2) and $NO_3^-$, formed from nitric acid, $HNO_3$.

(e) Aluminum sulfate is $Al_2(SO_4)_3$ formed from $Al^{3+}$ (group 3) and $SO_4^{2-}$ derived from sulfuric acid, $H_2SO_4$

(f) Calcium carbonate is $CaCO_3$ formed from $Ca^{2+}$ (group 2) and $CO_3^{2-}$ derived from carbonic acid, $H_2CO_3$

(g) Magnesium hydroxide is $Mg(OH)_2$ formed from $Mg^{2+}$ (group 2) and $OH^-$ derived from water, $H_2O$

27. First deduce the charge on each of the ions in the binary ionic compound: Cations are formed from metals when they lose all of their valence electrons; anions are formed from the nonmetals when they

gain sufficient electrons to complete an octet in their valence shells.

(a) Ca (group 2), $Ca^{2+}$; iodine (group 7), $I^-$; $CaI_2$ calcium iodide

(b) Be (group 2), $Be^{2+}$, O (group 6), $O^{2-}$; BeO, beryllium oxide

(c) Al (group 3), $Al^{3+}$; S (group 6), $S^{2-}$; $Al_2S_3$, aluminum sulfide

(d) Mg (group 2), $Mg^{2+}$; Br (group 7), $Br^-$; $MgBr_2$, magnesium bromide

(e) Rb (group 1), $Rb^+$; Se (group 6), $Se^{2-}$; $Rb_2Se$, rubidium selenide

(f) Ba (group 2), $Ba^{2+}$; O (group 6), $O^{2-}$; BaO, barium oxide

28. The increasing order as oxidizing agents is $Cl_2 > I_2$, thus $Cl_2$ oxidizes $I^-$, but $I_2$ will not oxidize $Cl^-$,

$Cl_2 + 2e^- \longrightarrow 2Cl^-$ reduction

$2I^- \longrightarrow I_2 + 2e^-$ oxidation

$Cl_2 + 2I^- \longrightarrow 2Cl^- + I_2$ overall

$Cl_2$ is reduced to $Cl^-$ and is the oxidizing agent; $I^-$ is oxidized to $I_2$ and is the reducing agent.

29. (a) KI is fully ionized as $K^+(aq)$ and $I^-(aq)$ in solution; $Cl_2(g)$ oxidizes $I^-(aq)$ to $I_2(s)$, but in the presence of excess $I^-(aq)$ $I_2$ remains in solution as the brown complex ion $I_3^-(aq)$.

$Cl_2(g) + 3KI(aq) \longrightarrow KI_3(aq) + 2KCl(aq)$

(b) The order of oxidizing strengths is $I_2 < Cl_2$, so $I_2$ cannot oxidize $Cl^-$ to $Cl_2$ - NR

(c) The order of oxidizing strengths is $Br_2 > I_2$, thus

$Br_2(l) + 3I^-(aq) \longrightarrow I_3^- + 2Br^-$ (c.f. part (a))

or $Br_2(l) + 3NaI(aq) \longrightarrow NaI_3(aq) + 2NaBr(aq)$

(d) $F_2(g)$ oxidizes water to $O_2(g)$,

$2[F_2(g) + 2e^- \longrightarrow 2F^-(aq)]$ reduction

$2H_2O(l) \longrightarrow O_2(g) + 4H^+(aq) + 4e^-$ oxidation

$2F_2(g) + 2H_2O(l) \longrightarrow O_2(g) + 4HF(aq)$ overall

30. $I^-$ is the most easily oxidized of the halide ions; traces of oxidizing impurities in the NaI(s) would produce some $I_2(s)$, which would impart a color to the the otherwise colorless NaI(s).

31. When a halogen, $X_2$, oxidizes $H_2S$ to sulfur, the reaction is

$X_2 + 2e^- \longrightarrow 2X^-$ reduction

$H_2S \longrightarrow 2H^+ + S + 2e^-$ oxidation

$X_2 + H_2S \longrightarrow 2HX + S$ overall

When a halogen oxidizes $HNO_2$ to $HNO_3$ the reaction is

$X_2 + 2e^- \longrightarrow 2X^-$ reduction

$HNO_2 + H_2O \longrightarrow HNO_3 + 2H^+ + 2e^-$ oxidation

$X_2 + HNO_2 + H_2O \longrightarrow 2HX + HNO_3$ overall

When $Cl_2(g)$ oxidizes $Br^-(aq)$, the reaction is

$Cl_2 + 2Br^- \longrightarrow 2Cl^- + Br_2$

Thus, for all of the reactions, we have

$H_2S(g) + Cl_2(g) \longrightarrow 2HCl(g) + S(s)$

$H_2S(g) + Br_2(l) \longrightarrow 2HBr(g) + S(s)$

$H_2S(g) + I_2(s) \longrightarrow 2HI(g) + S(s)$

$Cl_2 + HNO_2 + H_2O \longrightarrow 2HCl + HNO_3$

$Br_2 + HNO_2 + H_2O \longrightarrow 2HBr + HNO_3$ Thus, $Cl_2$ & $Br_2 > I_2$

$Cl_2 + 2Br^- \longrightarrow 2Cl^- + Br_2$ Thus, $Cl_2 > Br_2$

Hence, the order of oxidizing strengths is $Cl_2 > Br_2 > I_2$

32. (a) Rb is oxidized to $Rb^+$ and $I_2$ is reduced to $I^-$,

$2[Rb \longrightarrow Rb^+ + e^-]$ oxidation

$I_2 + 2e^- \longrightarrow 2I^-$ reduction

$2Rb + I_2 \longrightarrow 2RbI$ overall

Rb is oxidized and is the reducing agent
$I_2$ is reduced and is the oxidizing agent

(b) Al loses electrons to give $Al^{3+}$ and is oxidized; $O_2$ gains electrons and is reduced.

$4[Al \longrightarrow Al^{3+} + 3e^-]$ oxidation

$3[O_2 + 4e^- \longrightarrow 2O^{2-}]$ reduction

$4Al + 3O_2 \longrightarrow Al_2O_3$ overall

Al is oxidized to $Al^{3+}$ and is the reducing agent
$O_2$ is reduced to $O^{2-}$ and is the oxidizing agent

(c) $I^-$ loses electrons to give $I_2$ and is oxidized; $Cu^{2+}$ gains an electron to give $Cu^+$ and is reduced.

$2[Cu^+ + e^- \longrightarrow Cu^+]$ reduction

$2I^- \longrightarrow I_2 + 2e^-$ oxidation

$2Cu^{2+} + 2I^- \longrightarrow 2Cu^+ + I_2$ overall

and adding $3I^-$ to each side of the equation gives:

$2Cu^{2+} + 5I^- \rightarrow 2CuI + I_3^-$

$I^-$ is oxidized and is the reducing agent

$Cu^{2+}$ is reduced and is the oxidizing agent

(d) Zinc is oxidized to $Zn^{2+}$ and S is reduced to $S^{2-}$.

$$Zn \longrightarrow Zn^{2+} + 2e^- \quad \text{oxidation}$$

$$S + 2e^- \longrightarrow S^{2-} \quad \text{reduction}$$

$$Zn + S \longrightarrow ZnS \quad \text{overall}$$

Zn is oxidized and is the reducing agent
S is reduced and is the oxidizing agent

(e) $MgCl_2$ is $Mg^{2+}(Cl^-)_2$, so Mg loses electrons and is oxidized to $Mg^{2+}$, and, consisdering HCl as $H^+Cl^-$, $2H^+$ gain 2 electrons and are reduced to $H_2$.

$$Mg \longrightarrow Mg^{2+} + 2e^- \quad \text{oxidation}$$

$$2[H^+ + e^- \longrightarrow H_2] \quad \text{reduction}$$

$$Mg + 2H^+ \longrightarrow Mg^{2+} + H_2 \quad \text{overall}$$

and adding $2Cl^-$ to each side gives

$$Mg + 2HCl \longrightarrow MgCl_2 + H_2$$

Mg is oxidized and is the reducing agent

$H^+$ is reduced and is the oxidizing agent

33. Magnesium chloride is $MgCl_2$ or $Mg^{2+}(Cl^-)_2$, so that when Mg reacts with chlorine, Mg loses electrons to give $Mg^{2+}$ and $Cl_2$ gains electrons to give $2Cl^-$.

$$Mg(s) + Cl_2(g) \longrightarrow MgCl_2(s)$$

Mg is oxidized and $Cl_2$ is reduced, and the mass of $MgCl_2$ is

$$(2.45 \text{ g Mg})\left(\frac{1 \text{ mol Mg}}{24.31 \text{ g Mg}}\right)\left(\frac{1 \text{ mol } MgCl_2}{1 \text{ mol Mg}}\right)\left(\frac{95.21 \text{ g } MgCl_2}{1 \text{ mol } MgCl_2}\right) = 9.60 \text{ g } MgCl_2$$

34. The characteristic analytical test for halide ion is to add a solution of silver nitrate, $AgNO_3(aq)$, to the test solution. When AgCl(s), AgBr(s), or AgI(s) are precipitated, they can be identified from their colors: AgCl, white; AgBr, pale yellow, and AgI, dark yellow. They may also be distinguished in terms of their solubilities in $NH_3(aq)$: AgCl(s) is soluble, AgBr(s) is sparingly soluble, and AgI(s) is insoluble, due to the different abilities of the silver halides to form the soluble $Ag(NH_3)_2^+(aq)$ ion.

$$AgX(s) + 2NH_3(aq) \rightleftharpoons Ag(NH_3)_2^+(aq) + X^-(aq)$$

Alternatively, chlorine water, $Cl_2(aq)$, could be added to each of the test solutions. $Cl_2$ oxidizes $Br^-(aq)$ to $Br_2$ and $I^-(aq)$ to $I_2$, but does not oxidize $Cl^-$ ion. Thus, the solution with no reaction is the chloride solution. For the two solutions that react. they can be differentiated after reaction by adding 1 to 2 mL of carbon tetrachloride and shaking. Iodine and bromine are extracted into the organic solvent; bromine gives a brown solution and iodine gives a violet colored solution.

$$Cl_2(aq) + 2Br^-(aq) \longrightarrow 2Cl^-(aq) + Br_2(aq)$$
$$Cl_2(aq) + 3I^-(aq) \longrightarrow 2Cl^-(aq) + I_3^-(aq)$$

35. (a) Sulfates are usually soluble, except for those of $Sr^{2+}$, $Ba^{2+}$, $Pb^{2+}$, $Ag^+$, of which lead is an example; $PbSO_4$ is insoluble.

(b) Iodides are usually soluble but AgI(s) is an exception; AgI is insoluble.

(c) Salts containing alkali metal ions are generally soluble; $Na_2CO_3$ is soluble, even though carbonates are generally insoluble.

(d) Sulfides are generally insoluble, except those of group 1 and group 2 metals. FeS is insoluble.

(e) All nitrates are soluble; $AgNO_3$ is soluble

(f) Only the alkali metal hydroxides and those of $NH_4^+$ and $Ba^{2+}$ are soluble, $Ca(OH)_2(s)$ and $Sr(OH)_2(s)$ are sparingly soluble. $Cu(OH)_2$ is insoluble.

36. (a) Only $Ag^+$, $Pb^{2+}$, and $Hg_2^{2+}$ chlorides are insoluble; all others are soluble. $AlCl_3$ is soluble.

(b) $CaSO_4$ is one of the sparingly soluble sulfates

(c) $CuSO_4$ is soluble.

(d) LiOH (an alkali metal hydroxide) is soluble

(e) $BaCO_3$ is an insoluble carbonate; only the carbonates of group 1 metals and $NH_4^+$ are soluble.

(f) The soluble sulfides are those of group 1 and group 2 metals and $(NH_4)_2S$; $Na_2S$ is soluble.

37. All the reactant salts in this problem are soluble and will be present in solution as a mixture of the respective cations and anions. Consider if any possible combination of a cation and an anion gives an insoluble salt, and hence a precipitate when the solutions are mixed.

(a) The solution contains $Fe^{3+}(aq)$, $Cl^-(aq)$, $Na^+(aq)$ and $OH^-(aq)$. The only combination that gives an insoluble salt is

$$Fe^{3+}(aq) + 3OH^-(aq) \longrightarrow Fe(OH)_3(s)$$

NaCl is soluble and the $Na^+(aq)$ and $Cl^-(aq)$ ions remain in solution as spectator ions.

(b) All perchlorates and all nitrates are soluble; no precipitate is formed.

(c) Both of the possible products, $Ba(OH)_2$ and KCl are soluble; no precipitate is formed.

(d) $PbSO_4$ is insoluble and will form a precipitate

$$Pb(NO_3)_2(aq) + H_2SO_4(aq) \longrightarrow PbSO_4(s) + 2HNO_3(aq)$$

or $$Pb^{2+}(aq) + SO_4^{2-}(aq) \longrightarrow PbSO_4(s)$$

(e) $Ag_2S$ is <u>insoluble</u> and will be precipitated

$$2AgNO_3(aq) + Na_2S(aq) \longrightarrow Ag_2S(s) + 2NaNO_3(aq)$$

<u>or</u> $$2Ag^+(aq) + S^{2-}(aq) \longrightarrow Ag_2S(s)$$

38. This problem is similar to Problem 37.

(a) $CaCO_3$ is insoluble and will be precipitated

$$Na_2CO_3(aq) + CaCl_2(aq) \longrightarrow CaCO_3(s) + 2NaCl(aq)$$

<u>or</u> $$Ca^{2+}(aq) + CO_3^{2-}(aq) \longrightarrow CaCO_3(s)$$

(b) Neither NaBr nor $Ca(NO_3)_2$ is insoluble; no precipitate forms.

(c) AgI is insoluble and will be precipitated

$$AgNO_3(aq) + NaI(aq) \longrightarrow AgI(s) + NaNO_3(aq)$$

<u>or</u> $$Ag^+(aq) + I^-(aq) \longrightarrow AgI(s)$$

(d) $BaSO_4$ is insoluble and will be precipitated

$$BaCl_2(aq) + MgSO_4(aq) \longrightarrow BaSO_4(s) + MgCl_2(aq)$$

<u>or</u> $$Ba^{2+}(aq) + SO_4^{2-}(aq) \longrightarrow BaSO_4(s)$$

(e) HCl(aq) contains $H_3O^+(aq)$ and $Cl^-(aq)$ ions; $PbCl_2$ is insoluble and will be precipitated

$$2HCl(aq) + Pb(NO_3)_2(aq) \longrightarrow PbCl_2(s) + 2HNO_3(aq)$$

<u>or</u> $$Pb^{2+}(aq) + 2Cl^-(aq) \longrightarrow PbCl_2(s)$$

39. The balanced equation for the reaction is

$$2AgNO_3(aq) + BaCl_2(aq) \longrightarrow 2AgCl(s) + Ba(NO_3)_2(aq)$$

Since $AgNO_3(aq)$ is the limiting reactant

$$\text{mass AgCl(s)} = (2.50\text{ g }AgNO_3)\left(\frac{1\text{ mol }AgNO_3}{169.9\text{ g }AgNO_3}\right)\left(\frac{2\text{ mol AgCl}}{2\text{ mol }AgNO_3}\right)\left(\frac{143.4\text{ g AgCl}}{1\text{ mol AgCl}}\right)$$

$$= \underline{2.11\text{ g AgCl}}$$

40. The difficulty cannot be due to the solubility of $AgNO_3$, which greatly exceeds 0.1 M. However, most tap water contains a trace of $Cl^-(aq)$ because chlorine is often used in the purification of the domestic water supply,

$$Cl_2(aq) + 2H_2O(l) \longrightarrow H_3O^+(aq) + Cl^-(aq) + HOCl(aq)$$

Thus, the cloudiness of the solution was undoubtably due to a precipitate of AgCl(s), which has a very small solubility,

$$Ag^+(aq) + Cl^-(aq) \longrightarrow AgCl(s)$$

41. The balanced equation for the reaction is

$$Ba(OH)_2(aq) + H_2SO_4(aq) \longrightarrow BaSO_4(s) + 2H_2O(l)$$

Thus, for complete reaction

$$\text{mol } Ba(OH)_2 = (32.12 \text{ mL } H_2SO_4)\left(\frac{0.112 \text{ mol } H_2SO_4}{1 \text{ L}}\right)\left(\frac{1 \text{ L}}{10^3 \text{ mL}}\right)\left(\frac{1 \text{ mol } Ba(OH)_2}{1 \text{ mol } H_2SO_4}\right)$$

$$= \underline{3.60 \times 10^{-3} \text{ mol}}$$

(a) Let the molarity of the $Ba(OH)_2(aq)$ be X mol $L^{-1}$, then

$$(X \text{ mol } L^{-1})(35.00 \text{ mL})\left(\frac{1 \text{ L}}{10^3 \text{ mL}}\right) = 3.60\text{x}10^{-3} \text{ mol}; \ X = \underline{0.103 \text{ M}}$$

More simply, because 1 mol $Ba(OH)_2$ reacts with 1 mol $H_2SO_4$, we could have written

$$\text{mol } Ba(OH)_2 = \text{mol } H_2SO_4$$

i.e.,

$$(35.00 \text{ mL})(X \text{ mol } L^{-1}) = (32.12 \text{ mL})(0.112 \text{ mol } L^{-1})$$

$$\underline{X = 0.103 \text{ M}}$$

(b) From the balanced equation

$$\text{Mass } BaSO_4 = (3.60\text{x}10^{-3} \text{ mol } H_2SO_4)\left(\frac{1 \text{ mol } BaSO_4}{1 \text{ mol } H_2SO_4}\right)\left(\frac{233.4 \text{ g } BaSO_4}{1 \text{ mol } BaSO_4}\right)$$

$$= \underline{0.840 \text{ g}}$$

42. (a) $$Ca(OH)_2(aq) + 2HNO_3(aq) \longrightarrow Ca(NO_3)_2(aq) + 2H_2O(l)$$

2 mol $HNO_3$ react with 1 mol $Ca(OH)_2$. Thus, following the second method of Problem 41(a),

$$[V \text{ mL } HNO_3(aq)](0.012 \text{ mol } L^{-1}) = 2(25.1 \text{ mL } Ca(OH)_2)(0.021 \text{ mol } L^{-1})$$

$$\underline{V = 87.9 \text{ mL}}$$

(b) $$CaCO_3(s) + 2HCl(aq) \longrightarrow CaCl_2(aq) + CO_2(g) + H_2O(l)$$

$$\text{mol HCl} = (1.234 \text{ g } CaCO_3)\left(\frac{1 \text{ mol } CaCO_3}{100.1 \text{ g } CaCO_3}\right)\left(\frac{2 \text{ mol HCl}}{1 \text{ mol } CaCO_3}\right) = \underline{0.02466 \text{ mol}}$$

$$0.02466 \text{ mol HCl} = (V \text{ mL HCl})(0.112 \text{ mol } L^{-1})\left(\frac{1 \text{ L}}{10^3 \text{ mL}}\right)$$

$$\underline{V = 220 \text{ mL}}$$

(c) $$Ca(OH)_2(aq) + 2HCl(aq) \longrightarrow CaCl_2(aq) + 2H_2O(l)$$

and $$3Ca(OH)_2(aq) + 2H_3PO_4(aq) \longrightarrow Ca_3(PO_4)_2(aq) + 6H_2O(l)$$

From the first equation, calculate the molarity of the $Ca(OH)_2(aq)$,

$$(X \text{ mol } L^{-1})(25.0 \text{ mL } Ca(OH)_2) = (24.1 \text{ mL HCl})\left(\frac{0.056 \text{ mol}}{1 \text{ L}}\right)\left(\frac{1 \text{ mol } Ca(OH)_2}{2 \text{ mol HCl}}\right)$$

$$\underline{X = 0.027 \text{ mol } L^{-1}}$$

and using the second equation,

$$(28.2 \text{ mL } Ca(OH)_2)\left(\frac{0.027 \text{ mol}}{1 \text{ L}}\right)\left(\frac{2 \text{ mol } H_3PO_4}{3 \text{ mol } Ca(OH)_2}\right) = (V \text{ mL } H_3PO_4)\left(\frac{0.032 \text{ mol}}{1 \text{ L}}\right)$$

$$\underline{V = 15.9 \text{ mL}}$$

43. If a given aqueous solution is acidic it will

(i) act on an indicator such as litmus (red in acid);

(ii) dissolve a solid carbonate,

e.g., $CaCO_3(s) + 2H_3O^+(aq) \longrightarrow Ca^{2+}(aq) + CO_2(g) + 3H_2O(l)$

forming $CO_2(g)$, which can be detected by bubbling through $Ca(OH)_2(aq)$

$$Ca(OH)_2(aq) + CO_2(g) \longrightarrow CaCO_3(s) + H_2O\text{; and}$$

(iii) dissolve a reactive metal, such as magnesium or zinc,

$$Mg(s) + 2H_3O^+(aq) \longrightarrow Mg^{2+}(aq) + 2H_2O(l) + H_2(g)$$

forming $H_2(g)$ which can be tested by igniting, when it explodes with a pop and burns with a pale blue flame to give water,

$$2H_2(g) + O_2(g) \longrightarrow 2H_2O(l)$$

The first test given in the text should not be used even on a dilute solution of an acid. It is very dangerous to taste any unknown chemical.

44. (a) $HNO_3(aq) + H_2O(l) \longrightarrow H_3O^+(aq) + NO_3^-(aq)$ strong acid

(b) $H_3PO_4$, phosphoric acid, is a triprotic acid that ionizes in three stages:

$H_3PO_4(aq) + H_2O(l) \rightleftharpoons H_3O^+(aq) + H_2PO_4^-(aq)$ weak acid

$H_2PO_4^-(aq) + H_2O(l) \rightleftharpoons H_3O^+(aq) + HPO_4^{2-}(aq)$ weak acid

$HPO_4^{2-}(aq) + H_2O(l) \rightleftharpoons H_3O^+(aq) + PO_4^{3-}(aq)$ weak acid

(c) $HOCl(aq) + H_2O(l) \rightleftharpoons H_3O^+(aq) + OCl^-(aq)$ weak acid

(d) $H_2SO_4$ is a diprotic acid that ionizes in two stages:

$H_2SO_4(aq) + H_2O(l) \longrightarrow H_3O^+(aq) + HSO_4^-(aq)$ strong acid

$HSO_4^-(aq) + H_2O(l) \rightleftharpoons H_3O^+(aq) + SO_4^{2-}(aq)$ weak acid

(e) $HF(aq) + H_2O(l) \rightleftharpoons H_3O^+(aq) + F^-(aq)$ weak acid

(f) $HClO_4(aq) + H_2O(l) \longrightarrow H_3O^+(aq) + ClO_4^-(aq)$ strong acid

Remember that the common strong acids are $HClO_4$, $H_2SO_4$, $HNO_3$, HCl, HBr, and HI; all the other common acids are weak acids.

45. (a) $Na_2O(s)$ is ionic, $[Na^+]_2O^{2-}$, and forms the hydrated ions in aqueous solution:

$$Na_2O(s) \longrightarrow 2Na^+(aq) + O^{2-}(aq)$$

and the oxide ion behaves as a strong base

$$O^{2-}(aq) + H_2O(l) \longrightarrow 2OH^-(aq)$$

Thus, the overall reaction is

$Na_2O(s) + H_2O(l) \longrightarrow 2Na^+(aq) + 2OH^-(aq)$ strong base

(b) KOH(s) simply dissociates into its component ions

$KOH(s) \longrightarrow K^+(aq) + OH^-(aq)$ strong base

(c) Ammonia accepts a proton from a water molecule and is a weak base

$NH_3(aq) + H_2O(l) \rightleftharpoons NH_4^+(aq) + OH^-(aq)$, weak base

(d) LiH dissociates into $Li^+$ and $:H^-$ ions, and the $:H^-$ ion accepts a proton from a water molecule, to form $H_2(g)$ - strong base,

$LiH(aq) + H_2O(l) \longrightarrow Li^+(aq) + OH^-(aq) + H_2(g)$

46. (a) $KOH(s) \longrightarrow K^+(aq) + OH^-(aq)$ strong base

(b) $2H_2O(l) \rightleftharpoons H_3O^+(aq) + OH^-(aq)$ weak acid and weak base

Proton transfer occurs between one water molecule and another to a limited extent, so that water behaves both as a weak acid and a weak base, and is described as amphoteric or amphiprotic

(c) $CH_3CO_2H(aq) + H_2O(l) \rightleftharpoons H_3O^+(aq) + CH_3CO_2^-(aq)$ weak acid

(d) $H_2SO_4(aq) + H_2O(l) \longrightarrow H_3O^+(aq) + HSO_4^-(aq)$ strong acid

& $HSO_4^-(aq) + H_2O(l) \rightleftharpoons H_3O^+(aq) + SO_4^{2-}(aq)$ weak acid

(e) $NH_3(aq) + H_2O(l) \rightleftharpoons NH_4^+(aq) + OH^-(aq)$ weak base

(f) $KI(s) \longrightarrow K^+(aq) + I^-(aq)$ salt

Neither the alkali metal cation $K^+$ nor the anion $I^-$ react with water; $I^-$ is the anion of the strong acid HI and has no basic properties in aqueous solution.

(g) $CaCl_2(s) \longrightarrow Ca^{2+}(aq) + 2Cl^-(aq)$ salt

Neither the alkaline earth metal cation $Ca^{2+}$ nor the anion $Cl^-$ react with water; $Cl^-$ is the anion of the strong acid HCl and has no basic properties in water.

(h) $H_2CO_3(aq) + H_2O(l) \rightleftharpoons H_3O^+(aq) + HCO_3^-(aq)$ weak acid

$HCO_3^-(aq) + H_2O(l) \rightleftharpoons H_3O^+(aq) + CO_3^{2-}(aq)$ weak acid

(i) $Na_2O(s) \longrightarrow 2Na^+(aq) + O^{2-}(aq)$

and the oxide ion, $O^{2-}(aq)$ behaves as a strong base,

$O^{2-}(aq) + H_2O(l) \longrightarrow 2OH^-(aq)$ strong base

Note that like all soluble oxides, $Na_2O$ is a salt that behaves as a strong base in aqueous solution.

(j) $CH_3CO_2K(s) \longrightarrow K^+(aq) + CH_3CO_2^-(aq)$ salt

$K^+$ does not react further with water, but the acetate ion, $CH_3CO_2^-$, the anion of a weak acid, behaves as a weak base.

$CH_3CO_2^-(aq) + H_2O(l) \rightleftharpoons CH_3CO_2H(aq) + OH^-(aq)$ weak base

47. Here consider the component ions that constitute the salt; the relevant base is the hydroxide (or oxide) formed by the metal cation, and the acid is the parent acid of the anion.

(a) $Ca(OH)_2(s) + H_2SO_4(aq) \longrightarrow CaSO_4(aq) + 2H_2O(l)$

(b) $LiOH(aq) + HF(aq) \longrightarrow LiF(aq) + H_2O(l)$

(c) $2NaOH(aq) + H_2S(g) \longrightarrow Na_2S(aq) + 2H_2O(l)$

(d) $NH_3(aq) + HNO_3(aq) \longrightarrow NH_4NO_3(aq)$

(e) $Mg(OH)_2(s) + 2HClO_4(aq) \longrightarrow Mg(ClO_4)_2(aq) + 2H_2O(l)$

(f) $Al(OH)_3(s) + 3HCl(aq) \longrightarrow AlCl_3(aq) + 3H_2O(l)$

48. The acid-base properties of a salt depend on whether the cation and anion have acidic or basic properties.

(a) basic, RbF is fully dissociated into $Rb^+(aq)$ and $F^-(aq)$ ions. $Rb^+(aq)$ has no acidic properties, but $F^-(aq)$, the anion of the weak acid HF(aq), behaves as a weak base,

$$F^-(aq) + H_2O(l) \rightleftharpoons HF(aq) + OH^-(aq)$$

(b) neutral, $CaCl_2$ is fully dissociated into $Ca^{2+}(aq)$ and $Cl^-(aq)$, neither of which has acid or base properties in water

(c) neutral, $KNO_3$ is fully dissociated into $K^+(aq)$ and $NO_3^-(aq)$, neither of which has acid or basic properties

(d) acidic, $NH_4Cl$ is fully dissociated into $NH_4^+(aq)$ and $Cl^-(aq)$. $Cl^-$ has no basic properties, but $NH_4^+$ behaves as a weak acid, because $NH_3$ is a weak base.

$$NH_4^+(aq) + H_2O(l) \rightleftharpoons H_3O^+(aq) + NH_3(aq)$$

49. A conjugate acid-base pair differ only by a proton,

$$HA \longrightarrow H^+ + A^-$$

acid conjugate base

| | Acid | Conjugate Base | Name |
|---|---|---|---|
| (a) | HF | $F^-$ | fluoride ion |
| (b) | $HNO_3$ | $NO_3^-$ | nitrate ion |
| (c) | $HClO_4$ | $ClO_4^-$ | perchlorate ion |
| (d) | $H_2O$ | $OH^-$ | hydroxide ion |
| (e) | $H_3O^+$ | $H_2O$ | water |

50. $$B + H^+ \longrightarrow BH^+$$

base conjugate acid

| | Base | Conjugate Acid | Name |
|---|---|---|---|
| (a) | $NH_3$ | $NH_4^+$ | ammonium ion |
| (b) | $F^-$ | HF | hydrofluoric acid |
| (c) | $OH^-$ | $H_2O$ | water |
| (d) | $H_2O$ | $H_3O^+$ | hydronium ion |
| (e) | $H^-$ | $H_2$ | hydrogen |

51. Each of the conjugate bases is formed from the parent acid by the loss of a proton.

$OH^-$ $NH_2^-$ $Cl^-$ $F^-$ $NH_3$

52. Each of the conjugate acids is formed from the parent base by the gain of a proton.

$H_3O^+$ $NH_4^+$ $H_2O$ $NH_3$ $HS^-$

53. (a) HCl and $HNO_3$ are fully ionized and are strong acids, HF and $NH_4^+$ are incompletely ionized and are weak acids, $CH_4$ has no acid or base properties in water.

(b) $Cl^-$ and $NO_3^-$ are the conjugate bases of the strong acids HCl and $HNO_3$, respectively, and have no basic properties in water; $F^-$ is the conjugate base of the weak acid HF and behaves as a weak base in water; $NH_3$ is a weak base in water and $O^{2-}$ is quantitatively converted to $OH^-$ in aqueous solution and is a strong base.

54. (a) All acids give $H_3O^+$ ions in aqueous solution and all bases give $OH^-$ ions, and an acid-base reaction in water is simply the reaction of these two ions,

$$H_3O^+(aq) + OH^-(aq) \longrightarrow 2H_2O(l)$$

The strong acids HCl(aq) and $HNO_3$(aq) are fully ionized in aqueous solution,

$$HCl(aq) + H_2O(l) \longrightarrow H_3O^+(aq) + Cl^-(aq)$$

$$HNO_3(aq) + H_2O(l) \longrightarrow H_3O^+(aq) + NO_3^-(aq)$$

and the strong base NaOH(aq) is also completely dissociated into its component ions in aqueous solution,

$$NaOH(aq) \longrightarrow Na^+(aq) + OH^-(aq)$$

Thus, for the reactions of each of the strong acids with NaOH(aq), we can write,

$$H_3O^+(aq) + Cl^-(aq) + Na^+(aq) + OH^-(aq) \longrightarrow Na^+(aq) + Cl^-(aq) + 2H_2O(l)$$

and

$$H_3O^+(aq) + NO_3^-(aq) + Na^+(aq) + OH^-(aq) \longrightarrow Na^+(aq) + NO_3^-(aq) + 2H_2O(l)$$

In the first equation, $Na^+$(aq) and $Cl^-$(aq) are spectator ions, and in the second equation, $Na^+$(aq) and $NO_3^-$(aq) are spectator ions. Thus, in each case the overall reaction is simply

$$H_3O^+(aq) + OH^-(aq) \longrightarrow 2H_2O(l)$$

Similarly, $Ca(OH)_2$(aq) is fully dissociated into $Ca^{2+}$ and $OH^-$ ions, and we can write,

$$2H_3O^+(aq) + 2Cl^-(aq) + Ca^{2+}(aq) + 2OH^-(aq) \longrightarrow Ca^{2+}(aq) + 2Cl^-(aq) + 4H_2O(l)$$

and

$$2H_3O^+(aq) + 2NO_3^-(aq) + Ca^{2+}(aq) + 2OH^-(aq) \longrightarrow Ca^{2+}(aq) + 2NO_3^-(aq) + 4H_2O(l)$$

Again, $Ca^{2+}(aq)$ and $Cl^-(aq)$, or $NO_3^-(aq)$, ions are spectator ions and the overall reactions are simply

$$2H_3O^+(aq) + 2OH^-(aq) \longrightarrow 4H_2O(l)$$

(b) Although HF(aq) is a weak acid and incompletely ionized

$$HF(aq) + H_2O(l) \rightleftharpoons H_3O^+(aq) + F^-(aq),$$

addition of $OH^-$ ions from the base removes $H_3O^+$ ions, which causes more of the HF(aq) to ionize, giving more $H_3O^+$ ions, which in turn are removed as more $OH^-$ is added, until all of the HF(aq) is neutralized.

$$HF(aq) + NaOH(aq) \longrightarrow Na^+(aq) + F^-(aq) + H_2O(l)$$

55.

| | Name | Formula | Lewis Structure | Type | Geometry |
|---|---|---|---|---|---|
| (a) | Hydronium ion | $H_3O^+$ | +<br>H-Ö-H<br>\|<br>H | $AX_3E$ | trigonal pyramid |
| (b) | Ammonium ion | $NH_4^+$ | H<br>\| +<br>H-N-H<br>\|<br>H | $AX_4$ | tetrahedral |
| (c) | Hydroxide ion | $OH^-$ | ⁻:Ö-H (with lone pairs) | $AXE_3$ | (linear) |

56. In principle each solution contains two acid-base conjugate pairs. When the acid is strong, its conjugate base has no detectable basic properties in water; when it is a weak acid, its conjugate base behaves as a weak base in water:

(a) $\underset{acid_1}{HCl(aq)} + \underset{base_2}{H_2O(l)} \longrightarrow \underset{acid_2}{H_3O^+(aq)} + \underset{base_1}{Cl^-(aq)}$

HCl is a strong acid, so $Cl^-$ has no basic properties in water

(b) $\underset{acid_1}{CH_3CO_2H(aq)} + \underset{base_2}{H_2O(l)} \rightleftharpoons \underset{acid_2}{H_3O^+(aq)} + \underset{base_1}{CH_3CO_2^-(aq)}$

(c) $\underset{acid_1}{HClO_4(aq)} + \underset{base_2}{H_2O(l)} \longrightarrow \underset{acid_2}{H_3O^+(aq)} + \underset{base_1}{ClO_4^-(aq)}$

$HClO_4(aq)$ is a strong acid, so $ClO_4^-$ has no basic properties in water.

(d) $H_2SO_4$ is a diprotic acid and has two stages of ionization in water:

(i) $\underset{acid_1}{H_2SO_4(aq)} + \underset{base_2}{H_2O(l)} \longrightarrow \underset{acid_2}{H_3O^+(aq)} + \underset{base_1}{HSO_4^-(aq)}$

$H_2SO_4$ is a strong acid, so $HSO_4^-$ has no basic properties in water.

(ii) $\underset{acid_1}{HSO_4^-(aq)} + \underset{base_2}{H_2O(l)} \rightleftharpoons \underset{acid_2}{H_3O^+(aq)} + \underset{base_1}{SO_4^{2-}(aq)}$

(e) $\underset{acid_1}{HOCl(aq)} + \underset{base_2}{H_2O(l)} \rightleftharpoons \underset{acid_2}{H_3O^+(aq)} + \underset{base_1}{OCl^-(aq)}$

(f) $NH_4^+(aq) + H_2O(l) \rightleftharpoons H_3O^+(aq) + NH_3(aq)$

$\quad\text{acid}_1 \qquad \text{base}_2 \qquad \text{acid}_2 \qquad \text{base}_1$

57. (a) The salt NaOCl, sodium hypochlorite, is fully dissociated into ions,

$$NaOCl(aq) \longrightarrow Na^+(aq) + OCl^-(aq)$$

The $Na^+(aq)$ ion, like the other alkali metal ions, has no acidic or basic properties, but the hypochlorite ion, $OCl^-$, is the conjugate base of a weak acid, and is therefore a weak base,

$$OCl^-(aq) + H_2O(l) \rightleftharpoons HOCl(aq) + OH^-(aq) \quad \underline{\text{basic solution}}$$

(b) The salt $NH_4Cl$, ammonium chloride, is fully dissociated into ions

$$NH_4Cl(aq) \longrightarrow NH_4^+(aq) + Cl^-(aq)$$

$Cl^-$ is the conjugate base of the strong acid HCl, and therefore has no basic properties in water, but $NH_4^+$ is the conjugate acid of the weak base $NH_3$, and therefore behaves as a weak acid in water,

$$NH_4^+(aq) + H_2O(l) \rightleftharpoons H_3O^+(aq) + NH_3(aq) \quad \underline{\text{acidic solution}}$$

58. The strongest acid species that can exist in aqueous solution is $H_3O^+$, the hydronium ion; all intrinsically stronger acids are quantitatively converted in aqueous solution to $H_3O^+(aq)$. The strongest base that can exist in water is $OH^-$, the hydroxide ion; all intrinsically stronger bases in water are quantitatively converted to $OH^-(aq)$.

Both $H^-$ and $O^{2-}$ are strong bases in water and are quantitatively converted to $OH^-(aq)$:

$$H^-(aq) + H_2O(l) \longrightarrow H_2(g) + OH^-(aq)$$
$$O^{2-}(aq) + H_2O(l) \longrightarrow 2OH^-(aq)$$

Thus, the reactions of LiH, $(Li^+H^-)$; $CaH_2$, $[Ca^{2+}(H^-)_2]$; $Li_2O$, $[(Li^+)_2O^{2-}]$; and CaO, $(Ca^{2+}O^{2-})$, are those of their anions behaving as strong bases, the cations remaining in solution as spectator ions:

$$LiH(s) + H_2O(l) \longrightarrow Li^+(aq) + OH^-(aq) + H_2(g)$$
$$CaH_2(s) + 2H_2O(l) \longrightarrow Ca^{2+}(aq) + 2OH^-(aq) + 2H_2(g)$$
$$Li_2O(s) + H_2O(l) \longrightarrow 2Li^+(aq) + 2OH^-(aq)$$
$$CaO(s) + H_2O(l) \longrightarrow Ca^{2+}(aq) + 2OH^-(aq)$$

59. As we saw in Problem 54, the simplest way of expressing an acid-base neutralization reaction in water is

$$H_3O^+(aq) + OH^-(aq) \longrightarrow 2H_2O(l)$$

For all stoichiometry problems involving aqueous solutions of acids and bases, first write the balanced equation for the reaction, which in this case is

$$HCl(aq) + NaOH(aq) \longrightarrow NaCl(aq) + H_2O(l)$$

then calculate the initial moles of reactants,

$$\text{mol } HCl(aq) = (250\ mL)\left(\frac{1\ L}{10^3\ mL}\right)\left(\frac{1.00\ mol}{1\ L}\right) = \underline{0.250\ mol}$$

$$\text{mol } NaOH(aq) = (500\ mL)\left(\frac{1\ L}{10^3\ mL}\right)\left(\frac{0.500\ mol}{1\ L}\right) = \underline{0.250\ mol}$$

Thus, we have

| | $HCl(aq)$ + | $NaOH(aq)$ $\longrightarrow$ | $NaCl(aq)$ + | $H_2O(l)$ |
|---|---|---|---|---|
| initially | 0.250 | 0.250 | 0 | mol |
| after reaction | 0 | 0 | 0.250 | mol |

In the solution after reaction we have 0.250 mol NaCl(aq) in a total volume of (250+500) mL = 750 mL of solution, and the concentration of NaCl is

$$\frac{0.250\ mol\ NaCl}{(750\ mL)\left(\frac{1\ L}{10^3\ mL}\right)} = \underline{0.333\ M}$$

$$\text{Mass of NaCl} = (0.250\ mol\ NaCl)\left(\frac{58.44\ g\ NaCl}{1\ mol\ NaCl}\right) = \underline{14.6\ g}$$

60. From the information given, calculate the moles of $Ca(OH)_2$,

$$5.00\ g\ Ca(OH)_2 = (5.00\ g\ Ca(OH)_2)\left(\frac{1\ mol\ Ca(OH)}{74.10\ g\ Ca(OH)_2}\right) = \underline{0.0675\ mol}$$

and write the balanced equation for the reaction,

| | $2HCl(aq)$ + | $Ca(OH)_2(s)$ $\longrightarrow$ | $CaCl_2(aq)$ + | $2H_2O(l)$ |
|---|---|---|---|---|
| initially | x | 0.0675 mol | 0 | 0 |

Thus, mol HCl(aq) for complete reaction is given by

$$x = (0.0675\ mol\ Ca(OH)_2)\left(\frac{2\ mol\ HCl}{1\ mol\ Ca(OH)_2}\right) = \underline{0.135\ mol}$$

and if the required volume of 0.100 M HCl(aq) is V L, then

$$[V\ L\ HCl(aq)](0.100\ mol\ L^{-1}) = 0.135\ mol$$

$$\underline{V = 1.35\ L}$$

61. First calculate the moles of HCl(aq),

$$\text{mol } HCl(aq) = (2.00\ L)(0.120\ mol\ L^{-1}) = \underline{0.240\ mol}$$

and write the balanced equation for the reaction,

| | $NaAl(OH)_2CO_3(s)$ + | $4HCl(aq)$ $\longrightarrow$ | $NaCl(aq)$ + | $AlCl_3(aq)$ + | $CO_2(g)$ + | $3H_2O$ |
|---|---|---|---|---|---|---|
| initially | x | 0.240 mol | 0 | 0 | 0 | |

$$x = (0.240\ mol\ HCl)\left(\frac{1\ mol\ NaAl(OH)_2CO_3}{4\ mol\ HCl}\right) = \underline{0.060\ mol}$$

$$\text{Mass of } NaAl(OH)_2CO_3(s) = (0.060\ mol)(144.0\ g\ mol^{-1}) = \underline{8.64\ g}$$

62. Mol NaOH = (25.00 mL)$(\frac{1\ L}{10^3\ mL})$(0.107 mol $L^{-1}$) = $\underline{2.675 \times 10^{-3}\ mol}$

and the balanced equation for the reaction is

| | HBr(aq) | + NaOH(aq) ⟶ | NaBr(aq) | + $H_2O(l)$ | |
|---|---|---|---|---|---|
| initially | x | $2.675 \times 10^{-3}$ mol | 0 | 0 | mol |

For complete reaction x = ($2.675 \times 10^{-3}$ mol NaOH)$(\frac{1\ mol\ HBr}{1\ mol\ NaOH})$

= $\underline{2.675 \times 10^{-3}\ mol\ HBr}$

Thus, if the required volume of HBr(aq) is V L,

(V L HBr)(0.124 mol $L^{-1}$) = $2.675 \times 10^{-3}$ mol HBr

V = $\underline{2.16 \times 10^{-2}\ L}$ or $\underline{21.6\ mL}$

63. Mol HF = (100.0 mL)$(\frac{1\ L}{10^3\ mL})$(0.211 mol $L^{-1}$) = $\underline{2.11 \times 10^{-2}\ mol}$

and the balanced equation for the reaction is

| | HF(aq) | + KOH(aq) ⟶ | KF(aq) | + $H_2O(l)$ | |
|---|---|---|---|---|---|
| initially | $2.11 \times 10^{-2}$ | x | 0 | 0 | mol |

from which, x = $2.11 \times 10^{-2}$ mol KOH, and if the volume is V L,

(V L KOH)(0.115 mol $L^{-1}$) = $2.11 \times 10^{-2}$ mol; $\underline{V = 0.183\ L}$

64. (a) A simple compound of bromine that behaves as an acid is HBr(aq). Thus, for example

$$HBr(aq) + NaOH(aq) \longrightarrow NaBr(aq) + H_2O(l)$$

acid base salt

(b) $Br_2(l)$ is a suitable reagent to oxidize, for example, $I^-(aq)$ to $I_2(s)$. Thus in aqueous solution for example

$$Br_2(l) + 3NaI(aq) \longrightarrow NaI_3(aq) + 2NaBr(aq)$$

$I^-(aq)$ is oxidized to $I_2(s)$, which dissolves in excess $I^-(aq)$ to give the $I_3^-(aq)$ ion; $Br_2(l)$ is reduced to $Br^-(aq)$ ion.

(c) One of the insoluble salts containing bromine that has been mentioned in this chapter is silver bromide, AgBr(s) which is precipitated when a soluble silver salt, typically $AgNO_3(s)$, is reacted in aqueous solution with a soluble bromide, such as NaBr(aq).

$$NaBr(aq) + AgNO_3(aq) \longrightarrow AgBr(s) + NaNO_3(aq)$$

65. (a) $2AgNO_3(aq) + BaCl_2(aq) \longrightarrow Ba(NO_3)_2(aq) + 2AgCl(s)$ <u>precipitation</u>

(b) $2NH_3(aq) + H_2SO_4(aq) \longrightarrow (NH_4)_2SO_4(aq)$ <u>acid-base</u>

(c) $Na_2O(s) + H_2O(l) \longrightarrow 2NaOH(aq)$ <u>acid-base</u>

(d) $2Al(s) + 3Br_2(l) \longrightarrow 2AlBr_3(s)$ <u>oxidation-reduction</u>

$Al(s)$ is oxidized to $Al^{3+}$ and $Br_2$ is reduced to $Br^-$.

(e) $Ca(OH)_2(aq) + CO_2(g) \longrightarrow CaCO_3(s) + H_2O(l)$

This reaction is both an <u>acid-base</u> reaction and a <u>precipitation</u> reaction. Insoluble $CaCO_3(s)$ is precipitated, and that it is an acid-base reaction is clear if we first allow the $CO_2(g)$ to react with water to give carbonic acid

$$CO_2(g) + H_2O(l) \longrightarrow H_2CO_3(aq)$$

and then we can write

$$Ca(OH)_2(aq) + H_2CO_3(aq) \longrightarrow CaCO_3(s) + 2H_2O(l)$$

66. In general salts can be obtained by a number of different reactions, such as

(i) reaction of a metal oxide or hydroxide with the requisite acid

(ii) reaction of a metal with an acid

or in the case of halide salts by

(iii) direct reaction of metal with the halogen or hydrogen halide

and in the case of insoluble salts by

(iv) a precipitation reaction.

(a) $CaCl_2$ is a soluble salt, so preparation by precipitation is ruled out. Suitable methods would include:

$CaO(s) + 2HCl(aq) \longrightarrow CaCl_2(aq) + H_2O(aq)$ <u>acid-base</u>

$Ca(OH)_2(s) + 2HCl(aq) \longrightarrow CaCl_2(aq) + 2H_2O(l)$ <u>acid-base</u>

$Ca(s) + 2HCl(aq) \longrightarrow CaCl_2(aq) + H_2(g)$ <u>oxidation-reduction</u>

$Ca(s) + 2HCl(g) \longrightarrow CaCl_2(s) + H_2(g)$ <u>oxidation-reduction</u>

$Ca(s) + Cl_2(g) \longrightarrow CaCl_2(s))$ <u>oxidation-reduction</u>

(b) $MgSO_4$ is a soluble salt; preparation by a precipitation reaction is ruled out.

$MgO(s) + H_2SO_4(aq) \longrightarrow MgSO_4(aq) + H_2O(l)$ <u>acid-base</u>

$Mg(OH)_2(s) + H_2SO_4(aq) \longrightarrow MgSO_4(aq) + 2H_2O(l)$ <u>acid-base</u>

$Mg(s) + H_2SO_4(aq) \longrightarrow MgSO_4(aq) + H_2(g)$ <u>oxidation-reduction</u>

($MgSO_4$ also results from the reaction of magnesium with concentrated $H_2SO_4$, but reaction with gaseous $H_2SO_4$ is not feasible, because of the very high boiling point of sulfuric acid).

(c) $Na_2SO_4$ is soluble; preparation by a precipitation reaction is ruled out.

$Na_2O(s) + H_2SO_4(aq) \longrightarrow Na_2SO_4(aq) + H_2O(l)$ <u>acid-base</u>

$2NaOH(aq) + H_2SO_4(aq) \longrightarrow Na_2SO_4(aq) + 2H_2O(l)$ <u>acid-base</u>

$2Na(s) + H_2SO_4(aq) \longrightarrow Na_2SO_4(aq) + H_2(g)$ <u>oxidation-reduction</u>

(d) $AgCl(s)$ is an insoluble salt.

$AgNO_3(aq) + HCl(aq) \longrightarrow AgCl(s) + HNO_3(aq)$ <u>precipitation</u>

$Ag_2O(s) + 2HCl(aq) \longrightarrow 2AgCl(s) + H_2O(l)$ acid-base

$2Ag(s) + 2HCl(g) \longrightarrow 2AgCl(s) + H_2(g)$ oxidation-reduction

67. (a) Oxidation-reduction in which $Cl_2$ gains electrons and is reduced, and is the oxidizing agent, and $I^-$ loses an electron and is oxidized, and is the reducing agent.

(b) Acid-base in which HCl(aq) transfers a proton to $H_2O$; the acids in the system are HCl and $H_3O^+$ and the bases are $Cl^-$ and $H_2O$.

(c) Oxidation-reduction. The product $ZnCl_2$ is $Zn^{2+}(aq)$ and $2Cl^-(aq)$ in solution and the reactant HCl(aq) is present in solution as $H_3O^+(aq)$ and $Cl^-(aq)$. $Cl^-(aq)$ is a spectator ion in this reaction, which is more simply written as

$$Zn(s) + 2H_3O^+(aq) \longrightarrow Zn^{2+}(aq) + 2H_2O(l) + H_2(g)$$

Zn loses electrons and is oxidized, and is the reducing agent; $H_3O^+$ gains an electron and forms the unstable $H_3O$, which decomposes to $H_2(g)$ and water; $H_3O^+$ is reduced and is the oxidizing agent.

(d) Acid-base in which the acid $H_3O^+(aq)$ transfers a proton to the base $HCO_3^-$, to give $H_2CO_3(aq)$, carbonic acid, which decomposes to give $CO_2(g)$ and water. The acid-base conjugate pairs are $H_3O^+$ and $H_2O$, and $H_2CO_3$ and $HCO_3^-$.

68. The high melting point and boiling point of AB and the fact that AB when molten is an electrical conductor suggest that AB is an ionic compound. Since A is a metal, B must be a nonmetal. AB must be an ionic salt since its aqueous solution conducts electricity. Reaction of AB(aq) with $AgNO_3(aq)$ gives a pale yellow precipitate. This must be an insoluble silver salt since all nitrates are soluble. That the color is pale yellow strongly suggests that the precipitate is AgBr(s). That AB is a bromide salt is confirmed by the reaction of its aqueous solution with $Cl_2$, and by the fact that bromine is the only liquid nonmetal.

$$2Br^-(aq) + Cl_2(g) \longrightarrow Br_2(aq) + 2Cl^-(aq)$$

Thus, for the reaction of AB(aq) with $AgNO_3(aq)$,

$$ABr(aq) + AgNO_3(aq) \longrightarrow ANO_3(aq) + AgBr(s), \text{ and}$$

$$\text{mol ABr} = (0.857 \text{ g AgBr})\left(\frac{1 \text{ mol AgBr}}{187.8 \text{ g AgBr}}\right)\left(\frac{1 \text{ mol ABr}}{1 \text{ mol AgBr}}\right) = \underline{4.56\times10^{-3} \text{ mol}}$$

$$\text{Thus molar mass of ABr} = \left(\frac{0.543 \text{ g ABr}}{4.56\times 10^{-3} \text{ mol ABr}}\right) = \underline{119 \text{ g mol}^{-1}}$$

and molar mass of ABr = [(molar mass of A) + 79.90] g $mol^{-1}$

= 119 g $mol^{-1}$

Molar mass of A = 39.1 g $mol^{-1}$

and from the periodic table, we identify A as potassium (K).

A is potassium, B is bromine, and AB is potassium bromide, KBr

69. From the information that $MX_2(s)$ is an ionic halide salt, the ionic empirical formula is $M^{2+}(X^-)_2$, so that M must be a group 2 metal. Since $MX_2(aq)$ is not oxidized by $Br_2$, X cannot be iodine, but since it does react with $Cl_2$, X is most likely bromine, which is confirmed by the brown color,

$$MBr_2(aq) + Cl_2(g) \longrightarrow M^{2+}(aq) + 2Cl^-(aq) + Br_2(aq)$$

From the data concerning $Cl_2(g)$, calculate the moles of $Cl_2$ that react with the aqueous solution containing 5.35 g $MBr_2$.

$$\text{mol } Cl_2 = \frac{PV}{RT} = \frac{(1.00 \text{ atm})(710 \text{ mL } Cl_2)\left(\frac{1 \text{ L}}{10^3 \text{ mL}}\right)}{(0.0821 \text{ atm L mol}^{-1} \text{ K}^{-1})(298 \text{ K})} = \underline{0.0290 \text{ mol}}, \text{ and}$$

$$\text{mol } MBr_2 = (0.0290 \text{ mol } Cl_2)\left(\frac{1 \text{ mol } MBr_2}{1 \text{ mol } Cl_2}\right) = \underline{0.0290 \text{ mol}}$$

$$\text{Thus, molar mass of } MBr_2 = \left(\frac{5.35 \text{ g } MBr}{0.0290 \text{ mol } MBr_2}\right) = \underline{184 \text{ g mol}^{-1}}$$

$$= [(\text{molar mass of M}) + 2(79.90)] \text{ g mol}^{-1}$$

$$\text{Molar mass of M} = \underline{24 \text{ g mol}^{-1}}$$

which, from the periodic table, identifies M as magnesium.

The salt $MX_2$ is magnesium bromide, $MgBr_2$

70. (a) Electronegativity increases from left to right across any period of the periodic table because the core charge increases, for example, in period 2, from +1 for Li to +7 for F. It decreases in going down any group from top to bottom because as the number of filled inner shells of electrons increases the valence shell increases its distance from the nucleus. Thus, fluorine in group 7 and period 2 combines a high core charge of +7 with a small size, which makes it the most electronegative of the elements.

(b) Because of its very high electronegativity, fluorine is also a very powerful oxidizing agent, as exemplified by the following:

(i) $2F_2(g) + 2H_2O(l) \longrightarrow 4HF(aq) + O_2(g)$

(ii) $2K(s) + F_2(g) \longrightarrow 2KF(s)$

(iii) $2Al(s) + 3F_2(g) \longrightarrow 2AlF_3(s)$

(iv) $P_4(s) + 6F_2(g) \longrightarrow 4PF_3(g)$, and

$P_4(s) + 10F_2(g) \longrightarrow 2PF_5(g)$

71. As is shown by our discussion in the text, there are trends in the characteristic properties of the elements of group 7 in descending the group, and it is reasonable that these will continue in going from iodine to astatine.

(a) Physical state. The trend from $F_2$ and $Cl_2$ which are gases, to $Br_2$ (liquid) and $I_2$(solid) suggest that astatine should be a solid.

(b) Ionization energy and size. Astatine has one more filled inner shell of electrons than iodine but the same core charge of +7. It should be a larger atom than iodine and have a smaller ionization energy. Approximate values could be estimated by extrapolation.

(c) The simplest reaction between two different halogens gives a diatomic mixed halogen molecule, $X_2 + Y_2 \longrightarrow 2XY$, so we would expect

$$Br_2(l) + At_2(s) \longrightarrow 2AtBr \quad \text{Lewis structure } :\ddot{\underset{..}{At}}-\ddot{\underset{..}{Br}}:$$

(d) Like the other hydrogen halides HAt should be a strong acid in water

$$HAt(g) + H_2O(l) \longrightarrow H_3O^+(aq) + At^-(aq)$$

(e) KAt is a compound between a metal and a nonmetal and should be ionic with the Lewis structure

$$K^+ \; :\ddot{\underset{..}{At}}:^-$$

and containing an ionic bond

(f) $:PAt_3$ will be an $AX_3E$ type molecule with trigonal pyramidal geometry

(g) $At_2$ should behave as an oxidizing agent in all of these reactions:

(i) $2Na + At_2 \longrightarrow 2NaAt$

(ii) $Ca + At_2 \longrightarrow CaAt_2$

(iii) $P_4 + 6At_2 \longrightarrow 4PAt_3$

(iv) $H_2 + At_2 \longrightarrow 2HAt$

* Note that astatine has in fact a melting point of 302°C and a boiling point of 335°C.

72. (a) $Ca(OCl)_2(aq) + CO_2(g) + H_2O(l) \longrightarrow CaCO_3(s) + 2HOCl(aq)$

(b) To calculate the concentration of HOCl in the final solution, calculate moles of HOCl and use the known volume of 200 mL.

$$\text{mol HOCl} = (8.00\text{ g } Ca(OCl)_2)\left(\frac{1\text{ mol } Ca(OCl)}{143.0\text{ g } Ca(OCl)_2}\right)\left(\frac{2\text{ mol HOCl}}{1\text{ mol } Ca(OCl)_2}\right)$$

$$= \underline{0.112\text{ mol}}$$

$$\text{concentration of HOCl} = \left(\frac{0.112\text{ mol HOCl}}{200\text{ mL}}\right)\left(\frac{10^3\text{ mL}}{1\text{ L}}\right) = \underline{0.560\text{ M}}$$

73. Group 2 elements are metals and react with HCl(aq) to give solutions of ionic chlorides of empirical formula $MCl_2$. Thus,

$$Ra(s) + 2HCl(aq) \longrightarrow RaCl_2(aq) + H_2(g)$$

The high melting point of $RaCl_2(s)$ of 1000°C is consistent with its formulation as $Ra^{2+}(Cl^-)_2$, and in terms of the molar mass of Ra and that of $RaCl_2$, we can write

$$\left(\frac{1\text{ mol Ra}}{1\text{ mol } RaCl_2}\right) \times 100\% = 76.1\% = \left[\frac{100(\text{Molar mass of Ra})}{\text{Molar mass of Ra} + 2(35.45\text{ g mol}^{-1})}\right]$$

i.e., if molar mass of Ra is X g mol$^{-1}$,

$$\frac{X}{X + 2(35.45)} = 0.761 \; ; \; \underline{X = 226} \; ; \; \underline{\text{Lewis Structure}} \; Ra^{2+}[:\ddot{\underset{..}{Cl}}:^-]_2$$

74. According to the Brønsted-Lowry definition, an <u>acid</u> is a proton <u>donor</u> and a <u>base</u> is a proton <u>acceptor</u>.

A <u>conjugate acid-base pair</u> comprises two species that are related by the transfer of a proton between them. For example, HF and $F^-$, $H_2O$ and $OH^-$, $H_3O^+$ and $H_2O$, and $NH_4^+$ and $NH_3$, are conjugate acid-base pairs.

75. (a) (i) Carbon (group 4, valence 4) forms $CH_4$; (ii) Nitrogen (group 5, valence 3) forms $NH_3$; (iii) Oxygen (group 6, valence 2) forms $H_2O$, and (iv) Fluorine (group 7, valence 1) forms HF.

(b) (i) No reaction except at high temperature when synthesis gas is formed

$$CH_4(g) + H_2O(g) \longrightarrow CO(g) + 3H_2(g)$$

(ii) $NH_3$ behaves as a weak base

$$NH_3(aq) + H_2O(l) \rightleftharpoons NH_4^+(aq) + OH^-(aq)$$

(iii) Liquid water is amphiprotic and is slightly ionized to form $H_3O^+(aq)$ and $OH^-(aq)$

$$2H_2O(l) \rightleftharpoons H_3O^+(aq) + OH^-(aq)$$

(iv) HF behaves as a weak acid

$$HF(aq) + H_2O(l) \rightleftharpoons H_3O^+(aq) + F^-(aq)$$

(c) (i) Neither acid nor base
(ii) Weak base
(iii) Weak acid and weak base
(iv) Weak acid

76. Since both X and Y are gases, they must be nonmetals. The nonmetal gaseous elements are $H_2$, $N_2$, $O_2$, $F_2$, and $Cl_2$. Molecular $N_2$ can be ruled out on the basic that it does not react except under extreme conditions.

At 100°C and 1.00 atm pressure, 1 L of each gas must contain

$$n = \frac{PV}{RT} = \frac{(1.00 \text{ atm})(1 \text{ L})}{(0.0821 \text{ atm L mol}^{-1} \text{ K}^{-1})(373 \text{ K})} = \underline{0.0327 \text{ mol}}$$

$$\text{Thus, molar mass of X} = \left(\frac{0.0658 \text{ g X}}{0.0327 \text{ mol X}}\right) = \underline{2.01 \text{ g mol}^{-1}}$$

$$\text{molar mass of Y} = \left(\frac{2.315 \text{ g Y}}{0.0327 \text{ mol Y}}\right) = \underline{70.8 \text{ g mol}^{-1}}$$

From their molar masses, X is identified as $H_2$ and Y as $Cl_2$.

On this basis, for the reaction of X with Y, we have

$$H_2(g) + Cl_2(g) \longrightarrow 2HCl(g)$$

1 volume + 1 volume ⟶ 2 volumes (Avogadro's law)

| | | | | |
|---|---|---|---|---|
| initially | 50 | 100 | 0 | mL |
| after reaction | 0 | 50 | 100 | mL |

Which is consistent with the given data, and we can conclude that the mixture of gases after complete reaction has a volume of 150 mL and should contain 100 mL HCl(g) and an excess of 50 mL of $Cl_2(g)$, of which only the former is appreciably soluble in water

$$HCl(g) + H_2O(l) \longrightarrow H_3O^+(aq) + Cl^-(aq)$$

to give an acidic solution (turns blue litmus red). The remaining gas is $Cl_2(g)$, which does not support combustion but vigorously oxidizes hot sodium to give white solid NaCl(s),

$$2Na(l) + Cl_2(g) \longrightarrow 2Na^+Cl^-(s)$$

<u>X is $H_2(g)$ and Y is $Cl_2(g)$</u>

77.

| | Formula | Name | Type | Lewis Structure |
|---|---|---|---|---|
| (a) | $MgCl_2$ | magnesium chloride | ionic | $Mg^{2+}[:\ddot{\underset{..}{Cl}}:^-]_2$ |
| (b) | $SCl_2$ | sulfur dichloride | covalent | $\overset{\delta-}{:\ddot{\underset{..}{Cl}}}-\overset{2\delta+}{\ddot{\underset{..}{S}}}-\overset{\delta-}{\ddot{\underset{..}{Cl}}:}$ |
| (c) | $PCl_3$ | phosphorus trichloride | covalent | $\overset{\delta-}{:\ddot{\underset{..}{Cl}}}-\overset{3\delta+}{\ddot{P}}(-\overset{\delta-}{:\ddot{\underset{..}{Cl}}:})-\overset{\delta-}{\ddot{\underset{..}{Cl}}:}$ |
| (d) | HF | hydrogen fluoride | covalent | $\overset{\delta+}{H}-\overset{\delta-}{\ddot{\underset{..}{F}}:}$ |
| (e) | $OCl_2$ | oxygen dichloride | covalent | $\overset{\delta+}{:\ddot{\underset{..}{Cl}}}-\overset{2\delta-}{\ddot{\underset{..}{O}}}-\overset{\delta+}{\ddot{\underset{..}{Cl}}:}$ |
| (f) | $CS_2$ | carbon disulfide | covalent | $\overset{\delta+}{:\ddot{S}}=\overset{2\delta-}{C}=\overset{\delta+}{\ddot{S}:}$ |
| (g) | $NF_3$ | nitrogen trifluoride | covalent | $\overset{\delta-}{:\ddot{\underset{..}{F}}}-\overset{3\delta+}{\ddot{N}}(-\overset{\delta-}{:\ddot{\underset{..}{F}}:})-\overset{\delta-}{\ddot{\underset{..}{F}}:}$ |
| (h) | LiH | lithium hydride | ionic | $Li^+:H^-$ |

78. From the empirical formula XCl, the valence of X must be 1, and from the gas data we can calculate mol $Cl_2$ and hence mol XCl formed from 4.315 g of X. The balanced equation is

$$2X(s) + Cl_2(g) \longrightarrow 2XCl(s)$$

$$\text{mol XCl} = (\text{mol } Cl_2)\left(\frac{2 \text{ mol XCl}}{1 \text{ mol } Cl_2}\right) = 2\left[\frac{(1.00 \text{ atm})(0.481 \text{ L})}{(0.0821 \text{ atm L mol}^{-1} \text{ K}^{-1})(293 \text{ K})}\right]$$

$$= \underline{0.0400 \text{ mol}}$$

and mol X = mol XCl = 0.0400 mol, so that

$$\text{molar mass of X} = \left(\frac{4.315 \text{ g X}}{0.0400 \text{ mol X}}\right) = \underline{107.9 \text{ g mol}^{-1}}$$

Reference to the periodic table identifies X as silver, Ag, and XCl(s) as silver chloride, AgCl(s), which is insoluble in water.

Although insoluble AgCl(s) is formed in the reaction between silver and chlorine, this <u>is not</u> a precipitation reaction. A precipitation reaction is a reaction between a cation and an anion <u>in solution</u> to give the insoluble salt.

79. First calculate the volume of the cube, then the mass and the number of mol of NaCl, and finally the number of NaCl formula units.

$$\text{Volume of cube} = (1\ \text{mm})^3\left(\frac{1\ \text{cm}}{10\ \text{mm}}\right)^3$$

$$\text{Mass of cube} = (1\ \text{mm})^3\left(\frac{1\ \text{cm}}{10\ \text{mm}}\right)^3\left(\frac{2.17\ \text{g NaCl}}{1\ \text{cm}^3\ \text{NaCl}}\right) = \underline{2.17\text{x}10^{-3}\ \text{g NaCl}}$$

$$\text{NaCl formula units} = (2.17\text{x}10^{-3}\ \text{g NaCl})\left(\frac{1\ \text{mol NaCl}}{58.44\ \text{g NaCl}}\right)\left(\frac{6.022\text{x}10^{23}\ \text{NaCl}}{1\ \text{mol NaCl}}\right)$$

$$= \underline{2.24\text{x}10^{19}\ \text{NaCl formula units}}$$

Since each formula unit contains 1 $Na^+$ and 1 $Cl^-$ ion, this is also the number of $Na^+$ ions, and the number of $Cl^-$ ions.

80. (a) $HBr(g) + H_2O(l) \longrightarrow H_3O^+(aq) + Br^-(aq)$

(b) $CO_2(g) + H_2O(l) \rightleftharpoons H_2CO_3(aq)$

$H_2CO_3(aq) + H_2O(l) \rightleftharpoons H_3O^+(aq) + HCO_3^-(aq)$

$HCO_3^-(aq) + H_2O(l) \rightleftharpoons H_3O^+(aq) + CO_3^{2-}(aq)$

(c) $NH_3(aq) + H_2O(l) \rightleftharpoons NH_4^+(aq) + OH^-(aq)$

(d) $Cl_2(g) + H_2O(l) \rightarrow HCl(aq) + HOCl(aq)$

(e) $2F_2(g) + 2H_2O(l) \longrightarrow 4HF(aq) + O_2(g)$

81. According to the text (page 227) the annual U.S. and Canadian production of $F_2(g)$ is about 5000 tons. For the formation of $UF_6(g)$ we have

$$U(s) + 3F_2(g) \longrightarrow UF_6(g)$$

$$\text{Mass of } UF_6 = \left(\frac{70\ \text{ton F}}{100\ \text{ton } F_2}\right)(5000\ \text{ton } F_2)\left(\frac{1\ \text{mol } F_2}{38.00\ \text{g } F_2}\right)\left(\frac{1\ \text{mol } UF_6}{3\ \text{mol } F_2}\right)$$

$$\text{x}\left(\frac{352.0\ \text{g } UF_6}{1\ \text{mol } UF_6}\right) = \underline{1.08\text{x}10^4\ \text{ton } UF_6}$$

82. $Cl_2(aq)$ oxidizes $KI(aq)$ to $I_2(s)$ which dissolves in excess $KI(aq)$ to give a brown solution containing the brown $I_3^-(aq)$ ion and $KCl(aq)$,

$$Cl_2(aq) + 3KI(aq) \longrightarrow KI_3(aq) + 2KCl(aq)$$

On shaking with an equal volume of $CCl_4(l)$ the iodine is extracted into the organic layer, which settles to the bottom of the test tube and is colored purple by the iodine, leaving a colorless upper aqueous layer containing $KCl(aq)$ and excess $KI(aq)$.

$$KI_3(aq) \longrightarrow KI(aq) + I_2(sol'n)$$

# CHAPTER 6

1. (a) $CO(g) + H_2(g) \xrightarrow[\text{high T}]{\text{catalyst}} H_2CO(g) \xrightarrow{H_2} H_3COH(l)$

methanal — methanol

(b) $2CO(g) + O_2(g) \xrightarrow{\text{combustion}} 2CO_2(g)$

(c) $CO(g) + H_2O(g) \xrightarrow{\text{high T}} CO_2(g) + H_2(g)$

(d) $Fe_2O_3(s) + 3CO(g) \xrightarrow{\text{high T}} 2Fe(s) + 3CO_2(g)$

2. (a) Laboratory production (burning C in a limited supply of air),

$2C(s) + O_2(g) \longrightarrow 2CO(g)$

Industrial production,

$CH_4(g) + H_2O(g) \longrightarrow CO(g) + 3H_2(g)$ - synthesis gas production

(b) In the laboratory, a common preparation method for $CO_2(g)$ is addition of a dilute aqueous acid to a metal carbonate,

$Na_2CO_3(s) + 2HCl(aq) \longrightarrow 2NaCl(aq) + CO_2(g) + H_2O(l)$

Industrially, $CO_2$ is obtained by heating calcium carbonate,

$CaCO_3(s) \longrightarrow CaO(s) + CO_2(g)$

or on burning virtually any hydrocarbon in excess air,

$C_3H_8(g) + 5O_2(g) \longrightarrow 3CO_2(g) + 4H_2O(g)$

(c) Industrially, carbon disulfide is prepared directly from the elements,

$C(s) + 2S(s) \longrightarrow CS_2(l)$

3. For a molecule with atoms each of which obey the <u>octet rule</u>, a Lewis structure must have 8 electrons around each atom. Thus, counting the electrons around each of <u>n</u> atoms gives <u>8n</u> electrons. Since in the molecule we count the electrons forming bonds <u>twice</u> by this procedure, once for each atom forming the bonds, then the number of electrons forming bonds must be 8n - x, where x is the total number of electrons available (which we can obtain by counting up the number of electrons each atom contributes from its valence shell to form the molecule).

(a) <u>Carbon dioxide</u>, $CO_2$

For octets around each atom 3x8 = 24 electrons are required, but the number available is 4 from C (group 4) and 6 from each O atom (group 6). i.e., 4 + 2(6) = 16 electrons

Thus, the number forming bonds must be 24-16 = 8, or, in other words, there must be a total of $\underline{4}$ electron pair bonds in $CO_2$. In $CO_2$, the central atom is the unique C atom, hence we have

$$:\ddot{O}=C=\ddot{O}:$$

when we complete the octets for each atom. (Note, that there is in fact another possibility for four bonds, which is

$$^{+}:O\equiv C-\underset{\cdot\cdot}{\ddot{O}}:^{-} \quad )$$

(b) Carbon monoxide, CO

Following the method of part (a),

Bonding electrons = 2(8) - (4 + 6) = $\underline{6}$, or three bonds

$$^{-}:C\equiv O:^{+}$$

(c) Cyanide ion, $CN^-$

Bonding electrons = 2(8) - (4+5+1) = $\underline{6}$, or three bonds

$$^{-}:C\equiv N:$$

(d) Carbide ion, $C_2^{2-}$

Bonding electrons = 2(8) - [2(4) + 2] = $\underline{6}$, or three bonds

$$^{-}:C\equiv C:^{-}$$

(e) Hydrogen cyanide, HCN

The structure is readily obtained from that of $CN^-$ by adding $H^+$

$$H-C\equiv N:$$

4. lime, CaO(s); soda water, $CO_2(aq)$; natural gas, mainly $CH_4(g)$ with some $C_2H_6(g)$ and $C_3H_8(g)$; coke, mainly C(s) (graphite) produced by heating coal in the absence of air; carbon black, mainly C(s) (graphite) produced by burning hydrocarbons in a limited supply of air, and chalk, $CaCO_3(s)$.

5. (a) $BaCO_3(s) \xrightarrow{\text{heat}} BaO(s) + CO_2(g)$
   (b) $H_2CO_3(aq) \xrightarrow{\text{heat}} H_2O(l) + CO_2(g)$
   (c) $2HI(g) \xrightarrow{\text{heat}} H_2(g) + I_2(g)$
   (d) $C_2H_6(g) \xrightarrow{\text{heat}} C_2H_4(g) + H_2(g)$

6. (a) $C(s) + CuO(s) \longrightarrow Cu(s) + CO(g)$
   (b) $3C(s) + CaO(s) \longrightarrow CaC_2(s) + CO(g)$
   (c) $C(s) + 2S(s) \longrightarrow CS_2(l)$
   (d) $C(s) + O_2(g) \longrightarrow CO_2(g)$

7. (a) $CH_4(g) + H_2O(g) \xrightarrow{\text{catalyst}} CO(g) + 3H_2(g)$
   (b) $CH_4(g) + NH_3(g) \xrightarrow[\text{catalyst}]{1200^\circ C} HCN(g) + 3H_2(g)$

(c) $CH_4(g) + 2O_2(g) \xrightarrow{\text{combustion}} CO_2(g) + 2H_2O(l)$

8. (a) limestone, $CaCO_3(s)$; (b) hydrocyanic acid, $HCN(aq)$;
(c) acetylene, $C_2H_2(g)$; (d) calcium carbide, $CaC_2(s)$;
(e) carborundum, $SiC(s)$

9. In diamond each C atom is at the center of a tetrahedral arrangement of four other C atoms and has $AX_4$ geometry, with a CCC bond angle of exactly 109° 28'. In graphite, each C atom is part of a series of linked hexagonal arrangement of six carbon atoms in which the geometry around each C atom is $AX_3$, triangular planar, with a CCC bond angle of 120°. Diamond is very hard because it has an infinite three-dimensional network structure. In contrast, graphite consists of infinite planes of carbon atoms with strong bonds in the planes but only weak intermolecular forces linking the planes together. As a result the planes of carbon atoms in graphite slide over each other rather easily, which gives graphite its softness and lubricating properties.

10. For the method of obtaining Lewis structures, see Problem 3.

$SCN^-$ Bonding electrons = 3(8) - (6+4+6+1) = 8, or four bonds, which can be arranged as follows:

$$^{-}:\underset{..}{\ddot{S}}-C\equiv N: \qquad :\ddot{S}=C=\ddot{N}:^{-} \qquad \text{or} \qquad ^{+}:S\equiv C-\underset{..}{\ddot{N}}:^{2-}$$

I II III

III is ruled out on the basis of the large formal charge separation. Of I and II, I is preferred, since N forms multiple bonds more readily than does S. This ion is an example of a general rule, that second period elements, particularly C, O, and N, form multiple bonds more readily than third period elements, such as S. In fact the Lewis structure I agrees with the observed structure.

11. From the mass % composition, calculate the empirical formula.

| | | C | Br | O | |
|---|---|---|---|---|---|
| grams in 100 g | = | 6.4 | 85.0 | 8.6 | |
| moles in 100 g | = | $\frac{6.4}{12.01}$ | $\frac{85.0}{79.90}$ | $\frac{8.6}{16.00}$ | |
| | = | 0.53 | 1.06 | 0.54 | |
| ratio of moles (atoms) | = | $\frac{0.53}{0.53}$ : | $\frac{1.06}{0.53}$ : | $\frac{0.54}{0.53}$ | = 1:2:1 |

i.e., Empirical formula = $CBr_2O$ (Empirical formula mass = 187.8 u)

From the gas data and the mass, calculate the molar mass.

$$\text{Mol compound} = \frac{PV}{RT} = \frac{(122\ \text{mL})(\frac{1\ \text{L}}{10^3\ \text{mL}})(1.00\ \text{atm})}{(0.0821\ \text{atm L mol}^{-1}\ \text{K}^{-1})(298\ \text{K})} = 4.99\times10^{-3}\ \text{mol}$$

and molar mass = $\left(\frac{0.940 \text{ g}}{4.99\text{x}10^{-3} \text{ mol}}\right)$ = $\underline{188 \text{ g mol}^{-1}}$

Thus, the molecular formula is the same as the empirical formula,

$\underline{CBr_2O}$

Since the compound is formed from $CBr_4$, it must contain two C-Br bonds, and has the Lewis structure

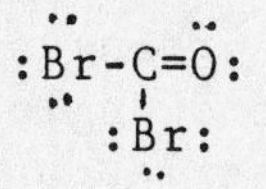

with $AX_3$ triangular planar geometry

12. The carbide contains only Al and C, and the data for $CH_4$ gives the moles of $CH_4$,

$$n = \frac{PV}{RT} = \frac{(1.02 \text{ atm})(250 \text{ mL})(\frac{1 \text{ L}}{10^3 \text{ mL}})}{(0.0821 \text{ atm L mol}^{-1} \text{ K}^{-1})(298 \text{ K})} = \underline{0.0104 \text{ mol}}$$

Thus, the mass of C in 0.500 g of carbide is

$$(0.0104 \text{ mol } CH_4)\left(\frac{1 \text{ mol C}}{1 \text{ mol } CH_4}\right)\left(\frac{12.01 \text{ g C}}{1 \text{ mol C}}\right) = \underline{0.125 \text{ g C}}$$

and the mass of Al = (0.500-0.125) g = $\underline{0.375 \text{ g Al}}$

$$\text{Hence, mol ratio Al : C} = \frac{0.375}{26.98} : \frac{0.125}{12.01} = 0.0139 : 0.0104$$

$$= \frac{0.0139}{0.0104} : \frac{0.0104}{0.0104} = 1.34 : 1.00$$

Thus the ratio of Al atoms : C atoms = 1.34 : 1.00 = $\underline{4.0 : 3.0}$

and the empirical formula of the carbide is $\underline{Al_4C_3}$, which is the expected empirical formula in terms of the common valences (Al,3, and C,4). The balanced equation for the reaction is thus,

$$Al_4C_3(s) + 6H_2O(l) \longrightarrow 2Al_2O_3(s) + 3CH_4(g)$$

and the mol $CH_4$ in 20.0 L $CH_4$(g) at 25°C and 1.00 atm pressure is

$$n = \frac{PV}{RT} = \frac{(1.00 \text{ atm})(20.0 \text{ L})}{(0.0821 \text{ atm L mol}^{-1} \text{ K}^{-1})(298 \text{ K})} = \underline{0.817 \text{ mol } CH_4}$$

$$\text{Mass of } Al_4C_3 = (0.817 \text{ mol } CH_4)\left(\frac{1 \text{ mol carbide}}{3 \text{ mol } CH_4}\right)\left(\frac{144.0 \text{ g carbide}}{1 \text{ mol carbide}}\right)$$

$$= \underline{39.2 \text{ g}}$$

13. For the oxide of carbon, moles of oxide in 0.200 g is given by

$$n = \frac{PV}{RT} = \frac{(740 \text{ mm Hg})(\frac{1 \text{ atm}}{760 \text{ mm Hg}})(74.3 \text{ mL})(\frac{1 \text{ L}}{10^3 \text{ mL}})}{(0.0821 \text{ atm L mol}^{-1} \text{ K}^{-1})(300 \text{ K})} = \underline{2.937\text{x}10^{-3} \text{ mol}}$$

Hence, molar mass of oxide = $\left(\frac{0.200 \text{ g}}{2.937\text{x}10^{-3} \text{ mol}}\right)$ = $\underline{68.1 \text{ g mol}^{-1}}$

And, from the composition, calculate the empirical formula:

| | | C | | O |
|---|---|---|---|---|
| Grams in 100 g | = | 53.0 | | 47.0 |
| Moles in 100 g | = | $\frac{53.0}{12.01}$ | | $\frac{47.0}{16.00}$ |
| Ratio of moles (atoms) | = | 4.41 | : | 2.94 |
| | = | $\frac{4.41}{2.94}$ | : | $\frac{2.94}{2.94}$ |
| | = | 1.50 | : | 1.00 or 3.0 : 2.0 |

Thus, the empirical formula is $C_3O_2$ (formula mass = 68.03 u)

(a) Since the molecular mass is the same as the empirical formula mass,

Molecular formula = $C_3O_2$

(b) The oxide is formed from malonic acid according to the equation,

$$CH_2(CO_2H)_2 \longrightarrow C_3O_2 + 2H_2O$$

(c) The structure of malonic acid is

```
 ..    ..
:O=C-O-H
   |  ..
 H-C-H
   |  ..
:O=C-O-H
 ..    ..
```

which suggests $:\ddot{O}=C=C=C=\ddot{O}:$ for the Lewis structure of the oxide.

14. (a) From the composition data:

| | | C | | N |
|---|---|---|---|---|
| 100 g of compound contains | | 46.2 g | | 53.8 g |
| Mol in 100 g | = | $\frac{46.2}{12.01}$ | | $\frac{53.8}{14.01}$ |
| Ratio of moles (atoms) | = | 3.85 | : | 3.85 = 1.0 : 1.0 |

Thus, the empirical formula is CN (empirical formula mass 26.02 u)

From the gas data:

$$\text{mol of gas} = \frac{PV}{RT} = \frac{(0.950 \text{ atm})(126 \text{ mL})(\frac{1 \text{ L}}{10^3 \text{ mL}})}{(0.0821 \text{ atm L mol}^{-1} \text{ K}^{-1})(373 \text{ K})} = 3.908\text{x}10^{-3} \text{ mol}$$

$$\text{Molar mass of X} = \left(\frac{0.208 \text{ g}}{3.908\text{x}10^{-3} \text{ mol}}\right) = 53.2 \text{ g mol}^{-1}$$

Hence, molecular formula = 2(empirical formula) = $C_2N_2$

(b) Following the method of Problem 3,

Bonding electrons = 4(8) - [2(4)+2(5)] = 14, or seven bonds,

so the Lewis structure is $:N\equiv C-C\equiv N:$

Each C atom is $AX_2$, so the molecule must be linear

15. The evidence all points to a gas containing C, S, and O, which suggests carbon oxosulfide, COS. Thus, when burned in oxygen

$$2COS(g) + 3O_2(g) \longrightarrow 2CO_2(g) + 2SO_2(g)$$

From the gas data,

$$\text{mol COS} = \frac{PV}{RT} = \frac{(1.00 \text{ atm})(100 \text{ mL})(\frac{1 \text{ L}}{10^3 \text{ mL}})}{(0.0821 \text{ atm L mol}^{-1} \text{ K}^{-1})(298 \text{ K})} = \underline{4.087\text{x}10^{-3} \text{ mol}}$$

and $\underline{\text{molar mass}} = (\frac{0.246 \text{ g}}{4.087\text{x}10^{-3} \text{ mol}}) = \underline{60.2 \text{ g mol}^{-1}}$ (which is the molar mass of COS)

According to Avogadro's law (constant P and T), we have for

| | $2COS(g)$ + | $3O_2(g)$ $\longrightarrow$ | $2CO_2(g)$ + | $2SO_2(g)$ | |
|---|---|---|---|---|---|
| before reaction | 100 | 200 | 0 | 0 | mL |
| after reaction | 0 | $[200 - \frac{3}{2}(100)]$ = 50 mL | 100 | 100 | mL |

Both $CO_2(g)$ and $SO_2(g)$ are soluble in water and will dissolve, leaving only the 50 mL of unreacted $O_2(g)$, which is insoluble, undissolved, which agrees with observation.

For the decomposition of 100 mL of COS(g) by a heated Pt wire,

| | $COS(g)$ $\longrightarrow$ | $CO(g)$ + | $S(s)$ | |
|---|---|---|---|---|
| before reaction | 100 | 0 | - | mL |
| after reaction | 0 | 100 | (solid) | mL |

and the volume after decomposition is expected to be 100 mL, as observed.

Finally, the reaction of COS(g) with KOH(aq) is

$$COS(g) + KOH(aq) \longrightarrow K^+(aq) + CO_3^{2-}(aq) + S^{2-}(aq) \quad \underline{\text{unbalanced}}$$

i.e. $COS(g) + 4KOH(aq) \longrightarrow 4K^+(aq) + CO_3^{2-}(aq) + S^{2-}(aq) + 2H_2O(l)$

for the balanced equation, which requires a 4:1 mol ratio of KOH to COS for complete reaction. This is in agreement with the given data, since

$$50.00 \text{ mL } 0.200 \text{ M KOH(aq)} = (50.00 \text{ mL})(\frac{1 \text{ L}}{10^3 \text{ mL}})(0.200 \text{ mol L}^{-1})$$
$$= \underline{1.00 \times 10^{-2} \text{ mol KOH}}$$

<u>and</u>

$$25.00 \text{ mL } 0.100 \text{ M COS(aq)} = (25.00 \text{ mL})(\frac{1 \text{ L}}{10^3 \text{ mL}})(0.100 \text{ mol L}^{-1})$$
$$= \underline{2.50\text{x}10^{-3} \text{ mol COS}}$$

i.e., $\frac{\text{mol KOH}}{\text{mol COS}} = (\frac{1.00\text{x}10^{-2} \text{ mol}}{2.50\text{x}10^{-3} \text{ mol}}) = \underline{4}$

16. Ethyne is prepared from CaO(s), coke (carbon), and water in two stages:

(i) Reaction of CaO(s) with coke at high temperature gives <u>calcium carbide</u>

$$CaO(s) + 3C(s) \longrightarrow CaC_2(s) + CO(g)$$

(ii) Reaction of calcium carbide with water at room temperature
gives acetylene.

$$CaC_2(s) + 2H_2O(1) \longrightarrow Ca(OH)_2(s) + C_2H_2(g)$$

Calcium carbide contains the $^-:C \equiv C:^-$ ion which behaves as a
strong base in water.

$$C_2^{2-}(aq) + 2H_2O(1) \longrightarrow C_2H_2(g) + 2OH^-(aq)$$

17. The alkanes with 1, 2, and 3 carbon atoms, respectively, are methane,
$CH_4$, ethane, $C_2H_6$, and propane, $C_3H_8$, with the Lewis structures:

$$
\begin{array}{ccc}
\text{H} & \text{H H} & \text{H H H} \\
\text{H-C-H} & \text{H-C-C-H} & \text{H-C-C-C-H} \\
\text{H} & \text{H H} & \text{H H H}
\end{array}
$$

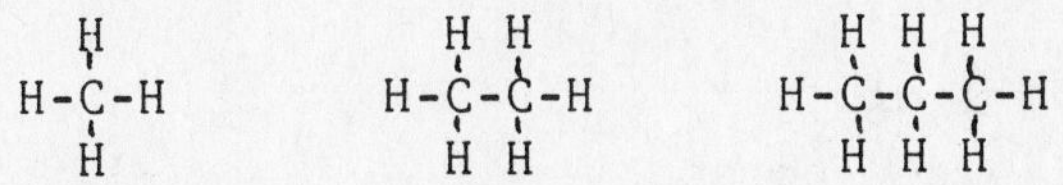

18. The Lewis structures of $C_2F_4$ and $C_2F_2$ are

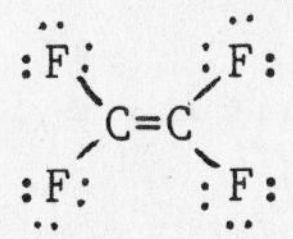

$$
\begin{array}{cc}
:\!\ddot{F}\!: \quad :\!\ddot{F}\!: \\
\phantom{:}\diagdown\phantom{x}\diagup \\
\phantom{:::}C=C \\
\phantom{:}\diagup\phantom{x}\diagdown \\
:\!\ddot{F}\!: \quad :\!\ddot{F}\!:
\end{array}
\qquad \text{and} \qquad :\!\ddot{F}\!-C \equiv C -\ddot{F}\!:
$$

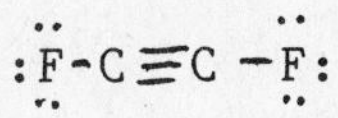

In $C_2F_4$, the bond arrangement around each C atom is $AX_3$, planar, and

in $C_2F_2$, the bond arrangement around each C atom is $AX_2$, linear.

19. When a system expands it does work by pushing the atmosphere away.
If the change in the volume of the system is $\Delta V$ at a constant pressure
P, the work done by the system is $P\Delta V$.

In this type of problem, $P\Delta V$ is evaluated using the ideal gas law.
1 mol of water is changed to 1 mol gas, and neglecting the volume of
water, the change in the number of moles, $\Delta n$, is 1 mol. Thus, in general

$$P\Delta V = \Delta nRT = (n \text{ mol})(8.314 \text{ J K}^{-1} \text{ mol}^{-1})(T \text{ K})$$

$$= \underline{8.314(nT) \text{ J}}$$

Note that the units of work are joules, so $8.314 \text{ J K}^{-1} \text{ mol}^{-1}$ is the
appropriate value to use for the gas constant R.

Hence, $P\Delta V = (1 \text{ mol})(8.314 \text{ J K}^{-1} \text{ mol}^{-1})(373 \text{ K}) = \underline{3.1 \times 10^3 \text{ J}}$

$$\text{or} \quad \underline{3.1 \text{ kJ}}$$

20. The reaction of zinc with HCl(aq) is

$$Zn(s) + 2HCl(aq) \longrightarrow Zn^{2+}(aq) + 2Cl^-(aq) + H_2(g)$$

In this case, the work done is due to the hydrogen gas that is evolved
expanding against a constant pressure of 1 atm. First we calculate the
number of moles of $H_2(g)$ evolved,

$$\text{mol } H_2(g) = (1 \text{ g Zn})(\frac{1 \text{ mol Zn}}{65.38 \text{ g Zn}})(\frac{1 \text{ mol } H_2}{1 \text{ mol Zn}}) = \underline{1.53 \times 10^{-2} \text{ mol}}$$

Then, $P\Delta V = \Delta nRT = (1.53 \times 10^{-2} \text{ mol})(8.314 \text{ J K}^{-1} \text{ mol}^{-1})(298 \text{ K})$

$$= \underline{37.9 \text{ J}}$$

21. Assuming that the heat capacity of the flame calorimeter is negligible, all of the heat produced by the reaction is transferred to the 1.500 kg of water, at constant pressure. Thus, the heat evolved is

$$(1.500 \text{ kg } H_2O)\left(\frac{10^3 \text{ g}}{1 \text{ kg}}\right)(4.184 \text{ J g}^{-1} \text{ K}^{-1})(26.386-25.246) \text{ K} = \underline{7155 \text{ J}}$$

i.e., (mass of water)(specific heat capacity of water)(temperature change)

Since heat is evolved, the reaction is exothermic, and q = -7155 J, or -7.155 kJ, and this is $\Delta H$ for the combustion of 0.150 g of liquid octane.

Thus, for 1 mol of liquid octane, $C_8H_{18}(l)$,

$$\Delta H^o = \left(\frac{-7.155 \text{ kJ}}{0.150 \text{ g octane}}\right)\left(\frac{114.2 \text{ g octane}}{1 \text{ mol octane}}\right) = \underline{-5450 \text{ kJ mol}^{-1}}$$

The standard enthalpy of combustion of octane is $\underline{-5450 \text{ kJ mol}^{-1}}$

22. First we have to decide what reaction occurs, so we calculate initial moles of the reactants,

Initial mol NaOH(aq) = (50.0 mL)(0.400 mol $L^{-1}$) = 20.0 mmol

Initial mol $H_2SO_4$(aq) = (20.0 mL)(0.500 mol $L^{-1}$) = 10.0 mmol

and the balanced equation for the reaction is

| | $H_2SO_4(aq)$ + | $2NaOH(aq) \longrightarrow$ | $Na_2SO_4(aq)$ + $2H_2O(l)$ | |
|---|---|---|---|---|
| initially | 10.0 | 20.0 | 0 | mmol |
| after reaction | 0 | 0 | 10.0 | mmol |

Hence, the reaction is the complete neutralization of 10.0 mmol $H_2SO_4$(aq) by 20.0 mmol NaOH(aq).

The total amount of heat evolved by the reaction is the sum of that absorbed by the calorimeter and by the (50.0+20.0) mL = 70.0 mL of solution. Assuming that the solution has a density of 1.00 g $mL^{-1}$, we have 70.0 g of solution, and we also assume that its heat capacity is the same as that of water (4.184 J $K^{-1}$ $g^{-1}$), since the solution is a dilute aqueous solution. Then,

q = (Heat absorbed by calorimeter) + (Heat absorbed by solution)

$$= [(39.0 \text{ J K}^{-1}) + (70.0 \text{ g})(4.184 \text{ J K}^{-1} \text{ g}^{-1})](3.60 \text{ K}) = \underline{1195 \text{ J}}$$

This is the heat evolved when 10.0 mmol of $H_2SO_4$ is completely neutralized. Thus, heat evolved when 1 mol $H_2SO_4$ is neutralized is given by

$$\left(\frac{1 \text{ mol } H_2SO_4}{10.0 \text{ mmol } H_2SO_4}\right)\left(\frac{10^3 \text{ mmol}}{1 \text{ mol}}\right)(1195 \text{ J})\left(\frac{1 \text{ kJ}}{10^3 \text{ J}}\right) = \underline{119.5 \text{ kJ}}$$

$$\underline{\Delta H^o = -119.5 \text{ kJ mol}^{-1}} \quad \text{(exothermic)}$$

23. Since we have equimolar amounts of HCl(aq) and KOH(aq), the reaction is the complete neutralization of HCl(aq) by KOH(aq), for which $\Delta H^o$ = -56.02 kJ $mol^{-1}$. The amount of heat evolved is given by the mass of solution multiplied by the heat capacity multiplied by the rise in

temperature, which is

$$(50.0\ mL)(\frac{1\ g}{1\ mL})(\frac{1\ mol}{18.02\ g})(75.4\ J\ K^{-1}\ mol^{-1})(1.60\ K) = \underline{334.7\ J}$$

Thus, the initial HCl(aq) solution must have contained

$$(334.7\ J)(\frac{1\ mol\ HCl}{56.02\ kJ})(\frac{1\ kJ}{10^3\ J}) = \underline{5.97\text{x}10^{-3}\ mol\ HCl}$$

and since the initial volume of HCl(aq) was 25.00 mL, the concentration must have been

$$(\frac{5.97\text{x}10^{-3}\ mol\ HCl}{25.00\ mL\ HCl})(\frac{10^3\ mL}{1\ L}) = \underline{0.239\ M}$$

24. The heat released by the reaction of 5.40 g CaO(s) is obtained from the temperature rise and the heat capacity of the calorimeter plus that of the solution (assuming a density of 1.00 g $mL^{-1}$ and that the heat capacity of the solution is the same as that of water).

$$q = -[(350\ J\ K^{-1})+(505.4\ g)(4.184\ J\ K^{-1}\ g^{-1})](2.60\ K) = \underline{-6408\ J}$$

$$\Delta H^o = (\frac{-6408\ J}{5.40\ g\ CaO})(\frac{56.08\ g\ CaO}{1\ mol\ CaO})(\frac{1\ kJ}{10^3\ J}) = \underline{-66.5\ kJ\ mol^{-1}}$$

25. Assuming that the heat capacity of the calorimeter is negligible, the heat released is that required to raise the temperature of the 1.00 kg of water by 5.4°C (5.4 K).

(a) $-q = (1.00\ kg)(\frac{10^3\ g}{1\ kg})(4.184\ J\ K^{-1}\ g^{-1})(5.4\ K)(\frac{1\ kJ}{10^3\ J}) = \underline{22.6\ kJ}$

The reaction is $S(s) + O_2(g) \longrightarrow SO_2(g)$, so that q per mol of $SO_2$ formed is

$$(\frac{-22.6\ kJ}{2.50\ g\ S})(\frac{32.06\ g\ S}{1\ mol\ S})(\frac{1\ mol\ S}{1\ mol\ SO_2}) = \underline{-290\ kJ\ mol^{-1}}$$

(b) $\Delta H$ for a reaction is q for the reaction carried out at <u>constant pressure</u>. In this reaction 1 mol $O_2(g)$ initially present is replaced by 1 mol $SO_2(g)$, so that the pressure remains constant and

$$\underline{\Delta H = -290\ kJ\ mol^{-1}}$$

26. The reaction is

$$C_{10}H_8(s) + 12O_2(g) \longrightarrow 10CO_2(g) + 4H_2O(l)$$

at <u>constant volume</u>. Thus, q, the heat evolved must be equal to $\Delta E$ for the reaction.

$$\underline{\Delta E = -40.10\ kJ}$$

For 1 mol $C_{10}H_8(s)$,

$$\Delta E^o = (\frac{-40.10\ kJ}{1.000\ g})(\frac{128.2\ g}{1\ mol}) = \underline{-5141\ kJ\ mol^{-1}}$$

27. The reaction is $C_2H_4(g) + H_2(g) \longrightarrow C_2H_6(g)$

For which

$$\Delta H^o = \Delta H_f^o(C_2H_6,g) - [\Delta H_f^o(C_2H_4,g) + \Delta H_f^o(H_2,g)]$$
$$= [-84.7 - (52.3 + 0)] \text{ kJ} = \underline{-137 \text{ kJ}}$$

However, the hydrogenation is carried out at constant volume, for which

$$\Delta H^o = \Delta E^o + \Delta(PV)$$

and $\Delta(PV) = \Delta nRT = [(1-2) \text{ mol}]RT = -(1 \text{ mol})(8.314 \text{ J K}^{-1} \text{ mol}^{-1})(298 \text{ K})$

$$= -2.48\text{x}10^3 \text{ J, } \underline{\text{or}} \ -2.48 \text{ kJ}$$

Thus, $\Delta E^o = (-137+2) \text{ kJ} = \underline{-135 \text{ kJ}}$

and the heat released on hydrogenation of 1.500 g of ethylene is

$$(\frac{-135 \text{ kJ}}{1 \text{ mol}})(1.500 \text{ g})(\frac{1 \text{ mol}}{28.05 \text{ g}}) = \underline{-7.22 \text{ kJ}}$$

If the temperature rise is $\Delta T$, then $(\Delta T \text{ K})(1.500 \text{ kJ K}^{-1}) = 7.22 \text{ kJ}$

$$\underline{\Delta T = 4.81 \text{ K}}$$

The observed temperature rise should be $\underline{4.81 \text{ K}}$

28. To apply Hess's law in general the equations for which the $\Delta H^o$ values are given are added in such a way as to obtain the equation for the reaction whose $\Delta H^o$ is sought. In this problem, reverse the first equation and add it to the second equation, remembering that when a reaction is reversed the sign of its $\Delta H^o$ changes to the reverse sign.

$$CuCl_2(s) \longrightarrow Cu(s) + Cl_2(g) \qquad \Delta H^o = +206 \text{ kJ}$$
$$2Cu(s) + Cl_2(g) \longrightarrow 2CuCl(s) \qquad \Delta H^o = -36 \text{ kJ}$$
$$CuCl_2(s) + Cu(s) \longrightarrow 2CuCl(s) \qquad \Delta H^o = [+206 + (-36)] \text{ kJ}$$
$$= \underline{+170 \text{ kJ}}$$

29. The strategy is similar to that employed in Problem 28. In the final equation, $F_2(g)$ appears as 2 mol, and in the first equation it appears as 1 mol. Thus, multiply the first equation by 2 and add it to the reverse of the second equation to obtain the required equation.

$$2H_2(g) + 2F_2(g) \longrightarrow 4HF(g) \qquad \Delta H^o = 2(-542 \text{ kJ}) = -1084 \text{ kJ}$$
$$2H_2O(l) \longrightarrow 2H_2(g) + O_2(g) \qquad \Delta H^o = -(-572 \text{ kJ}) = +572 \text{ kJ}$$
$$2F_2(g) + 2H_2O(l) \longrightarrow 4HF(g) + O_2(g) \quad \Delta H^o = (-1084+572) \text{ kJ} = \underline{-512 \text{ kJ}}$$

30. Using the strategy used in Problems 28 and 29, multiply the first equation by 2 and add it to the reverse of the second equation, which eliminates C(graphite) from both sides of the equation.

$2C(graphite) + 2O_2(g) \longrightarrow 2CO_2(g)$ $\Delta H^\circ = 2(-393.5) = -787$ kJ

$2CO(g) \longrightarrow 2C(graphite) + O_2(g)$ $\Delta H^\circ = -(-221.0) = +221.0$ kJ

$2CO(g) + O_2(g) \longrightarrow 2CO_2(g)$ $\Delta H^\circ = (-787+221.0)$ kJ $= \underline{-566.0 \text{ kJ}}$

31. First write the balanced equation for the required reaction.

$FeO(s) + CO(g) \longrightarrow Fe(s) + CO_2(g)$

Add the reverse of the first equation in the problem to the second.

$2FeO(s) + CO_2(g) \longrightarrow Fe_2O_3(s) + CO(g)$ $\Delta H^\circ = -38$ kJ

$Fe_2O_3(s) + 3CO(g) \longrightarrow 2Fe(s) + 3CO_2(g)$ $\Delta H^\circ = -28$ kJ

$2FeO(s) + 2CO(g) \longrightarrow 2Fe(s) + 2CO_2(g)$ $\Delta H^\circ = [-38+(-28)]$ kJ $= \underline{-66 \text{ kJ}}$

This is twice the required equation, for which

$$\Delta H^\circ = \frac{1}{2}(-66 \text{ kJ}) = \underline{-33 \text{ kJ}}$$

32. $NO_2(g)$ appears only in the first equation; reverse it and multiply by 3/2. $HNO_3(aq)$ appears only in the second equation; divide this equation by 2 and add the equations.

$3NO_2(g) \longrightarrow 3NO(g) + \frac{3}{2} O_2(g)$ $\Delta H^\circ = -\frac{3}{2}(-173 \text{ kJ}) = +259.5$ kJ

$N_2(g) + \frac{5}{2} O_2(g) + H_2O(l) \longrightarrow 2HNO_3(aq)$ $\Delta H^\circ = \frac{1}{2}(-255 \text{ kJ}) = -127.5$ kJ

$3NO_2(g) + N_2(g) + O_2(g) + H_2O(l) \longrightarrow 2HNO_3(aq) + 3NO(g)$

$\Delta H^\circ = [+259.5+(-127.5)]$ kJ $= \underline{+132.0 \text{ kJ}}$

and now the required equation results from adding the reverse of the third equation to the above equation.

$2NO(g) \longrightarrow N_2(g) + O_2(g)$ $\Delta H^\circ = -(+181 \text{ kJ}) = -181$ kJ

$3NO_2(g) + H_2O(l) \longrightarrow 2HNO_3(aq) + NO(g)$ ; $\Delta H^\circ = (+132-181)$ kJ $= \underline{-49 \text{ kJ}}$

This could have been calculated in one step, but for more complicated problems, such as this, you will probably find it easier to do the calculation in more than one step.

33. Reverse the first equation and divide by 2; multiply the second equation by 2 and add the resulting equations.

$$2CO_2(g) + H_2O(l) \longrightarrow C_2H_2(g) + \frac{5}{2}O_2(g) \quad \Delta H^\circ = -\frac{1}{2}(-2600) = +1300 \text{ kJ}$$
$$2C(s) + 2O_2(g) \longrightarrow 2CO_2(g) \quad \Delta H^\circ = 2(-390 \text{ kJ}) = -780 \text{ kJ}$$

$$2C(s) + H_2O(l) \longrightarrow C_2H_2(g) + \frac{1}{2}O_2(g) \quad \Delta H^\circ = (+1300-780) = \underline{520 \text{ kJ}}$$

Now add the last equation, divided by 2, to give the required equation.

$$H_2(g) + \frac{1}{2}O_2(g) \longrightarrow H_2O(l) \quad \Delta H^\circ = \frac{1}{2}(-572) = -286 \text{ kJ}$$

$$2C(s) + H_2(g) \longrightarrow C_2H_2(g) \quad \Delta H^\circ = [520+(-286)] = \underline{234 \text{ kJ}}$$

34. Double the reverse of the last equation; double the reverse of the second equation, and add the two.

$$2H_2O_2(l) \longrightarrow 2H_2(g) + 2O_2(g) \quad \Delta H^\circ = -2(-187.6) \text{ kJ} = 375.2 \text{ kJ}$$
$$2H_2O(g) \longrightarrow 2H_2O(l) \quad \Delta H^\circ = -2(+44.0) \text{ kJ} = -88.0 \text{ kJ}$$

$$2H_2O_2(l) + 2H_2O(g) \longrightarrow 2H_2(g) + O_2(g) + 2H_2O(l) \quad \Delta H^\circ = (375.2-88.0) = \underline{287.2 \text{ kJ}}$$

Finally, add the above equation to the last equation.

$$2H_2(g) + O_2(g) \longrightarrow 2H_2O(g) \quad \Delta H^\circ = -483.6 \text{ kJ}$$

$$2H_2O_2(l) \longrightarrow 2H_2O(l) + O_2(g) \quad \Delta H^\circ = [287.2+(-483.6)] \text{ kJ} = \underline{-196.4 \text{ kJ}}$$

35. Halve the reverse of the second equation and add it to the first equation.

$$HO(g) \longrightarrow \frac{1}{2}H_2O_2(g) \quad \Delta H^\circ = -\frac{1}{2}(134 \text{ kJ}) = -67 \text{ kJ}$$
$$Cl_2(g) \longrightarrow 2Cl(g) \quad \Delta H^\circ = 242 \text{ kJ}$$

$$HO(g) + Cl_2(g) \longrightarrow \frac{1}{2}H_2O_2(g) + 2Cl(g) \quad \Delta H^\circ = (-67+242) \text{ kJ} = \underline{175 \text{ kJ}}$$

Finally, halve the final equation and add it to the above equation.

$$\frac{1}{2}H_2O_2(g) + Cl(g) \longrightarrow HOCl(g) \quad \Delta H^\circ = \frac{1}{2}(-209 \text{ kJ}) = \underline{-104.5 \text{ kJ}}$$

$$HO(g) + Cl_2(g) \longrightarrow HOCl(g) + Cl(g) \quad \Delta H^\circ = [175+(-104.5)] \text{ kJ} = \underline{70.5 \text{ kJ}}$$

36. First write the balanced equations from the given data:

$$P(s,\text{white}) + \frac{3}{2}Cl_2(g) \longrightarrow PCl_3(g) \quad \Delta H^\circ = -306 \text{ kJ}$$
$$P(s,\text{red}) + \frac{3}{2}Cl_2(g) \longrightarrow PCl_3(g) \quad \Delta H^\circ = -288 \text{ kJ}$$

Reversing the first equation and adding it to the second gives,

$$P(s,\text{red}) \longrightarrow P(s,\text{white}) \quad \Delta H^\circ = [-306+(-288)] \text{ kJ} = \underline{-18 \text{ kJ}}$$

37. The balanced equation for the formation of $N_2O_4(g)$ from its elements in their standard states is

$$N_2(g) + 2O_2(g) \longrightarrow N_2O_4(g)$$

Adding the last equation to the second gives,

$$2NO_2(g) \longrightarrow N_2O_4(g) \qquad \Delta H^o = -58.0 \text{ kJ}$$

$$2NO(g) + O_2(g) \longrightarrow 2NO_2(g) \qquad \Delta H^o = -114.2 \text{ kJ}$$

$$2NO(g) + O_2(g) \longrightarrow N_2O_4(g) \qquad \Delta H^o = [-58+(-114.2)] \text{ kJ}$$

$$= \underline{-172.2 \text{ kJ}}$$

Now add this equation to the first equation,

$$N_2(g) + O_2(g) \longrightarrow 2NO(g) \qquad \Delta H^o = +180.6 \text{ kJ}$$

$$N_2(g) + 2O_2(g) \longrightarrow N_2O_4(g) \qquad \Delta H^o = [-172.2+(+180.6)] \text{ kJ}$$

$$= \underline{+8.4 \text{ kJ}}$$

38. By definition, the standard enthalpy of formation of a substance is $\Delta H^o$ for the reaction where 1 mole of the substance is formed from its elements in their standard states. Thus, for $H_2O(l)$, the appropriate reaction is

$$H_2(g) + \frac{1}{2} O_2(g) \longrightarrow H_2O(l); \; \Delta H^o = \Delta H^o_f = \underline{-285.8 \text{ kJ}} \; .... \; (1)$$

For the formation of $H_2O(g)$ from its elements in their standard states

$$H_2(g) + \frac{1}{2} O_2(g) \longrightarrow H_2O(g)$$

which results from adding $H_2O(l) \longrightarrow (H_2O(g)$, and equation 1. Reversing the second equation, we obtain

$$H_2O(l) \longrightarrow H_2O(g) \qquad \Delta H^o = +44.0 \text{ kJ}$$

$$H_2(g) + \frac{1}{2} O_2(g) \longrightarrow H_2O(l) \qquad \Delta H^o = -285.8 \text{ kJ}$$

$$H_2(g) + \frac{1}{2} O_2(g) \longrightarrow H_2O(g) \qquad \Delta H^o = \underline{-241.8 \text{ kJ}} = \Delta H^o_f(H_2O,g)$$

The equation for the formation of $NH_3(g)$ from its elements is the reverse of the third equation

$$N_2(g) + 3H_2(g) \longrightarrow 2NH_3(g) \qquad \underline{\Delta H^o = -92.4 \text{ kJ}}$$

which is the $\Delta H^o$ for the formation of 2 moles of $NH_3(g)$.

Thus, $\Delta H^o_f(NH_3,g) = \underline{-46.2 \text{ kJ mol}^{-1}}$

$$\Delta H^o_f(H_2O,l) = \underline{-285.8 \text{ kJ}} \; ; \; \Delta H^o_f(H_2O,g) = \underline{-241.8 \text{ kJ}} \; ;$$

$$\Delta H^o_f(NH_3,g) = \underline{-46.2 \text{ kJ}}$$

39. The standard enthalpy of formation of an element is defined as 0 kJ $mol^{-1}$ for the element in its standard state, i.e., in the stable form in which it exists under standard conditions of 25°C and 1 atm pressure. Under standard conditions $H_2$, $N_2$, $O_2$, $F_2$, and $Cl_2$, are gases, while bromine is a liquid and iodine is a solid. In other words the standard enthalpies of formation of $Br_2(l)$ and $I_2(s)$ are zero, rather than the standard enthalpies of formation of $Br_2(g)$ and $I_2(g)$.

40. First write the balanced equation for the combustion reaction.

$$C_7H_{16}(l) + 11\, O_2(g) \longrightarrow 7CO_2(g) + 8H_2O(l)$$

The $\Delta H_f^o$ of any reactant or product can be calculated provided the $\Delta H_f^o$ values of the remainder of the reactants and products, and $\Delta H^o$ for the reaction are known, since

$$\Delta H^o = \sum \Delta H_f^o(\text{products}) - \sum \Delta H_f^o(\text{reactants})$$

In this case, we have

$$\Delta H = [7\,\Delta H_f^o(CO_2,g) + 8\,\Delta H_f^o(H_2O,l)] - [\Delta H_f^o(C_7H_{16},l) + 11\,\Delta H_f^o(O_2,g)]$$

Using the given $\Delta H^o$ for the reaction and $\Delta H_f^o$ values from Appendix B,

$$-4816.9 \text{ kJ} = [7(-393.5)+8(-285.8)]-[\Delta H_f^o(C_7H_{16},l)+11(0)] \text{ kJ}$$

where each $\Delta H_f^o$ value is multiplied by the number of moles of reactant or product in the balanced equation for the reaction.

Hence, $\Delta H_f^o(C_7H_{16},l) = +4816.9 - 2754.5 - 2286.4 = \underline{-224 \text{ kJ}}$

41. For the method, see Problem 40.

$$SO_3(g) + H_2O(g) \longrightarrow H_2SO_4(l)$$

$$\Delta H^o = [\Delta H_f^o(H_2SO_4,l)] - [\Delta H_f^o(SO_3,g) + \Delta H_f^o(H_2O,g)]$$
$$= -814.0 - [-395.7 + (-241.8)] \text{ kJ} = \underline{-176.5 \text{ kJ}}$$

42. For the method, see Problem 40.

$$C_2H_2(g) + 2H_2(g) \longrightarrow C_2H_6(g)\ ;\ \Delta H^o = -312 \text{ kJ}$$

$$\Delta H^o = [\Delta H_f^o(C_6H_6,g)] - [\Delta H_f^o(C_2H_2,g) + 2\,\Delta H_f^o(H_2,g)]$$
$$-312 \text{ kJ} = -84.7 - \Delta H_f^o(C_2H_2,g) - 2(0)$$
$$\Delta H_f^o(C_2H_2,g) = \underline{+227 \text{ kJ}}$$

43. The balanced equation for the formation of $Mg(OH)_2(s)$ from its elements is

$$Mg(s) + O_2(g) + H_2(g) \longrightarrow Mg(OH)_2(s)$$

Thus, it is the $\Delta H^o$ for this reaction that is required, which is obtained as follows:

Add the reverse of the third reaction to the first equation

$$2H_2(g) + O_2(g) \longrightarrow 2H_2O(l) \qquad \Delta H^o = -571.6 \text{ kJ}$$
$$2Mg(s) + O_2(g) \longrightarrow 2MgO(s) \qquad \Delta H^o = -1203.7 \text{ kJ}$$

$$2Mg(s) + 2H_2(g) + 2O_2(g) \longrightarrow 2MgO(s) + H_2O(l)\ ;\ \Delta H^o = \underline{-1775.3 \text{ kJ}}$$

Now twice the second equation is added to this equation, to give

$$2Mg(s) + 2H_2(g) + 2O_2(g) \longrightarrow 2Mg(OH)_2(s)$$

for which $\Delta H^o = -1775.3 + 2(-36.7) = \underline{-1848.7 \text{ kJ}}$, which is for the formation of 2 mol of $Mg(OH)_2(s)$.

$$\Delta H_f^o[Mg(OH)_2,s] = \underline{-924.4 \text{ kJ}}$$

44. The balanced equation for the reaction is

$$2C_4H_{10}(l) + 13\,O_2(g) \longrightarrow 8CO_2(g) + 10\,H_2O(g)$$

and the reaction is carried out under standard conditions. Thus,

$$\Delta H^\circ = [8\,\Delta H_f^\circ(CO_2,g) + 10\,\Delta H_f^\circ(H_2O,g)] - [2\,\Delta H_f^\circ(C_4H_{10},l) + 13\,\Delta H_f^\circ(O_2,g)]$$
$$= [8(-393.5) + 10(-241.8)] - [2(-127) + 13(0)] = \underline{-5312\ kJ}$$

which is the standard enthalpy of combustion for 2 mol liquid butane. For 1.00 g of butane

$$\Delta H^\circ = \left(\frac{-5312\ kJ}{2\ mol\ C_4H_{10}}\right)\left(\frac{1\ mol\ C_4H_{10}}{58.14\ g\ C_4H_{10}}\right) = \underline{-45.7\ kJ\ g^{-1}}$$

45. $C_2H_5OH(l) + 3O_2(g) \longrightarrow 2CO_2(g) + 3H_2O(l);\ \Delta H^\circ = -1370\ kJ$

$$\Delta H^\circ = [2\,\Delta H_f^\circ(CO_2,g) + 3\,\Delta H_f^\circ(H_2O,l)] - [\,\Delta H_f^\circ(C_2H_5OH,l) + 3\,\Delta H_f^\circ(O_2,g)]$$
$$-1370\ kJ = [2(-393.5) + 3(-285.8)] - [\,\Delta H_f^\circ(C_2H_5OH,l) + 3(0)]\ kJ$$
$$\Delta H_f^\circ(C_2H_5OH,l) = \underline{-274\ kJ}$$

46. $C_3H_8(g) + 5O_2(g) \longrightarrow 3CO_2(g) + 4H_2O(l);\ \Delta H^\circ = -2044\ kJ$

$$\Delta H^\circ = [3\,\Delta H_f^\circ(CO_2,g) + 4\,\Delta H_f^\circ(H_2O,l)] - [\,\Delta H_f^\circ(C_3H_8,g) + 5\,\Delta H_f^\circ(O_2,g)]$$
$$-2044\ kJ = [3(-393.5) + 4(-285.8)] - [\,\Delta H_f^\circ(C_3H_8,g) + 5(0)]\ kJ$$
$$\Delta H_f^\circ(C_3H_8,g) = \underline{-280\ kJ}$$

47. The data in Appendix B gives $\Delta H_f^\circ(C_6H_6,l)$ but not $\Delta H_f^\circ(C_6H_6,g)$. However, the latter can be found since we are given the amount of heat required to convert 1.00 g of $C_6H_6(l)$ to $C_6H_6(g)$.

$$C_6H_6(l) \longrightarrow C_6H_6(g)\ ;\ \Delta H^\circ = 434.5\ J\ g^{-1}$$

Note that heat is absorbed on conversion of liquid to gas, so that the process is endothermic.

For 1 mol $C_6H_6(l)$, $\Delta H^\circ_{vap} = \left(\frac{434.5\ J}{1.00\ g}\right)\left(\frac{78.11\ g\ C_6H_6}{1\ mol\ C_6H_6}\right)\left(\frac{1\ kJ}{10^3\ J}\right) = \underline{33.9\ kJ}$

Thus, for the vaporization of benzene

$$\Delta H^\circ = 33.9\ kJ = [\,\Delta H_f^\circ(C_6H_6,g)] - [\,\Delta H_f^\circ(C_6H_6,l)$$
$$= \Delta H_f^\circ(C_6H_6,g) - (49.0\ kJ)$$
$$\Delta H_f^\circ(C_6H_6,g) = \underline{+82.9\ kJ}$$

<u>Note</u> that in general $\Delta H_f^\circ$ values for gases are <u>less</u> than those for the corresponding liquids; heat is absorbed in converting liquid to gas.

The reaction in question is

$$3C_2H_2(g) \longrightarrow C_6H_6(g)$$

for which

$$\Delta H^\circ = \Delta H_f^\circ(C_6H_6,g) - 3\,\Delta H_f^\circ(C_2H_2,g) = [+82.9\ ] - [3(228.0]\ kJ$$
$$= \underline{-601.1\ kJ}$$

48. $$H_2(g) + I_2(s) \rightarrow 2HI(g) \; ; \; \Delta H^\circ = +51.9 \text{ kJ}$$

and since $H_2(g)$ and $I_2(s)$ are the standard states for these elements, $\Delta H^\circ = 2[\Delta H_f^\circ(HI,g)] = +51.9$ kJ.

$$\Delta H_f^\circ(HI,g) = \underline{+26.0 \text{ kJ mol}^{-1}}$$

49. This problem is very similar to Problem 40.

$$C_8H_{18}(l) + \frac{25}{2}O_2(g) \longrightarrow 8CO_2(g) + 9H_2O(l)$$

$$\Delta H^\circ = [8\,\Delta H_f^\circ(CO_2,g) + 9\,\Delta H_f^\circ(H_2O,l)] - [\Delta H_f^\circ(C_8H_{18},l) + \frac{25}{2}\Delta H_f^\circ(O_2,g)]$$
$$= [8(-393.5) + 9(-285.8)] - [(-224) + \frac{25}{2}(0)] \text{ kJ} = \underline{-5496 \text{ kJ}}$$

50. $$CH_3CO_2H(l) + 2O_2(g) \longrightarrow 2CO_2(g) + 2H_2O(l)$$

First calculate $\Delta H^\circ$ for the above reaction for the combustion of 1 mol $CH_3CO_2H(l)$ (molar mass 60.05 g $mol^{-1}$).

$$\Delta H^\circ = \left(\frac{-72.62 \text{ kJ}}{4.766 \text{ mL acid}}\right)\left(\frac{1 \text{ mL acid}}{1.049 \text{ g acid}}\right)\left(\frac{60.05 \text{ g acid}}{1 \text{ mol acid}}\right) = \underline{-872.2 \text{ kJ mol}^{-1}}$$

Then use the information on combustion of acetic acid to calculate $\Delta H_f^\circ(CH_3CO_2H,l)$.

$$\Delta H^\circ = [2\,\Delta H_f^\circ(CO_2,g) + 2\,\Delta H_f^\circ(H_2O,l)] - [\Delta H_f^\circ(CH_3CO_2H,l) + 2\,\Delta H_f^\circ(O_2,g)]$$

$$-872.2 = [2(-393.5) + 2(-285.8)] - [\Delta H_f^\circ(CH_3CO_2H,l) + 2(0)] \text{ kJ}$$

$$\Delta H_f^\circ(CH_3CO_2H,l) = \underline{-486.4 \text{ kJ}}$$

Now the $\Delta H^\circ$ can be calculated for the reaction

$$C_2H_6(g) + \frac{3}{2}O_2(g) \longrightarrow CH_3CO_2H(l) + H_2O(l)$$

$$\Delta H^\circ = [\Delta H_f^\circ(CH_3CO_2H,l) + \Delta H_f^\circ(H_2O,l)] - [\Delta H_f^\circ(C_2H_6,g) + \frac{3}{2}\Delta H_f^\circ(O_2,g)]$$
$$= [(-486.4) + (-285.8)] - [(-84.7) + \frac{3}{2}(0)] \text{ kJ} = \underline{-687.5 \text{ kJ}}$$

51. The reaction is

$$C(\text{graphite},s) \longrightarrow C(\text{diamond},s)$$

Write the balanced equations for the combustion of graphite and diamond, and reverse the latter to obtain the required equation.

$$C(\text{graphite},s) + O_2(g) \longrightarrow CO_2(g) \qquad \Delta H^\circ = -393.7 \text{ kJ}$$
$$CO_2(g) \longrightarrow C(\text{diamond},s) + O_2(g) \qquad \Delta H^\circ = -(-395.6) = +395.6 \text{ kJ}$$

$$C(\text{graphite},s) \longrightarrow C(\text{diamond},s) \qquad \Delta H^\circ = (-393.7) + (395.6) \text{ kJ}$$
$$= \underline{+1.9 \text{ kJ}}$$

Thus, $\Delta H_f^\circ(\text{diamond},s) = \underline{+1.9 \text{ kJ mol}^{-1}}$

Heat is absorbed when diamond is formed from graphite; diamond is less stable than graphite.

52. (a) The dissociation energy of $O_2(g)$ is the $\Delta H^o$ for

$$O_2(g) \longrightarrow 2O(g)$$

For which, $\Delta H^o = 2\,\Delta H_f^o(O,g) - \Delta H_f^o(O_2,g) = [2(249) - 0]$ kJ

$= \underline{498 \text{ kJ mol}^{-1}}$

(b) Similarly, for $N_2(g) \longrightarrow 2N(g)$

$\Delta H^o = 2\,\Delta H_f^o(N,g) + \Delta H_f^o(N_2,g) = [2(473) - 0]$ kJ $= \underline{946 \text{ kJ mol}^{-1}}$

53. The bond dissociation energy of ClF is the $\Delta H^o$ for the reaction

$$ClF(g) \longrightarrow Cl(g) + F(g)$$

For which $\Delta H^o = [\Delta H_f^o(Cl,g) + \Delta H_f^o(F,g)] - \Delta H_f^o(ClF,g)$

As we saw in Problem 52, $\Delta H_f^o(X,g)$ is half of the dissociation energy of $X_2(g)$, so that

$\Delta H_f^o(Cl,g) = \dfrac{239 \text{ kJ}}{2} = \underline{119.5 \text{ kJ mol}^{-1}}$

and $\Delta H_f^o(F,g) = \dfrac{155 \text{ kJ}}{2} = \underline{77.5 \text{ kJ}}$

Thus, $\Delta H^o = [(119.5) + (77.5)]$ kJ $= \underline{253 \text{ kJ mol}^{-1}}$

54. In bond energy problems, $\Delta H^o$ for the complete dissociation of a gaseous compound into its gaseous atoms is the sum of the energies of all of the bonds. Assuming that all of the bonds are the same, $\Delta H^o$ divided by the number of bonds gives the average bond energy.

(i) $PCl_3(g) \longrightarrow P(g) + 3Cl(g)$

$\Delta H^o = [\Delta H_f^o(P,g) + 3\,\Delta H_f^o(Cl,g)] - \Delta H_f^o(PCl_3,g)$

$= [(316.2) + 3(119.5)] - (-287)$ kJ $= \underline{962 \text{ kJ}}$

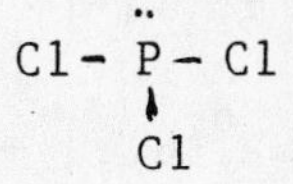

i.e., $3BE(P\text{-}Cl) = 962$ kJ; $BE(P\text{-}Cl) = \underline{321 \text{ kJ mol}^{-1}}$

(ii) $PCl_5(g) \longrightarrow P(g) + 5Cl(g)$

$\Delta H^o = [\Delta H_f^o(P,g) + 5\,\Delta H_f^o(Cl,g)] - \Delta H_f^o(PCl_5,g)$

$= [(316.2) + 5(119.5)] - (-374.7)$ kJ $= \underline{1288 \text{ kJ}}$

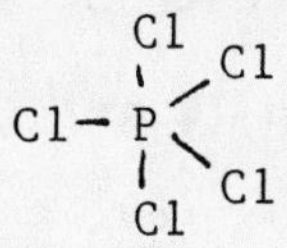

i.e., $5BE(P\text{-}Cl) = 1288$ kJ; $BE(P\text{-}Cl) = \underline{258 \text{ kJ mol}^{-1}}$

The energy of the two additional bonds in $PCl_5$ is given by the difference between the $\Delta H^o$ values in (ii) and (i).

i.e. $2BE(P\text{-}Cl) = (1288 - 962)$ kJ $= 326$ kJ

$BE(P\text{-}Cl) = \underline{163 \text{ kJ mol}^{-1}}$

55. (a) For the reaction $H_2O(g) \longrightarrow 2H(g) + O(g)$, $\Delta H^o = 2BE(O\text{-}H)$.

$\Delta H^o = [2\,\Delta H_f^o(H,g) + \Delta H_f^o(O,g)] - \Delta H_f^o(H_2O,g)$

$= [2(218) + (249.1)] - (-241.8)$ kJ $= \underline{926.9 \text{ kJ}}$



$2BE(O\text{-}H) = 926.9$ kJ; $BE(O\text{-}H) = \underline{463.5 \text{ kJ mol}^{-1}}$

(b) $H_2O_2(g) \longrightarrow 2H(g) + 2O(g)$;  $\Delta H^o = 2BE(O\text{-}H) + BE(O\text{-}O)$

$$\Delta H^o = [2\,\Delta H_f^o(H,g) + 2\,\Delta H_f^o(O,g)] - [\,\Delta H_f^o(H_2O_2,g)]$$
$$= [2(218) + 2(249.1)] - [(-136.4)] = \underline{1070.6\ kJ}$$

Assuming BE(O-H) = 463.5 kJ,

$$\Delta H^o = 1070.6\ kJ = 2(463.5\ kJ) + BE(O\text{-}O)$$

H-Ö-Ö-H

$$BE(O\text{-}O) = \underline{143.6\ kJ\ mol^{-1}}$$

(c) The single O-O bond in $H_2O_2$ is much weaker than the average C-C bond energy of 348 kJ $mol^{-1}$ .

56. Carbon monoxide   $CO(g) \longrightarrow C(g) + O(g)$;  $\Delta H^o = BE(C{\equiv}O)$

$$\Delta H^o = [\Delta H_f^o(C,g) + \Delta H_f^o(O,g)] - [\Delta H_f^o(CO,g)]$$

$^-:C{\equiv}O:^+$

$$BE(C{\equiv}O) = \Delta H^o = [(716.7) + (249.1)] - [(-110.5)] = \underline{1076.3\ kJ\ mol^{-1}}$$

Carbon dioxide   $CO_2(g) \longrightarrow C(g) + 2O(g)$;  $\Delta H^o = 2BE(C{=}O)$

$$\Delta H^o = [\,\Delta H_f^o(C,g) + 2\,\Delta H_f^o(O,g)] - [\,\Delta H_f^o(CO_2,g)]$$

:Ö=C=Ö:

$$2BE(C{=}O) = \Delta H^o = [(716.7) + 2(249.1)] - [(-393.5)] = \underline{1608.4\ kJ}$$
$$BE(C{=}O) = \underline{804.2\ kJ\ mol^{-1}}$$

As expected, the CO double bond in $CO_2$ is weaker than the triple bond in carbon monoxide.

57. Methanal, $H_2C{=}O$, has two C-H bonds and one C=O bond, and for

$$H_2CO(g) \longrightarrow 2H(g) + C(g) + O(g),\ \Delta H^o = 2BE(C\text{-}H) + BE(C{=}O)$$

$$\Delta H^o = [2\,\Delta H_f^o(H,g) + \Delta H_f^o(C,g) + \Delta H_f^o(O,g)] - [\Delta H_f^o(H_2CO,g)]$$
$$= [2(218.0) + (716.7) + (249.1)] - [(-115.9)] = \underline{1517.7\ kJ}$$

Assuming that BE(C-H) = 413 kJ $mol^{-1}$

$$2BE(C\text{-}H) + BE(C{=}O) = [2(413) + BE(C{=}O)]\ kJ = 1517.7\ kJ$$

$BE(C{=}O) = \underline{692\ kJ\ mol^{-1}}$ , which is <u>weaker</u> than C=O in $CO_2$.

58. The standard enthalpy of formation , $\Delta H_f^o(C_2H_6,g)$, is the $\Delta H^o$ for

$$2C(s) + 3H_2(g) \longrightarrow C_2H_6(g)$$

which can be achieved through the intermediate step of first converting C(s) to C(atomic,g), and $H_2(g)$ to H(atomic,g), and then combining the atoms to form the product, $C_2H_6(g)$.

Thus,

$$\Delta H^o = \sum[\text{bond energies reactants}] - \sum[\text{bond energies products}]$$

```
  H H
  | |
H-C-C-H
  | |
  H H
```

$$\Delta H_f^o(C_2H_6,g) = [2BE(C,s) + 3BE(H_2,g)] - [BE(C\text{-}C) + 6BE(C\text{-}H)]$$
$$= [2(716.7) + 3(436)] - [(348) + 6(413)] \text{ kJ}$$
$$= \underline{-84.6 \text{ kJ mol}^{-1}}$$

59. Use the method given in Problem 58.

$\Delta H^o = \sum$[bond energies reactants] - $\sum$[bond energies products]

(a) $H_2S(g) + Cl_2(g) \longrightarrow SCl_2(g) + H_2(g)$

<u>bonds</u> H-S-H Cl-Cl Cl-S-Cl H-H

$$\Delta H^o = [2BE(H\text{-}S) + BE(Cl\text{-}Cl)] - [2BE(S\text{-}Cl) + BE(H\text{-}H)]$$
$$= [2(364) + (239)] - [2(276) + (436)] \text{ kJ} = \underline{-21 \text{ kJ}}$$

(b) $CH_4(g) + 2F_2(g) \longrightarrow CH_2F_2(g) + 2HF(g)$

<u>bonds</u> H-C-H (with H above and below C) 2 F-F H-C-H (with F above and below C) 2 H-F

$$\Delta H^o = [4BE(C\text{-}H) + 2BE(F\text{-}F)] - [2BE(C\text{-}H) + 2BE(C\text{-}F) + 2BE(H\text{-}F)]$$
$$= [4(413) + 2(155)] - [2(413) + 2(485) + 2(565)] \text{ kJ} = \underline{-964 \text{ kJ}}$$

(c) $CH_2Cl_2(g) + CH_4(g) \longrightarrow 2CH_3Cl(g)$

<u>bonds</u> Cl-C-Cl (with H above and below C) H-C-H (with H above and below C) H-C-Cl (with Cl above and below C)

In this case, we have 6(C-H) and 2(C-Cl) bonds in both the reactants and products, so that $\underline{\Delta H^o = 0 \text{ kJ}}$

60. This problem is similar to Problem 59.

(a) $C_2H_2(g) + C_2H_6(g) \longrightarrow 2C_2H_4(g)$

<u>bonds</u> H-C≡C-H H-C-C-H (with H above and below each C) H-C=C-H (with H below each C)

i.e., <u>reactants</u>: 8(C-H), C≡C, and C-C

<u>products</u>: 8(C-H) and 2(C=C)

$$\Delta H^o = [8BE(C\text{-}H) + BE(C{\equiv}C) + BE(C\text{-}C)] - [8BE(C\text{-}H) + 2BE(C{=}C)]$$
$$= BE(C{\equiv}C) + BE(C\text{-}C) - 2BE(C{=}C) = [(812) + (348) - 2(619)] \text{ kJ}$$
$$= \underline{-78 \text{ kJ}}$$

(b) $2H_2O_2(g) \longrightarrow 2H_2O(g) + O_2(g)$

<u>bonds</u> 2 H-O-O-H 2 H-O-H O=O

i.e., <u>reactants</u>: 4(O-H) and 2(O-O); <u>products</u>: 4(O-H) and O=O

$$\Delta H^o = [4BE(O\text{-}H) + 2BE(O\text{-}O)] - [4BE(O\text{-}H) + BE(O{=}O)]$$

$\Delta H^\circ = 2BE(O\text{-}O) - BE(O{=}O) = [2(138) - (494)]$ kJ = $\underline{-218 \text{ kJ}}$

(c) $CO(g) + H_2(g) \longrightarrow H_2CO(g)$

<u>bonds</u> C≡O   H-H   H-C=O (with H on C)

i.e., <u>reactants:</u> C≡O and H-H; <u>products</u>: 2(C-H) and C=O

$$\Delta H^\circ = [BE(C{\equiv}O) + BE(H\text{-}H)] - [2BE(C\text{-}H) + BE(C{=}O)]$$
$$= (1070) + (436) - 2(413) - (707) = \underline{-27 \text{ kJ}}$$

61. The equation for the standard enthalpy of formation is

$$4C(s) + 4H_2(g) \xrightarrow{\Delta H_f^\circ} H_2C{=}C(CH_3)_2(g)$$

$\Delta H_1^\circ \downarrow$ $\Delta H_2^\circ \downarrow$ $\Delta H_3^\circ \nearrow$

$$4C(g) + 8H(g)$$

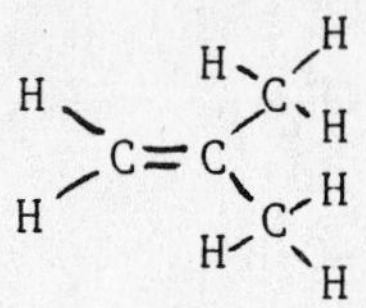

$$\Delta H_f^\circ(C_4H_8,g) = \Delta H_1^\circ + \Delta H_2^\circ + \Delta H_3^\circ$$
$$= 4\Delta H_f^\circ(C,g) + 4BE(H\text{-}H) - [8BE(C\text{-}H) + 2BE(C\text{-}C) + BE(C{=}C)]$$
$$= [4(716.7) + 4(436)] - [8(413) + 2(348) + (619)$$
$$= \underline{-8 \text{ kJ}}$$

<u>Note</u> that this is essentially the bond energy method, except that we are not given the average bond energy in C(graphite,s), so, instead, we use $\Delta H_f^\circ(C,g)$.

62. (a) $H_2(g) + \frac{1}{2}O_2(g) \longrightarrow H_2O(g)$

$$\Delta H_f^\circ(H_2O,g) = [BE(H\text{-}H) + \tfrac{1}{2}BE(O{=}O)] - [2BE(O\text{-}H)]$$
$$= [(436) + \tfrac{1}{2}(494)] - [2(463)] = \underline{-243 \text{ kJ}}$$

(b) $H_2(g) + \frac{1}{8}S_8(g) \longrightarrow H_2S(g)$

$$\Delta H_f^\circ(H_2S,g) = [BE(H\text{-}H) + BE(S\text{-}S)] - [2BE(H\text{-}S)]$$
$$= [(436) + (255)] - [2(364)] = \underline{-37 \text{ kJ}}$$

(c) $\frac{1}{2}N_2(g) + \frac{3}{2}H_2(g) \longrightarrow NH_3(g)$

$$\Delta H_f^\circ(NH_3,g) = [\tfrac{1}{2}BE(N\quad N) + \tfrac{3}{2}BE(H\text{-}H)] - [3BE(N\text{-}H)]$$
$$= [\tfrac{1}{2}(941) + \tfrac{3}{2}(436)] - [3(389)] = \underline{-43 \text{ kJ}}$$

(d) $\frac{1}{4}P_4(g) + \frac{3}{2}H_2(g) \longrightarrow PH_3(g)$

$$\Delta H_f^\circ(PH_3,g) = [\tfrac{3}{2}BE(P\text{-}P) + \tfrac{3}{2}BE(H\text{-}H)] - [3BE(P\text{-}H)]$$
$$= [\tfrac{3}{2}(172) + \tfrac{3}{2}(436)] - [3(318)] = \underline{-42 \text{ kJ}}$$

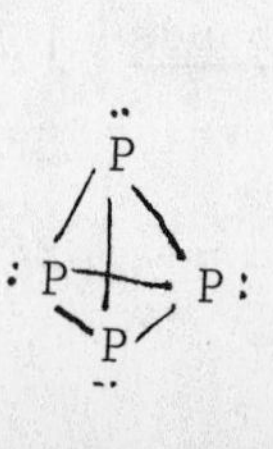

(e) $C(s,gr) + 2H_2(g) \longrightarrow CH_4(g)$ =C

$$\Delta H_f^o(CH_4,g) = BE(C-C) + \frac{1}{2}BE(C=C) + 2BE(H-H) - 4BE(C-H)$$
$$= (348) + \frac{1}{2}(619) + 2(436) - 4(413) = \underline{-122\ kJ}$$

63. The standard enthalpy of dissociation of HCl(g) is BE(H-Cl), so we can use the bond energy method to calculate $\Delta H_f^o(HCl,g)$.

$$\frac{1}{2}H_2(g) + \frac{1}{2}Cl_2(g) \longrightarrow HCl(g)$$

$$\Delta H_f^o(HCl,g) = [\frac{1}{2}BE(H-H) + \frac{1}{2}BE(Cl-Cl)] - [BE(H-Cl)]$$
$$= \frac{1}{2}(436) + \frac{1}{2}(239) - (431) = \underline{-94\ kJ}$$

64. (a) The standard enthalpy change, $\Delta H^o$, of a reaction is the difference between the enthalpy of the final and initial states of the system, when they are both at a pressure of 1 atmosphere and 25°C. It is equal to the heat absorbed, q, at 1 atm and 25°C.
(b) The standard enthalpy of formation, $\Delta H_f^o$, of a substance is the $\Delta H^o$ for the reaction where the substance is formed from its elements in their standard states at 25°C and 1 atm pressure.
(c) The standard enthalpy of neutralization is the enthalpy change for the reaction where 1 mol $H_3O^+(aq)$ reacts with 1 mol $OH^-(aq)$ under standard conditions.
(d) The standard state of an element is the form in which it normally exists at 25°C and 1 atm pressure.
(e) The average bond energy of a bond of a particular type in a polyatomic molecule is the average energy needed to break all of the bonds of that type.

65. (a) See Problem 64. $\Delta H_f^o(MgCO_3,s)$ is the $\Delta H^o$ for the reaction in which 1 mol $MgCO_3(s)$ is formed from Mg(s), C(graphite,s) and $O_2(g)$ under standard conditions;

$$Mg(s) + C(graphite,s) + \frac{3}{2}O_2(g) \longrightarrow MgCO_3(s)$$

(b) The heat capacity of the calorimeter and its contents is obtained from the results of the second experiment.

Heat capacity of calorimeter = $(\frac{506\ J}{1.02\ K}) = \underline{496\ J\ K^{-1}}$

Thus, the heat released in the first experiment, q, is

$q = (496\ J\ K^{-1})(8.61\ K) = \underline{4270\ J}$ $\quad (q = -4270\ J)$.

and for 1 mol Mg

$$\Delta H = (\frac{-4270\ J}{0.203\ g\ Mg})(\frac{24.31\ g\ Mg}{1\ mol\ Mg})(\frac{1\ kJ}{1000\ J}) = \underline{-511\ kJ\ mol^{-1}}$$

(c) $MgCO_3(s) + 2HCl(aq) \longrightarrow MgCl_2(aq) + CO_2(g) + H_2O(l)$; $\Delta H = \underline{-90.4\ kJ}$

and from part (b),

$Mg(s) + 2HCl(aq) \longrightarrow MgCl_2(aq) + H_2(g)$; $\Delta H = -511\ kJ$

We require $\Delta H$ for the reaction given in part (a). Reversing the first equation above and adding it to the second equation gives,

$Mg(s) + CO_2(g) + H_2O(l) \longrightarrow MgCO_3(s) + H_2(g)$, for which

$\Delta H^\circ = [(-511) - (-90.4)]\ kJ = \underline{-421\ kJ}$

This equation contains reactants and products whose $\Delta H_f^\circ$ values, except that for $MgCO_3(s)$, are given in Appendix B. Thus,

$$\Delta H^\circ = -421\ kJ = [\ \Delta H_f^\circ(MgCO_3,s) + \Delta H_f^\circ(H_2,g)]$$
$$- [\ \Delta H_f^\circ(Mg,s) + \Delta H_f^\circ(CO_2,g) + \Delta H_f^\circ(H_2O,l)]$$

$$-421\ kJ = [\Delta H_f^\circ(MgCO_3,s) + (0)] - [(0) + (-393.5) + (-285.8)]$$

$$\Delta H_f^\circ(MgCO_3,s) = \underline{-1100\ kJ}$$

66. (a) $SO_2(g)$ is reduced to S(s), so $H_2S(g)$ must be oxidized to S(s). $SO_2(g)$ is the <u>oxidizing agent</u> and $H_2S(g)$ is the <u>reducing</u> agent.

(b) All of the standard enthalpies of formation are given in Appendix B.

$$\Delta H^\circ = [3\,\Delta H_f^\circ(S,s) + 2\,\Delta H_f^\circ(H_2O,g)] - [2\,\Delta H_f^\circ(H_2S,g) + \Delta H_f^\circ(SO_2,g)]$$
$$= [3(0) + 2(-241.8)] - [2(-20.6) + (-296.8)]\ kJ$$
$$= \underline{-145.6\ kJ}$$

67. For the combustions, first write the balanced equations:

$C(graphite,s) + O_2(g) \longrightarrow CO_2(g)$ $\qquad \Delta H^\circ = -393.5\ kJ$

$H_2(g) + \frac{1}{2}O_2(g) \longrightarrow H_2O(l)$ $\qquad \Delta H^\circ = -285.8\ kJ$

$C_2H_6(g) + \frac{7}{2}\ O_2(g) \longrightarrow 2CO_2(g) + 3H_2O(l)$ $\qquad \Delta H^\circ = -1559.8\ kJ$

$C_3H_8(g) + 5O_2(g) \longrightarrow 3CO_2(g) + 4H_2O(l)$ $\qquad \Delta H^\circ = -2219.9\ kJ$

By definition, $\Delta H_f^\circ(C_2H_6,g)$ is the $\Delta H^\circ$ for

$2C(graphite,s) + 3H_2(g) \longrightarrow C_2H_6(g)$

Using Hess's law, twice the first combustion equation above, plus three times the second equation, plus the reverse of the third equation, gives the required equation. Thus,

$$\Delta H_f^\circ(C_2H_6,g) = 2(-393.5) + 3(-285.8) - (-1559.8) = \underline{-84.6\ kJ}$$

Similarly, $\Delta H_f^\circ(C_3H_8,g)$ is the $\Delta H$ for

$3C(graphite,s) + 4H_2(g) \longrightarrow C_3H_8(g)$

Using Hess's law, three times the first equation, plus four times the second equation, plus the reverse of the fourth equation, gives the required equation. Thus,

$$\Delta H_f^\circ(C_3H_8,g) = 3(-393.5) + 4(-285.8) - (-2219.9) = \underline{-103.8\ kJ}$$

```
  H          H H        H H H        H H H H
  |          | |        | | |        | | | |
H-C-H      H-C-C-H    H-C-C-C-H    H-C-C-C-C-H
  |          | |        | | |        | | | |
  H          H H        H H H        H H H H
```

In terms of bond energies, $C_2H_6$ has 1 C-C and 6 C-H bonds, while $C_3H_8$ has 2 C-C and 8 C-H bonds, and $C_4H_{10}$ has 3 C-C and 10 C-H bonds. In going from $C_2H_6$ to $C_3H_8$, there is an additional C-C bond and 2 more C-H bonds, and in going from $C_3H_8$ to $C_4H_{10}$ there is one more C-C bond and 2 more C-H bonds. Thus, to a good approximation we would expect

$\Delta H_f^o(C_4H_{10},g) - \Delta H_f^o(C_3H_8,g) = \Delta H_f^o(C_3H_8,g) - \Delta H_f^o(C_2H_6,g)$, or

$$\Delta H_f^o(C_4H_{10},g) = 2\,\Delta H_f^o(C_3H_8,g) - \Delta H_f^o(C_2H_6,g) = 2(-103.8) - (-84.6)$$
$$= \underline{-123\ kJ}$$

For the combustion of $C_4H_{10}(g)$,

$$C_4H_{10}(g) + \frac{13}{2}O_2(g) \longrightarrow 4CO_2(g) + 5H_2O(l)$$

and Hess's law tells us that the change in $\Delta H^o$(combustion) in going from $C_3H_8(g)$ to $C_4H_{10}(g)$ should be approximately the same as that obtained in going from $C_2H_6(g)$ to $C_3H_8(g)$; i.e., (-2219.8) - (-1559.8) $= \underline{-660.0\ kJ}$

i.e., $\Delta H^o(\text{combustion}, C_4H_{10},g) = (-2219.8) + (-660.0) = \underline{-2879.9\ kJ}$

The same result is obtained using $\Delta H_f^o$ values and the balanced equation for the combustion of $C_4H_{10}(g)$.

$$\Delta H^o = [4\,\Delta H_f^o(CO_2,g) + 5\,\Delta H_f^o(H_2O,l)] - [\Delta H_f^o(C_4H_{10},g) - \frac{13}{2}\Delta H_f^o(O_2,g)]$$
$$= 4(-393.5) + 5(-285.8) - (-123) - (0) = \underline{-2880\ kJ}$$

68. The reaction is exothermic. First calculate the enthalpy change when 1 mol $C_3H_8(g)$ is burned.

$$\Delta H^o = \left(\frac{-46.3\ kJ}{1\ g\ C_3H_8}\right)\left(\frac{44.09\ g\ C3H8}{1\ mol\ C_3H_8}\right) = \underline{-2040\ kJ}$$

i.e., $C_3H_8(g) + 5O_2(g) \longrightarrow 3CO_2(g) + 4H_2O(g);\ \Delta H^o = -2040\ kJ$

$$\Delta H^o = -2040\ kJ = [3\,\Delta H_f^o(CO_2,g) + 4\,\Delta H_f^o(H_2O,g)] - [\Delta H_f^o(C_3H_8,g) + 5\,\Delta H_f^o(O_2,g)]$$
$$= [3(-393.5) + 4(-241.8)] - [\Delta H_f^o(C_3H_8,g) + 5(0)]\ kJ$$
$$\Delta H_f^o(C_3H_8,g) = \underline{-108\ kJ}$$

69. $6CO_2(g) + 6H_2O(l) \longrightarrow C_6H_{12}O_6(s) + 6O_2(g)$

$$\Delta H^o = [\Delta H_f^o(C_6H_{12}O_6,s) + 6\,\Delta H_o^f(O_2,g)] - [6\,\Delta H_o^f(CO_2,g) + 6\,\Delta H_f^o(H_2O,l)]$$
$$= [(-1273) + 6(0)] - [6(-393.5) + 6(-285.8)] = \underline{+2803\ kJ\ mol^{-1}}$$

For 1 gram, $\Delta H^o = \left(\frac{+2803\ kJ}{1\ mol\ glucose}\right)\left(\frac{1\ mol\ glucose}{180.2\ g\ glucose}\right) = \underline{15.55\ kJ\ g^{-1}}$

70. For the combustion of graphite (assuming complete conversion to $CO_2(g)$),

$$C(s) + O_2(g) \longrightarrow CO_2(g);\ \Delta H^o = \Delta H_f^o(CO_2,g) = \underline{-393.5\ kJ\ mol^{-1}}$$

Similarly, for the combustion of methane

$$CH_4(g) + 2O_2(g) \longrightarrow CO_2(g) + 2H_2O(g)$$

$$\Delta H^o = [\Delta H_f^o(CO_2,g) + 2\Delta H_f^o(H_2O,g)] - [\Delta H_f^o(CH_4,g) + 2\Delta H_f^o(O_2,g)]$$
$$= [(-393.5) + 2(-241.8)] - [(-74.5) + 2(0)] = \underline{-802.6 \text{ kJ mol}^{-1}}$$

(a) The heat evolved per gram is

$$\underline{\text{for coke}} \quad \Delta H^o = \left(\frac{-393.5 \text{ kJ}}{1 \text{ mol C}}\right)\left(\frac{1 \text{ mol C}}{12.01 \text{ g C}}\right) = \underline{-32.8 \text{ kJ g}^{-1}}$$

$$\underline{\text{for methane}} \quad \Delta H^o = \left(\frac{-802.6 \text{ kJ}}{1 \text{ mol } CH_4}\right)\left(\frac{1 \text{ mol } CH_4}{16.04 \text{ g } CH_4}\right) = \underline{-50.0 \text{ kJ g}^{-1}}$$

For identical costs, methane is the more economical.

(b) For a doubling of the cost of methane, only (-50.0 kJ)/2 = -25.0 kJ would be produced per unit cost; the coke would be the more economical.

71. The daily intake in kJ is

$$(2500 \text{ k cal})\left(\frac{4.184 \text{ kJ}}{1 \text{ kcal}}\right) = \underline{1.046 \times 10^4 \text{ kJ}}$$

(a) Pounds of cake $= (2.500\text{x}10^3 \text{ kcal})\left(\frac{1 \text{ oz}}{130 \text{ kcal}}\right)\left(\frac{1 \text{ lb}}{16 \text{ oz}}\right) = \underline{1.20 \text{ lb}}$

(b) $\Delta H^o$ for bread is $-73 \text{ kcal oz}^{-1}$; $\Delta H^o$ for butter is $-221 \text{ kcal oz}^{-1}$;

for 1 oz of each, $\Delta H^o = \underline{-294 \text{ kcal}}$.

lbs of bread&butter $= (2.500\text{x}10^3 \text{ kcal})\left(\frac{2 \text{ oz}}{294 \text{ kcal}}\right)\left(\frac{1 \text{ lb}}{16 \text{ oz}}\right) = \underline{1.06 \text{ lb}}$

or 0.53 lb bread, and 0.53 lb butter

72. $\Delta H^o$(combustion) for hexane is the $\Delta H^o$ for the reaction

$$C_6H_{14}(l) + \frac{19}{2} O_2(g) \longrightarrow 6CO_2(g) + 7H_2O(l)$$

$$\Delta H^o = [6\Delta H_f^o(CO_2,g) + 7\Delta H_f^o(H_2O,l)] - [\Delta H_f^o(C_6H_{14},l) + \frac{19}{2}\Delta H_f^o(O_2,g)]$$
$$= [6(-393.5) + 7(-285.8)] - [(-199) + \frac{19}{2}(0)] \text{ kJ} = \underline{-4162.6 \text{ kJ}}$$

Thus, 4162.6 kJ of heat energy comes from the combustion of 1 mol of $C_6H_{14}(l)$. Convert this to heat per liter (of gasoline).

$$\Delta H^o = \left(\frac{4162.6 \text{ kJ}}{1 \text{ mol } C_6H_{14}}\right)\left(\frac{1 \text{ mol } C_6H_{14}}{86.17 \text{ g } C_6H_{14}}\right)\left(\frac{1.38 \text{ g}}{1 \text{ mL}}\right)\left(\frac{1000 \text{ mL}}{1 \text{ L}}\right) = \underline{6.67\text{x}10^4 \text{ kJ L}^{-1}}$$

If you know the local gasoline cost in dollars per liter, you can deduce the kJ of energy that can be purchased for 1 dollar. (If the price is in dollars per gallon, convert to cost per liter using the conversion factor 1 gallon(U.S) = 3.7854 L. For example, if the cost is \$0.43 per liter, $1.55 \times 10^5$ kJ could be purchased for \$1.00.

For electricity, one kilowatt hour corresponds to

$$(1 \text{ kW h})\left(\frac{1000 \text{ W}}{1 \text{ kW}}\right)\left(\frac{1 \text{ J s}^{-1}}{1 \text{ W}}\right)\left(\frac{3600 \text{ s}}{1 \text{ h}}\right) = \underline{3.600 \times 10^3 \text{ kJ}}$$

Dividing this amount of energy by the dollar cost of one kW h of electricity gives kJ of electricity per dollar.

73. This is a Hess's law problem. First double the fourth equation and add the result to the reverse of the second equation

$$2H_2(g) + O_2(g) \longrightarrow 2H_2O(l) \qquad \Delta H^\circ = 2(-285.8) = -571.6\ kJ$$

$$CO(g) \longrightarrow C(graphite,s) + \frac{1}{2}O_2(g) \qquad \Delta H^\circ = -(-110.5) = +110.5\ kJ$$

$$2H_2(g) + \frac{1}{2}O_2(g) + CO(g) \longrightarrow C(s) + 2H_2O(l) \qquad \Delta H^\circ = (-571.6) + 110.5 = \underline{-461.1\ kJ}$$

Add this equation to the reverse of the first equation

$$CO_2(g) + 2H_2O(l) \longrightarrow CH_3OH(l) + \frac{3}{2}O_2(g) \qquad \Delta H^\circ = -(-726.6) = +726.6\ kJ$$

$$2H_2(g) + CO(g) + CO_2(g) \longrightarrow C(s) + CH_3OH(l) + O_2(g)$$

$$\Delta H^\circ = (-461.1) + 726.6 = \underline{265.5\ kJ}$$

and the required equation results when we add this equation to the third equation

$$C(s) + O_2(g) \longrightarrow CO_2(g) \qquad \Delta H^\circ = -393.5\ kJ$$

$$2H_2(g) + CO(g) \longrightarrow CH_3OH(l) \qquad \Delta H^\circ = 265.5 + (-393.5) = \underline{-128.0\ kJ}$$

# CHAPTER 7

1. Frequency, $\nu$, is the number of waves per second, and if the wavelength is $\lambda$, then the distance traveled in one second is $\nu\lambda$, which is the same as the velocity per second, which for all electromagnetic radiation is the speed of light, c. i.e., for all electromagnetic radiation: $\nu\lambda = c$

or $\lambda = \frac{c}{\nu}$

For the FM signal:

$$\lambda = \left(\frac{3.00 \times 10^8 \text{ m s}^{-1}}{94.1 \text{ MHz}}\right)\left(\frac{1 \text{ MHz}}{10^6 \text{s}^{-1}}\right) = 3.19 \text{ m}$$

For the AM signal:

$$\lambda = \left(\frac{3.00 \times 10^8 \text{ m s}^{-1}}{740 \text{ kHz}}\right)\left(\frac{1 \text{ kHz}}{10^3 \text{s}^{-1}}\right) = 4.05 \times 10^2 \text{ m}$$

2. This questions calls for the selection of the appropriate unit conversions factors, using the relationships:

(a) 1 nm = $10^{-9}$m; (b) 1 μm = $10^{-6}$m; (c) 1 Å = $10^{-10}$m

(a) $(435.8 \text{ nm})\left(\frac{10^{-9} \text{m}}{1 \text{ nm}}\right) = 4.358 \times 10^{-7} \text{ m}$

(b) $(435.8 \text{ nm})\left(\frac{10^{-9} \text{ m}}{1 \text{ nm}}\right)\left(\frac{1 \mu\text{m}}{10^{-6}\text{m}}\right) = 4.358 \times 10^{-1} \mu\text{m}$

(c) $(435.8 \text{ nm})\left(\frac{10^{-9} \text{m}}{1 \text{ nm}}\right)\left(\frac{1 \text{ Å}}{10^{-10}\text{m}}\right) = 4.358 \times 10^{3} \text{ Å}$

and $\nu = c/\lambda$

$$= \frac{3.00\times10^8 \text{ m s}^{-1}}{4.378\times10^{-7} \text{ m}}$$

$$= 6.88\times10^{14} \text{ s}^{-1} \text{ (Hz)}$$

3. This problem is similar to Problem 1:

$$\lambda = \frac{c}{\nu} = \left(\frac{3.00 \times 10^8 \text{ m s}^{-1}}{27.3 \text{ MHz}}\right)\left(\frac{1 \text{ MHz}}{10^6 \text{s}^{-1}}\right) = 11.0 \text{ m}$$

4. Again we use the relationship $\lambda\nu = c$, in the form $\nu = c/\lambda$, and employ the conversion factors:

1 cm = $10^{-2}$ m; 1 mm = $10^{-3}$m; 1 μm = $10^{-6}$m; 1 nm = $10^{-9}$m; 1 pm = $10^{-12}$m, and 1 km = $10^3$m, to give:

| Type of Radiation | Wavelength | Frequency (Hz) |
|---|---|---|
| Radio frequency | 1 km to 30 cm | $3.0\times10^5$ to $1.0 \times 10^9$ |
| Microwave | 30 cm to 2 mm | $1.0 \times 10^9$ to $1.5 \times 10^{11}$ |
| Far infra-red | 2 mm to 30 μm | $1.5 \times 10^{11}$ to $1.0 \times 10^{13}$ |
| Near infra-red | 30 μm to 710 nm | $1.0 \times 10^{13}$ to $4.2 \times 10^{14}$ |
| Visible | 710 nm to 400 nm | $4.2 \times 10^{14}$ to $7.5 \times 10^{14}$ |
| Ultraviolet | 400 nm to 4 nm | $7.5 \times 10^{14}$ to $7.5 \times 10^{16}$ |

| | | |
|---|---|---|
| X-rays | 4 nm to 30 pm | $7.5 \times 10^{16}$ to $1.0 \times 10^{19}$ |
| $\gamma$-rays | 30 pm to 0.1 pm | $1.0 \times 10^{19}$ to $3.0 \times 10^{21}$ |

5. $\nu = \frac{c}{\lambda} = \left(\frac{3.00\text{x}10^{8}\ \text{m s}^{-1}}{633\ \text{nm}}\right)\left(\frac{1\ \text{nm}}{10^{-9}\ \text{m}}\right) = \underline{4.74\text{x}10^{14}\ \text{s}^{-1}}$ (Hz)

   This corresponds to light in the red part of the visible spectrum.

6. $\nu = \frac{c}{\lambda} = \left(\frac{3.00\text{x}10^{8}\ \text{m s}^{-1}}{589.2\ \text{nm}}\right)\left(\frac{1\ \text{nm}}{10^{-9}\ \text{m}}\right) = \underline{5.09\text{x}10^{14}\ \text{s}^{-1}}$ (Hz)

   This corresponds to light in the yellow part of the visible spectrum.

7. The energy of one photon is given by $E = h\nu$, where h is Planck's constant = $6.63\text{x}10^{-34}$ J s, and $\nu = c/\lambda$

$$E = \frac{hc}{\lambda} = \frac{(6.63\text{x}10^{-34}\ \text{J s})(3.00\text{x}10^{8}\ \text{m s}^{-1})}{670.8\ \text{nm}}\left(\frac{1\ \text{nm}}{10^{-9}\ \text{m}}\right) = \underline{2.97\text{x}10^{-19}\ \text{J}}$$

1 mol of photons = $6.022\text{x}10^{23}$ photons,

$$E = (2.97\text{x}10^{-19}\ \text{J photon}^{-1})\left(\frac{6.022 \times 10^{23}\ \text{photon}}{1\ \text{mol}}\right)\left(\frac{1\ \text{kJ}}{1000\ \text{J}}\right)$$

$$= \underline{1.79\text{x}10^{2}\ \text{kJ mol}^{-1}}$$

8. Given the energy per mole, calculate the energy per potassium atom, which will be equal to the energy of the photon with just enough energy to ionize one potassium atom; this energy is then expressed as the corresponding maximum wavelength of light to ionize the atom, using $\lambda = c/\nu$ . With practice this can be done in one step, but first do the calculation in two steps.

$$E\ \text{atom}^{-1} = \frac{E\ \text{mol}^{-1}}{N} = \left(\frac{0.42\ \text{MJ}}{1\ \text{mol}}\right)\left(\frac{10^{6}\ \text{J}}{1\ \text{MJ}}\right)\left(\frac{1\ \text{mol}}{6.022\text{x}10^{23}\ \text{atom}}\right)$$

$$= \underline{6.97\text{x}10^{-19}\ \text{J atom}^{-1}}$$

$$E = h\nu\ ;\ \lambda = \frac{hc}{E} = \frac{(6.63\text{x}10^{-34}\ \text{J s})(3.00\text{x}10^{8}\ \text{m s}^{-1})}{6.97\text{x}10^{-19}\ \text{J}}\left(\frac{1\ \text{nm}}{10^{-9}\ \text{m}}\right) = \underline{285\ \text{nm}}$$

9. $\nu = \frac{c}{\lambda}\ ;\ \nu_1 = \left(\frac{2.998\text{x}10^{8}\ \text{m s}^{-1}}{589.6\ \text{nm}}\right)\left(\frac{1\ \text{nm}}{10^{-9}\ \text{m}}\right) = \underline{5.085\text{x}10^{14}\ \text{s}^{-1}}$

$$\nu_2 = \left(\frac{2.998\text{x}10^{8}\ \text{m s}^{-1}}{589.0\ \text{nm}}\right)\left(\frac{1\ \text{nm}}{10^{-9}\ \text{m}}\right) = \underline{5.090 \times 10^{14}\ \text{s}^{-1}}$$

and the energy difference is given by $E = E_2 - E_1 = h(\nu_2 - \nu_1)$, i.e.,

$$E = (6.63\text{x}10^{-34}\ \text{J s})(0.005\text{x}10^{14}\ \text{s}^{-1}) = \underline{3.32\text{x}10^{-22}\ \text{J}}$$

10. (a) For one photon, $E = h\nu = \frac{hc}{\lambda}$

$$= \frac{(6.63\times10^{-34}\ \text{J s})(3.00\times10^{8}\ \text{m s}^{-1})}{546.1\ \text{nm}}\left(\frac{1\ \text{nm}}{10^{-9}\ \text{m}}\right) = \underline{3.64\times10^{-19}\ \text{J}}$$

(b) For one mole of photons,

$$E = \left(\frac{3.64\times10^{-19}\ \text{J}}{1\ \text{photon}}\right)\left(\frac{6.022\times10^{23}\ \text{photon}}{1\ \text{mol}}\right)\left(\frac{1\ \text{kJ}}{10^{3}\ \text{J}}\right) = \underline{219\ \text{kJ}}$$

11. In the photoelectric effect, $E_{photon} = \emptyset + KE$, where $E_{photon}$ is the energy of the incident photon and $\emptyset$ is the energy required to just ionize an electron. $\emptyset$ is calculated from the longest wavelength of light that just ionizes an electron, in this case 520 nm. The difference between the energy of the photons of irradiation (of wavelength 360 nm) and the energy of the maximum wavelength photon equals the kinetic energy of the emitted electrons.

$$KE = E_{photon} - \emptyset, = hc\left(\frac{1}{360\ \text{nm}} - \frac{1}{520\ \text{nm}}\right)$$

$$= (6.63\times10^{-34}\ \text{J s})(3.00\times10^{8}\ \text{m s}^{-1})\left[\left(\frac{520-360}{(520\times360)\ \text{nm}}\right)\left(\frac{1\ \text{nm}}{10^{-9}\ \text{m}}\right)\right]$$

$$E\ \text{mol}^{-1} = (1.70\times10^{-19}\ \text{J electron}^{-1})(6.022\times10^{23}\ \text{electron mol}^{-1})$$

$$= \underline{1.02\times10^{5}\ \text{J mol}^{-1},\ \text{or}\ 102\ \text{kJ mol}^{-1}}$$

12. $$Na(g) + 496\ \text{kJ mol}^{-1} \longrightarrow Na^{+}(g)$$

For one photon, $E = \left(\frac{496\ \text{kJ mol}^{-1}}{N\ \text{mol}^{-1}}\right) = h\nu$

$$\nu = \frac{E}{h} = \left(\frac{496\ \text{kJ mol}^{-1}}{6.022\times10^{23}\ \text{mol}^{-1}}\right)\left(\frac{10^{3}\ \text{J}}{1\ \text{kJ}}\right)\left(\frac{1}{6.63\times10^{-34}\ \text{J s}}\right) = \underline{1.24\times10^{15}\ \text{s}^{-1}}$$

Light of this frequency is in the ultraviolet part of the electromagnetic spectrum, i.e., at lower wavelength than visible light.

From the difference between 600 kJ mol$^{-1}$ and 496 kJ mol$^{-1}$ (i.e. 104 kJ mol$^{-1}$) we have the kinetic energy of 1 mol of emitted electrons (see Problem 11). Thus for 1 electron,

$$KE\ \text{electron}^{-1} = \left(\frac{104\ \text{kJ}}{1\ \text{mol}}\right)\left(\frac{10^{3}\ \text{J}}{1\ \text{kJ}}\right)\left(\frac{1\ \text{mol}}{6.022\times10^{23}\ \text{electron}}\right) = \underline{1.73\times10^{-19}\ \text{J}}$$

and $KE = \frac{1}{2}mv^2$, where m is the electron mass and v is its velocity. From the data on the back cover of the text, 1 J = 1 kg m$^2$ s$^{-2}$, and the electron mass is m = $9.11\times10^{-31}$ kg, hence,

$$KE = (1.73\times10^{-19}\ \text{J})\left(\frac{1\ \text{kg m}^2\ \text{s}^{-2}}{1\ \text{J}}\right) = \frac{1}{2}(9.11\times10^{-31}\ \text{kg})v^2$$

$$v = (0.38\times10^{12}\ \text{m}^2\ \text{s}^{-2})^{1/2} = \underline{6.2\times10^{5}\ \text{m s}^{-1}}$$

13. (a) The kinetic energy of the emitted electrons is $\frac{1}{2}mv^2$, where m is the electron mass of $9.11\times10^{-31}$ kg and v = $6.4\times10^{4}$ m $s^{-1}$. Again (as in Problem 12), use the conversion factor

$$1 \text{ J} = 1 \text{ kg m}^2 \text{ s}^{-2}$$

$$\text{Thus, KE} = \frac{1}{2}mv^2 = \frac{1}{2}(9.11\times10^{-31}\text{ kg})(6.4\times10^{4}\text{ m s}^{-1})^2\left(\frac{1 \text{ J}}{1 \text{ kg m}^2 \text{ s}^{-2}}\right)$$

$$= \underline{1.9\times10^{-21}\text{ J}}$$

(b) For 470 nm photons

$$E = h\nu = \frac{hc}{\lambda} = \frac{(6.63\times10^{-34}\text{ J s})(3.00\times10^{8}\text{ m s}^{-1})}{470.0\text{ nm}}\left(\frac{1\text{ nm}}{10^{-9}\text{ m}}\right)$$

$$= \underline{4.23\times10^{-19}\text{ J photon}^{-1}}$$

(c) The minimum energy needed to remove an electron from potassium metal is the difference between the energy of a 470.0 nm photon and the KE of the emitted electron,

$$= [(4.23\times10^{-19})-(1.9\times10^{-21})]\text{ J} = \underline{4.21\times10^{-19}\text{ J}}$$

14. This problem is similar to Problem 13. The energy of the photon that will just ionize cesium metal is the difference between the energy of the photon of frequency $1.30\times10^{15}$ Hz ($s^{-1}$) and the energy of a photo-electron.

$$E_{photon} = h\nu = (6.63\times10^{-34}\text{ J s})(1.30\times10^{15}\text{ s}^{-1}) = \underline{8.62\times10^{-19}\text{ J}}$$

and the energy of the photon of longest wavelength is

$$(8.62-5.20)\times10^{-19}\text{ J} = \underline{3.42\times10^{-19}}\text{ J}$$

which corresponds to a photon of wavelength given by

$$E = \frac{hc}{\lambda};\ \lambda = \frac{hc}{E} = \frac{(6.63\times10^{-34}\text{ J s})(3.00\times10^{8}\text{ m s}^{-1})}{3.42\times10^{-19}\text{ J}}\left(\frac{10^{9}\text{ nm}}{1\text{ m}}\right)$$

$$= \underline{582\text{ nm}}$$

15. From the lowest frequency, calculate the maximum wavelength of light that will activate the alarm.

$$\lambda = \frac{c}{\nu} = \left(\frac{3.00\times10^{8}\text{ m s}^{-1}}{8.95\times10^{14}\text{ s}^{-1}}\right)\left(\frac{10^{9}\text{ nm}}{1\text{ m}}\right) = \underline{335\text{ nm}}$$

This corresponds to the wavelength of ultraviolet light, so that magnesium would be unsuitable, except in bright sunlight.

16. Calculate the energy of 439 nm photons, per mol of photons.

$$E = \frac{hcN}{\lambda} = \frac{(6.63\times10^{-34}\text{ J s})(3.00\times10^{8}\text{ m s}^{-1})(6.022\times10^{23}\text{ mol}^{-1})}{(439\text{ nm})\left(\frac{1\text{ m}}{10^{9}\text{ nm}}\right)}$$

$$= (2.73\times10^{5}\text{ J})\left(\frac{1\text{ kJ}}{10^{3}\text{ J}}\right) = \underline{273\text{ kJ mol}^{-1}}$$

273 kJ $mol^{-1}$ > 190 kJ $mol^{-1}$, is more than sufficient energy to dissociate 1 mol of $Br_2$ molecules into Br atoms.

For photons of energy 190 kJ $mol^{-1}$, calculate energy per photon and the corresponding wavelength.

$$E' = \frac{E}{N} = h\nu = \frac{hc}{\lambda};$$

$$\lambda = \frac{hcN}{\nu} = \frac{(6.63\times10^{-34}\ J\ s)(3.00\times10^{8}\ m\ s^{-1})(6.022\times10^{23}\ mol^{-1})}{(190\ kJ\ mol^{-1})(\frac{10^3\ J}{1\ kJ})}$$

$$= (6.30\times10^{-7}\ m)(\frac{10^{9}\ nm}{1\ m}) = \underline{630\ nm}$$

17. First calculate the energy per mol of photons for each type of radiation.

For 1 photon, $E_{photon} = h\nu = \frac{hc}{\lambda}$

For 1 mol of photons,

$$E = \frac{(6.63\times10^{-34}\ J\ s)(3.00\times10^{8}\ m\ s^{-1})(6.022\times10^{23}\ mol^{-1})}{\lambda}$$

| Type of radiation | Wavelength | Energy (kJ $mol^{-1}$) |
|---|---|---|
| Radiofrequency | 1 km to 30 cm | $1.20\times10^{-7}$ to $3.99\times10^{-4}$ |
| Microwave | 30 cm to 2 mm | $3.99\times10^{-4}$ to $5.98\times10^{-2}$ |
| Far infrared | 2 mm to 30 μm | $5.98\times10^{-2}$ to 3.99 |
| Near infrared | 30 μm to 710 nm | 3.99 to $1.69\times10^{2}$ |
| Visible | 710 nm to 400 nm | $1.69\times10^{2}$ to $3.00\times10^{2}$ |
| Ultraviolet | 400 nm to 4 nm | $3.00\times10^{2}$ to $3.00\times10^{4}$ |
| X-rays | 4 nm to 30 pm | $3.00\times10^{4}$ to $3.99\times10^{6}$ |
| γ-rays | 30 pm to 0.1 pm | $3.99\times10^{6}$ to $1.20\times10^{9}$ |

For each of the $X_2$(g) molecules in question, the energy required for dissociation is $2\Delta H_f^o(X,g)$.

(a) $2\Delta H_f^o(H,g)$ = 436 kJ $mol^{-1}$ corresponding to ultraviolet light.

(b) $2\Delta H_f^o(O,g)$ = 494 kJ $mol^{-1}$ corresponding to ultraviolet light.

(c) $2\Delta H_f^o(Cl,g)$ = 240 kJ $mol^{-1}$ corresponding to visible light.

(d) $2\Delta H_f^o(F,g)$ = 156 kJ $mol^{-1}$ corresponding to near infrared radiation.

(e) $2\Delta H_f^o(N,g)$ = 942 kJ $mol^{-1}$ corresponding to ultraviolet light.

18. (a) Use the formula developed in Problem 17, $E = \frac{hcN}{\lambda}$, to calculate the wavelength, $\lambda$.

$$\lambda = \frac{(6.63\times10^{-34}\ J\ s)(3.00\times10^{8}\ m\ s^{-1})(6.022\times10^{23}\ mol^{-1})}{(305\ kJ\ mol^{-1})(\frac{1000\ J}{1\ kJ})}(\frac{10^{9}\ nm}{1\ m})$$

$$= \underline{393\ nm}$$

(b) Light of wavelength 393 nm is just beyond violet light, in the ultraviolet spectrum.

(c) Since the energy of light is inversely proportional to its wavelength, it can be deduced without calculation that light of wavelength 320 nm has an energy greater than that of light of wavelength 393 nm, which has an energy of 305 kJ $mol^{-1}$, from part (b), so that 320 nm light will cause the dissociation of $NO_2(g)$ to NO(g) and O(g) at sea level.

19. The energy emitted when an electron moves from an initial $n_i$ level to a final $n_f$ level ($n_i > n_f$) is given by

$$\Delta E = 1312\left(\frac{1}{n_f^2} - \frac{1}{n_i^2}\right) \text{ kJ mol}^{-1}$$

which, substituting $n_i = 6$ and $n_f = 2$, gives

$$\Delta E = 1312\left(\frac{1}{(2)^2} - \frac{1}{(6)^2}\right) \text{ kJ mol}^{-1} = \underline{291.6 \text{ kJ mol}^{-1}}$$

corresponding to the energy of 1 mol of photons.

For 1 photon, $E = \frac{\Delta E}{N} = \frac{hc}{\lambda}$ ; $\lambda = \frac{hcN}{\Delta E}$

$$= \frac{(6.63\text{x}10^{-34} \text{ J s})(3.00\text{x}10^{8} \text{ m s}^{-1})(6.022\text{x}10^{23} \text{ mol}^{-1})}{(291.6 \text{ kJ mol}^{-1})\left(\frac{1000 \text{ J}}{1 \text{ kJ}}\right)}\left(\frac{10^9 \text{ nm}}{1 \text{ m}}\right)$$

$= \underline{411 \text{ nm}}$

which is a wavelength in the violet visible spectrum.

20. The calculation is similar to that in Problem 19.

$\Delta E = \underline{-246.0 \text{ kJ mol}^{-1}}$ (energy absorbed); corresponding to a photon of wavelength $\underline{487 \text{ nm}}$.

21. In the Lyman series of lines of the hydrogen spectrum, photons are emitted when electrons fall from an energy level with n > 1 to the n = 1 energy level. First calculate the energy per mol corresponding to the wavelength of 103 nm, for which

$$\Delta E = \frac{hcN}{\lambda} = \frac{(6.63\text{x}10^{-34} \text{ Js})(3.00\text{x}10^{8} \text{ m s}^{-1})(6.022\text{x}10^{23} \text{ mol}^{-1})}{(103 \text{ nm})\left(\frac{1 \text{ m}}{10^9 \text{ nm}}\right)}\left(\frac{1 \text{ kJ}}{10^3 \text{ J}}\right)$$

$= \underline{1162 \text{ kJ mol}^{-1}}$

Use the formula given in Problem 19, with $n_f = 1$,

$$1162 \text{ kJ mol}^{-1} = (1312 \text{ kJ mol}^{-1})\left(\frac{1}{(1)^2} - \frac{1}{n_i^2}\right)$$

$$\frac{1132}{n_i^2} = (1312-1162); \quad n_i^2 = \frac{1312}{150} \; ; \quad n_i = 3 \text{ (an integer)}$$

The excited electrons are in the $\underline{n = 3}$ energy level

22. (a) n = 1 for the ground state and for ionization n = ∞ Using the formula from Problem 19 gives $\underline{E = -1312 \text{ kJ mol}^{-1}}$. The wavelength of photons of this energy is given by $\lambda = \frac{hcN}{\Delta E}$ and $\underline{\lambda = 91.2 \text{ nm}}$

(b) The calculation is similar to that of part (a), except that the electron transition is from the n = 2 to the n = $\infty$ level:

$\Delta E = -328.0\ \text{kJ mol}^{-1}$ ; $\lambda = 365\ \text{nm}$

23. For the Balmer series the transitions are from higher energy levels to the n = 2 level. The longest wavelength emission will correspond to the minimum energy (wavelength inversely proportional to energy), i.e., to the transition from n = 3 to n = 2, for which

$$\Delta E = 1312\left(\frac{1}{(2)^2} - \frac{1}{(3)^2}\right) = 182.2\ \text{kJ mol}^{-1}$$

$$\text{and } \lambda = \frac{hcN}{\Delta E} = \frac{(6.63\text{x}10^{-34}\ \text{J s})(3.00\text{x}10^{8}\ \text{m s}^{-1})(6.022\text{x}10^{23}\ \text{mol}^{-1})}{(182.2\ \text{kJ mol}^{-1})\left(\frac{1000\ \text{J}}{1\ \text{kJ}}\right)}$$

$$= 6.57 \times 10^{-7}\ \text{m} = \underline{657\ \text{nm}}$$

24. Emission transitions are possible from:

n = 2,3,4, or 5 to n = 1 (Lyman series); 4 transitions (U.V)

n = 3,4,or 5 to n = 2 (Balmer series); 3 transitions (Visible)

n = 4 or 5 to n = 3 (Paschen series); 2 transitions (infrared)

n = 5 to n = 4; 1 transition (infrared)

The total number of possible transitions is 4+3+2+1 = 10

4 of these give spectral lines in the ultraviolet; 3 give spectral lines in the visible spectrum.

25. This problem is similar to Problem 21. For the Paschen series, the transitions are from higher energy levels to the n = 3 energy level.

A wavelength of 1094 nm gives $\Delta E = 109.5\ \text{kJ mol}^{-1}$

$$109.5\ \text{kJ mol}^{-1} = -1312\left(\frac{1}{n^2} - \frac{1}{(3)^2}\right)\ \text{kJ mol}^{-1}\ ;\ \underline{n = 6}$$

26. In problems to do with the de Broglie postulate, $\lambda = \frac{h}{p}$ , where p is the momentum of the particle ( = mass x velocity, mv). In this problem we have to use the electron mass ($9.11 \times 10^{-31}$ kg) and the given velocity:

$$\lambda = \frac{h}{mv} = \frac{(6.63 \times 10^{-34}\ \text{J s})}{(9.11 \times 10^{-31}\ \text{kg})(0.20)(3.00 \times 10^{8}\ \text{m s}^{-1})}\left(\frac{1\ \text{kg m}^2\ \text{s}^{-2}}{1\ \text{J}}\right)$$

$$= 1.21 \times 10^{-11}\ \text{m} = \underline{12.1\ \text{pm}}$$

For an Li atom moving at 20% the velocity of light, the mass of 7.02 u must be converted to SI units of kg, (1 u = $1.661 \times 10^{-27}$ kg):

$$\lambda = \frac{h}{mv} = \frac{(6.63\times10^{-34}\ \text{J s})}{(7.02\ \text{u})(\frac{1.661\times10^{-27}\ \text{kg}}{1\ \text{u}})(0.20)(3.00\times10^{8}\ \text{m s}^{-1})}(\frac{1\ \text{kg m}^2\ \text{s}^{-2}}{1\ \text{J}})$$

$$= 9.48\times10^{-16}\ \text{m} = \underline{9.48\times10^{-4}\ \text{pm}}$$

27. This problem is similar to Problem 26.

$$\lambda = \frac{h}{mv} = \frac{(6.63\times10^{-34}\ \text{J s})}{(9.11\times10^{-31}\ \text{kg})(5.00\times10^{7}\ \text{m s}^{-1})}(\frac{1\ \text{kg m}^2\ \text{s}^{-2}}{1\ \text{J}})$$

$$= 1.46\times10^{-11}\ \text{m} = \underline{14.6\ \text{pm}}$$

28. This problem is also similar to Problem 26.

$$\lambda = \frac{h}{mv} = \frac{(6.63\times10^{-34}\ \text{J s})}{(1.67\times10^{-27}\ \text{kg})(1.00\times10^{2}\ \text{m s}^{-1})}(\frac{1\ \text{kg m}^2\ \text{s}^{-2}}{1\ \text{J}})$$

$$= 3.97\times10^{-9}\ \text{m} = \underline{3.97\ \text{nm}}$$

29. $\lambda = \frac{h}{mv}$ ; $v = \frac{h}{m\lambda}$

$$v = \frac{(6.63\times10^{-34}\ \text{J s})}{(1.67\times10^{-27}\ \text{kg})(3.00\times10^{-10}\ \text{m})}(\frac{1\ \text{kg m}^2\ \text{s}^{-2}}{1\ \text{J}}) = \underline{1.32\times10^{3}\ \text{m s}^{-1}}$$

30. In this problem, convert the velocity in mph to SI units of $\text{m s}^{-1}$, and the mass in oz to kg.

$$(\frac{95\ \text{mile}}{1\ \text{h}})(\frac{1\ \text{h}}{3600\ \text{s}})(\frac{1\ \text{km}}{0.6214\ \text{mile}})(\frac{1000\ \text{m}}{1\ \text{km}}) = \underline{42.5\ \text{m s}^{-1}}$$

$$(5\ \text{oz})(\frac{1\ \text{lb}}{16\ \text{oz}})(\frac{1\ \text{kg}}{2.205\ \text{lb}}) = \underline{0.142\ \text{kg}}$$

Applying the de Broglie relationship, as before,

$$\lambda = \frac{(6.63\times10^{-34}\ \text{J s})}{(0.142\ \text{kg})(42.5\ \text{m s}^{-1})}(\frac{1\ \text{kg m}^2\ \text{s}^{-2}}{1\ \text{J}}) = \underline{1.1\times10^{-34}\ \text{m}}$$

which is a very short wavelength indeed.

31. This problem is similar to Problem 26, but first convert the mass to kg, and calculate the velocity from the kinetic energy.

$$\text{Mass} = (6.65\times10^{-24}\ \text{g})(\frac{1\ \text{kg}}{1000\ \text{g}}) = \underline{6.65\times10^{-27}\ \text{kg}}$$

$$\text{KE} = \frac{1}{2}mv^2 = (4.0\times10^{-12}\ \text{J})(\frac{1\ \text{kg m}^2\ \text{s}^{-2}}{1\ \text{J}}) = 4.0\times10^{-12}\ \text{kg m}^2\ \text{s}^{-2}$$

$$v = (2[\frac{4.0\times10^{-12}\ \text{kg m}^2\ \text{s}^{-2}}{6.65\times10^{-27}\ \text{kg}}])^{1/2} = \underline{3.46\times10^{7}\ \text{m s}^{-1}}$$

$$\lambda = \frac{h}{mv} = \frac{(6.63\times10^{-34}\ \text{J s})}{(6.65\times10^{-27}\ \text{kg})(3.46\times10^{7}\ \text{m s}^{-1})}(\frac{1\ \text{kg m}^2\ \text{s}^{-2}}{1\ \text{J}})$$

i.e., $\lambda = 2.88\text{x}10^{-15}$ m = $\underline{2.88\text{x}10^{-3}\ \text{pm}}$

Expressed as a percentage of the velocity of light, the velocity of the particle is

$$\left(\frac{3.46\text{x}10^{7}\ \text{m s}^{-1}}{3.00\text{x}10^{8}\ \text{m s}^{-1}}\right) \text{x } 100\% = \underline{11.5\%}$$

32. (a) $\lambda = \frac{h}{mv}$ ; $v = \frac{h}{m\lambda}$ ; $m = 9.11\text{x}10^{-31}$ kg.

$$v = \frac{(6.63\text{x}10^{-34}\ \text{J s})}{(9.11\text{x}10^{-31}\ \text{kg})(0.100\ \text{nm})\left(\frac{1\ \text{m}}{10^{9}\ \text{nm}}\right)} \left(\frac{1\ \text{kg m}^{2}\ \text{s}^{-2}}{1\ \text{J}}\right)$$

$$= \underline{7.28\text{x}10^{6}\ \text{m s}^{-1}}$$

(b) The wavelength of 0.100 nm = $(0.100\ \text{nm})\left(\frac{1\ \text{m}}{10^{9}\ \text{nm}}\right)\left(\frac{10^{12}\ \text{pm}}{1\ \text{m}}\right) = \underline{100\ \text{pm}}$ which is in the X-ray region of the electromagnetic spectrum.

33. Ionization energy increases across any period, with increasing core charge, and decreases down any group as the valence shell becomes increasingly more distant from the nucleus. Thus, only (c) is correct.

Li < Si < C < Ne

34. (a) Li (b) Mg (c) Cs (d) P (e) $O^{2-}$

35. The order of increasing first ionization energy is

K < Ca < Cl < $Ca^{2+}$

K has an n = 4 valence shell and a core charge of +1, while Ca has an n = 4 valence shell and a core charge of +2; Both $Ca^{2+}$ and Cl have an n = 3 valence shell, with core charges of +7 and +10, respectively.

36. The noble gas atoms all have core charge +8 but He has an n = 1, Ne has an n = 2, and Ar has an n = 3 valence shell. Thus,

He > Ne > Ar

$Li^{+}$ and $Be^{2+}$ both have a $1s^{2}$ valence shell configuration, as does He, but the core charges are +3, +4, and +2, respectively. Thus

$Be^{2+}$ > $Li^{+}$ > He > Ne > Ar

37. Use the general trends for ionization energies given in Problem 33.

(a) C and Si are both in group 4 and have core charges of +4; C has an n = 2 valence shell, while Si has an n = 3 valence shell.

C > Si

(b) Ca and Mg are both in group 2 with core charges of +2; Ca has an n = 4 valence shell while Mg has an n = 3 valence shell.

Mg > Ca

(c) Be and B are both in period 2; Be has a core charge of +2 and that of B is +3, so that B should have the higher IE. However, the electron configurations are Be $1s^2 2s^2$ and B $1s^2 2s^2 2p^1$ with the extra 2p electron of B at a greater distance from the nucleus than the outer 2s electron of Be.

Be > B

(d) Br is in group 7 (core charge +7), and S is in group 6 (core charge +6), but because Br is in period 4 and S in period 3

S > Br

(e) Ar and $K^+$ are isoelectronic, so the determining factor is their relative core charges of +8 and +9, respectively.

$K^+$ > Ar

38. $K^+$ and $Ca^{2+}$ both have the electron configuration $[Ne]3s^2 3p^6$, but the core charge of $K^+$ is +9, while that of $Ca^{2+}$ is +10. Thus, the ionization energy of $Ca^{2+}$ is greater than that of $K^+$.

39. (a) See the description in the text, page 335--338.

(b) The electron configuration of Ar is $1s^2 2s^2 2p^6 3s^2 3p^6$, with the most easily removed electrons in the outermost 3p sub-shell. The peaks in the photoelectron spectrum of Ar are identified with each of these energy levels. In order of increasing energy

| Peak | -1.52 | -2.82 | -24.1 | -31.5 | -309 | $MJ\ mol^{-1}$ |
|---|---|---|---|---|---|---|
| Energy Level | 3p | 3s | 2p | 2s | 1s | |

The spectrum shows that there is an outermost n = 3 shell containing 3p and 3s energy levels, an inner n = 2 shell containing 2p and 2s subshells, and an innermost n = 1 shell consisting of a single energy level. The order of energy levels is

$$3p < 3s << 2p < 2s << 1s$$

40. A maximum wavelength photon corresponds to the energy to just remove a 2s electron, which is given by

$$E = h\nu = \frac{hc}{\lambda} = \frac{(6.63\text{x}10^{-34}\ J\ s)(3.00\text{x}10^{8}\ m\ s^{-1})}{(3.80\ nm)(\frac{1\ m}{10^9\ nm})} = \underline{5.23\text{x}10^{-17}\ J}$$

The energy of a higher energy photon must be equal to the energy to just remove a 2s electron plus the kinetic energy of the ejected electron,

$$E = (5.23\text{x}10^{-17}\ J) + (1.12\text{x}10^{-16}\ J) = \underline{1.64\text{x}10^{-16}\ J}$$

and the wavelength of a higher energy photon is given by

$$\lambda = \frac{hc}{E} = \frac{(6.63\text{x}10^{-34}\ J\ s)(3.00\text{x}10^{8}\ m\ s^{-1})}{(1.64\text{x}10^{-16}\ J)}\left(\frac{10^9\ nm}{1\ m}\right) = \underline{1.21\ nm}$$

41. First calculate the ionization energy per atom, and the energy of an incident photon.

$$\text{IE per atom} = \left(\frac{2.37\ \text{MJ}}{1\ \text{mol}}\right)\left(\frac{10^6\ \text{J}}{1\ \text{MJ}}\right)\left(\frac{1\ \text{mol}}{6.022\text{x}10^{23}\ \text{atom}}\right) = \underline{3.94\text{x}10^{-18}\ \text{J}}$$

$$E_{\text{photon}} = \frac{hc}{\lambda} = \frac{(6.63\text{x}10^{-34}\ \text{J s})(3.00\text{x}10^{8}\ \text{m s}^{-1})}{(40.0\ \text{nm})\left(\frac{1\ \text{m}}{10^9\ \text{nm}}\right)} = \underline{4.97\text{x}10^{-18}\ \text{J}}$$

The kinetic energy of the photoelectrons produced is the difference between the energy of an incident photon and the IE per atom.

$$\text{KE} = [(4.97-3.94)\ \text{x}\ 10^{-18}\ \text{J}) = \underline{1.03\text{x}10^{-18}\ \text{J}}$$

This is the KE of 1 photoelectron. For 1 mol of photoelectrons,

$$\text{KE} = (1.03\text{x}10^{-18}\ \text{J})(6.022\text{x}10^{23}\ \text{mol}^{-1})\left(\frac{1\ \text{kJ}}{1000\ \text{J}}\right) = \underline{620\ \text{kJ mol}^{-1}}$$

42. Each electron has to have a unique combination of the allowed values of the n, $\ell$, m, and $m_s$ quantum numbers:

| n | $\ell$ | m | $m_s$ | | Number of Electrons ($2n^2$) |
|---|---|---|---|---|---|
| 1 | 0 | 0 | ±1/2 | | 2 |
| 2 | 0 | 0 | ±1/2 | 2 | 8 |
| | 1 | -1,0,+1 | ±1/2 | 6 | |
| 3 | 0 | 0 | ±1/2 | 2 | 18 |
| | 1 | -1,0,+1 | ±1/2 | 6 | |
| | 2 | -2,-1,0,+1,+2 | ±1/2 | 10 | |

43. For a given value of the quantum number n, the $\ell$ quantum number can have any value up to n-1; s,p, and d designated $\ell$= 0, 1, and 2.

(a) n = 6, $\ell$ = 0; possible

(b) n = 1, $\ell$ = 1, <u>not possible</u>

(c) n = 4, $\ell$ = 2, possible

(d) n = 2, $\ell$ = 2, <u>not possible</u>

(e) n = 3, $\ell$ = 1, possible

(f) n = 4, $\ell$ = 2, possible

(b) n = 5, $\ell$ = 1, possible

(h) n = 2, $\ell$ = 0, possible

44. (a) The quantum numbers designate an allowed 2s orbital.

(b) The quantum numbers designate a 1p orbital - <u>not allowed</u>.

(c) The quantum numbers designate an allowed 2p orbital.

(d) The quantum numbers designate an allowed orbital, because $\ell$ is less than n, and the allowed values of m for n = 6 are n = 0,±1,±2,±3,±4, and ±5.

45. A 1s orbital can accommodate a maximum of two electrons, so that the ground state configuration of beryllium is $1s^2\ 2s^2$. A configuration of $1s^4$ would suggest that two electrons with the same spin can occupy the same orbital.

46. (a) allowed
(b) <u>not allowed</u>, because the 3s orbital is restricted to $3s^2$
(c) allowed - although it is not the ground state
(d) <u>not allowed</u>, because the three 3p orbitals are restricted to a maximum of <u>two</u> electrons each, for a total of 6 electrons, $3p^6$

47. s,p, and d are shorthand for $\ell = 0$, 1, and 2, respectively:
(a) $\ell = 0$, (b) $\ell = 1$, (c) $\ell = 2$, (d) $\ell = 0$, (e) $\ell = 2$, and (f) $\ell = 1$

48. (a) <u>not</u> ground state (2s follows 1s); (b) <u>not</u> ground state (2p follows 2s); (c) ground state; (d) <u>not</u> ground state (4s follows 3p).

49. The ground state electron configurations of (a), (b), and (c) are deduced from their positions in the periodic table.
(a) K is in group 1 and period 4; the outermost electron must be $4s^1$ and all the inner orbitals must be filled: $1s^2 2s^2 2p^6 3s^2 3p^6 4s^1$.
(b) Al is in group 3 and period 3; the three valence electrons must be $3s^2 3p^1$; all the inner orbitals must be filled: $1s^2 2s^2 2p^6 3s^2 3p^1$.
(c) Cl is in group 7 and period 3; the seven valence electrons must be $3s^2 3p^5$; all the inner orbitals must be filled: $1s^2 2s^2 2p^6 3s^2 3p^5$.
(d) Ti (Z = 22) must be: $1s^2 2s^2 2p^6 3s^2 3p^6 4s^2 3d^2$, (3d follows 4s).
(e) Zn (Z = 30) must be: $1s^2 2s^2 2p^6 3s^2 3p^6 4s^2 3d^{10}$.
(f) As (Z = 33) must be $1s^2 2s^2 2p^6 3s^2 3p^6 4s^2 3d^{10} 4p^3$

50. (a) <u>Be</u>, group 2, period 2; $1s^2 2s^2$.
(b) <u>N</u>, group 5, period 2; $1s^2 2s^2 2p^3$.
(c) <u>F</u>, group 7, period 2; $1s^2 2s^2 2p^5$.
(d) <u>Mg</u>, group 2, period 3; $1s^2 2s^2 2p^6 3s^2$.
(e) <u>$Cl^+$</u>, one electron less than Cl in group 7 and period 3;
$$1s^2 2s^2 2p^6 3s^2 3p^4$$
(f) <u>$Ne^+$</u>, one electron less than Ne in group 8 and period 2;
$$1s^2 2s^2 2p^5$$
(g) <u>$Al^{3+}$</u>, three electrons less than Al in group 3 and period 3;
$$1s^2 2s^2 2p^6$$

51. In each case the number of electrons in the orbitals with the largest n quantum number gives the group number (number of valence electrons), and this n quantum number corresponds to the period to which the atom belongs.
(a) Group 1, period 2: <u>Li</u> (b) Group 5, period 2: <u>N</u>
(c) Group 2, period 4: <u>Ca</u>
(d) The second transition element in period 4: <u>Ti</u>
(e) Group 5, period 4: <u>As</u> (f) Group 7, period 5: <u>I</u>
(g) Group 2, period 6: <u>Ba</u>

52. In each case, deduce the electron configuration and count up the number of electrons in singly occupied orbitals.

$\underline{O}$ $1s^2 2s^2 2p_x^2 2p_y^1 2p_z^1$ — $\underline{\text{two}}$ unpaired electrons

$\underline{O^-}$ $1s^2 2s^2 2p_x^2 2p_y^2 2p_z^1$ — $\underline{\text{one}}$ unpaired electron

$\underline{O^{2-}}$ $1s^2 2s^2 2p_x^2 2p_y^2 2p_z^2$ — $\underline{\text{no}}$ unpaired electrons

$\underline{S}$ $1s^2 2s^2 2p^6 3s^2 3p_x^2 3p_y^1 3p_z^1$ — $\underline{\text{two}}$ unpaired electrons

$\underline{F}$ $1s^2 2s^2 2p_x^2 2p_y^2 2p_z^1$ — $\underline{\text{one}}$ unpaired electron

$\underline{Ar}$ $1s^2 2s^2 2p^6 3s^2 3p^6$ — $\underline{\text{no}}$ unpaired electrons

$\underline{Fe}$ $[Ar]4s^2 3d_{xy}^2 3d_{xz}^1 3d_{yz}^1 3d_{z^2}^1 3d_{x^2-y^2}^1$ — $\underline{\text{four}}$ unpaired electrons

53. (a) The energy of an electron in a one-electron atom or ion depends only on the value of the principal quantum number n.

(b) In a multi-electron atom or ion the energy depends on the values of both the n and $\ell$ quantum numbers, because of the different shapes of the s, p, and d orbitals.

54. (a) n = 2, and the symbol for $\ell = 1$ is p; $\underline{2p}$

(b) n = 4 and the symbol for $\ell = 0$ is s; $\underline{4s}$

(c) n =3 and the symbol for $\ell = 2$ is d; $\underline{3d}$

(d) n = 2 and the symbol for $\ell = 0$ is s; $\underline{2s}$

55. (a) The main group element with the lowest ionization energy in any period is that with the smallest core charge of +1; i.e., the group 1 alkali metal with an $ns^1$ valence shell.

(b) The main group element with the highest ionization energy in any period is that with the largest possible core charge of +8; i.e., the group 8 noble gas with an $ns^2np^6$ valence shell.

56. (a) $\underline{P}$ (group 5 and period 3) has the ground state valence shell

| 3s | 3p | | |
|---|---|---|---|
| ↑↓ | ↑ | ↑ | ↑ |

(b) $\underline{Ca}$ (group 2 and period 4) has the ground state valence shell

| 4s |
|---|
| ↓↑ |

(c) $\underline{V}$ is the third transition element of the first series in period $\underline{4}$ with the ground state valence shell

| 4s | 3d | | | | |
|---|---|---|---|---|---|
| ↑↓ | ↑ | ↑ | ↑ | | |

(d) O (group 6 and period 2) has the ground state valence shell

2s: [↑↓]  2p: [↑↓|↑|↑]

(e) Br (group 7 and period 4) has the valence shell

4s: [↑↓]  4p: [↑↓|↑↓|↑]

57. Hund's rule states that for orbitals in the same p or d subshell, the ground state configurations have their electrons distributed as far as is possible in separate orbitals with parallel spins. Here, (b) and (e) are not ground states, because

(b) has two electrons in the same 2p orbital but there is still an empty p orbital, and (e) has two pairs of electrons in filled 3p orbitals but has an empty 3p orbital.

The elements are: (a) N, (b) N, (c) N, (d) O, and (e) S.

58. Only the valence shell can contain unpaired electrons in the ground state; only the valence shell needs to be considered.

(a) P is in group 5 and period 3; [Ne] 3s: [↑↓] 3p: [↑|↑|↑]

(b) Si is in group 4 and period 3; [Ne] 3s: [↑↓] 3p: [↑|↑| ]

(c) I is in group 7 and period 5; [Kr] $4d^{10}$ 5s: [↑↓] 5p: [↑↓|↑↓|↑]

(d) Se is in group 6 and period 4; [Ar] $3d^{10}$ 4s: [↑↓] 4p: [↑↓|↑|↑]

(e) Scandium is the first transition element in period 4;

[Ar] 4s: [↑↓] 3d: [↑| | | | ]

Unpaired electrons: (a) 3, (b) 2, (c) 1, (d) 2, (e) 1

59. (a) $1s^2 2s^2$ (Z = 4), ground state of Be

(b) $1s^2 3s^1$ (Z = 3), an excited state of Li

(c) Not possible; there are no 2d orbitals

(d) $[Ne]3s^2 3d^1$ (Z = 13), an excited state of Al

(e) $[Ar]4s^2 3d^2$ (Z = 22), ground state of Ti

(f) $1s^2 2s^2 2p^6 3s^1$ (Z = 11), ground state of Na

(g) $[Ne]3s^2 3d^{12}$, not possible because five 3d orbitals cannot accommodate twelve electrons

(h) $[Ar]3s^2 3p^1$, not possible because within the [Ar] core the 3s and three 3p orbitals are already filled

60. Arsenic is in period 4 and below N and P in group 5, thus its electron configuration is $1s^2 2s^2 2p^6 3s^2 3p^6 4s^2 3d^{10} 4p^3$, with 15 fully occupied orbitals and an electron in each of three 4p orbitals, for a total of 18 occupied orbitals. The electron configurations of N and P are

$$N\ 1s^2 2s^2 2p^3;\quad P\ 1s^2 2s^2 2p^6 3s^2 3p^3$$

61. In general, for a given value of the n quantum number, the total number of nodes (spherical and planar) is n-1, and the number of planar nodes is equal to the value of the quantum number $\ell$.

(a) n = 3

| $\ell$ | spherical nodes | planar nodes | total nodes |
|---|---|---|---|
| 0 | 2 | 0 | 2 |
| 1 | 1 | 1 | 2 |
| 2 | 0 | 2 | 2 |

(b) No. The maximum possible value for $\ell$ is n-1.

(c) Orbitals are designated by their n quantum number, followed by the value of the $\ell$ number, using the symbols: s for $\ell = 0$, p for $\ell = 1$, d for $\ell = 2$, and f for $\ell = 3$.

Hence, n = 3, $\ell = 0$, is designated 3s

n = 3, $\ell = 1$, is designated 3P

62. The number of planar nodes in an atomic orbital is given by the value of its $\ell$ quantum number; s orbitals, 0 planar nodes; p orbitals, 1 planar node; d orbitals, 2 planar nodes, etc.

63. For n = 3, $\ell$ = 0, 1, or 2, so that the orbitals in question are the 3s, the three 3p, and the five 3d orbitals. The 3s orbital is spherical with two spherical nodes; each 3p orbital has a nodal plane dividing it into two equal "lobes", which are intersected by a spherical node. Each 3d orbital is divided into four "lobes" by two planar nodes.

64. (a) For the 1s and 2s orbitals, see Figure 7.24, (a) and (b). The 3s orbital has a similar spherical shape with two spherical nodes.

(b) For the 2p orbital, see Figure 7.24(c). A 3p orbital is similar, except that it has a spherical node in addition to the planar node.

65. As the number of nodes increases (with the increasing value of the quantum number n), the energy increases. An atomic orbital with a principal quantum number n has (n-1) nodes.

66. (a) An atomic orbital is the standing wave that describes the behaviour of an electron in one of the energy levels of an atom.
(b) A molecular orbital is an orbital that extends over all the nuclei in a molecule. It is the standing wave that describes the behaviour of an electron in one of the energy levels of a molecule.
(c) A localized orbital is an orbital that is confined to only one or two nuclei in a molecule. It may be a bonding orbital or a nonbonding (lone pair) orbital.
(d) A bonding orbital is the orbital occupied by the two electrons of a single bond.
(e) A nonbonding orbital is an orbital that is associated with only one nucleus in a molecule. It is also called a lone-pair orbital.

67. When two hydrogen atoms are brought together the nucleus of one atom begins to attract the electron of the other atom. Each H atom has space in its 1s orbital for a second electron, provided the two electrons have opposite spin, and both electrons are drawn towards the region between the two nuclei where the two 1s orbitals overlap. This increases the electron density in the region between the two nuclei and the two nuclei are held together by their attractions for this internuclear electron density, which we describe as an electron pair, or covalent, bond. The two 1s orbitals are said to form a bonding molecular orbital that is occupied by two electrons with opposite spin. In contrast, a helium atom has a filled, $1s^2$, atomic orbital, and when two He atoms are brought together they cannot attract the electrons of each other into their 1s orbitals because these are already filled with two electrons of opposite spin. Therefore, the He atoms repel each other, and formation of an $He_2$ molecule does not occur. In other words, the formation of an $He_2$ molecule would violate the Pauli exclusion principle.

68. The light emitted has an energy of $3.62 \times 10^{-19}$ J photon$^{-1}$, which corresponds to the wavelength

$$\lambda = \frac{hc}{E} = \frac{(6.63 \times 10^{-34}\ \text{J s})(3{,}00 \times 10^{8}\ \text{m s}^{-1})}{(3.62 \times 10^{-19}\ \text{J})}\left(\frac{10^{9}\ \text{nm}}{\text{m}}\right) = \underline{549\ \text{nm}}$$

Light of this wavelength is in the green region of the visible spectrum

69. The energy difference between the 3d and 4s energy levels is given by the difference between the energy of a photon of wavelength 319 nm and a photon of wavelength 395 nm, per atom.

$$E = \frac{hc}{\lambda}\ ;\quad E_{319} - E_{395} = hc\left(\frac{1}{\lambda_{319}} - \frac{1}{\lambda_{395}}\right) = hc\left(\frac{\lambda_{395} - \lambda_{319}}{\lambda_{395}\lambda_{319}}\right)$$

$$= (6.63 \times 10^{-34}\ \text{J s})(3.00 \times 10^{8}\ \text{m s}^{-1})\left[\frac{(395-319)\ \text{nm}}{(395\ \text{nm})(319\ \text{nm})}\right]\left(\frac{1\ \text{nm}}{10^{-9}\ \text{nm}}\right)$$

$$= (1.20 \times 10^{-19}\ \text{J atom}^{-1})(6.022 \times 10^{23}\ \text{atom mol}^{-1})\left(\frac{1\ \text{kJ}}{1000\ \text{J}}\right)$$

$$= \underline{72.3\ \text{kJ mol}^{-1}}$$

The energy of a photon is inversely proportional to its wavelength, so the difference between the 3d energy level and the ground state is greater than the difference between the 4s energy level and the ground state by 72.3 kJ mol-1; $E_{3d} > E_{4s}$.

70. Silicon is in group 4 and period 3, so its ground state electron configuration is

$$1s^2 2s^2 2p^6 3s^2 3p_x^1 3p_y^1$$

with two unpaired electrons. Thus, the expected valence is 2, in contrast to the common valence state of 4 in both $SiH_4$ and $SiCl_4$, which must be formed from the excited state of Si with the electron configuration

$$1s^2 2s^2 2p^6 3s^1 3p_x^1 3p_y^1 3p_z^1$$

71. Boron is in group 3 and period 2 & has the ground state configuration

$$[He]2s^2 2p^1$$

with one unpaired electron, corresponding to a valence of 1. To form molecules such as $BF_3$, boron requires three unpaired electrons, which corresponds to the excited state

$$[He]2s^1 2p_x^1 2p_y^1$$

Boron in this excited state can form three bonds to fluorine to form $BF_3$, and still has an empty $2p_z$ orbital, so that $BF_3$ can accept an electron pair from the fluoride ion, $:\ddot{\underset{..}{F}}:^-$, to form the $BF_4^-$ ion.

72. AgCl(s) is dissociated by 310 kJ $mol^{-1}$ energy, which corresponds to that of a photon whose minimum wavelength is

$$\lambda = \frac{hc}{E_{photon}}$$ , where $E_{photon}$ is the energy of a single photon.

$$E_{photon} = \frac{(310 \text{ kJ mol}^{-1})}{(6.022\text{x}10^{23} \text{ mol}^{-1}}$$

$$\lambda = \frac{(6.63\text{z}10^{-34} \text{ J s})(3.00\text{x}10^{8} \text{ m s}^{-1})(6.022\text{x}10^{23} \text{ mol}^{-1})}{(310 \text{ kJ mol}^{-1})(\frac{1000 \text{ J}}{1 \text{ kJ}})}$$

$$= 3.86\text{x}10^{-7} \text{ m} = \underline{386 \text{ nm}}$$

This corresponds to light in the ultraviolet spectrum, an important constituent of bright sunlight.

73. Ionization energy is the energy required for the reaction

$$A(g) \longrightarrow A(g)^+ + e^-(g)$$

while electron affinity is the energy released in the reaction

$$A(g) + e^- \longrightarrow A^-(g)$$

Thus, provided there is room in its valence shell for another electron, an atom with a high ionization energy (high core charge and small radius) will also have a high electron affinity. Exceptions to this rule are the noble gases, with filled valence shells.

## CHAPTER 8

1. Sulfur is one of the few elements that occur in nature as the element. Its recovery from natural deposits depends on the fact that the melting point of orthorhombic sulfur, its most stable form, is 113 °C, so that it can be melted with superheated steam, with which it does not react. In the Frasch process, a tube made up of three concentric pipes is driven down to the subterranean sulfur deposit. Superheated steam is passed down the outer tube to melt the sulfur, which is then forced up through the middle pipe by compressed air forced down the inner tube. The steam in the outer tube keeps the sulfur molten. Sulfur is also obtained in large amounts by the reaction of $SO_2(g)$, from the smelting of sulfides and other industrial processes, with $H_2S(g)$ from natural gas wells.

$$2H_2S(g) + SO_2(g) \longrightarrow 3S(s) + 2H_2O(l)$$

2.

| Allotrope | Properties |
|---|---|
| Orthorhombic | Brilliant yellow rhombic shaped crystals containing $S_8$ molecules melting at 113°C. Insoluble in water but soluble in organic solvents such as carbon disulfide. It is the most stable form of sulfur. |
| Monoclinic | Yellow long needle-shaped crystals containing $S_8$ molecules. Stable above 119°C and boiling at 445°C. It slowly reverts to the orthorhomic form on standing or on recrystallization from carbon disulfide. It is made by cooling molten sulfur. |
| Plastic | Brown rubbery elastic solid made by pouring molten sulfur into cold water. It consists of long polymer chains of sulfur atoms and slowly reverts to the stable orthorhombic form on standing. |

When orthorhombic sulfur is heated, it melts at 113 °C. On further heating, the liquid becomes darker in color and its viscosity increases as long polymeric chains of S atoms are formed. Eventually it becomes very mobile again before boiling at 445 °C. (See text).

3. In its S(II) ground state, sulfur has a valence of 2; the 4 and 6 valence states correspond to excited states in which one and two electrons, respectively, are promoted from (i) a filled 3p orbital to a 3d orbital, and (ii) from a filled 3s orbital and a filled 3p orbital to two 3d orbitals.

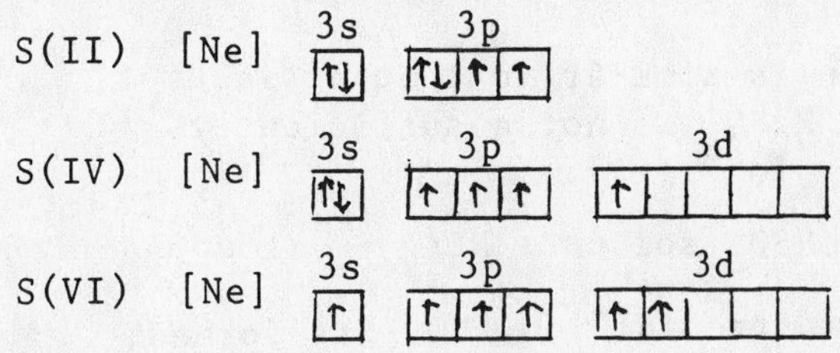


Sulfur cannot have a valence of 8 because its valence shell contains only 6 electrons; the only other possibility would entail the promotion of an electron from a 2p orbital to a 3d orbital, which is energetically impossible in a chemical reaction.

The ground state of oxygen is $[He]2s^2 2p_x^{\,2} 2p_y^{\,1} 2p_z^{\,1}$, corresponding to a valence of 2. Valences of 4 and 6 requires promotion of electrons to higher energy orbitals but since oxygen is in period 2, its valence shell is limited to a maximum of eight electrons (octet rule) and there are no 2d orbitals. Thus, no compounds are known in which oxygen has valences of 4 or 6.

4. (a) The industrial process for the manufacture of sulfuric acid is called the contact process because its first step involves the reaction of $SO_2(g)$ with $O_2(g)$ when they come into contact on the surface of a vanadium pentoxide catalyst to form $SO_3(g)$.

(b) The enthalpy of solution of sulfuric acid is very large and exothermic, so that when water is added to sulfuric acida large amount of heat is evolved, to the extent that it is sufficient to boil the water that comes into contact with the sulfuric acid. This can result in drops of very corrosive hot sulfuric acid being thrown violently out of the mixing vessel, which is dangerous.

5. Sulfuric acid has a very high boiling point (> 340°C, decomposes), which makes it suitable for the preparation of more volatile acids, such as HCl (b.p. -65°C) and $HNO_3$ (b.p. 86°C):

$$NaCl(s) + H_2SO_4(l) \longrightarrow NaHSO_4(sol'n) + HCl(g)$$

$$NaNO_3(s) + H_2SO_4(l) \longrightarrow NaHSO_4(sol'n) + HNO_3(g)$$

However, since sulfuric acid is also an oxidizing agent, it is not suitable for preparing acids that are readily oxidized, such as HBr or HI, which are oxidized to the corresponding halogen.

$$2NaX(s) + 5H_2SO_4(l) \longrightarrow X_2 + SO_2(g) + 2H_3O^+ + 2Na^+ + 4HSO_4^-$$

6. (a) $H_2SO_4 + NaOH \longrightarrow NaHSO_4 + H_2O$

$H_2SO_4 + 2NaOH \longrightarrow Na_2SO_4 + 2H_2O$

These acid-base reaction occur at room temperature with both concentrated and dilute acid, although the first reaction dominates in concentrated acid, and the second in dilute acid.

(b) $BaCl_2(aq) + H_2SO_4(aq) \longrightarrow BaSO_4(s) + 2HCl(aq)$

This reaction occurs readily at room temperature to give a precipitate of insoluble $BaSO_4(s)$. When concentrated $H_2SO_4$ is used the $BaSO_4(s)$ dissolves as $Ba(HSO_4)_2$ and HCl(g) is evolved.

(c) With concentrated acid, the reaction is similar to that of salts of other volatile acids. (Note that $H_2SO_4$ is not a sufficiently strong oxidizing agent to oxidize $F^-$ to $F_2(g)$).

$$NaF(s) + H_2SO_4(l) \longrightarrow NaHSO_4(sol'n) + HF(g) \quad \text{(slow reaction)}$$

(d) With concentrated acid in the cold, nitronium ion, $NO_2^+$ is formed.

$$NaNO_3 + 3H_2SO_4 \longrightarrow Na^+ + NO_2^+ + H_3O^+ + 3HSO_4^-$$

On heating, volatile nitric acid (b.p. 86°C) is driven off.

$$NaNO_3 + H_2SO_4 \longrightarrow NaHSO_4 + HNO_3(g)$$

(e) With concentrated sulfuric acid, on heating, carbon is oxidized to $CO_2(g)$.

$$C(s) + 4H_2SO_4 \longrightarrow CO_2(g) + 2SO_2(g) + 2H_3O^+ + 2HSO_4^-$$

7. (a) Reactions in which $H_2SO_4$ reacts as an oxidizing agent are those in which some substance is oxidized and $H_2SO_4$ is reduced to $SO_2(g)$. Examples in the text include:

$$C(s) + 4H_2SO_4 \longrightarrow CO_2(g) + 2SO_2(g) + 2H_3O^+ + 2HSO_4^-$$
$$2Br^- + 5H_2SO_4 \longrightarrow Br_2 + SO_2 + 2H_3O^+ + 4HSO_4^-$$
$$2I^- + 5H_2SO_4 \longrightarrow I_2 + SO_2 + 2H_3O^+ + 4HSO_4^-$$
$$Cu + 5H_2SO_4 \longrightarrow Cu^{2+} + SO_2 + 2H_3O^+ + 4HSO_4^-$$
$$2Ag + 5H_2SO_4 \longrightarrow 2Ag^+ + SO_2 + 2H_3O^+ + 4HSO_4^-$$

Note that in each of these reactions the oxidizing agent is molecular $H_2SO_4$, which only exists in concentrated acid.

(b) Sulfuric acid is a diprotic acid, so that when it behaves as an acid it donates either one or two protons to a base, to give either $HSO_4^-$ ion, or $SO_4^{2-}$ ion, depending on the conditions. For example, with a limited amount of water

$$H_2O + H_2SO_4 \rightarrow H_3O^+ + HSO_4^-$$

but with excess water

$$HSO_4^- + H_2O \rightleftharpoons H_3O^+ + SO_4^{2-}$$ (which is substantially complete).

Other common bases with which sulfuric acid reacts include the alkali metal or alkaline earth metal oxides and hydroxides and weak bases such as ammonia, $NH_3$, or the anions of weak acids:

| | |
|---|---|
| $NaOH + 2H_2SO_4 \longrightarrow Na^+ + H_3O^+ + 2HSO_4^-$ | concentrated acid |
| $Ca(OH)_2 + 2H_2SO_4 \longrightarrow Ca^{2+} + 2H_3O^+ + 4HSO_4^-$ | concentrated acid |
| $NH_3 + H_2SO_4 \longrightarrow NH_4^+ + HSO_4^-$ | concentrated acid |
| $F^- + H_2SO_4 \longrightarrow HF(g) + HSO_4^-$ | concentrated acid |
| $Cl^- + H_2SO_4 \longrightarrow HCl(g) + HSO_4^-$ | concentrated acid |
| $2NH_3 + H_2SO_4 \longrightarrow 2NH_4^+ + SO_4^{2-}$ | dilute acid |
| $2NaOH + H_2SO_4 \longrightarrow 2Na^+ + SO_4^{2-} + 2H_2O$ | dilute acid |
| $Ca(OH)_2 + H_2SO_4 \longrightarrow Ca^{2+} + SO_4^{2-} + 2H_2O$ | dilute acid |

(c) As a dehydrating agent, concentrated sulfuric acid removes the elements of water from substances, usually on heating, and this water reacts with the sulfuric acid to form $H_3O^+$ and $HSO_4^-$. Examples from the text include:

$$CuSO_4.5H_2O \longrightarrow CuSO_4 + 5H_2O$$
$$C_{12}H_{22}O_{11} \longrightarrow 12C(s) + 11H_2O$$
$$HCO_2H \longrightarrow CO(g) + H_2O$$

8. The Lewis structures for all of these species are given in the text. To deduce the structure of all but the last, $S_2O_3{}^{2-}$, thiosulfate ion, recollect that $:\ddot{O}\cdot$ forms two bonds, and $\cdot\ddot{O}:{}^-$ and $\cdot\ddot{O}$-H each form one bond. Thus, $SO_2$ and $SO_3{}^{2-}$ have four bonds to a central S atom, so that each of these species contains S(IV), with one unshared (lone) pair of electrons in its valence shell, as we saw in Problem 3. While in $SO_3$, $H_2SO_4$, and $SO_4{}^{2-}$, there are six bonds to a central S atom, so that each of these species contains S(VI) with no lone pairs on the S atom. $S_2O_3{}^{2-}$ also contains S(VI) as the central atom and is related to $SO_4{}^{2-}$ in that one of the singly bonded $-\ddot{O}:{}^-$ atoms is replaced by $-\ddot{S}:{}^-$, in which sulfur is in the S(II) valence state. Thus,

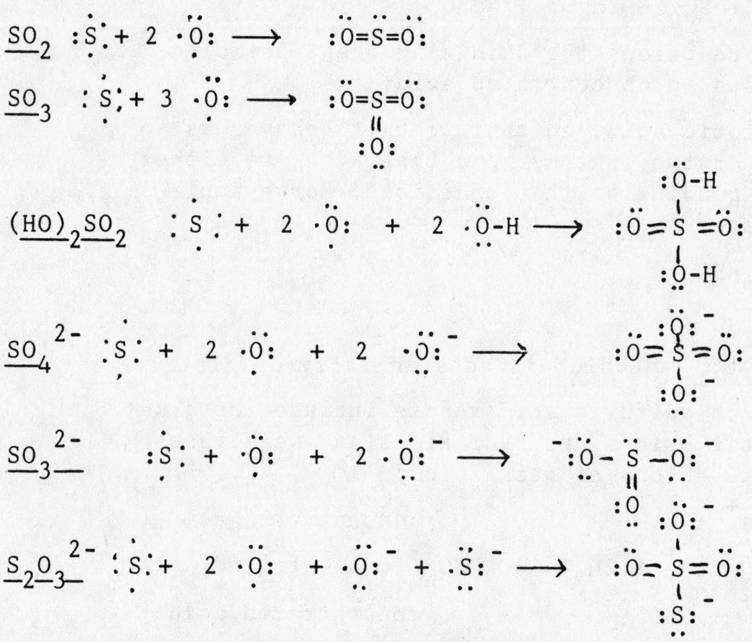

$\underline{SO_2}$ $:\dot{\ddot{S}}\cdot + 2 \cdot\ddot{O}: \longrightarrow :\ddot{O}=\ddot{S}=\ddot{O}:$

$\underline{SO_3}$ $:\dot{S}\cdot + 3 \cdot\ddot{O}: \longrightarrow :\ddot{O}=S=\ddot{O}:$
$$\overset{\|}{\underset{\ddot{O}:}{}}$$

$\underline{(HO)_2SO_2}$ $:\dot{S}\cdot + 2 \cdot\ddot{O}: + 2 \cdot\ddot{O}\text{-H} \longrightarrow$
$$\begin{array}{c}:\ddot{O}\text{-H}\\ | \\ :\ddot{O}=\overset{}{S}=\ddot{O}:\\ | \\ :\ddot{O}\text{-H}\end{array}$$

$\underline{SO_4{}^{2-}}$ $:\dot{S}\cdot + 2 \cdot\ddot{O}: + 2 \cdot\ddot{O}:{}^- \longrightarrow$
$$\begin{array}{c}:\ddot{O}:{}^-\\ | \\ :\ddot{O}=S=\ddot{O}:\\ | \\ :\ddot{O}:{}^-\end{array}$$

$\underline{SO_3{}^{2-}}$ $:\dot{S}\cdot + \cdot\ddot{O}: + 2 \cdot\ddot{O}:{}^- \longrightarrow$
$$\begin{array}{c}{}^-:\ddot{O}-\overset{}{S}-\ddot{O}:{}^-\\ \overset{\|}{\ddot{O}}\end{array}$$

$\underline{S_2O_3{}^{2-}}$ $:\dot{S}\cdot + 2 \cdot\ddot{O}: + \cdot\ddot{O}:{}^- + \cdot\ddot{S}:{}^- \longrightarrow$
$$\begin{array}{c}:\ddot{O}:{}^-\\ | \\ :\ddot{O}=S=\ddot{O}:\\ | \\ :\ddot{S}:{}^-\end{array}$$

| Molecule | $\underline{AX_nE_m}$ Structure | Shape |
|---|---|---|
| $SO_2$ | $AX_2E$ | angular |
| $SO_3$ | $AX_3$ | trigonal planar |
| $H_2SO_4$ | $AX_4$ | tetrahedral |
| $SO_4{}^{2-}$ | $AX_4$ | tetrahedral |
| $SO_3{}^{2-}$ | $AX_3E$ | trigonal pyramidal |
| $S_2O_3{}^{2-}$ | $AX_4$ | tetrahedral |

9. (a) $Zn(s) + H_2SO_4(aq) \longrightarrow Zn^{2+}(aq) + SO_4{}^{2-}(aq) + H_2(g)$

(b) $2NaI(s) + 5H_2SO_4(conc) \longrightarrow I_2(s) + SO_2(g) + 2Na^+(soln)$
$+ 2H_3O^+(soln) + 4HSO_4{}^-(soln)$

(c) $2Ag(s) + 5H_2SO_4(conc) \longrightarrow 2Ag^+(soln) + SO_2(g) + 2H_3O^+(soln)$
$+ 4HSO_4{}^-(soln)$

(d) $Mg(OH)_2(s) + H_2SO_4(aq) \longrightarrow Mg^{2+}(aq) + SO_4{}^{2-}(aq) + 2H_2O(l)$

10. By bubbling $SO_2(g)$ into NaOH(aq), a solution of $Na_2SO_3(aq)$ may be prepared

$$SO_2(g) + 2NaOH(aq) \longrightarrow Na_2SO_3(aq) + H_2O(l)$$

and when this solution is boiled with S(s) the thiosulfate results

$$SO_3^{2-}(aq) + S(s) \longrightarrow S_2O_3^{2-}(aq)$$

11. From the volume of $SO_2$ produced on combustion, calculate mol $SO_2$.

$$\text{mol } SO_2 = \frac{PV}{RT} = \frac{(750 \text{ mm Hg})(\frac{1 \text{ atm}}{760 \text{ mm Hg}})(173.6 \text{ mL})(\frac{1 \text{ L}}{1000 \text{ mL}})}{(0.0821 \text{ atm L mol}^{-1} \text{ K}^{-1})(298 \text{ K})}$$

$$= \underline{7.00\times10^{-3} \text{ mol}}$$

Thus, the mass of sulfur in 0.4203 g of the Fe/S compound is

$$(7.00\times10^{-3} \text{ mol } SO_2)(\frac{1 \text{ mol S}}{1 \text{ mol } SO_2})(\frac{32.06 \text{ g S}}{1 \text{ mol S}}) = \underline{0.224 \text{ g S}}$$

Therefore, mass of Fe = (0.4203-0.224) = <u>0.1963 g Fe</u>

$$\text{and mol Fe in sample} = (0.1963 \text{ g Fe})(\frac{1 \text{ mol Fe}}{55.85 \text{ g Fe}}) = \underline{3.51\times10^{-3} \text{ mol}}$$

Mol Fe : mol S = $3.51\times10^{-3}$ mol : $7.00\times10^{-3}$ mol S <u>or</u> 1.00 : 1.99

which is the same as the ratio of atoms, so that

the <u>empirical formula</u> is <u>$FeS_2$</u>

Since the compound is formed between a metal and a nonmetal, it is most probably an <u>ionic</u> compound.

$$Fe^{2+} \; {}^{-}:\ddot{\underset{..}{S}}-\ddot{\underset{..}{S}}:^{-}$$

12. The characteristic test for $SO_4^{2-}(aq)$ ion is to add $BaCl_2(aq)$ to a solution acidified with HCl(aq). A white precipitate of $BaSO_4(s)$ indicates the presence of $SO_4^{2-}$(ion). ($BaSO_3$, like all sulfites, is soluble).

$$BaCl_2(aq) + SO_4^{2-}(aq) \longrightarrow BaSO_4(s) + 2Cl^-(aq)$$

A number of tests could be used to detect the presence of $SO_3^{2-}(aq)$ ion, which contains S(IV) and is a good reducing agent since it is readily oxidized to $SO_4^{2-}$ containing S(VI). Addition of dilute acid liberates $SO_2(g)$, because $SO_3^{2-}$ is a weak base,

$$SO_3^{2-}(aq) + 2H_3O^+(aq) \longrightarrow SO_2(g) + 3H_2O(l)$$

$SO_2(g)$ is readily detected by its characteristic odor and its ability to bleach moistened litmus paper. $SO_3^{2-}(aq)$ readily oxidizes $Br_2(aq)$, for example, to $Br^-(aq)$.

$$Br_2(aq) + SO_3^{2-}(aq) + 3H_2O(l) \longrightarrow 2Br^-(aq) + SO_4^{2-}(aq) + 2H_3O^+(aq)$$

13. $$SO_2(g) + 2H_2S(g) \longrightarrow 3S(s) + 2H_2O(g)$$

First calculate the moles of sulfur in $10^6$ g of coal containing 4.00 mass % S.

$$\text{Mol S} = (10^6 \text{ g coal})(\frac{4.00 \text{ g S}}{100 \text{ g coal}})(\frac{1 \text{ mol S}}{32.06 \text{ g S}}) = \underline{1.25\times10^3 \text{ mol S}}$$

and since the balanced equation for the combustion of S is

$$S(s) + O_2(g) \longrightarrow SO_2(g)$$

the mol $SO_2(g)$ from combustion of $10^6$ g coal is also <u>$1.25\times10^3$ mol</u>, and this amount of $SO_2$ reacts with

$$(1.25\times10^3 \text{ mol } SO_2)\left(\frac{2 \text{ mol } H_2S}{1 \text{ mol } SO_2}\right) = \underline{2.50\times10^3 \text{ mol}}\ H_2S$$

Finally, calculate the volume of this $H_2S$ at 25°C and 750 mm Hg.

$$V = \frac{nRT}{P} = \frac{(2.50\times10^3 \text{ mol})(0.0821 \text{ atm L mol}^{-1} \text{ K}^{-1})(298 \text{ K})}{(750 \text{ mm Hg})\left(\frac{1 \text{ atm}}{760 \text{ mm Hg}}\right)}$$

$$= \underline{6.20\times10^4 \text{ L } H_2S}$$

From the balanced equation for the reaction of $SO_2$ with $H_2S$,

$$\text{Mass of S} = (1.25\times10^3 \text{ mol } SO_2)\left(\frac{3 \text{ mol S}}{1 \text{ mol } SO_2}\right)\left(\frac{32.06 \text{ g S}}{1 \text{ mol S}}\right)\left(\frac{1 \text{ kg}}{1000 \text{ kg}}\right)$$

$$= \underline{1.20\times10^2 \text{ kg of S}}$$

14. $H_2S$ is oxidized to S(s) and $Cl_2(aq)$ must be reduced to $Cl^-(aq)$,

$$Cl_2(aq) + H_2S(aq) \longrightarrow S(s) + 2HCl(aq)$$

The $H_2S$ content of 5000 L of spring water is

$$(5000 \text{ L water})\left(\frac{1000 \text{ g water}}{1 \text{ L water}}\right)\left(\frac{22 \text{ g } H_2S}{10^6 \text{ g } H_2O}\right) = \underline{110 \text{ g } H_2S}$$

Convert this mass of $H_2S$ to mol $H_2S$, then to mol $Cl_2$, and finally to grams of $Cl_2$.

$$(110 \text{ g } H_2S)\left(\frac{1 \text{ mol } H_2S}{34.08 \text{ g } H_2S}\right)\left(\frac{1 \text{ mol } Cl_2}{1 \text{ mol } H_2S}\right)\left(\frac{70.90 \text{ g } Cl_2}{1 \text{ mol } Cl_2}\right) = \underline{229 \text{ g } Cl_2}$$

15. First calculate the empirical formula of the compound.

| | | S | Cl | O |
|---|---|---|---|---|
| Grams in 100 g | = | 23.7 | 52.6 | 23.7 |
| Mol in 100 g | = | $\frac{23.7}{32.06}$ | $\frac{52.6}{35.45}$ | $\frac{23.7}{16.00}$ |
| Ratio of mol (atoms) | = | 0.739 : | 1.48 : | 1.48 |
| | = | $\frac{0.739}{0.739}$ : | $\frac{1.48}{0.739}$ : | $\frac{1.48}{0.739}$ |
| | = | 1.00 : | 2.01 : | 2.01 |

The <u>empirical formula</u> is <u>$SCl_2O_2$</u> (formula mass 135.0 u)

From the gas data, calculate mol in 0.337 g, and hence the molar mass.

$$\text{Mol of gas} = \frac{PV}{RT} = \frac{(770 \text{ mm Hg})\left(\frac{1 \text{ atm}}{760 \text{ mm Hg}}\right)(75.6 \text{ mL})\left(\frac{1 \text{ L}}{1000 \text{ mL}}\right)}{(0.0821 \text{ atm L mol}^{-1} \text{ K}^{-1})(373 \text{ K})}$$

$$= \underline{2.50\times10^{-3} \text{ mol}}$$

$$\text{Molar mass} = \left(\frac{0.337 \text{ g}}{2.50 \times 10^{-3} \text{ mol}}\right) = \underline{135 \text{ g mol}^{-1}}$$

Thus, the molecular formula is also $\underline{SCl_2O_2}$

The relatively low boiling point of $69^\circ C$ suggests that the compound is a $\underline{\text{covalent}}$ substance. The oxidation state of sulfur in the compound is $S(VI)$, suggesting the $\underline{\text{Lewis structure}}$

$$\begin{array}{c} \ddot{O}: \\ \| \\ :\ddot{C}l-S-\ddot{C}l: \\ \| \\ \ddot{O}: \end{array} \qquad AX_4, \text{ tetrahedral}$$

16. (a) $H_2S$ is prepared in the laboratory by the reaction of a metal sulfide with a dilute aqueous acid, such as $HCl(aq)$.

e.g., $\quad FeS(s) + 2HCl(aq) \longrightarrow H_2S(g) + FeCl_2(aq)$

(b) $2H_2S(g) + 3O_2(g) \longrightarrow 2SO_2(g) + 2H_2O(g) \qquad$ (in air)

$2H_2S(g) + O_2(g) \longrightarrow 2S(s) + 2H_2O(g) \qquad$ (limited air)

(c) $H_2S(aq) + H_2O(l) \rightleftharpoons H_3O^+(aq) + HS^-(aq)$

$HS^-(aq) + H_2O(l) \rightleftharpoons H_3O^+(aq) + S^{2-}(aq)$

(d) When $H_2S$ behaves as a reducing agent, the sulfur must be oxidized to an oxidation state higher than -2. Examples include:

$H_2S(aq) + Cl_2(aq) \longrightarrow 2HCl(aq) + S(s)$

$2H_2S(g) + 3O_2(g) \longrightarrow 2H_2O(l) + 2SO_2(g)$

$H_2S(g) + O_2(g) \longrightarrow 2S(s) + H_2O(l)$

17. (a) potassium sulfate, $K_2SO_4$

(b) calcium hydrogen sulfate, $Ca(HSO_4)_2$

(c) calcium sulfite, $CaSO_3$

(d) potassium tetrasulfide, $K_2S_4$

(e) sodium disulfate, $Na_2S_2O_7$

(f) aluminum sulfate, $Al_2(SO_4)_3$

18. See Problem 7, parts (a), (b), and (c), and Problem 12.

19. The $\underline{\text{ionic}}$ compounds are those formed between a metal and a nonmetal; the $\underline{\text{covalent}}$ compounds are those formed between two nonmetals.

(a) $\underline{\text{ionic}}$  $[Na^+]_2 \ :\ddot{S}:^{2-}$

(b) $\underline{\text{ionic}}$  $[Na^+]_2 \ ^-:\ddot{S}-\ddot{S}:^-$

(c) $\underline{\text{ionic}}$  $Mg^{2+} \ :\ddot{S}:^{2-}$

(d) $\underline{\text{covalent}}$  $\begin{array}{c} :\ddot{S} \quad \ddot{S}: \quad \ddot{S}: \\ \quad \ddot{S} \quad \ddot{S} \quad S \\ \quad :\ddot{S} \quad \ddot{S}: \end{array}$

(e) $\underline{\text{covalent}}$  $:\ddot{S}=C=\ddot{S}:$

(f) $\underline{\text{covalent}}$  $:\ddot{O}=\ddot{S}=\ddot{O}:$

(g) $\underline{\text{covalent}}$  $\begin{array}{c} :\ddot{O}=S=\ddot{O}: \\ \| \\ \ddot{O}: \end{array}$

(h) $\underline{\text{covalent}}$  $:\ddot{C}l-\ddot{S}-\ddot{C}l:$

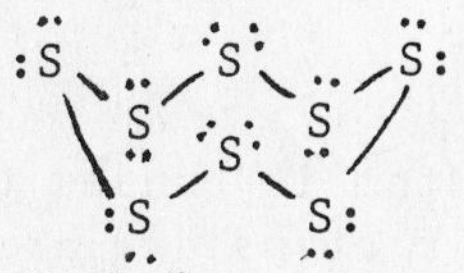

Species with double bonds: $CS_2$, $SO_2$, $SO_3$

Species with polar bonds: $CS_2$, $SO_2$, $SO_3$, $SCl_2$

Names: (a) sodium sulfide (b) sodium disulfide (c) magnesium sulfide
(d) orthorhombic or monoclinic sulfur (e) carbon disulfide
(f) sulfur dioxide (e) sulfur trioxide (h) sulfur dichloride

20. (a) White phosphorus consists of $P_4$ molecules and is a molecular solid. To melt it, energy has to be supplied to overcome only rather weak intermolecular forces to allow $P_4$ molecules to move relative to each other. In contrast, black phosphorus is an infinite covalent network solid composed of infinite two-dimensional buckled sheets of phosphorus atoms. To melt black phosphorus, many strong P-P have to be broken.

(b) The relatively small $P_4$ molecules in white phosphorus go into solution quite readily in the covalent carbon disulfide solvent. In contrast, red phosphorus, which has a structure very similar to that of black phosphorus, is insoluble in carbon disulfide. In order for it to be dissolved many P-P bonds would have to be broken.

21. $4PH_3(g) + 8O_2(g) \longrightarrow P_4O_{10}(s) + 6H_2O(g)$

22. Phosphate rock is calcium phosphate, $Ca_3(PO_4)_3(s)$.

(a) $CaSO_4.2H_2O(s)$ is insoluble. When phosphate rock is treated with aqueous sulfuric acid of an appropriate concentration the reaction

$$Ca_3(PO_4)_2(s) + 3H_2SO_4(aq) + 6H_2O \longrightarrow 2H_3PO_4(aq) + 3CaSO_4.2H_2O(s)$$

occurs, and the $CaSO_4.2H_2O(s)$ is then filtered off.

(b) Sodium phosphate, $Na_3PO_4$, could be obtained by neutralizing the phosphoric acid formed from phosphate rock according to part (a), with sodium hydroxide, according to the equation

$$H_3PO_4(aq) + 3NaOH(aq) \longrightarrow Na_3PO_4(aq) + 3H_2O(l)$$

(c) Reaction of $Ca_3(PO_4)_3(l)$ with silica, $SiO_2(l)$, in an electric furnace gives $P_4O_{10}(g)$, which is then reduced with carbon.

$$2Ca_3(PO_4)_2(l) + 6SiO_2(l) \longrightarrow P_4O_{10}(g) + 6CaSiO_3(l)$$

$$P_4O_{10}(g) + 10C(s) \longrightarrow P_4(g) + 10CO\ (g)$$

(d) Superphosphate-of-lime fertilizer is a mixture of $CaSO_4.2H_2O$, gypsum, and hydrated calcium dihydrogen phosphate, $Ca(H_2PO_4)_2.H_2O$ and results from the reaction of phosphate rock with sulfuric acid and an appropriate amount of water, according to the equation

$$Ca_3(PO_4)_2(s) + 2H_2SO_4(l) + 5H_2O(l) \longrightarrow Ca(H_2PO_4)_2.H_2O(s) + 2CaSO_4.2H_2O(s)$$

23. In $P_4$ each P atom is bonded to three other P atoms by P-P bonds; in $P_4O_6$ the four P atoms are arranged at the corners of a tetrahedron, as they are in $P_4$, but the six O atoms form P-O-P bridges between each pair of P atoms (bridging each of the six edges of the $P_4$ tetrahedron,

so that each P atom forms three P-O bonds. Presumably, the three S atoms in $P_4S_3$ can form P-S-P bridge bonds, thereby bonding one P atom to the other three P atoms, that remain joined by P-P bonds as they are in $P_4$.

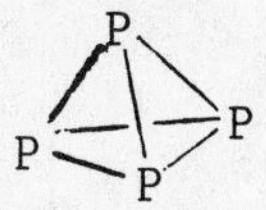

$P_4$

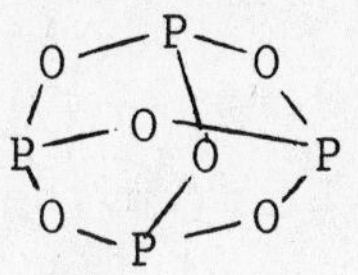

$P_4O_6$

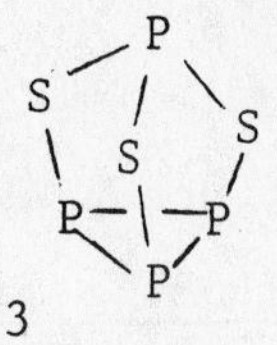

$P_4S_3$

24. $PF_5$ and $PCl_5$ are formed from the P(V) excited state of phosphorus, in which phosphorus has the electron configuration

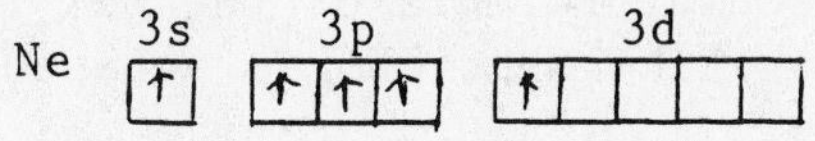


Energy is required to form this valence state from the ground state, and this is possible provided that strong bonds are formed, as is the case for P-F and P-Cl bonds. $PH_5$ is unknown because P-H bonds have insufficient energy to bring about the required promotion of a 3s electron to a 3d orbital, and the only simple hydride of phosphorus is $PH_3$. $NCl_5$ is unknown because nitrogen is a second period element, for which the occupancy of the valence shell is confined to an octet of electrons. Alternatively, nitrogen cannot expand its valence shell because it has no low lying orbitals in addition to the 2s and 2p orbitals into which electrons can be promoted.

25. $$P_4S_3(s) + 8O_2(g) \longrightarrow P_4O_{10}(g) + 3SO_2(g)$$

From the given mass of $P_4S_3$, calculate the moles of $P_4S_3$ and the moles of $SO_2$ formed on its complete combustion.

$$\text{Mol } SO_2 = (0.157 \text{ g } P_4S_3)\left(\frac{1 \text{ mol } P_4S_3}{220.1 \text{ g } P_4S_3}\right)\left(\frac{3 \text{ mol } SO_2}{1 \text{ mol } P_4S_3}\right) = \underline{2.14\text{x}10^{-3} \text{ mol}}$$

Then, from PV = nRT, $V = \frac{nRT}{P}$

$$V_{SO_2} = \frac{(2.14\text{x}10^{-3} \text{ mol})(0.0821 \text{ atm L mol}^{-1} \text{ K}^{-1})(293 \text{ K})}{(772 \text{ mm Hg})\left(\frac{1 \text{ atm}}{760 \text{ mm Hg}}\right)}$$

$$= \underline{5.07\text{x}10^{-2} \text{ L}}, \underline{\text{or}} \quad \underline{50.7 \text{ mL}}$$

26. First calculate the empirical formula of the compound from the mass % composition, by the normal method. The ratio of moles (atoms) is P:F:O = 0.962:2.88:0.963, or 1 : 3 : 1, and the empirical formula is $PF_3O$ (formula mass = 104.0 u). Using the relationship that 1 mol of any gas at STP has a volume equal to 22.4 L (assuming ideal behaviour).

$$\text{Molar mass} = (4.64 \text{ g L}^{-1})\left(\frac{22.4 \text{ L}}{1 \text{ mol}}\right) = \underline{104 \text{ g L}^{-1}}$$

Molecular formula = $PF_3O$

A possible Lewis structure has all of the atoms bonded to P.

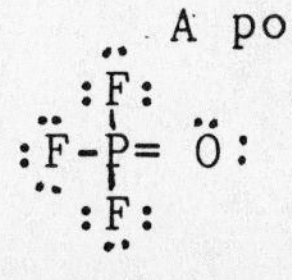

Oxidation numbers: P(+5), O(-2), F(-1)

27. From the information given, we have

$$AlP + H_2SO_4(aq) \longrightarrow Al,S,O \text{ salt } + \text{gas} \quad .... (1)$$

$$\text{gas} + HI(g) \longrightarrow P,H,I \text{ solid compound} \quad .... (2)$$

From (1), the gas must contain phosphorus, and from (2) it seems likely that it contains only P and H. First, let us calculate the molar mass of the gas.

At STP, 1 mole of gas occupies 22.4 L; from the density of 1.531 g $L^{-1}$

$$\text{Molar mass} = \left(\frac{1.531 \text{ g}}{1 \text{ L}}\right)\left(\frac{22.4 \text{ L}}{1 \text{ mol}}\right) = \underline{34.3 \text{ g mol}^{-1}}$$

For at least one P atom per molecule the molar mass has to be greater than 30.97 g $mol^{-1}$, the atomic mass of P. The difference of 3.3 g $mol^{-1}$ can only be due to hydrogen, and the molecular formula of the gas is evidently $PH_3$ (molar mass 33.99 g $mol^{-1}$). Like $NH_3$, $PH_3$ is a base and its reaction with HI(g) is expected to be

$$PH_3(g) + HI(g) \longrightarrow PH_4I(s)$$

That the product is the salt phosphonium iodide, $PH_4I(s)$, should be confirmed from the stoichiometry of the reaction and from the analytical data on the salt formed with HI.

Mass of $PH_4I$ expected from 0.158 g X is given by

$$(0.158 \text{ g } PH_3)\left(\frac{1 \text{ mol } PH_3}{33.99 \text{ g } PH_3}\right)\left(\frac{1 \text{ mol } PH_4I}{1 \text{ mol } PH_3}\right)\left(\frac{161.9 \text{ g } PH_4I}{1 \text{ mol } PH_4I}\right) = \underline{0.753 \text{ g } PH_4I}$$

which is in agreement with the observed mass of $\underline{0.752 \text{ g}}$.

From the analytical data:

| | | P | H | I |
|---|---|---|---|---|
| Grams in 100 g of compound | = | 19.1 | 2.5 | 78.4 |
| Moles in 100 g of compound | = | $\frac{19.1}{30.97}$ | $\frac{2.5}{1.008}$ | $\frac{78.4}{126.9}$ |
| Ratio of moles (atoms) | = | 0.617 : | 2.48 : | 0.618 |
| | = | 1.00 : | 4.01 : | 1.00 |

which confirms the empirical formula $PH_4I$

Finally, we can calculate the empirical formula of the aluminum salt:

| | | Al | S | O |
|---|---|---|---|---|
| Grams in 100 g of salt | = | 15.8 | 28.1 | 56.1 |
| Moles in 100 g of salt | = | $\frac{15.8}{26.98}$ | $\frac{28.1}{32.06}$ | $\frac{56.1}{16.00}$ |
| Ratio of moles (atoms) | = | 0.585 : | 0.876 : | 3.51 |
| | = | 1.00 : | 1.50 : | 6.00 |

Empirical Formula: $Al_2S_3O_{12}$ or $[Al^{3+}]_2[SO_4^{2-}]_3$

The gas is phosphine, $PH_3$, which is formed by the reaction

$$2AlP(s) + 3H_2SO_4(aq) \longrightarrow Al_2(SO_4)_3(aq) + 2PH_3(g)$$

28. (a) $P + HNO_3 \longrightarrow H_3PO_4 + NO_2 + H_2O$ unbalanced

Clearly this is an oxidation-reduction reaction in which phosphorus, (oxidation number 0) is oxidized to P(+5) in $H_3PO_4$, and N(+5) in $HNO_3$ is reduced to N(+4) in $NO_2$.

$$P + 4H_2O \longrightarrow H_3PO_4 + 5e^- + 5H^+ \quad \text{oxidation}$$
$$5[HNO_3 + e^- + H^+ \longrightarrow NO_2 + H_2O] \quad \text{reduction}$$
$$P + 5HNO_3 \longrightarrow H_3PO_4 + 5NO_2 + H_2O \quad \underline{\text{balanced overall}}$$

(b) $Ca_3(PO_4)_2 + C + SiO_2 \longrightarrow CaSiO_3 + CO + P_4$ unbalanced

or, more simply

$$PO_4^{3-} + C + SiO_2 \longrightarrow SiO_3^{2-} + CO + P_4 \quad \text{unbalanced}$$

in which P(+5) in $PO_4^{3-}$ is reduced to P(0) in $P_4$, and C(0) in carbon is oxidized to C(+2) in CO, and the oxidation number of Si is +4 in both $SiO_2$ and $SiO_3^{2-}$.

$$4PO_4^{3-} + 20e^- + 32H^+ \longrightarrow P_4 + 16H_2O \quad \text{reduction}$$
$$10[\ C + H_2O \longrightarrow CO + 2e^- + 2H^+] \quad \text{oxidation}$$
$$4PO_4^{3-} + 10C + 12H^+ \longrightarrow P_4 + 10CO + 6H_2O \quad \underline{\text{balanced}} \ldots\ldots (1)$$

Note that $H^+$ and $H_2O$ appear in this equation, which are not reactants or products, but these may be eliminated by using the water to form $SiO_3^{2-}$ ions from the $SiO_2$.

$$6SiO_2 + 6H_2O \longrightarrow SiO_3^{2-} + 12H^+ \quad \ldots. (2)$$

Adding equations (1) and (2) gives

$$4PO_4^{3-} + 10C + 6SiO_2 \longrightarrow P_4 + 10CO + 6SiO_3^{2-}$$

and the final balanced equation results from adding $6Ca^{2+}$ to both sides

$$2Ca_3(PO_4)_2 + 10C + 6SiO_2 \longrightarrow P_4 + 10CO + 6CaSiO_3$$

(c) Disodium hydrogen phosphate is $Na_2HPO_4$, phosphoric acid is $H_3PO_4$ and sodium hydroxide is NaOH; phosphorus is in the +5 oxidation state in both $H_3PO_4$ and the $HPO_4^{2-}$ ion, so that there is no oxidation or reduction. This is an acid-base reaction

$$H_3PO_4(aq) + 2NaOH(aq) \longrightarrow Na_2HPO_4(aq) + 2H_2O$$

(d) Phosphorous acid is $H_3PO_3$, and the unbalanced equation is

$$H_3PO_3 + Cu^{2+} \longrightarrow Cu(s) + H_3PO_4$$

in which P(+3) in $H_3PO_3$ is oxidized to P(+5) in $H_3PO_4$, and Cu(+2) in $Cu^{2+}$ is reduced to Cu(0) in Cu.

$$H_3PO_3 + H_2O \longrightarrow H_3PO_4 + 2e^- + 2H^+ \quad \text{oxidation}$$
$$Cu^{2+} + 2e^- \longrightarrow Cu \quad \text{reduction}$$
$$Cu^{2+} + H_3PO_3 + H_2O \longrightarrow Cu + H_3PO_4 + 2H^+ \quad \underline{\text{balanced}}$$

<u>or</u> $$Cu^{2+} + H_3PO_3 + 3H_2O \longrightarrow Cu + H_3PO_4 + 2H_3O^+$$

(e) The unbalanced equation is

$P_4 + NaOH \longrightarrow NaH_2PO_2 + PH_3$, or $P_4 + OH^- \longrightarrow H_2PO_2^- + PH_3$

in which P(0) in $P_4$ is reduced to P(-3) in $PH_3$ and oxidized to P(+1) in the $H_2PO_2^-$ ion.

$$P_4 + 12e^- + 12H_2O \longrightarrow 4PH_3 + 12OH^- \qquad \text{reduction}$$
$$3[P_4 + 8OH^- \longrightarrow 4H_2PO_2^- + 4e^-] \qquad \text{oxidation}$$
$$4P_4 + 12H_2O + 12OH^- \longrightarrow 4PH_3 + 12H_2PO_2^- \qquad \text{balanced}$$

Dividing throughout by 4, and adding $3Na^+$ to each side gives

$$P_4 + 3H_2O + 3NaOH \longrightarrow PH_3 + 3NaH_2PO_2$$

29. From the given data, 3.064 g of P reacts with (7.020-3.064) g, = 3.956 g, of oxygen to give 7.020 g of phosphorous oxide. First calculate the empirical formula of the oxide.

| | | P | O |
|---|---|---|---|
| Grams in 7.020 g oxide | = | 3.064 | 3.956 |
| Moles in 7.020 g oxide | = | $\frac{3.064}{30.97}$ | $\frac{3.956}{16.00}$ |
| Ratio of moles (atoms) | = | 0.0989 : | 0.2473 |
| | = | 1.00 : | 2.50 or 2:5 |

Empirical formula of oxide = $P_2O_5$ (formula mass = 141.9 u)

Also from the given data, 7.020 g oxide reacts with (9.691-7.020) g, = 2.671 g water

$$P_2O_5 + xH_2O \longrightarrow \text{phosphorous oxoacid}$$

$$\text{Mol } P_2O_5 = (7.020 \text{ g } P_2O_5)\left(\frac{1 \text{ mol } P_2O_5}{141.9 \text{ g } P_2O_5}\right) = 4.95\text{x}10^{-2} \text{ mol}$$

$$\text{Mol } H_2O = (2.671 \text{ g } H_2O)\left(\frac{1 \text{ mol } H_2O}{18.02 \text{ g } H_2O}\right) = 0.148 \text{ mol}$$

$$\text{In oxoacid, ratio of mol } P_2O_5\text{: mol } H_2O = 1.00 : \frac{0.148}{0.0495} = 2.99$$

i.e., x in the above equation for the reaction with water is 3,

$$P_2O_5 + 3H_2O \longrightarrow H_6P_2O_8$$

and the empirical formula of the oxoacid is $H_3PO_4$

30. The oxidation state of P in $P_4$ is 0, and it is +5 in $Na_3PO_4$. As we saw in Problem 29, the anhydride of $H_3PO_4$ is P(V) oxide, which has the molecular formula $P_4O_{10}$(s). Thus, to prepare $Na_3PO_4$(aq) from $P_4$, the phosphorus is first burned in excess air, or oxygen

$$P_4(s) + 5O_2(g) \longrightarrow P_4O_{10}(s)$$

and the resulting P(V) oxide is dissolved in water to give an aqueous solution of phosphoric acid, $H_3PO_4$.

$$P_4O_{10}(s) + 6H_2O(l) \longrightarrow 4H_3PO_4(aq)$$

Addition of NaOH(aq) to this solution, according to the equation

$$H_3PO_4(aq) + 3NaOH(aq) \longrightarrow Na_3PO_4(aq) + 3H_2O(l)$$

gives the required solution of $Na_3PO_4(aq)$.

31. First calculate the empirical formula.

| | | P | O | Na |
|---|---|---|---|---|
| Grams in 100 g of salt | = | 25.3 | 43.5 | 31.2 |
| Mol in 100 g of salt | = | $\frac{25.3}{30.97}$ | $\frac{43.5}{16.00}$ | $\frac{31.2}{22.99}$ |
| Mol (atom) ratio | = | 0.817 : | 2.72 : | 1.36 |
| | = | 1.00 : | 3.33 : | 1.66 |
| | = | 3.0 : | 10.0 : | 5.0 |

The empirical formula is $Na_5P_3O_{10}$

Formulated as ions, the salt is $[Na^+]_3\ P_3O_{10}^{5-}$, containing the triphosphate ion, with the Lewis structure

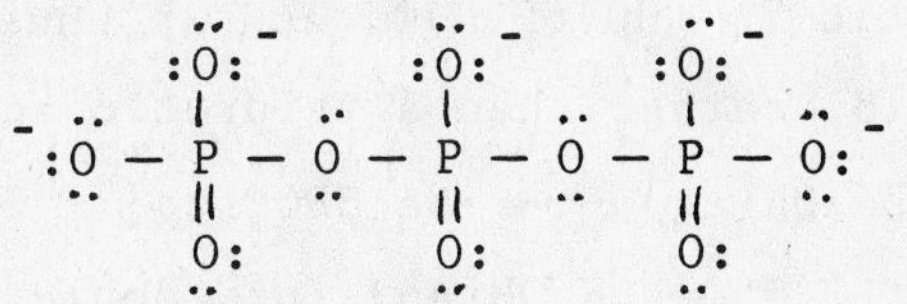

32. (a)

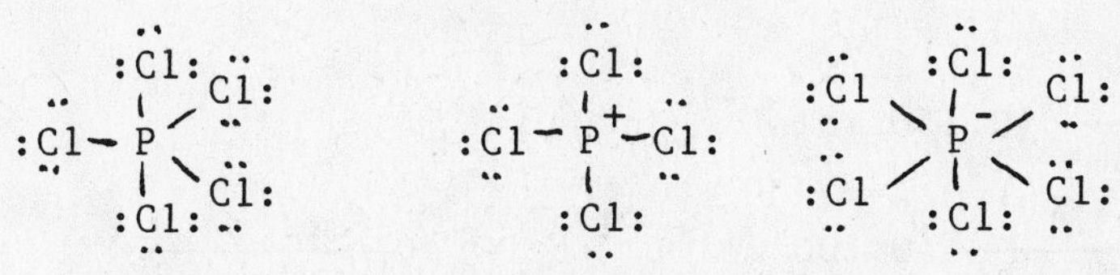

$PCl_5$ $\qquad$ $PCl_4^+$ $\qquad$ $PCl_6^-$

(b) $P + \frac{3}{2}Cl_2 \longrightarrow PCl_3$ ; $P + \frac{5}{2}Cl_2 \longrightarrow PCl_5$

The total moles of $PCl_3$ and $PCl_5$ must be the same as the initial moles of phosphorus

Mol P = $(5.43\ g\ P)(\frac{1\ mol\ P}{30.97\ g\ P}) = \underline{0.175\ mol\ P}$

and the molar masses are: $PCl_3$, 137.3 g $mol^{-1}$; $PCl_5$, 208.2 g $mol^{-1}$.

Let mol $PCl_3$ = x, then mol $PCl_5$ = (0.175-x).

Thus, $(x\ mol\ PCl_3)(\frac{137.3\ g\ PCl_3}{1\ mol\ PCl_3}) + [(0.175-x)\ mol\ PCl_5](\frac{208.2\ g\ PCl_5}{1\ mol\ PCl_5})$

$= 28.05$ g

$137.3x + 36.43 - 208.2x = 36.43 - 70.9x = 28.05$; $70.9x = 8.38$

$\underline{x = 0.118}$

Mass of $PCl_3$ = $(0.118\ mol\ PCl_3)(\frac{137.3\ g\ PCl_3}{1\ mol\ PCl_3}) = \underline{16.3\ g}$

Mass % $PCl_3$ = $(16.3\ g\ PCl_3)(\frac{100\ g}{28.05\ g}) = \underline{58.1\ mass\ \%}$

Mass % $PCl_5$ = $\underline{41.9\ mass\ \%}$

33. (a) $P_4O_6(s)$ (b) $Ca(H_2PO_4)_2(s)$ (c) $Ca_3P_2(s)$ (d) $H_3PO_3$
(e) $PH_4I(s)$ (f) $NaPO_3(s)$

34. $$CaCO_3(s) + H_3PO_4(aq) \longrightarrow CaHPO_4(aq) + CO_2(g) + H_2O(l)$$

To calculate the mass of $CaCO_3(s)$ required, first calculate the initial mol of $H_3PO_4$.

$$\text{mol } H_3PO_4 = (48.0 \text{ mL})\left(\frac{1.70 \text{ g}}{1 \text{ mL}}\right)\left(\frac{85.5 \text{ g } H_3PO_4}{100 \text{ g solution}}\right)\left(\frac{1 \text{ mol } H_3PO_4}{97.99 \text{ g } H_3PO_4}\right)$$
$$= \underline{0.712 \text{ mol}}$$

$$\text{Mass } CaCO_3 = (0.712 \text{ mol } H_3PO_4)\left(\frac{1 \text{ mol } CaCO_3}{1 \text{ mol } H_3PO_4}\right)\left(\frac{100.1 \text{ g } CaCO_3}{1 \text{ mol } CaCO_3}\right) = \underline{71.3 \text{ g}}$$

$$\text{Mass } CaHPO_4 = (0.712 \text{ mol } H_3PO_4)\left(\frac{1 \text{ mol } CaHPO_4}{1 \text{ mol } H_3PO_4}\right)\left(\frac{136.1 \text{ g } CaHPO_4}{1 \text{ mol } CaHPO_4}\right)$$
$$= \underline{96.9 \text{ g}}$$

35. $H_3PO_3$, phosphorus(III) acid is $(HO)_2P(H)O$, and is unusual in that it is a <u>diprotic</u> acid, rather than a triprotic acid.

$$H_3PO_3(aq) + 2NaOH(aq) \longrightarrow Na_2HPO_3(aq) + 2H_2O(l)$$

If the required volume of 0.150 M NaOH(aq) for complete neutralization is V mL,

$$\text{mol NaOH} = (V \text{ mL})\left(\frac{1 \text{ L}}{1000 \text{ mL}}\right)(0.150 \text{ mol L}^{-1} \text{ NaOH})$$
$$= (25.00 \text{ mL})\left(\frac{1 \text{ L}}{1000 \text{ mL}}\right)(0.200 \text{ mol L}^{-1}\ H_3PO_3)\left(\frac{2 \text{ mol NaOH}}{1 \text{ mol } H_3PO_3}\right)$$

$$\underline{V = 66.7 \text{ mL}}$$

36. (a) <u>Oxidation</u> is the <u>loss</u> of electrons, and <u>reduction</u> is the <u>gain</u> of electrons.

For example: $2Cl^- \longrightarrow Cl_2 + 2e^-$ <u>oxidation</u>

$Cu^{2+} + 2e^- \longrightarrow Cu$ <u>reduction</u>

(b) <u>Oxidation</u> involves an <u>increase</u> in oxidation number, and <u>reduction</u> involves a <u>decrease</u> in oxidation number.

For example: $Cl^-(-1) \longrightarrow \frac{1}{2}Cl_2(0)$ <u>oxidation</u>

$Cu^{2+}(+2) + 2e^- \longrightarrow Cu(0)$ <u>reduction</u>

37. The range in oxidation states of phosphorus is from -3 to +5.

For example: <u>-3</u> $PH_3$, $P^{3-}$ (phosphide ion), $PH_4^+$

<u>+5</u> $P_4O_{10}$, $H_3PO_4$, $H_2PO_4^-$, $HPO_4^{2-}$, $PO_4^{3-}$, $PCl_5$, $PF_5$, $POCl_3$, $POF_3$

Other possible oxidation states include:

<u>+3</u> $P_4O_6$, $PCl_3$, $PF_3$, $H_3PO_3$, $H_2PO_3^-$, $HPO_3^{2-}$

<u>0</u> the allotropes of elemental phosphorus

38. (a) P(-3), H(+1) (b) As(-3), H(+1) (c) P(+3), F(-1)
(d) K(+1), Mn(+7), O(-2) (e) Si(+4), O(-2) (f) P(-3),H(+1)

39. (a) 0 (b) 0 (c) +2 (d) -1 (e) -1 (f) -2

40. (a) C(-4), H(+1) (b) C(+4), O(-2) (c) C(+3), O(-2)
(d) C(0), H(+1), O(-2) (e) S(+6), O(-2) (f) Cr(+6), O(-2)

41. (a) +2 (b) +4 (c) +5 (d) +6 (e) -1 (f) +3

42. (a) N(-2), H(+1) (b) Li(+1), O(-2) (c) P(-3), H(+1), I(-1)
(d) Na(+1), Cl(+7), O(-2) (e) Na(+1), O(-2), Cl(+1) (f) Ca(+2), S(-1)
(g) Al(+3), P(-3) (h) S(0) (i) Ba(+2), S(+4), O(-2)

43. (a) N(0) (b) H(+1),C(+4),O(-2) (c) N(-3),H(+1) (d) P(-3),H(+1)
(e) V(+3), O(-2) (f) V(+5), O(-2) (g) Mn(+4), O(-2)
(h) H(+1), N(+5), O(-2) (i) H(+1), N(+3), O(-2)

44. (a) $\overset{-2}{H_2S} \longrightarrow \overset{0}{S} + 2e^- + 2H^+$

(b) $\overset{+4}{SO_2} + 2H_2O \longrightarrow \overset{+6}{SO_4^{2-}} + 2e^- + 4H^+$

(c) $\overset{+3}{H_3PO_3} + H_2O \longrightarrow \overset{+5}{H_3PO_4} + 2e^- + 2H^+$

45. (a) The oxidation number increases.
(b) The oxidation number decreases.
(c) $Cu^+ \longrightarrow Cu^{2+} + e^-$

$S_8 + 16e^- \longrightarrow 8S^{2-}$

$NO_3^- + 3e^- + 4H^+ \longrightarrow NO + 2H_2O$

$S_2O_3^{2-} + 5H_2O \longrightarrow 2SO_4^{2-} + 8e^- + 10H^+$

46. For the first reaction, the unbalanced equation is

$Br^- + H_2SO_4 \longrightarrow Br_2 + SO_2$

in which Br(-1) in $Br^-$ is oxidized to Br(0) in $Br_2$, and S(+6) in $H_2SO_4$ is reduced to S(+4) in $SO_2$.

| | |
|---|---|
| $2Br^- \longrightarrow Br_2 + 2e^-$ | oxidation |
| $H_2SO_4 + 2e^- + 2H^+ \longrightarrow SO_2 + 2H_2O$ | reduction |
| $H_2SO_4 + 2Br^- + 2H^+ \longrightarrow Br_2 + SO_2 + 2H_2O$ | balanced |

The water produced in this reaction will be protonated by $H_2SO_4$

$H_2O + H_2SO_4 \longrightarrow H_3O^+ + HSO_4^-$

For the second reaction, the unbalanced equation is

$$Br_2 + SO_2 \longrightarrow Br^- + SO_4^{2-}$$

in which Br(0) in $Br_2$ is reduced to Br(-1) in $Br^-$, and S(+4) in $SO_2$ is oxidized to S(+6) in $SO_4^{2-}$.

$$Br_2 + 2e^- \longrightarrow 2Br^- \qquad \underline{\text{reduction}}$$

$$SO_2 + 2H_2O \longrightarrow SO_4^{2-} + 2e^- + 4H^+ \qquad \underline{\text{oxidation}}$$

$$Br_2 + SO_2 + 2H_2O \longrightarrow 2Br^- + SO_4^{2-} + 4H^+ \qquad \underline{\text{balanced}}$$

The first reaction is carried out in concentrated acid, which contains molecular $H_2SO_4$, which is the strongest oxidizing agent present. The second reaction is carried out in dilute acid, which contains no $H_2SO_4$ but only $H_3O^+$ and $SO_4^{2-}$ ions. The direction of reaction under given conditions can be predicted using Le Châtelier's principle. For

$$Br_2 + SO_2 + 2H_2O \rightleftharpoons 2Br^- + H_2SO_4 + 2H^+$$

when little or no $H_2SO_4$ is present, as is the case in dilute acid, the reaction will proceed from left to right; if a large amount of $H_2SO_4$ is present, and particularly if the $SO_2$(g) is allowed to escape, the reaction will proceed from right to left.

47. The oxidation number of S in $SO_2$ is +4; when it behaves as an oxidizing agent it must be reduced to a compound in which the S has an oxidation number less than +4, and when it behaves as a reducing agent it must be oxidized to a compound in which the S has an oxidation number greater than +4. For example:

(a) $SO_2$ can oxidize $H_2S$ to S, a reaction in which it is itself reduced to S.

$$2[\overset{-2}{H_2S} \longrightarrow \overset{0}{S} + 2e^- + 2H^+] \qquad \underline{\text{oxidation}}$$

$$\overset{+4}{SO_2} + 4e^- + 4H^+ \longrightarrow \overset{0}{S} + 2H_2O \qquad \underline{\text{reduction}}$$

$$2H_2S + SO_2 \longrightarrow 3S + 2H_2O \qquad \underline{\text{overall}}$$

(b) The only oxidation state of S greater than +4 is +6. In the reaction of $SO_2$ with $Br_2$ in aqueous solution, to give $SO_4^{2-}$ and $Br^-$, $SO_2$ is oxidized and $Br_2$ is reduced:

$$\overset{0}{Br_2}(aq) + 2e^- \longrightarrow 2\overset{-1}{Br^-}(aq) \qquad \underline{\text{reduction}}$$

$$\overset{+4}{SO_2}(aq) + 2H_2O(l) \longrightarrow \overset{+6}{SO_4^{2-}}(aq) + 2e^- + 4H^+ \qquad \underline{\text{oxidation}}$$

$$Br_2(aq) + SO_2(aq) + 2H_2O(l) \longrightarrow 2Br^-(aq) + SO_4^{2-}(aq) + 4H^+(aq)$$

48. In writing Lewis structures, it is useful to determine the valence of the central atom in a species, and thus the number of bonds that it forms. In species (a) and (c) the oxidation state of S is +6, so that in each case the total number of bonds to S is 6. In (b) and (d), the oxidation state of S is +4, with 4 bonds to S in each case, and in (e) the oxidation state of S is +2, with two bonds to S. S(II) has two unshared pairs of electrons in its valence shell, S(IV) has one unshared pair, and S(VI) has no unshared pairs. It is also useful to remember that when an O atom forms one bond, as in $HO^-$, it has a formal charge of -1, when it forms two bonds, as in $H_2O$, it has a formal charge of 0, and when it forms three bonds, as in $H_3O^+$, it has a formal charge of +1. Thus:

(a) 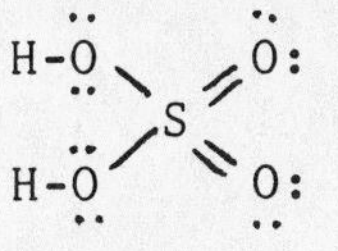

$AX_4$

tetrahedral

(b) 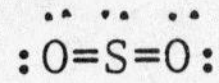

$AX_2E$

angular

(c) 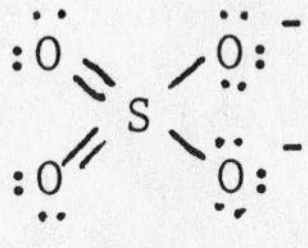

$AX_4$

tetrahedral

(d) 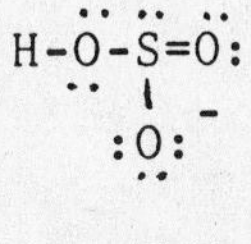

$AX_3E$

trigonal pyramid

(d) 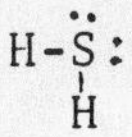 $AX_2E_2$ angular

49. (a) P(+5), 5 bonds to P

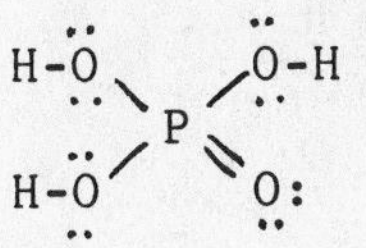

$AX_4$

tetrahedral

(b) P(+3), 3 bonds to P, which would give $:P(OH)_3$, but the structure is actually

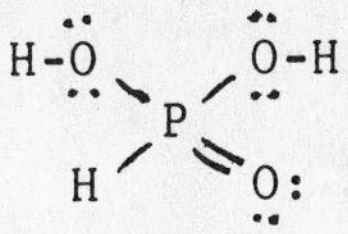

$AX_4$

tetrahedral

(c) P(+5), 5 bonds to P and the structure is derived from that of $H_3PO_4$ (part a), by removing a proton from each of 3 O-H groups

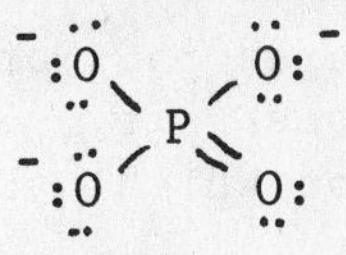

$AX_4$

tetrahedral

(d) See part (b), and the structure is derived from that of $H_3PO_3$ by removing a proton from each of two O-H groups

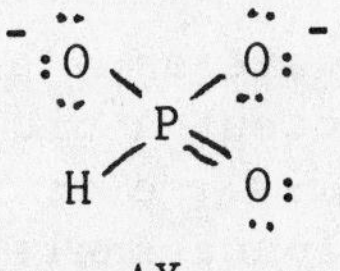

$AX_4$

tetrahedral

50. S(+6), 6 bonds to each S

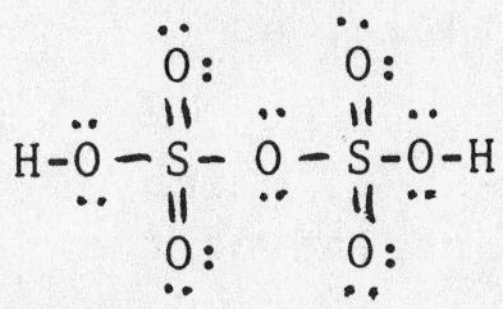

$H_2S_2O_7$

(b) P(+5), five bonds to each P

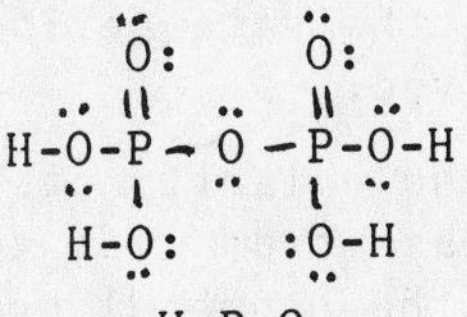

$H_4P_2O_7$

(c) S(+6), 6 bonds to S

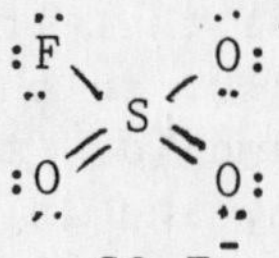

$SO_3F^-$

(d) P(+5), 5 bonds to S

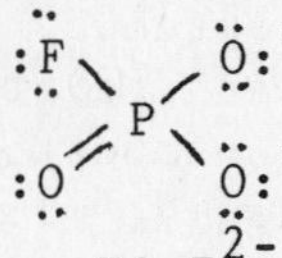

$PO_3F^{2-}$

51. (a) S(+6), six bonds to S

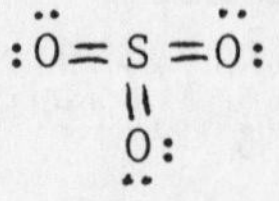

$SO_3$

(b) S(+4), 4 bonds to S

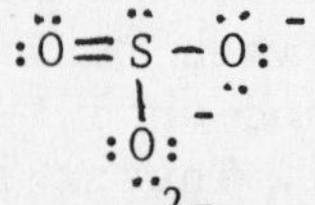

$SO_3^{2-}$

(c) The central S atom is S(+6)

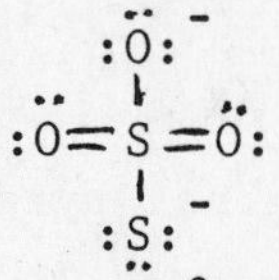

$S_2O_3^{2-}$

(d) S(+6), 6 bonds to each S

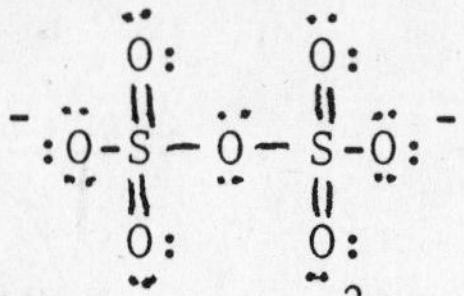

$S_2O_7^{2-}$

52. (a) P(+5), 5 bonds to each P

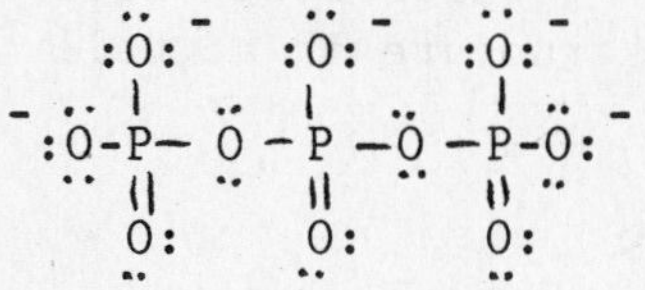

$P_3O_{10}^{5-}$

(b) S(+6), 6 bonds to each S

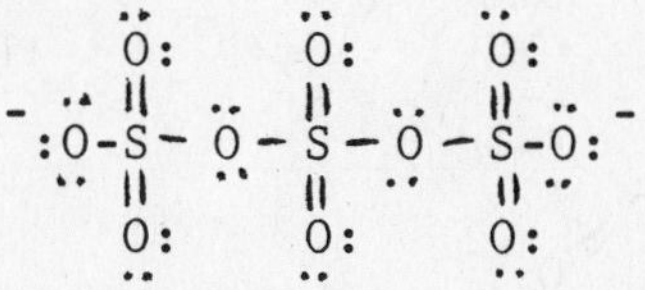

$S_3O_{10}^{2-}$

53. Write each acid in the form $XO_m(OH)_n$ in order to classify it as weak or strong:

(a) $B(OH)_3$ $X(OH)_3$ weak

(b) $HPO(OH)_2$ $XO(OH)_2$ weak

(c) $SeO(OH)_2$ $XO(OH)_2$ weak

(d) $NO_2(OH)$ $XO_2(OH)$ strong

(e) $PO(OH)_3$ $XO(OH)_3$ weak

54. For a compound to behave as a base, it must have an unshared pair of electrons that can accept a proton; as the size of the lone pair increases with decreasing electronegativity of the atom of which it is part of the valence shell, the basicity increases.

$CH_4$ - no unshared electron pairs, no base properties

$NH_3$, $H_2O$, HF - all have unshared electron pairs on their central atoms; the basicity is expected to decrease from left to right along the same period of the periodic table, as observed.

Acidity depends on the polarity of the X-H bonds ($^{\delta-}X\text{-}H^{\delta+}$), which is expected to increase from left to right across any period of the periodic table, for the nonmetal hydrides.

The C-H bond of $CH_4$ is essentially nonpolar; $CH_4$ has no acidic properties. The polarities of the X-H bonds in $NH_3$, $H_2O$, and HF, increase in the same order, with increasing electronegativity of X, and the acidity increases in the same order.

55. This problem is similar to Problem 54. The trends in acidity and basicity for the period 3 hydrides are expected to parallel those of the period 2 hydrides, except that because period 3 atoms are larger than those of period 2 and the valence shell electron pairs are not as strongly held, the X-H bonds will be weaker, and therefore more easily broken.
For the basic behaviour, $SiH_4$ has no lone pairs and cannot behave as a base. The lone pair of $PH_3$ is more diffuse than that of $NH_3$ and forms a weaker P-H bond, so that $PH_4^+$ is a stronger acid than $NH_4^+$, or $PH_3$ is a weaker base than $NH_3$. Similar considerations apply to $H_2S$ and HCl in comparison to $H_2O$ and HF, which are weaker bases than $PH_3$ and have no detectable base properties in water.

The acidities of the 3rd period nonmetal hydrides will increase across the period from left to right as the polarity of the X-H bond increases from left to right with increased electronegativity of X. $SiH_4$ and $PH_3$ have no acid properties in water, $H_2S$ is a weak acid and HCl is a strong acid, which is consistent with the increasing polarity of the X-H bonds. That $H_2S$ is a stronger acid than $H_2O$, and HCl a stronger acid than HF, is due to the decrease in X-H bond strength in descending any group.

56. The acidity of a hydride is expected to depend on the polarity of the X-H bonds and their strength. The former is related to the electronegativity difference between X and H and the latter to the size of X, which is much greater for a third period element than it is for a second period element in the same group. Thus, (a) the H-F bond is more polar than the H-Cl bond but the H-Cl bond is much weaker than the H-F bond and the latter effect apparently dominates, so that HCl is a strong acid and HF is a weak acid in water, and (b) $H_2S$ is a stronger acid than $H_2O$ for similar reasons - H-O more polar than H-S but H-S much weaker than H-O.

57. In comparison to the parent acid, a $H_{m-1}XO_n^-$ ion is a weaker acid than its parent acid, $H_mXO_n$ in aqueous solution, because it carries a negative charge, which makes it more difficult to remove a proton from the anion than from the neutral parent acid. Thus $HSO_4^-$ is a weaker acid in water than $H_2SO_4$, and for a triprotic acid such as $H_3PO_4$, the trend to decreased acidity increases with increased formal charge. i.e., $H_3PO_4 > H_2PO_4^- > HPO_4^{2-}$. Alternatively, the basicity of the anion could be considered. A negatively charged ion will attract a proton more strongly than than a neutral species. Thus, for example, $SO_4^{2-}$, the conjugate base of $HSO_4^-$, is expected to be a stronger base than $HSO_4^-$, the conjugate base of $H_2SO_4$. The stronger is a conjugate base, the weaker is its conjugate acid. Thus $HSO_4^-$ is a weaker acid than $H_2SO_4$.

58. (a) $B(OH)_3$ is the known structure of boric acid, $H_3BO_3$.

(b) Carbonic acid is $(HO)_2C{=}O$, formed by the elimination of a $H_2O$ molecule from $C(OH)_4$, but even $(HO)_2C{=}O$ is unstable and readily eliminates another water molecule to give $:\ddot{O}{=}C{=}\ddot{O}:$

(c) Nitric acid is $HONO_2$ rather than $N(OH)_5$ , which would violate the octet rule, and is formally formed from it by elimination of two $H_2O$.
(d) $Si(OH)_4$ is the known structure of silicic acid, $H_4SiO_4$.
(e) Phosphoric acid is $(HO)_3PO$, rather than $P(OH)_5$ , from which it is formed formally by the elimination of $H_2O$. (f) Sulfuric acid is $(HO)_2SO_2$, rather than $S(OH)_6$, from which it is formed formally by the elimination of $2H_2O$.

The tendency of an atom to form X-OH single bonds is apparently related to its electronegativity and size. Large atoms of low electronegativity tend to form the maximum number of -OH bonds; small atoms of relatively high electronegativity tend to replace some of their -OH bonds by =O bonds by the elimination of water molecules.

59. Whether a compound with a A-O-H bonding arrangement is acidic or basic depends on which bond, A-O or O-H, is the more easily broken. In the cases of KOH and $Ca(OH)_2$, the electronegativity difference between the metal atom and oxygen is large and causes these compounds to be ionic, releasing $OH^-$ in water. In $Si(OH)_4$, the Si-O bond is covalent and the O-H bond is covalent with a polarity $^{\delta-}O\text{-}H^{\delta+}$ and can transfer a proton to water to give $H_3O^+$.

60. (a) Hydrofluoric acid, HF(aq). Note that the term applies to an aqueous solution. HF(g) is called hydrogen fluoride.
(b) Phosphoric acid, $H_3PO_4$. (c) Phosphorous acid, $H_3PO_3$.
(d) Hypochlorous acid, HOCl(aq). (e) Sulfuric acid, $H_2SO_4$.
(f) The aqueous solution called sulfurous acid consists almost entirely of solvated $SO_2$ molecules, $SO_2(aq)$, and contains very little, if any, $H_2SO_3$, although the ions $HSO_3^-$ and $SO_3^{2-}$ are present.

61. (a) $CaSO_4(s)$ (b) $PBr_5(s)$ (c) $PI_3(s)$ (d) $(NH_4)_2HPO_4(s)$
(e) $Ca_3N_2(s)$ (f) $SF_4(g)$ (g) $CrCl_3(s)$

62. (a) Potassium phosphate (b) Carbon tetrafluoride (tetrafluoromethane)
(c) Zinc carbonate (d) Calcium hypochlorite (e) Calcium sulfite
(f) Phosphonium iodide (g) Sodium disulfate (h) Sodium triphosphate

63. (a) Potassium selenate (b) Hydrogen telluride (c) Sodium tetrasulfide
(d) Iron(II) disulfide (e) Rubidium hydrogen sulfate
(f) Phosphorus(III) oxide* (g) Sodium hydrogen phosphate
(h) Sodium hydrogen phosphite (* <u>or</u> tetraphosphorous hexaoxide).

64. (a) Sulfuric acid (b) Phosphoric acid (c) Phosphorous acid
(d) Carbonic acid (e) Nitric acid (f) Sulfurous acid
(g) Silicic acid (h) Hypochlorous acid (i) Perchloric acid

65. Write each acid to indicate the number of -OH groups, and thus the number of ionizable protons. Successive removal of these protons by reaction with $OH^-(aq)$ gives each of the possible products.

(a) <u>Nitric acid</u>, $HONO_2$ (monoprotic acid)

$$HNO_3(aq) + NaOH(aq) \longrightarrow \underset{\text{sodium nitrate}}{NaNO_3(aq)} + H_2O(l)$$

(b) Sulfuric acid $(HO)_2SO_2$ (diprotic acid)

$H_2SO_4(aq) + NaOH(aq) \longrightarrow Na^+(aq) + HSO_4^-(aq) + H_2O(l)$

but the $HSO_4^-(aq)$ so formed is a relatively strong acid and $NaHSO_4$, sodium hydrogen sulfate cannot be obtained from the solution.

$H_2SO_4(aq) + 2NaOH(aq) \longrightarrow Na_2SO_4(aq) + 2H_2O(l)$
sodium sulfate

(c) Phosphoric acid $(HO)_3PO$ (triprotic acid)

$H_3PO_4(aq) + NaOH(aq) \longrightarrow NaH_2PO_4(aq) + H_2O(l)$
sodium dihydrogen phosphate

$H_3PO_4(aq) + 2NaOH(aq) \longrightarrow Na_2HPO_4(aq) + 2H_2O$
sodium hydrogen phosphate

$H_3PO_4(aq) + 3NaOH(aq) \longrightarrow Na_3PO_4(aq) + 3H_2O(l)$
sodium phosphate

(d) Phosphorous acid $(HO)_2PO(H)$ (diprotic acid)

$H_3PO_3(aq) + NaOH(aq) \longrightarrow NaH_2PO_3(aq) + H_2O(l)$
sodium dihydrogen phosphite

$H_3PO_3(aq) + 2NaOH(aq) \longrightarrow Na_2HPO_3(aq) + 2H_2O(l)$
sodium hydrogen phosphite

66. Consider the ionization of the compound in water, and any further reaction of the species formed as acids or bases.

(a) $H_2SO_4(aq) + H_2O(l) \longrightarrow H_3O^+(aq) + HSO_4^-(aq)$
hydronium ion hydrogen sulfate ion

$HSO_4^-(aq) + H_2O(l) \rightleftharpoons H_3O^+(aq) + SO_4^{2-}(aq)$
hydronium ion sulfate ion

(b) $H_2O(l) + SO_2(g) \rightleftharpoons H_2SO_3(aq)$
sulfur dioxide sulfurous acid

$H_2SO_3(aq) + H_2O(l) \rightleftharpoons H_3O^+(aq) + HSO_3^-(aq)$
hydronium ion hydrogen sulfite ion

$HSO_3^-(aq) + H_2O(l) \rightleftharpoons H_3O^+(aq) + SO_3^{2-}(aq)$
hydronium ion sulfite ion

(c) $H_3PO_4(aq) + H_2O(l) \rightleftharpoons H_3O^+(aq) + H_2PO_4^-(aq)$
phosphoric acid hydronium ion dihydrogen phosphate ion

$H_2PO_4^-(aq) + H_2O(l) \rightleftharpoons H_3O^+(aq) + HPO_4^{2-}(aq)$
hydronium ion hydrogen phosphate ion

$HPO_4^{2-}(aq) + H_2O(l) \rightleftharpoons H_3O^+(aq) + PO_4^{3-}(aq)$
hydronium ion phosphate ion

(d) $H_2CO_3(aq) + H_2O(l) \rightleftharpoons H_3O^+(aq) + HCO_3^-(aq)$

carbonic acid — hydronium ion — hydrogen carbonate ion

$HCO_3^-(aq) + H_2O(l) \rightleftharpoons H_3O^+(aq) + CO_3^{2-}(aq)$

hydronium ion — carbonate ion

$H_2CO_3(aq) \rightleftharpoons CO_2(g) + H_2O(l)$

carbon dioxide

(e) $Na_3PO_4(s) \longrightarrow 3Na^+(aq) + PO_4^{3-}(aq)$

sodium ion — phosphate ion

$PO_4^{3-}(aq) + H_2O(l) \rightleftharpoons HPO_4^{2-}(aq) + OH^-(aq)$

hydrogen phosphate ion — hydroxide ion

and in principle we might also expect to find very small amounts of the $H_2PO_4^-(aq)$, dihydrogen phosphate ion.

(f) $CaHPO_4(s) \longrightarrow Ca^{2+}(aq) + HPO_4^{2-}(aq)$

calcium ion — hydrogen phosphate ion

$HPO_4^{2-}(aq) + H_2O(l) \rightleftharpoons PO_4^{3-}(aq) + H_3O^+(aq)$

phosphate ion — hydronium ion

and in principle we might also expect to find very small amounts of the dihydrogen phosphate ion, $H_2PO_4^-(aq)$.

(g) $NH_4NO_3(s) \longrightarrow NH_4^+(aq) + NO_3^-(aq)$

ammonium ion — nitrate ion

$NH_4^+(aq) + H_2O(l) \rightleftharpoons NH_3(aq) + H_3O^+(aq)$

ammonia — hydronium ion

(h) $(NH_4)_2SO_4(s) \longrightarrow 2NH_4(aq) + SO_4^{2-}(aq)$

ammonium ion — sulfate ion

$NH_4^+(aq) + H_2O(l) \rightleftharpoons NH_3(aq) + H_3O^+(aq)$

ammonia — hydronium ion

$SO_4^{2-}(aq) + H_2O(l) \rightleftharpoons HSO_4^-(aq) + OH^-(aq)$

hydrogen sulfate ion — hydroxide ion

67. The acid-base reactions are those where reactants and products are related by the transfer of protons, and the oxidation-reduction reactions are those in which there are changes in oxidation numbers between reactants and products. Thus, (a), (b), (e), and (g) are acid-base reactions, and (c), (d), and (f) are oxidation-reduction reactions.

(a) $CO_3^{2-} + H_2SO_4 \longrightarrow HSO_4^- + HCO_3^-$ — balanced

$\text{base}_1$ $\text{acid}_2$ $\text{base}_2$ $\text{acid}_1$

(b) $HSO_4^- + H_2O \longrightarrow H_3O^+ + SO_4^{2-}$ — balanced

$\text{acid}_1$ $\text{base}_2$ $\text{acid}_2$ $\text{base}_1$

(c) Cu(0) in Cu(s) is oxidized to Cu(2+) in $Cu^{2+}$ & S(+6) in $H_2SO_4$ is reduced to S(+4) in $SO_2$,

$$Cu \longrightarrow Cu^{2+} + 2e^- \quad \text{oxidation}$$

$$H_2SO_4 + 2e^- + 2H^+ \longrightarrow SO_2 + 2H_2O \quad \text{reduction}$$

$$Cu + H_2SO_4 + 2H^+ \longrightarrow Cu^{2+} + SO_2 + 2H_2O \quad \underline{\text{overall}}$$

and add $SO_4^{2-}$ to both sides of the equation to obtain the balanced equation in the appropriate form,

$$Cu + 2H_2SO_4 \longrightarrow Cu^{2+} + SO_4^{2-} + SO_2 + 2H_2O \quad \underline{\text{balanced}}$$

(d) Br(-1) in $Br^-$ is oxidized to Br(0) in $Br_2$, and S(+6) in $H_2SO_4$ is reduced to S(+4) in $SO_2$.

$$2Br^- \longrightarrow Br_2 + 2e^- \quad \text{oxidation}$$

$$H_2SO_4 + 2e^- + 2H^+ \longrightarrow SO_2 + 2H_2O \quad \text{reduction}$$

$$2Br^- + H_2SO_4 + 2H^+ \longrightarrow Br_2 + SO_2 + 2H_2O \quad \underline{\text{overall}}$$

Adding $SO_4^{2-}$ to both sides of the equation gives

$$2Br^- + 2H_2SO_4 \longrightarrow Br_2 + SO_2 + SO_4^{2-} + 2H_2O \quad \underline{\text{balanced}}$$

(e) $CaCO_3$ is $Ca^{2+}CO_3^{2-}$, so that $Ca^{2+}$ appears on both sides of the equation, which is more simply written as

$$\underset{\text{base}_1}{CO_3^{2-}} + \underset{\text{acid}_2}{H_3O^+} \rightarrow \underset{\text{acid}_1}{HCO_3^-} + \underset{\text{base}_2}{H_2O}$$

and adding $Ca^{2+}$ to both sides gives

$$CaCO_3 + H_3O^+ \longrightarrow Ca^{2+} + HCO_3^- + H_2O \quad \underline{\text{balanced}}$$

(f) Mg(0) in Mg is oxidized to Mg(+2) in $Mg^{2+}$, and H(+1) in $H_3O^+$ is reduced to H(0) in $H_2$.

$$Mg \longrightarrow Mg^{2+} + 2e^- \quad \text{oxidation}$$

$$2[H_3O^+ + e^- \longrightarrow 1/2\ H_2 + H_2O] \quad \text{reduction}$$

$$Mg + 2H_3O^+ \longrightarrow Mg^{2+} + H_2 + 2H_2O \quad \underline{\text{balanced}}$$

(g) $Ca_3P_2$ is $(Ca^{2+})_3(P^{3-})_2$ and $Ca(OH)_2$ is $Ca^{2+}(OH^-)_2$, so in this reaction, $P^{3-}$ adds three protons to give $PH_3$, and these come from three $H_2O$ molecules, leaving three $OH^-$ ions.

$$\underset{\text{base}_1}{P^{3-}} + \underset{\text{acid}_2}{3H_2O} \longrightarrow \underset{\text{acid}_1}{PH_3} + \underset{\text{base}_2}{3OH^-}$$

Double this equation and add three $Ca^{2+}$ ions to each side,

$$Ca_3P_2 + 6H_2O \longrightarrow 3Ca(OH)_2 + 2PH_3 \quad \underline{\text{balanced}}$$

68. $Cl^-$ is a very weak base and the equilibrium reaction in which HCl is formed by protonating it with an acid lies far to the left.

$$Cl^- + H^+ \rightleftharpoons HCl$$

On heating a mixture of NaCl and an acid, HCl(g) is driven off, so that according to Le Châtelier's principle, more HCl is formed, which in turn is removed by heating, so that eventually the equilibrium goes to completion. The success of the method depends on removing the volatile HCl(g), so that only acids with high boiling points, such as $H_2SO_4$ and $H_3PO_4$ are suitable. In contrast, nitric acid is relatively volatile (b.p. 86 $^\circ$C) and is unsuitable.

69. (a) $P_4$, $P_4O_6$, and $P_4O_{10}$ all contain a tetrahedral arrangement of four P atoms (see text)

(b) Both $(HPO_3)_3$ and $(SO_3)_3$ contain a cyclic six membered -X-O-X-O-X-O- ring:

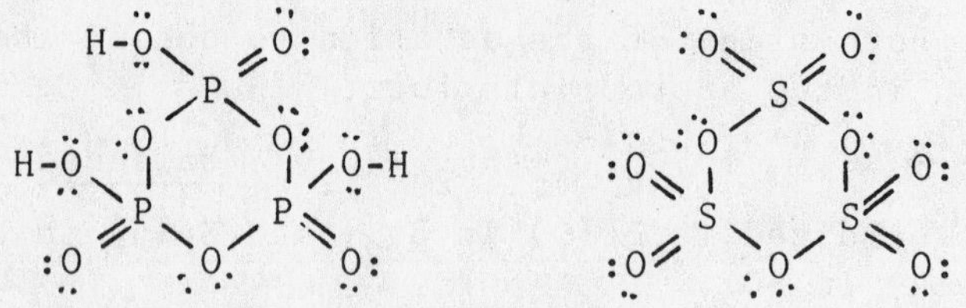

(c) Each of these species contains X-O-X bridge bonds:

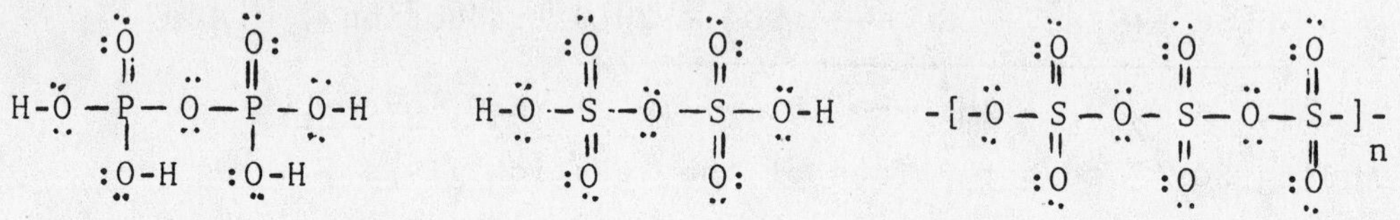

(d) Each of these species has $AX_4$ tetrahedral geometry around a P atom or an S atom:

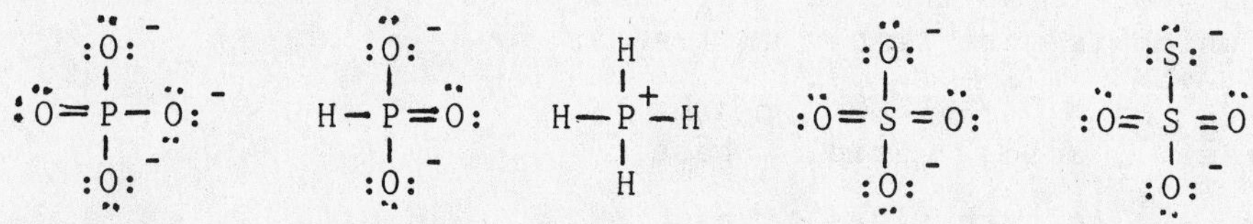

phosphate ion | hydrogen phosphate ion | phosphonium ion | sulfate ion | thiosulfate ion

70. As and Se are both in period 4; As is in group 5 and Se is in group 6. Thus the ground state configurations are:

As $[Ar]3d^{10}$ 4s [↑↓] 4p [↑|↑|↑] Se $[Ar]3d^{10}$ 4s [↑↓] 4p [↑↓|↑|↑]

So that the expected hydrides are $AsH_3$ and $H_2Se$

$AsCl_5$ contains As(V), $SeF_4$ contains Se(IV), and $SeF_6$ contains Se(VI), which must be derived from valence states with 5, 4, and 6 unpaired electrons, respectively:

As(V) $[Ar]3d^{10}$ 4s [↑] 4p [↑|↑|↑] 4d [↑| | | | ]

Se(IV) $[Ar]3d^{10}$ 4s [↑↓] 4p [↑|↑|↑] 4d [↑| | | | ]

Se(VI) $[Ar]3d^{10}$ 4s [↑] 4p [↑|↑|↑] 4d [↑|↑| | | ]

71. (a) $NaH_2PO_2$ contains the $H_2PO_2^-$ ion (see Problem 69(e)) so the parent acid must be $H_3PO_2$, with the structure $H_2P(O)OH$

(b) Hypophosphorous acid (c)

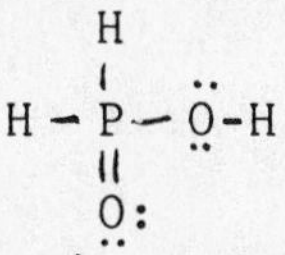

(d) $P_4$ contains P(0), $PH_3$ contains P(-3), and $H_2PO_2^-$ contains P(+1)

(e) $P_4$ is reduced to $PH_3$, and oxidized to $H_2PO_2^-$

(f) Molecular nitrogen, $N_2$, is very unreactive because it contains a strong triple bond, :N≡N:

(g) Calculate mol $PH_3$ from the gas data.

$$\text{mol } PH_3 = \frac{PV}{RT} = \frac{(0.95 \text{ atm})(100 \text{ mL})(\frac{1 \text{ L}}{1000 \text{ mL}})}{(0.0821 \text{ atm L mol}^{-1} \text{ K}^{-1})(298 \text{ K})} = \underline{3.883\text{x}10^{-3} \text{ mol}}$$

$$\text{Minimum mass of P} = (3.883\text{x}10^{-3} \text{ mol } PH_3)(\frac{1 \text{ mol } P_4}{1 \text{ mol } PH_3})(\frac{123.9 \text{ g } P_4}{1 \text{ mol } P_4})$$
$$= \underline{0.481 \text{ g}}$$

72. $SO_2$ reacts with $O_2$ from air to give $SO_3$, which combines with water to give $H_2SO_4$,

$$2SO_2(g) + O_2(g) \longrightarrow 2SO_3(g)$$
$$SO_3(g) + H_2O(l) \longrightarrow H_2SO_4(l)$$

$10^3$ kg coal contains $(\frac{5 \text{ g S}}{100 \text{ g coal}})(10^3 \text{ kg coal}) = 50$ kg S, which gives

$$(50 \text{ kg S})(\frac{1000 \text{ g}}{1 \text{ kg}})(\frac{1 \text{ mol S}}{32.06 \text{ g S}})(\frac{1 \text{ mol } H_2SO_4}{1 \text{ mol S}})(\frac{98.08 \text{ g } H_2SO_4}{1 \text{ mol } H_2SO_4})(\frac{1 \text{ kg}}{1000 \text{ g}})$$
$$= \underline{153 \text{ kg}} \; H_2SO_4$$

For neutralization of $H_2SO_4$ with $Ca(OH)_2(s)$, we have

$$H_2SO_4 + Ca(OH)_2 \longrightarrow CaSO_4 + 2H_2O$$

Mass of $Ca(OH)_2(s)$ to neutralize 153 kg $H_2SO_4$ =

$$(153 \text{ kg } H_2SO_4)(\frac{1 \text{ mol } H_2SO_4}{98.08 \text{ g } H_2SO_4})(\frac{1 \text{ mol } Ca(OH)_2}{1 \text{ mol } H_2SO_4})(\frac{74.10 \text{ g } Ca(OH)_2}{1 \text{ mol } Ca(OH)_2})$$
$$= \underline{116 \text{ kg}} \; Ca(OH)_2(s)$$

73. Among the yellow solids encountered in this chapter, the most important was sulfur, which fits the observations described.

A is sulfur, which when heated in air burns to give sulfur dioxide gas, B, according to

$$S(s) + O_2(g) \longrightarrow SO_2(g)$$

and $SO_2(g)$ is soluble in water, to give a solution of sulfurous acid, $H_2SO_3(aq)$, C,

$$SO_2(g) + H_2O(l) \longrightarrow H_2SO_3(aq)$$

Sulfur, A, combines with iron to give iron(II) sulfide, FeS(s), D, which is a black solid,

$$Fe(s) + S(s) \longrightarrow FeS(s)$$

and FeS(s) reacts with dilute acid to give hydrogen sulfide gas, $H_2S(g)$, E.

$$FeS(s) + H_2SO_4(aq) \longrightarrow FeSO_4(aq) + H_2S(g)$$

$H_2S(g)$ in aqueous solution is oxidized to yellow sulfur, A, by $H_2SO_3(aq)$,

$$2H_2S(g) + H_2SO_3(aq) \longrightarrow 3S(s) + 3H_2O(l)$$

74. X has the properties described in the text for white phosphorus, $P_4$, which is confirmed by calculating the molar mass from the gas data.

$$\text{mol of X} = \frac{PV}{RT} = \frac{(1\ \text{atm})(1\ \text{L})}{(0.0821\ \text{atm L mol}^{-1}\ \text{K}^{-1})(673\ \text{K})} = \underline{1.81\text{x}10^{-2}\ \text{mol}}$$

$$\text{Molar mass of X} = \left(\frac{2.242\ \text{g X}}{0.0181\ \text{mol X}}\right) = \underline{124\ \text{g mol}^{-1}}\ \ (\text{c.f. } P_4,\ 123.9\ \text{g mol}^{-1})$$

For the gaseous reaction with $Cl_2(g)$,

$$\begin{array}{ccccc} P_4(g) & + & 6Cl_2(g) & \longrightarrow & 4PCl_3(g) \\ 1\ \text{vol} & & 6\ \text{vol} & & 4\ \text{vol} \end{array}$$

which is consistent with the observed reaction of 25 mL $P_4(g)$ with 150 mL $Cl_2(g)$ to give 100 mL $PCl_3(g)$, (Avogadro's law of combining volumes). The formation of $PCl_3(g)$ is confirmed by the elemental composition.

$$\text{mass \% P in } PCl_3 = \left(\frac{106.4\ \text{g}}{137.3\ \text{g}}\right) \times 100\% = 77.5\%\ (\text{c.f. } 77.4\%\ \text{experimental})$$

The expected molar mass of $PCl_3$ is 137.3 g $mol^{-1}$ (c.f. 137.3 g $mol^{-1}$), and at room temperature it is a liquid (b.p. 74°C).

Lewis structures:

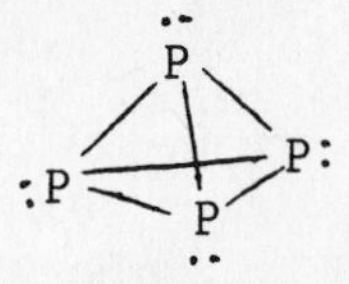

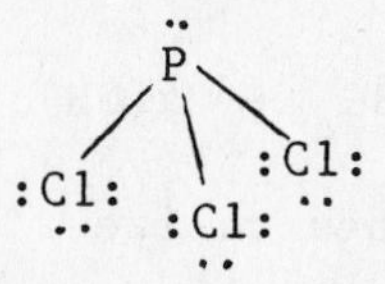

75. First calculate the empirical formula: 4.00 g P react with (14.35-4.00) = 10.35 g S.

| | | P | | S |
|---|---|---|---|---|
| Moles | = | $\frac{4.00\ \text{g}}{30.97\ \text{g}}$ | | $\frac{10.35\ \text{g}}{32.06\ \text{g}}$ |
| | = | 0.129 | | 0.323 |
| Ratio of moles (atoms) | = | $\frac{0.129}{0.129}$ | : | $\frac{0.323}{0.129}$ |
| | = | 1.00 | : | 2.50 |
| | = | 2.00 | : | 5.00 |

Thus, the empirical formula is $P_2S_5$, which by analogy with phosphorus(V) oxide, $P_4O_{10}$, suggests that the molecular formula is $P_4S_{10}$, with the structure

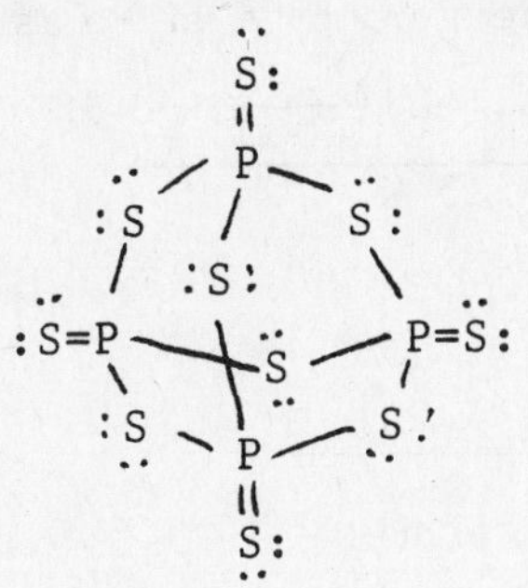

76. From the given data, calculate the composition of the compound.

25.00 mL of the 250 mL of solution containing 0.9209 g X reacts with 32.47 mL 0.0500 M $AgNO_3(aq)$ to precipitate all of the Cl in the sample as AgCl(s).

$$AgNO_3(aq) + Cl^-(aq) \longrightarrow AgCl(s) + NO_3^-(aq)$$

Mol Cl in sample is given by

$$(32.47 \text{ mL } AgNO_3)\left(\frac{0.0500 \text{ mol}}{1 \text{ L}}\right)\left(\frac{1 \text{ mol } Cl^-}{1 \text{ mol } AgNO_3}\right)\left(\frac{250 \text{ mL sol'n}}{25 \text{ mL sol'n}}\right)\left(\frac{1 \text{ L}}{1000 \text{ mL}}\right)$$

$$= \underline{1.624\text{x}10^{-2} \text{ mol } Cl^-}$$

0.4437 g X gives 0.2932 g $Mg_2P_2O_7(s)$, Y, which contains

$$(0.2932 \text{ g Y})\left(\frac{1 \text{ mol Y}}{222.6 \text{ g Y}}\right)\left(\frac{2 \text{ mol P}}{1 \text{ mol Y}}\right) = \underline{2.634\text{x}10^{-3} \text{ mol P}}$$

Now convert mol Cl and mol P to mass %, to obtain mass % S by difference.

$$\text{mass Cl} = (1.624\text{x}10^{-2} \text{ mol})\left(\frac{35.45 \text{ g Cl}}{1 \text{ mol Cl}}\right) = \underline{0.5757 \text{ g Cl}}$$

$$\text{mass P} = (2.534\text{x}10^{-3} \text{ mol})\left(\frac{30.97 \text{ g P}}{1 \text{ mol P}}\right) = \underline{0.08157 \text{ g P}}$$

$$\text{mass \% Cl} = \left(\frac{0.5757 \text{ g}}{0.9209 \text{ g}}\right) \times 100 = \underline{62.52\%}$$

$$\text{mass \% P} = \left(\frac{0.08157 \text{ g}}{0.4437 \text{ g}}\right) \times 100 = \underline{18.39\%}$$

and, by difference, mass % S = (100.0-62.5-18.4) = <u>19.1%</u>

Now calculate the empirical formula.

| | | P | S | Cl |
|---|---|---|---|---|
| Grams in 100g | = | 18.4 | 19.1 | 62.5 |
| Mol in 100 g | = | $\frac{18.4}{30.97}$ | $\frac{19.1}{32.06}$ | $\frac{62.5}{35.45}$ |
| Ratio of moles (atoms) | = | 0.594 : | 0.596 : | 1.763 |
| | = | $\frac{0.594}{0.594}$ : | $\frac{0.596}{0.594}$ : | $\frac{1.763}{0.594}$ |
| | = | 1.00 : | 1.00 : | 2.97 |

The <u>empirical formula</u> is $PSCl_3$ (formula mass 169.4 u)

The molar mass of X is calculated from the gas data and mass of X.

$$\text{Mol X} = \frac{PV}{RT} = \frac{(444.5\ \text{mm Hg})\left(\frac{1\ \text{atm}}{760\ \text{mm Hg}}\right)(300\ \text{mL})\left(\frac{1\ \text{L}}{1000\ \text{mL}}\right)}{(0.0821\ \text{atm L mol}^{-1}\ \text{K}^{-1})(523\ \text{K})}$$

$$= 4.10\text{x}10^{-3}\ \text{mol}$$

$$\text{Molar mass of X} = \left(\frac{0.6775\ \text{g X}}{4.10\text{x}10^{-3}\ \text{mol X}}\right) = 165\ \text{g mol}^{-1}$$

The molar mass = the empirical formula mass.

Thus, the molecular formula is $PSCl_3$ (containing P(V)).

Lewis structure:

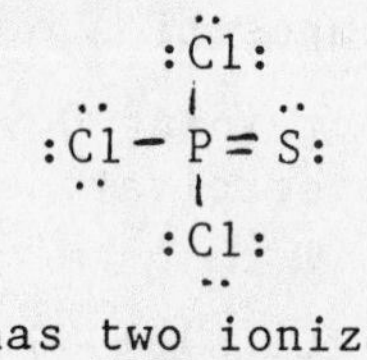

77. Sulfuric acid, $(HO)_2SO_2$, has two ionizable protons and is referred to as a diprotic acid; phosphoric acid, $(HO)_3PO$, has three ionizable protons and is referred to as a triprotic acid.

In terms of the $XO_n(OH)_m$ formulation, $H_2SO_4$ has n = m = 2, and $H_3PO_4$ has n = 1, and m =3. Thus, sulfuric acid is a strong acid and phosphoric acid is a weak acid, in terms of their first stages of ionization in water.

The products of dehydration of these acids are their anhydrides, which are the acidic oxides $SO_3$ and $P_4O_{10}$, respectively.

$$H_2SO_4 \longrightarrow SO_3 + H_2O$$

$$4H_3PO_4 \longrightarrow P_4O_{10} + 6H_2O$$

78. All the substances are gases. Helium (c) is characterized by its complete lack of reactivity. Only hydrogen (a) is flammable in air, and burns with a pale blue flame to give water, which can be condensed on a cold surface and identified by its reaction with colorless $CuSO_4$(s), which turns blue. Alone among the gases, oxygen, (b) ignites a glowing splint. Sulfur dioxide (e), ammonia (g), chlorine (d), and hydrogen chloride (h) have distinct characteristic odors. Chlorine (d) and sulfur dioxide (e) bleach moistened litmus paper. Sulfur dioxide (e), carbon dioxide (f), and hydrogen chloride (h) are acids and change the color of moistened blue litmus paper to red. Ammonia (g) is the only base, and turns the color of moistened litmus paper from red to blue. Among the acidic gases, hydrogen chloride reacts with $NH_3$(g) to give a white cloud of $NH_4Cl$(s).

# CHAPTER 9

1. In writing Lewis structures we follow the following procedures:

   1. Connect all of the atoms to the central atom by single bonds
   2. Count up all of the available valence electrons, obtaining the number for each atom from its group in the periodic table, and adding or subtracting the requisite number of electrons to allow for any charge on the species
   3. Convert the total number of electrons to electron pairs and subtract the number of pairs already used to form single bonds
   4. Use these electron pairs to complete the octets (four electron pairs) around each atom, starting with the most electronegative atoms
   5. If any octets remain incomplete, delocalize unshared electron pairs to form double (or triple) bonds until all of the octets are complete, unless this generates greater formal charges on any of the atoms
   6. Delocalize additional electron pairs to exceed an octet and minimize the formal charges if there are atoms from the third period and beyond, such as S or P.  Thus:

   (a) $S_2^{2-}$  1. S-S  2. S is in group 6, so $S_2^{2-}$ has (6+6+2) = 14 electrons, or 7 pairs

   3. 7-1 = 6 pairs of electrons remain. Completing the octet on each S atom gives:

   $$:\overset{..}{\underset{..}{S}}-\overset{..}{\underset{..}{S}}:$$

   and assigning formal charges gives  $^-:\overset{..}{\underset{..}{S}}-\overset{..}{\underset{..}{S}}:^-$

   (b) $SO_3^{2-}$  1. O-S-O  2. Both S and O are in group 6, so $SO_3^{2-}$ has [4(6)+2]
   |
   O
   = 26 electrons, or 13 pairs

   3. 13-3 = 10 pairs of electrons remain to be distributed to give

   4.  $^-:\overset{..}{\underset{..}{O}}-\overset{..}{\underset{}{S}}^+-\overset{..}{\underset{..}{O}}:^-$
   |
   $:\overset{..}{\underset{..}{O}}:^-$

   5. Since S is a third period atom it can exceed the octet; delocalization of a lone pair from one of the O atoms to share with S gives

   $$:\overset{..}{O}=\overset{..}{\underset{}{S}}-\overset{..}{\underset{..}{O}}:^-$$
   |
   $:\overset{}{\underset{..}{O}}:^-$

   This is the required Lewis structure; delocalization of further lone pairs would produce unacceptable - formal charges on the S atom.

   (c)  $:\overset{..}{\underset{..}{Cl}}-\overset{..}{\underset{..}{O}}:^-$   (d)

   $^-:\overset{..}{\underset{..}{O}}-\overset{\overset{:\overset{..}{O}:^-}{|}}{\underset{\underset{:\overset{..}{O}:^-}{|}}{Cl}}^{3+}-\overset{..}{\underset{..}{O}}:^-$  $\longrightarrow$  $:\overset{..}{O}=\overset{\overset{:O}{||}}{\underset{\underset{:O}{||}}{Cl}}-\overset{..}{\underset{..}{O}}:^-$

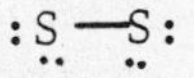

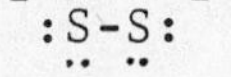

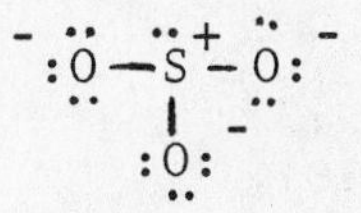

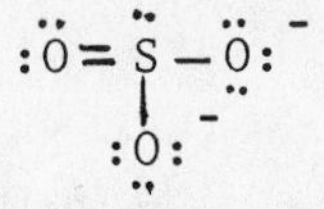

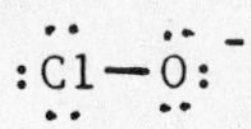

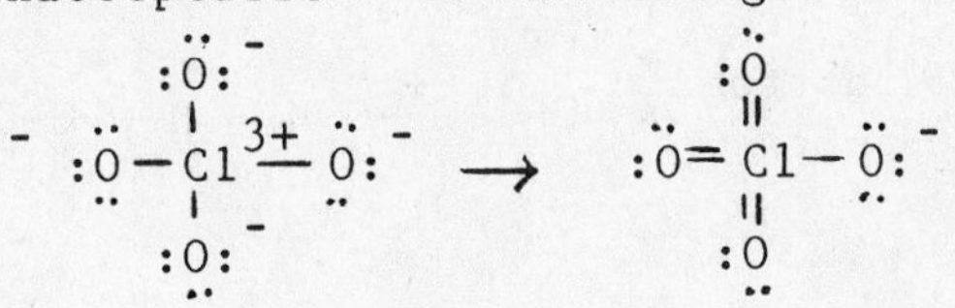

2.

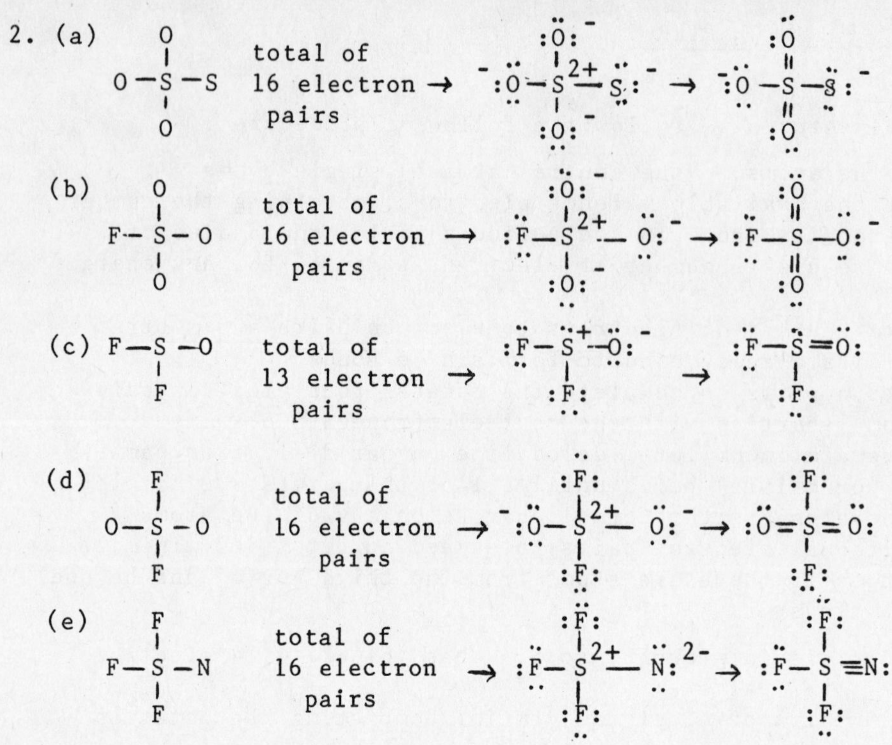


3. We should first write the formulas to indicate the -OH groups in these oxoacids:

HO-NO  $HO\text{-}NO_2$  $(HO)_2SO_2$  $HO\text{-}ClO_2$  $(HO)_2CO$

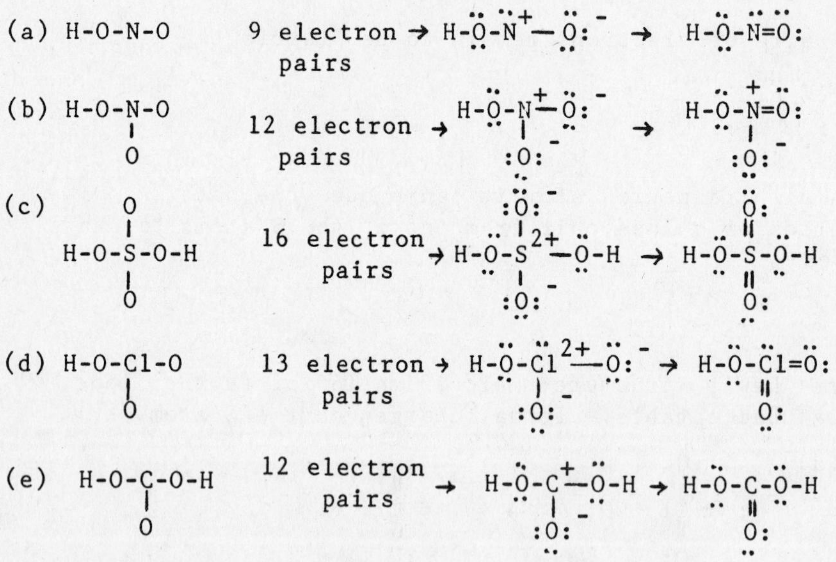


4. $\underline{CN^-}$ has 10 valence electrons or 5 electron pairs: $:C\equiv N:^-$

$\underline{NO_2}^-$ 18 valence electrons or 9 pairs: O-N-O →

$\underline{O_3}$ 18 valence electrons or 9 pairs: O-O-O →

5. 16 valence electrons, or 8 electron pairs:

<u>NON</u>

$\text{N-O-N} \rightarrow {}^{2-}:\ddot{\text{N}}-\text{O}{}^{4+}\text{—}\ddot{\ddot{\text{N}}}:{}^{2-} \rightarrow {}^{-}:\ddot{\text{N}}=\text{O}{}^{2+}=\text{N}:{}^{-}$ <u>or</u> $:\text{N}{\equiv}\text{O}{}^{2+}-\ddot{\ddot{\text{N}}}:{}^{2-}$

<u>NNO</u>

$\text{N-N-O} \rightarrow {}^{2-}:\ddot{\text{N}}-\text{N}{}^{3+}\text{—}\ddot{\ddot{\text{O}}}:{}^{-} \rightarrow {}^{-}:\ddot{\text{N}}=\overset{+}{\text{N}}=\ddot{\text{O}}:$ <u>or</u> $:\text{N}{\equiv}\overset{+}{\text{N}}-\ddot{\ddot{\text{O}}}:{}^{-}$ <u>or</u> ${}^{2-}:\ddot{\text{N}}-\overset{+}{\text{N}}{\equiv}\overset{+}{\text{O}}:$

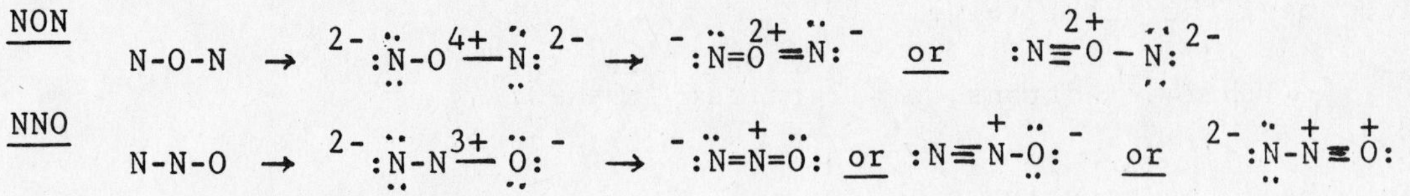

Of these structures, $:\ddot{\text{N}}=\overset{+}{\text{N}}=\ddot{\text{O}}:$ and $:\text{N}{\equiv}\overset{+}{\text{N}}-\ddot{\text{O}}:^{-}$ with the smallest formal charges are the most probable in terms of their contribution to the observed structure. In other words, the arrangement NNO is preferred. Generally the less electronegative atom is the central atom.

6. Since H has a valence of only one, these molecules must contain bonds between the two N atoms:

$\underline{\text{N}_2\text{H}_4}$    H-N-N-H with 14 valence electrons, or 7 pairs:   H-$\ddot{\text{N}}$-$\ddot{\text{N}}$-H
             | |                                              | |
             H H                                              H H

$\underline{\text{N}_2\text{H}_2}$    H-N-N-H    with 12 valence electrons, or 6 pairs:   H-$\ddot{\text{N}}$=$\ddot{\text{N}}$-H

7. (a) Cl-C-Cl with 24 valence
        |
        O       electrons <u>or</u> 12 pairs

     $:\ddot{\text{C}}\text{l}$               $:\ddot{\text{C}}\text{l}$
         $\overset{+}{\diagdown}\text{C}-\ddot{\text{O}}:^{-} \longrightarrow$    $\diagdown\text{C}=\ddot{\text{O}}:$
     $:\ddot{\text{C}}\text{l}$               $:\ddot{\text{C}}\text{l}$

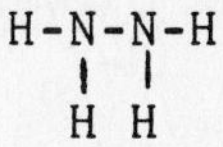

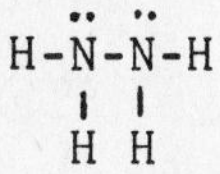

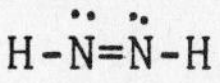

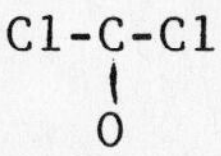

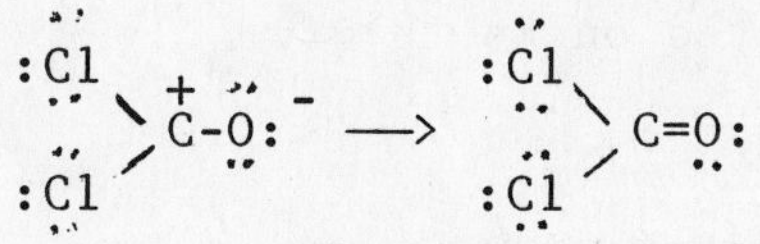

<ClCCl ~ 110°; <ClCO ~ 125°

(b)        $\ddot{\text{N}}=\ddot{\text{N}}$
      $:\ddot{\text{F}}:$    $\ddot{\text{F}}$    <FNF ~ 120°

(c) $:\text{O}=\text{C}=\ddot{\text{S}}:$   <OCS = 180°

(d)   H $\diagdown$      H     <HCH <120°
         C=C $\diagdown$        <HCC >120°
    H $\diagup$      C'      <CCN = 180°
                  N:

(e)       S
        ‖
   $:\ddot{\text{F}}:$ | $\ddot{\text{O}}:$    <FSF <110°; <FSO >110°
      $:\ddot{\text{F}}:$

8. (a) The middle carbon atom has $AX_2$ geometry; thus, C=C=C is linear.

   (b) The two end $CH_2$ groups form planes that are perpendicular to each other, as can be seem by depicting each double bond as two bent bonds:

     H $\diagdown$                  H
        C    C    C
   H $\diagup$                H

   Each $CH_2$ group is in a plane perpendicular to the plane formed by the two bent bonds to the same C atom - thus the two double bonds are in perpendicular planes, and the two $CH_2$ groups are in perpendicular planes.

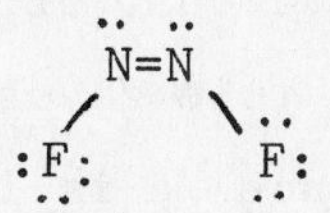

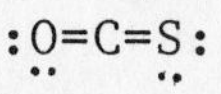

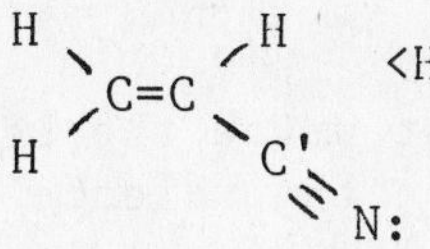

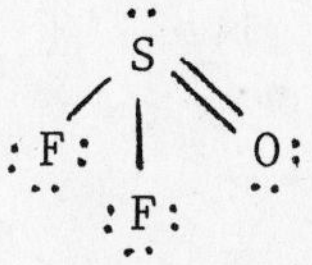

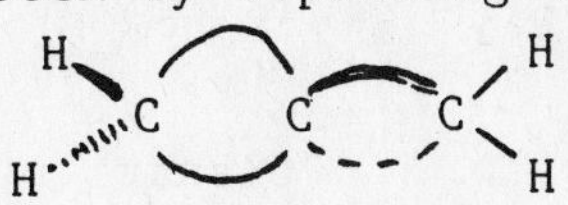

9. (a)     Cl                                        $:\ddot{\text{C}}\text{l}:$
         |                                              |
   Cl — B     with 24 valence electrons, or 12 pairs:   $:\ddot{\text{C}}\text{l}$ — B
         |                                              |
        Cl                                        $:\ddot{\text{C}}\text{l}:$

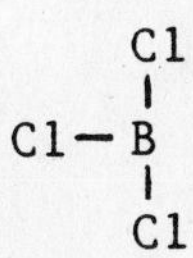

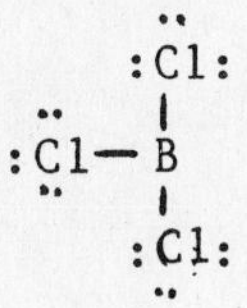

which is an exception to the octet rule.

(b) O-N-O with 16 valence electrons, or 8 pairs. $:\ddot{O}=\overset{+}{N}=\ddot{O}:$

which obeys the octet rule.

(c) O-N-O with 24 electrons, or 12 pairs: $:O=\overset{+}{N}-\ddot{O}:^-$ (with $:\ddot{O}:^-$ bonded to N)

which obeys the octet rule.

(d) $[O\text{-}Br\text{-}O]^-$ with 22 valence electrons, or 11 pairs. Applying the octet rule gives

$^-:\ddot{O}-\ddot{Br}^+-\ddot{O}:^-$, but since Br is in period 4 it can <u>exceed</u> the octet rule, to give $:\ddot{O}=\ddot{Br}-\ddot{O}:^-$

(e) F-P-F with 26 valence electrons, or 13 pairs, $:\ddot{F}-\ddot{P}-\ddot{F}:$ (with $:\ddot{F}:$ bonded to P), which obeys the octet rule

(f) F-Xe-F with 22 electrons, or 11 electron pairs.

Applying the octet rule gives $:\ddot{F}-\ddot{Xe}-\ddot{F}:$, which requires only 10 electron pairs. Since Xe is in period 5, it can exceed the octet rule and the 11th electron pair can be accommodated on Xe to give

$:\ddot{F}-\dot{\ddot{Xe}}-\ddot{F}:$

10. In each case we first write the Lewis structure, and then express the electron pairs in the valence shell of the central atom A in the $AX_nE_m$ nomenclature. The arrangement of the n+m electron pairs gives the arrangement of electron pairs, and the arrangement of the n bonding pairs gives the molecular shape.

| Molecule | Lewis Structure | $AX_nE_m$ | Molecular Shape |
|---|---|---|---|
| $BH_3$ | H-B-H (with H bonded to B) | $AX_3$ | planar triangular |
| $BO_3^{3-}$ | $^-:\ddot{O}-B-\ddot{O}:^-$ (with $:\ddot{O}:^-$ bonded to B) | $AX_3$ | planar triangular |
| $SnCl_3^-$ | $:\ddot{Cl}-\ddot{Sn}^--\ddot{Cl}:$ (with $:\ddot{Cl}:$ bonded to Sn) | $AX_3E$ | trigonal pyramidal |
| $H_2Se$ | $H-\ddot{Se}-H$ | $AX_2E_2$ | angular |
| $SiF_5^-$ | $Si^-$ bonded to five $:\ddot{F}:$ | $AX_5$ | trigonal bipyramidal |

| Molecule | Lewis Structure | $AX_nE_m$ | Molecular Shape |
|---|---|---|---|
| $BrF_5$ | 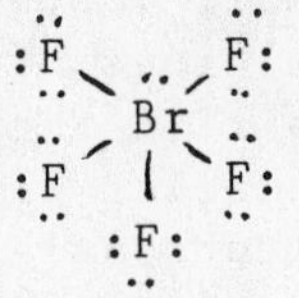 | $AX_5E$ | square pyramidal |

11. The arrangement of electron pairs is determined by the total number of bonding electron pairs and unshared (lone) pairs of electrons in the valence shell of a central atom A, but the geometric shape is determined only by the positions of the atom A and the ligands X. Thus, $AX_4$, $AX_3E$, and $AX_2E_2$ molecules all have a tetrahedral arrangement of electron pairs in the valence shell of A. The shape of a $AX_4$ molecule is tetrahedral because it is determined by the tetrahedral arrangement of four X ligands around A. However, an $AX_3E$ molecule has a shape determined by the positions of the A atom and three X ligands, i.e., that of a tetrahedron with three corners occupied by X ligands, which is described as triangular pyramidal, and an $AX_2E_2$ molecule has a shape determined by the positions of the A atom and two X ligands, i.e., that of a tetrahedron with two corners occupied by X ligands, which is described as angular.

12. First write the Lewis structures and then count up the number of bonds, n, to the central atom A, and its number of unshared electron pairs, m. The arrangement of bonds and unshared electron pairs is determined by their total number (n+m); the geometric shape is determined by the positions of the central atom A and the n ligands:

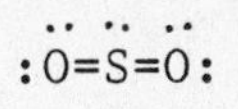 $AX_2E$ angular

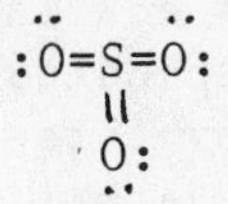 $AX_3$ triangular planar

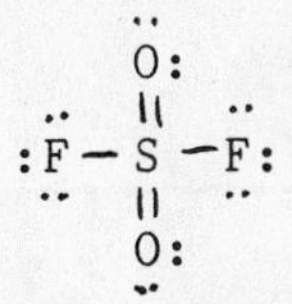 $AX_4$ tetrahedral

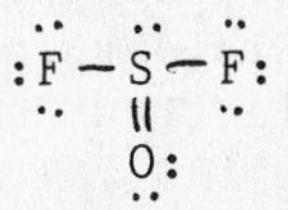 $AX_3E$ triangular pyramidal

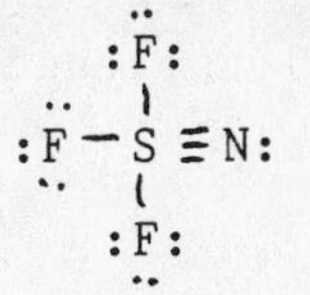 $AX_4$ tetrahedral

13. In an $AX_4$ molecule, with all identical X ligands, all of the bonds are equivalent, so the shape is exactly regular tetrahedral, with all the bond angles XAX equal to 109° 28', (109.5°). The Lewis structures of $NH_3$ and $H_2O$ are:

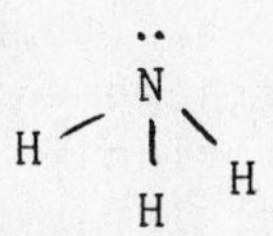

$AX_3E$

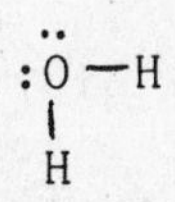

$AX_2E_2$

Because they are attracted by a single nucleus, unshared (lone) pairs of electrons occupy more room in the valence shell of the atom A than do bonding pairs (which are attracted away from A towards the ligand atom X). The consequence of this is to make lone pair : bond pair repulsions greater than bond pair : bond pair repulsions. In $NH_3$ the bonding electrons are repelled by a single lone pair in the valence shell of the N atom; thus, the HNH bond angle is smaller than the tetrahedral angle of $109.5^{\circ}$. In $H_2O$ the bonding electrons are repelled by two lone pairs in the valence shell of the O atom; thus, the HOH bond angle is also smaller than the tetrahedral angle of $109.5^{\circ}$, and even smaller than the HNH bond angle of $NH_3$, due to the greater repulsions of two lone pairs as compared to that of one lone pair.

14. The important factor affecting the bond angles here is the relative electronegativities of the atoms involved in bonding; for bonds to a common atom A, the bond to an atom X occupies less and less space in the valence shell of A the greater the electronegativity of X, because with increasing electronegativity of X the bonding electron pair is increasingly strongly drawn into the space between A and X. When two molecules have central atoms from different periods and lone pairs of electrons in their valence shells, the bond angles are primarily influenced by the amount of space occupied by these lone pairs, which, for example is much greater for a third period atom than it is for a second period atom.

    (a) F is more electronegative than H; <FOF < <HOH.

    (b) N is in period 2 and P is in period 3; <HNH > <HPH

    (c) F is more electronegative than H; <FNF < <HNH

15. The shape of an $AX_5$ molecule is trigonal bipyramidal, with three bonds directed towards the corners of an equilateral triangular (equatorial), and two colinear (axial) bonds perpendicular to the equatorial plane. Each axial bond pair is repelled by three equatorial bond pairs making angles of $90^{\circ}$ to it, while each equatorial bond pair is repelled by only two bond pairs at $90^{\circ}$ to it, and two bond pairs at $120^{\circ}$ to it. Because inter electron pair repulsions fall off rapidly with increasing distance (Coulomb's law), electron pair repulsions at $90^{\circ}$ are far more important than those at $120^{\circ}$. Thus, axial bonds experience relatively much greater repulsions with other electron pairs than do equatorial bonds, and axial bonds are lengthened relative to equatorial bonds.

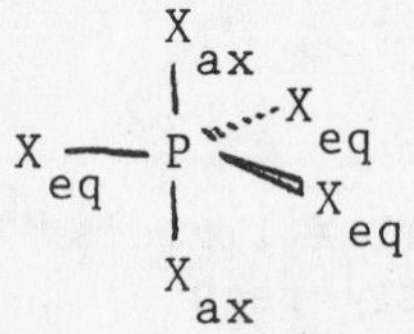


16. Lone pairs occupy more space in a valence shell than do bonding pairs, because they are attracted by only one nucleus rather than two nuclei. In molecules based on the trigonal bipyramidal arrangement of five electron pairs, the three equatorial positions provide more space to accomodate lone pairs than do the axial positions. Thus, lone pairs are always found to preferentially occupy these equatorial positions, rather than the axial positions. Examples include:

| $\underline{AEX_4}$ | $\underline{AE_2X_3}$ | $\underline{AE_3X_2}$ |
|---|---|---|
| $:SF_4$ disphenoid | $:\ddot{Cl}F_3$ T-shaped | $:\ddot{Xe}F_2$, $:\ddot{I}-\ddot{I}-\ddot{I}:$ linear |

17. The Lewis structures for $CO_3^{2-}$ and $SO_3^{2-}$ are, respectively:

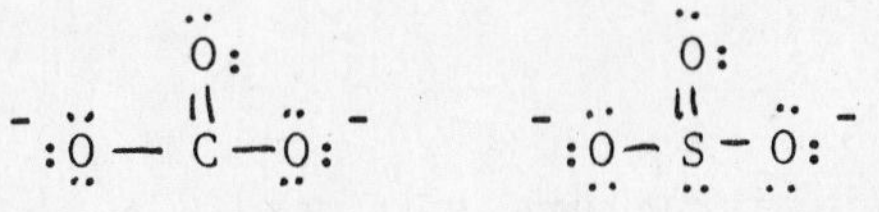

which imply that each species should have a double bond and two single bonds to a central atom, which are of substantially different lengths. However, experimentally it is found that all the CO bonds in $CO_3^{2-}$ are of equivalent lengths, as are the SO bonds in $SO_3^{2-}$, and in each case the experimental bond lengths are intermediate in length between those expected for a single bond and a double bond. Thus, a single Lewis structure is inadequate to depict such species. When we draw the Lewis structures, we first write

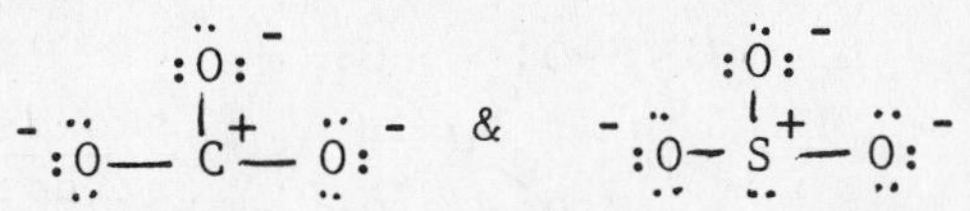

and then complete the valence shell of the central atom by delocalizing a pair of electrons from an oxygen into the bonding region. However, for species such as these with a number of O atoms singly bonded, there is no reason to suppose that any particular O atom will delocalize a lone pair of electrons in this way, in preference to any of the other "equivalent" O atoms; in fact all of the equivalent O atoms participate to the same extent in delocalizing electrons to complete the valence shell of the central atom, but there is no simple way to depict such a phenomenon. In each of these species, each O atom delocalizes one third of an electron pair, but since no one Lewis structure is adequate to represent this, we achieve the same result by writing all of the possible Lewis structures, none of which represents the bonding in the actual species, and then average them:

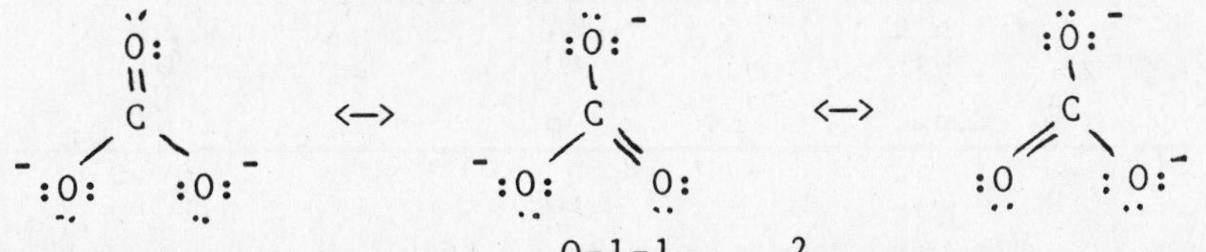

Average charge on each O atom $= \frac{0-1-1}{3} = -\frac{2}{3}$

Average CO bond order $= \frac{2+1+1}{3} = 1\frac{1}{3}$

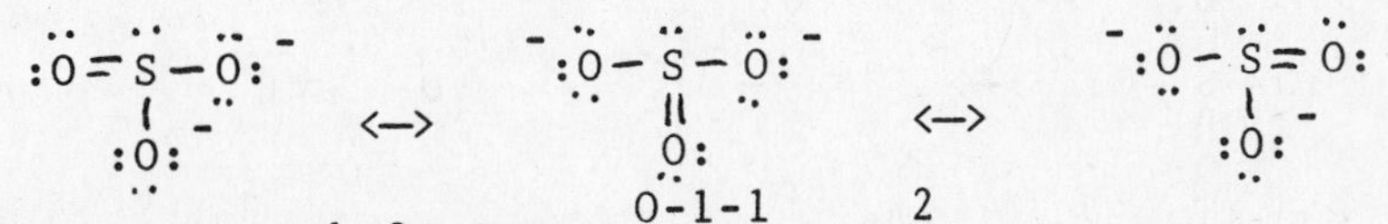

Average charge on each O atom $= \frac{0-1-1}{3} = -\frac{2}{3}$

Average SO bond order $= \frac{2+1+1}{3} = 1\frac{1}{3}$

In each case, the observed bond length is consistent with a bond order of $1\frac{1}{3}$, giving a bond intermediate in length between the lengths of the respective single and double bonds.

18. $\underline{CH_3CO_2^-}$ 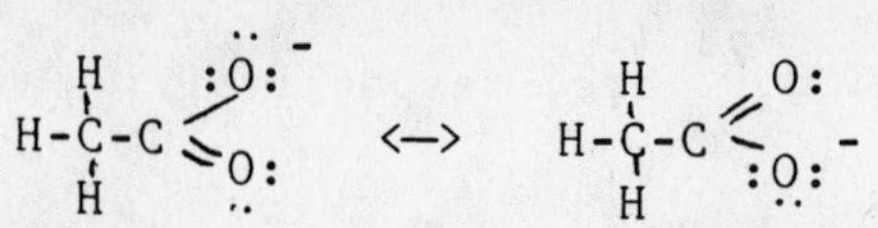 $\underline{\text{two resonance structures}}$

$\underline{CH_3OH}$ 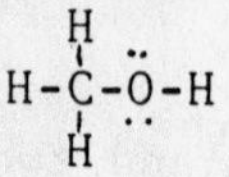 only $\underline{\text{one}}$ structure required

$\underline{ClO_3^-}$ 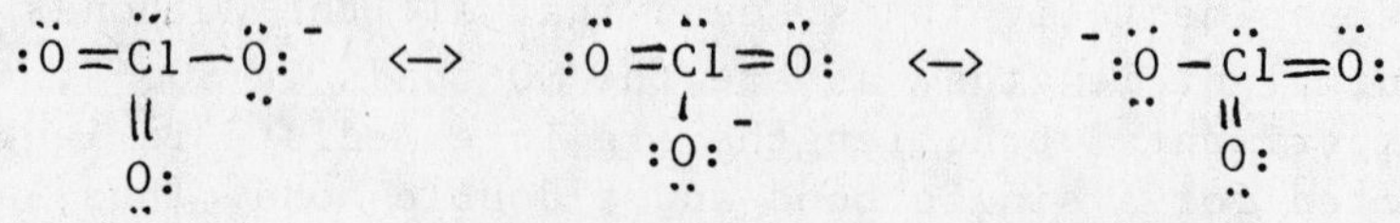

$\underline{\text{three resonance structures}}$

$\underline{(HO)_2PO_2^-}$ 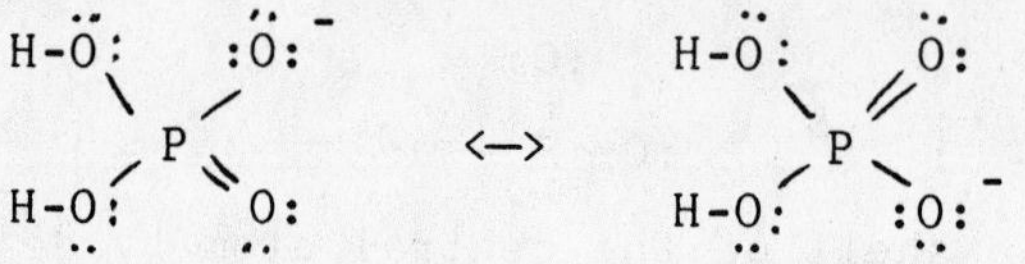 $\underline{\text{two resonance structures}}$

19. (a) 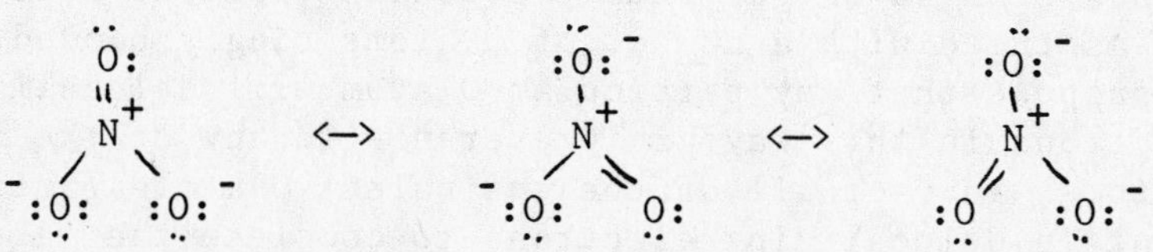

Formal charge on N = +1; formal charge on each O atom = $\frac{0-1-1}{3} = -\frac{2}{3}$

NO bond order = $\frac{2+1+1}{3} = 1\frac{1}{3}$

(b) 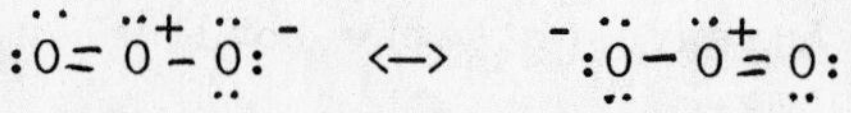

Formal charges: Central O atom, +1; terminal O atoms, $\frac{0-1}{2} = -\frac{1}{2}$

(c) 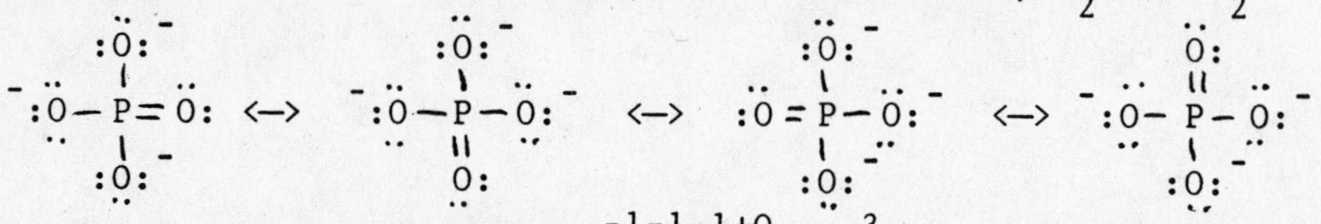

Formal charge on each O atom = $\frac{-1-1-1+0}{4} = \frac{3}{4}$

PO bond order = $\frac{1+1+1+2}{4} = 1\frac{1}{4}$

(d) 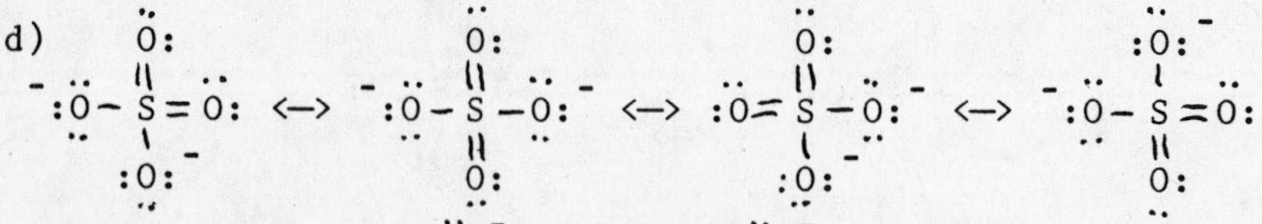

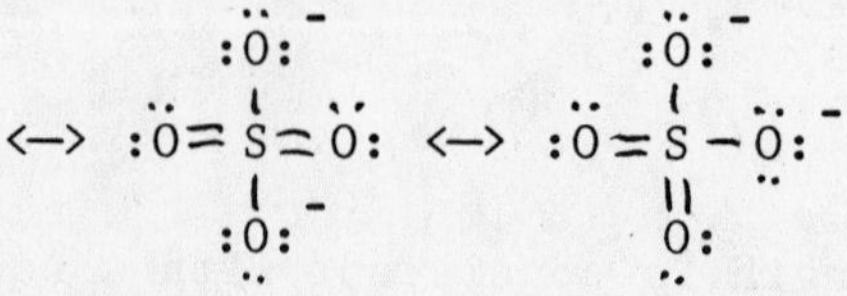

Formal charge on each O atom = $\frac{0+0+0-1-1-1}{6} = -\frac{1}{2}$

SO bond order = $\frac{2+2+2+1+1+1}{6} = 1\frac{1}{2}$

(e)

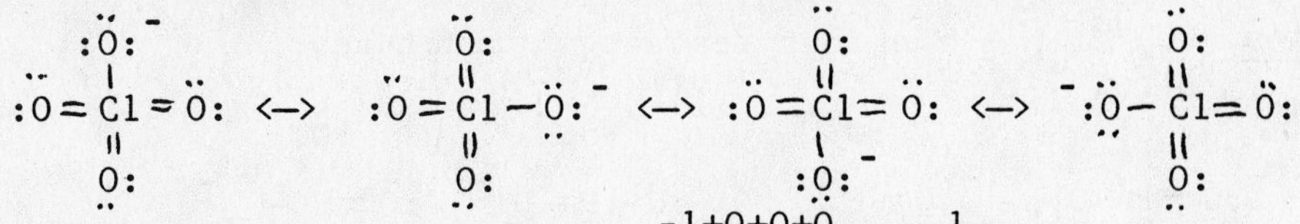

Formal charge on each O atom $= \frac{-1+0+0+0}{4} = -\frac{1}{4}$

ClO bond order $= \frac{1+2+2+2}{4} = 1\frac{3}{4}$

Note that the formal charge and the bond order is calculated selecting the top O atom in each case; the same result is obtained using any other O atom.

20. $ClO^-$ has the single Lewis structure $:\ddot{C}l-\ddot{O}:^-$, for a ClO bond order of 1.00, while $ClO_4^-$ has four possible resonance structures (see Problem 19(e)), with each ClO bond of order 1.75. Since bond length decreases as bond order increases, the ClO bond in $Cl-O^-$ is longer than the ClO bond in $ClO_4^-$.

21. Nitric acid, $HONO_2$, has the two important resonance structures:

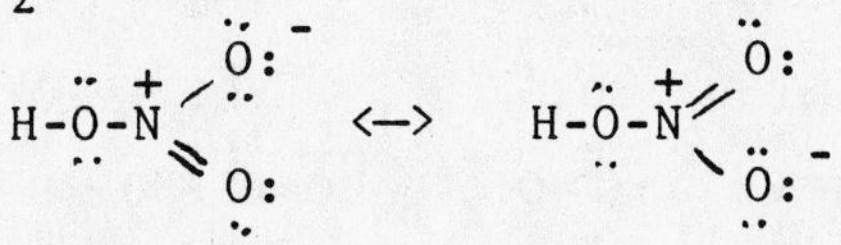

so that the N-OH bond is expected to have a bond order of 1.00, and the two NO bonds are expected to have bond orders of 1.50. Thus, the NO bond of length 136 pm is that of the N-OH bond, and the bond length of 121 pm is that of the NO bonds.

The nitrate ion has the three resonance structures

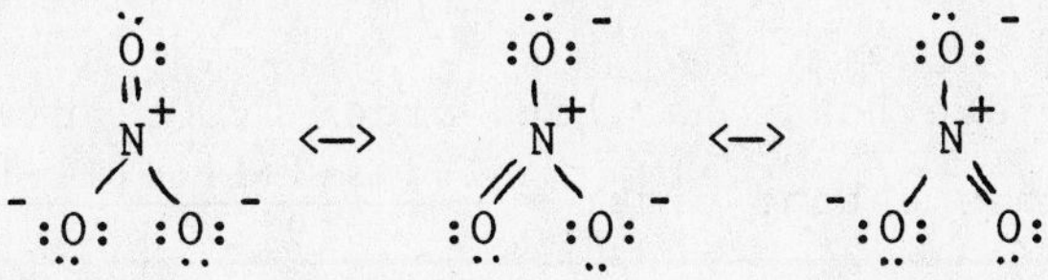

with three NO bonds of order 1.33, corresponding to the observed bond length of 126 pm, and the nitrite ion has the two resonance structures

with the two NO bonds each of order 1.50 and a bond length of 121 pm. Thus, for these species:

| NO bond order: | 1.00 | 1.33 | 1.50 |
|---|---|---|---|
| Bond length (pm) | 136 | 126 | 121 |

As expected, NO bond length decreases with increasing bond order.

22. $P_4O_{10}$ has a cage structure with single P-O bonds and each P atom also has a terminal P=O bond

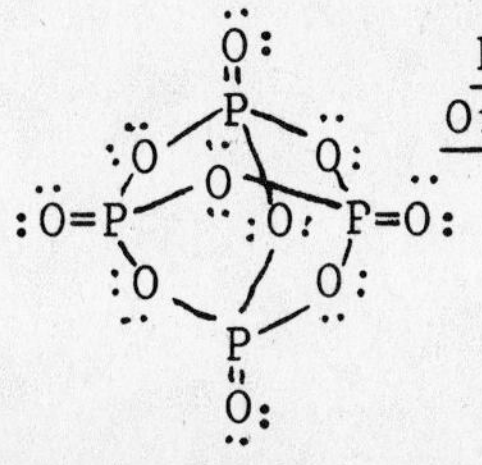

| Bond Order | Bond Length (pm) |
|---|---|
| 1.00 | 160 |
| 2.00 | 140 |

Phosphate ion, $PO_4^{3-}$, has the four resonance structures:

$$
\begin{array}{cccc}
\ddot{O}: & :\ddot{O}:^- & :\ddot{O}:^- & :\ddot{O}:^- \\
\| & | & \setminus & \setminus \\
{}^-:\ddot{O}-P-\ddot{O}:^- & {}^-:\ddot{O}-P=\ddot{O}: & {}^-:\ddot{O}-P-\ddot{O}:^- & :\ddot{O}=P-\ddot{O}:^- \\
| & | & \| & | \\
:\ddot{O}:^- & :\ddot{O}:^- & \ddot{O}: & :\ddot{O}:^-
\end{array}
$$

and a PO bond order of $\dfrac{2+1+1+1}{4} = 1.25$

Hydrogen phosphite ion has the three resonance structures:

$$
\begin{array}{ccc}
\ddot{O}: & :\ddot{O}:^- & :\ddot{O}:^- \\
\| & \setminus & \setminus \\
H-P-\ddot{O}:^- & H-P=\ddot{O}: & H-P-\ddot{O}:^- \\
| & | & \| \\
:\ddot{O}:^- & :\ddot{O}:^- & \ddot{O}:
\end{array}
$$

and a PO bond order of $\dfrac{2+1+1}{3} = 1.33$

Diphosphate ion, $P_2O_7^{4-}$, has the nine resonance structures:

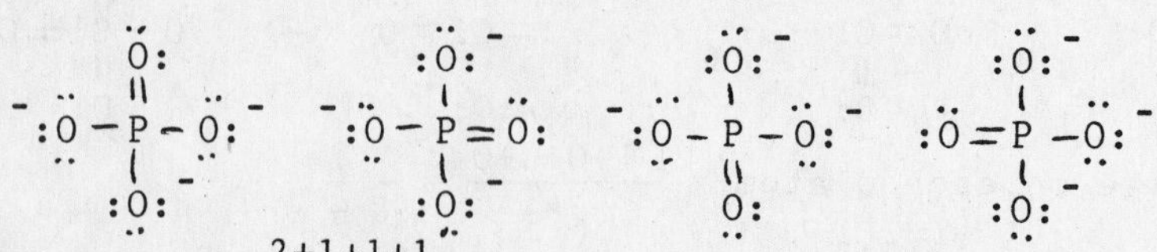

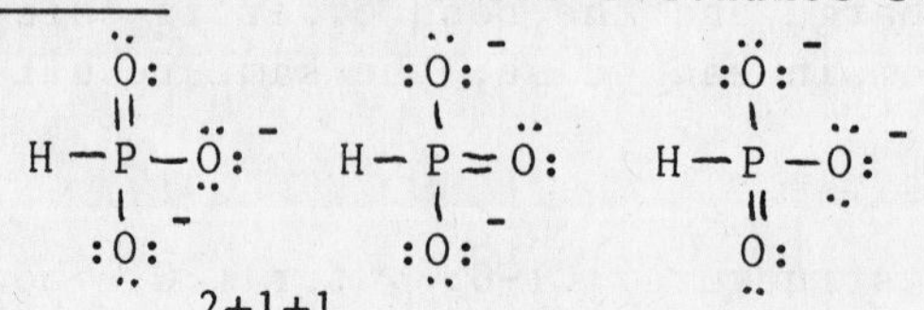

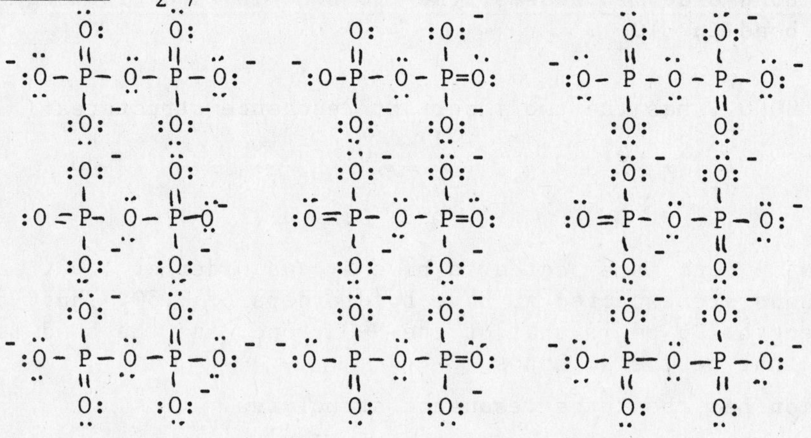

Thus, each of the P-O-P bridge bonds has bond order 1.00, and each of the six terminal PO bonds has a bond order $= \dfrac{2+2+2+1+1+1+1+1+1}{9} = 1.33$
and summarizing, we have:

| PO Bond Order | Bond Length (pm) | Species |
|---|---|---|
| 1.00 | 160 | $P_4O_{10}$ |
| 1.00 | 161 | $P_2O_7^{4-}$ |
| 1.25 | 154 | $PO_4^{3-}$ |
| 1.33 | 152 | $P_2O_7^{4-}$ |
| 1.33 | 151 | $HPO_3^{2-}$ |
| 2.00 | 140 | $P_4O_{10}$ |

As we have seen previously, there is a progressive decrease in bond length as the bond order increases.

23. The two Lewis structures are:

$$:\overset{..}{\underset{..}{Cl}} - \overset{+}{N} \overset{\overset{..}{O}:}{\underset{:\overset{..}{O}:^-}{\diagdown}} \quad \& \quad :\overset{..}{\underset{..}{Cl}} - \overset{+}{N} \overset{:\overset{..}{O}:^-}{\underset{\overset{..}{O}:}{\diagdown}}$$

with an NO bond order of $\dfrac{2+1}{2} = 1.50$, for each of the two NO bonds. $NO_2Cl$ is an $AX_3$ triangular planar molecule. For all the bonds the same, the expected bond angle is $120^o$. Here, there are two NO bonds of order 1.50 and an NCl bond of order 1.00. Thus, there will be a greater repulsion between the NO bonds than between an NO bond and the NCl bond, to give >ONO greater than $120^o$ and two >ClNO less than $120^o$. (The observed ONO bond angle is in fact $135^o$C).

24. (a) $\underline{CF_4}$ is an $AX_4$ tetrahedral molecule and will utilize the 2s and three 2p orbitals of carbon for overlap with 2p orbitals of F for bonding. Thus, the appropriate set of hybrid orbitals on C is $sp^3$.

(b) $\underline{SF_6}$ contains S(VI) with a $3s^1 3p^3 3d^2$ valence state. $SF_6$ is an octahedral $AX_6$ molecule, so the appropriate set of orbitals on S is a set of six $sp^3d^2$ hybrid orbitals.

25. In each case, classify the species in terms of the $AX_nE_m$ nomenclature and use sufficient s, p, and d orbitals to accommodate n+m electron pairs:

(a) $H_2O$, $AX_2E_2$, $sp^3$;  (b) $H_3O^+$, $AX_3E$, $sp^3$;  (c) $CO_2$, $AX_2$, sp;

(d) $CO_3^{2-}$, $AX_3$, $sp^2$;  (e) $PF_5$, $AX_5$, $sp^3d$;  (f) $ClF_3$, $AX_3E_2$, $sp^3d$

26. Ethene is a planar molecule, $\overset{H}{\underset{H}{\diagup}}C=C\overset{\diagup H}{\diagdown_H}$, with HCH bond angles close to $116^o$C. According to the bent bond model, the molecule is formed from four $sp^3$ hybrid orbitals on each C atom, two of which overlap with hydrogen 1s orbitals to form two C-H bonds, leaving two others to overlap with similar orbitals on the other C atom, to form the two bent bonds that represent the double bond. According to the $\sigma - \pi$ model, the planar arrangement of three bonds around each C atom results from the formation of three $sp^2$ hybrid orbitals on each C atom, two of which overlap with hydrogen 1s orbitals to form two C-H $\sigma$ bonds, leaving the other $sp^2$ hybrid orbital to overlap with the similar orbital of the other C atom, to form a C-C $\sigma$ bond. The other bonding electron pair of the C=C bond is then accommodated in an orbital formed from overlap of a single electron in each of the unutilized p orbitals on each C atom

which are perpendicular to the σ bonded planar framework of the molecule, to form a CC π bond.

The bent bond model predicts a HCH bond angle of 109.5°, the tetrahedral angle, while the σ-π model predicts a HCH bond angle of 120°, neither of which is exactly the observed bond angle of 116°, so both models are approximate descriptions.

27. In propene, 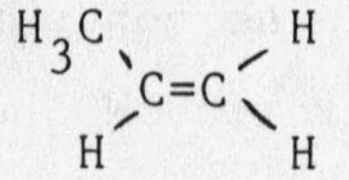 the first C atom is $AX_4$ tetrahedral, while each of the second and third C atoms are $AX_3$ triangular planar. Thus, the first carbon atom is described as forming four σ bonds utilizing a set of $sp^3$ hybrid orbitals. The remaining carbon atoms utilize sets of $sp^2$ hybrid orbitals to form three σ bonds, leaving a single electron on each C atom that is in the p orbital that is perpendicular to the σ bonded framework, and these p orbitals overlap to form the π component of the double bond.

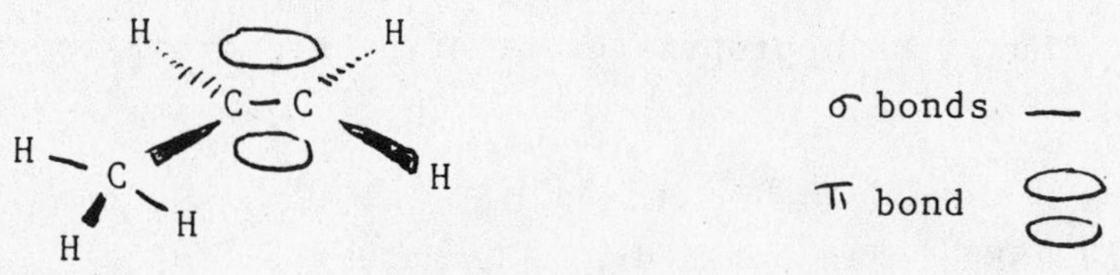


28. In propyne, $H_3C\text{-}C{\equiv}C\text{-}H$, the first C atom is $AX_4$ tetrahedral, and the second and third C atoms are $AX_2$, linear. The first C atom is described as forming four σ bonds utlizing a set of $sp^3$ hybrid orbitals, and the second and third C atoms as utilizing sets of sp hybrid orbitals to form two σ bonds each. This leaves two singly occupied p orbitals on each of the second and third C atoms in planes mutually perpendicular to the C-C σ bond, which overlap sideways to form two π bonds.

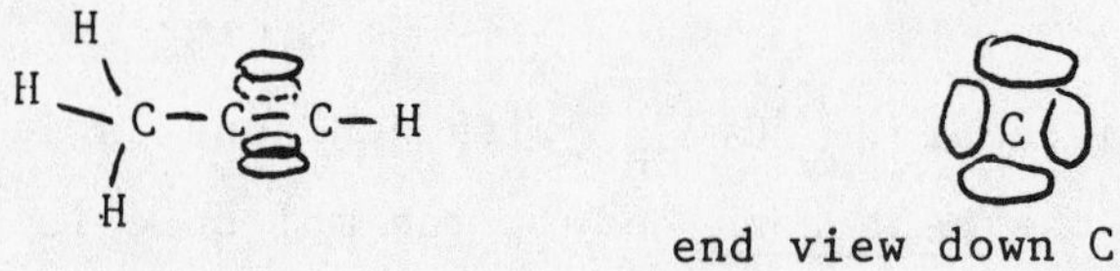


end view down $C_2C_3$ axis

29. (a) A <u>double bond</u> in a molecule such as ethene may be described as a σ bond and a π bond. Starting with three $sp^2$ hybrid orbitals on each C atom, two of these orbitals overlap with hydrogen 1s orbitals to form two C-H σ bonds, and the third is overlapped with a similar orbital on the other C atom to form a C-C σ bond. This leaves a single electron on each C atom in a p orbital that is perpendicular to the plane formed by the o bonds. Sideways overlap of these two p orbitals forms the π bond with its electron density concentrated in two lobes, one above the plane of the molecule and the other below.

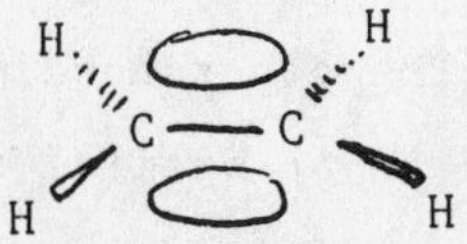

(b) A <u>triple bond</u> in a molecule such as ethyne consists of a σ bond and two π bonds. Starting with a set of sp hybrid orbitals on each C atom, one of these orbitals overlaps with a 1s orbital of H to form a σ bond, and the second is overlapped with a similar orbital on the other C atom to form a C-C σ bond. This leaves an electron in each of the unutilized p orbitals on each C atom which are situated in planes mutually perpendicular to the C-C axis. Sideways overlap of each pair of p orbital gives the two π bonds.

30. (a) In its ground state, As has the electron configuration $[Ar]3d^{10}4s^{2}4p^{3}$ with three unpaired 4p electrons and a valence of three. Thus $AsF_5$ containing five valent As must be formed from the

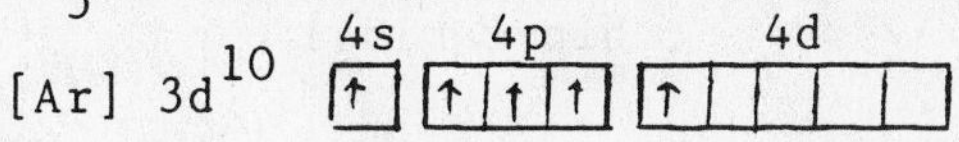


valence state of As, so that $AsF_5$ is an $AX_5$ type molecule with trigonal bipyramidal geometry.

(b) The appropriate set of hybrid orbital on As is five $sp^3d$ orbitals. Each of these singly occupied orbitals are overlapped with the singly occupied 2p orbital of a fluorine atom, to form five As-F σ bonds.

31. In each case, draw the Lewis structure and predict the $AX_nE_m$ arrangement of bonds and lone pairs on the central atom; then select the appropriate hybridization scheme to give this geometry.

(a) $NF_3$ is $AX_3E$ trigonal pyramidal, based on the tetrahedral arrangement of four pairs of electrons in the valence shell of N. Thus the four orbitals on N are a set of $sp^3$ hybrid orbitals.

(b) The N atom in $H_3C\text{-}C{\equiv}N$: has AXE linear geometry. Thus, two sp hybrid orbitals on N are required to accommodate the lone pair and to form a N-C σ bond, leaving unpaired electrons in each of two mutually perpendicular p orbitals to form the two π bonds.

(c) $NH_4^+$ has $AX_4$ tetrahedral geometry. The appropriate set of hybrid orbitals on N is a set of four $sp^3$ hybrid orbitals.

(d) $NO^+$ has the Lewis structure $:N{\equiv}\overset{+}{O}:$, in which the N atom has AXE geometry. As in (b) above, two sp hybrid orbitals are needed to accommodate the N lone pair and to form a N-O σ bond, leaving two singly occupied p orbitals on N to form the two π bonds.

(e) $NO_2^+$ has the Lewis structure $:\ddot{O}{=}\overset{+}{N}{=}\ddot{O}:$, with $AX_2$ geometry at N. Two sp hybrid orbitals are used to form the σ bonds to the two O atoms, leaving two singly occupied p orbitals on N to form π bonds with the two O atoms.

(f) $(CH_3)_4N^+$ is similar to $NH_4^+$ in that the four C atoms and the N have an $AX_4$ tetrahedral arrangement. The N-C bonds are formed by overlap of each of a set of $sp^3$ hybrid orbitals on N with an $sp^3$ hybrid orbital on each of the C atoms.

32.

| Molecule | Lewis Structure | Type | Shape | Hybrid Orbitals on A |
|---|---|---|---|---|
| $BrF_3$ | 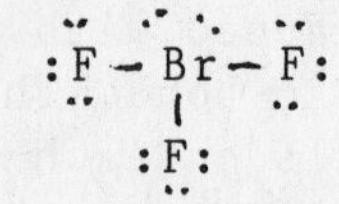 | $AX_3E_2$ | T-shaped | $sp^3d$ |
| $SbF_6^-$ | 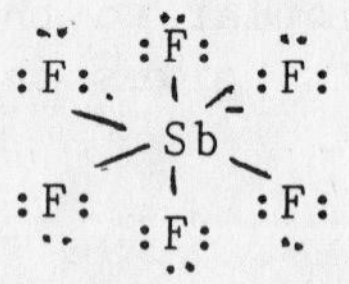 | $AX_6$ | octahedral | $sp^3d^2$ |
| $BrF_4^-$ | 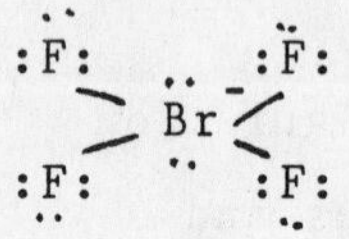 | $AX_4E_2$ | square planar | $sp^3d^2$ |
| $SiF_5^-$ | 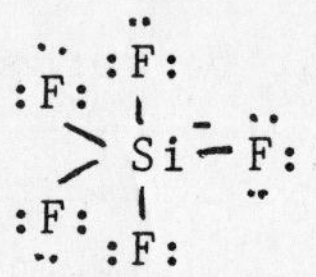 | $AX_5$ | trigonal bipyramidal | $sp^3d$ |
| $SiF_6^{2-}$ | 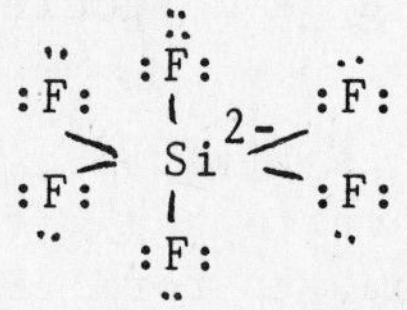 | $AX_6$ | octahedral | $sp^3d^2$ |

33. The atom A, with valence 3, must belong to either group 3, or to group 5. For A from group 3, $ACl_3$ would have $AX_3$ triangular planar geometry and no dipole moment. For A from group 5, $ACl_3$ would have $AX_3E$ triangular pyramidal geometry and would have a dipole moment. Thus, A must be an element from group 5.

34. A covalent bond is polar in the sense $\overset{\delta+}{A}\text{-}\overset{\delta-}{X}$ when the electronegativity of X is greater than that of A. A molecule has a dipole moment when its center of positive charge does not coincide with its center of negative charge, which depends on its geometry in most cases.

(a) $CO_2$ O is more electronegative than C.

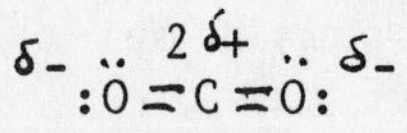

Since this is an $AX_2$ linear molecule, the center of negative charge is on the C atom, coincident with the center of positive charge. ( $\mu = 0$).

(b) $SO_2$ O is more electronegative than S.

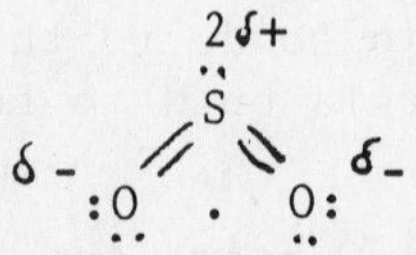

This is an $AX_2E$ angular molecule with the center of negative charge situated at a point midway between the two O atoms, which does not coincide with the center of positive charge (situated on the S atom). ($\mu \neq 0$)

(c) $\underline{H_2O}$ O is more electronegative than H.

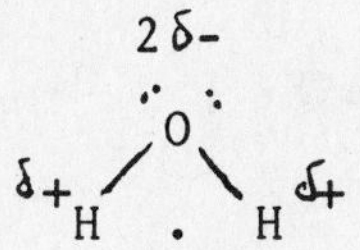

Since this is an $AX_2E_2$ angular molecule, the center of positive charge is at the midpoint between the two H atoms, and does not coincide with the center of negative charge on the O atom. ($\mu \neq 0$).

(d) $\underline{NH_3}$ N is more electronegative than H.

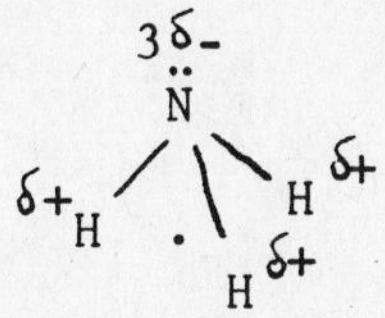

Since this is an $AX_3E$ triangular pyramidal molecule, the center of positive charge is at the center of the equilateral arrangement of the three H atoms, and does not coincide with the center of negative charge on the N atom. ($\mu \neq 0$).

(e) $\underline{SO_3}$ O is more electronegative than S.

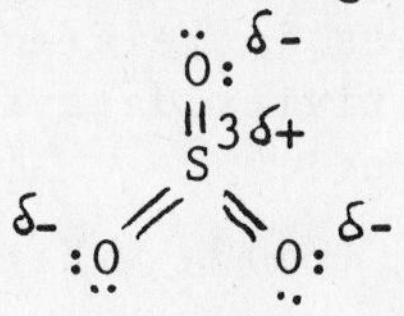

Since this is an $AX_3$ triangular planar molecule, the center of negative charge is at the center of the equilateral triangle formed by the O atom, i.e., coincides with the center of positive charge, which is on the S atom. ($\mu = 0$).

(f) $\underline{BeH_2}$ H is more electronegative than Be.

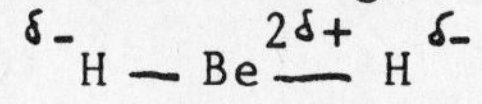

Since this is an $AX_2$ linear molecule, the center of negative charge coincides with the center of positive charge at the Be atom. ($\mu = 0$).

35. This problem is similar to Problem 34.

(a) 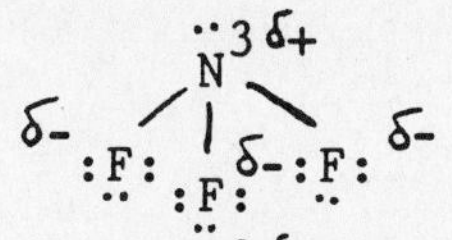 $AX_3E$ trigonal pyramidal geometry, $\mu \neq 0$.

(b) 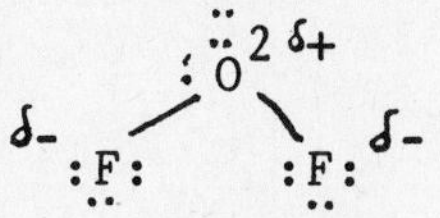 $AX_2E_2$ angular geometry, $\mu \neq 0$.

(c) 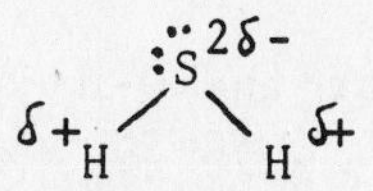 $AX_2E_2$ angular geometry, $\mu \neq 0$.

(d) 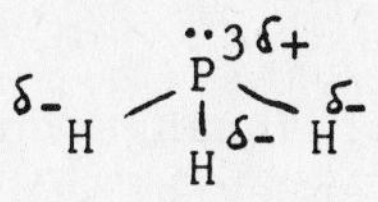 $AX_3E$ trigonal pyramidal geometry, $\mu \neq 0$.

(e) 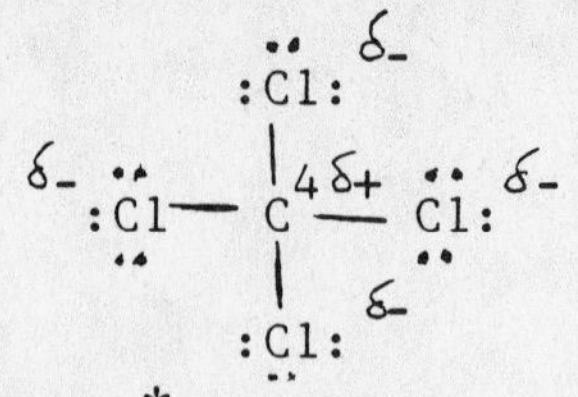 $AX_4$ tetrahedral geometry, $\mu = 0$.

(f) $\delta-$ :S=C=O: $\Delta-$ (with * on C) AXY linear geometry, $\mu \neq 0$.

*The charge on the C atom is $\delta+ + \Delta+$, but the CS bond is longer than the CO bond and the center of negative charge is not at the C atom, the center of positive charge, but somewhere between the C atom and the O atom.

(g) 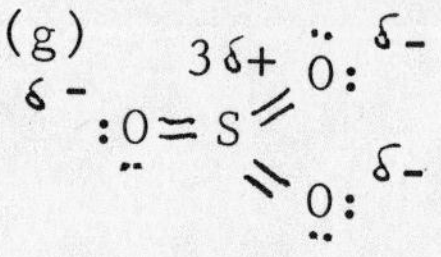 $AX_3$ triangular planar $\mu = 0$

(h) 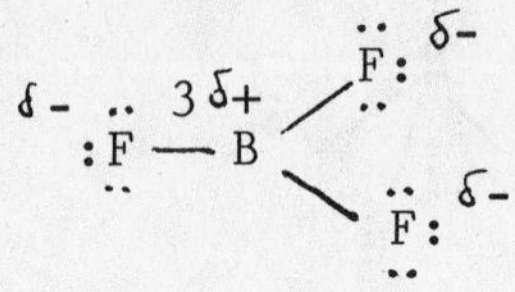 $AX_3$ triangular planar $\mu = 0$

36. The magnitude of a dipole moment is given by Qr, where Q is the charge at the positive and negative centers of charge, and r is their distance apart. Q is approximately proportional to the difference in the electronegativities of the atoms involved in bonding, and r is proportional to the bond length $r_{XY}$, provided the bond angles in related molecules do not differ significantly. Using electronegativity data from Figure 5.6 and covalent radii from Figure 4.9, we have:

| | | $\chi_X$ | $\chi_Y$ | $\Delta\chi$ | $r_X$ (pm) | $r_Y$ (pm) | $r_{XY}$ (pm) | $\Delta\chi \cdot r_{XY}$ |
|---|---|---|---|---|---|---|---|---|
| (a) | HF | 2.2 | 4.1 | 1.9 | 27 | 64 | 91 | 173 |
| | HCl | 2.2 | 2.8 | 0.6 | 27 | 99 | 126 | 76 |
| | HBr | 2.2 | 2.7 | 0.5 | 27 | 114 | 141 | 71 |
| | HI | 2.2 | 2.2 | 0.0 | 27 | 133 | 160 | 0 |

HF > HCl > HBr > HI

| | | $\chi_X$ | $\chi_Y$ | $\Delta\chi$ | $r_X$ (pm) | $r_Y$ (pm) | $r_{XY}$ (pm) | $\Delta\chi \cdot r_{XY}$ |
|---|---|---|---|---|---|---|---|---|
| (b) | $AsH_3$ | 2.2 | 2.2 | 0.0 | 121 | 27 | 148 | 0 |
| | $PH_3$ | 2.1 | 2.2 | 0.1 | 110 | 27 | 137 | 14 |
| | $NH_3$ | 3.1 | 2.2 | 0.9 | 70 | 27 | 97 | 87 |

$NH_3 > PH_3 > AsH_3$

| | | $\chi_X$ | $\chi_Y$ | $\Delta\chi$ | $r_X$ (pm) | $r_Y$ (pm) | $r_{XY}$ (pm) | $\Delta\chi \cdot r_{XY}$ |
|---|---|---|---|---|---|---|---|---|
| (c) | $Cl_2O$ | 2.8 | 3.5 | 0.7 | 99 | 66 | 165 | 116 |
| | $F_2O$ | 4.1 | 3.5 | 0.6 | 64 | 66 | 130 | 78 |
| | $H_2O$ | 2.2 | 3.5 | 1.3 | 27 | 66 | 93 | 121 |

$H_2O > Cl_2O > F_2O$

| | | $\chi_X$ | $\chi_Y$ | $\Delta\chi$ | $r_X$ (pm) | $r_Y$ (pm) | $r_{XY}$ (pm) | $\Delta\chi \cdot r_{XY}$ |
|---|---|---|---|---|---|---|---|---|
| (d) | $H_2O$ | 2.2 | 3.5 | 1.3 | 27 | 66 | 93 | 121 |
| | $H_2S$ | 2.2 | 2.4 | 0.2 | 27 | 104 | 131 | 26 |

$H_2O > H_2S$

37. (a) The polarity of the O-F bond is much smaller than that of the O-H bond (see Problem 36(c)). (b) For OCS, see Problem 35(f); for :S=C=S:, $\overset{\delta+}{S}=\overset{2\delta-}{C}=\overset{\delta+}{S}$, the centers of + and - charge coincide; $\mu = 0$.

38. The S-F bonds in $SF_6$ are polar $\overset{\delta+}{S}-\overset{\delta-}{F}$, but for an octahedral arrangement of bonds, the center of negative charge coincides with the center of positive charge, at the S atom. Thus, $\mu = 0$. This is easily seen by taking each pair of colinear F-S-F bonds in turn; for each the center of - charge is halfway between the F atoms, i.e., at the S atom.

39. (a) Tetrahedral $CCl_4$ has no dipole moment. Even though each C-Cl bond is polar, the centers of positive and negative charge are at the central C atom (see text).

In contrast, in tetrahedral $CCl_2F_2$ the center of negative charge is not at the C atom, because the C-F bonds are more polar than the C-Cl bonds.

(b) Each P-F bond is polar, $\overset{\delta+}{P}-\overset{\delta-}{F}$, and the center of negative charge is coincident with the center of positive charge, at the P atom, so that the molecule has no dipole moment. This is most clearly seen if the three equatorial bonds are considered as one group of bonds and the two axial bonds are considered as another group of bonds. The equatorial bonds point towards the corners of an equilateral triangle with the P atom at the center, and the axial bonds form a linear arrangement with the P atom at the center. For both an $AX_3$ and an $AX_2$ arrangement of bonds the center of negative charge coincides with the center of positive charge, at the A atom.

40. The dipole moment of a molecule is given by $\mu = Q.r$, with Q in coulombs, C, and r in meters, m, where Q is the magnitude of the charges at the positive and negative centers of charge separated by their distance apart, r.

<u>HF</u> The distance between the centers of positive and negative charge is the bond length 91.7 pm. Thus,

$$Q(91.7\ \text{pm})\left(\frac{1\ \text{m}}{10^{12}\ \text{pm}}\right) = 6.36\times10^{-30}\ \text{C m}\ ;\ Q = 6.94\times10^{-20}\,\text{C}$$

and in terms of the charge on the electron, $1.602\times10^{-19}$ C,

$$Q = (6.94\times10^{-20}\ \text{C})\left(\frac{1\ \text{e}}{1.602\times10^{-19}\ \text{C}}\right) = \underline{0.43\ \text{e}}$$

i.e., $\overset{+0.43\ e}{H}\!-\!\overset{-0.43\ e}{F}$

<u>$H_2O$</u>

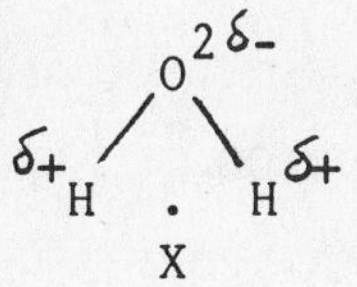


For $H_2O$, the center of positive charge is at X, the midpoint between the two H atoms, and the center of negative charge is at the O atom.

$$r_{OX} = (r_{OH})\left(\cos\frac{<HOH}{2}\right) = (95.7\ \text{pm})\left(\cos\frac{104.5^\circ}{2}\right)$$
$$= (95.7\ \text{pm})(0.6122) = \underline{58.6\ \text{pm}}$$

$$\mu = Q(58.6\ \text{pm})\left(\frac{1\ \text{m}}{10^{12}\ \text{pm}}\right) = 6.17\times10^{-30}\ \text{C m};\ Q = \underline{1.05\times10^{-19}\ \text{C}}$$

$$Q = (1.05\times10^{-19}\,\text{C})\left(\frac{1\ \text{e}}{1.602\times10^{-19}\ \text{C}}\right) = \underline{0.66\ \text{e}}$$

i.e.,

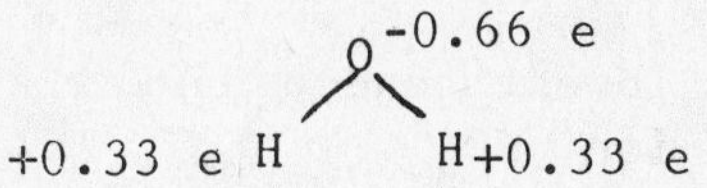

41. This problem is similar to Problem 40.

<u>HCl</u>

$\mu = Q(127.4 \text{ pm})(\frac{1 \text{ m}}{10^{12} \text{ pm}}) = 3.43 \times 10^{-30}$ C m; $Q = 2.69 \times 10^{-20}$ C

$Q = (2.69 \times 10^{-20}$ C$)(\frac{1 \text{ e}}{1.602 \times 10^{-19} \text{ C}}) = \underline{0.17 \text{ e}}$

$$+0.17 \text{ e}_{H} \text{——} Cl^{-0.17 \text{ e}}$$

<u>H$_2$S</u>  $r_{SX} = (135 \text{ pm})(\cos\frac{92^{\circ}}{2}) = (135 \text{ pm})(0.6947) = 93.8$ pm

$\mu = Q(93.8 \text{ pm})(\frac{1 \text{ m}}{10^{12} \text{ pm}}) = 3.12 \times 10^{-30}$ C m; $Q = 3.33 \times 10^{-20}$ C

$Q = (3.33 \times 10^{-20}$ C$)(\frac{1 \text{ e}}{1.602 \times 10^{-19} \text{ C}}) = \underline{0.208 \text{ e}}$

$$S^{-0.208 \text{ e}}$$
$$+0.104 \text{ e}_H \diagup \diagdown_H {}^{+0.104 \text{ e}}$$

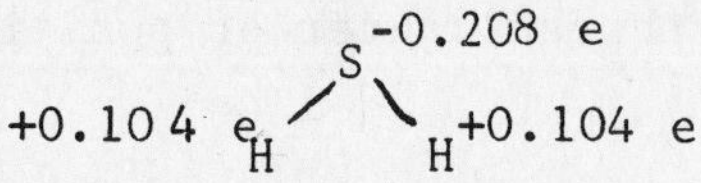

42. The usual Lewis structure for BF$_3$ is

$$:\ddot{F}—B\begin{smallmatrix}\diagup \ddot{F}: \\ \diagdown \ddot{F}: \end{smallmatrix}$$

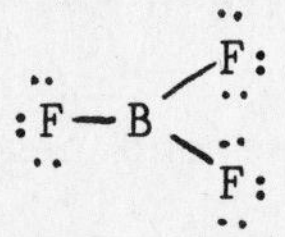

in which the boron atom has an incomplete valence shell of 3 pairs
of electrons. The octet can be achieved by moving a lone pair of
electrons from any one of the F atoms into the B-F bonding region,
to give the three possible resonance structures

$$^+:\ddot{F}\!\!=\!\!B\begin{smallmatrix}\diagup \ddot{F}: \\ \diagdown \ddot{F}: \end{smallmatrix} \quad \longleftrightarrow \quad :\ddot{F}—B\begin{smallmatrix}\diagup \ddot{F}:^+ \\ \diagdown \ddot{F}: \end{smallmatrix} \quad \longleftrightarrow \quad :\ddot{F}—B\begin{smallmatrix}\diagup \ddot{F}: \\ \diagdown \ddot{F}:^+ \end{smallmatrix}$$

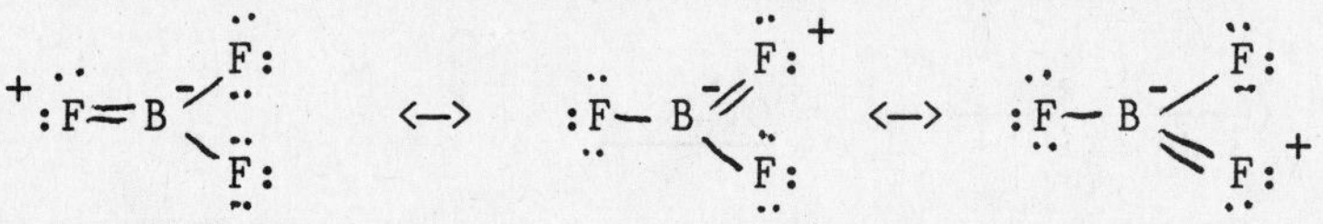

These structures are expected to have only a minor contribution to
the actual structure of BF$_3$ because of the separation of + and -
formal charges.

43. $:\ddot{O}\!\!=\!\!\overset{+}{N}\!\!=\!\!\ddot{O}:$                                     NO bond order 2.00

$$\begin{array}{ccccc} \ddot{O}: & & :\ddot{O}:^- & & :\ddot{O}:^- \\ \overset{+}{\underset{N}{\|}} & & \overset{+}{\underset{N}{|}} & & \overset{+}{\underset{N}{|}} \\ \diagup \diagdown & \longleftrightarrow & \diagup \diagdown & \longleftrightarrow & \diagup \diagdown \\ {}^-:\ddot{O}\, \quad \, :\ddot{O}:^- & & :\ddot{O}: \quad \ddot{O}: & & :\ddot{O} \quad :\ddot{O}:^- \end{array}$$  NO bond order 1.33

$:\ddot{O}\!\!=\!\!\ddot{N}\!\!-\!\!\ddot{O}:^- \longleftrightarrow {}^-:\ddot{O}\!\!-\!\!\ddot{N}\!\!=\!\!\ddot{O}:$                    NO bonder order 1.50

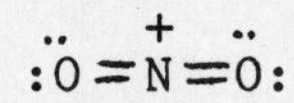

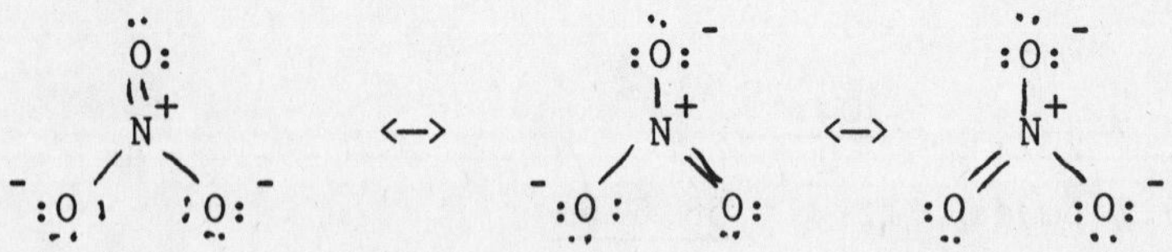

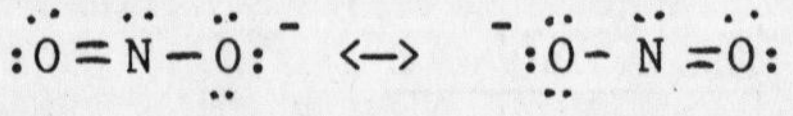

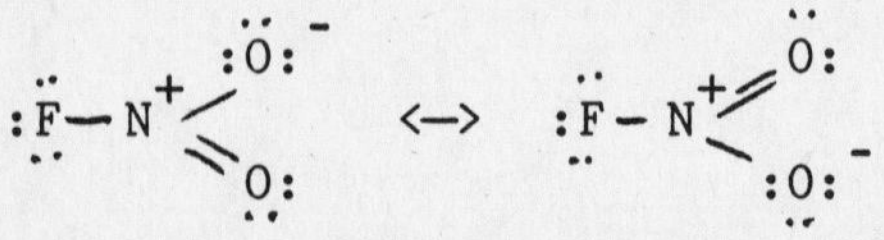

$$:\ddot{F}—\overset{+}{N}\begin{smallmatrix}\diagup :\ddot{O}:^- \\ \diagdown\!\!= \ddot{O}: \end{smallmatrix} \quad \longleftrightarrow \quad :\ddot{F}—\overset{+}{N}\begin{smallmatrix}\diagup\!\!= \ddot{O}: \\ \diagdown :\ddot{O}:^- \end{smallmatrix}$$  NO bond order 1.50

The longest NO bond will be that of lowest bond order, i.e.,
each of the NO bonds in the nitrate ion, NO$_3^-$

44. (a) The <u>bond order</u> of a bond is the number of electron pairs that participate in the bond. For molecules and ions that are described by resonance structures, the bond order may be fractional.

   (b) As the number of electron pairs involved in a bond increases, the strength of the bond increases; the atoms are held more closely together and the <u>bond length</u> decreases.

45. (a) The valence state electron configurations of Xe in each of the fluorides of xenon are

   $$\underline{XeF_2} \quad Xe(II) \quad [Kr]4d^{10} \quad \begin{array}{c} 5s \\ \boxed{\uparrow\downarrow} \end{array} \quad \begin{array}{c} 5p \\ \boxed{\uparrow\downarrow|\uparrow\downarrow|\uparrow} \end{array} \quad \begin{array}{c} 5d \\ \boxed{\uparrow|\;|\;|\;|\;} \end{array}$$

   $$\underline{XeF_4} \quad Xe(IV) \quad [Kr]4d^{10} \quad \boxed{\uparrow\downarrow} \quad \boxed{\uparrow\downarrow|\uparrow|\uparrow} \quad \boxed{\uparrow|\uparrow|\;|\;|\;}$$

   $$\underline{XeF_6} \quad Xe(VI) \quad [Kr]4d^{10} \quad \boxed{\uparrow\downarrow} \quad \boxed{\uparrow|\uparrow|\uparrow} \quad \boxed{\uparrow|\uparrow|\uparrow|\;|\;}$$

   (b) $XeF_2$ has three lone pairs of electrons and two bond pairs in the valence shell of Xe; it has $AX_2E_3$ <u>linear</u> geometry. $XeF_4$ has two lone pairs of electrons and four bond pairs in the valence shell of Xe; it has $AX_4E_2$ <u>square planar</u> geometry.

   (c) $XeF_6$ has one lone pair of electrons and six bond pairs in the valence shell of Xe. It has $AX_6E$ geometry in which the Xe atom has <u>seven</u> electron pairs in its valence shell. Since there are more than six electron pairs, the geometry cannot be octahedral.

46. (a) A <u>polar bond</u> is a covalent bond in which the electron pair is unequally shared between two atoms of different electronegativities, so that there are partial formal charges on the bonded atoms.
   e.g., $^{\delta+}H-\overset{\cdot\cdot}{\underset{\cdot\cdot}{F}}:^{\delta-}$ <u>or</u> $^{\delta+}H-\overset{\cdot\cdot}{O}:^{2\delta-}$
   $\qquad\qquad\qquad\qquad\qquad\quad \underset{\overset{|}{H}^{\delta+}}{}$

   (b) <u>Bond dipole</u> refers to the dipole moment of a specific polar bond, which is given by the product of the partial charges on the atoms multiplied by the bond length.
   e.g., Bonds such as $^{\delta+}H-\overset{\cdot\cdot}{\underset{\cdot\cdot}{F}}:^{\delta-}$ and $^{\delta+}C-\overset{\cdot\cdot}{\underset{\cdot\cdot}{Cl}}:^{\delta-}$ in $CCl_4$ will have a bond dipole.

   (c) <u>Dipole moment</u> is the net effect of all of the bond dipoles of a molecule on its overall polarity. For example, in a molecule such as $CCl_4$ the individual bond dipoles cancel each other, and overall the molecule has no net dipole moment, while in a molecule such as water, $H_2O$, the individual bond dipoles reinforce each other and there is a net dipole moment. The magnitude of a dipole moment is given by the net charges at the centers of positive and negative charge multiplied by their distance apart.

   (d) <u>Bent bond</u> is used to describe the individual components of a double bond or a triple bond, in which there are two and three bent bonds, respectively. The concept arises from the model in which double and triple bonds are formed by the overlap of, for example, two and three $sp^3$ hybrid orbitals, respectively, on the atoms forming these bonds. In such bonds the electron density is mainly outside of the straight line joining the nuclei. $H_2C{=}CH_2$ and $HC{\equiv}CH$ are examples.

(e) Ligand is a general term used to describe the entities that are bonded to a central atom. They may be single atoms or groups of atoms. For example, each of the F atoms in $SF_6$ may be described as a ligand, as may each of the water molecules in a hydrated ion such as $Cu(H_2O)_4{}^{2+}$.

(f) A resonance structure is one of a number of possible Lewis structures that can be drawn for a molecular species that has no single unique Lewis structure. None of the resonance structures represent the actual structure, in which some of the electron pairs are more delocalized than can be depicted by a single Lewis structure. The average of the Lewis resonance structures gives a more accurate description of the bonding in the molecular species than does any single Lewis structure. For example

$CO_3{}^{2-}$

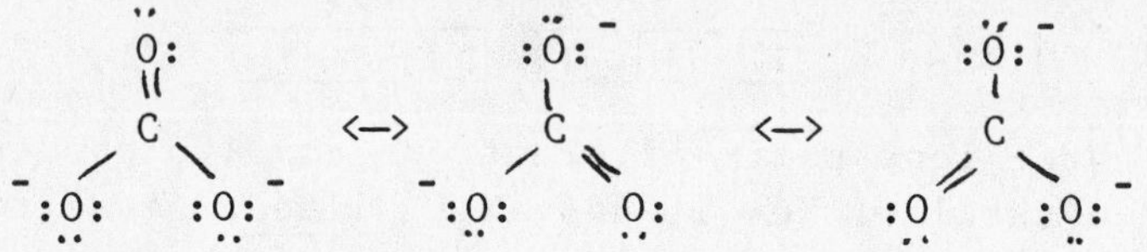

$NO_2{}^-$ :Ö=N̈–Ö:⁻ <-> ⁻:Ö–N̈=Ö:

(g) Bond order is the number of electron pairs that bond two atoms together, which in the case of species that are represented by a number of resonance structures may be a fractional number. For example, the CC bonds in ethane, $H_3C{-}CH_3$, ethene, $H_2C{=}CH_2$, and ethyne, $HC{\equiv}CH$, have bond orders of 1, 2, and 3, respectively, while those in the carbonate ion, $CO_3{}^{2-}$ and the nitrite ion, $NO_2{}^-$, are 1.33 and 1.50, respectively.

47. (a) The octet rule states that when an atom forms a compound, it gains, loses, or shares electrons to obtain eight electrons in its valence shell. It only applies without exception to C, N, O, and F.

(b) A Lewis structure is a structure which represents the arrangement of bonds (bonding electron pairs) and unshared (lone) electron pairs in a molecular species, assuming that they are localized, either in bonds or on individual atoms. A complete Lewis structure also shows the individual formal charges on individual atoms of the molecular species.

(c) See Problem 46(f).

(d) Molecules in which the central atom obeys the octet rule are those in which the central atom has four pairs of electrons in its valence shell. Examples include hydrocarbons such as $CH_4$, water, $H_2\ddot{O}$:, and ammonia, $:NH_3$.

(e) Molecules in which the central atom has less than an octet of electrons are those in which the central atom is from group 2 or 3 of the periodic table. Examples include covalent species such as $BeH_2$, $BCl_3$, and $AlCl_3$.

(f) Molecules in which the central atom exceeds the octet rule are those in which the central atom is from period 3 (or 4,5, and 6) and is in a valence state greater than its ground state, involving the use of valence shell d orbitals in addition to s and p. Examples include molecular species such as $SF_4$, $SF_6$, $XeF_2$, $H_2SO_4$, $H_3PO_4$, etc.

48. (a) $H_2C{=}CH_2$, $H_2C{=}O$, O=O, $(HO)_3P{=}O$, etc.

(b) $N_2$, CO, HCN, HC≡CH, etc.

(c) Linear geometry is possessed by $AX_2$ molecules and $AX_2E_3$ molecules. Examples of $AX_2$ molecules include $BeH_2$, HCN, and H-C≡C-H; $AX_2E_3$ species include $XeF_2$, $I_3^-$, and $ICl_2^-$.

(d) Trigonal bipyramidal geometry is possessed by any species of the type $AX_5$. Examples include $PCl_5$, $PF_5$, $AsF_5$, and $SiF_5^-$.

(e) Octahedral geometry is possessed by species of the $AX_6$ type. For example, $SF_6$, $SeF_6$, $PCl_6^-$ and $SiF_6^{2-}$.

(f) A formal positive charge on A occurs when a new molecular species is formed by utilizing a lone pair of electrons on A to form an additional bond, as in $NH_4^+$ or $PH_4^+$ or $H_3O^+$. Alternatively, there is a formal negative charge on A if it forms an additional bond by accepting into its valence shell an additional pair of electrons from a ligand, as in $BF_4^-$ or $SiF_5^-$.

49. (a) The Lewis structure for $NO_3^-$ is

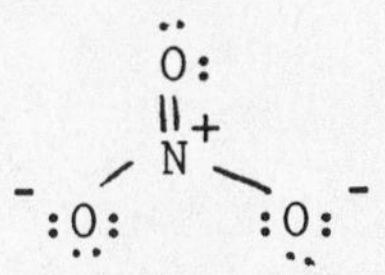

with trigonal planar $AX_3$ geometry and approximate bond angles <O-N=O > 120° and <O-N-O < 120°, due to the greater space taken up in the valence shell of the N atom by a double bond in comparison to that occupied by a single bond.

(b) In order to account for the exact equilateral triangular shape of the $NO_3^-$ ion, the concept of resonance is used. We write all of the likely Lewis structures, in this case

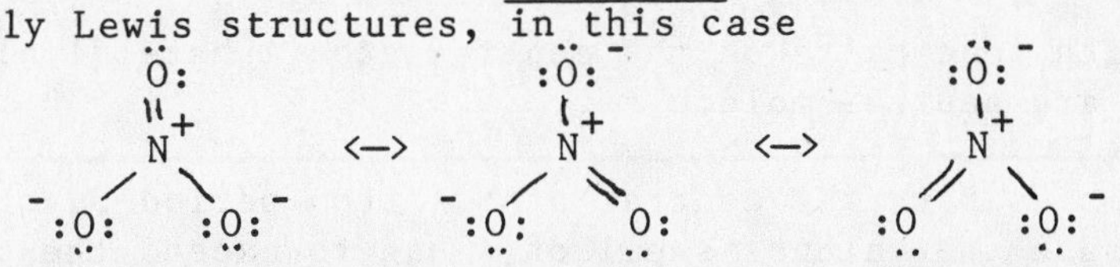

and mix them together to give the average structure. For the $NO_3^-$ ion this leads to a structure in which the formal charge on the N atom is +1 and those on each of the O atoms is $\frac{2}{3}$ -, and the NO bond order of each bond is $1\frac{1}{3}$, consistent with the observed bond length of 126 pm, intermediate in length between a NO single bond (136 pm) and an NO double bond (116 pm). All the <ONO = 120°.

50. (a) The two Lewis structures are 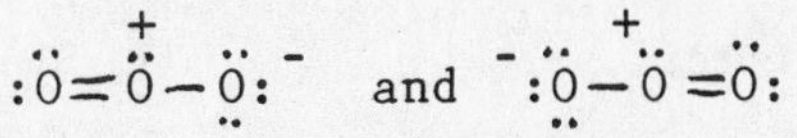

(b) (i) $O_3$ is an $AX_2E$ type molecule and is therefore angular.

(ii) From the two resonance structures in (a), the OO bond order in $O_3$ is $\frac{2+1}{2}$ = 1.50, and from the Lewis structures for $O_2$ and H-O-O-H, we have:

:O=O: , bond order 2.00 (121 pm) & 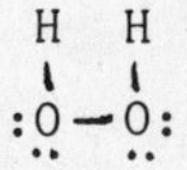 , bond order 1.00 (147 pm).

Thus, as expected the observed bond lengths in $O_3$ are intermediate between those of the double bond in $O_2$ and the single bond in $H_2O_2$.

51.

| | Species | Molecular Shape | Hybrid Orbitals On Central Atom | Approximate Bond Angles |
|---|---|---|---|---|
| (a) | $CH_4$ | $AX_4$, tetrahedral | $sp^3$ | $109.5^{o}$* |
| (b) | $PH_4^+$ | $AX_4$, tetrahedral | $sp^3$ | $109.5^{o}$* |
| (c) | $NF_3$ | $AX_3E$, trigonal pyramidal | $sp^3$ | $< 109.5^{o}$ |
| (d) | $F_2O$ | $AX_2E_2$, angular | $sp^3$ | $< 109.5^{o}$ |
| (e) | $H_3O^+$ | $AX_3E$, trigonal pyramidal | $sp^3$ | $< 109.5^{o}$ |

*exactly

52. In each case, examples can be cited where AX bonds are either single bonds or multiple bonds, such as

| | Type | Shape | Examples |
|---|---|---|---|
| (a) | $AX_2$ | linear | H-Be-H F-B=O O=C=O H-C≡N |
| (b) | $AX_3$ | triangular planar | $BF_3$ $SO_3$ $H_2C{=}O$ |
| (c) | $AX_2E$ | angular | $:SnCl_2$ $:CH_2$ $:SO_2$ |
| (d) | $AX_4$ | tetrahedral | $CH_4$ $NH_4^+$ $SO_4^{2-}$ $PO_4^{3-}$ $ClO_4^-$ $F_2SO_2$ |
| (e) | $AX_3E$ | triangular pyramidal | $:NH_3$ $:PH_3$ $:SO_3^{2-}$ $F\ddot{S}O_2$ $H_3O:^+$ |
| (f) | $AX_2E_2$ | angular | $H_2\ddot{O}:$ $F_2\ddot{O}:$ $:\ddot{C}lO_2^-$ $:\ddot{B}rO_2^-$ |
| (g) | $AXE_3$ | linear | $H\ddot{F}:$ $H\ddot{C}l:$ $H\ddot{B}r:$ $H\ddot{I}:$ $^-:\ddot{O}H$ $^-:\ddot{S}H$ |

All the species with lone pairs on the central atom A have dipole moments when they are neutral molecules.

53. All of these examples have the central atom A from period 3, or beyond, because in each case the valence shell of A has to exceed the octet.

| | Type | Shape | Examples |
|---|---|---|---|
| (a) | $AX_5$ | trigonal bipyramidal | $PF_5$ $PCl_5$ |
| (b) | $AX_4E$ | disphenoidal | $:SF_4$ $:SeF_4$ |
| (c) | $AX_3E_2$ | T-shaped | $:\ddot{C}lF_3$ $:\ddot{I}F_3$ |
| (d) | $AX_2E_3$ | linear | $:\ddot{X}eF_2$ $I_3^-$ $:\ddot{I}Cl_2^-$ |
| (e) | $AX_6$ | octahedral | $SF_6$ $PF_6^-$ $SiF_6^{2-}$ |
| (f) | $AX_5E$ | square pyramidal | $:BrF_5$ $:IF_5$ |
| (g) | $AX_4E_2$ | square planar | $:\ddot{X}eF_4$ $:\ddot{I}Cl_4^-$ |

All the neutral species with lone pairs in the valence shell of A, except those of the $AX_3E_2$ and $AX_4E_2$ type, have dipole moments.

54.

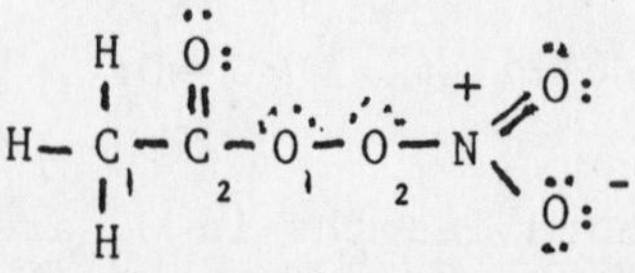

(a)

| Atom | Type | Bond Arrangement | Approximate Bond Angle | Hybrid Orbitals |
|---|---|---|---|---|
| $C_1$ | $AX_4$ | tetrahedral | 109.5° | $sp^3$ |
| $C_2$ | $AX_3$ | trigonal planar | 120° | $sp^2$ |
| $O_1$ | $AX_2E_2$ | angular | < 109.5° | $sp^3$ |
| $O_2$ | $AX_2E_2$ | angular | < 109.5° | $sp^3$ |
| N | $AX_3$ | trigonal planar | 120° | $sp^2$ |

(b) The above table shows the hybrid orbitals used on each atom to form σ bonds:

$\underline{C_1}$ 3 σ bonds to H atoms and 1 σ bond to $C_2$

$\underline{C_2}$ σ bonds to $C_1$, O, and $O_1$ and 1π bond to O

$\underline{O_1}$ a σ bond to C and a σ bond to $O_2$

$\underline{O_2}$ a σ bond to $O_1$ and a σ bond to N

$\underline{N}$ three σ bonds, one to $O_2$ and one to each of the terminal O atoms, and a π bond to one of the terminal O atoms*

*There are two similar Lewis structures, giving the terminal NO bonds bond orders of 1.50 each.

55. Iodine is in period 5 and can expand its valence shell beyond the octet by utilizing 5d orbitals. Thus the Lewis structure of $I_3^-$ is

:Ï—Ï—Ï:

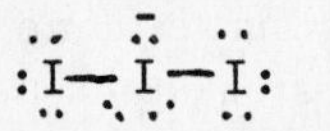

which is an $AX_2E_3$ linear species. $F_3^-$ would have to have a similar Lewis structure, which is not possible because F in period 2 has no low lying orbitals (there are no 2d orbitals) and obeys the octet rule.

56. The Kekulé structure for benzene is

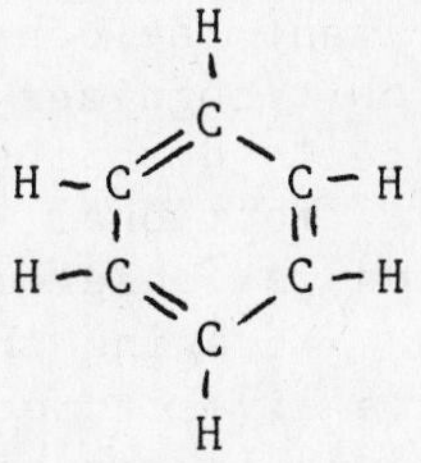

For the formation of $C_6H_6(g)$ from its elements,

$$6C(s,\text{graphite}) + 3H_2(g) \xrightarrow{\Delta H_f^o} C_6H_6(g)$$

$$\downarrow \Delta H_1 \qquad \downarrow \Delta H_2 \qquad \nearrow \Delta H_3$$

$$6C(g) \quad + \quad 6H(g)$$

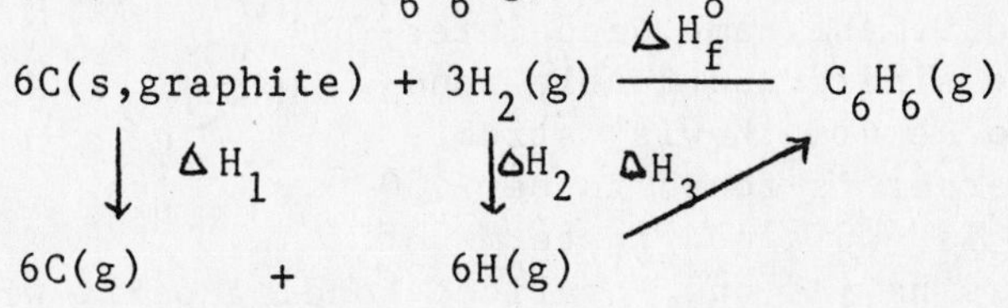

$$\Delta H_f^o(C_6H_6,g) = \Delta H_1 + \Delta H_2 + \Delta H_3$$

where, $\Delta H_1 = 6\,\Delta H_f^o(C,g)$; $\Delta H_2 = 3BE(H\text{-}H)$

and $\Delta H_3 = -[6BE(C\text{-}H) + 3BE(C\text{-}C) + 3BE(C{=}C)$

Thus, $\Delta H_f^o(C_6H_6,g) = [6(716.7) + 3(436)] - [6(413)+3(348)+3(619)]$ kJ

$= (5608.2-5379)$ kJ $=$ +229 kJ

Thus, on the basis of the Kekulé structure, $\Delta H_f^o(C_6H_6,g) = +229\ kJ\ mol^{-1}$ which is to be compared with the experimental value of $83\ kJ\ mol^{-1}$

Benzene is more stable by $(229-83)\ kJ\ mol^{-1} = 146\ kJ\ mol^{-1}$ than expected in terms of the Kekulé structure.

In fact, the Kekulé structure above is just one of the important resonance structures of benzene, which is better represented by

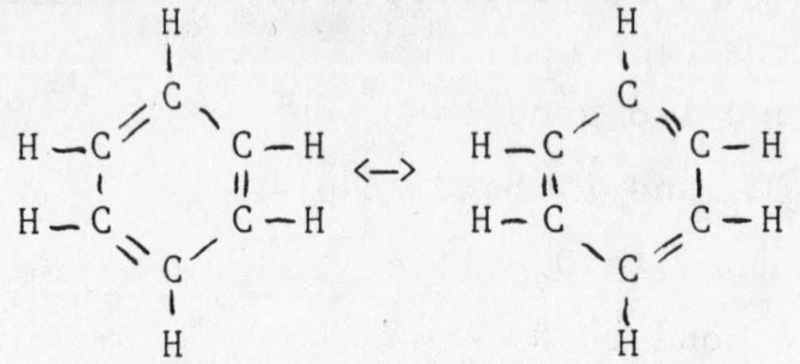

indicating that electrons are more delocalized in benzene than either of the Kekulé structures imply, and therefore more stable by $146\ kJ\ mol^{-1}$ Hence, the use of the term resonance or delocalization energy.

57. Relative to a Lewis structure with nonbonding (lone) electron pairs assigned entirely to individual atoms, and bonding pairs assigned to bonds between pairs of atoms, some molecules have lone pairs that are partially bonding and partially nonbonding, and bond pairs that are spread over more than two nuclei; in both cases the electron pairs are said to be delocalized. For example, while the Lewis structure of the carbonate ion, $CO_3^{2-}$, has one CO double bond and two CO single bonds, the experimental structure shows that all of the CO bonds are of equal length, intermediate between the length of a double bond and the length of a single bond. To achieve a description that is consistent with experiment the concept of resonance was introduced, which recognizes that no one Lewis structure is adequate to describe molecular species with bonds that have fractional bond order, but a better representation is achieved by writing all of the possible Lewis structures and then averaging them. Another example is benzene, for which a Lewis structure gives three CC single bonds and three CC double bonds in the six membered carbon ring, yet experimentally all of the six CC distances are found to be the same, and intermediate in length between the single and double bond lengths. By averaging the two Kekulé (Lewis) structures of benzene the CC bond order is found to be 1.50 for each of the bonds, which is explained in terms of each CC bond being depicted as having one localized bonding pair, with an additional three electron pairs being spread over all six carbon atoms, rather than being localized between specific pairs of carbon atoms to give three CC double bonds.

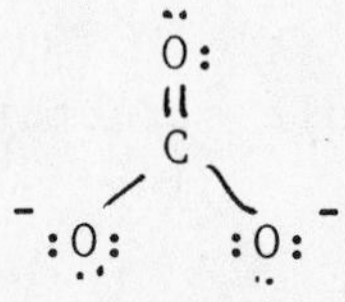

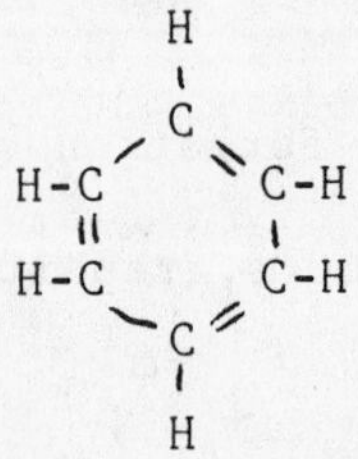

58. The carbon atom in methane forms four bonds and must therefore be in the C(IV) valence state

C(IV) [He] 2s [↑] 2p [↑|↑|↑]

Formation of methane from this valence state, by overlap of the atomic orbitals of carbon with hydrogen 1s orbitals predict a geometry for methane in which the C-H bonds formed by overlap of the three 2p carbon orbitals would make angles of $90^{o}$ with each other for maximum overlap. However, since the overlap of a carbon 2s orbital with a hydrogen 1s orbital is the same in all directions (both 1s and 2s orbitals are spherical), the direction of the fourth C-H bond would be indeterminate with no prefered direction. Experimentally, methane is found to have an exactly tetrahedral geometry with all the CH bonds equal in length and with all the HCH bond angles equal to the tetrahedral angle of $109.5^{o}$, whereas the structure based on the use of carbon atomic orbitals would have three equivalent CH bonds with bond angles of $90^{o}$ and a fourth CH bonds of different length and making no specific bond angle with the other CH bonds. In order to describe each bond in terms of a localized orbital, the 2s orbital and the three 2p orbitals of carbon are combined to give four equivalent orbitals, each with 1/4 s and 3/4 p character, which point towards the corners of a tetrahedron. This process is referred to as the hybridization of atomic orbitals, and the resulting set of orbitals is called a set of four $sp^3$ hybrid orbitals. These can be used, by maximum overlap with four hydrogen 1s atomic orbitals, to form four equivalent C-H bonds. (Formation of four $sp^3$ orbitals does not change the electron density distribution around the carbon atom).

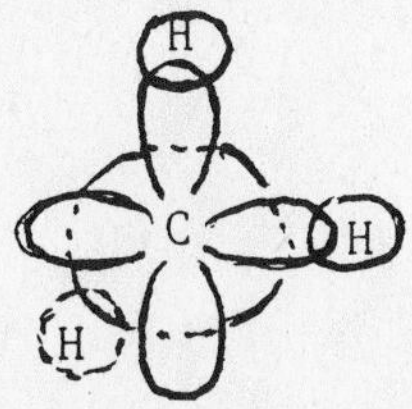


For clarity the third p orbital is omitted

59. In terms of the two Kekulé Lewis structures of benzene, the resonance model takes the average of the two structures

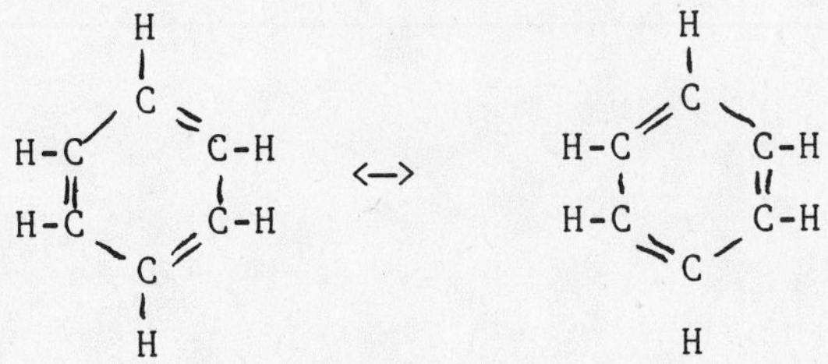

to represent the actual structure, giving each of the six CC bonds a bond order of 1.50, consistent with a six-membered carbon ring with regular hexagonal geometry. In the alternative model, a σ bonded framework is constructed using a set of $sp^2$ hybrid orbitals on each C atom, to form a C-H bond by overlap with a H is orbital and two C-C σ bonds, by overlap with two $sp^2$ hybrid orbitals on adjacent C atoms. This leaves a single electron in each of six carbon p orbitals that are perpendicular to the σ bonded carbon ring, that are overlapped sideways to form three π bonds delocalized over all six carbon atoms. In this description, each CC bond has an average of three electrons, and a bond order of 1.50, and the six C atoms form a regular hexagon.

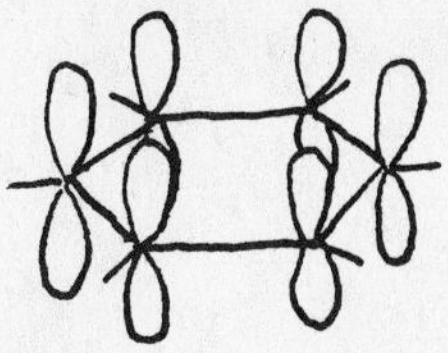

# CHAPTER 10

1. (a) Suitable reducing agents are those that are more readily oxidized than the metal cations that they are used to reduce. Common examples include: carbon (coke), carbon monoxide, hydrogen, reactive metals, such as the alkali metals and aluminium, and electrons.

(b) (i) $Cu_2O + CO \longrightarrow 2Cu + CO_2$

(ii) $Fe_2O_3 + 2Al \xrightarrow{\text{heat}} 2Fe + Al_2O_3$ (thermite process)

(iii) $PbO + C \longrightarrow Pb + CO$

(iv) $AlCl_3 + 3K \longrightarrow Al + 3KCl$

In this case, the reaction given was used before the Hall process was discovered:

Hall process:

$$2Al^{3+} + 6e^- \longrightarrow 2Al$$
$$3O^{2-} + 3C \longrightarrow 3CO_2 + 6e^- \quad \text{(Electrolysis)}$$
$$Al_2O_3 + 3C \longrightarrow 2Al + 3CO_2$$

The carbon in this case comes from the graphite anode.

2.

$10^3$ kg ore contains $(10^3 \text{ kg})(\frac{1000 \text{ g}}{1 \text{ kg}})(\frac{1.60 \text{ g Cu}}{100 \text{ g}}) = 1.60 \times 10^4$ g Cu

The balanced equation for the reaction is:

$$Cu_2S(s) + O_2(g) \longrightarrow 2Cu(l) + SO_2(g)$$

Moles of $SO_2$ obtained $= (1.60\text{x}10^4 \text{ g Cu})(\frac{1 \text{ mol Cu}}{63.54 \text{ g Cu}})(\frac{1 \text{ mol } SO_2}{2 \text{ mol Cu}})$

$= \underline{126 \text{ mol } SO_2}$

$$V_{SO_2} = \frac{nRT}{P} = \frac{(126 \text{ mol})(0.0821 \text{ atm L mol}^{-1} \text{ K}^{-1})(293 \text{ K})}{1.00 \text{ atm}} = \underline{3.03\text{x}10^3 \text{ L}}$$

3. The reaction between $CaCO_3$ and $SiO_2$ in the blast furnace is:

$$CaCO_3(s) + SiO_2(l) \longrightarrow CaSiO_3(l) + CO_2(g)$$

2.00 metric tons of iron ore contains

$$(2.00 \text{ ton})(\frac{10^6 \text{ g}}{1 \text{ ton}})(\frac{12.2 \text{ g } SiO_2}{100 \text{ g}})(\frac{1 \text{ mol } SiO_2}{60.09 \text{ g } SiO_2}) = 4.06\text{x}10^3 \text{ mol } SiO_2$$

Minimum mass of $CaCO_3 = (4.06\text{x}10^3 \text{ mol } SiO_2)(\frac{1 \text{ mol } CaCO_3}{1 \text{ mol } SiO_2})(\frac{100.1 \text{ g } CaCO_3}{1 \text{ mol } CaCO_3})$

$= 4.06 \times 10^5$ g pure $CaCO_3$

Since the limestone used is only 96% pure:

Minimum mass of limestone $= (4.06\text{x}10^5 \text{ g})(\frac{100 \text{ g limestone}}{96 \text{ g}})(\frac{1 \text{ ton}}{10^6 \text{ g}})$

$= \underline{0.423 \text{ ton limestone}}$

4. A mineral is a naturally occurring substance containing one or more metallic compounds. An ore is a mineral deposit which is economical to use for the extraction of one or more metals. The common ores are oxides, sulfides, and carbonates, due to the abundance of oxygen and sulfur in the earth's crust, and the formation of carbonates from carbon containing plants and animals in ancient seas, and due to the fact that most metals occur in the crust as positive metal ions, that is in their oxidized form. Extraction of metals from their ores is by a reduction reaction.

5. (a) Oxides are commonly reduced using a reducing agent such as carbon:

$$Cu_2O + C \longrightarrow 2Cu + CO$$

$$Fe_2O_3 + 3CO \longrightarrow 2Fe + 3CO_2$$

Sulfides are normally roasted first in air, to oxidize $S^{2-}$ to $SO_2$, and the metal oxide is then reduced:

$$2PbS + 3O_2 \longrightarrow 2PbO + 2SO_2$$

$$PbO + C \longrightarrow Pb + CO$$

Carbonates of heavy metals decompose to the oxide and $CO_2$ on heating, and the oxide is then reduced. An example is the reduction of the ore malachite, $Cu_2(CO_3)(OH)_2$, copper(II) carbonate hydroxide:

$$Cu_2(CO_3)(OH)_2 + C \longrightarrow 2Cu + 2CO_2 + H_2O$$

(b) (i) Reduction of an oxide with a metal

$$Fe_2O_3 + 2Al \longrightarrow 2Fe + Al_2O_3 \quad \text{thermite process}$$

$$AlCl_3 + 3K \longrightarrow Al + 3KCl$$

(ii) Reduction of an oxide with a nonmetal

Carbon (coke) is the commonest nonmetal reducing agent for the preparation of metals

$$Fe_2O_3 + 3C \longrightarrow 2Fe + 3CO$$

$$PbO + C \longrightarrow Pb + CO$$

6. (a) 100 g ore contains 30 g of $Fe_2O_3$, or in terms of moles

$$\text{Mol of Fe from 100 g ore} = (30\text{ g }Fe_2O_3)\left(\frac{1\text{ mol }Fe_2O_3}{159.7\text{ g }Fe_2O_3}\right)\left(\frac{2\text{ mol Fe}}{1\text{ mol }Fe_2O_3}\right)$$

$$= 0.376\text{ mol Fe};\quad (0.376\text{ mol Fe})\left(\frac{55.85\text{ g Fe}}{1\text{ mol Fe}}\right) = \underline{21\text{ g Fe}}$$

Thus, for the production of $10^6$ ton of steel

$$\text{Mass of ore} = \left(\frac{10^6\text{ ton steel}}{21\text{ g Fe}}\right)(100\text{ g ore}) = \underline{4.8 \times 10^6\text{ ton}}$$

(b) 100 g of ore contains 21 g of Fe

$$\text{1 ton of ore contains } (1\text{ ton})\left(\frac{10^6\text{ g}}{1\text{ ton}}\right)\left(\frac{21\text{ g Fe}}{100\text{ g ore}}\right)\left(\frac{1\text{ kg}}{1000\text{ g}}\right) = \underline{210\text{ kg Fe}}$$

(c) The balanced equation for the reaction is

$$Fe_2O_3 + 3C \longrightarrow 2Fe + 3CO$$

1 ton of ore produces a maximum of 210 kg Fe,

$$\text{Mass of C required} = (210 \text{ kg Fe})\left(\frac{1 \text{ mol Fe}}{55.85 \text{ g Fe}}\right)\left(\frac{3 \text{ mol C}}{1 \text{ mol Fe}}\right)\left(\frac{12.01 \text{ g C}}{1 \text{ mol C}}\right)$$

$$= \underline{68 \text{ kg coke}}$$

7. (a) $\underline{Fe_3O_4}$ is a component of iron ore, an impure mixture of $Fe_2O_3$ and $Fe_3O_4$, which is the feedstock for the furnace.

(b) Coke is relatively pure carbon which at relatively high temperature under the conditions in the blast furnace is oxidized to carbon monoxide gas

$$2C(s) + O_2(g) \longrightarrow 2CO(g)$$

(c) Carbon monoxide is produced by the reaction in part (b) and is the reducing agent that reduces the $Fe_3O_4/Fe_2O_3$ mixture to iron,

$$Fe_3O_4(s) + 4CO(g) \longrightarrow 3Fe(l) + 4CO_2(g)$$

$$Fe_2O_3(s) + 3CO(g) \longrightarrow 2Fe(l) + 3CO_2(g)$$

(d) Calcium oxide is formed in the blast furnace by the decomposition of calcium carbonate, the principal constituent of limestone, which is part of the blast furnace charge.

$$CaCO_3(s) \longrightarrow CaO(s) + CO_2(g)$$

It is a basic oxide that combines with silica, $SiO_2$, impurities in the iron ore, to form a liquid layer of molten slag, calcium silicate, $CaSiO_3(l)$, which covers the liquid iron formed and prevents its reoxidation.

$$CaO(s) + SiO_2(l) \longrightarrow CaSiO_3(l)$$

8. (a) A compound such as $AlCl_3$ can be reduced using a reactive metal, such as an alkali metal.

$$AlCl_3 + 3K \longrightarrow 3KCl + Al$$

(b) Galena, PbS(s), is first roasted in air to give lead(II) oxide and sulfur dioxide gas. The PbO(s) is then reduced with carbon.

$$2PbS + 3O_2 \longrightarrow 2PbO + 2SO_2$$

$$PbO + C \longrightarrow Pb + CO$$

(c) The limestone decomposes at the temperature of the furnace to give the basic oxide CaO and $CO_2$ gas. The basic oxide CaO then combines with the acidic oxide $SiO_2$, silica, found as an impurity in iron ore, to give the salt calcium silicate, $CaSiO_3$, which at the temperature of the furnace forms a liquid layer on top of the iron that is produced, and protects it from oxidation by air. The $CaSiO_3$ is known as slag.

$$CaCO_3(s) \longrightarrow CaO(s) + CO_2(g)$$

$$CaO(s) + SiO_2(l) \longrightarrow CaSiO_3(l)$$

9. In the manufacture of steel in the blast furnace, a mixture of impure iron oxide, coke, and limestone is fed in at the top of the furnace, while preheated air or oxygen is blown in at the bottom. The iron ore is reduced to iron by carbon monoxide formed by incomplete oxidation of the coke, according to the overall equation

$$Fe_2O_3(s) + 3CO(g) \longrightarrow 2Fe(l) + 3CO_2(g)$$

The function of the limestone, $CaCO_3(s)$, is to form lime, $CaO(s)$ at the temperature of the furnace,

$$CaCO_3(s) \longrightarrow CaO(s) + CO_2(g)$$

which combines with silica impurities, $SiO_2$, to form calcium silicate, $CaSiO_3$, (slag), which floats on top of the molten iron, protecting it from reoxidation. The molten iron that is tapped off at the bottom of the blast furnace is called pig iron, and contains a number of impurities, notably carbon, silicon, and sulfur. Purification of the pig iron is accomplished by blowing air through the molten iron, which converts most of the impurities to their oxides. The final product is steel, which depends for its properties on the residual small amount of carbon in the iron alloy (carbon steel)

10. See Problem 9. Calcium silicate slag is the product of the reaction of the basic oxide CaO with the acidic oxide $SiO_2$.

11. The majority of the elements are metals; they are found mainly on the left-hand side of the periodic table, in groups 1, 2, and 3, in the series of transition elements in periods 4, 5, and 6, and the lanthanoids and actinoids. The metals and nonmetals are divided by a diagonal band of elements, starting at beryllium in group 2, and extending to polonium in group 6, which exhibit both metallic and nonmetallic properties and are called semimetals (or metalloids). The nonmetals are found on the right-hand side of the periodic table (mainly in groups 4, 5, 6, 7, and 8). In group 4, carbon is a true nonmetal, silicon is a nonmetal in its reactions, although solid element is a semiconductor of electricity and is metallic in appearance and can be described as a semimetal. Germanium is a semimetal, and tin and lead are both metals. In the third period, sodium, magnesium, and aluminum are metals.

12. (a)

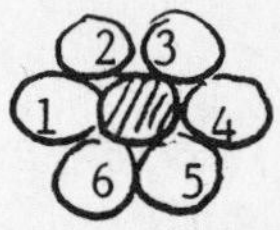


(hexagonal)
close-packing

Each atom is surrounded by six atoms in the same plane

Three-dimensions - six atoms in the same plane, three in the plane above and three in the plane below for a total of 12 atoms

ABA (or ABC)

(b)

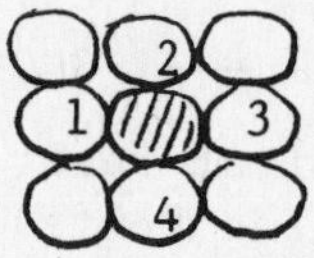


square-packing

Each atom is surrounded by four atoms in the same plane

Three-dimensions - four atoms in the same plane, one in the plane above and one in the plane below, for a total of 6 atoms

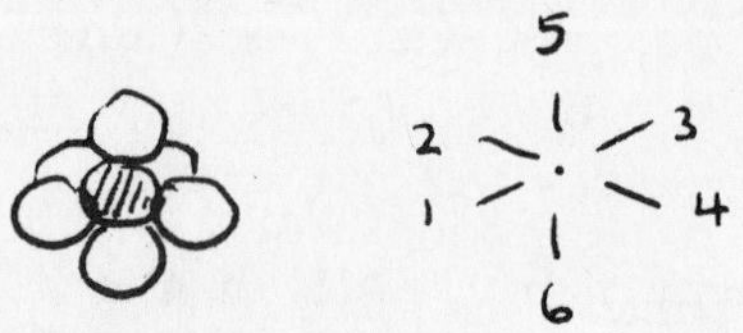

13. (a) Each Na atom has a single $3s^1$ valence electron and $Na_2(g)$ is formed in the gas phase by the overlap of two singly occupied 3s orbitals (in the same way that $H_2$ is formed by the overlap of two singly occupied 1s hydrogen orbitals). Thus, the bonding in $Na_2$ is a single covalent bond between the two atoms.

(b) In sodium metal the 3s valence orbital of any particular Na atom can overlap with the 3s orbitals of any of the surrounding Na atoms. Thus, one model to describe the bonding in the solid metal is to draw a very large number of resonance structures illustrating all of the ways in which single bonds can be formed between a very large number of Na atoms. Bearing in mind that any Na atom can form only one bond, many of the Na atoms in any structure remain unbonded. The actual structure is obtained by averaging all of the possible resonance structures. This is the same as describing the structure as having one valence electron per Na atom delocalized over all of the atoms. Alternatively, the structure can be described as a large number of $Na^+$ ions surrounded by a sea of electrons. It is the attraction between the positive ions and the delocalized electrons that constitute the metallic bond. The essential difference between an $Na_2$ molecule and sodium metal is that in the former an electron pair is localized in a covalent bond joining the atoms, while in the latter one electron per atom is delocalized over all of the atoms. The Na-Na distance in $Na_2$ (bond order 1) should be less than that in the metal (bond order $< 1$).

14. (a) The delocalized electrons in a metal are free to move between the positively charged metal ions and they migrate freely under the influence of an electric potential. When a battery is connected to metal wire, the electrons from the battery replace electrons in the wire as they are removed from the wire at the other battery terminal, and a current flows through the metal. When one part of the metal is heated, the heat energy is transferred to the electrons and metal ions - which increases their kinetic energy and the relatively light electrons move at relatively high speeds to other parts of the metal where their energy is transferred to other metal ions and the metal heats up.

(b) In electric conduction, electrons flow freely through the metal and are impeded only by collisions with the vibrating metal ions With decrease in temperature, the amplitude of the vibrations of the ions is decreased, which decreases their impedance of the electron flow, and the electrical conductivity increases.

15. <u>Body-centered cubic structure</u>

The selected atom is at the center of a cube. The eight atoms at the corners of the cube (A) touch the central atom and are the <u>nearest</u> neighbors. The six <u>next-nearest</u> neighbors are the (B) atoms located at the centers of the six cubes that share common faces with the cube shown.

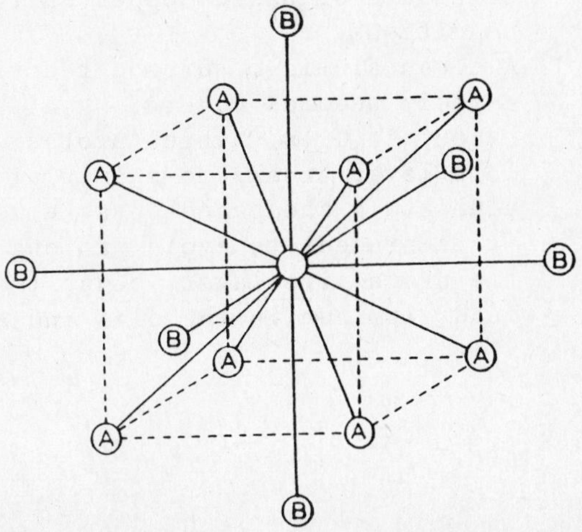

16. The biblical metals are all found as the free metals in nature, or occur as ores that are relatively easily reduced to the metal, by, for example heating an ore with charcoal - a reaction that probably was discovered when a fire was built over a rock containing the ore. In contrast, none of the alkali metals or aluminum occur as the free metals in nature, and their compounds are difficult to reduce to the metals; only very strong reducing agents, such as electrons in electrolysis, or other reactive metals are suitable. Thus none of these metals could have been discovered accidently and their first preparation was only possible after the science of metallurgy was reasonably well advanced. For example, sodium was first prepared electrolytically by Humphrey Davy in 1807, who used the same method to discover aluminum in 1808.

17. (a) Ideal properties for a space vehicle would include a low density, to minimize its mass, high tensile strength, a high melting point, and low chemical reactivity. A good choice would be aluminum, or even better one of its alloys, such as duralumin (an alloy of Al with Mg, Cu, and Mn). Although aluminum is a relatively reactive metal, it is protected by a thin hard layer of its oxide, $Al_2O_3$, at all but very high temperatures (otherwise it could not be used for articles such as cooking pots).
(b) For a statue, the density would not be a consideration, but desirable properties would include the ease of casting, resistance to weathering, and aesthetic appeal. Copper has traditionally been the metal of choice, or bronze (an alloy of copper and tin that has been known since ancient times) which is harder than copper and not so easily damaged but nevertheless easy to cast. Gold would be an even better choice, but few towns are sufficiently prosperous to bear the expense or would risk the chance of theft, although in Russia, thin layers of gold have traditionally been used to decorate church spires and even statues. Copper is relatively difficult to oxidize and therefore quite resistant to corrosion, although it rapidly acquires a brown protective tarnish of oxide or sulfide. Exposure to town air forms a green patina of basic copper sulfate, $Cu_2(OH)_2SO_4(s)$, which is generally considered to be visually pleasing, especially on copper roofs.

18. (a) Copper is found as the metal in nature and many of its ores are readily reduced with, for example, charcoal. The earliest cast copper articles from Egypt date back to about 4000 B.C.
(b) Copper covered roofs acquire a green patina due to the slow reaction of the copper to give copper oxide or sulfide, and then copper carbonate or basic copper sulfate, depending on the atmospheric conditions.
(c) A clean aluminum surface reacts readily with oxygen in air and rapidly becomes covered by a thin film of the very hard, insoluble, oxide, $Al_2O_3(s)$, which protects the metal from further reaction. This layer breaks down only at very high temperature, which makes aluminum a cheap and durable material for cooking utensils, although it is prudent to avoid prolonged contact with acids, which dissolve the protective layer, since there well may be a link between too much aluminum in the diet and Alzheimer's disease.

(d) Prior to the discovery of theelectrolytic (Hall) process in 1886, which utlized the electrolysis of the relatively low melting cryolyte/bauxite mixture, aluminum was difficult and expensive to prepare and was therefore a rare and expensive metal. The principal method of preparation was by the reduction of anhydrous aluminum chloride, $AlCl_3$, with potassium metal (a process first discovered by Oersted in 1825).

$$AlCl_3 + 3K \longrightarrow 3KCl + Al$$

19. The melting points of metals are related to their strengths of metallic bonding, which depend on the number of electrons per atom that are available for metallic bonding and the sizes of the ions that attract these electrons. Na (group 1), Ca (group 2), and Al(group 3), have, respectively 1, 2, and 3, relatively easily ionized valence electrons, and the sizes of the ions decrease in the order $Na^+ > Ca^{2+} > Al^{3+}$. Thus, the strength of electrostatic attraction for an ever increasing number of delocalized metallic bonding electrons is expected to increase in the order Na < Ca < Al, which is the order of increasing melting points. The transition metal iron has the valence shell electron configuration $4s^2 3d^6$ and therefore in all probability has available an even greater number of metallic bonding electrons than Al, which accounts for its greater melting point.

20. Diamond is a covalent network substance in which the bonding electrons are localized in C-C bonds, which accounts for its properties such as hardness. It is not easily deformed and fractures only under high stress when a large number of C-C bonds are broken, which requires a large amount of energy. In contrast, in a metal such as copper, the metallic bonding electrons are delocalized and bond individual copper atoms relatively weakly, so that under stress copper atoms move rather easily and the bonding electrons can easily flow to assume the new shape. Thus copper is relatively soft and readily deforms.

21. (a) Common metals that dissolve in dilute HCl(aq) include reactive metals, such as Mg, Al, Fe, and Zn (but not metals such as Cu or Ag). In dilute acid the oxidizing agent is $H_3O^+(aq)$, which is incapable of oxidizing the less reactive metals.

$$Mg(s) + 2H_3O^+(aq) \longrightarrow Mg^{2+}(aq) + H_2(g) + 2H_2O(l)$$

$$2Al(s) + 6H_3O^+(aq) \longrightarrow 2Al^{3+}(aq) + 3H_2(g) + 6H_2O(l)$$

$$Fe(s) + 2H_3O^+(aq) \longrightarrow Fe^{2+}(aq) + H_2(g) + 2H_2O(l)$$

$$Zn(s) + 2H_3O^+(aq) \longrightarrow Zn^{2+}(aq) + H_2(g) + 2H_2O(l)$$

In such reactions $Cl^-(aq)$ ions take no part in the reactions and remain in solution as spectator ions.

(b) The less reactive metals, such as Cu, Ag, and Au do not react with HCl(aq). The solution contains only $H_3O^+(aq)$ and $Cl^-(aq)$, and $H_3O^+(aq)$ is an insufficiently strong oxidizing agent to oxidize these metals to their cations.

(c) Dilute $HNO_3(aq)$ contains $H_3O^+(aq)$ and $NO_3^-(aq)$ ions. The latter contains nitrogen in the +5 oxidation state and is a stronger oxidizing agent than $H_3O^+(aq)$, since it is readily reduced to lower oxidation state species such as $NO_2(g)$ (N, +4) and NO(g) (N, +2).

22. (a) $2Al(s) + 3H_2SO_4(aq) \longrightarrow Al_2(SO_4)_3(aq) + 3H_2(g)$

As in Problem 21, $H_3O^+(aq)$ is the oxidizing agent, and the reaction in its simplest form is

$$2Al(s) + 6H_3O^+(aq) \longrightarrow 2Al^{3+}(aq) + 6H_2O(l) + 3H_2(g)$$

(b) Silver is not oxidized by $H_3O^+(aq)$ but can be oxidized by hot concentrated $H_2SO_4(l)$. The latter contains $H_2SO_4$ molecules containing sulfur in the +6 oxidation state, which is readily reduced to $SO_2(g)$, with sulfur in the +4 oxidation state.

$2[Ag \longrightarrow Ag^+ + e^-]$ oxidation

$H_2SO_4 + 2e^- + 2H^+ \longrightarrow SO_2 + 2H_2O$ reduction

$2Ag + H_2SO_4 + 2H^+ \longrightarrow 2Ag^+ + SO_2 + 2H_2O$ overall

In concentrated acid, $H_2O$ is ionized as $H_3O^+HSO_4^-$ & adding $2HSO_4^-$ and $2H_2SO_4$ to each side of the equation gives

$$2Ag(s) + 5H_2SO_4(l) \longrightarrow 2Ag^+(soln) + SO_2(g) + 2H_3O^+(soln) + 4HSO_4^-(soln)$$

23. The order of reactivity with respect to $H_2O$ is

$$Na > Mg > Al > Cu$$

Sodium reacts vigorously with cold water, magnesium reacts slowly with hot water. To study the reactivity of aluminum the protective $Al_2O_3$ film must first be removed with dilute acid, or by forming an amalgam with mercury. It is then found that the aluminum reacts very slowly with hot water and more vigorously with steam. In contrast, copper does not react with water under any conditions. Recall that copper is not even oxidized by dilute acid but only by nitric acid or hot concentrated sulfuric acid.

24. Using the rule that the sum of the oxidation numbers of all the atoms is equal to the overall charge on the species, and remembering that in all but a few compounds the oxidation numbers of O and H are -2 and +1, respectively, and that of N is normally -3, we have:

(a) H +1, O, -2, Al +3 (b) Cl -1, Al +3 (c) O -2, Si +4

(d) O -2, Si +4 (e) O -2, H +1, Pb +2 (f) Cl -1, Pb +4.

25. See Problem 24.

(a) H +1, O -2, Al +3 (b) Cl -1, Al +3 (c) O -2, S +6, Fe +3

(d) Cl -1, Cu +1 (e) $Ca^{2+}CO_3^{2-}$; Ca +2, O -2, C +4

(f) $K^+Al^{3+}(SO_4^{2-})_3$; K +1, Al +3, O -2, S +6. $H_2O$ ; H +1, O -2.

(g) $Fe^{2+}[^-:S\text{-}S:^-]$; Fe +2, S -1 (h) $(Zn^{2+})_2CO_3^{2-}(OH^-)_2$; Zn +2, O -2, C +4.

26. In terms of the transfer of electrons, oxidation involves the loss of electrons from a species, and reduction involves the gain of electrons by a species.

(a) Zn(0) in Zn(s) is oxidized to Zn(+2) in $Zn^{2+}$ in $ZnSO_4$(aq); H(+1) in $H_2SO_4$ is reduced to H(0) in $H_2$(g), and S(+6) remains unchanged in $ZnSO_4$(aq) and $H_2SO_4$(aq).
The equation is balanced as written.

(b) I(-1/3) in $I_3^-$ is reduced to I(-1) in $I^-$; S(+2) in $S_2O_3^{2-}$ is oxidized to S(+5/2) in $S_4O_6^{2-}$.

$I_3^- + 2e^- \longrightarrow 3I^-$ reduction

$2S_2O_3^{2-} \longrightarrow S_4O_6^{2-} + 2e^-$ oxidation

$I_3^- + 2S_2O_3^{2-} \longrightarrow 3I^- + S_4O_6^{2-}$ overall

27. (a) S(+4) in $SO_3^{2-}$ is oxidized to S(+6) in $SO_4^{2-}$; Mn(+7) in $MnO_4^-$ is reduced to Mn(+2) in $Mn^{2+}$.

$5[SO_3^{2-} + H_2O \longrightarrow SO_4^{2-} + 2e^- + 2H^+]$ oxidation

$2[MnO_4^- + 5e^- + 8H^+ \longrightarrow Mn^{2+} + 4H_2O]$ reduction

$5SO_3^{2-} + 2MnO_4^- + 6H^+ \longrightarrow 5SO_4^{2-} + 2Mn^{2+} + 3H_2O$ overall

(b) O(-1) in $H_2O_2$ is oxidized to O(0) in $O_2$; Mn(+7) in $MnO_4^-$ is reduced to Mn(+2) in $Mn^{2+}$.

$5[H_2O_2 \longrightarrow O_2 + 2e^- + 2H^+]$ oxidation

$2[MnO_4^- + 8H^+ + 5e^- \longrightarrow Mn^{2+} + 4H_2O]$ reduction

$5H_2O_2 + 2MnO_4^- + 6H^+ \longrightarrow 2Mn^{2+} + 5O_2 + 8H_2O$ overall

(c) Zn(0) in Zn(s) is oxidized to Zn(+2) in $Zn^{2+}$; N(+5) in $NO_3^-$ is reduced to N(+1) in $N_2O$.

$4[Zn \longrightarrow Zn^{2+} + 2e^-]$ oxidation

$2NO_3^- + 8e^- + 10H^+ — N_2O + 5H_2O$ reduction

$4Zn + 2NO_3^- + 10H^+ \longrightarrow 4Zn^{2+} + N_2O + 5H_2O$ overall

(d) P(0) in $P_4$ is oxidized to P(+5) in $H_2PO_4^-$; N(+5) in $NO_3^-$ is reduced to N(+2) in NO.

$3[P_4 + 16H_2O \longrightarrow 4H_2PO_4^- + 20e^- + 24H^+]$ oxidation

$20[NO_3^- + 3e^- + 4H^+ \longrightarrow NO + 2H_2O]$ reduction

$3P_4 + 20NO_3^- + 8H_3O^+ \longrightarrow 12H_2PO_4^- + 20\ NO$ overall

(e) N(+5) in $NO_3^-$ is reduced to N(+4) in $NO_2$; I(0) in $I_2$ is oxidized to I(+5) in $IO_3^-$.

$10[NO_3^- + e^- + 2H^+ \longrightarrow NO_2 + H_2O]$ reduction

$I_2 + 6H_2O \longrightarrow 2IO_3^- + 10e^- + 12H^+$ oxidation

$10NO_3^- + I_2 + 8H^+ \longrightarrow 2IO_3^- + 10NO_2 + 4H_2O$ overall

28. (a) Mn(+7) in $MnO_4^-$ is reduced to Mn(+2) in $Mn^{2+}$; Fe(+2) in $Fe^{2+}$ is oxidized to Fe(+3) in $Fe^{3+}$.

$MnO_4^- + 5e^- + 8H^+ \longrightarrow Mn^{2+} + 2H_2O$ reduction

$5[Fe^{2+} \longrightarrow Fe^{3+} + e^-]$ oxidation

$MnO_4^- + 5Fe^{2+} + 8H^+ \longrightarrow Mn^{2+} + 5Fe^{3+} + 4H_2O$ overall

(b) Cl(+5) in $ClO_3^-$ is reduced to Cl(+4) in $ClO_2$; Cl(-1) in $Cl^-$ is oxidized to Cl(0) in $Cl_2$.

$2[ClO_3^- + 2H^+ + e^- \longrightarrow ClO_2 + H_2O]$ reduction

$2Cl^- \longrightarrow Cl_2 + 2e^-$ oxidation

$2ClO_3^- + 2Cl^- + 4H^+ \longrightarrow Cl_2 + 2ClO_2 + 2H_2O$ overall

(c) Cu(0) in Cu(s) is oxidized to Cu(+2) in $Cu^{2+}$; N(+5) in $HNO_3$ is reduced to N(+2) in NO.

$3[Cu \longrightarrow Cu^{2+} + 2e^-]$ oxidation

$2[HNO_3 + 3e^- + 3H^+ \longrightarrow NO + 2H_2O]$ reduction

$3Cu + 2HNO_3 + 6H^+ \longrightarrow 3Cu^{2+} + 2NO + 4H_2O$ overall

(d) Mn(+7) in $MnO_4^-$ is reduced to Mn(+2) in $Mn^{2+}$; S(+4) in $SO_2$ is oxidized to S(+6) in $SO_4^{2-}$.

$2[MnO_4^- + 5e^- + 8H^+ \longrightarrow Mn^{2+} + 4H_2O]$ reduction

$5[SO_2 + 2H_2O \longrightarrow SO_4^{2-} + 2e^- + 4H^+]$ oxidation

$2MnO_4^- + 5SO_2 + 2H_2O \longrightarrow 2Mn^{2+} + 5SO_4^{2-} + 4H^+$ overall

(e) I(-1) in HI is oxidized to I(0) in $I_2$; N(+5) in $HNO_3$ is reduced to N(+2) in NO.

$3[2HI \longrightarrow I_2 + 2e^- + 2H^+]$ oxidation

$2[HNO_3 + 3e^- + 3H^+ \longrightarrow NO + 2H_2O]$ reduction

$6HI + 2HNO_3 \longrightarrow 3I_2 + 2NO + 4H_2O$ overall

29. (a) Mn(+7) in $MnO_4^-$ is reduced to Mn(+4) in $MnO_2$; I(-1) in $I^-$ is oxidized to I(+1) in $IO^-$.

$2[MnO_4^- + 3e^- + 2H_2O \longrightarrow MnO_2 + 4OH^-]$ reduction

$3[I^- + 2OH^- \longrightarrow IO^- + 2e^- + H_2O]$ oxidation

$2MnO_4^- + 3I^- + H_2O \longrightarrow 2MnO_2 + 3IO^- + 2OH^-$ overall

(b) Br(0) in $Br_2$ is reduced to Br(-1) in $Br^-$ and oxidized to Br(+5) in $BrO_3^-$.

$5[Br_2 + 2e^- \longrightarrow 2Br^-]$ reduction

$Br_2 + 12OH^- \longrightarrow 2BrO_3^- + 10e^- + 6H_2O$ oxidation

$6Br_2 + 12OH^- \longrightarrow 10Br^- + 2BrO_3^- + 6H_2O$ overall

i.e., $3Br_2 + 6OH^- \longrightarrow 5Br^- + BrO_3^- + 3H_2O$

(c) I(-1) in $I^-$ is oxidized to I(+5) in $IO_3^-$; Cl(+1) in $ClO^-$ is reduced to Cl(-1) in $Cl^-$:

$$I^- + 6OH^- \longrightarrow IO_3^- + 6e^- + 3H_2O \qquad \text{oxidation}$$
$$3[ClO^- + 2e^- + H_2O \longrightarrow Cl^- + 2OH^-] \qquad \text{reduction}$$
$$I^- + 3ClO^- \longrightarrow IO_3^- + 3Cl^- \qquad \text{overall}$$

(d) S(+4) in SO is oxidized to S(+6) in $SO_4^{2-}$; Mn(+7) in $MnO_4^-$ is reduced to Mn(+4) in $MnO_2$:

$$3[SO_3^{2-} + 2OH^- \longrightarrow SO_4^{2-} + 2e^- + H_2O] \qquad \text{oxidation}$$
$$2[MnO_4^- + 3e^- + 2H_2O \longrightarrow MnO_2 + 4OH^-] \qquad \text{reduction}$$
$$3SO_3^{2-} + 2MnO_4^- + H_2O \longrightarrow 3SO_4^{2-} + 2MnO_2 + 2OH^- \qquad \text{overall}$$

(e) Mn(+2) in $Mn^{2+}$ is oxidized to Mn(+4) in $MnO_2$; the other product must be water, so O(-1) in $H_2O_2$ must be reduced to O(-2) in $H_2O$:

$$Mn^{2+} + 4OH^- \longrightarrow MnO_2 + 2e^- + 2H_2O \qquad \text{oxidation}$$
$$H_2O_2 + 2e^- \longrightarrow 2OH^- \qquad \text{reduction}$$
$$Mn^{2+} + H_2O_2 + 2OH^- \longrightarrow MnO_2 + 2H_2O \qquad \text{overall}$$

30. (a) C(+2) in $CN^-$ is oxidized to C(+4) in $CNO^-$; Mn(+7) in $MnO_4^-$ is reduced to Mn(+4) in $MnO_2$:

$$3[CN^- + 2OH^- \longrightarrow CNO^- + 2e^- + H_2O] \qquad \text{oxidation}$$
$$2[MnO_4^- + 3e^- + 2H_2O \longrightarrow MnO_2 + 4OH^-] \qquad \text{reduction}$$
$$3CN^- + 2MnO_4^- + H_2O \longrightarrow 3CNO^- + 2MnO_2 + 2OH^- \qquad \text{overall}$$

(b) Cr(+3) in $Cr^{3+}$ is oxidized to Cr(+6) in $CrO_4^{2-}$; Cl(+1) in $OCl^-$ is reduced to Cl(-1) in $Cl^-$:

$$2[Cr^{3+} + 8OH^- \longrightarrow CrO_4^{2-} + 3e^- + 4H_2O] \qquad \text{oxidation}$$
$$3[OCl^- + 2e^- + H_2O \longrightarrow Cl^- + 2OH^-] \qquad \text{reduction}$$
$$2Cr^{3+} + 3OCl^- + 10\,OH^- \longrightarrow 2CrO_4^{2-} + 3Cl^- + 5H_2O \qquad \text{overall}$$

(c) I(-1) in $I^-$ is oxidized to I($-\frac{1}{3}$) in $I_3^-$; Cl(+5) in $ClO_3^-$ is reduced to Cl(-1) in $Cl^-$:

$$3[3I^- \longrightarrow I_3^- + 2e^-] \qquad \text{oxidation}$$
$$ClO_3^- + 6e^- + 3H_2O \longrightarrow Cl^- + 6OH^- \qquad \text{reduction}$$
$$9I^- + ClO_3^- + 3H_2O \longrightarrow 3I_3^- + Cl^- + 6OH^- \qquad \text{overall}$$

(d) N(-3) in $NH_3$ is oxidized to N(-2) in $N_2H_4$; Cl(+1) in $ClO^-$ is reduced to Cl(-1) in $Cl^-$:

$$2NH_3 + 2OH^- \longrightarrow N_2H_4 + 2e^- + 2H_2O \qquad \text{oxidation}$$
$$OCl^- + 2e^- + H_2O \longrightarrow Cl^- + 2OH^- \qquad \text{reduction}$$
$$2NH_3 + OCl^- \longrightarrow N_2H_4 + Cl^- + H_2O \qquad \text{overall}$$

(e) Mn(+4) in $MnO_2$ is oxidized to Mn(+6) in $MnO_4^{2-}$; O(0) in $O_2$ is reduced to O(-2) in $OH^-$.

$$2[MnO_2 + 4OH^- \longrightarrow MnO_4^{2-} + 2e^- + 2H_2O] \qquad \text{oxidation}$$

$$O_2 + 4e^- + 2H_2O \longrightarrow 4OH^- \qquad \text{reduction}$$

$$2MnO_2 + 4OH^- + O_2 \longrightarrow 2MnO_4^{2-} + 2H_2O \qquad \text{overall}$$

31. In each case the anhydride results from removing all the H from the parent acid in the form of water molecules.

(a) $H_2SO_4 \longrightarrow H_2O + SO_3$ (b) $2HNO_3 \longrightarrow H_2O + N_2O_5$

(c) $H_2CO_3 \longrightarrow H_2O + CO_2$ (d) $2HClO_4 \longrightarrow H_2O + Cl_2O_7$

(a) sulfur trioxide; (b) dinitrogen pentaoxide; (c) carbon dioxide
(d) dichlorine heptaoxide

32. Most metal oxides are basic oxides because they consist of metal cations and oxide ions, $O^{2-}$. A soluble oxide dissociates into its ions in aqueous solution, and the $O^{2-}$(aq) ion behaves as a strong base:

$$O^{2-}(aq) + H_2O(l) \longrightarrow 2OH^-(aq)$$

Nonmetal oxides are acidic oxides; they are the anhydrides of oxoacids and react with water to give oxoacids.

(a) $K_2O(s) + H_2O(l) \longrightarrow 2K^+(aq) + 2OH^-(aq)$ basic oxide

(b) $SrO(s) + H_2O(l) \longrightarrow Sr^{2+}(aq) + 2OH^-(aq)$ basic oxide

(c) $SO_2(g) + H_2O(l) \longrightarrow H_2SO_3(aq)$ acidic oxide

(d) $SO_3(g) + H_2O(l) \longrightarrow H_2SO_4(aq)$ acidic oxide

(e) $CO_2(g) + H_2O(l) \longrightarrow H_2CO_3(aq)$ acidic oxide

(f) $P_4O_6(s) + 6H_2O(l) \longrightarrow 4H_3PO_3(aq)$ acidic oxide

(g) $Cl_2O_7(l) + H_2O(l) \longrightarrow 2HClO_4(aq)$ acidic oxide

The oxoacids above will of course be ionized in aqueous solution.

33. Al is more metallic than Si, which is classified as a semimetal. Thus, $Al_2O_3$ is more basic than $SiO_2$. In fact $Al_2O_3$ behaves as an amphoteric oxide while $SiO_2$ is an acidic oxide. Both $SiO_2$ and $SO_2$ are nonmetal oxides and are therefore acidic oxides. $SiO_2$ reacts with water very slowly to give the very weak acid $Si(OH)_4$, silicic acid, while $SO_2$ reacts to give a solution that behaves as an aqueous solution of $(HO)_2SO$, sulfurous acid. In terms of the $XO_m(OH)_n$ formulation of oxoacids, silicic acid is a very weak acid and sulfurous acid is a weak acid. $SO_2 > SiO_2$.

34. The reactions are all reactions between a basic (metal) oxide and an acidic (nonmetal) oxide to give a salt. Since all basic oxides contain metal cations and oxide ions, the reactions are all between the $O^{2-}$ ions and an acidic oxide, with the metal ions acting as spectator ions.

(a) $O^{2-} + SiO_2 \longrightarrow SiO_3^{2-}$ Product: $Li_2SiO_3$, lithium silicate
(b) $O^{2-} + N_2O_5 \longrightarrow 2NO_3^{-}$ Product: $NaNO_3$, sodium nitrate
(c) $6O^{2-} + P_4O_{10} \longrightarrow 4PO_4^{3-}$ Product: $Ca_3(PO_4)_2$, calcium phosphate

35. (a) A Brønsted-Lowry acid is a proton donor and a Brønsted-Lowry base is a proton acceptor. A Lewis acid is an electron pair acceptor and a Lewis base is an electron pair donor.

Bases are the same according to both concepts, since a base has to have a (lone) electron pair in order to accept a proton. In other words, a Brønsted base donates an electron pair to a proton In terms of the acid concepts, the proton is a specific example of a Lewis acid, but it is unique in that it has no separate existence but is always found in combination with a base. For example, as an acid, HA, or as the $H_3O^+(aq)$ ion.

(b) CaO is the basic metal oxide $Ca^{2+}O^{2-}$ and $SiO_2$ is an acidic oxide. Thus, the reaction involves the Lewis acid $SiO_2$ (electron pair acceptor) and the $O^{2-}$ ion - a Lewis base (electron pair donor), and is an acid-base reaction.

$$SiO_2 + :\ddot{\underset{\cdot\cdot}{O}}:^{2-} \longrightarrow SiO_3^{2-}$$

| $SiO_2$ | :O:$^{2-}$ | $SiO_3^{2-}$ |
|---|---|---|
| Lewis acid | Lewis base | silicate ion |

36. (a) Acidic oxides are the nonmetal oxides found on the right-hand side of the periodic table; amphoteric oxides have both acid and base properties are are formed by the elements that are on the borderline between the metals and the nonmetals.

| | | | | | | |
|---|---|---|---|---|---|---|
| Period 2 | BeO | $B_2O_3$ | $CO_2$ | $N_2O_3$, $N_2O_5$ | | ($O_2$ $F_2O$)* |
| Period 3 | $Al_2O_3$ | $SiO_2$ | $P_4O_6$, $P_4O_{10}$ | $SO_2$, $SO_3$ | $Cl_2O_7$** | |
| | amphoteric | | acidic | | | |

*no acid or base properties in water
**see Chapter 20 for other examples of chlorine oxides

(b) Basic oxides (any oxide from groups 1 or 2), e.g. $Na_2O$. CaO

Acidic oxides (any oxide from groups 5, 6, or 7), e.g. $SO_3$, $N_2O_5$

Amphoteric oxides e.g. BeO, $Al_2O_3$

(c) $Na_2O(s) + H_2O(l) \longrightarrow 2Na^+(aq) + 2OH^-(aq)$

$CaO(s) + H_2O(l) \longrightarrow Ca^{2+}(aq) + 2OH^-(aq)$

Both BeO(s) and $Al_2O_3(s)$ are very insoluble in water

$SO_3(g) + H_2O(l) \longrightarrow H_2SO_4(aq)$

$N_2O_5(s) + H_2O(l) \longrightarrow 2HNO_3(aq)$

37. A Lewis acid is an electron pair acceptor. Thus the reactions in question must be a reaction between a Lewis acid and a Lewis base.

$AlCl_3 + :\ddot{\underset{..}{Cl}}:^- \longrightarrow AlCl_4^-$ ; $AlCl_3 + 3H_2O \longrightarrow Al(OH)_3(s) + 3HCl$

$AlBr_3 + :\ddot{\underset{..}{Br}}:^- \longrightarrow AlBr_4^-$ ; $AlBr_3 + 3H_2O \longrightarrow Al(OH)_3(s) + 3HBr$

$BF_3 + :\ddot{\underset{..}{F}}:^- \longrightarrow BF_4^-$; $BF_3 + :NH_3 \longrightarrow F_3\overset{-}{B}-\overset{+}{N}H_3$

$BCl_3 + :\ddot{\underset{..}{Cl}}:^- \longrightarrow BCl_4^-$ ; $BCl_3 + 3H_2O \longrightarrow B(OH)_3 + 3HCl$

38. A Lewis acid is an electron pair acceptor and a Lewis base is an electron pair donor.

$:NH_3$ Lewis base; $Cu^{2+}$ Lewis acid; $Al^{3+}$ Lewis acid;

$SiO_2$ Lewis acid; $:\ddot{\underset{..}{Cl}}:^-$ Lewis base.

Metal ions in general are Lewis acids; in water for example they accept electron pairs from water molecules to form hydrated ions. Nonmetal (acidic) oxides are also Lewis acids; for example, they react with water to give oxoacids and with oxide ions to give the anions of oxoacids.

39. All the compounds are soluble in water to give solutions of the salts. Thus, initially each solution contains the ions from the salt and $Na^+(aq)$ and $OH^-(aq)$ ions from the NaOH. These are all examples where the combination of the metal ion from the salt with $OH^-$ give an insoluble hydroxide. Thus, they are all precipitation reactions.

(a) $Al^{3+}(aq) + 3OH^-(aq) \longrightarrow Al(OH)_3$ - a gelatinous white ppt*

(b) $Cu^{2+}(aq) + 2OH^-(aq) \longrightarrow Cu(OH)_2(s)$ - a pale-blue ppt

(c) $Fe^{2+}(aq) + 2OH^-(aq) \longrightarrow Fe(OH)_2(s)$ - a white ppt that rapidly turns brown**

(d) $Pb^{2+}(aq) + 2OH^-(aq) \longrightarrow Pb(OH)_2(s)$ - a white ppt*

*$Al(OH)_3(s)$ and $Pb(OH)_2(s)$ are both amphoteric hydroxides that dissolve in excess of NaOH(aq):

$$Al(OH)_3(s) + OH^-(aq) \longrightarrow Al(OH)_4^-(aq)$$

$$Pb(OH)_2(s) + 2OH^-(aq) \longrightarrow Pb(OH)_4^{2-}(aq)$$

**$Fe(OH)_2(s)$ is rapidly oxidized by air to brown $Fe(OH)_3(s)$.

40. The hydroxides of Al(III), Fe(III), and Cu(II), all decompose on heating to give the oxides:

$$2Al(OH)_3(s) \longrightarrow Al_2O_3(s) + 3H_2O(g)$$

$$2Fe(OH)_3(s) \longrightarrow Fe_2O_3(s) + 3H_2O(g)$$

$$Cu(OH)_2(s) \longrightarrow CuO(s) + H_2O(g)$$

41. Some of the most important reactions that could be used to distinguish these solutions are shown in the following table:

| Test | $CuSO_4(aq)$ | $Al_2(SO_4)_3(aq)$ | $Pb(NO_3)_2(aq)$ | $FeSO_4(aq)$ |
|---|---|---|---|---|
| Color | blue | colorless | colorless | colorless |
| HCl(aq) | - | - | white ppt of $PbCl_2(s)$ | - |
| NaOH(aq) | Pale blue ppt $Cu(OH)_2$ insoluble in excess | White ppt $Al(OH)_3(s)$ soluble in excess | White ppt of $Pb(OH)_2(s)$ soluble in excess | White ppt of $Fe(OH)_2(s)$ insoluble in excess. Turns brown in air |
| $NH_3(aq)$ | Blue ppt $Cu(OH)_2(s)$ soluble in excess | White ppt $Al(OH)_3(s)$ slightly soluble in excess | White ppt $Pb(OH)_2(s)$ insoluble in excess | White ppt of $Fe(OH)_2(s)$ insoluble in excess |

42. (a) Al(0) in Al(s) is oxidized to Al(+3) in $Al(OH)_4^-$; H(+1) in $H_2O$ is reduced to H(0) in $H_2$:

$$2[Al + 4OH^- \longrightarrow Al(OH)_4^- + 3e^-] \quad \text{oxidation}$$
$$3[2H_2O + 2e^- \longrightarrow H_2 + 2OH^-] \quad \text{reduction}$$
$$2Al(s) + 6H_2O(l) + 2OH^-(aq) \longrightarrow 2Al(OH)_4^- + 3H_2(g) \quad \text{overall}$$

or $2Al(s) + 2NaOH(aq) + 6H_2O(l) \longrightarrow 2NaAl(OH)_4(aq) + 3H_2(g)$

(b) $Al_2O_3(s) + C(s) + Cl_2(g) \longrightarrow AlCl_3 + CO$ unbalanced

C(0) in C(s) is oxidized to C(+2) in CO; Cl(0) in $Cl_2$ is reduced to Cl(-1) in $AlCl_3$; Al(+3) is unchanged in $Al_2O_3$ and $AlCl_3$. In ionic form:

$$C + O^{2-} \longrightarrow CO + 2e^- \quad \text{oxidation}$$
$$Cl_2 + 2e^- \longrightarrow 2Cl^- \quad \text{reduction}$$
$$C + O^{2-} + Cl_2 \rightarrow CO + 2Cl^- \quad \text{overall}$$

$$2Al^{3+} + 3O^{2-} + 3C + 3Cl_2 \longrightarrow 2Al^{3+} + 6Cl^- + 3CO$$

i.e., $Al_2O_3(s) + 3C(s) + 3Cl_2(g) \longrightarrow 2AlCl_3(s) + 3CO(g)$

(c) $2(NH_4)Al(SO_4)_2 \cdot 12H_2O(s) \longrightarrow 2NH_3(g) + 4H_2SO_4(l) + Al_2O_3(s) + 21H_2O(g)$

43. (a) <u>$HNO_3$</u>

| | | |
|---|---|---|
| (a) $HNO_3$ | $Cu(NO_3)_2$ | Copper(II) nitrate |
| | $Fe(NO_3)_2$ | Iron(II) nitrate * |
| | $Fe(NO_3)_3$ | Iron(III) nitrate |
| (b) $H_2SO_4$ | $CuSO_4$ | Copper(II) sulfate |
| | $Cu(HSO_4)_2$ | Copper(II) hydrogen sulfate |
| | $FeSO_4$ | Iron (II) sulfate |
| | $Fe(HSO_4)_2$ | Iron (II) hydrogen sulfate |

| | | |
|---|---|---|
| | $Fe_2(SO_4)_3$ | Iron(III) sulfate |
| | $Fe(HSO_4)_3$ | Iron (III) hydrogen sulfate |
| (b) $\underline{H_3PO_4}$ | $Cu_3(PO_4)_2$ | Copper(II) phosphate |
| | $CuHPO_4$ | Copper(II) hydrogen phosphate |
| | $Cu(H_2PO_4)_2$ | Copper (II) dihydrogen phosphate |
| | $Fe_3(PO_4)_2$ | Iron(II) phosphate |
| | $FeHPO_4$ | Iron(II) hydrogen phosphate |
| | $Fe(H_2PO_4)_2$ | Iron(II) dihydrogen phosphate |
| | $FePO_4$ | Iron(III) phosphate |
| | $Fe_2(HPO_4)_3$ | Iron(III) hydrogen phosphate |
| | $Fe(H_2PO_4)_3$ | Iron(III) dihydrogen phosphate |

*$Fe^{2+}$ would be oxidized to $Fe^{3+}(aq)$ in $HNO_3(aq)$

44. (a) Limestone - calcium carbonate, $CaCO_3(s)$

(b) Alumina - Aluminum oxide, $Al_2O_3(s)$ (Corundum or emery)

(c) Magnetite - Iron(II) iron(III) oxide, $Fe_3O_4(s)$ or $Fe(II)(Fe(III))_2O_4$

(d) Pyrite - Iron(II) disulfide, $FeS_2(s)$

(e) Coke - Carbon (graphite), C

(f) Red Lead - Lead(II) lead(IV) oxide, $Pb_3O_4$ or $(Pb(II))_2Pb(IV)O_4$

(g) Rust - Hydrated iron(III) oxide, $Fe_2O_3.xH_2O(s)$

(h) Ammonium alum - Hydrated ammonium aluminum sulfate, $(NH_4)Al(SO_4)_2.12H_2O$

45. Iron rusts (corrodes) in the presence of air and moisture. The first stage of oxidation gives $Fe^{2+}(aq)$ , according to the equation

$$2Fe(s) + 4H^+(aq) + O_2(g) \longrightarrow 2Fe^{2+}(aq) + 2H_2O(l)$$

and then $Fe^{2+}(aq)$ is oxidized by $O_2(g)$ to insoluble iron(III) oxide

$$4Fe^{2+}(aq) + O_2(g) + 4H_2O(l) \longrightarrow 2Fe_2O_3(s) + 8H^+(aq)$$

Adding the two reactions gives the overall reaction

$$4Fe(s) + 3O_2(g) \longrightarrow 2Fe_2O_3(s)$$

46. Clay, $H_2Al_2(SiO_4)_2.H_2O$ has a molar mass of

$$[2(1.008) + 2(26.98) + 2[28.09 + 4(16.00)] + 2(1.008) + 16.00]\ g$$

$$= \underline{258.2\ g\ mol^{-1}}$$

$$\text{Mass \% Al} = \left(\frac{1 \text{ mol clay}}{258.2 \text{ g clay}}\right)\left(\frac{2 \text{ mol Al}}{1 \text{ mol clay}}\right)\left(\frac{26.98 \text{ g Al}}{1 \text{ mol Al}}\right) \times 100 = \underline{20.90\%}$$

$$\text{Mass \% Si} = \left(\frac{1 \text{ mol clay}}{258.2 \text{ g clay}}\right)\left(\frac{2 \text{ mol Si}}{1 \text{ mol clay}}\right)\left(\frac{28.09 \text{ g Si}}{1 \text{ mol Si}}\right) \times 100 = \underline{21.76\%}$$

$$\underline{20.90 \text{ mass \% Al}; \quad 21.76 \text{ mass \% Si}}$$

47. From the given data,

0.250 g Al(s) combines with (1.236-0.250) g = 0.986 6 g $Cl_2(g)$

| | | Al | | Cl |
|---|---|---|---|---|
| Masses of Al and Cl (g) | = | 0.250 | | 0.986 |
| Ratio of moles (atoms) | = | $\frac{0.250}{26.98}$ | : | $\frac{0.986}{35.45}$ |
| | = | $9.27 \times 10^{-3}$ | : | $2.78 \times 10^{-2}$ |
| | = | $\frac{9.27 \times 10^{-3}}{9.27 \times 10^{-3}}$ | : | $\frac{2.78 \times 10^{-2}}{9.27 \times 10^{-3}}$ |
| | = | 1.00 | : | 3.00 |

Thus, the empirical formula is $AlCl_3$ (formula mass 133.3 u)

From the gas data, calculate the moles of compound,

$$n = \frac{PV}{RT} = \frac{(720 \text{ torr})\left(\frac{1 \text{ atm}}{760 \text{ torr}}\right)(210 \text{ mL})\left(\frac{1 \text{ L}}{1000 \text{ mL}}\right)}{(0.0821 \text{ atm L mol}^{-1} \text{ K}^{-1})(523 \text{ K})} = \underline{4.63 \times 10^{-3} \text{ mol}}$$

and the molar mass is given by

$$\left(\frac{1.236 \text{ g}}{4.63 \times 10^{-3} \text{ mol}}\right) = \underline{267 \text{ g mol}^{-1}}$$

i.e., molecular mass = 2(formula mass) = 266.6 u, and the

Molecular formula = $Al_2Cl_6$

Lewis structure:

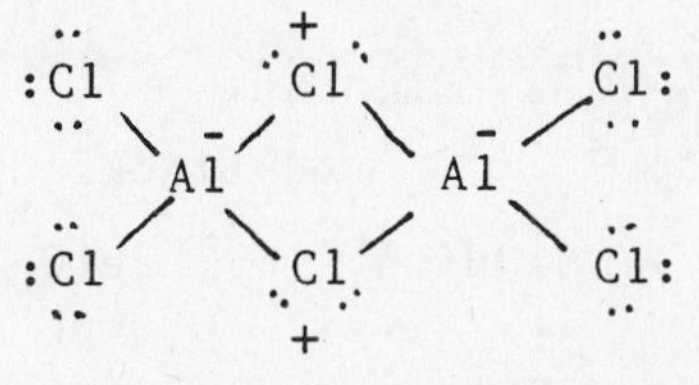

48. Calculate the mass of tin in the sample from the mass of $SnO_2 \cdot 2H_2O$, for which the molar mass is

$$[118.7+2(16.00)]+2[2(1.008)+16.00] = \underline{186.7 \text{ g mol}^{-1}}$$

$$\text{Mass Sn} = (0.778 \text{ g oxide})\left(\frac{1 \text{ mol oxide}}{186.7 \text{ g oxide}}\right)\left(\frac{1 \text{ mol Sn}}{1 \text{ mol oxide}}\right)\left(\frac{118.7 \text{ g Sn}}{1 \text{ mol Sn}}\right)$$

$$= \underline{0.495 \text{ g}}$$

$$\text{Mass \% tin} = \left(\frac{0.495 \text{ g}}{1.00 \text{ g}}\right) \times 100 = \underline{49.5} \text{ ; mass \% Pb} = \underline{50.5\%}$$

49. When lead is strongly heated in oxygen, the red powder produced is red lead, $Pb_3O_4$, containing Pb(II) and Pb(IV), which may be formulated as $(Pb(II)O)_2$, $Pb(IV)O_2$. When red lead is treated with concentrated nitric acid, only the Pb(II)O reacts, to form $Pb(NO_3)_2(aq)$, leaving $Pb(IV)O_2$ as the brown powder that is filtered off. The filtrate contains $Pb^{2+}(aq)$ and $NO_3^-(aq)$, which when treated with KI(aq) gives a precipitate of yellow $PbI_2(s)$:

$$3Pb(s) + 2O_2(g) \longrightarrow Pb_3O_4 \text{ (red lead)} \quad \text{.... (1)}$$

$$Pb_3O_4(s) + 4HNO_3(aq) \longrightarrow \underset{\text{lead(II) nitrate}}{2Pb(NO_3)_2(aq)} + \underset{\text{lead(IV) oxide}}{PbO_2(s)} \quad \text{.... (2)}$$

$$Pb(NO_3)_2(aq) + 2KI(aq) \longrightarrow \underset{\text{lead(II) iodide}}{PbI_2(s)} + \underset{\text{potassium nitrate}}{2KNO_3(aq)} \quad \text{.... (3)}$$

From (1), 3 mol Pb gives 1 mol $Pb_3O_4$

From (1) and (2), mol $PbO_2$ = mol $Pb_3O_4$

From (1), (2), and (3), mol $PbI_2$ = mol $Pb(NO_3)_2$ = 2(mol $Pb_3O_4$)

Thus

$$\text{Mass } PbO_2 = (2.00 \text{ g Pb})\left(\frac{1 \text{ mol Pb}}{207.2 \text{ g Pb}}\right)\left(\frac{1 \text{ mol } PbO_2}{3 \text{ mol Pb}}\right)\left(\frac{239.2 \text{ g } PbO_2}{1 \text{ mol } PbO_2}\right) = \underline{0.770 \text{ g}}$$

$$\text{Mass } PbI_2 = (2.00 \text{ g Pb})\left(\frac{1 \text{ mol Pb}}{207.2 \text{ g Pb}}\right)\left(\frac{2 \text{ mol } PbI_2}{3 \text{ mol Pb}}\right)\left(\frac{461.0 \text{ g } PbI_2}{1 \text{ mol } PbI_2}\right) = \underline{2.97 \text{ g}}$$

50. Iron is oxidized to $FeCl_3$, iron(III) chloride, by $Cl_2(g)$, which dissolves in water to give a solution containing $Fe^{3+}(aq)$ and $Cl^-(aq)$ ions. When NaOH(aq) is added to the $FeCl_3(aq)$ solution, a brown precipitate of $Fe(OH)_3(s)$ is formed. When $Fe(OH)_3(s)$ is strongly heated it forms the oxide, $Fe_2O_3(s)$:

$$2Fe(s) + 3Cl_2(g) \longrightarrow \underset{\text{red-black solid}}{2FeCl_3(s)}$$

$$FeCl_3(aq) + 3NaOH(aq) \longrightarrow \underset{\text{brown Fe(III) hydroxide}}{Fe(OH)_3(s)} + 3NaCl(aq)$$

$$2Fe(OH)_3(s) \xrightarrow{\text{heat}} \underset{\text{red-brown Fe(III) oxide}}{Fe_2O_3(s)} + 3H_2O(g)$$

$$\text{Mass of } Fe_2O_3(s) = (1.50 \text{ g Fe})\left(\frac{1 \text{ mol Fe}}{55.85 \text{ g Fe}}\right)\left(\frac{1 \text{ mol } Fe_2O_3}{2 \text{ mol Fe}}\right)\left(\frac{159.7 \text{ g } Fe_2O_3}{1 \text{ mol } Fe_2O_3}\right)$$

$$= \underline{2.14 \text{ g}}$$

51. Aluminum metal reacts rapidly with oxygen of the air at ordinary temperatures and soon becomes covered with a thin film of $Al_2O_3(s)$ which protects the metal from further oxidation in air. $Al_2O_3(s)$ dissolves

readily in both dilute acid and dilute base because it is an amphoteric oxide:

$$Al_2O_3(s) + 6H_3O^+(aq) \longrightarrow 2Al^{3+}(aq) + 9H_2O(l)$$

$$Al_2O_3(s) + 2OH^-(aq) + 3H_2O(l) \longrightarrow \underset{\text{aluminate ion}}{2Al(OH)_4^-}$$

In the highly exothermic thermite reaction, once the reaction is initiated, the heat produced is sufficient to melt the aluminum, which then reacts vigorously, for example

$$2Al(s) + Fe_2O_3(s) \longrightarrow 2Fe(l) + Al_2O_3(s)$$

as is also the case for the reaction with $Cl_2(g)$

$$2Al(s) + 3Cl_2(g) \longrightarrow 2AlCl_3(g)$$

At high temperature, particularly if the aluminum melts, the protective coating of $Al_2O_3(s)$ breaks down and exposes the metal to further reaction, as is the case, for example, when warships are attacked by missiles, or aircraft are involved in crashes in which their fuel ignites. Application of water to douse such fires is useless because the exposed Al metal reacts vigorously with water at high temperature.

52. Of the metals in the alloy, only Zn and Al react with dilute $H_2SO_4$, but all of the metals react with hot concentrated $H_2SO_4$ to give $SO_2(g)$

For the reaction with dilute $H_2SO_4$:

$$\text{moles of } H_2 = \frac{PV}{RT} = \frac{(1\text{ atm})(149.3\text{ mL})(\frac{1\text{ L}}{1000\text{ mL}})}{(0.0821\text{ atm L mol}^{-1}\text{ K}^{-1})(298\text{ K})} = \underline{6.102\text{x}10^{-3}\text{ mol}}$$

For the reaction with concentrated $H_2SO_4$

$$\text{moles of } SO_2 = \frac{PV}{RT} = \frac{(1\text{ atm})(411.1\text{ mL})(\frac{1\text{ L}}{1000\text{ mL}})}{(0.0821\text{ atm L mol}^{-1}\text{ K}^{-1})(298\text{ K})} = \underline{1.680\text{x}10^{-2}\text{ mol}}$$

In each case the moles of gas is related to the moles of the metals:

$$\left\{\begin{array}{l} Zn(s) + H_2SO_4(aq) \longrightarrow Zn^{2+}(aq) + SO_4^{2-}(aq) + H_2(g) \\ 2Al(s) + 3H_2SO_4(aq) \longrightarrow 2Al^{3+}(aq) + 3SO_4^{2-}(aq) + 3H_2(g) \end{array}\right. \quad \underline{\text{dilute acid}}$$

$$\left\{\begin{array}{l} Cu(s) + 2H_2SO_4(conc) \longrightarrow Cu^{2+} + SO_4^{2-} + 2H_2O + SO_2(g) \\ Zn(s) + 2H_2SO_4(conc) \longrightarrow Zn^{2+} + SO_4^{2-} + 2H_2O + SO_2(g) \\ 2Al(s) + 6H_2SO_4(conc) \longrightarrow 2Al^{3+} + 3SO_4^{2-} + 6H_2O + 3SO_2(g) \end{array}\right. \quad \underline{\text{concentrated acid}}$$

Thus, for $n_{Zn}$ moles Zn, $n_{Al}$ moles of Al, and $n_{Cu}$ moles of Cu

$$n_{H_2} = n_{Zn} + \frac{3}{2}n_{Al} = 6.102 \times 10^{-3}\text{ mol} \quad \text{..... (1)}$$

$$n_{SO_2} = n_{Cu} + n_{Zn} + \frac{3}{2}n_{Al} = 1.680 \times 10^{-2}\text{ mol} \quad \text{..... (2)}$$

Subtracting (1) from (2) gives the moles of Cu:

$$n_{Cu} = (1.680\text{x}10^{-2}) - (6.10\text{x } 10^{-3}) = \underline{1.070 \times 10^{-2}\text{ mol}}$$

$$\text{Mass of Cu} = (1.070 \times 10^{-2}\text{ mol Cu})(\frac{63.55\text{ g Cu}}{1\text{ mol Cu}}) = \underline{0.680\text{ g}}$$

By difference, mass Zn + mass Al = (1.000 - 0.680) = 0.320 g

If the mass of Zn is x g, then mass Al = (0.320-x)g, and from (1):

$$n_{Zn} + \frac{3}{2}n_{Al} = (x\text{ g Zn})\left(\frac{1\text{ mol Zn}}{65.38\text{ g Zn}}\right) + \frac{3}{2}[(0.320-x)\text{ g Al}]\left(\frac{1\text{ mol Al}}{26.98\text{ g Al}}\right)$$

$$= (1.530\times10^{-2})x + (1.779\times10^{-2}) - (5.560\times10^{-2})x$$

$$= (1.779\times10^{-2}) - (4.03\times10^{-2})x = \underline{6.102\times10^{-3}}$$

i.e., $(4.03\times10^{-2})x = 1.169\times10^{-2}$ ; $\underline{x = 0.290\text{ g}}$

Thus, mass of Zn = 0.290 g; mass of Al = (0.320 - 0.290) = 0.030 g

and 1.000 g aluminum brass contains:

0.680 g Cu, 0.290 g Zn, and 0.030 g Al

Composition of aluminum brass is:

68.0% Cu, 29.0% Zn, and 3.00% Al

53. (a) Copper(I) phosphide is $(Cu^+)_3P^{3-}$. Thus, P(0) in $P_4$ is reduced to P(-3) in $P^{3-}$. Simultaneously, P(0) in $P_4$ is oxidized to P(+3) in $H_3PO_3$, and Cu(+2) in $Cu^{2+}$ is reduced to Cu(+1) in $Cu^+$. Thus we can write the following two equations:

1. $P_4 + 12e^- \longrightarrow 4P^{3-}$ reduction

$P_4 + 12H_2O \longrightarrow 4H_3PO_3 + 12e^- + 12H^+$ oxidation

$2P_4 + 12H_2O \rightarrow 4P^{3-} + 4H_3PO_3 + 12H^+$ overall ... (1)

2. $12(Cu^{2+} + e^- \longrightarrow Cu^+\quad]$ reduction

$P_4 + 12H_2O \longrightarrow 4H_3PO_3 + 12e^- + 12H^+$ oxidation

$12Cu^{2+} + P_4 + 12H_2O \longrightarrow 12Cu^+ + 4H_3PO_3 + 12H^+$ overall .(2)

Adding (1) and (2) gives:

$$3P_4 + 24H_2O + 12Cu^{2+} \longrightarrow 4Cu_3P + 8H_3PO_4 + 24H^+$$

and adding $12SO_4^{2-}$ to both sides gives the final balanced equation:

$$\underline{3P_4 + 24H_2O + 12CuSO_4 \longrightarrow 4\,Cu_3P + 8H_3PO_3 + 12H_2SO_4}$$

(b) P(0) in $P_4$ is oxidized to P(+5) in $H_3PO_4$; Cu(+2) in $CuSO_4$ is reduced to Cu(0) in Cu metal, and S is unchanged in $CuSO_4$ and $H_2SO_4$:

$P_4 + 16H_2O \longrightarrow 4H_3PO_4 + 20e^- + 20\ H^+$ oxidation

$10[Cu^{2+} + 2e^- \longrightarrow Cu\quad]$ reduction

$P_4 + 10Cu^{2+} + 16H_2O \longrightarrow 4H_3PO_4 + 10Cu + 20\ H^+$ overall

and adding $10SO_4^{2-}$ to both sides gives:

$$\underline{P_4 + 10CuSO_4 + 16H_2O \longrightarrow 4H_3PO_4 + 10Cu + 10H_2SO_4}$$

54. Heavy metal nitrates such as lead nitrate, $Pb(NO_3)_2(s)$, decompose on heating to give $PbO(s)$, $NO_2(g)$ and $O_2(g)$, according to the balanced equation

$$2Pb(NO_3)_2(s) \longrightarrow 2PbO(s) + 4NO_2(g) + O_2(g)$$

On heating, brown fumes of $NO_2(g)$ are observed.

55. When dissolved in water an aqueous solution of ammonium alum, $(NH_4)Al(SO_4)_2 . 12H_2O$ will give a solution containing $NH_4^+(aq)$, $Al^{3+}(aq)$, and $SO_4^{2-}(aq)$, each of which can be tested for using standard analytical tests:

(i) $\underline{NH_4^+(aq)}$ On addition of NaOH(aq) and warming, $NH_3(g)$ would be evolved, which can be detected by its characteristic odor, action on moist red litmus paper, which turns blue, and the formation of a white cloud of $NH_4Cl(s)$ when the test tube is placed near another test tube containing concentrated HCl(aq).

$$NH_4^+(aq) + OH^-(aq) \xrightarrow{\text{heat}} NH_3(g) + H_2O(l)$$

$$NH_3(g) + HCl(g) \longrightarrow NH_4Cl(s)$$

(ii) $\underline{Al^{3+}(aq)}$ On addition of NaOH(aq) a gelatinous white precipitate of $Al(OH)_3(s)$ forms, which is soluble in excess of the reagent.

$$Al^{3+}(aq) + 3OH^-(aq) \longrightarrow Al(OH)_3(s)$$

$$Al(OH)_3(s) + OH^-(aq) \longrightarrow Al(OH)_4^-(aq)$$

(iii) $\underline{SO_4^{2-}(aq)}$ Addition of dilute HCl(aq), followed by $BaCl_2(aq)$ gives a white insoluble precipitate of $BaSO_4(s)$.

$$Ba^{2+}(aq) + SO_4^{2-}(aq) \longrightarrow BaSO_4(s)$$

(iv) $\underline{H_2O}$ On gentle heating of the solid compound, the water of hydration is lost, which could be condensed on a cold object. Water could be confirmed by its action on anhydrous copper(II) sulfate, $CuSO_4(s)$, which is hydrated by the water to give blue $CuSO_4 . 5H_2O$.

$$\underset{\text{white}}{CuSO_4(s)} + H_2O(l) \longrightarrow \underset{\text{blue}}{CuSO_4 . 5H_2O(s)}$$

The blue color is due to the hydrated $Cu(H_2O)_4^{2+}$ ion.

56. From the data in (i) the moles of Fe in 0.7840 g of salt is

$$\text{mol Fe} = (0.1600\text{ g }Fe_2O_3)\left(\frac{1\text{ mol }Fe_2O_3}{159.7\text{ g }Fe_2O_3}\right)\left(\frac{2\text{ mol Fe}}{1\text{ mol }Fe_2O_3}\right) = \underline{2.00\times10^{-3}\text{ mol}}$$

From the data in (ii) the moles of $SO_4^{2-}$ in 0.7840 g of salt is

$$\text{mol }SO_4^{2-} = (0.9336\text{ g }BaSO_4)\left(\frac{1\text{ mol }BaSO_4}{233.4\text{ g }BaSO_4}\right)\left(\frac{1\text{ mol }SO_4^{2-}}{1\text{ mol }BaSO_4}\right)$$
$$= \underline{4.00\times10^{-3}\text{ mol}}$$

The reactions involved in (iii) are:

$$NH_4^+(aq) + OH^-(aq) \rightleftharpoons NH_3(g) + H_2O(l)$$
$$NH_3(g) + HCl(aq) \longrightarrow NH_4Cl(aq)$$
$$HCl(aq) + NaOH(aq) \longrightarrow NaCl(aq) + H_2O(l)$$

Since the HCl(aq) was in excess, and the excess acid required 30.0 mL 0.10 M NaOH(aq) for neutralization, (50.0 - 30.0) = 20.0 mL of the original 50.0 mL 0.10 M HCl(aq) must have been neutralized by the $NH_3(g)$ liberated from the salt. i.e.,

$$\text{mol }NH_3 = (20.0\text{ mL HCl})(0.10\text{ mol L}^{-1})\left(\frac{1\text{ L}}{1000\text{ mL}}\right)\left(\frac{1\text{ mol }NH_3}{1\text{ mol HCl}}\right) = \underline{2.00\times10^{-3}\text{ mol}}$$

Thus, mol of $NH_4^+$ in 0.3920 g of salt = $2.00\times10^{-3}$ mol, <u>or</u> mol of $NH_4^+$ in 0.7840 g of salt = $\underline{4.00\times10^{-3}\text{ mol}}$

Thus the masses of the ions in 0.7840 g of salt are

$$\text{Mass of Fe} = (2.00\times10^{-3}\text{ mol Fe})\left(\frac{55.85\text{ g Fe}}{1\text{ mol Fe}}\right) = 0.1117\text{ g Fe}$$

$$\text{Mass of }SO_4^{2-} = (4.00\times10^{-3}\text{ mol }SO_4^{2-})\left(\frac{96.06\text{ g }SO_4^{2-}}{1\text{ mol }SO_4^{2-}}\right) = 0.3842\text{ g }SO_4^{2-}$$

$$\text{Mass of }NH_4^+ = (4.00\times10^{-3}\text{ mol }NH_4^+)\left(\frac{18.04\text{ g }NH_4^+}{1\text{ mol }NH_4^+}\right) = 0.0722\text{ g }NH_4^+$$

i.e., (0.1117 + 0.3842 + 0.0722) = $\underline{0.5681\text{ g}}$

and the mass of $H_2O$ is (0.7840 - 0.5681) = $\underline{0.2159\text{ g}}$, by difference

$$(0.2159\text{ g }H_2O)\left(\frac{1\text{ mol }H_2O}{18.02\text{ g }H_2O}\right) = \underline{12.00\times10^{-3}\text{ mol}}$$

Thus in 0.7840 g of sample there is

| | Fe | | $SO_4^{2-}$ | | $NH_4^+$ | | $H_2O$ |
|---|---|---|---|---|---|---|---|
| moles x $10^3$ | 2.00 | : | 4.00 | : | 4.00 | : | 12.00 |
| <u>or</u> | 1 | : | 2 | : | 4 | : | 6 |

so that the empirical formula is $\underline{Fe(NH_4)_2(SO_4)_2\cdot6H_2O}$

57. 1.000 g of salt contains 0.2703 g $H_2O$

$$\text{moles }H_2O\text{ in 1.000 g} = (0.2703\text{ g }H_2O)\left(\frac{1\text{ mol }H_2O}{18.02\text{ g }H_2O}\right) = \underline{0.0150\text{ mol}}$$

1.000 g of salt gives 0.1989 g CuO(s)

$$\text{mol Cu in 1.000 g} = (0.1989\ \text{g CuO})\left(\frac{1\ \text{mol CuO}}{79.55\ \text{g CuO}}\right)\left(\frac{1\ \text{mol Cu}}{1\ \text{mol CuO}}\right)$$
$$= \underline{2.500 \times 10^{-3}\ \text{mol}}$$

The salt contains $Cu^{2+}$, $NH_4^+$, and $SO_4^{2-}$ ions and water of hydration. Thus, its empirical formula must be of the form:

$$(CuSO_4)_x[(NH_4)_2SO_4]_y \cdot zH_2O$$

For 1.000 g of salt we have:

$$(2.500\text{x}10^{-3}\ \text{mol Cu})\left(\frac{1\ \text{mol}\ CuSO_4}{1\ \text{mol Cu}}\right)\left(\frac{159.6\ \text{g}\ CuSO_4}{1\ \text{mol}\ CuSO_4}\right) = \underline{0.3990\ \text{g}}\ CuSO_4$$

Thus, 1.000 g of salt contains 0.3990 g $CuSO_4$ and 0.2703 g $H_2O$, so, by difference:

$$\text{Mass of}\ (NH_4)_2SO_4\ \text{in 1.000 g} = (1.000-0.3990-0.2703) = \underline{0.3307\ \text{g}}$$

$$\text{Mol}\ (NH_4)_2SO_4\ \text{in 1.000 g salt} = (0.3307\ \text{g})\left(\frac{1\ \text{mol}\ (NH_4)_2SO_4}{132.1\ \text{g}\ (NH_4)_2SO_4}\right)$$
$$= \underline{2.500 \times 10^{-3}\ \text{mol}}$$

| i.e., | $CuSO_4$ | | $(NH_4)_2SO_4$ | | $H_2O$ |
|---|---|---|---|---|---|
| ratio of moles ($\text{x}10^3$) | 2.500 | : | 2.500 | : | 15.0 |
| <u>or</u> | 1 | : | 1 | : | 6 |

<u>Empirical formula</u> is $\underline{CuSO_4 \cdot (NH_4)_2SO_4 \cdot 6H_2O}$ <u>or</u> $\underline{(NH_4)_2Cu(SO_4)_2 \cdot 6H_2O}$

58. (a) $Cu^{2+}$(aq) in a solution of $CuSO_4 \cdot 5H_2O$ may be reacted with NaOH(aq) to give a precipitate of $Cu(OH)_2$(s), which when filtered off and heated strongly gives CuO(s), which may be reduced to Cu metal with a reducing agent such as $H_2$(g), or carbon.

$$Cu^{2+}(aq) + 2OH^-(aq) \longrightarrow Cu(OH)_2(s) \xrightarrow{\text{heat}} CuO(s) \xrightarrow{H_2} Cu(s) + H_2O$$

(b) CuO(s), prepared as in part (a) could be dissolved in HCl(aq) to give a solution from which $CuCl_2 \cdot 2H_2O$(s) could be crystallized after careful concentration by evaporating off most of the water.

$$CuO(s) + 2HCl(aq) \longrightarrow CuCl_2(aq) + H_2O(l)$$

(c) CuCl(s) cannot be obtained from aqueous solution, but can be prepared by passing HCl(g) over heated copper (obtained from $CuSO_4 \cdot 5H_2O$ as described in part (a))

$$2Cu(s) + 2HCl(g) \longrightarrow 2CuCl(s) + H_2(g)$$

(d) $Cu(NH_3)_4SO_4 \cdot H_2O$(s) can be crystallized from a solution obtained by adding the appropriate amount of concentrated $NH_3$(aq) to an aqueous solution of $CuSO_4 \cdot 5H_2O$, according to the equation

$$Cu^{2+}(aq) + 4NH_3(aq) \longrightarrow Cu(NH_3)_4^{\ 2+}(aq)$$

59. (i) $Al_2(SO_4)_3(aq)$ could be prepared most simply by dissolving the metal in dilute $H_2SO_4(aq)$.

$$2Al(s) + 3H_2SO_4(aq) \longrightarrow 2Al^{3+}(aq) + 3SO_4^{2-}(aq) + 3H_2(g)$$

(ii) Copper could be dissolved in $HNO_3(aq)$ to give a solution of $Cu(NO_3)_2(aq)$, which on evaporation and heating strongly gives $CuO(s)$, which when dissolved in $H_2SO_4(aq)$ gives $CuSO_4(aq)$.

$$Cu(s) \xrightarrow{HNO_3(aq)} Cu(NO_3)_2(aq) \xrightarrow{heat} CuO(s) \xrightarrow{H_2SO_4} CuSO_4(aq)$$

Alternatively, Cu(s) could be heated strongly in air to give CuO(s) and this could be dissolved in $H_2SO_4(aq)$. (See text, page 527).

(iii) Lead could be heated strongly in air to give PbO(s). Dissolving PbO(s) in dilute nitric acid gives $Pb(NO_3)_2(aq)$, which on adding a solution of a soluble sulfate, such as $Na_2SO_4(aq)$, precipitates insoluble $PbSO_4$ (see text, page 519).

$$Pb(s) \xrightarrow{O_2} PbO(s) \xrightarrow{HNO_3} Pb(NO_3)_2(aq) \xrightarrow{SO_4^{2-}} PbSO_4(s)$$

60. Both of these oxides are soluble in HCl(aq). Addition of NaOH(aq) in excess to solutions in HCl(aq) would give precipitates of the respective hydroxides.

$$Fe_2O_3(s) + 6HCl(aq) \longrightarrow 2Fe^{3+}(aq) + 6Cl^- + 3H_2O(l)$$

$$Fe^{3+}(aq) + 3OH^-(aq) \longrightarrow Fe(OH)_3(s)$$

$$CuO(s) + 2HCl(aq) \longrightarrow Cu^{2+}(aq) + 2Cl^-(aq) + H_2O(l)$$

$$Cu^{2+}(aq) + 2OH^-(aq) \longrightarrow Cu(OH)_2(s)$$

The insoluble hydroxides could be filtered off, washed with distilled water and carefully dried.

61. As Problem 60 demonstrates, the compounds that precipitate from aqueous solution are the hydroxides, rather than the oxides. However, the oxides are readily obtained from the hydroxides by heating them. Thus, Starting with $CuSO_4(aq)$, addition of excess NaOH(aq) gives $Cu(OH)_2(s)$, which on heating, even in the solution, decomposes to black CuO(s), which can be filtered off and dried.

$$CuSO_4(aq) + 2NaOH(aq) \longrightarrow 2Na^+(aq) + SO_4^{2-}(aq) + Cu(OH)_2(s)$$

$$Cu(OH)_2(s) \longrightarrow CuO(s) + H_2O(l)$$

$FeSO_4(aq)$ contains Fe(II), while $Fe_2O_3(s)$ contains Fe(III). A suitable oxidizing agent to oxidize Fe(II) to Fe(III) is $H_2O_2(aq)$ in acid.

$$2[Fe^{2+}(aq) \longrightarrow Fe^{3+}(aq) + e^-] \qquad \text{oxidation}$$

$$H_2O_2(aq) + 2H^+(aq) + 2e^- \longrightarrow 2H_2O(l) \qquad \text{reduction}$$

$$2Fe^{2+}(aq) + H_2O_2(aq) + 2H^+(aq) \longrightarrow 2Fe^{3+}(aq) + 2H_2O(l)$$

Then, addition of NaOH(aq) precipitates $Fe(OH)_3(s)$, which can be filtered off and heated to give $Fe_2O_3(s)$.

$$Fe^{3+}(aq) + 3OH^-(aq) \longrightarrow Fe(OH)_3(s)$$

$$2Fe(OH)_3(s) \longrightarrow Fe_2O_3(s) + 3H_2O(g)$$

62. Silver iodide, AgI(s), is very insoluble in water, while lead iodide, $PbI_2(s)$, while not very soluble in cold water is considerably more soluble in hot water, from which it crystallizes as golden-yellow spangles on cooling. Taking a small sample of the unknown solid in a test tube with a few mL of water, and then heating to boiling would be a suitable test. A clear yellow solution from which yellow crystals separate on cooling would confirm $PbI_2(s)$. If the solid does not dissolve, then AgI(s) is indicated.

63. When KI(aq) is added to $CuSO_4(aq)$, a precipitate of $CuI_2(s)$ is first formed, which rapidly decomposes to give white insoluble copper(I) iodide, CuI(s), and iodine. The iodine dissolves in excess KI(aq) to give a solution containing the brown $I_3^-(aq)$ ion:

$$2[Cu^{2+}(aq) + e^- \longrightarrow Cu^+(aq)] \qquad \text{reduction}$$
$$2I^-(aq) \longrightarrow I_2(s) + 2e^- \qquad \text{oxidation}$$
$$2Cu^{2+}(aq) + 2I^-(aq) \longrightarrow 2Cu^+(aq) + I_2(s)$$

or $2CuSO_4(aq) + 4KI(aq) \longrightarrow 2CuI(s) + 2K_2SO_4(aq) + I_2(s)$

and $I_2(s) + KI(aq) \longrightarrow KI_3(aq)$

Adding these two equations gives the overall reaction,

$$2CuSO_4(aq) + 5KI(aq) \longrightarrow 2CuI(s) + K_2SO_4(aq) + KI_3(aq)$$

and the solution changes from its initial blue color due to $Cu^{2+}(aq)$ to give a brown solution ($I_3^-(aq)$) containing a white precipitate of CuI(s).

64. Addition of dilute acid to an insoluble sulfide liberates $H_2S(g)$, which can be detected by its characteristic odor and the blackening of filter paper soaked in an aqueous solution of a soluble lead salt. In contrast when copper oxide is dissolved in dilute acid it dissolves without evolution of a gas to give a solution with the characteristic blue color of the $Cu^{2+}(aq)$ ion.

$$CuS(s) + H_2SO_4(aq) \longrightarrow CuSO_4(aq) + H_2S(g)$$
$$Pb^{2+}(aq) + H_2S(g) + 2H_2O(l) \longrightarrow PbS(s) + 2H_3O^+(aq)$$
$$CuO(s) + H_2SO_4(aq) \longrightarrow CuSO_4(aq) + H_2O(l)$$

65. The balanced equation for the reaction is

$$CuCl_2.xCu(OH)_2(s) + 2xHCl(aq) \longrightarrow (x+1)CuCl_2(aq) + xH_2O(l)$$

From the given data, mol HCl(aq) that react with 0.6217 g atacamite

$$= (21.45\text{ mL HCl(aq)})\left(\frac{0.4071\text{ mol}}{1\text{ L}}\right)\left(\frac{1\text{ L}}{1000\text{ mL}}\right) = \underline{8.732\text{x}10^{-3}\text{ mol}}$$

and the mol of $Cu(OH)_2$ that react with this amount of acid is

$$(8.732\text{x}10^{-3}\text{ mol HCl})\left(\frac{x\text{ mol Cu(OH)}_2}{2x\text{ mol HCl}}\right) = \underline{4.366\text{x}10^{-3}\text{ mol}}$$

which corresponds to a mass of $Cu(OH)_2(s)$ of

$$(4.366\text{x}10^{-3}\text{ mol Cu(OH)}_2)\left(\frac{97.57\text{ g Cu(OH)}_2}{1\text{ mol Cu(OH)}_2}\right) = \underline{0.4260\text{ g}}$$

Thus, by difference, the mass of $CuCl_2$ in 0.6217 g atacamite is (0.6127-0.4260) g = <u>0.1957 g</u>, and

$$\text{mol } CuCl_2 \text{ in sample} = (0.1957 \text{ g } CuCl_2)\left(\frac{1 \text{ mol } CuCl_2}{135.4 \text{ g } CuCl_2}\right)$$

$$= \underline{1.455\times10^{-3} \text{ mol}}$$

| | $CuCl_2$ | | $Cu(OH)_2$ | |
|---|---|---|---|---|
| Ratio of moles = | $1.455\times10^{-3}$ | : | $4.336\times10^{-3}$ | |
| = | $\frac{1.455\times10^{-3}}{1.455\times10^{-3}}$ | : | $\frac{4.336\times10^{-3}}{1.455\times10^{-3}}$ | = <u>1.00 : 2.98</u> |

Thus, the <u>empirical formula</u> is <u>$CuCl_2.3Cu(OH)_2$</u> ; <u>x = 3</u>

66. (a) 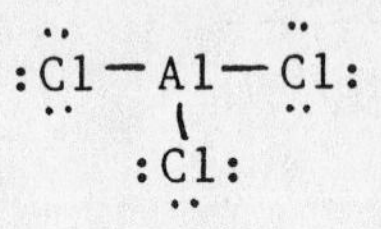 (b) 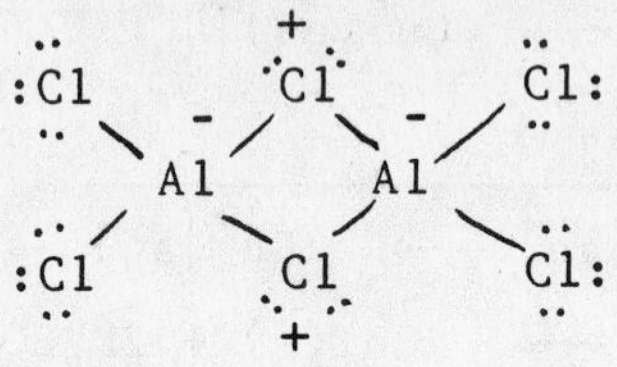

(c) 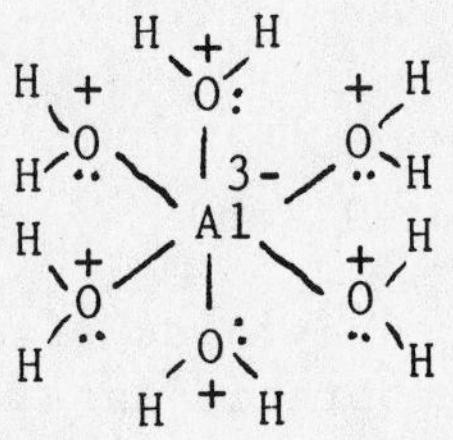 (d) 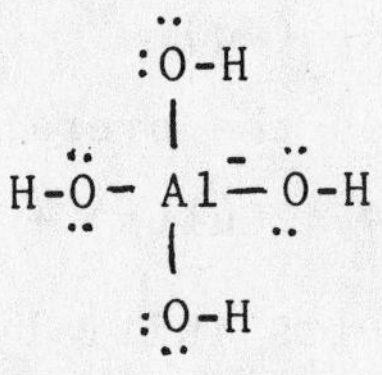

67. $Fe_2O_3(s) + 2Al(s) \longrightarrow Al_2O_3(s) + 2Fe(s)$ (thermite reaction)

$$\Delta H^\circ = [\Delta H_f^\circ(Al_2O_3,s) + 2\,\Delta H_f^\circ(Fe,s)] - [\Delta H_f^\circ(Fe_2O_3,s) + 2\,\Delta H_f^\circ(Al,s)]$$

$$= [(-1676) + 2(0)] - [(-824) + 2(0)] \text{ kJ} = \underline{-852 \text{ kJ}}$$

68. $3Al(s) + 3NH_4ClO_4(s) \longrightarrow Al_2O_3(s) + AlCl_3(s) + 3NO(g) + 6H_2O(g)$

$$\Delta H^\circ = [\Delta H_f^\circ(Al_2O_3,s) + \Delta H_f^\circ(AlCl_3,s) + 3\,\Delta H_f^\circ(NO,g) + 6\,\Delta H_f^\circ(H_2O,g)]$$
$$- [\Delta H_f^\circ(Al,s) + 3\,\Delta H_f^\circ(NH_4ClO_4,s)]$$

$$= [(-1676) + (-704.2) + 3(90.3) + 6(-241.8)] - [3(0) + 3(-295)]$$

$$= \underline{-2675 \text{ kJ}}$$

## CHAPTER 11

1. Carbon, both as graphite and as diamond, is a covalent <u>network</u> solid.
   $S_8$ molecules are found in both orthorhombic and monoclinic <u>sulfur</u> which
   are <u>molecular</u> solids
   Solid <u>carbon dioxide</u> consists of $CO_2$ molecules; it is a <u>molecular</u> solid.
   <u>Phosphorus(III) oxide</u> consists of $P_4O_6$ molecules; it is a <u>molecular</u> solid.
   Sodium chloride, NaCl, consists of an infinite array of $Na^+$ and $Cl^-$ ions;
   it is a <u>network</u> solid
   Magnesium oxide, MgO, consists of an infinite array of $Mg^{2+}$ and $O^{2-}$ ions;
   it is a <u>network solid</u>
   <u>Aluminum</u> consists of an infinite array of Al atoms; it is a <u>network solid</u>

2. When a molecular solid melts, only the weak intermolecular forces between
   its individual covalent molecules need to be overcome. When a network
   solid melts, strong covalent or ionic bonds must be broken in order for
   the atoms to become sufficiently mobile to slide by each other. Thus,
   molecular solids melt at much lower temperatures than network solids

3. (a) In orthorhombic sulfur, $S_8$ molecules are held together by weak inter-
       molecular forces.
   (b) Black phosphorus consists of infinite two-dimensional layers of
       covalently bonded phosphorus atoms.
   (c) In magnesium oxide the $Mg^{2+}$ and $O^{2-}$ ions are held together by ionic
       bonds.
   (d) In silica, $SiO_2$, each Si atom is surrounded by four covalently bonded
       O atoms, to form an infinite three-dimensional network.
   (e) In copper metal the atoms are held together by metallic bonds.

4. (a) Polymeric sulfur trioxide consists of long chains of $SO_3$ groups linked
       through S-O bonds:

$$\cdots-O-\overset{\overset{O}{\|}}{\underset{\underset{O}{\|}}{S}}-O-\overset{\overset{O}{\|}}{\underset{\underset{O}{\|}}{S}}-O-\overset{\overset{O}{\|}}{\underset{\underset{O}{\|}}{S}}-O-\overset{\overset{O}{\|}}{\underset{\underset{O}{\|}}{S}}-O-\overset{\overset{O}{\|}}{\underset{\underset{O}{\|}}{S}}-\cdots$$

   The molecules are long and thin and attract each other by intermolecular forces, which are optimized when the molecules are aligned
   parallel to each other or wind around each other, giving needle-like
   crystals.

   (b) Plastic sulfur consists of long chains of sulfur atoms which are
       weakly attracted by intermolecular forces, and the molecules become
       intertangled forming a soft rubbery solid.

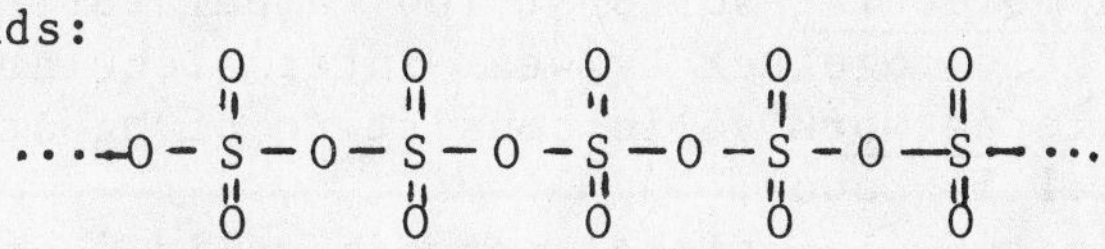

5. A crystalline solid has a regular arrangement of atoms, between which
   equivalent bonds are all of the same strength and therefore take exactly
   the same amount of energy to break them. Thus, when a crystalline solid
   is heated, all the equivalent bonds break at the same temperature and the
   solid has a sharp melting point. In contrast, the atoms in an amorphous
   solid have a range of lengths and strengths and do not all require the
   same amount of energy to break them. Thus, as the temperature is increased,
   some bonds break at a lower temperature than others and the solid softens
   and gradually changes to a liquid without exhibiting a sharp melting point.

6. (a) Silica, $SiO_2$, is a covalent network solid in which each Si atom is covalently bonded to four O atoms, to form a three-dimensional infinite array. In contrast, carbon dioxide consists of discrete $CO_2$ molecules, each having two CO double bonds, that are weakly attracted to each other in $CO_2(s)$ by weak intermolecular forces.
(b) Oxygen consists of discrete $O_2$ molecules (with OO double bonds) which attract each other through only weak London forces, so that under normal conditions oxygen is a gas. Sulfur consists of $S_8$ molecules in which the atoms are linked by single S-S bonds. Again, the intermolecular forces are London forces, but because the molecules are relatively large and have a large number of electrons, they have a relatively large polarizability so that the London forces are fairly strong and sulfur is therefore a solid with a relatively low melting point.
(c) Diamond has an infinite three-dimensional array of carbon atoms in which each C atom is covalently bonded to four others in a tetrahedral $AX_4$ arrangement, while in graphite the C atoms are arranged in planes with each C atom bonded to three others, with the infinite planes of C atoms weakly attracted to each other through London forces. Thus, diamond is a very hard substance while graphite is soft and is a good lubricant because the planes of C atoms can easily slide over each other.

7. A two-dimensional network solid is one in which the strongly bonded network extends indefinitely only in two-dimensions, and the planes of atoms attract each other only by weak intermolecular forces. Examples include graphite, red and black phosphorus, and magnesium chloride.

8. Carbon monoxide, CO, consists of individual CO molecules held together in the low melting solid by weak intermolecular forces, and below the melting point is a molecular solid. At room temperature it is a gas.
Silicon carbide, SiC, is an infinite network solid with a structure similar to that of diamond and containing an infinite network of Si-C bonds.
Chlorine, $Cl_2$, forms a molecular solid at low temperatures, in which the $Cl_2$ molecules are held together by weak intermolecular forces.
Magnesium is an infinite network solid consisting of Mg atoms held together by strong metallic bonds.
Magnesium chloride is a two-dimensional network solid containing $Mg^{2+}$ and $Cl^-$ ions held together by strong ionic bonds.

9. Barium oxide, BaO, is an infinite network solid composed of $Ba^{2+}$ and $O^{2-}$ ions. For BaO(s) to melt, strong ionic bonds must be broken.
Diamond, C, is an infinite covalent network solid consisting of C atoms strongly covalently bonded in three dimensions. For diamond to melt covalent bonds have to be broken.
Phosphorus(V) oxide is a molecular solid composed of $P_4O_{10}$ molecules. When it melts, weak intermolecular forces between $P_4O_{10}$ molecules have to be overcome.
Copper consists of Cu atoms held together by strong metallic bonds. For copper to melt, strong metallic bonds have to be broken.
Iodine, $I_2$, consists of $I_2$ molecules held together in the solid by weak intermolecular forces. For $I_2$ to melt only the weak forces have to be overcome.
Graphite, C, is a two-dimensional infinite network solid. For it to melt strong covalent CC bonds have to be broken.

10. Molecular solids consist of relatively small individual molecules held together by weak intermolecular forces. They are relatively low melting and are soluble in water only if they have polar bonds that can interact with polar water molecules.

(b) Ionic solids are network solids composed of positive ions and negative ions. They have relatively high melting points because the ions have to be given sufficient kinetic energy to be able to move relative to each other. Ionic solids are soluble in water if the interaction between the ions and polar water molecules is sufficiently strong to overcome the strong ionic forces in the solid.

(c) Metallic solids are network solids in which the atoms are held together by metallic bonds. The melting points range from close to room temperature to very high temperatures, depending on the strength of the metallic bonds, which depends on the number of bonding electrons and the sizes and charges of the metal ions. The forces between neutral metal atoms and water molecules are much weaker than the forces between metal atoms (metallic bonds), so that metals are insoluble in water (although some metals react with water).

(d) Covalent network solids have infinite arrays of atoms bonded together by covalent bonds. A relatively large number of these bonds have to be broken for the solid to melt, so that they have relatively high melting points. Also, they are generally insoluble in water because their interactions with water molecules are very weak and insufficiently strong to break any of their covalent bonds.

11. The melting points of ionic solids are related to the sizes of their ions and the charges on these ions. NaCl is composed of $Na^+$ ions and $Cl^-$ ions, while MgO is composed of $Mg^{2+}$ ions and $O^{2-}$ ions. The $Mg^{2+}$ ion is smaller than the $Na^+$ ion and the $O^{2-}$ ion is smaller than the $Cl^-$ ion. Thus, the ionic bonds in MgO are much stronger than those in NaCl, since electrostatic forces are proportional to the attracting charges and inversely proportional to their distance apart. It takes a much higher temperature to break MgO ionic bonds than to break NaCl ionic bonds.

12. The large electronegativity difference between Na and Cl results in a strong tendency to form $Na^+$ and $Cl^-$ ions by transfer of an electron from Na to Cl

$$\mathrm{Na}\cdot + \cdot\ddot{\underset{\cdot\cdot}{\mathrm{Cl}}}: \longrightarrow \mathrm{Na}^+ + :\ddot{\underset{\cdot\cdot}{\mathrm{Cl}}}:^-$$

In the gas phase, Na and Cl form highly polar Na-Cl molecules, but in the solid there is an infinite array of $Na^+$ ions and $Cl^-$ ions in which each $Na^+$ ion is surrounded by six $Cl^-$ ions and each $Cl^-$ ion is surrounded by six $Cl^-$ ions; in other words, each ion is surrounded by as many ions of opposite charge that can be packed around it. Thus, NaCl(s) is an infinite network ionic solid in which no individual NaCl molecules can be detected, with ionic bonds between the $Na^+$ and $Cl^-$ ions. The formula NaCl is its empirical formula.

13. A space lattice is a regular three-dimensional array of points in space.

14. A unit cell is obtained by joining together four points in a 2D lattice, or eight points in a 3D lattice, to give a parallelogram or a parallelopiped. The most convenient cell, and the one that is usually chosen, is the cell that shows the full symmetry of the lattice. By moving the unit cell repeatedly in directions parallel to the cell edges the complete lattice is generated.

15. The primitive cubic unit cell consists of a cube with a point at each of the eight corners. The body-centered cubic unit cell consists of a cube with a point at each of the eight corners and an additional point at the center of the cube. The face-centered cubic unit cell consists of a cube with a point at each of the eight corners and a point at the center of each of six faces:

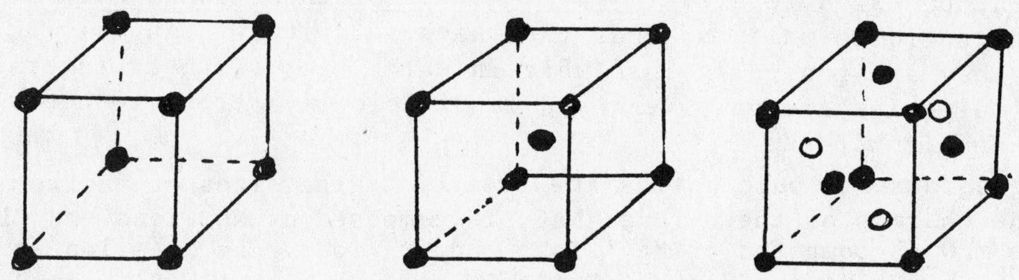

(a) primitive cube (b) body-centered cube (c) face-centered cube

16. (a) Primitive cubic unit cell (refer to the above figures)

Eight cubes intersect at each corner, so that each point is shared between eight unit cells and contributes one-eighth of a point to any one unit cell. There are eight corners.

Lattice points $= 8(\frac{1}{8}) = \underline{1}$

(b) Body-centered cubic unit cell

As for the primitive cell, each corner point contributes $\frac{1}{8}$ th of a point to the unit cell, and the central point belongs entirely to the unit cell.

Lattice points $= 8(\frac{1}{8}) + 1 = \underline{2}$

(c) Face-centered cubic unit cell

Each corner point contributes $\frac{1}{8}$ th of a point to the unit cell. Each of the six face-centered points is shared between two unit cells and contributes one-half of a point to the unit cell.

Lattice points $= 8(\frac{1}{8}) + 6(\frac{1}{2}) = \underline{4}$

17. The three common structures for metals are hexagonal close-packed, face-centered cubic, and body-centered cubic. Since all of the atoms are identical the motif is a single atom at a lattice point

18. (a) Copper has the face-centered cubic (cubic close-packed) structure.

(b) Sodium has the body-centered cubic structure.
(c) Diamond is based on the face-centered cubic cell, but with four additional C atoms located one-fourth of the way along each of the body diagonals. Each C atom is arranged at the center of a tetrahedral arrangement of four other C atoms.

19. (a) Potassium chloride has the sodium chloride structure, with the $K^+$ ions arranged in a face-centered lattice, and the $Cl^-$ ions arranged in a face-centered lattice, arranged so that the $Cl^-$ ions are half-way along the lattice formed by the $K^+$ ions, and vice-versa.
(b) Barium oxide, composed of $Ba^{2+}$ and $O^{2-}$ ions, also has the NaCl structure.
(c) Copper(I) chloride, composed of $Cu^+$ ions and $Cl^-$ ions has the sphalerite (ZnS) structure based on the face-centered cubic lattice, with $Cl^-$ ions at the lattice points and $Cu^+$ ions located one-quarter of the distance along each body diagonal.

20. In both of these close-packed structures, each atom is surrounded by six atoms in the same plane with three atoms touching it in the plane of atoms above, and three touching it in the plane below. There are two ways in which this can occur - either with the successive planes of atoms arranged in the sequence ABABAB.. , so that every third plane of atoms repeats the pattern of the first, or with the planes of atoms arranged in the sequence ABCABCABC..., so that the pattern of any plane of atoms is repeated with each successive fourth layer. The ABCABC... pattern corresponds to face-centered cubic packing, and the ABABABAB.... pattern corresponds to hexagonal close-packing:

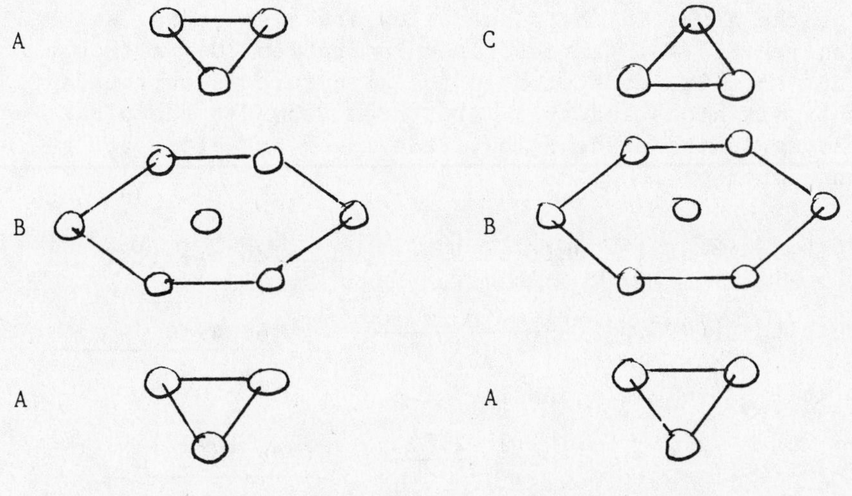


hexagonal close-packing            face-centered cubic packing

21. The cubic close-packed structure corresponds to the face-centered cubic unit cell, with atoms at each of the eight corners of the cube and at the centers of each of the six faces. The diagram on the following page shows this arrangement for two adjacent unit cells. If we focus our attention on the atom in the center of the diagram, that is in the center of the face that is common to both unit cells, we can see that it is surrounded by six atoms in the same plane, which form a hexagon around

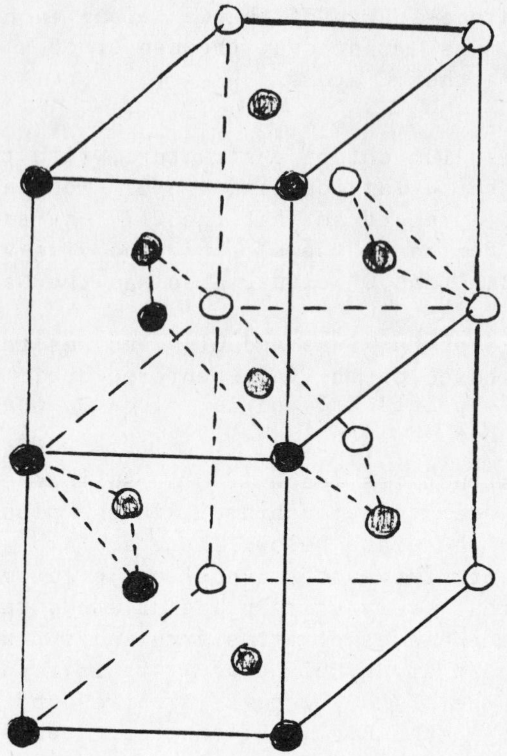

Face-centered cubic packing showing the relationship to ABC hexagonal close packing

Looking from the front, the nearest atoms are shown as ●, the next nearest atoms are shown as ◍, and the atoms farthest away are shown as ○

In the actual structure the atoms are close-packed; here for clarity the structure has been expanded

it in the same plane (joined by dotted lines in the diagram). In the plane above this plane just three atoms are shown (joined by dotted lines) and three of the atoms in the plane below are also shown. We see that the arrangement is ABC (as described in Problem 20). Although the diagram does not show it, each atom in the structure is surrounded in the same plane by six atoms and there are three atoms in the plane above, and three atoms in the plane below, that touch any atom, for a total coordination number of 12

22. From the length of the edge of the unit cell, we can calculate its volume in $cm^3$:

$$\text{Volume of unit cell} = [(405\ \text{pm})(\frac{1\ \text{m}}{10^{12}\text{pm}})(\frac{100\ \text{cm}}{1\ \text{m}})]^3 = \underline{6.64\text{x}10^{-23}\ \text{cm}^3}$$

Thus, mass of unit cell = (volume x density)

$$= (6.64\text{x}10^{-23}\ \text{cm}^3)(\frac{2.70\ \text{g}}{1\ \text{cm}^3}) = \underline{1.79\text{x}10^{-22}\ \text{g}}$$

and this must be the mass of the Al atoms in the unit cell.

The mass of one Al atom is $(\frac{26.98\ \text{g}}{1\ \text{mol}})(\frac{1\ \text{mol}}{6.022\ \text{x}\ 10^{23}\ \text{atoms}}) = \underline{4.480\text{x}10^{-23}\ \text{g}}$

$$\text{Thus, number of atoms per unit cell} = \frac{\text{mass of unit cell}}{\text{mass of 1 atom}}$$

$$= \frac{1.79\ \text{x}\ 10^{-22}\ \text{g}}{4.480\ \text{x}\ 10^{-23}\ \text{g}} = \underline{4.00}$$

and the lattice must be <u>face-centered cubic</u>

23. For the cubic close-packed (face-centered cubic) structure, the number of atoms per unit cell is 4; thus, we can calculate the volume of 4 atoms from the length of the cell edge.

$$\text{Volume of 4 Pt atoms} = [(392\ \text{pm})(\frac{1\ \text{m}}{10^{12}\ \text{pm}})(\frac{100\ \text{cm}}{1\ \text{m}})]^3$$
$$= \underline{6.02 \times 10^{-23}\ \text{cm}^3}$$

Thus, the volume of 1 mol of Pt atoms is given by:

$$V = (\frac{6.022 \times 10^{23}\ \text{atoms}}{1\ \text{mol}})(\frac{6.02 \times 10^{-23}\ \text{cm}^3}{4\ \text{atoms}}) = \underline{9.06\ \text{cm}^3\ \text{mol}^{-1}}$$

and the mass of 1 mol of Pt atoms is

$$(9.06\ \text{cm}^3\ \text{mol}^{-1})(21.5\ \text{g cm}^{-3}) = \underline{195\ \text{g mol}^{-1}}$$

Thus, the atomic mass of Pt is $\underline{195\ u}$

24. Considering the face of the unit cell of the cubic close-packed (face-centered cubic) structure. The length of the face-diagonal is four times the radius of an atom in the structure.

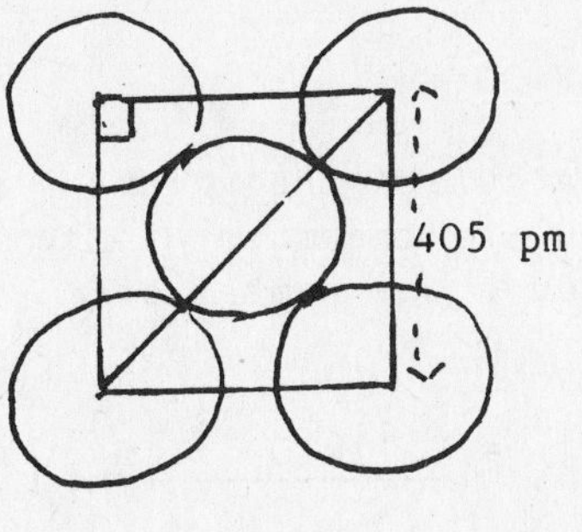


Unit Cell Face

The length of the face-diagonal, b, is related to the length of the cell edge, a, by Pythagoras's theorem

$$b^2 = a^2 + a^2 = 2a^2$$

i.e., $b = \sqrt{2}\ a$

In this case. a = 405 pm

Thus:

$$4r_{Al} = b = \sqrt{2}(405\ \text{pm}) = 573\ \text{pm}$$
$$r_{Al} = \frac{573\ \text{pm}}{4} = \underline{143\ \text{pm}}$$

25. Copper(I) chloride has the sphalerite (ZnS) structure, with four formula units of CuCl per unit cell.

(a) 4CuCl formula mass units occupy the volume of the unit cell; thus, the density of CuCl is given by

$$\text{density} = \frac{\text{mass of 4 CuCl}}{\text{volume of unit cell}} = 3.41\ \text{g cm}^{-3}$$

and one mole of CuCl units has a mass of (63.55+35.45) = 99.00 g

$$\text{i.e., } V_{\text{unit cell}} = (\frac{99.00\ \text{g CuCl}}{1\ \text{mol CuCl}})(\frac{1\ \text{mol CuCl}}{6.022\times10^{23}\ \text{CuCl}})(4\ \text{CuCl})(\frac{1\ \text{cm}^3}{3.41\ \text{g}})$$
$$= \underline{1.93 \times 10^{-22}\ \text{cm}^3}$$

Thus, if $\underline{a}$ is the length of the edge of the unit cell

$$V_{\text{unit cell}} = a^3 = 1.93\times10^{-22}\ \text{cm}^3$$

$$a = (1.93\times10^{-22}\ \text{cm}^3)^{1/3}(\frac{1\ \text{m}}{100\ \text{cm}})(\frac{10^{12}\ \text{pm}}{1\ \text{m}}) = \underline{578\ \text{pm}}$$

(b) In the unit cell, the $Cu^+$ ions are at the lattice points and the $Cl^-$ ions are one-quarter of the length of a body-diagonal away. In Problem 24 we used Pythagoras's theorem to show that the length of the face-diagonal of the cube is $b = \sqrt{2}a$.

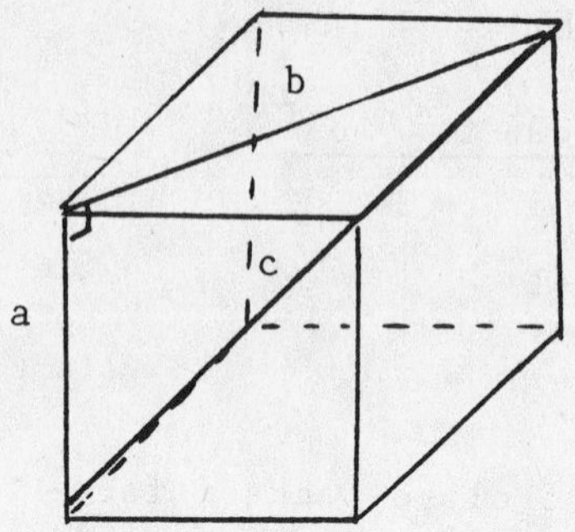


Thus the length of the body-diagonal, c, is given by

$$c^2 = a^2 + b^2 = a^2 + 2a^2 = 3a^2$$

$$c = \sqrt{3}a$$

In this case a = 578 pm

$$c = \sqrt{3}a = \sqrt{3}(578 \text{ pm}) = 1000 \text{ pm}$$

$$\frac{c}{4} = \left(\frac{1000 \text{ pm}}{4}\right) = \underline{250 \text{ pm}}$$

(c) The distance between the centers of a $Cu^+$ ion and a $Cl^-$ ion is 250 pm, which is the sum of the radii of the two ions.

$$r_{Cu^+} + r_{Cl^-} = 250 \text{ pm} = r_{Cu^+} + 180 \text{ pm}$$

$$r_{Cu^+} = \underline{70 \text{ pm}}$$

26. This problem is similar to problem 22; we first calculate the volume of the unit cell, then we use the density to calculate the mass of the unit cell, and from the mass of one Ne atom we deduce the number of atoms per unit cell:

$$\text{Volume of unit cell} = \left[(450 \text{ pm})\left(\frac{1 \text{ m}}{10^{12} \text{ pm}}\right)\left(\frac{100 \text{ cm}}{1 \text{ m}}\right)\right]^3 = \underline{9.11\text{x}10^{-23} \text{ cm}^3}$$

$$\text{Mass of unit cell} = (9.11\text{x}10^{-23} \text{ cm}^3)\left(\frac{1.45 \text{ g}}{1 \text{ cm}^3}\right) = 1.32 \text{ x } 10^{-22} \text{ g}$$

$$\text{Number of atoms per unit cell} = (1.32\text{x}10^{-22} \text{ g})\left(\frac{1 \text{ mol Ne}}{20.18 \text{ g Ne}}\right)\left(\frac{6.022\text{x}10^{23}}{1 \text{ mol Ne}}\right)$$

$$= \underline{3.94}$$

There are four Ne atoms per unit cell, so Ne must have the face-centered cubic lattice. As in Problem 24, the length of the face diagonal is $\sqrt{2}(450 \text{ pm}) = 636$ pm, which is four times the radius of a Ne atom.

$$r_{Ne} = \frac{636 \text{ pm}}{4} = \underline{159 \text{ pm}}$$

27. The unit cell is face-centered cubic (cubic close packed).

(a)

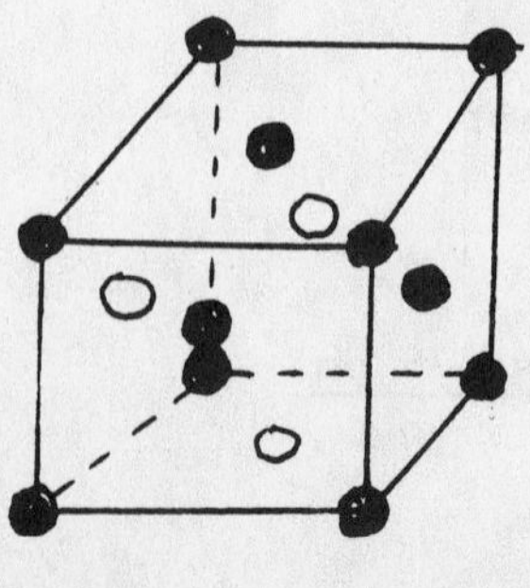

Each of the 6 faces has 4 atoms at the corners & 1 atom in the center of the face

(b) There are 4 Kr atoms per unit cell

(c) Density = $\frac{\text{mass of 4 Kr atoms}}{\text{volume of unit cell}}$

$$= \frac{4\left(\frac{83.80\ \text{g Kr}}{1\ \text{mol Kr}}\right)\left(\frac{1\ \text{mol Kr}}{6.022\times10^{23}\ \text{Kr atoms}}\right)}{\left[(559\ \text{pm})\left(\frac{1\ \text{m}}{10^{12}\ \text{pm}}\right)\left(\frac{100\ \text{cm}}{1\ \text{m}}\right)\right]^3} = \frac{5.566 \times 10^{-22}\ \text{g}}{1.747 \times 10^{-22}\ \text{cm}^3}$$

$$= \underline{3.18\ \text{g cm}^{-3}}$$

(d) The length of the face-diagonal is $\sqrt{2}a$, where a is the length of the cell edge, and it is equal to four times the radius of a Kr atom.

$$r_{Kr} = \frac{\sqrt{2}(559\ \text{pm})}{4} = \underline{198\ \text{pm}}$$ (see Problem 24)

(e) The covalent radius of Kr given in Figure 4.9 is 111 pm, which will have been obtained from dividing up the length of a covalent bond in a molecule such as $KrF_2$ between Kr and F. The 198 pm obtained above is one-half of the distance between nonbonded Kr atoms in the solid, where they are attracted by only weak intermolecular forces. In other words, 111 pm is the radius of Kr when it forms covalent bonds to other atoms, and 396 pm is the distance between nonbonded Kr atoms in the solid.

28. From the atomic mass of gold (197.0 g $mol^{-1}$) we know the mass of 1 mol of Au atoms. Since there are 4 Au atoms in the cubic close-packed unit cell, the cell dimensions and the density enable us to calculate the mass of 4 Au atoms. Thus, we can calculate the number of Au atoms in 1 mol of Au atoms to give a value for Avogadro's constant.

Mass of 4 Au atoms = (density of Au)(volume of unit cell)

$$= (19.329\ \text{g cm}^{-3})\left[(407\ \text{pm})\left(\frac{1\ \text{m}}{10^{12}\ \text{pm}}\right)\left(\frac{100\ \text{cm}}{1\ \text{m}}\right)\right]^3$$

$$= \underline{1.30 \times 10^{-21}\ \text{g}}$$

Number of Au atoms in 1 mol = $\left(\frac{4\ \text{Au atoms}}{1.30\times10^{-21}\ \text{g}}\right)(197.0\ \text{g mol}^{-1})$

$$= \underline{6.06 \times 10^{23}\ \text{atom mol}^{-1}}$$

29. Since NaCl has the cubic close-packed (face-centered cubic) lattice, there are 4 NaCl formula units per unit cell. For any cell edge there are $Na^+$ ion at each end - and a $Cl^-$ ion in the middle (Figure 11.21); the length of the cell edge is twice the NaCl distance = 2(281 pm), or 562 pm. As in Problem 28, we calculate the mass of 4 NaCl formula units from the volume of the unit cell and the density of NaCl and can then calculate a value for the Avogradro constant from the molar mass of NaCl = (22.99 + 35.45) = 58.44 g $mol^{-1}$.

Mass of 4 NaCl formula units = (density of NaCl)(volume of unit cell)

$$= (2.165\ \text{g cm}^{-3})\left[(562\ \text{pm})\left(\frac{1\ \text{m}}{10^{12}\ \text{pm}}\right)\left(\frac{100\ \text{cm}}{1\ \text{m}}\right)\right]^3 = \underline{3.843 \times 10^{-22}\ \text{g}}$$

Number of NaCl formula units in 1 mol = $\left(\frac{4\text{NaCl units}}{3.843\times10^{-22}\ \text{g}}\right)(58.44\ \text{g mol}^{-1})$

$$= \underline{6.08 \times 10^{23}\ \text{units mol}^{-1}}$$

30. The methodology is similar to that of Problems 28 and 29. Copper has the cubic close-packed structure and, thus, four Cu atoms per unit cell

$$\text{Avogadro constant} = \underline{6.03 \times 10^{23}\ \text{mol}^{-1}}$$

31. As in Problem 22, we first calculate the volume of the unit cell, and then its mass, using the density. From the molar mass and Avogadro's constant we can then calculate the number of $CaF_2$ formula units with a mass equal to that of the unit cell.

$$\text{Mass of unit cell} = (3.180\ \text{g cm}^{-3})[(546.3\ \text{pm})(\frac{1\ \text{m}}{10^{12}\ \text{pm}})(\frac{100\ \text{cm}}{1\ \text{m}})]^3$$
$$= 5.185 \times 10^{-22}\ \text{g}$$

Number of $CaF_2$ units per unit cell

$$= (5.185 \times 10^{-22}\ \text{g})(\frac{1\ \text{mol CaF}_2}{78.08\ \text{g CaF}_2})(\frac{6.022\text{x}10^{23}}{1\ \text{mol}}) = \underline{4.00}$$

32. There are 4 KF units per unit cell.

(a)

$$\text{Volume of unit cell} = (\frac{\text{mass of 4 KF formula units}}{\text{density}})$$

$$= \frac{(\frac{58.10\ \text{g KF}}{6.022\text{x}10^{23}\ \text{KF units}})(4\ \text{KF units})}{2.481\ \text{g cm}^{-3}} = 1.555 \times 10^{-22}\ \text{cm}^3$$

$$\text{Length of cell edge} = (\text{Volume})^{1/3} = (155.5 \times 10^{-24}\ \text{cm}^3)^{1/3}$$

$$= (5.377\text{x}10^{-8}\ \text{cm})(\frac{1\ \text{m}}{100\ \text{cm}})(\frac{10^{12}\ \text{pm}}{1\ \text{m}}) = \underline{537.7\ \text{pm}}$$

(b) As in NaCl (Figure 11.21), the length of the cell edge is twice the K-F distance.

$$\text{KF distance} = \frac{537.7\ \text{pm}}{2} = \underline{268.9\ \text{pm}}$$

33. The unit cell is similar to that of diamond, with Si atoms replacing the C atoms, but with O atoms arranged so that every Si atom is surrounded by four tetrahedrally disposed O atoms, each of which is bonded to another Si atom. Thus, the lattice is face-centered cubic with Si atoms at the lattice points and at one-fourth of the distance along each body diagonal, and with O atoms, between each pair of Si atoms

34. An octahedral hole is formed in the center of four spherical ions packed in a square, with two other ions resting in the "hole in the middle", one above, and the other below, the plane of the four square packed ions:

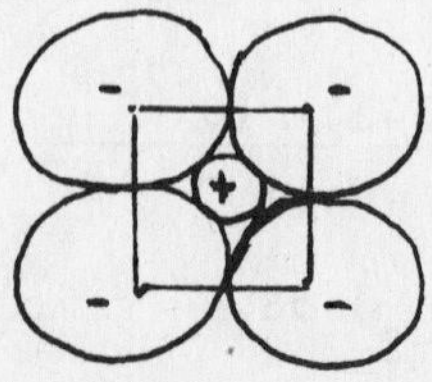

The cation will just fit into the hole formed by the anions, in other words, into the hole formed by the square packed ions, whose centers form a square of side $2r_-$. The length of the diagonal of this square is $\sqrt{2}(2r_-)$ - see Problem 24. Thus.

$$\sqrt{2}(2r_-) = 2r_- + 2r_+$$

i.e., $2r_+ = (1.414)[2(195\text{ pm})] - 2(195\text{ pm}) = 162\text{ pm}$

$$r_+ = \underline{81\text{ pm}}$$

The maximum radius of cations that can fit into the octahedral holes is 81 pm.

35. This is similar to Problem 31.

Mass of unit cell = (density) x (volume)

$$= (16.6\text{ g cm}^{-3})[(328\text{ pm})(\frac{1\text{ m}}{10^{12}\text{ pm}})(\frac{100\text{ cm}}{1\text{ m}})]^3$$

$$= \underline{5.86 \times 10^{-22}\text{ g}}$$

Number of Ta atoms per unit cell $= (5.86\text{x}10^{-22}\text{ g})(\frac{1\text{ mol Ta}}{180.9\text{ g Ta}})(\frac{6.022\text{x}10^{23}}{1\text{ mol}})$

$= \underline{1.95}$ (approximately $\underline{\text{two}}$).

There are 2 Ta atoms per unit cell, which corresponds to the expected number for a body-centered cubic structure.

36. Cesium chloride has a primitive cubic lattice with eight $Cl^-$ ions at the lattice points and the $Cs^+$ ion situated at the center of the cube. Thus the number of CsCl formula units per unit cell is one. Sodium chloride has the face-centered cubic lattice with an $Na^+$ ion at each lattice point and an equal number of $Cl^-$ ions situated halfway between each pair of closest $Na^+$ ions. Thus the number of formula units is four.

Thus, we have to compare the mass of CsCl with that of 4NaCl.

$$\frac{\text{mass of CsCl}}{\text{mass of 4NaCl}} = \frac{1\text{ mol CsCl}}{4\text{ mol NaCl}} = \frac{168.4\text{ g}}{233.8\text{ g}} = 0.720$$

The unit cell of sodium chloride has the higher mass.

37. The monatomic body-centered unit cell contains two atoms, and we can use the given data to calculate first the volume of the unit cell:

$$\text{Volume} = \frac{\text{mass}}{\text{density}} = 2(\frac{95.94\text{ g Mo}}{1\text{ mol Mo}})(\frac{1\text{ mol Mo}}{6.022\text{x}10^{23}})(\frac{1\text{ cm}^3}{10.22\text{ g}}) = \underline{3.12\text{x}10^{-23}\text{ cm}^3}$$

Thus the length of the cell edge, $\underline{a}$, is $V^{1/3} = \underline{3.15 \times 10^{-8}\text{ cm}} = \underline{315\text{ pm}}$

In the body-centered unit cell the atom at the center just touches two other atoms at opposite corners, i.e., the length of the body diagonal is 4r, where r is the radius of the atom.

As we saw in Problem 25, the length of the body diagonal is $\sqrt{3}a$

i.e., $4r = \sqrt{3}(315\text{pm})$ ; $r = \underline{136\text{ pm}}$

The radius of the molybdenum atom is 136 pm.

38. HgS has the sphalerite structure in which $Hg^{2+}$ ions are situated at the lattice points and the $S^{2-}$ ions are situated one-quarter of the distance along each body diagonal. i.e., the length of the body diagonal is 4(253 pm) = 1012 pm, and there are four HgS formula units per unit cell.

As we saw in Problem 25, the length of the body-diagonal is $\sqrt{3}a$, where a is the length of the edge of the cubic unit cell, so that

$$\sqrt{3}a = 1012 \text{ pm} \; ; \quad a = \underline{584 \text{ pm}}$$

$$\text{Thus, density} = \frac{\text{mass of 4 HgS formula units}}{\text{volume of unit cell}}$$

$$= \frac{4\left(\frac{232.7 \text{ g HgS}}{1 \text{ mol HgS}}\right)\left(\frac{1 \text{ mol}}{6.022\times10^{23}}\right)}{\left[(584 \text{ pm})\left(\frac{1 \text{ m}}{10^{12} \text{ pm}}\right)\left(\frac{100 \text{ cm}}{1 \text{ m}}\right)\right]^3} = \underline{7.76 \text{ g cm}^{-3}}$$

39. Cesium bromide has the CsCl structure, which is body-centered cubic with the $Cs^+$ ion at the central lattice point and $Br^-$ ions at the lattice points situated at the eight corners of the cube. There is one CsBr formula unit per unit cell and the distance between the centers of the $Cs^+$ ion and a $Br^-$ ion is one-half of the length of the body diagonal of the cube.

As we saw in Problem 25, the length of the body-diagonal is $\sqrt{3}a$, where a is the length of the edge of the cubic unit cell, so that

$$\sqrt{3}a = 2(371 \text{ pm}) \; ; \quad a = \underline{428 \text{ pm}}$$

$$\text{Thus, density} = \frac{\text{mass of 1 CsBr formula unit}}{\text{volume of unit cell}}$$

$$= \frac{\left(\frac{212.8 \text{ g CsBr}}{1 \text{ mol CsBr}}\right)\left(\frac{1 \text{ mol}}{6.022\times10^{23}}\right)}{\left[(428 \text{ pm})\left(\frac{1 \text{ m}}{10^{12} \text{ pm}}\right)\left(\frac{100 \text{ cm}}{1 \text{ m}}\right)\right]^3} = \underline{4.51 \text{ g cm}^{-3}}$$

40. (a) Barium oxide has the same structure as sodium chloride, which is face-centered cubic with $Ba^{2+}$ ions at the lattice points and an $O^{2-}$ ion to the right of each $Ba^{2+}$ ion at a distance equal to one-half of the length of the cell edge. Each of the face-centered $Ba^{2+}$ ions is associated with two unit cells, for a total of $6 \times \frac{1}{2} = 3Ba^{2+}$ ions, and each of the eight $Ba^{2+}$ ions at the corners of the cube is shared between eight unit cells for a total contribution of $8\times\frac{1}{8} = 1Ba^{2+}$ ion; i.e., the total number of $Ba^{2+}$ ions associated with the unit cell is 4. There is one $O^{2-}$ ion at the center of the cube that is associated entirely with the unit cell, and 12 more $O^{2-}$ ions situated on the cell edges and associated with a total of four unit cells each, for a total of $12\times\frac{1}{4} = 3O^{2-}$ ions; i.e., the total number of $O^{2-}$ ions associated with the unit cell is 1+3 = 4:

$4Ba^{2+}$ ions and $4O^{2-}$ ions are associated with the unit cell

Cesium iodide has the cesium chloride structure with a primitive cubic lattice with eight $I^-$ ions located at the corners of the cube and a one $Cs^+$ ion situated at the center of the cube. Each $I^-$ ion is associated with eight unit cells, for a total of $8\times\frac{1}{8}$ = one $I^-$ ion, and the $Cs^+$ ion belongs entirely to the unit cell.

$1Cs^+$ ion and 1 $I^-$ ion is associated with the unit cell

Lithium sulfide has the antifluorite structure, which is face-centered cubic with $S^{2-}$ ions at the lattice points (for a total of $4S^{2-}$ ions associated with the unit cell. Eight $Li^{+}$ ions are situated entirely within the unit cell in the tetrahedral holes formed by the $S^{2-}$ ions, for a total of $8Li^{+}$ ions for the unit cell.

$8Li^{+}$ ions and $4S^{2-}$ ions are associated with the unit cell

(b) Barium oxide Face centered cubic with 8+6 = 14 lattice points
Cesium iodide Primitive cubic with eight lattice points
Lithium sulfide Face centered cubic with 14 lattice points

41. Zinc sulfide, ZnS, has the sphalerite structure. The lattice is face-centered cubic with sulfide ions at the lattice points and $Zn^{2+}$ ions situated one-fourth of the distance along each body-diagonal, so that each $Zn^{2+}$ ion is at the center of a tetrahedral arrangement of $S^{2-}$ ions, for a coordination number of 4. Each $S^{2-}$ ion is also surrounded by a tetrahedral arrangement of 4 $Zn^{2+}$ ions. Thus, both ions have a coordination number of 4.

The arrangement of ions in the sphalerite structure is the same as that for the carbon atoms in the diamond structure. The $S^{2-}$ ions are at the lattice points of a face-centered cubic unit cell, and there are four $Zn^{2+}$ ions replacing the four C atoms in diamond that are situated one-quarter of the distance along each body-diagonal.

42. (a) In the fluorite structure, the $Ca^{2+}$ ions are located at the face--centered cubic lattice points. Each face-centered $Ca^{2+}$ ion is at the center of a cubic arrangement of eight $F^-$ ions, so the coordination number is 8. (b) All the tetrahedral holes in the cubic close-packed arrangement of $Ca^{2+}$ ions are occuppied by $F^-$ ions, so each $F^-$ ion has a coordination number of 4. (see Figure 11.25)
(c) The arrangement of the $Ca^{2+}$ ions in the fluorite structure is the same as the arrangement of the $S^{2-}$ ions in the ZnS structure, but in $CaF_2$ there are twice as many $F^-$ ions as there are $Zn^{2+}$ ions in ZnS. The $F^-$ ions in $CaF_2$ occupy all of the tetrahedral holes, whereas the $Zn^{2+}$ ions in ZnS occupy only one-half of the tetrahedral holes.

43. In the monatomic body-centered cubic structure the number of Fe atoms per unit cell is 2, while in the face-centered cubic structure the number of Fe atoms per unit cell is 4. As in Problems 38 and 39.

$$\text{Density} = \frac{\text{mass of unit cell}}{\text{volume of unit cell}} = \frac{\text{mass of 4 Fe atoms}}{\text{volume of unit cell}}$$

$$= \frac{4\left(\frac{55.85\text{ g Fe}}{1\text{ mol Fe}}\right)\left(\frac{1\text{ mol}}{6.022\times10^{23}}\right)}{\left[(363\text{ pm})\left(\frac{1\text{ m}}{10^{12}\text{ pm}}\right)\left(\frac{100\text{ cm}}{1\text{ m}}\right)\right]^3} = \underline{7.76\text{ g cm}^{-3}}$$

44. The positions of the $Na^{+}$ ions and the $Cl^{-}$ ions in the unit cell are discussed in Problem 40(a). The $Na^+$ ions are at the lattice points and thus none of them belong entirely to any one unit cell. Of the $Cl^-$ ions, only one, that situated at the center of the cube, belongs entirely to the unit cell.
Of the $Na^+$ ions, each of the eight ions situated at the corners of the cube is shared between eight unit cells, and each of the six ions situated at the centers of the six faces are each shared between two

unit cells. Of the thirteen $Cl^-$ ions, one is at the center of the unit cell, and the the others are arranged in pairs halfway along each of the 12 edges of the cube, and each is shared between four unit cells. For any unit cell, there is a total of 4 NaCl formula units associated with the cell.

$Na^+$ ions $= 8(\frac{1}{8}) + 6(\frac{1}{2}) = \underline{4}$ ; $Cl^-$ ions $= 1 + 12(\frac{1}{4}) = \underline{4}$

45. As in Problems 38, 39, and 43, Density = mass/volume, and for the body-centered cubic structure, there are 2 Ba atoms per unit cell.

$$\text{Density} = \frac{2(\frac{137.3\text{ g Ba}}{1\text{ mol Ba}})(\frac{1\text{ mol}}{6.022\times10^{23}})}{[(502\text{ pm})(\frac{1\text{ m}}{10^{12}\text{ pm}})(\frac{100\text{ cm}}{1\text{ m}})]^3} = \underline{3.60\text{ g cm}^{-3}}$$

46. (a) In the fluorite structure of $BaF_2$ there are 14 atoms at each of the lattice points in a cubic close packed arrangement. i.e. $4Ba^{2+}$ ions per unit cell, and $8F^-$ ions occupying the 8 tetrahedral holes within the unit cell, for a total of <u>four</u> $BaF_2$ formula units.

(b) Each $F^-$ ion is one-quarter of the length of the body diagonal away from a $Ba^{2+}$ ion, and the length of the body diagonal is $\sqrt{3}a$, where a is the length of the edge of the unit cell (see Problem 25). Thus,

$$b = \sqrt{3}(618\text{ pm}) = 1070\text{ pm}$$

and the sum of the radii of the ions is given by

$$r_{Ba^{2+}} + r_{F^-} = \frac{1070\text{ pm}}{4} = \underline{268\text{ pm}}$$

Thus, for $r_{F^-} = 135$ pm, $r_{Ba^{2+}} = (268\text{-}135)$ pm $= \underline{133\text{ pm}}$

(c) The value for $r_{Ba^{2+}}$ obtained in part (b) is close to the value of 136 pm given in Table 11.5.

47. In the face-centered cubic structure of silver there are 4 Ag atoms per unit cell, and from the density of 10.50 g $cm^{-3}$, we can calculate the volume of the unit cell, and hence the number of Ag atoms in a cube with an edge 1 mm in length, i.e., a volume of 1 $mm^3$.

$$\text{Volume of unit cell} = \frac{\text{mass of unit cell}}{\text{density}}$$

$$= (4\text{ Ag atoms}(\frac{107.9\text{ g Ag}}{1\text{ mol Ag}})(\frac{1\text{ mol}}{6.022\times10^{23}\text{ Ag atoms}})(\frac{1\text{ cm}^3}{10.50\text{ g}})$$

$$= \underline{6.826\times10^{-23}\text{ cm}^3}$$

$$\text{Number of atoms in }(1\text{ mm})^3 = (1\text{ mm})^3(\frac{1\text{ cm}}{10\text{ mm}})^3(\frac{4\text{ Ag atoms}}{6.826\times10^{-23}\text{ cm}^3})$$

$$= \underline{5.86\times10^{19}\text{ Ag atoms}}$$

48. (a) $F^-$ and $I^-$ have the same +7 core charge but I is in period 5 below F in period 2; $\underline{I^- > F^-}$. (215 pm versus 135 pm).

(b) $Na^+$ and $K^+$ have the same core charge of +9 but K is one period below Na in the periodic table. $K^+ > Na^+$ (138 pm versus 102 pm).

(c) O and F are in the same period but the core charge of O is +6 while that of F is +7. $O^{2-} > F^-$ (140 pm versus 135 pm).

(d) $Be^{2+}$ has the same electronic configuration as He, and $Cl^-$ has the same configuration as Ar. Thus, $Cl^-$ has two more filled shells than $Be^{2+}$ and is expected therefore to be much larger. $Cl^- >> Be^{2+}$. (180 pm versus 35 pm).

(e) Both $K^+$ and $Cl^-$ have the Ar configuration, but the core charge of $K^+$ is +9, versus +7 for $Cl^-$; $Cl^- > K^+$ (180 pm versus 138 pm).

49. Diamond is an allotrope of carbon that forms a three-dimensional network solid (giant molecule) in which each C atom is at the center of a tetrahedral arrangement of 4 other C atoms, to which it is bonded by four strong covalent bonds. Each carbon atom may be described as using a set of tetrahedral $sp^3$ hybrid orbitals, each of which forms a C-C bond by overlap with a similar orbital on each of 4 C atoms. The unit cell of diamond is face-centered cubic with a C atom at each lattice point and at one-quarter of the distance along each body diagonal. Thus the unit cell contains 8 C atoms.

50. Graphite is an allotrope of carbon that forms a two-dimensional network solid (giant molecule) in which each C atom is bonded to three others in the same plane, so that the carbon atoms are arranged in interconnected hexagons. A single Lewis (resonance) structure has each C atom forming two C-C covalent bonds and one C=C bond. Averaging all of the resonance structures gives identical CC bonds of order $1\frac{1}{3}$, which is consistent with the observed CC bond length of 142 pm, compared to a bond length of 154 pm for the single bonds in diamond, and the arrangement of the C atoms in planar regular hexagons. Each C atom may also be described as using a set of $sp^2$ orbitals which overlap with similar orbitals on three adjacent C atoms to form a $\sigma$ bonded framework. This leaves a single electron on each C atom in the p orbital that is perpendicular to the planar sheets of C atoms. These orbitals overlap to form a delocalized system of $\pi$ bonds. The distance between the planar layers of C atoms is 334 pm, so that they may be regarded as being held together only by weak intermolecular forces, which enables the sheets of C atoms to slide over each other rather easily. Thus, carbon in the form of graphite is soft and behaves as a lubricant.

51. Diamond is the hardest mineral known while graphite is relatively soft. Diamond is a nonconductor of electricity, while carbon conducts electricity rather well in the direction of its sheets of C atoms but not in other directions. Diamond has a much higher density than graphite (3.53 g $cm^{-3}$ versus 2.25 g $cm^{-3}$). Chemically, graphite reacts at room temperature with sulfuric acid and other oxidizing agents and burns readily in air and oxygen. In contrast, diamond is chemically inert at ordinary temperatures and must be heated to 800°C in order to react even with oxygen.
All of these differences in properties are readily explained in terms of the different structures of diamond and graphite (see Problems 49 & 50). All the bonds in diamond are localized, while graphite has delocalized electrons between the planes of C atoms. This accounts for the metal-like electrical conductivity of graphite along the planes of C

atoms, for its shiny black "metallic" appearance, and for its chemical reactivity.

52. See Problem 50. The CC bond order in graphite is 1.33, consistent with the CC bond length of 142 pm, intermediate in length between the length of the CC single bond in diamond of 154 pm and the CC double bond length of 134 pm.

53. NaCl has a structure in which the $Na^+$ ions form a face-centered cubic lattice, with one $Cl^-$ ion at the center of the cube and 12 $Cl^-$ ions arranged each at the center of one of the 12 edges of the cube, half-way between two $Na^+$ ions. In contrast, the structure of CsCl has the $Cl^-$ ion occupying the lattice points of a primitive cubic lattice, with a $Cs^+$ ion at the center of the cube. Thus, the coordination number of the $Na^+$ ion in NaCl is <u>six</u> and that of the $Cs^+$ ion in CsCl is <u>eight</u>. The differences in the structures is due to the different cation:anion radius ratios. The ionic radii are $Na^+$ 102 pm, $Cs^+$ 170 pm and $Cl^-$ 180 pm. Thus, the radius ratio in NaCl is $Na^+ : Cl^- = 0.567$, and the radius ratio in CsCl is $Cs^+ : Cl^- = 0.944$. Consideration of the limiting radius ratios for cations with different anion coordination numbers gives the range 0.414 to 0.732 for six coordination, and the range 0.732 to 1.00 for eight coordination. Thus, a coordination number of 6 is expected for the $Na^+$ ions in NaCl, and a coordination number of 8 is expected for the $Cs^+$ ions in CsCl, as observed. In the simplest terms, the different structures of NaCl and CsCl is due to the much larger size of $Cs^+$ relative to the size of $Na^+$.

54. When the structures of ionic crystals are determined by means of X-ray diffraction, the data that can be obtained is the distances between ions (interionic distances). The problem then is to divide up these interionic distances to obtain a set of useful radii for each of the individual ions, which can be used, for example, to accurately predict interionic distances in other crystals. The underlying assumption is that a particular ion has a constant size, independent of any other ions in the crystal or their coordination numbers. Such an assumption is only approximate for several reasons. For example, the same ion may occupy holes of different sizes in different structures so that the ions are not necessarily in contact, the same ion may be compressed to different extents in different structures, and the assumption that the bonds are pure ionic bonds and have no covalent character may be incorrect in many cases.

55. An anion has one or more electrons than the corresponding neutral atom. The core charge remains the same but addition of more electrons to the neutral atom leads to greater repulsions between the valence shell electrons than is the case for the neutral atom. Thus, anions are larger than the corresponding neutral atoms. A cation has one or more electrons less than the corresponding neutral atom, which leads to diminished repulsions in the cation compared to those in the neutral atom. Thus, cations are generally smaller than the corresponding neutral atoms.

# CHAPTER 12

1. In problems concerned with a change from one kind of concentration unit to another, it is important to remember the definitions of the concentration units involved - in this case:

$$\text{Mole fraction of X} = \frac{\text{moles of X}}{\text{total moles of all the components}}$$

In particular for a two component solution,

$$\text{Mole fraction of X} = \frac{\text{moles of X}}{\text{moles of X + moles of Y}}$$

$$\text{Mole fraction of Y} = \frac{\text{moles of Y}}{\text{moles of X + moles of Y}}$$

Mole fraction of X + mole fraction of Y = 1

and

Molality of solute X in a solvent Y = moles of X in 1 kg of solvent Y

Distinguish carefully between molality, m, and molarity, M. The molarity of a solute X in a solvent Y = moles of X in 1 L of solution

Thus:

Mole fraction of $H_2O$ = 0.925; mole fraction of $C_2H_5OH$ = 1.000 - 0.925 = 0.075

and the solution contains 0.075 mol $C_2H_5OH$ and 0.925 mol $H_2O$

$$\text{Molality} = \frac{\text{moles of } C_2H_5OH}{1 \text{ kg } H_2O} = \frac{(0.075 \text{ mol } C_2H_5OH)}{(0.925 \text{ mol } H_2O)\left(\frac{18.02 \text{ g } H_2O}{1 \text{ mol } H_2O}\right)\left(\frac{1 \text{ kg } H_2O}{1000 \text{ g } H_2O}\right)}$$

$$= \underline{4.5 \text{ m}}$$

2. The solution contains 9.65 g NaCl and (100 - 9.65) = 90.35 g $H_2O$

i.e., $(9.65 \text{ g NaCl})\left(\frac{1 \text{ mol NaCl}}{58.44 \text{ g NaCl}}\right) = 0.165 \text{ mol NaCl}$, and

$$(90.35 \text{ g } H_2O)\left(\frac{1 \text{ mol } H_2O}{18.02 \text{ g } H_2O}\right) = 5.014 \text{ mol } H_2O$$

$$\text{Mole fraction } H_2O = \frac{5.014 \text{ mol } H_2O}{0.165 \text{ mol NaCl} + 5.014 \text{ mol } H_2O} = \underline{0.968}$$

3. For definitions of concentration units, see Problem 1.

100 g 69.0 mass% $HNO_3$ contains 69.0 g $HNO_3$ and 31.0 g $H_2O$

i.e., $(69.0 \text{ g } HNO_3)\left(\frac{1 \text{ mol } HNO_3}{63.02 \text{ g } HNO_3}\right) = 1.095 \text{ mol } HNO_3$ and

$$(31.0 \text{ g } H_2O)\left(\frac{1 \text{ mol } H_2O}{18.02 \text{ g } H_2O}\right) = 1.720 \text{ mol } H_2O$$

To calculate the mole fraction we need to know mol $HNO_3$ and mol $H_2O$; molarity requires mol $HNO_3$ and the volume of the solution, and molality requires mol $HNO_3$ and the mass of water. All of these data are known <u>except</u> for the volume of the solution, which we can obtain via the given density, since

$$\text{Volume of 100 g solution} = (100\text{ g})\left(\frac{1\text{ mL}}{1.41\text{ g}}\right) = 70.9\text{ mL}$$

Thus

$$\text{Molarity} = \frac{1.095\text{ mol HNO}_3}{(70.9\text{ mL})\left(\frac{1\text{ L}}{1000\text{ mL}}\right)} = \underline{15.4\text{ mol L}^{-1}}$$

$$\text{Molality} = \frac{1.095\text{ mol}}{(31.0\text{ g H}_2\text{O})\left(\frac{1\text{ kg}}{1000\text{ g}}\right)} = \underline{35.3\text{ m}}$$

$$\text{Mole fraction HNO}_3 = \frac{1.095\text{ mol}}{(1.095\text{ mol} + 1.720\text{ mol})} = \underline{0.389}$$

4. This problem is similar to problem 3.

100 g solution contains 37.0 g HCl and 63.0 g $H_2O$

$$= (37.0\text{ g HCl})\left(\frac{1\text{ mol HCl}}{36.45\text{ g HCl}}\right) = 1.015\text{ mol HCl} \quad \text{and}$$

$$(63.0\text{ g H}_2\text{O})\left(\frac{1\text{ mol H}_2\text{O}}{18.02\text{ g H}_2\text{O}}\right) = 3.496\text{ mol H}_2\text{O}$$

and the volume of the solution is $(100\text{ g})\left(\frac{1\text{ mL}}{1.18\text{ g}}\right) = 84.7\text{ mL}$

$$\text{Molarity} = \frac{1.015\text{ mol HCl}}{(84.7\text{ mL})\left(\frac{1\text{ L}}{1000\text{ mL}}\right)} = \underline{12.0\text{ mol L}^{-1}}$$

$$\text{Molality} = \frac{1.015\text{ mol HCl}}{(63.0\text{ g H}_2\text{O})\left(\frac{1\text{ kg}}{1000\text{ g}}\right)} = \underline{16.1\text{ m}}$$

$$\text{Mole fraction} = \frac{1.015\text{ mol}}{(1.015 + 3.496)\text{ mol}} = \underline{0.225}$$

5. $22.40\text{ g MgCl}_2 = (22.40\text{ g})\left(\frac{1\text{ mol}}{95.21\text{ g}}\right) = 0.2353\text{ mol MgCl}_2$

$200\text{ mL H}_2\text{O} = (200\text{ mL})\left(\frac{1.00\text{ g}}{1\text{ mL}}\right)\left(\frac{1\text{ mol}}{18.02\text{ g}}\right) = 11.1\text{ mol H}_2\text{O}$

Mass of solution = (200 + 22.40) g = 222.4 g

$\text{Volume of solution} = (222.4\text{ g})\left(\frac{1\text{ mL}}{1.089\text{ g}}\right) = 204.2\text{ mL}$

Thus

$$\text{Mole fraction MgCl}_2 = \frac{0.2353\text{ mol}}{(0.2353 + 11.1)\text{ mol}} = \underline{0.0208}$$

$$\text{Molality} = \frac{0.2353\text{ mol MgCl}_2}{(200\text{ g H}_2\text{O})\left(\frac{1\text{ kg}}{1000\text{ g}}\right)} = \underline{1.18\text{ m}}$$

$$\text{Molarity} = \frac{0.2353 \text{ mol MgCl}_2}{(204.2 \text{ mL})(\frac{1 \text{ L}}{1000 \text{ mL}})} = \underline{1.15 \text{ mol L}^{-1}}$$

6. Mass of solution = (95.94 + 10.66) g = 106.6 g

$$\text{Density} = \frac{\text{mass}}{\text{volume}} = \frac{106.6 \text{ g}}{100.0 \text{ mL}} = \underline{1.066 \text{ g mL}^{-1}}$$

$$\text{Moles of } H_2SO_4 = (10.66 \text{ g})(\frac{1 \text{ mol}}{98.08 \text{ g}}) = 0.1087 \text{ mol}$$

$$\text{Moles of } H_2O = (95.94 \text{ g})(\frac{1 \text{ mol}}{18.02 \text{ g}}) = 5.324 \text{ mol}$$

$$\text{Molality} = \frac{0.1087 \text{ mol } H_2SO_4}{(106.6 \text{ g})(\frac{1 \text{ kg}}{95.94 \text{ g}})} = \underline{1.133 \text{ mol kg}^{-1}}$$

$$\text{Molarity} = \frac{0.1087 \text{ mol } H_2SO_4}{(100 \text{ mL })(\frac{1 \text{ L}}{1000 \text{ mL}})} = \underline{1.087 \text{ mol L}^{-1}}$$

$$\text{Mole fraction of } H_2O = \frac{5.342 \text{ mol } H_2O}{(0.1087 + 5.324) \text{ mol}} = \underline{0.9800}$$

7. In the process of sublimation, a substance passes directly from the solid phase to the vapor (gaseous phase), which is equivalent to going from solid to liquid and then from liquid to gas. The enthalpy of sublimation must be the sum of the enthalpy of fusion and the enthalpy of vaporization. i.e.,

$$\Delta H_{subl} > \Delta H_{vap}$$

8. (a) With increase in temperature, the average kinetic energy of the molecules increases, so that a greater proportion of them have sufficient energy to escape from the surface of the liquid into the vapor. Thus, vapor pressure increases with increase in temperature.
(b) A liquid boils when its vapor pressure becomes equal to atmospheric pressure. Since benzene has a higher vapor pressure than toluene at 20$^{\circ}$C, and we expect this difference to be maintained, the vapor pressure of benzene will become equal to the pressure of the atmosphere at a lower temperature than that at which the vapor pressure of toluene becomes equal to that of the atmosphere. Thus, toluene is expected to have the higher boiling point.
(c) Assuming that the relative differences in the vapor pressures of the liquids is maintained as the temperature is lowered, the liquid with the lowest boiling point will be expected to have the highest vapor pressure at 25$^{\circ}$C, and the liquid with the highest boiling point will be expected to have the lowest vapor pressure at 25$^{\circ}$C. That is, diethyl ether will have the highest vapor pressure, and methanol will have the lowest vapor pressure at 25$^{\circ}$C. At all temperatures,

$$P_{CH_3OH} < P_{(CH_3)_2CO} < P_{(CH_3)_2O}$$

9. The porous clay pot standing in water absorbs the water, which evaporates on a hot day. Since the molar enthalpy of vaporization of water is large, this evaporation removes a large amount of heat from the clay pot, which keeps it and its contents cool, even on the hottest day.

10. $\Delta H^\circ$ for the reaction is calculated from the balanced equation for the combustion and the standard enthalpies of formation of the products and the reactants:

$$CH_4(g) + 2O_2(g) \longrightarrow CO_2(g) + 2H_2O(l)$$

$$\Delta H^\circ = (\Sigma \Delta H_f^\circ \text{ products}) - (\Sigma \Delta H_f^\circ \text{ reactants})$$

$$\Delta H^\circ = [\Delta H_f^\circ(CO_2,g) + 2\Delta H_f^\circ(H_2O,l)] - [\Delta H_f^\circ(CH_4,g) + 2\Delta H_f^\circ(O_2,g)]$$

$$= [-393.5 + 2(-285.8)] - [-74.5 + 2(0)] = \underline{-890.6 \text{ kJ}}$$

where the data is taken from Appendix B.

The enthalpy of vaporization of water is 40.7 kJ $mol^{-1}$. Thus, the amount of methane that has to be burned to vaporize 1000 kg of water is

$$(1000 \text{ kg } H_2O)\left(\frac{1000 \text{ g}}{1 \text{ kg}}\right)\left(\frac{1 \text{ mol } H_2O}{18.02 \text{ g } H_2O}\right)\left(\frac{40.7 \text{ kJ}}{1 \text{ mol}}\right)\left(\frac{1 \text{ mol } CH_4}{890.6 \text{ kJ}}\right)\left(\frac{16.04 \text{ g } CH_4}{1 \text{ mol } CH_4}\right)$$

$$= \underline{4.07 \times 10^4 \text{ g } CH_4} \quad \underline{\text{or}} \quad \underline{40.7 \text{ kg } CH_4}$$

11. (a) Condensation is the formation of a liquid when a vapor (gas) is cooled.
(b) Evaporation is the gradual disappearance of a liquid when the equilibrium between the liquid and its vapor is disturbed by allowing the vapor to escape into the atmosphere.
(c) Sublimation is the process in which a solid is converted directly to vapor.
(d) The boiling point of a liquid is the temperature at which its vapor pressure is the same as the external pressure.

12. (a) No effect; (b) no effect; (c) the vapor pressure increases with increase in temperature, because an increased number of molecules have sufficient energy to escape from the liquid into the vapor; (d) the greater the intermolecular forces between molecules in the liquid, the smaller is the vapor pressure; and (e) no effect.

13. (a) The sample collected over water contains water vapor as well as $O_2(g)$. Thus,

$$P_{total} = P_{O_2} + P_{H_2O} = 745 \text{ torr}$$

the pressure of the atmosphere, since the water levels inside and outside the jar were adjusted to be equal. $P_{H_2O}$ is the vapor pressure of water at 21°C,

which from the data in Table 12.4 is 18.6 mm Hg, or 18.6 torr. Thus,

$$P_{O_2} = P_{total} - P_{H_2O} = (745 - 18.6) = \underline{726 \text{ torr}}$$

(b) Since mercury has a negligible vapor pressure at ordinary temperatures, the pressure of $O_2(g)$ is equal to the atmospheric pressure of 745 torr, and, according to Boyle's law,

$$PV = \text{constant (at constant n and T)}$$

i.e., $P_1V_1 = (726 \text{ torr})(365 \text{ mL}) = P_2V_2 = (745 \text{ torr})V_2$

and $\underline{V_2 = 356 \text{ mL}}$

(c) Initially, $V_1 = 356$ mL ; $P_1 = 745$ torr, and $T_1 = 294$ K; the new conditions are $P_2 = 1$ atm (760 torr), and $T_2 = 273$ K (STP). Thus:

$$\frac{P_1V_1}{T_2} = \frac{P_2V_2}{T_2}; \frac{(745 \text{ torr})(356 \text{ mL})}{(294 \text{ K})} = \frac{(760 \text{ torr})V_2}{(273 \text{ K})}$$

$$V_2 = \underline{324 \text{ mL}}$$

(d) The balanced equation for the decomposition of $KClO_3(s)$ is

$$2KClO_3(s) \rightarrow 3KCl(s) + 3O_2(g)$$

From the gas data, we can calculate the mol of $O_2$ produced

$$\text{mol of } O_2 = \frac{PV}{RT} = \frac{(1 \text{ atm})(324 \text{ mL})(\frac{1 \text{ L}}{1000 \text{ mL}})}{(0.0821 \text{ atm L mol}^{-1} \text{ K}^{-1})(273 \text{ K})}$$

$$= \underline{1.45 \times 10^{-2} \text{ mol}}$$

$$\text{Mass of } KClO_3 = (1.45\text{x}10^{-2} \text{ mol } O_2)(\frac{2 \text{ mol } KClO_3}{3 \text{ mol } O_2})(\frac{122.6 \text{ g } KClO_3}{1 \text{ mol } KClO_3})$$

$$= \underline{1.19 \text{ g}}$$

14. The increase in the mass of the sulfuric acid must be the mass of water that saturates 5.00 L of air at 25°C.

$$0.115 \text{ g } H_2O = (0.115 \text{ g})(\frac{1 \text{ mol}}{18.02 \text{ g}}) = 6.38 \times 10^{-3} \text{ mol } H_2O$$

The vapor pressure of water is the pressure exerted by this amount of water in a volume of 5.00 L at 25°C.

$$P_{H_2O} = \frac{nRT}{V} = \frac{(6.38\text{x}10^{-3} \text{ mol})(0.0821 \text{ atm L mol}^{-1} \text{ K}^{-1})(298 \text{ K})}{5.00 \text{ L}}$$

$$= \underline{0.0312 \text{ atm}}$$

$\underline{\text{or}}$ $(0.0312 \text{ atm})(\frac{760 \text{ mm Hg}}{1 \text{ atm}}) = \underline{23.7 \text{ mm Hg}}$

15. The reaction is $LiH(s) + H_2O(l) \longrightarrow LiOH(aq) + H_2(g)$

$$\text{mol } H_2 = (0.540 \text{ g LiH})(\frac{1 \text{ mol LiH}}{7.949 \text{ g LiH}})(\frac{1 \text{ mol } H_2}{1 \text{ mol LiH}}) = \underline{6.79\text{x}10^{-2} \text{ mol}}$$

The total pressure of 754 torr = $P_{H_2} + P_{H_2O}$, and $P_{H_2O}$ at 25°C is 23.8 torr (Table 12.4).

Thus, for the $H_2$ gas we have. P = 730 torr, n = $6.79\text{x}10^{-2}$ mol, and T = 298 K, so that

$$V = \frac{nRT}{P} = \frac{(6.79\text{x}10^{-2} \text{ mol})(0.0821 \text{ atm L mol}^{-1} \text{ K}^{-1})(298 \text{ K})}{(730 \text{ torr})(\frac{1 \text{ atm}}{760 \text{ torr}})}$$

$$= \underline{1.73 \text{ L}}$$

The number of moles of LiOH resulting from the reaction is the same as the moles of $H_2(g)$

$$\text{Molar concentration LiOH} = \left(\frac{6.79\times10^{-2}\ \text{mol LiOH}}{50.5\ \text{mL}}\right)\left(\frac{1000\ \text{mL}}{1\ \text{L}}\right) = \underline{1.34\ \text{M}}$$

Finally, the volume of dry $H_2(g)$ at STP is given by

$$\frac{P_1V_1}{T_1} = \frac{P_2V_2}{T_2}, \quad \text{or } V_2 = \frac{P_1V_1T_2}{T_1P_2} = \frac{(730\ \text{torr})(1.73\ \text{L})(273\ \text{K})}{(298\ \text{K})(760\ \text{torr})}$$

$$= \underline{1.52\ \text{L}}$$

16. The reaction is $2Al(s) + 3H_2SO_4(aq) \longrightarrow Al_2(SO_4)_3(aq) + 3H_2(g)$

$$\text{mol } H_2 = (0.1022\ \text{g Al})\left(\frac{1\ \text{mol Al}}{26.98\ \text{g Al}}\right)\left(\frac{3\ \text{mol } H_2}{2\ \text{mol Al}}\right) = \underline{5.682\times10^{-3}\ \text{mol}}$$

The total pressure of 740 mm = $P_{H_2} + P_{H_2O}$, and $P_{H_2O}$ at 27°C is (by interpolation from Table 12.4) 30 mm Hg. Thus, $P_{H_2}$ = 710 mm Hg

$$V_{H_2} = \frac{nRT}{P} = \frac{(5.682\times10^{-3}\ \text{mol})(0.0821\ \text{atm L mol}^{-1}\ \text{K}^{-1})(300\ \text{K})}{(710\ \text{mm Hg})\left(\frac{1\ \text{atm}}{760\ \text{mm Hg}}\right)}$$

$$= \underline{0.150\ \text{L}}$$

17. On the basis of the information given, boron must be a covalent network solid. The lack of electric conductivity indicates that it is not a metal and since only one kind of atom is present it cannot be an ionic solid. The fact that it resembles diamond in its hardness and its very high melting point is consistent with a covalent network structure. It is not likely to be soluble in water because this would involve the breaking of many covalent bonds.

18. The properties of $SnCl_4$ are consistent with its formulation as a low melting molecular solid, composed of $AX_4$ tetrahedral $SnCl_4$ molecules which attract each other only by weak intermolecular forces. Although $SnCl_2$ contains fewer atoms than $SnCl_4$, its melting point is almost 300°C higher than that of $SnCl_4$, which suggests an ionic structure in which $Sn^{2+}$ ions and $Cl^-$ ions are held together by ionic bonds. The relatively low melting point for an ionic solid is due to the relatively large sizes of the ions. Recollect that a similar difference between the structures of $PbCl_4$ (covalent) and $PbCl_2$ (ionic) was described in Chapter 10. (Although covalent $SnCl_4$ would be expected to be insoluble in water, it would also be expected to react with it, as do other covalent halides such as $BCl_3$ and $SiCl_4$, to give a solution of stannic acid, $Sn(OH)_4$. In contrast, $SnCl_2$ would be expected to be soluble and dissolve to give a solution containing $Sn^{2+}(aq)$ and $Cl^-(aq)$ ions.)

19. (a) <u>Nitrogen</u> consists of nonpolar $N_2$ molecules and N atoms have a low polarizability. Thus, $N_2(s)$ would be expected to be a low melting molecular solid with only weak intermolecular (London) forces.
(b) <u>Hydrogen sulfide</u> consists of slightly polar $H_2S$ molecules although, because of the relatively low electronegativity of sulfur, it would not be expected to form hydrogen bonds. Thus, it would be expected to be a low melting molecular solid with weak London forces.
(c) <u>Chromium</u> consists of metal atoms and would be expected to be a metallic network solid with metallic bonds.
(d) <u>Calcium oxide</u> is composed of $Ca^{2+}$ and $O^{2-}$ ions. Thus, it would be

expected to be an ionic network solid with ionic bonds.

(e) Silane, $SiH_4$, is a covalent hydride. It would be expected to form a covalent molecular solid with only weak intermolecular (London) forces between the $SiH_4$ molecules.

(f) Silica, $SiO_2$, is a covalent network solid with strong covalent bonds between its Si and O atoms which are arranged so that each pair of Si atoms are linked by two Si-O bonds and each Si is surrounded by a tetrahedral arrangement of O atoms.

(g) Potassium hydroxide is an ionic solid composed of an infinite array of $K^+$ and $OH^-$ ions bonded by ionic bonds.

(h) Sulfuric acid, $H_2SO_4$, is composed of covalent $(HO)_2SO_2$ molecules and the solid is a covalent solid. The major intermolecular forces are O-H... O hydrogen bonds.

20. (a) BrF should have the higher boiling point. The electronegativity difference for Br and F is greater than that for Cl and F and BrF is more polar than ClF. The Br atom in BrF is more polarizable than the Cl atom in ClF. Both the dipole-dipole and the London forces are greater for BrF than for ClF.

(b) BrCl should have the higher boiling point. Not only is Br more polarizable than Cl, so that the London forces between BrCl molecules should be stronger than those between $Cl_2$, but BrCl will have a small dipole moment whereas the $Cl_2$ molecule is nonpolar.

(c) KBr will have the higher b.p. Its $K^+$ and $Br^-$ ions are held by strong ionic forces, whereas BrCl will form a molecular solid with only weak London forces and dipole-dipole interactions between its molecules

(d) Potassium would be expected to have the higher boiling point. In order to vaporize potassium the metallic bonds have to be broken, while to vaporize $Br_2$ only the weak London forces between $Br_2$ molecules have to be overcome.

21. Chlorine is larger & more polarizable than the very small H atom. Thus, replacement of a H atom in benzene by a Cl atom, to give chlorobenzene, increases the polarizability of the molecule and hence the strength of the intermolecular forces (London forces) between the molecules. The effect is even more pronounced when all of the H atoms in benzene are replaced by Cl atoms, to give hexachlorobenzene, which therefore has a higher boiling point than chlorobenzene, and a much higher boiling point than benzene.

22. (a) The most important intermolecular forces between the $I_2$ molecules in solid iodine are London forces between the relatively large and highly polarizable iodine molecules.

(b) Calcium oxide is an ionic solid composed of $Ca^{2+}$ and $O^{2-}$ ions. Thus, the most important interactions in this ionic network solid are the electrostatic forces between the ions.

(c) Carbon dioxide is a covalent molecule. In the gas there will be only very weak London forces between the nonpolar molecules.

(d) Methylchloride, $CH_3Cl$, consists of slightly polar covalent molecules but the most important intermolecular attractions in the liquid will be the London forces between the molecules.

(e) Liquid hydrogen fluoride consists of HF molecules that are bonded together through hydrogen bonding.

23. LiF and $BeF_2$ contain a metal and a nonmetal and are therefore ionic substances, with strong ionic bonds, which accounts for their very high

boiling points. The remainder of the fluorides of the second period elements are small covalent molecules. In order of increasing boiling points, $F_2 < OF_2 < CF_4 < NF_3 < BF_3$. The boiling points for these covalent substances are expected to depend primarily on their polarizabilities, (which is a function of their numbers of atoms and the atom polarizabilities), their shapes (planar molecules pack with more atoms in closer proximity than do trigonal pyramidal or tetrahedral molecules), and the magnitude of their dipole moments. Atom polarizability decreases in any period with increasing core charge (e.g., from B to F in the second period), and of the molecules in question, only $OF_2$ and $NF_3$ have dipole moments. Thus, in terms of the numbers of atoms and their polarizabilities, the expected increase in boiling pints is $F_2 < OF_2 < NF_3 < BF_3 < CF_4$. However, although C is more polarizable than N, because of its tetrahedral shape, only the F atoms of $CF_4$ interact closely, so that the b.p. of $CF_4$ is actually lower than that of trigonal pyramidal $NF_3$, which is also a consequence of the polarity of $NF_3$ versus that of nonpolar $CF_4$. Thus, the boiling points of the second period covalent fluorides increase in the order $F_2 < OF_2 < CF_4 < NF_3 < BF_3$.

24. Gasoline is a mixture of nonpolar hydrocarbon molecules with six to nine carbon atoms and with the general formula $C_nH_{2n+2}$. Thus, the only intermolecular forces between gasoline molecules are London forces. In contrast, water molecules are polar and also form strong hydrogen bonds between molecules. Among the intermolecular forces between water molecules the hydrogen bonding dominates and accounts for its unusual properties. Thus, gasoline molecules interact through a relatively large number of very weak forces while water molecules form strong hydrogen bonds, so that at any given temperature the number of water molecules with sufficient energy to escape into the vapor phase will be very small compared to the number of gasoline molecules that can escape. Thus, the concentration of gaseous water molecules will be very small compared to the concentration of gasoline molecules under similar conditions and water will exert a much smaller vapor pressure than gasoline.

25. (a) $\underline{H_2O}$ will have the higher enthalpy of vaporization. For water to boil relatively strong hydrogen bonds have to be broken, whereas in the vaporization of $Cl_2O$ the strongest intermolecular forces that have to be overcome are the London forces between the molecules.

(b) $\underline{CBr_4}$ will have the higher enthalpy of vaporization. Both $CCl_4$ and $CBr_4$ are nonpolar tetrahedral molecules and the dominant forces are the London forces between the Cl and Br atoms, respectively. Since Br is a larger atom than Cl, and is more polarizable, the London forces will be stronger in $CBr_4$ than they are in $CCl_4$.

(c) Ar will have the higher enthalpy of vaporization. Both substances consist of atoms but Ar is a larger atom than He and therefore more polarizable, so that the London forces between Ar atoms in the liquid will be much greater than those between He atoms in liquid He.

26. Of the properties listed, (a), (b), (d), and (e) depend on the strength of the intermolecular forces.

(a) For a liquid to boil the intermolecular forces have to be overcome so that a vapor (gas) can be formed, in which the molecules are far apart and have only very weak intermolecular attractions.

(b) When a liquid vaporizes the intermolecular forces between the molecules in the liquid have to be overcome for the gas to be formed. Thus the enthalpy of vaporization is expected to be closely related to the strength of the intermolecular forces in the liquid.
(c) Molar mass depends only on the masses of the atoms of which a substance is composed and not on the nature and strength of any intermolecular forces.
(d) One factor that affects the solubility of a substance is the difference between the magnitude of the intermolecular forces between the solute molecules, those between solute molecules and solvent molecules, and those between solvent molecules.
(e) Viscosity is a measure of the ease with which a liquid flows; in other words it depends on how easily the molecules of a liquid can slide over each other, which depends on the magnitude of the intermolecular forces.
(f) The covalent radius of an atom is related to the lengths of the covalent bonds that it forms in molecules, which could only be affected in a very minor way by any intermolecular forces. The fact that the covalent radius of an atom is almost constant in all of the covalent molecules that it forms shows that the influence of the intermolecular forces is negligible.
(g) The electronegativity of an atom in a molecule depends primarily on its size and core charge and is unaffected in any important way by any intermolecular forces between molecules.
(h) Bond energy is related to the strength of a covalent bond and is unaffected by any intermolecular forces. Note also that in any event bond energies are measured for molecules in the gas phase, where intermolecular forces are negligible.

27. (a) HCl(g) is a polar molecule and will be soluble in water which is composed of polar $H_2O$ molecules. Not only is it very soluble but it also transfers a proton to $H_2O$ to give $H_3O^+$ and $Cl^-$ in solution. In contrast, it has a low solubility in nonpolar pentane, $C_5H_{12}(l)$, which is incapable of forming strong intermolecular interactions with polar HCl molecules.
(b) Water is very soluble in liquid HF because it can interact with it to form strong hydrogen bonds, and because HF is a stronger acid than $H_2O$ it reacts to give $H_3O^+$ and $F^-$. In contrast, the intermolecular attractions between $H_2O$ molecules and the hydrocarbons of which gasoline is composed will be very small, because hydrogen bonds are not formed and polar water molecules interact only weakly with nonpolar molecules of gasoline.
(c) Chloroform is insoluble in water because its slightly polar molecules cannot form hydrogen bonds with water molecules to replace those between water molecules that would have to be broken in order for it to dissolve. However, it is very soluble in tetrachloromethane because the intermolecular (London) forces are similar in both substances.
(d) Naphthalene is an aromatic hydrocarbon consisting of nonpolar molecules and is insoluble in water because its nonpolar molecules are incapable of forming strong intermolecular forces with polar water molecules. However, it is soluble in nonpolar benzene because of the similarity of the London forces in naphthalene, and in benzene, and in a solution of the two substances.

(e) Nitrogen, $N_2(g)$, is negligibly soluble in water since it is a small nonpolar molecule of low polarizability and incapable of interacting sufficiently strongly with the hydrogen bonded water to disrupt the water structure. In contrast, HCN(g) is composed of polar molecules which can interact strongly with polar $H_2O$ molecules and, moreover, because HCN is a weak carbon acid it forms hydrogen bonds with water molecules and is slightly ionized to $H_3O^+$ and $CN^-$ ions.

(f) Nonpolar benzene is soluble in nonpolar toluene because of the similarity of the London forces of attraction between the molecules. It cannot however interact strongly with the hydrogen bonded molecules of water and is insoluble in this solvent.

28. (a) Solid argon contains argon atoms held together by weak London forces.

(b) $Cl_2(g)$ molecules interact only very weakly as the result of London forces.

(c) In molten LiF the strongest intermolecular forces are the ionic forces (bonds) between $Li^+$ and $F^-$ ions.

(d) $AlCl_3.6H_2O$ is composed of $Al(H_2O)_6{}^{3+}$ ions and $Cl^-$ ions that attract each other by ionic forces and the water molecules are held to the $Al^{3+}$ ion by polar covalent bonds.

(e) In liquid methanol, $CH_3OH$, the strongest intermolecular forces are the hydrogen bonds formed between the O-H groups of the $CH_3OH$ molecules.

(f) Liquid $H_2SO_4$ has strong hydrogen bonds between the two O-H groups of each molecule and two oxygen atoms on adjacent molecules.

(g) HCl(g) molecules interact only very weakly through London forces and dipole-dipole attractions.

(h) $C_2F_6(g)$ molecules are attracted only by very weak London forces.

29. (a) $CO_2$ is a nonpolar molecule; the most important intermolecular forces are induced dipole-induced dipole (London) forces which are very weak at room temperature, so that $CO_2$ is a gas under these conditions.

(b) Water is a polar molecule that can form hydrogen bonds. At room temperature the most important intermolecular forces are these hydrogen bonds, and since each $H_2O$ molecule can participate in a maximum of four hydrogen bonds, water is a liquid at room temperature.

(c) HF is a polar molecule that can form a maximum of two hydrogen bonds per molecule, and these are the dominant intermolecular forces so that HF forms a liquid at a temperature just below normal room temperature.

(d) Iodine consists of nonpolar $I_2$ molecules and the most important intermolecular forces are London forces, which are relatively strong because the iodine atom is a relatively large atom with a high polarizability, so that $I_2$ forms a solid at room temperature.

(e) ICl is a polar molecule and the most important intermolecular forces are dipole-dipole forces and London forces. Although Cl is not as polarizable as I, the London forces are relatively strong and ICl is a solid at room temperature (m.p. $27^{\circ}C$).

(f) Helium is composed of He atoms that have a very small polarizability, so that the only important intermolecular forces are very weak London forces and He is a gas at room temperature.

30. (a) Benzene, $C_6H_6$, consists of planar nonpolar molecules that attract each other in the solid by London forces between a relatively large number of atoms. These London forces have to be overcome to some extent in order for solid benzene to melt.
(b) Ethanol, $C_2H_5OH$, molecules are associated in the solid through hydrogen bonds, some of which have to be broken when solid ethanol melts.
(c) Ethane, $C_2H_6$, molecules are attracted by only very weak London forces, so that ethane is a solid only at very low temperatures. For ethane to melt these weak London forces have to be overcome.
(d) Barium oxide consists of $Ba^{2+}$ and $O^{2-}$ ions and contains strong ionic bonds, some of which have to be broken in order for the solid to melt.
(e) Chlorine consists of nonpolar $Cl_2$ molecules that are attracted in the solid by weak London forces. When low melting $Cl_2(s)$ melts, some of these weak forces have to be overcome so that $Cl_2$ molecules can slide past each other.
(f) Hydrogen chloride consists of polar HCl molecules that attract each other most importantly by London forces and by dipole-dipole attractions, which have to be overcome in order for HCl(s) to melt.

31. (a) Both calcium and sodium are metals in which $Ca^{2+}$ and $Na^+$ ions, respectively, are held together by a cloud of delocalized electrons. Since the metallic bonds in calcium consist of two electrons per atom, and those in metallic sodium consist of only one electron per atom, and the $Ca^{2+}$ ion is much smaller than the $Na^+$ ion, calcium has a much higher boiling point than sodium.
(b) Both silane, $SiH_4$, and methane, $CH_4$, are tetrahedral $AX_4$ nonpolar molecules and the dominant intermolecular forces in the low boiling liquids are London forces. Since $SiH_4$ is a larger and more polarizable molecule than $CH_4$, $SiH_4$ has the higher boiling point.
(c) Ethane, $C_2H_6$, and methane, $CH_4$, are both nonpolar molecules and the intermolecular forces in the liquids are London forces. Because ethane has more atoms than $CH_4$ it has the higher boiling point.
(d) Ammonia, $NH_3$, and phosphine, $PH_3$, are both $AX_3E$ polar molecules with dipole moments. Although $PH_3$ is a larger molecule than $NH_3$, and thus more polarizable than $NH_3$, the dominant intermolecular attractions in liquid $NH_3$ are hydrogen bonds, which do not form between $PH_3$ molecules, and $NH_3$ has the higher boiling point.
(e) Both $F_2$ and $Cl_2$ consist of nonpolar diatomic molecules. The dominant intermolecular forces in the liquids are London forces but since Cl is a larger and more polarizable atom than F, $Cl_2(l)$ has the stronger London forces and has a higher boiling point than $F_2(l)$.
(f) Liquid $SO_2$ consists of discrete polar $SO_2$ molecules while much of the infinite covalent network structure of $SiO_2(s)$ must be retained in $SiO_2(l)$. For $SO_2(l)$ to boil the principal intermolecular forces that have to be overcome are London forces and dipole-dipole forces, but for $SiO_2(l)$ to boil a large number of Si-O covalent bonds have to be broken, so that $SiO_2(l)$ has a much higher boiling point than $SO_2(l)$.

32. For the diatomic molecules, the ones containing atoms of different electronegativities, CO and HBr, are polar and have dipole-dipole interactions between their molecules. Of the triatomic molecules, $H_2S$ with $AX_2E_2$ angular geometry and $SO_2$ with $AX_2E$ angular geometry have dipole moments and exhibit dipole-dipole interactions. Of the tetrahedral molecules, only $CH_2Cl_2$ has a dipole moment, because of the different polarities of the C-H and the C-Cl bonds, and its molecules interact via dipole-dipole interactions. The remaining molecules - $Cl_2$, $CCl_4$, and $CH_4$ are all nonpolar and interact only through London forces. Thus, the substances with dipole-dipole interactions between their molecules are:

$$CO \quad H_2S \quad CH_2Cl_2 \quad HBr \quad SO_2$$

33. (a) Hydrogen peroxide, H-O-O-H, forms hydrogen bonds with water and is miscible with water in all proportions, while nonpolar benzene is insoluble in polar water. A better solvent for benzene is some other nonpolar liquid hydrocarbon, such as toluene, hexane, or cyclohexane.
   (b) Ethane diol, $HOCH_2CH_2OH$, forms hydrogen bonds with water and is miscible in all proportions, while nonpolar ethane is insoluble. A better solvent for ethane is any nonpolar liquid hydrocarbon.
   (c) Sugar is a carbohydrate containing -OH groups that form hydrogen bonds with water, in which it is quite soluble. Nonpolar hexane, $C_6H_{14}(l)$, is insoluble. A better solvent for hexane is another nonpolar liquid hydrocarbon.
   (d) Magnesium chloride is appreciably soluble in water because it dissolves to form hydrated $Mg(H_2O)_6^{2+}$ and $Cl^-$ ions. Slightly polar chloroform, $CHCl_3$, has a very small solubility in water. A better solvent for slightly polar $CHCl_3$ is a solvent such as $CCl_4$ or a liquid hydrocarbon such as hexane.
   (e) Hydrogen iodide, HI(g), is a polar molecule that forms strong hydrogen bonds with water, in which it dissolves as a strong acid to give $H_3O^+(aq)$ and $I^-(aq)$ ions. Iodine consists of nonpolar $I_2$ molecules which are insoluble in polar water. A better solvent for iodine is a nonpolar liquid, such as tetrachloromethane, $CCl_4$.
   (f) Lithium chloride, LiCl(s), is very soluble in water due to the formation of highly solvated $Li^+(aq)$ and $Cl^-(aq)$ ions in solution. Nonpolar $CCl_4$ is very insoluble in polar water. A better solvent for $CCl_4$ is another nonpolar liquid, such as chloroform or hexane.
   (g) Methanol, $CH_3OH$, is miscible in all proportions because it forms hydrogen bonds with water. Ethane is insoluble because it is a nonpolar molecule and is incapable of forming hydrogen bonds with water. A better solvent for ethane is another nonpolar liquid hydrocarbon, such as hexane.

34. Deviations from ideality are due to the fact that ideal behaviour assumes that gas molecules have negligible volumes and that there are no intermolecular interactions between them. Thus, the greater deviations from ideal behaviour are expected for the larger molecules particularly when their intermolecular attractions are significant.
   (a) $Cl_2 > F_2$, both on account of its larger size and its greater polarizability.
   (b) $BrCl > Cl_2$, on account of its larger size, the greater polarizability of Br compared to that of Cl, and the fact that BrCl is a polar molecule while $Cl_2$ is nonpolar.

(c) $CF_4 > CH_4$, on account of its larger size compared to that of $CH_4$ and its greater polarizability.

(d) $SO_2 > CO_2$, on account of its greater size, its greater polarizability and the fact that it is a polar molecule while $CO_2$ is nonpolar.

35. (a) Carbon readily forms multiple bonds with the second period element oxygen and $CO_2$ consists of small linear $AX_2$ nonpolar molecules and is a gas under ordinary conditions. At low temperature it forms a molecular solid in which the only intermolecular forces are weak London forces between $CO_2$ molecules. In contrast, silicon is incapable of forming double bonds with oxygen and silica, empirical formula $SiO_2$, is a covalent network solid or giant polymeric molecule with each Si atom bonded in an $AX_4$ tetrahedral arrangement to four O atoms, to which it is bonded by covalent bonds.

(b) All the halogens form nonpolar diatomic $X_2$ molecules that attract each other in the solids by London forces and they are molecular covalent solids. The melting points increase as the size and the polarizability of X increases, so that under ordinary conditions $F_2$ and $Cl_2$ are gases, $Br_2$ is a liquid, and $I_2$ is a solid.

(c) Salts are soluble in water primarily when their ions are capable of strong solvation by polar water molecules. In contrast, nonpolar - tetrachloromethane is incapable of solvating any ion, and no ionic solid is soluble in this solvent.

(d) In both pairs of these hydrides, $H_2O$ and $H_2S$, and $NH_3$ and $PH_3$, the significant difference in melting points is due to the capacity of the first member of each pair to form intermoelcular hydrogen bonds, which is not possible for the hydrides of the third period elements. $H_2O$ has a more anomalous behaviour than $NH_3$ because it can form four hydrogen bonds per $H_2O$ molecule to other $H_2O$ molecules, while $NH_3$, with only a single lone pair on nitrogen, can form only two hydrogen bonds per $NH_3$ molecule.

36. Tetrachloromethane, $CCl_4$, is an $AX_4$ tetrahedral molecule whose physical properties are largely determined by four Cl atoms of relatively high polarizability. Chloroform, $CHCl_3$, is also a tetrahedral molecule. In contrast to nonpolar $CCl_4$ it has a small dipole moment. Nevertheless it boils at a lower temperature than $CCl_4$ because it contains only three Cl atoms and a H atom of low polarizability. As is the case with many other substances, it is the London forces that are dominant in determining physical properties - such as boiling point.

37. (a) For a molecule to form intermolecular hydrogen bonds it has to have highly polar X-H bonds, which is the case when X is a highly electronegative second period element, such as N, O, or F. In addition the atom X must have highly localized lone pairs with which the H atoms of other molecules can interact to form hydrogen bonds, which is also the case when X is a second period element, such as N, O, or F.

(b) (i) Ammonia, $:NH_3$, has a highly electronegative N atom with one highly localized lone pair. Its polar N-H bonds can form hydrogen bonds with the nitrogen lone pair of another $NH_3$ molecule.

(ii) Sodium chloride contains no hydrogen and forms an ionic melt, $Na^+Cl^-(l)$ in which there is no possibility of hydrogen bond formation.

(iii) Hydrogen fluoride, H-F:, satisfies all the criteria for the formation of hydrogen bonds and forms zig-zag chains of hydrogen bonded molecules in the liquid.

(iv) Hydrogen, $H_2$, has no lone electron pairs and cannot form hydrogen bonds.

(v) Methane has no lone electron pairs and cannot form hydrogen bonds.

(vi) Lithium hydride, $Li^+$ $:H^-$, is an ionic solid and cannot form hydrogen bonds.

(vii) Methanol, $H_3C$-OH, contains a highly polar O-H bond and the O atom has two highly localized lone electron pairs. It readily forms hydrogen bonds in the liquid.

(viii) Acetic acid, $H_3C$-C(O)OH, contains a highly polar O-H bond and has two O atoms with highly localized lone pairs. It is strongly hydrogen bonded even in the gas phase, where it forms hydrogen bonded dimers.

38. The stronger are the forces between the molecules or ions of a substance the higher is its expected boiling point. The forces operating between the substances in question are ion-ion forces (ionic bonds), hydrogen bonds, dipole-dipole forces, and London forces which operate between all molecules and depend on the number and the polarizability of the atoms in the molecule. The strongest of these forces are the ion-ion forces, followed by hydrogen bonds. On this basis, the expected order of increase of boiling point is

$$Ne < Ar < O_2 < HCl < Cl_2 < HF < H_2O < NaCl$$

The boiling points of the first five substances are related to their relative polarizabilities, which are expected to increase in the order indicated. Both HF and $H_2O$ are hydrogen bonded in their liquids but HF can form only two hydrogen bonds per molecule, while $H_2O$ can form four per molecule. NaCl is the only ionic substance among the group and will have the highest boiling point. Note that although the HCl molecule is polar and therefore there are dipole-dipole forces and London forces between HCl molecules, there are two polarizable Cl atoms in $Cl_2$ and because London forces between such relatively large atoms are rather strong, $Cl_2$ is expected to have a higher boiling point than HCl, despite the fact that the latter is a polar molecule.

39. (a) The higher boiling point of ethanol compared to that of dimethyl ether, is due to the hydrogen bonding between molecules of the former, which are not possible for the latter.

(b) The higher boiling point of HF is due to its hydrogen bonded structure, which is not present in HCl(l).

(c) LiCl >> $CCl_4$ because LiCl is an ionic melt whereas $CCl_4$ is a molecular liquid.

(d) LiCl >> HCl because LiCl is an ionic melt while HCl(l) is a molecular liquid.

40. (a) Polarity refers to the nature of a covalent bond. A covalent bond X-Y is polar when X and Y have different electronegativities, which gives a partial negative charge on the more electronegative atom and a partial positive charge on the less electronegative atom. Polarizability is a measure of the ease with which the electrons of an atom can be deformed (polarized), which decreases with increasing core charge of an atom and increases as its size increases.

(b) London (dispersion) forces are a consequence of the instantaneous dipoles (induced dipoles) that arise when nonbonded atoms approach each other closely, while dipole-dipole forces arise from the interaction of the permanent dipoles of polar molecules. London forces are also called induced dipole-induced dipole forces.

(c) The van der Waals radius of an atom is one-half of the distance between identical nonbonded atoms that touch each other in a covalent molecular solid, while the covalent radius is one-half the bond length between identical atoms forming a covalent bond. To a good approximation, the distance between nonbonded atoms in molecular solids is given by the sum of the van der Waals radii of the atoms, and bond lengths in covalent molecules are given by the sum of the covalent radii of the atoms forming the bond.

(d) A polar covalent bond is an electron pair bond formed between two atoms with different electronegativities. A hydrogen bond is the strong attraction between the H atom of an X-H bond (X = N, O, or F) and a lone pair of another of the same highly electronegative atoms.

41. The smaller density of ice at the freezing point of water compared to that of water at the same temperature is a consequence of the different extents to which $H_2O$ molecules form hydrogen bonds in the solid and the liquid. Ice has an open cage-like structure in which each O atom is at the center of an approximately tetrahedral arrangement of four H atoms, two of which form O-H covalent bonds with the oxygen and the other two are on other water molecules and form hydrogen bonds with each of the two lone pairs of the oxygen. When ice melts, some of the hydrogen bonds between water molecules are broken and the ice structure partially collapses to give an increase in density, because now more $H_2O$ molecules are packed into the same volume. In contrast, the intermolecular (London) forces between $Br_2$ molecules have their maximum strength in solid bromine and hold the $Br_2$ molecules in a regular array in the solid. When bromine melts, the intermolecular forces decrease to allow $Br_2$ molecules to slide past each other in liquid and the molecules are farther apart in the liquid than they are in the solid. There is a smaller number of molecules in a given volume of the liquid at the melting point than there are in the solid at the melting point. Hence, the density of liquid $Br_2$ is less than that of solid $Br_2$ at the melting point.

42. (a) The kinetic molecular theory of gases defines an ideal gas as having molecules of negligible volume compared to the volume in which they move, and having no intermolecular forces between them, so that molecular collisions are elastic. In real gases, neither assumption is exactly true because the molecules take up part of the volume in which they are contained and the intermolecular attractions are not entirely negligible. These effects are small at low pressures and high temperatures when the volume of the molecules is small compared to the volume that they occupy and the intermolecular forces are very small because on average the molecules are far apart and have large kinetic energies. As the pressure is increased the molecules are forced closer together as the gas density increases and the intermolecular forces increase in magnitude. The observed pressure becomes less than the ideal pressure, since the molecules now have a tendency to stick to each other, which slows them down, particularly at low temperature where their average kinetic energies are not very large. Also under these conditions the actual volume of the molecules is no longer negligible compared to the total volume and the measured volume is

greater than the actual volume in which the molecules are free to move. Both factors contribute to the nonideal behaviour of real gases.
(b) (i). Bromine is expected to be more nonideal than $F_2$ both on account of its greater molar volume and stronger intermolecular forces due to the greater polarizability of $Br_2$ molecules in comparison to that of $F_2$ molecules.
(ii) CO and $N_2$ are expected to have rather similar molar volumes and polarizabilities but CO is a polar molecule while $N_2$ is nonpolar. Thus CO is expected to show the greater deviation from ideal behaviour because of dipole-dipole attractions between its molecules.
(iii) Both molecules are polar are should not have too dissimilar molar volumes, but acetic acid is capable of forming intermolecular hydrogen bonds, which will make it more nonideal than acetyl chloride.

43. In solving problems concerned with colligative properties such as boiling point elevation and freezing point depression, the appropriate unit of concentration is <u>molality</u>. For a boiling point elevation $\Delta T$,

$$\Delta T = K_b m$$

where $K_b$ is the boiling point elevation constant ($0.52^\circ C\ kg\ mol^{-1}$ for water) and $\underline{m}$ is the molal concentration of solute species (moles per kg of <u>solvent</u>). Thus,

$$\Delta T = (101.0-100.0) = 1.0^\circ C = (0.52^\circ C\ kg\ mol^{-1})m$$

$$m = \left(\frac{1.00^\circ C}{0.52^\circ C\ kg\ mol^{-1}}\right) = \underline{1.92\ mol\ kg^{-1}}$$

Assuming that the density of water is $1.00\ g\ mL^{-1}$,

$$\text{mass of sugar} = \left(\frac{1.92\ mol}{1\ kg}\right)(250\ mL)\left(\frac{1\ g}{1\ mL}\right)\left(\frac{1\ kg}{1000\ g}\right)\left(\frac{342.3\ g\ sugar}{1\ mol\ sugar}\right)$$

$$= \underline{164\ g\ sugar}$$

44. The method is given in Problem 43.

$$\Delta T = (78.89-78.41) = 0.48^\circ C$$

$$m = \frac{\Delta T}{K_b} = \left(\frac{0.48^\circ C}{1.19^\circ C\ kg\ mol^{-1}}\right) = 0.40\ mol\ kg^{-1}$$

$$\text{i.e., mol of mothballs in 100.0 g ethanol} = \left(\frac{0.40\ mol}{1\ kg}\right)(100.0\ g)\left(\frac{1\ kg}{1000\ g}\right)$$

$$= \underline{0.040\ mol}$$

$$\text{Thus, molar mass of mothballs} = \left(\frac{5.00\ g}{0.040\ mol}\right) = \underline{125\ g\ mol^{-1}}$$

Comparing the experimentally determined molar mass with the actual molar masses of camphor, $C_{10}H_{16}O$, of $152.2\ g\ mol^{-1}$, and naphthalene, $C_{10}H_8$, of $128.2\ g\ mol^{-1}$, shows clearly that these mothballs are composed of <u>naphthalene</u>.

45. From the boiling point elevation data calculate the molality of urea, and hence its molar mass,

$$m = \frac{\Delta T}{K_b} = \left(\frac{0.65^\circ C}{0.52^\circ C\ kg\ mol^{-1}}\right) = 1.25\ mol\ kg^{-1}$$

mol of urea in 17.0 g $H_2O$ = $(\frac{1.25 \text{ mol}}{1 \text{ kg}})(17.0 \text{ g})(\frac{1 \text{ kg}}{1000 \text{ g}})$ = <u>0.0213 mol</u>

molar mass of urea = $(\frac{1.25 \text{ g urea}}{0.0213 \text{ mol urea}})$ = <u>58.7 g mol$^{-1}$</u>

From the mass % composition, calculate the empirical formula:

| | | C | N | H | O |
|---|---|---|---|---|---|
| Grams in 100 g urea | = | 20.0 | 46.7 | 6.7 | 26.6 |
| Mol in 100 g urea | = | $\frac{20.0}{12.01}$ | $\frac{46.7}{14.01}$ | $\frac{6.7}{1.01}$ | $\frac{26.6}{16.00}$ |
| Ratio of mol (atoms) | = | 1.67 : | 3.33 : | 6.63 : | 1.66 |
| | = | $\frac{1.67}{1.66}$ : | $\frac{3.33}{1.66}$ : | $\frac{6.63}{1.66}$ : | $\frac{1.66}{1.66}$ |
| | = | 1.01 : | 2.01 : | 3.99 : | 1.00 |

and the <u>empirical formula</u> of urea is $CN_2H_4O$ (formula mass 60.1 u) and from the molar mass above, the <u>molecular formula</u> is also $CN_2H_4O$. Assuming that carbon is the central atom in the molecule and that urea is a covalent substance (since it dissolves in organic solvents), then the only likely Lewis structure, remembering that carbon should have a valence of 4, is

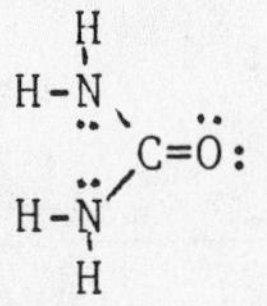

For <u>ammonium cyanate</u>, with the empirical formula $CN_2H_4O$, one component ion is $NH_4^+$, and the other must be $CNO^-$.

<u>$CNO^-$</u> has 16 valence electrons, or eight pairs, and all of its atoms must obey the octet rule. Hence, the following Lewis structures can be written:

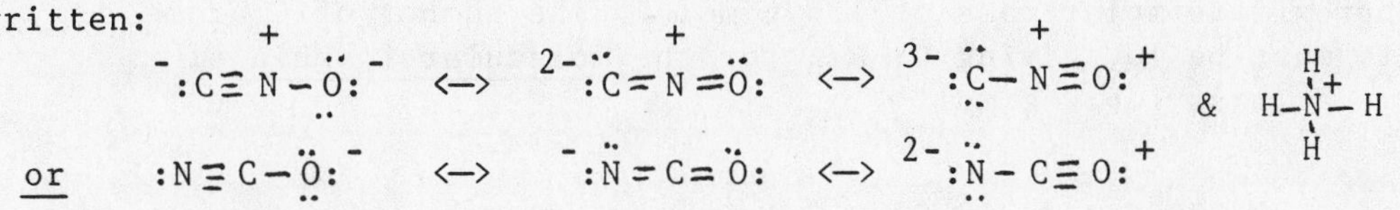

of which $:N\equiv C-\ddot{O}:^-$ <-> $^-:\ddot{N}=C=\ddot{O}:$, are the most probable resonance structures in terms of the distribution of formal charges.

Ammonium cyanate, $NH_4^+NCO^-$ would give an aqueous solution containing $NH_4^+(aq)$ and $NCO^-(aq)$, so that a given molal concentration of this salt would give <u>twice</u> the boiling point elevation of the same concentration of urea.

46. $m = \frac{\Delta T}{K_b} = (\frac{0.105^\circ C}{2.10^\circ C \text{ kg mol}^{-1}}) = 0.050 \text{ mol kg}^{-1}$

and mol S = $(\frac{0.050 \text{ mol sulfur}}{1 \text{ kg ether}})(200 \text{ g ether})(\frac{1 \text{ kg}}{1000 \text{ g}})$ = <u>0.010 mol</u>

Molar mass of sulfur = $(\frac{2.60 \text{ g sulfur}}{0.010 \text{ mol sulfur}})$ = $\underline{260 \text{ g mol}^{-1}}$

The empirical formula of sulfur is S (formula mass 32.06 u).

Thus, the molecular formula of sulfur is $\underline{S_8}$ (molar mass 256.5 g $mol^{-1}$)

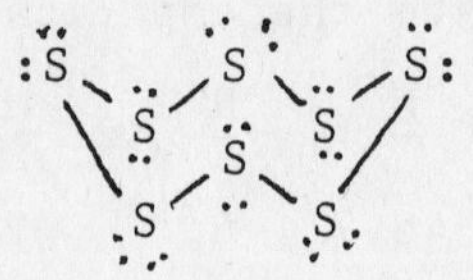

47. The vapor pressure of the solvent in a solution is proportional to the mole fraction of the solvent in the solution.

$$p = p^o X_{solvent} \quad \underline{\text{or}} \quad \frac{p}{p_o} = X_{solvent}$$

For a two component system, $X_{solvent} + X_{solute} = 1$,

i.e., $p = p^o(1 - X_{solute})$

For the solution of limonene,

$$\frac{p}{p_o} = \frac{90.6 \text{ mm Hg}}{95.2 \text{ mm Hg}} = 0.952 = (1 - X_{solute}) \; ; \; X_{solute} = \underline{0.048}$$

and moles of benzene = $(78.1 \text{ g})(\frac{1 \text{ mol}}{78.11 \text{ g}})$ = $\underline{1.00 \text{ mol}}$

If mol of limonene = n, $\frac{n}{1.00+n} = 0.048$ ; n = $\underline{0.050 \text{ mol}}$

and molar mass of limonene = $(\frac{6.80 \text{ g}}{0.050 \text{ mol}})$ = $\underline{136 \text{ g mol}^{-1}}$

From the approximate molar mass of 136 g $mol^{-1}$, the number of C atoms per molecule must be 10, giving $C_{10}H_{16}$ for the molecular formula of limonene (molar mass 136.2 g $mol^{-1}$).

48. $\Delta T = K_f m$ ; $0.099^oC = (5.12^oC \text{ kg mol}^{-1})m$

$$m = (\frac{0.099^oC}{5.12^oC \text{ kg mol}^{-1}}) = \underline{0.019 \text{ mol kg}^{-1}}$$

and moles of bromide = $(\frac{0.019 \text{ mol}}{1 \text{ kg}})(100 \text{ g})(\frac{1 \text{ kg}}{1000 \text{ g}})$ = $\underline{0.0019 \text{ mol}}$

Thus, molar mass of bromide = $(\frac{1.00 \text{ g}}{0.0019 \text{ mol}})$ = $\underline{530 \text{ g mol}^{-1}}$

Al is a group 3 element, therefore the empirical formula of the bromide is $AlBr_3$ (formula mass 266.7 u), and the experimental molar mass of 530 g $mol^{-1}$ is consistent with the molecular formula $Al_2Br_6$ (molar mass 533.4 g $mol^{-1}$) with the Lewis structure,

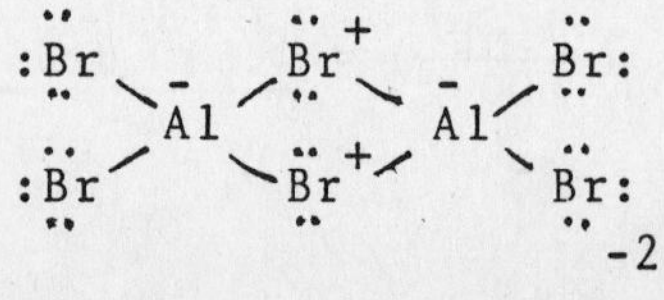

49. NaCl is fully dissociated into $Na^+(aq)$ and $Cl^-(aq)$ ions in aqueous solution,

$$NaCl(s) \longrightarrow Na^+(aq) + Cl^-(aq)$$

Therefore, the molal concentration of ions in solution is 2(5.8 m) = 11.6 m, and the approximate freezing point depression is,

$$T = iK_f m = (1.86^\circ C\ kg\ mol^{-1})(11.6\ mol\ kg^{-1}) = \underline{21.6^\circ C}$$

Salt is effective in melting ice down to a temperature of approximately $\underline{-22^\circ C}$.

50. $\Delta T = K_f m\ ;\ m = \frac{\Delta T}{K_f} = (\frac{0.48^\circ C}{1.86^\circ C\ kg\ mol^{-1}}) = \underline{0.26\ m}$

Mol sucrose in 1.00 kg solution = $(\frac{0.26\ mol}{1\ kg})(1\ kg) = \underline{0.26\ mol}$

Molar mass of sucrose = $(\frac{89.0\ g}{0.26\ mol}) = \underline{340\ g\ mol^{-1}}$

(The actual molar mass of sucrose, $C_{12}H_{22}O_{11}$) is $\underline{342.3\ g\ mol^{-1}}$)

51. Calculate the molality of ethylene glycol using,

$$\Delta T = K_f m\ ;\ m = \frac{\Delta T}{K_f} = (\frac{30^\circ C}{1.86^\circ C\ kg\ mol^{-1}}) = \underline{16\ m}$$

52. In each case calculate the molal concentrations of the solutes and use the formulas,

$$\Delta T = iK_f m \quad \underline{and} \quad \Delta T = iK_b m$$

(a) molality of urea = $(50.0\ g)(\frac{1\ mol}{60.06\ g})(\frac{1}{750\ g\ H_2O})(\frac{1000\ g}{1\ kg}) = \underline{1.11\ m}$

$$\Delta T_f = (1.86^\circ C\ kg\ mol^{-1})(1.11\ mol\ kg^{-1}) = \underline{2.06^\circ C}$$
$$\Delta T_b = (0.52^\circ C\ kg\ mol^{-1})(1.11\ mol\ kg^{-1}) = \underline{0.58^\circ C}$$

Thus the freezing point of the solution is $\underline{-2.06^\circ C}$ and the boiling point is $\underline{100.58^\circ C}$.

(b) molality of sucrose = $(\frac{17.1\ g}{500\ g\ H_2O})(\frac{1\ mol}{342.3\ g})(\frac{1000\ g}{1\ kg}) = \underline{0.100\ m}$

$$\Delta T_f = (1.86^\circ C\ kg\ mol^{-1})(0.100\ kg\ mol^{-1}) = \underline{0.186^\circ C}$$
$$\Delta T_b = (0.52^\circ C\ kg\ mol^{-1})(0.100\ mol\ kg^{-1}) = \underline{0.052^\circ C}$$

Thus the freezing point of the solution is $\underline{-0.186^\circ C}$ and the boiling point is $\underline{100.52^\circ C}$.

(c) molality of glycerol = $(\frac{25.0\ g}{100\ g\ H_2O})(\frac{1\ mol}{92.09\ g})(\frac{1000\ g}{1\ kg}) = \underline{2.71\ m}$

$$\Delta T_f = (1.86^\circ C\ kg\ mol^{-1})(2.71\ mol\ kg^{-1}) = \underline{5.04^\circ C}$$
$$\Delta T_b = (0.52^\circ C\ kg\ mol^{-1})(2.71\ mol\ kg^{-1}) = \underline{1.4^\circ C}$$

Thus the freezing point of the solution is $\underline{-5.04^\circ C}$ and the boiling point is $\underline{101.4^\circ C}$.

53. Sucrose is a nonelectrolyte and dissolves in water as sucrose molecules.

$$\Delta T = (1.86^\circ C\ kg\ mol^{-1})(0.10\ mol\ kg^{-1}) = \underline{0.19^\circ C}$$

The freezing point is $\underline{-0.19^\circ C}$.

Sodium nitrate dissociates in solution completely to give $Na^+(aq)$ and $NO_3^-(aq)$ ions,

$$NaNO_3(s) \longrightarrow Na^+(aq) + NO_3^-(aq), \quad \underline{i = 2}$$

$$\Delta T = (1.86^\circ C\ kg\ mol^{-1})[2(0.10\ mol\ kg^{-1})] = \underline{0.37^\circ C}$$

Calcium nitrate dissociates completely to give $Ca^{2+}(aq)$ and $NO_3^-(aq)$,

$$Ca(NO_3)_2(s) \longrightarrow Ca^{2+}(aq) + 2NO_3^-(aq), \quad \underline{i = 3}$$

$$\Delta T = (1.86^\circ C\ kg\ mol^{-1})[3(0.10\ mol\ kg^{-1})] = \underline{0.56^\circ C}$$

(a) $\underline{-0.19^\circ C}$; (b) $\underline{-0.37^\circ C}$; (c) $\underline{-0.56^\circ C}$.

54. The depression of freezing point is proportional to the total concentration of solute species, i.e., (m)(i).

(a) $NaCl(s) \longrightarrow Na^+(aq) + Cl^-(aq)$; m (ions) = 0.20 molal

(b) $C_6H_{12}O_6(s) \longrightarrow C_6H_{12}O_6(aq)$; m (ions) = 0.40 molal

(c) $BaCl_2(s) \longrightarrow Ba^{2+}(aq) + 2Cl^-(aq)$ m (ions) = 0.30 molal

(d) $C_{12}H_{22}O_{11}(s) \longrightarrow C_{12}H_{22}O_{11}(aq)$; m (ions) = 0.10 molal

Thus, in order of <u>decreasing</u> freezing point:

$$H_2O > \text{sucrose} > NaCl(aq) > BaCl_2(aq) > \text{urea}$$

55. (a) This equation has the same form as the ideal gas equation,

$$\pi V = nRT \quad \underline{\text{or}} \quad \pi = \frac{n}{V} RT = MRT$$

where $\pi$ is the osmotic pressure, M is the molarity of the solute species, R is the gas constant, and T is the absolute temperature.

(b) For a substance that dissolves unchanged (a nonelectrolyte),

$$\pi = MRT = (0.001\ mol\ L^{-1})(0.0821\ atm\ L\ mol^{-1}\ K^{-1})(298\ K)$$

$$= \underline{0.0245\ atm}$$

(c) For the aromatic hydrocarbon, use the osmotic pressure equation to give the moles of hydrocarbon.

$$n = \frac{\pi V}{RT} = \frac{(74.7\ mm\ Hg)(\frac{1\ atm}{760\ mm\ Hg})(25.00\ mL)(\frac{1\ L}{1000\ mL})}{(0.0821\ atm\ L\ mol^{-1}\ K^{-1})(298\ K)} = \underline{1.00\times10^{-4}\ mol}$$

$$\text{Molar mass of hydrocarbon} = (\frac{0.01279\ g}{1.00\times10^{-4}\ mol}) = \underline{128\ g\ mol^{-1}}$$

56. This problem is similar to Problem 55(c).

$$n = \frac{\pi V}{RT} = \frac{(0.427\ \text{atm})(50\ \text{mL})(\frac{1\ \text{L}}{1000\ \text{mL}})}{(0.0821\ \text{atm L mol}^{-1}\ \text{K}^{-1})(298\ \text{K})} = \underline{8.73\text{x}10^{-4}\ \text{mol}}$$

$$\text{Molar mass of insulin} = (\frac{5.00\ \text{g}}{8.73\text{x}10^{-4}\ \text{mol}}) = \underline{5.73\text{x}10^{3}\ \text{g mol}^{-1}}$$

57. This problem is similar to Problems 55 and 56. 1.0 mass % is 1.0 g in 100 g of solution (or in 100 mL of solution, assuming a density of 1.00 g $mL^{-1}$).

$$n = \frac{\pi V}{RT} = \frac{(7.22\ \text{mm Hg})(\frac{1\ \text{atm}}{760\ \text{mm Hg}})(100\ \text{mL})(\frac{1\ \text{L}}{1000\ \text{mL}})}{(0.0821\ \text{atm L mol}^{-1}\ \text{K}^{-1})(298\ \text{K})} = \underline{3.88\text{x}10^{-5}\ \text{mol}}$$

$$\text{Molar mass} = (\frac{1.0\ \text{g}}{3.88\text{x}10^{-5}\ \text{mol}}) = \underline{2.6\text{x}10^{4}\ \text{g mol}^{-1}}$$

The molar mass of the monomer, $C_{12}H_{22}O_{11}$, is 342.3 g $mol^{-1}$.

Thus, average number of monomer units per polymer molecule

$$= (\frac{2.6\text{x}10^{4}\ \text{g}}{342.3\ \text{g}}) = \underline{76}$$

58. (a) Calcium carbide, $CaC_2$ is the salt $Ca^{2+}\ ^{-}{:}C{\equiv}C{:}^{-}$, containing the anion of the very weak carbon acid H-C≡C-H, ethyne. In solution $CaC_2$ dissociates into its ions and the $C_2^{2-}$ ion behaves as a moderately strong base and is protonated to give ethyne, $C_2H_2(g)$, which bubbles off, so that the reaction goes to completion.

$$CaC_2(s) + 2H_2O(l) \longrightarrow Ca^{2+}(aq) + 2OH^-(aq) + C_2H_2(g)$$

This is an acid-base reaction that occurs at 25°C.

(b) Sulfur dioxide, $SO_2(g)$, is a nonmetal oxide and therefore behaves as an acidic oxide, reacting with water to give a solution that behaves as a solution of the weak acid $H_2SO_3(aq)$, sulfurous acid.*

$$SO_2(g) + H_2O(l) \rightleftharpoons H_2SO_3(aq)$$

$$H_2SO_3(aq) + H_2O(l) \rightleftharpoons H_3O^+(aq) + HSO_3^-(aq)$$

$$HSO_3^-(aq) + H_2O(l) \rightleftharpoons H_3O^+(aq) + SO_3^{2-}(aq)$$

weak diprotic acid

The primary reaction with water to give $H_2SO_3(aq)$ is a Lewis acid-base reaction; the ensuing reactions are Brønsted-Lowry acid-base reactions. All the reactions occur at 25°C.

(c) Sulfur trioxide, $SO_3(g)$, is an acidic nonmetal oxide, which reacts with water to give a solution of the strong diprotic acid $H_2SO_4$ at room temperature.

$$SO_3(g) + H_2O(l) \longrightarrow H_2SO_4(aq)$$

$$H_2SO_4(aq) + H_2O(l) \longrightarrow H_3O^+(aq) + HSO_4^-(aq)$$

$$HSO_4^-(aq) + H_2O(l) \rightleftharpoons H_3O^+(aq) + SO_4^{2-}(aq)$$

The first reaction is a Lewis acid-base reaction, and the ensuing stages of ionization of $H_2SO_4$ are Brønsted-Lowry acid-base reactions.

*Note that there is no direct evidence for the parent acid, but sulfite and hydrogen sulfite salts are well known.

(d) Magnesium nitride, $Mg_3N_2(s)$, is the ionic compound $(Mg^{2+})_3(N^{3-})_2$ and the nitride ion behaves as a strong base in water and reacts to give $NH_3(aq)$ at room temperature.

$$Mg_3N_2(s) + 6H_2O(l) \longrightarrow 3Mg(OH)_2(s) + 2NH_3(aq)$$

This is an acid-base reaction.

(e) Carbon monoxide, $CO(g)$, is in principle an acidic nonmetal oxide that is the anhydride of $HCO_2H$, methanoic (formic) acid but it does not react with water as an acidic oxide. At room temperature no reaction occurs with water but at high temperature, using a catalyst, steam oxidizes CO to $CO_2$ and is itself reduced to $H_2(g)$.

$$CO(g) + H_2O(g) \longrightarrow CO_2(g) + H_2(g)$$

This is an oxidation-reduction reaction.

(f) Sodium amide, $NaNH_2(s)$, is an ionic salt that dissociates in water into $Na^+(aq)$ and $NH_2^-$ ions. The latter is the conjugate base of ammonia, $NH_3$, and reacts with water as a strong base at room temperature.

$$NaNH_2(s) + H_2O(l) \longrightarrow Na^+(aq) + OH^-(aq) + NH_3(aq)$$

This is a Brønsted acid-base reaction that occurs at 25°C.

(g) White phosphorus, $P_4(s)$, is a reactive allotrope of phosphorus that ignites in oxygen or air. It is stored under water as a safety precaution. There is no reaction at 25°C.

(h) Phosphorus trichloride, $PCl_3(l)$, is a nonmetal halide that is readily hydrolyzed by water to give a solution of the weak diprotic acid $H_3PO_3$, phosphorous acid, and the strong acid HCl(aq). The reaction is initiated by the Lewis base $H_2O$ donating a pair of electrons to the P atom of $PCl_3$ (Lewis acid), and then HCl is eliminated.

$$PCl_3(l) + 6H_2O(l) \longrightarrow H_3PO_3(aq) + 3H_3O^+(aq) + 3Cl^-(aq)$$

This is a Lewis acid-base reaction that occurs at 25°C.

59. (a) Phosphorus pentachloride, $PCl_5(s)$, is a nonmetal halide that is readily hydrolyzed by water to give a solution of the weak acid $H_3PO_4$, phosphoric acid and the strong acid HCl(aq). The reaction is initiated by the Lewis base $H_2O$ donating a pair of electrons to the P atom of $PCl_5$ (Lewis acid), which is followed by the elimination of HCl(aq). The reaction occurs readily at 25°C.

$$PCl_5(s) + 9H_2O(l) \longrightarrow H_3PO_4(aq) + 5H_3O^+(aq) + 5Cl^-(aq)$$

This is a Lewis acid-base reaction.

(b) Calcium, Ca(s), is a reactive group 2 metal that reduces water to $H_2(g)$ at room temperature.

$$Ca(s) + 2H_2O(l) \longrightarrow Ca^{2+}(aq) + 2OH^-(aq) + H_2(g)$$

This is an oxidation-reduction reaction that occurs relatively slowly at 25°C.

(c) Sodium oxide, $Na_2O(s)$, is an ionic basic metal oxide that ionizes to give $Na^+(aq)$ and $O^{2-}(aq)$ ions. The latter behaves as a strong base in water.

$Na_2O(s) + H_2O(l) \longrightarrow 2Na^+(aq) + 2OH^-(aq)$

This is an acid-base reaction that occurs at room temperature.

(d) Fluorine, $F_2(g)$, is the strongest oxidizing agent among the halogens. It oxidizes water to $O_2(g)$ at room temperature.

$2F_2(g) + 2H_2O(l) \longrightarrow 4HF(aq) + O_2(g)$

This is an oxidation-reduction reaction.

(e) Chlorine, $Cl_2(g)$, is not as strong an oxidizing agent as fluorine and the analogous reaction to that in part (d) occurs only very slowly at room temperature. Rather, Cl(0) in $Cl_2$ is reduced to Cl(-1) in HCl(aq) and oxidized to Cl(+1) in HOCl(aq), hypochlorous acid.

$Cl_2(g) + H_2O(l) \longrightarrow HCl(aq) + HOCl(aq)$

This is an oxidation-reduction (disproportionation) reaction.

(f) Methane, $CH_4$, reacts with steam at high temperatures and is oxidized to CO(g). The products of this oxidation-reduction reaction are called synthesis gas. No reaction at room temperature.

$CH_4(g) + H_2O(g) \longrightarrow CO(g) + 3H_2(g)$

(g) Magnesium hydride, $MgH_2(s)$, is an ionic hydride composed of $Mg^{2+}$ and $H^-$ (hydride) ions. $H^-$ is the conjugate base of $H_2$ and behaves as a strong base in water, accepting a proton from $H_2O$ to form $H_2$.

$MgH_2(s) + 2H_2O(l) \longrightarrow Mg(OH)_2(s) + H_2(g)$

The reaction occurs at room temperature. It is an acid-base reaction, and also an oxidation-reduction reaction (because H(-1) in $H^-$ is oxidized to H(0) in $H_2$, and H(+1) in $H_2O$ is reduced to H(0) in $H_2$.

60. Water behaves as a Lewis base in any reaction in which it donates and electron pair to some other species (Lewis acid). Examples include:

(i) Hydration of metal ions: $Al^{3+} + 6H_2O \longrightarrow Al(H_2O)_6^{3+}$

(ii) Adduct formation: $BF_3 + NH_3 \longrightarrow F_3\overset{-}{B}\text{-}\overset{+}{N}H_3$

(iii) Reaction with an acidic oxide:

$SO_3 + H_2O \longrightarrow H_2SO_4$

(iv) Hydrolysis of a nonmetal chloride:

$BCl_3 + 3H_2O \longrightarrow H_3BO_3 + 3HCl$

61. (a) When water behaves as an oxidizing agent it is reduced to $H_2(g)$,

$2H_2O + 2e^- \longrightarrow H_2(g) + 2OH^-$

(b) When water behaves as a reducing agent it is oxidized to $O_2(g)$,

$2H_2O \longrightarrow 4H^+ + O_2(g) + 4e^-$

62. Water is described as an amphoteric substance because it can behave either as an acid (proton donor) or as a base (proton acceptor). This property is exemplified by its self-ionization (autoprotolysis) reaction.

$H_2O + H_2O \rightleftharpoons H_3O^+ + OH^-$

63. (a) $2Al(s) + 3H_2O(g) \longrightarrow Al_2O_3(s) + 3H_2(g)$

$H_2O$ is reduced to $H_2(g)$ and behaves as an oxidizing agent.

(b) The equation is balanced as written.
$H_2O$ protonates the $NH_2^-$ ion and behaves as an acid.

(c) $2F_2(g) + 2H_2O(l) \longrightarrow 4HF(aq) + O_2(g)$

$H_2O$ is oxidized to $O_2(g)$ and behaves as a reducing agent.

(d) $P_4O_{10}(s) + 6H_2O(l) \longrightarrow 4H_3PO_4(aq)$

$H_2O$ molecules donate electron pairs to $P_4O_{10}$; water behaves as a Lewis base.

(e) $BaO_2(s) + 2H_2O(l) \longrightarrow Ba(OH)_2(aq) + H_2O_2(aq)$

Two water molecules each donate a proton to the $O_2^{2-}$ (peroxide) ion to give $H_2O_2$ (hydrogen peroxide); water behaves as an acid.

(f) $Cl_2O_7(l) + H_2O(l) \longrightarrow 2HClO_4(aq)$

Water donates an electron pair to $Cl_2O_7$ and the strong acid $HClO_4$ is formed; water behaves as a Lewis base.

(g) $CaH_2(s) + 2H_2O(l) \longrightarrow Ca(OH)_2(aq) + 2H_2(g)$

Water donates a proton to the hydride ion, $H^-$, to give $H_2(g)$; water behaves as an acid and as an oxidizing agent.

64. (a) Any salt containing a neutral metal cation and an anion that is the conjugate base of a weak monoprotic acid gives a basic solution in water, provided that it is soluble. Examples include:

$KF(aq) + H_2O(l) \rightleftharpoons KOH(aq) + HF(aq)$

$NaCN(aq) + H_2O(l) \rightleftharpoons NaOH(aq) + HCN(aq)$

$CH_3CO_2Na(aq) + H_2O(l) \rightleftharpoons NaOH(aq) + CH_3CO_2H(aq)$

(b) A salt that contains a neutral metal cation and a neutral anion (the conjugate base of a strong acid) will give a neutral solution in water. Suitable cations include those of the groups 1 and 2 metals. Examples include:

$NaCl$ $Ba(NO_3)_2$ $CaBr_2$ $NaClO_4$ $KI$

(c) Here there are two possibilities. Either the salt contains an acidic cation and a neutral anion, or it can contain a neutral cation and an acidic anion (of a polyprotic acid). Acidic cations are highly solvated small cations, such as $Al(H_2O)_6^{3+}$ or cations that are the conjugate acids of weak bases, the most common of which is the ammonium ion, $NH_4^+$. Acidic anions include $HSO_4^-$ and $H_2PO_4^-$. Examples include:

$AlCl_3 \cdot 6H_2O$ $BeCl_2 \cdot 4H_2O$ $NH_4Cl$ $NH_4NO_3$ $NaHSO_4$ $KH_2PO_4$

65. $SO_2$, $SO_3$, and $N_2O_5$ are examples of acidic nonmetal oxides that are the anhydrides of proton acids; each reacts with water to give the parent acid.

$SO_2(g) + H_2O(l) \rightleftharpoons H_2SO_3(aq)$

$$SO_3(g) + H_2O(l) \rightarrow H_2SO_4(aq)$$
$$N_2O_5(s) + H_2O(l) \rightarrow 2HNO_3(aq)$$

In each case the reaction is initiated by an $H_2O$ molecule donating a lone pair to the acidic oxide. In other words, water acts as a Lewis base and the anhydride behaves as a Lewis acid. For example, in detail, for the reaction of $SO_3$,

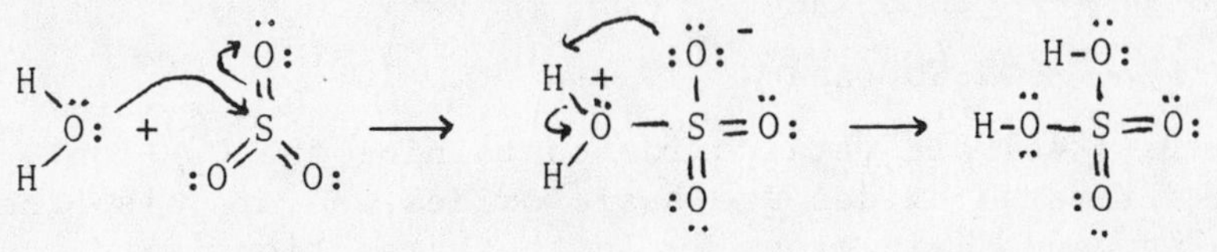

66. (a) The acidic behaviour of $FeCl_3.6H_2O$ in aqueous solution is due to the presence of the strongly hydrated $Fe(H_2O)_6^{3+}$ ion in the solution. This behaves as a weak acid,

$$Fe(H_2O)_6^{3+}(aq) + H_2O(l) \rightleftharpoons Fe(H_2O)_5OH^{2+}(aq) + H_3O^+(aq)$$

(b) (i) $Al_2O_3(s) + 6HCl(aq) \longrightarrow 2Al^{3+}(aq) + 6Cl^-(aq) + 3H_2O(l)$

(ii) $Al_2O_3(s) + 2NaOH(aq) + 3H_2O \longrightarrow 2Na^+(aq) + 2Al(OH)_4^-(aq)$

(c) $NH_4NO_3(s)$ is completely dissociated in aqueous solution into $NH_4^+$ and $NO_3^-$, of which the former is the conjugate acid of the weak base ammonia, $NH_3$, and the latter is the conjugate base of the strong acid nitric acid, $HNO_3$. Thus, nitrate ion is neutral in aqueous solution, while $NH_4^+$ behaves as a weak acid.

$$NH_4^+(aq) + H_2O(l) \rightleftharpoons H_3O^+(aq) + NH_3(aq)$$

67. (a) This is the reaction of a hydrocarbon with steam using a catalyst to give synthesis gas, which is then reacted with additional steam to convert carbon monoxide to carbon dioxide, which is soluble in water, which enables it to be separated from the hydrogen.

$$CH_4(g) + H_2O(g) \longrightarrow CO(g) + 3H_2(g) \xrightarrow{H_2O} CO_2(g) + 4H_2(g)$$

Alternatively, coke is reacted with steam to give water gas, and the CO(g) converted to $CO_2(g)$ by the above procedure.

$$C(s) + H_2O(g) \longrightarrow CO(g) + H_2(g)$$

(b) Pure oxygen is prepared in the laboratory by heating potassium chlorate with $MnO_2(s)$ as a catalyst.

$$2KClO_3(s) \longrightarrow 2KCl(s) + 3O_2(g)$$

(c) Water behaves as an oxidizing agent in any reaction in which it is reduced to $H_2(g)$. Examples include both reactions in (a) above, and,

$$2Na(s) + 2H_2O(l) \longrightarrow 2Na^+(aq) + 2OH^-(aq) + H_2(g)$$
$$Mg(s) + H_2O(g) \longrightarrow MgO(s) + H_2(g)$$
$$2Fe(s) + 3H_2O(g) \longrightarrow Fe_2O_3(s) + 3H_2(g)$$

(d) Water behaves as a reducing agent in any reaction in which it is oxidized to $O_2(g)$. For example,

$$2F_2(g) + 2H_2O(l) \longrightarrow 4HF(aq) + O_2(g)$$

(e) An acidic oxide is a nonmetal oxide which behaves as the anhydride of an oxoacid. Examples include $B_2O_3$, $CO_2$, $N_2O_3$, and $N_2O_5$ among the oxides of the period 2 elements, and $SiO_2$, $P_4O_6$, $P_4O_{10}$, $SO_2$, $SO_3$, and $Cl_2O_7$ among the oxides of the period 3 elements. Examples include,

$$B_2O_3(s) + 3H_2O(l) \longrightarrow 2H_3BO_3(aq)$$
$$N_2O_5(s) + H_2O(l) \longrightarrow 2HNO_3(aq)$$
$$SO_3(g) + H_2O(l) \longrightarrow H_2SO_4(aq)$$

(f) A basic oxide is a soluble metal oxide containing the $O^{2-}$ ion. All of the alkali metal oxides are basic oxides, as are also the oxides of Ca, Sr, and Ba in group 2. In their reactions with water the esential reaction that takes place is

$$O^{2-}(aq) + H_2O(l) \longrightarrow 2OH^-(aq)$$

Examples include,

$$Na_2O(s) + H_2O(l) \longrightarrow 2Na^+(aq) + 2OH^-(aq)$$
$$CaO(s) + H_2O(l) \longrightarrow Ca^{2+}(aq) + 2OH^-(aq)$$

68. (a) For a molecule to escape from the liquid phase into the gaseous phase it must have sufficient kinetic energy to overcome the intermolecular forces between the molecules in the liquid. Although the average kinetic energy of the molecules in the liquid is determined by the temperature, there is a distribution of kinetic energies, so that a fraction of the molecules have sufficient energy to enter the vapor even at temperatures below the boiling point. In a closed system an equilibrium is established between liquid and vapor, giving a vapor pressure that depends only on the temperature. In an open system the gaseous molecules escape as they are formed and eventually all of the liquid is converted to vapor at temperatures below the boiling point. This process is called evaporation.

(b) The explanation is similar to that given in part (a). Solids usually exert small vapor pressures that depend on the temperature. At any temperature below the melting point a small fraction of the molecules in the solid have sufficient energy to escape into the gaseous phase. Ice is normally in equilibrium with the water vapor in the atmosphere and on a very cold day the vapor pressure of the water in the air (humidity) is very small. Snow, a form of ice, absorbs heat energy from the sun and because the vapor pressure of snow is greater than that of the water vapor in the air, it is slowly converted to vapor without melting.

69. Use Graham's law (Chapter 3) to determine the molar mass of the acetic acid vapor.

i.e., $\dfrac{r_2}{r_1} = \sqrt{\dfrac{M_1}{M_2}}$ where $r_1$ and $r_2$ are the rates of diffusion and $M_1$ and $M_2$ are the molar masses of compounds 1 and 2, respectively.

The rates of diffusion are inversely proportional to the time that it takes for a given volume (number of moles) of gas to diffuse. Thus,

$$\frac{r_2}{r_1} = \frac{9.69\ \text{min}}{5.00\ \text{min}} = \left(\frac{M_2}{32.00\ \text{g mol}^{-1}}\right)^{1/2}$$

$M_2 = 3.75(32.00 \text{ g mol}^{-1}) = \underline{120 \text{ g mol}^{-1}}$

The molar mass of acetic acid, $CH_3CO_2H$, is 60.05 g $mol^{-1}$. Thus, in the vapor, acetic acid must be present as dimers of molecular formula $(CH_3CO_2H)_2$ (molar mass 120.1 g $mol^{-1}$).

The presence of dimers in the vapor is accounted for by the association of two acetic acid molecules through hydrogen bonds.

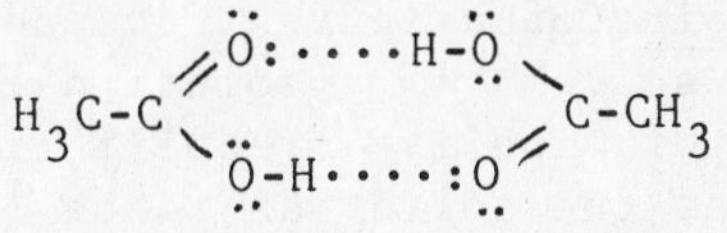

70. (a) Solids exert a vapor pressure. If when the solid is heated its vapor pressure exceeds the external pressure before it melts it is converted directly from solid to vapor in the process known as <u>sublimation</u>.

(b) The vapor pressure is proportional to the mole fraction of the solvent, which for a solution is less than that for the pure solvent. Thus the vapor pressure of a solution containing a non-volatile solute is less than that of the pure solvent. Alternatively the problem can be discussed in terms of <u>entropy</u>, which is a measure of the randomness of the molecules in a system. The vapor pressure measures the tendency for solvent molecules to escape into the vapor phase, which is related to the difference in entropy between the solvent molecules in the liquid and in the gas. This is greatest for the pure solvent and the vapor because the solvent molecules in a solution have a more random arrangement than those in the pure solvent. In other words, the difference in entropy between a pure solvent and vapor is greater than that between any solution and the vapor, so that in a solution there is less tendency for the conversion from solvent to gas than there is for the pure solvent. The solution has a smaller vapor pressure than does the solution.

(c) Ice, because of its hydrogen bonded structure, has an open cage-like structure in which each O atom is bonded to 4 H atoms, forming two O-H bonds and two hydrogen bonds. When ice melts, this structure collapses to some extent, so that there are more $H_2O$ molecules per unit volume in water at $0^{\circ}C$ than there are in ice at the same temperature. Thus liquid water has a greater density at $0^{\circ}C$ than does ice at the same temperature.

(d) For a pure solid to be in equilibrium with a solution, the vapor pressures of the solid and the liquid solvent must be the same. But the vapor pressure of a solution is less than that of the pure solvent, so that for equilibrium to be maintained the temperature has to be lowered below the freezing point of the pure solid to reduce its vapor pressure to that of the solvent in the solution. Thus, a solution has a lower freezing point than the pure solvent.

71. (a) <u>Fractional distillation</u> makes use of the property that at a given temperature different liquids have different vapor pressures. When a mixture of liquids is heated, the vapor is enriched in the more volatile components relative to the composition of the liquid mixture. When these vapors are condensed and vaporized again, the vapor becomes even more enriched in the more volatile components, and repetition of this process eventually gives a vapor containing only the most volatile component, which can be condensed to give the pure liquid. To avoid

having to perform a large number of successive distillations, use is made of a fractionating column in which the vapor can condense on the packing inside the column, so that within the column the liquid gradually becomes more enriched in the most volatile component as it climbs the column. At the top, the vapor contains essentially only one component which can be tapped off. An important application is in the distillation of petroleum (Chapter 6).

(b) Depression of freezing point of a solvent depends on the property that the vapor pressure of a solvent in a solution is less than that of the pure solvent. When a solution is in equilibrium with pure solid solvent, the vapor pressure of the solid has to be lower than it is at the melting point of the pure solvent, which can only be achieved by lowering the temperature. Thus, the freezing point of a solution is less than that of the pure solvent. Important applications include the use of salt to melt ice in the winter and the use of aqueous solutions of antifreeze to prevent automobile radiators from freezing at low temperatures.

(c) In osmosis, a pure solvent passes through a membrane into a solution thus creating an osmotic pressure. In reverse osmosis, pure solvent is separated from a solution, for example, pure water from seawater, by applying a pressure greater than the osmotic pressure to the seawater, so that pure water is forced through the membrane. This method is used to prepare pure water from seawater.

72. From the given freezing point depressions calculate the value of i, the number of solute particles formed in sulfuric acid solution when one solute "molecule" dissolves. Since each solution is 0.010 m, it is convenient to calculate the expected depression of freezing point for i = 1,

i.e., $\Delta T = (6.12^{\circ}C\ kg\ mol^{-1})(0.010\ mol\ kg^{-1}) = \underline{0.061^{\circ}C}$

Then i for each solute is given by the observed $\Delta T$ divided by this quantity.

| | $H_2O$ | $K_2SO_4$ | $HNO_3$ | $N_2O_5$ |
|---|---|---|---|---|
| $\Delta T$ ($^{\circ}C$) | 0.122 | 0.245 | 0.245 | 0.367 |
| i | $\frac{0.122}{0.061}$ | $\frac{0.245}{0.061}$ | $\frac{0.245}{0.061}$ | $\frac{0.367}{0.061}$ |
| i | 2 | 4 | 4 | 6 |

For $H_2O$ and $K_2SO_4$ the observed results are consistent with

$H_2O + H_2SO_4 \longrightarrow H_3O^+ + HSO_4^-$    i = 2

$K_2SO_4 + H_2SO_4 \longrightarrow 2K^+ + 2HSO_4^-$    i = 4

and since $NO_2^+$ is formed in sulfuric acid from both $HNO_3$ and $N_2O_5$, yielding 4 particles and 6 particles per molecule, respectively,

$HNO_3 + 2H_2SO_4 \longrightarrow NO_2^+ + H_3O^+ + 2HSO_4^-$    i = 4, and

$N_2O_5 + 3H_2SO_4 \longrightarrow H_3O^+ + 2NO_2^+ + 3HSO_4^-$    i = 6.

73. (a) Among the group 6 hydrides only $H_2O$ forms hydrogen bonds. Each water molecule forms a maximum of four hydrogen bonds to other water molecules, which accounts for the very anomalous boiling point of water.

(b) Ammonia molecules also form hydrogen bonds in liquid ammonia but since the number of hydrogen bonds per molecule is a maximum of two, since each $:NH_3$ molecule has only one unshared electron pair, the boiling point is expected to be less than that of water which can form a maximum of four hydrogen bonds per molecule, since there are two unshared electron pairs and two H atoms per molecule. Thus, water has an even more anomalous boiling point than ammonia.
(c) In most solids the molecules attract each other more strongly than is the case in the liquids, and thus for most substances the number of molecules per unit volume in the solid is greater than that in the liquid; the density of the solid is greater than that of the liquid. Water is unusual in that ice has a structure composed of cages of of water molecules joined by hydrogen bonds. When ice melts this open structure partially collapses so that the average number of water molecules surrounding any given molecule is slightly greater than four. Thus, at $0^{\circ}C$, the number of molecules per unit volume increases and the density increases.
(d) Above $0^{\circ}C$ the structures continues to collapse as more hydrogen bonds are broken; eventually the ensuing density increase is outweighed by expansion due to greater thermal vibrations of the $H_2O$ molecules and above $3.98^{\circ}C$ the molar volume increases, resulting in a decrease in density.
(e) Methanol is a hydrogen bonded substance, as is water. Methanol is soluble in water in all proportions (miscible) because it can replace its own hydrogen bonds, and those between water molecules, by new hydrogen bonds between $CH_3OH$ molecules and $H_2O$ molecules. Hexane can interact with its own molecules, or with water molecules, only through weak van der Waals forces (principally London forces), which are insufficiently strong to disrupt the hydrogen bonded structure of water, so that hexane is insoluble in water.
(f) Sodium oxide, $Na_2O$, and magnesium oxide, MgO, are both ionic solids. For an ionic substance to be soluble in water, ionic bonds have to be broken. Comparing the sizes of the ions in these oxides and their charges, $Na^+$ is a much larger ion than $Mg^{2+}$ and has only one-half the charge, so in $(Na^+)_2O^{2-}$ the electrostatic attraction of the ions is much weaker than that in $Mg^{2+}O^{2-}$. The difficulty of separating $Mg^{2+}$ and $O^{2-}$ ions is apparently so great that MgO(s) is insoluble, while $Na_2O$ , with very much weaker ionic bonds dissolves to give a solution of NaOH(aq).

74. (a) Phosphoric oxide, $P_4O_{10}$, is a nonmetal oxide and thus an acidic oxide. It reacts with water to give phosphorus(V) acid, of which it is the anhydride.

$$P_4O_{10}(s) + 5H_2O(l) \longrightarrow 4H_3PO_4(aq)$$

The product phosphoric acid is a weak triprotic acid; water behaves as a Lewis base.
(b) Phosphorus(III) chloride, $PCl_3(l)$, is a nonmetal chloride of a third period element that can expand its valence shell beyond the octet. Thus it behaves as a Lewis acid and accepts a lone pair from the Lewis base $H_2O$, which leads to the hydrolysis of $PCl_3$.

$$PCl_3(l) + 3H_2O(l) \longrightarrow H_3PO_3(aq) + 3HCl(aq)$$

(c) Calcium is a metal. The only possible reaction with water is its oxidation to $Ca^{2+}$ ions and the reduction of water to $H_2(g)$. This reaction occurs at ordinary temperatures because Ca is a reactive group 2 metal.

$$Ca(s) + 2H_2O(l) \longrightarrow Ca^{2+}(aq) + 2OH^-(aq) + H_2(g)$$

Water behaves as an oxidizing agent.

(d) $F_2(g)$ is the strongest oxidizing agent among the halogens. It oxidizes water to $O_2(g)$ and is itself reduced to $F^-(aq)$; water behaves as a reducing agent.

$$F_2(g) + 2H_2O(l) \longrightarrow 4HF(aq) + O_2(g)$$

(e) Carbonate ion, $CO_3^{2-}$, is the conjugate base of the weak acid $HCO_3^-$, hydrogen carbonate ion, and therefore behaves as a weak base. Water behaves as a Brønsted acid.

$$HCO_3^-(aq) + H_2O(l) \rightleftharpoons HCO_3^-(aq) + OH^-(aq)$$

(f) CaO(s), calcium oxide, is a metal oxide containing neutral $Ca^{2+}$ ions and strongly basic $O^{2-}$ ions. $O^{2-}$ reacts completely with water to give $OH^-$ ions. Water behaves as a Brønsted acid.

$$CaO(s) + H_2O(l) \longrightarrow Ca^{2+}(aq) + 2OH^-(aq)$$

(g) $HNO_3$, nitric acid, is a strong oxoacid that donates a proton to water to give hydronium ion, $H_3O^+$, and nitrate ion, $NO_3^-$. Water behaves as a Brønsted base (and as a Lewis base since it donates an electron pair to the hydrogen of $HNO_3$).

$$HNO_3(aq) + H_2O(l) \longrightarrow H_3O^+(aq) + NO_3^-(aq)$$

(h) $NH_4Br$, ammonium bromide, is an ionic salt composed of $NH_4^+$ ions and $Br^-$ ions and is fully dissociated into these ions in solution. $Br^-$ is the conjugate base of the strong acid HBr and has no basic properties in water, while $NH_4^+$ is the conjugate acid of the weak base $NH_3$ and behaves as a weak acid in aqueous solution.

$$NH_4^+(aq) + Br^-(aq) + H_2O(l) \rightleftharpoons NH_3(aq) + H_3O^+(aq) + Br^-(aq)$$

Water behaves as a Brønsted base (it is also a Lewis base because it donates an electron pair to a hydrogen of $NH_4^+$ to give $H_3O^+$).

# CHAPTER 13

1. (a) Equilibrium refers to any situation where a system has reached a steady state and there is no apparent change in the system. The system may involve the reactants and products of a reaction (a chemical change), or a simple phase change (physical change)

   (b) For a system at equilibrium, there is no apparent change in the system with time but the system is not static. A forward change is occurring at exactly the same rate as a backward (reverse) change, and the system is described as having achieved a state of dynamic equilibrium.

   (c) A physical equilibrium is one that involves only physical changes, such as a change in phase. A chemical equilibrium is one that occurs in a system undergoing a chemical change, where the atoms of the reactants are undergoing rearrangement to give products. Physical equilbria are exemplified by (i) a solid solute in equilibrium with its saturated solution (Chapter 1), and (ii) a liquid in equilibrium with its vapor in a closed system (Chapter 12). Chemical equilibrium is eventually achieved in any chemical reaction under conditions where none of the reactants or products are allowed to escape from the system. Examples of chemical equilibria include:

   (i) $HF(aq) + H_2O(l) \rightleftharpoons H_3O^+(aq) + F^-(aq)$

   (ii) $CaCO_3(s) \rightleftharpoons CaO(s) + CO_2(g)$

2. (a) The equilibrium constant, $K_{eq}$, is the value of the equilibrium constant expression when each of the concentrations has its equilibrium value. The reaction quotient, Q, is the value of the equilibrium constant expression when the concentrations are different from their equilibrium values.

   (b) For $Q > K_{eq}$, the reactant concentrations have to increase and the product concentrations have to decrease for equilibrium to be achieved. The system is at equilibrium when $Q = K_{eq}$.

3. (a) $K_c = \dfrac{[NO_2]^4[O_2]}{[N_2O_5]^2}$ ; $K_p = \dfrac{(p_{NO_2})^4(p_{O_2})}{(p_{N_2O_5})^2}$

   (b) $K_c = \dfrac{[SO_3]^2}{[SO_2]^2[O_2]}$ ; $K_p = \dfrac{(p_{SO_3})^2}{(p_{SO_2})^2(p_{O_2})}$

   (c) $K_c = \dfrac{[SO_3]}{[SO_2][O_2]^{1/2}}$ $K_p = \dfrac{(p_{SO_3})}{(p_{SO_2})(p_{O_2})^{1/2}}$

(d) $K_c = \frac{[P_4O_{10}]}{[P_4][O_2]^5}$ ; $K_p = \frac{(p_{P_4O_{10}})}{(p_{P_4})(p_{O_2})^5}$

(e) $K_c = \frac{[PCl_3][Cl_2]}{[PCl_5]}$ ; $K_p = \frac{(p_{PCl_3})(p_{Cl_2})}{(p_{PCl_5})}$

4. (a) units of $K_c = \frac{(mol\ L^{-1})^4(mol\ L^{-1})}{(mol\ L^{-1})^2} = \underline{mol^3\ L^{-3}}$

units of $K_p = \frac{(atm)^4(atm)}{(atm)^2} = \underline{atm^3}$

(b) units of $K_c = \frac{(mol\ L^{-1})^2}{(mol\ L^{-1})^2(mol\ L^{-1})} = \underline{mol^{-1}\ L}$

units of $K_p = \frac{(atm)^2}{(atm)^2(atm)} = \underline{atm^{-1}}$

(c) units of $K_c = \frac{(mol\ L^{-1})}{(mol\ L^{-1})(mol\ L^{-1})^{1/2}} = \underline{mol^{-1/2}\ L^{1/2}}$

units of $K_p = \frac{(atm)}{(atm)(atm)^{1/2}} = \underline{atm^{-1/2}}$

(d) units of $K_c = \frac{(mol\ L^{-1})}{(mol\ L^{-1})(mol\ L^{-1})^5} = \underline{mol^{-5}\ L^5}$

units of $K_p = \frac{(atm)}{(atm)(atm)^5} = \underline{atm^{-5}}$

(e) units of $K_c = \frac{(mol\ L^{-1})(mol\ L^{-1})}{(mol\ L^{-1})} = \underline{mol\ L^{-1}}$

units of $K_p = \frac{(atm)(atm)}{(atm)} = \underline{atm}$

5.
(a) $K_c = \frac{[NO]^2[Br_2]}{[NOBr]^2}$ ; $K_p = \frac{(p_{NO})^2(p_{Br_2})}{(p_{NOBr})^2}$

(b) $K_c = \frac{[CO_2]^3}{[CO]^3}$ ; $K_p = \frac{(p_{CO_2})^3}{(p_{CO})^3}$

(c) $K_c = \frac{[NH_3]^2[H_2O]}{[N_2O][H_2]^4}$ ; $K_p = \frac{(p_{NH_3})^2(p_{H_2O})}{(p_{N_2O})(p_{H_2})^4}$

(d) $K_c = [O_2]$ ; $K_p = (P_{O_2})$

(e) $K_c = [NO_2]^4[O_2]$ ; $K_p = (p_{NO_2})^4(p_{O_2})$

Note that for the heterogeneous equilibria (b), (d), and (e), the components in the solid phase do not appear in the equilibrium constant expression.

6. (a) mol $L^{-1}$; atm (b) no units for $K_c$ or for $K_p$
(c) $mol^{-2}$ $L^2$ ; $atm^{-2}$ (d) mol $L^{-1}$; atm (e) $mol^5$ $L^{-5}$ ; $atm^5$

7. Arrange PV = nRT in the form $P = \frac{n}{V}$ (RT), then with R with units of atm L $mol^{-1}$ $K^{-1}$, the units of $\frac{n}{V}$ are mol $L^{-1}$,

i.e., $P_A$ = [A]RT, and for the equilibrium

$$N_2O_4(g) \rightleftharpoons 2NO_2(g)$$

we can write,

$$K_p = \frac{(p_{NO_2})^2}{(p_{N_2O_4})} = \frac{([NO_2]RT)^2}{[N_2O_4]RT} = \frac{[NO_2]^2}{[N_2O_4]}(RT) = K_c(RT)$$

hence, $K_p$ = (0.212 mol $L^{-1}$)(0.0821 atm L $mol^{-1}$ $K^{-1}$)(373 K)
= <u>6.49 atm</u>

8. This problem is similar to Problem 7. For each pressure term in the $K_p$ expressions, we have

$$p_A = [A]RT$$

and we can use the conversion factor <u>atm = (mol $L^{-1}$)(RT)</u>

(a) $K_c$ has no units; $K_p = K_c = \underline{2.5 \times 10^{-3}}$ at 2100°C

(b) $$K_p = \frac{K_c}{(RT)^2} = \frac{300 \text{ mol}^{-2} \text{ L}^2}{[(0.0821 \text{ atm L mol}^{-1} \text{ K}^{-1})(698 \text{ K})]^2} = \underline{9.18\text{x}10^{-2} \text{ atm}^{-2}}$$

(c) $$K_p = K_c(RT) = (6.35\text{x}10^{-2} \text{ mol L}^{-1})(0.0821 \text{ atm L mol}^{-1} \text{ K}^{-1})$$
$$= \underline{2.73 \text{ atm}}$$

9. (a) $$K_c = \frac{[NO]^2}{[N_2][O_2]} \quad \text{or} \quad K_p = \frac{(P_{NO})^2}{(P_{N_2})(P_{O_2})}$$

(b) $K_p = K_c = 2.5\text{x}10^{-3}$ at 2100°C; regardless of whether we use concentrations in mol $L^{-1}$ or partial pressures in atm (or indeed any other units), the equilibrium constant has the same numerical value and is unitless.

10. For the reaction as written. $N_2O_4(g) \rightleftharpoons 2NO_2(g)$

$$K_c = \frac{[NO_2]^2}{[N_2O_4]} = 0.212 \text{ mol L}^{-1} \text{ at } 100°C$$

(a) $$K'_c = \frac{[N_2O_4]}{[NO_2]^2} = (K_c)^{-1} = (0.212 \text{ mol L}^{-1})^{-1} = \underline{4.72 \text{ mol}^{-1} \text{ L}}$$

(b) $$K''_c = \frac{[N_2O_4]^{1/2}}{[NO_2]} = K_c^{-1/2} = (0.212 \text{ mol L}^{-1})^{-1/2} = \underline{2.17 \text{ mol}^{-1/2} \text{ L}^{1/2}}$$

11. Initially

$$Q = \frac{[CH_3OH]}{[H_2]^2[CO]} = \frac{0.10 \text{ mol L}^{-1}}{(0.10 \text{ mol L}^{-1})^2(0.10 \text{ mol L}^{-1})} = \underline{1.0\text{x}10^2 \text{ mol}^{-2} \text{ L}^2}$$

$Q < K_c = 300 \text{ mol}^2 \text{ L}^{-2}$ ; initially the system is <u>not</u> at equilibrium. To attain equilibrium $[CH_3OH]$ would have to <u>increase</u>, so that it will be <u>greater</u> than 0.10 mol $L^{-1}$ at equilibrium.

12. $$K_c = \frac{[SO_3]^2}{[SO_2]^2[O_2]} = \frac{(0.100 \text{ mol L}^{-1})^2}{(0.010 \text{ mol L}^{-1})^2(0.20 \text{ mol L}^{-1})} = \underline{5.0\text{x}10^2 \text{ mol}^{-1} \text{ L}}$$

13. $$K_c = \frac{[H_2O][CO]}{[H_2][CO_2]} = \frac{(0.500 \text{ mol L}^{-1})(0.425 \text{ mol L}^{-1})}{(0.600 \text{ mol L}^{-1})(0.459 \text{ mol L}^{-1})} = \underline{0.772}$$

$$K_p = K_c = \underline{0.772}$$

14.

| | $H_2(g)$ + | $CO_2(g)$ $\rightleftharpoons$ | $H_2O(g)$ + | $CO(g)$ | |
|---|---|---|---|---|---|
| initially | 0.200 | 0.200 | 0 | 0 | mol $L^{-1}$ |
| at eq'm | 0.200-x | 0.200-x | x | x | mol $L^{-1}$ |

Note that initially we have 1.00 mol of $H_2(g)$ and 1.00 mol of $CO_2(g)$ in a volume of 5.00 L. Thus the initial concentration of each gas is 1.00 mol/5.00 L = 0.200 mol $L^{-1}$

For the equilibrium situation,

$$K_c = \frac{[H_2O][CO]}{[H_2][CO_2]} = \frac{(x)(x)}{(0.200-x)(0.200-x)} = [\frac{x}{0.200-x}]^2 = 0.772$$

$$\frac{x}{0.200-x} = 0.879 \;;\quad x = 0.176-0.879x \;;\quad \underline{x = 0.0937 \text{ mol } L^{-1}}$$

Thus the equilibrium concentrations are

$[H_2] = [CO_2] = 0.200-x = \underline{0.106 \text{ mol } L^{-1}}$; $[H_2O] = [CO] = \underline{0.0937 \text{ mol } L^{-1}}$

15.

| | $I_2(g) \rightleftharpoons$ | $2I(g)$ | |
|---|---|---|---|
| initially | 0.500 | 0 | mol $L^{-1}$ |
| at eq'm | 0.500-x | 2x | mol $L^{-1}$ |

$$K_c = \frac{[I]^2}{[I_2]} = \frac{4x^2}{0.500 - x} = 3.76 \times 10^{-5} \text{ mol } L^{-1}$$

Since the value of $K_c$ is small, we can assume x << 0.500

$$4x^2 = 1.88 \times 10^{-5} \;;\; x = 2.17 \times 10^{-3}$$

The equilibrium concentrations are $[I_2] = (0.500-x) = \underline{0.498 \text{ mol } L^{-1}}$

$[I^-] = 2x = \underline{0.00434 \text{ mol } L^{-1}}$

<u>Check</u>

$$K_c = \frac{[I]^2}{[I_2]} = \frac{(4.34 \times 10^{-3} \text{ mol } L^{-1})^2}{0.498 \text{ mol } L^{-1}} = 3.78\times10^{-5} \text{ mol } L^{-1}$$

the assumption x << 0.500 is justified

$$\% \text{ dissociation } I_2 = (\frac{0.00217 \text{ mol } L^{-1}}{0.500 \text{ mol } L^{-1}}) \times 100\% = \underline{0.43\%}$$

16.

| | $PCl_5(g) \rightleftharpoons$ | $PCl_3(g)$ + | $Cl_2(g)$ | |
|---|---|---|---|---|
| initially | 0.0400 | 0 | 0 | mol $L^{-1}$ |
| at eq'm | 0.0400-x | x | x | mol $L^{-1}$ |

$$(0.040 - x) \text{ mol } L^{-1} = 0.015 \text{ mol } L^{-1} \;;\quad x = \underline{0.025 \text{ mol } L^{-1}}$$

$$K_c = \frac{[PCl_3][Cl_2]}{[PCl_5]} = \frac{(0.025 \text{ mol } L^{-1})(0.025 \text{ mol } L^{-1})}{(0.015 \text{ mol } L^{-1})} = \underline{0.042 \text{ mol } L^{-1}}$$

17. Concentration of water at equilibrium = $(1.541 \text{ g})(\frac{1 \text{ mol}}{18.02 \text{ g}})/(2.00 \text{ L})$

$$[H_2O] = 0.0428 \text{ mol } L^{-1}$$

$$\begin{array}{lccccl} & H_2(g) & + \; CO_2(g) & \rightleftharpoons \; H_2O(g) & + \; CO(g) & \\ \text{initially} & 0.0500 & 0.100 & 0 & 0 & \text{mol L}^{-1} \\ \text{at eq'm} & 0.0500-x & 0.100-x & x & x & \text{mol L}^{-1} \end{array}$$

and, as we saw above, $x = 0.0428 \text{ mol L}^{-1}$, so that at equilibrium:

$$[H_2] = 0.0072 \;,\; [CO_2] = 0.0572,\; [H_2O] = [CO] = 0.0428 \text{ mol L}^{-1}$$

$$K_c = \frac{[H_2O][CO]}{[H_2][CO_2]} = \frac{(0.0428 \text{ mol L}^{-1})(0.0428 \text{ mol L}^{-1})}{(0.0072 \text{ mol L}^{-1})(0.0572 \text{ mol L}^{-1})} = \underline{4.45}$$

18. Initial [HI] = (2.00 mol)/4.3 L = 0.465 mol $L^{-1}$

$$\begin{array}{lcccl} & 2HI(g) & \rightleftharpoons \; H_2(g) & + \; I_2(g) & \\ \text{initially} & 0.465 & 0 & 0 & \text{mol L}^{-1} \\ \text{at eq'm} & 0.465-2x & x & x & \text{mol L}^{-1} \end{array}$$

$$K_c = \frac{[H_2][I_2]}{[HI]^2} = \frac{x^2}{(0.465-2x)^2} = 0.022$$

$$\text{i.e.,} \quad \frac{x}{0.465-2x} = 0.148 \;; \quad x = 0.0688 - 0.296\,x$$

$$\underline{x = 0.053}$$

$$[HI] = (0.465-0.106) = \underline{0.36 \text{ mol L}^{-1}}$$

$$[H_2] = [I_2] = \underline{0.053 \text{ mol L}^{-1}}$$

19.

$$\begin{array}{lcccl} & H_2(g) & + \; I_2(g) & \rightleftharpoons \; 2HI(g) & \\ \text{initially} & 0.10 & 0.10 & 1.0 \text{ mol L}^{-1} & \\ \text{at eq'm} & 0.10+x & 0.10+x & 1.0-2x & \text{mol L}^{-1} \end{array}$$

(a)

$$Q = \frac{[HI]^2}{[H_2][I_2]} = \frac{(1.0 \text{ mol L}^{-1})^2}{(0.10 \text{ mol L}^{-1})(0.10 \text{ mol L}^{-1})} = 100$$

$Q \neq K_c = 54.4$, mixture is <u>not</u> at equilibrium

(b) $Q > K_c$, so to achieve equilibrium, some HI must be converted to $H_2$ and $I_2$.

$$K_c = \frac{(1.0-2x)^2}{(0.10+x)^2} = 54.4 \;; \quad \frac{1.0-2x}{0.10+x} = 7.38$$

$$1.0 - 2x = 0.74 + 7.38x \;; \; 9.38x = 0.26 \;; \; \underline{x = 0.028}$$

At equilibrium, the concentrations are

$$[H_2] = [I_2] = \underline{0.13 \text{ mol L}^{-1}} \;; \quad [HI] = \underline{0.94 \text{ mol L}^{-1}}$$

(c) $$H_2(g) + I_2(g) \rightleftharpoons 2HI(g)$$

| | $H_2$ | $I_2$ | $HI$ |
|---|---|---|---|
| initially | 0 | 0 | 1.2 mol $L^{-1}$ |
| at eq'm | x | x | (1.2-2x) mol $L^{-1}$ |

$$K_c = \frac{[HI]^2}{[H_2][I_2]} = \frac{(1.2-2x)^2}{x^2} = 54.4$$

$$\frac{1.2-2x}{x} = 7.38 \;; \quad 9.38x = 1.2 \;; \quad \underline{x = 0.128}$$

and the equilibrium concentrations are

$[H_2] = [I_2] = \underline{0.13 \text{ mol } L^{-1}}$ ; $[HI] = \underline{0.94 \text{ mol } L^{-1}}$

20. (a)

$$Q = \frac{[NH_3]^2}{[N_2][H_2]^3} = \frac{(0.10 \text{ mol } L^{-1})^2}{(1.0 \text{ mol } L^{-1})(1.0 \text{ mol } L^{-1})^3} = \underline{1.0 \text{x} 10^{-2} \text{ M}^{-2}}$$

$Q = 1.0 \times 10^{-2}$ $M^{-2}$ < $K_c = 6.0 \times 10^{-2}$ $M^{-2}$

The system is not at equilibrium.

(b) $$N_2(g) + 3H_2(g) \rightleftharpoons 2NH_3(g)$$

| | $N_2$ | $H_2$ | $NH_3$ | |
|---|---|---|---|---|
| first system | 1.0 | 1.0 | 0.10 | mol $L^{-1}$ |
| second system | 0.99 | 0.97 | 0.12 | mol $L^{-1}$ |

(0.01 mol $N_2$ converted to $NH_3$ reacts with 0.03 mol $H_2$ to give 0.02 mol $NH_3$)

$$Q = \frac{(0.12 \text{ mol } L^{-1})^2}{(0.99 \text{ mol } L^{-1})(0.97 \text{ mol } L^{-1})^3} = \underline{1.5 \times 10^{-2} \text{ mol } L^{-1}}$$

Q is still less than $K_c$, so the system is still not at equilibrium (even more $N_2$ and $H_2$ would have to be converted to $NH_3$ to achieve equilibrium). The student's estimate was too small.

21.(a) Let the partial pressure of $N_2O_3$ be P atm.

$$K_p = \frac{P}{(1.00 \text{ atm})(0.500 \text{ atm})} = 2.00 \text{ atm}^{-1}$$

$\underline{P = 1.00 \text{ atm}}$

(b) $$NO(g) + NO_2(g) \rightleftharpoons N_2O_3(g)$$

| | NO | $NO_2$ | $N_2O_3$ | |
|---|---|---|---|---|
| initially | 1.00 | 0.500 | 0 | atm |
| at eq'm | 1.00-P | 0.500-P | P | atm |

$$K_p = \frac{P}{(1.00-P)(0.500-P)} = 2.00$$

i.e., $2[0.500 - 1.50P + P^2) = P$ or $2P^2 - 4P + 1 = 0$

Solving the quadratic equation gives P = 1.71 or 0.293 atm

The first value is inadmissable; the second gives

$P_{NO}$ = 0.707 atm, $P_{NO_2}$ = 0.207 atm, and $P_{N_2O_3}$ = 0.293 atm

We can check to show that these are correct by recalculating the value of $K_p$:

$$K_p = \frac{(0.293 \text{ atm})}{(0.707 \text{ atm})(0.207 \text{ atm})} = 2.00 \text{ atm}^{-1}$$

22. The initial concentration of HI is 1.00 mol/5.00 L = 0.200 mol $L^{-1}$

| | $2HI(g) \rightleftharpoons$ | $H_2(g)$ + | $I_2(g)$ | |
|---|---|---|---|---|
| initially | 0.200 | 0 | 0 | mol $L^{-1}$ |
| at eq'm | 0.200-2x | x | x | mol $L^{-1}$ |

$$K_c = \frac{[H_2][I_2]}{[HI]^2} = \frac{x^2}{(0.200-2x)^2} = 6.34 \times 10^{-4}$$

i.e., $\frac{x}{0.200-2x} = 2.52 \times 10^{-2}$ ; $x = (5.04 \times 10^{-3}) - (5.04 \times 10^{-2})x$

$$x = \frac{5.04 \times 10^{-3}}{1.050} = 4.80 \times 10^{-3}$$

The equilibrium concentrations are [HI] = 0.190 mol $L^{-1}$

$[H_2] = [I_2] = 4.80 \times 10^{-3}$ mol $L^{-1}$

% Dissociation of HI = $\left(\frac{9.60 \times 10^{-3} \text{ mol L}^{-1}}{0.200 \text{ mol L}^{-1}}\right) \times 100\% = 4.80\%$

23. (a)

$$Q = \frac{[NO_2]^2}{[N_2O_4]} = \frac{(0.12 \text{ mol L}^{-1})^2}{(0.10 \text{ mol L}^{-1})} = 0.14 \text{ mol L}^{-1}$$

(b) $Q < K_c = 0.212$ mol $L^{-1}$ ; the reaction is not at equilibrium.

(c) Q has to increase for equilibrium to be established, i.e., $[NO_2]$ has to increase.

(d)

| | $N_2O_4(g) \rightleftharpoons$ | $2NO_2(g)$ | |
|---|---|---|---|
| initially | 0.10 | 0.12 | mol $L^{-1}$ |
| at eq'm | 0.10-x | 0.12+2x | mol $L^{-1}$ |

$$K_c = \frac{(0.12+2x)^2}{0.10-x} = 0.212 \text{ mol L}^{-1}$$

$0.0144 + 0.48x + x^2 = 0.0212 - 0.212x$, or $x^2 + 0.69x - 0.0068 = 0$

Solving the quadratic equation for x gives x = 0.01.

At equilibrium: $[N_2O_4]$ = 0.09 mol $L^{-1}$; $[NO_2]$ = 0.14 mol $L^{-1}$

Check: $K_c = [NO_2]^2/[N_2O_4] = (0.14)^2/(0.09) = 0.022$ mol $L^{-1}$

(e) $PV = nRT;\quad P = \frac{n}{V}(RT) = [\ ](RT)$

$$K_p = \frac{(P_{NO_2})^2}{(P_{N_2O_4})} = \frac{([NO_2](RT))^2}{([N_2O_4](RT))} = K_c(RT)$$

$$K_p = (0.212\ \text{mol L}^{-1})(0.0821\ \text{atm L mol}^{-1}\ \text{K}^{-1})(373\ \text{K}) = \underline{6.49\ \text{atm}}$$

(f) From the concentrations in part (d), the partial pressures are

$$P_{NO_2} = [NO_2](RT) = (0.14\ \text{mol L}^{-1})(0.0821\ \text{L atm mol}^{-1}\ \text{K}^{-1})(373\ \text{K}) = \underline{4.3\ \text{atm}}$$

$$P_{N_2O_4} = [N_2O_4](RT) = (0.09\ \text{mol L}^{-1})(0.0821\ \text{L atm mol}^{-1}\ \text{K}^{-1})(373\ \text{K}) = \underline{2.8\ \text{atm}}$$

We can check these values by calculating the value of $K_p$.

$$K_p = (4.3\ \text{atm})^2/(2.8\ \text{atm}) = \underline{6.6\ \text{atm}}$$

24. According to Le Châtelier's principle, adding heat to a system by increasing the temperature will favor whichever of the forward or reverse reaction absorbs heat, i.e., is endothermic. If the value of the equilibrium constant increases with increase in temperature, the forward reaction is favored by increase in temperature, i.e., the forward reaction must be endothermic.

If the forward reaction is endothermic, then the reverse reaction must be exothermic; the enthalpy change is negative for the reverse reaction.

25. $$PCl_5(g) + \text{heat} \longrightarrow PCl_3(g) + Cl_2(g)$$

(a) When the mixture is compressed, the number of molecules per unit volume is increased, which is counteracted according to Le Châtelier by $PCl_3$ and $Cl_2$ recombining to give $PCl_5$, to diminish the pressure: percent dissociation of $PCl_5$ decreases.

(b) Increasing the volume decreases the number of molecules per unit volume, which is counteracted by the formation of more $PCl_3$ and $Cl_2$ to increase the concentration of gas molecules; percent dissociation of $PCl_5$ increases.

(c) Decreasing the temperature will favor the reaction that produces heat, i.e., the exothermic reaction, which in this case is the reverse reaction: percent dissociation of $PCl_5$ decreases.

(d) Adding $Cl_2(g)$ will be counteracted by the reaction that decreases the concentration of $Cl_2$ being favored; percent dissociation of $PCl_5$ decreases.

26. (a) $$Q = \frac{(P_{NOBr})^2}{(P_{NO})^2(P_{Br_2})} = \frac{(0.108\ \text{atm})^2}{(0.100\ \text{atm})^2(0.010\ \text{atm})} = 116.6\ \text{atm}^{-1}$$

This value is the same as that given for 25°C, so unless the enthalpy change for the reaction is very small, the reaction mixture would not be at equilibrium at 0°C. We are in fact told that the forward reaction is endothermic, so a decrease in temperature from 25°C to 0°C should favor the exothermic (reverse) reaction, in other words, the formation of NO and $Br_2$. This would decrease the value of $K_p$,

$$K_p^{0^\circ C} < K_p^{25^\circ C}$$

(b) $2NO(g) + Br_2(g) \rightleftharpoons 2NOBr(g)$

| | 2NO(g) | $Br_2$(g) | 2NOBr(g) | |
|---|---|---|---|---|
| initially | 0 | 0 | 5.00 | atm |
| at eq'm | 0.60 | 0.30 | 4.40 | atm |

$$K_p^{50^\circ C} = \frac{(4.40\text{ atm})^2}{(0.60\text{ atm})^2(0.30\text{ atm})} = 179\text{ atm}^{-1}$$

$K_p^{50^\circ C} > K_p^{25^\circ C}$ as expected; an increase in temperature should favor the endothermic (forward) reaction, which in this case would lead to an increase in $K_p$ with increased temperature. The change in $K_p$ from 25°C to 50°C is rather small, indicating that $\Delta H$ for the reaction is small in magnitude.

27. In each case, use the standard enthalpies of formation from Appendix B to calculate $\Delta H^\circ$ for the forward reaction,

$$\Delta H^\circ = \Sigma \Delta H_f^\circ(\text{products}) - \Sigma \Delta H_f^\circ(\text{reactants})$$

If in going from reactants to products there is an increase in moles of gases, the formation of products is favored by decreased pressure, if there is a decrease in moles, the formation of products is favored by increased pressure.

(a) $\Delta H^\circ = [2\,\Delta H_f^\circ(NOCl,g)] - [2\,\Delta H_f^\circ(NO,g) + \Delta H_f^\circ(Cl_2,g)]$

$= [2(51.7)] - [2(90.3) + 0] = -77.2\text{ kJ}$

and 2 mol NO + 1 mol $Cl_2$ ⟶ 2 mol NOCl (decrease in moles).

Thus, the reaction is exothermic and leads to a decrease in the number of moles of gases; low temperature and high pressure will favor the formation of products.

(b) $\Delta H^\circ = [2\,\Delta H_f^\circ(SO_3,g)] - [2\,\Delta H_f^\circ(SO_2,g) + \Delta H_f^\circ(O_2,g)]$

$= [(2(-395.7)] - [2(-296.8) + 0] = -197.8\text{ kJ}$

and 2 mol $SO_2$ + 1 mol $O_2$ ⟶ 2 mol $SO_3$ (decrease in moles).

Thus, as in (a), the reaction is exothermic and leads to a decrease in the number of moles of gases; low temperature and high pressure will favor the formation of products.

(c) $\Delta H^\circ = [2\,\Delta H_f^\circ(NH_3,g)] - [\Delta H_f^\circ(N_2,g) + 3\,\Delta H_f^\circ(H_2,g)]$

$= [2(-46.2)] - [0 + 3(0)] = -92.4\text{ kJ}$

and 1 mol $N_2$ + 3 mol $H_2$ ⟶ 2 mol $NH_3$ (decrease in moles).

The situation is the same as in (a) and (b). The reaction is exothermic and leads to a decrease in the number of moles of gases; low temperature and high pressure will favor products.

(d) $\Delta H^o = [\Delta H_f^o(CO_2,g) + \Delta H_f^o(H_2,g)] - [\Delta H_f^o(CO,g) + \Delta H_f^o(H_2O,g)]$
$= [-393.5 + 0] - [-110.5 + (-241.8)] = \underline{-41.2\ kJ}$

and 1 mol $CO_2$ + 1 mol $H_2$ $\longrightarrow$ 1 mol CO + 1 mol $H_2O$

Thus, the reaction is exothermic but there is no change in the moles of gases from reactants to products; low temperature would be the only condition for a high yield of products.

28. This problem is similar to Problems 25 and 27.

The forward reaction is exothermic and leads to a decrease in the number of moles of gases.

(a) Doubling the volume of the reaction vessel decreases the pressure, which would favor the reverse reaction; $[SO_3]$ decreases.

(b) Increase in the temperature would favor the reverse (endothermic) reaction; $[SO_3]$ decreases.

(c) When the concentration of $O_2$ is increased, the system attempts to remove the excess $O_2$; the forward reaction is favored; $[SO_3]$ increases

(d) Adding helium to the reaction vessel at constant volume would not affect the concentrations of any of $SO_2$, $O_2$, or $SO_3$; $[SO_3]$ would be unchanged.

29. This problem is similar to Problem 28.

This is a heterogeneous equilibrium and C(s) does not appear in the equilibrium constant expression.

(a) Increase in temperature favors the reverse (endothermic) reaction; the equilibrium shifts to the left.

(b) Increase in volume decreases the pressure, which favors the formation of more moles of gas, which occurs by the equilibrium shifting to the left.

(c) Increase in the pressure of $H_2$ is counteracted by the formation of more $CH_4$; the equilibrium shifts to the right.

(d) Addition of more carbon has no affect since it does not appear in the equilibrium constant expression. Alternatively, we can argue that any amount of solid carbon has a constant concentration (molar density).

30.

$$K_p = \frac{(P_{SO_3})}{(P_{SO_2})(P_{O_2})^{1/2}}$$

Since $K_p$ decreases with increase in temperature, the amount of $SO_3$ must also decrease with increase in temperature.

In other words, the reverse reaction is favored by increase in temper- ture; the reverse reaction must be endothermic, and the reaction as written must be exothermic (see Problem 27(b)).

31. See Problem 27 for method of calculating $\Delta H^o$ for a reaction

(a) $\Delta H^o = [\Delta H_f^o(N_2O_4,g)] - [2\,\Delta H_f^o(NO_2,g)]$
$= [9.3] - [2(33.2)] = \underline{-57.1\ kJ}$

The reaction is exothermic and gives a decrease in the number of moles of gas. Thus, increased temperature favors the reverse reaction (the formation of $NO_2$), and increased pressure favors the forward reaction (formation of $N_2O_4$).

(b) $\Delta H^\circ = [2\Delta H_f^\circ(NO,g) + \Delta H_f^\circ(O_2,g)] - [2\Delta H_f^\circ(NO_2,g)]$

$= [2(90.3) + 0] - [2(-33.2)] = \underline{114.2\ kJ}$

The forward reaction is endothermic and leads to an increase in the number of moles of gas. Thus, increased temperature favors the forward reaction (the formation of NO and $O_2$), and increased pressure favors the reverse reaction (formation of $NO_2$).

32. In writing the equilibrium constant expression for any heterogeneous equilibrium, omit any component that has a constant concentration, for example, any pure liquid or solid, (for which mol $L^{-1}$ is the density).

(a) $K_p = \dfrac{(p_{H_2})(p_{O_2})^{1/2}}{(p_{H_2O})}$ ; $K_c = \dfrac{[H_2][O_2]^{1/2}}{[H_2O]}$

(b) $K_p = (p_{H_2})(p_{O_2})^{1/2}$ ; $K_c = [H_2][O_2]^{1/2}$

(c) $K_p = (p_{H_2})^2(p_{O_2})$ ; $K_c = [H_2]^2[O_2]$

(d) $K_p = \dfrac{1}{(p_{H_2})(p_{O_2})}$ ; $K_c = \dfrac{1}{[H_2][O_2]}$

(e) $K_p = (p_{O_2})$ ; $K_c = [O_2]$

33. This problem is similar to Problem 33.

(a) $K_p = \dfrac{(p_{CO_2})}{(p_{O_2})}$ ; $K_c = \dfrac{[CO_2]}{[O_2]}$

(b) $K_p = (p_{CO_2})$ ; $K_c = [CO_2]$

(c) $K_p = (p_{CO_2})(p_{H_2O})$ ; $K_c = [CO_2][H_2O]$

(d) $K_p = \dfrac{(p_{CO_2})}{(p_{CO})}$ ; $K_c = \dfrac{[CO_2]}{[CO]}$

(e) $K_p = \dfrac{(p_{H_2})^4}{(p_{H_2O})^4}$ ; $K_c = \dfrac{[H_2]^4}{[H_2O]^4}$

34. (a) Graphite does not appear in the equilibrium constant expression; addition of more graphite will have no effect.

(b) Addition of $CO_2(g)$ will favor the formation of C(s) and $O_2(g)$.

(c) The concentration of $O_2$ increases, but the increase is less than that due to the added $O_2$ because more $CO_2(g)$ is formed.

(d) The reaction as written is exothermic; lowering the temperature will favor the forward (exothermic) reaction and produce more $CO_2$ at equilibrium.

(e) Addition of a catalyst will speed up the rate at which equilibrium is achieved but has no effect on the position of equilibrium.

35. PV = nRT, or $P = \frac{n}{V}(RT)$, so that in converting pressure units (atm) to molar concentration units (mol $L^{-1}$) every pressure is replaced by mol $L^{-1}$ times RT. In general, for a $K_p$ expression for r moles of reactants giving p moles of products in the balanced equation,

$$K_p = K_c(RT)^{p-r}$$

because RT appears p times in the numerator and r times in the denominator of the $K_p$ expression.

(a) $K_c = K_p(RT)^2 = (9.1\ \text{atm}^{-2})[(0.0821\ \text{atm L mol}^{-1}\ \text{K}^{-1})(298\ \text{K})]^2$
$= \underline{5.4 \times 10^3\ \text{mol}^{-2}\ \text{L}^2}$

(b) When the balanced equation for the reaction is multiplied throughout by 2, the equilibrium constant expression changes from

$$K_c = \frac{1}{[NH_3][H_2S]} \quad \text{to} \quad K'_c = \frac{1}{[NH_3]^2[H_2S]^2} = (K_c)^2$$

$$K'_c = (5.4\text{x}10^3\ \text{mol}^{-2}\ \text{L}^2)^2 = \underline{2.9\text{x}10^7\ \text{mol}^{-4}\ \text{L}^4}$$

(c)

| | $NH_3(g)$ + | $H_2S(g)$ ⇌ | $NH_4HS(s)$ | |
|---|---|---|---|---|
| initially | 0 | 0 | - | mol $L^{-1}$ |
| at eq'm | x | x | - | mol $L^{-1}$ |

$$K_c = \frac{1}{[NH_3][H_2S]} = \frac{1}{x^2} = 5.4\text{x}10^3\ \text{mol}^{-2}\ \text{L}^2\ ;\ x = \underline{0.014\ \text{mol L}^{-1}}$$

The total moles of gases at equilibrium is 2x = $\underline{0.028\ \text{mol L}^{-1}}$

$$PV = nRT;\quad P = \frac{n}{V}(RT) = (0.028\ \text{mol L}^{-1})(0.0821\ \text{atm L mol}^{-1}\text{K}^{-1})(298\ \text{K})$$
$$= \underline{0.69\ \text{atm}}$$

Note that this is the vapor pressure of $NH_4HS(s)$ at 25°C, which is independent of the amount of $NH_4HS(s)$.

36.

| | $NH_3(g)$ + | $H_2S(g)$ ⇌ | $NH_4HS(s)$ | |
|---|---|---|---|---|
| initially | 1.00 | 1.00 | - | mol $L^{-1}$ |
| at eq'm | 1.00-x | 1.00-x | - | mol $L^{-1}$ |

$$K_c = \frac{1}{[NH_3][H_2S]} = 5.4\text{x}10^3\ \text{mol}^{-2}\ \text{L}^2 = \frac{1}{(1.00-x)^2}$$

$$(1.00-x)^2 = 1.85 \times 10^{-4}; \quad 1.00-x = 0.014; \quad \underline{x = 0.99}$$

The amount of $NH_4HS(s)$ at equilibrium is $(1.00+0.99) = \underline{1.99 \text{ mol}}$

Note that the concentration of $NH_3(g)$ and $H_2S(g)$ in equilibrium with $NH_4HS(s)$ at $25°C$ (0.014 mol L$^{-1}$ each) is the same as in Problem 35. It is in fact the amount of these gases that is responsible for the vapor pressure of $NH_4HS(s)$ at $25°C$, which is independent of the amount of $NH_4HS(s)$.

37. The required equation is obtained by adding together the first equation , where the value of $K_c$ is given, and twice the second equation, where the value of $K_c$ is also given.

$$\begin{array}{ll} MnO_2(s) + 2H_2(g) \rightleftharpoons Mn(s) + 2H_2O(g) & K_c = 182 \\ 2[CO(g) + H_2O(g) \rightleftharpoons CO_2(g) + H_2(g)] & K_c = (0.052)^2 \\ \hline MnO_2(s) + 2CO(g) \rightleftharpoons Mn(s) + 2CO_2(g) & K_c = 182(0.052)^2 \end{array}$$

i.e., $K_c = 182(0.052)^2 = \underline{0.49}$

Since the $K_c$ for the reaction is unitless,

$$K_p = K_c = \underline{0.49}$$

38. The molecules involved in the equilibrium

$$CO(g) + H_2(g) \rightleftharpoons H_2CO(g)$$

are: $\phantom{xx}^-{:}C \equiv O{:}\phantom{xx}$ H-H $\phantom{xx}$ and $\phantom{x}$ H-C=Ö:
$$\phantom{xxxxxxxxxxxxxxxxxxxxxxxxxxxxxxxx}|$$
$$\phantom{xxxxxxxxxxxxxxxxxxxxxxxxxxxxxxxx}H$$

and we use,

$$\Delta H° = \sum [BE(\text{reactants})] - \sum[BE(\text{products})]$$
$$= [BE(C \equiv O) + BE(H-H)] - [2BE(C-H) + BE(C=O)]$$
$$= [(1070) + (436)] - [2(413) + (707)]$$
$$= \underline{-27 \text{ kJ}}$$

The reaction is slightly exothermic, so the formation of methanal is favored by low temperature, and since there is a decrease in moles of gas in going from reactants to products, high pressure would also be favorable to a high yield of methanal.

39. From the data given, calculate the total number of moles of $N_2O_4(g)$ and $NO_2(g)$.

$$PV = nRT \text{ ; } n = \frac{PV}{RT} = \frac{(1.00 \text{ atm})(1.00 \text{ L})}{(0.0821 \text{ atm L mol}^{-1} \text{ K}^{-1})(333 \text{ K})} = \underline{0.0366 \text{ mol}}$$

Then assume that there are x mol $N_2O_4$ and (0.0366-x) mol $NO_2$.

and since the total mass of gases is 2.50 g,

$$(x \text{ mol } N_2O_4)\left(\frac{92.02 \text{ g } N_2O_4}{1 \text{ mol } N_2O_4}\right) + [(0.0366-x) \text{ mol } NO_2]\left(\frac{46.01 \text{ g } NO_2}{1 \text{ mol } NO_2}\right)$$
$$= 2.500 \text{ g}$$

$$92.02x + 1.684 - 46.01x = 2.500 \;;\; 46.01x = 0.816$$
$$x = 0.0177 \text{ mol} \;;\; (0.0366-x) = 0.0189 \text{ mol}$$

i,e., $[N_2O_4] = 0.0177 \text{ mol L}^{-1}$ ; $[NO_2] = 0.0189 \text{ mol L}^{-1}$

(a) The equilibrium concentrations of $N_2O_4$ and $NO_2$ give the initial moles of $N_2O_4$ as $(0.0177 + 0.0189/2) = 0.0272$ mol

$$\% \text{ dissociation} = \left(\frac{(0.0272-0.0177) \text{ mol L}^{-1}}{0.0272 \text{ mol L}^{-1}}\right) \times 100 = \underline{34.9\%}$$

(b)
$$N_2O_4(g) \rightleftharpoons 2NO_2(g)$$

$$K_c = \frac{[NO_2]^2}{[N_2O_4]} = \frac{(0.0189 \text{ mol L}^{-1})^2}{(0.0177 \text{ mol L}^{-1})} = \underline{0.0202 \text{ mol L}^{-1}}$$

(c)
$$K_p = K_c(RT) = (0.0202 \text{ mol L}^{-1})(0.0821 \text{ atm L mol}^{-1} \text{ K}^{-1})(333 \text{ K})$$
$$= \underline{0.552 \text{ atm}}$$

40. If the initial pressure of $SO_2(g)$ is P, then the initial pressure of $O_2(g) = P/2$ since the pressure is proportional to the number of moles of gas.

One third of the $SO_2(g)$ is converted to $SO_3(g)$, thus the equilibrium pressure of $SO_2$ is $2P/3$, and the pressure of $SO_3(g)$ produced must be $P/3$, since one mol $SO_2$ gives one mol $SO_3$, and the pressure of $O_2$ to produce this amount of $SO_3$ must be

$$\frac{1}{2}\left(\frac{P}{3}\right) = \frac{P}{6}$$

since 2 mol of $SO_2$ reacts with 1 mole of $O_2$. Thus, we have

| | $2SO_2(g)$ | $+ \; O_2(g)$ | $\rightleftharpoons 2SO_3(g)$ | |
|---|---|---|---|---|
| initially | $P$ | $\frac{P}{2}$ | $0$ | atm |
| at eq'm | $P - \frac{P}{3}$ | $\frac{P}{2} - \frac{P}{6}$ | $\frac{P}{3}$ | atm |
| i.e., | $\frac{2P}{3}$ | $\frac{P}{3}$ | $\frac{P}{3}$ | atm |

and since the total pressure is 5.00 atm

$$\frac{2P}{3} + \frac{P}{3} + \frac{P}{3} = 5.00 \text{ atm} \;;\; P = \underline{3.75 \text{ atm}}$$

and we have at equilibrium $P_{SO_2} = \frac{2P}{3} = 2.50$ atm

$$P_{O_2} = \frac{P}{3} = 1.25 \text{ atm} = P_{SO_3}$$

Thus, $K_p = \dfrac{(P_{SO_3})^2}{(P_{SO_2})^2(P_{O_2})} = \dfrac{(1.25\ \text{atm})^2}{(2.50\ \text{atm})^2(1.25\ \text{atm})} = \underline{0.200\ \text{atm}^{-1}}$

41. (a)

| | $N_2O_4(g) \rightleftharpoons$ | $2NO_2(g)$ | |
|---|---|---|---|
| initially | 0.160 | 0 | mol $L^{-1}$ |
| at eq'm | 0.090 | 0.140 | mol $L^{-1}$ |

(Note 0.140 mol $L^{-1}$ $NO_2$ must have been formed from 0.070 mol $L^{-1}$ $N_2O_4$, so the concentration of $N_2O_4$ at equilibrium = (0.160-0.070) = 0.090 mol $L^{-1}$)

$$K_c = \frac{[NO2]^2}{[N_2O_4]} = \frac{(0.140\ \text{mol L}^{-1})^2}{(0.090\ \text{mol L}^{-1})} = \underline{0.218\ \text{mol L}^{-1}}$$

(b)

| | $N_2O_4(g) \rightleftharpoons$ | $2NO_2(g)$ | |
|---|---|---|---|
| initially | 0.080 | 0 | mol $L^{-1}$ |
| at eq'm | 0.080-x | 2x | mol $L^{-1}$ |

$$K_c = \frac{(2x)^2}{0.080-x} = 0.218\ \text{mol L}^{-1}$$

$4x^2 = 0.01744 - 0.218x$ <u>or</u> $4x^2 + 0.218x - 0.01744 = 0$

Solving the quadratic equation gives x = 0.0441 mol $L^{-1}$

Thus, at equilibrium: $[N_2O_4] = \underline{0.036\ \text{mol L}^{-1}}$

$[NO_2] = \underline{0.088\ \text{mol L}^{-1}}$

(c) (i) The reaction is endothermic as written. Thus, the effect of increasing the temperature would be to favor the forward reaction and <u>increase</u> the concentration of $NO_2$.

(ii) Decreasing the volume of the vessel increases the number of moles per unit volume, which would be counteracted by the reaction proceeding in the reverse direction to achieve a new equilibrium where more $N_2O_4$ is formed; decreasing the volume would lead to a decrease in the concentration of $NO_2$.

42. (a) We can use the density data to calculate the molar mass of aluminum chloride at the specified temperatures.

<u>At 200°C</u>, the density is $6.87\times10^{-3}$ g $mL^{-1}$ = 6.87 g $L^{-1}$, and for 1 L;

$$PV = nRT\ ;\quad n = \frac{PV}{RT} = \frac{(1\ \text{atm})(1\ \text{L})}{(0.0821\ \text{atm L mol}^{-1}\ \text{K}^{-1})(473\ \text{K})} = 2.58\times10^{-2}\ \text{mol}$$

molar mass at 200°C = $\left(\dfrac{6.87\ \text{g}}{2.58\times10^{-2}\ \text{mol}}\right) = \underline{266\ \text{g mol}^{-1}}$

Al is in group 3, so the empirical formula of the chloride is $AlCl_3$, with a formula mass of 133.3 u

Thus, at <u>200°C</u>, the molecular formula is $\underline{Al_2Cl_6}$ (molar mass = 266.6 g $mol^{-1}$)

<u>At 800°C</u>, the density is $1.51\times10^{-3}$ g $mL^{-1}$ = 1.51 g $L^{-1}$, and for 1 L;

$$PV = nRT \; ; \; n = \frac{PV}{RT} = \frac{(1\ \text{atm})(1\ \text{L})}{(0.0821\ \text{atm L mol}^{-1}\ \text{K}^{-1})(1073\ \text{K})} = 1.14\text{x}10^{-2}\ \text{mol}$$

$$\text{molar mass at } 800^{\circ}\text{C} = \left(\frac{1.51\ \text{g}}{1.14\text{x}10^{-2}\ \text{mol}^{-1}}\right) = \underline{132\ \text{g mol}^{-1}}$$

so that at 800°C, the molecular formula is $AlCl_3$ (molar mass 133.3 g $mol^{-1}$)

(b) At the intermediate temperature of 600°C the density is 2.65 g $mol^{-1}$, intermediate between the density at 200°C and 800°C, so that the apparent molecular formula is between that for $AlCl_3$ and that for $Al_2Cl_6$, consistent with the equilibrium:

$$Al_2Cl_6(g) \rightleftharpoons 2AlCl_3(g)$$

(c) At 600°C the density is $2.65\text{x}10^{-3}$ g $mL^{-1}$ = 2.65 g $L^{-1}$, and we can calculate the total number of moles of $AlCl_3$ and $Al_2Cl_6$ with this average density.

$$PV = nRT \; ; \; n = \frac{(1\ \text{atm})(1\ \text{L})}{(0.0821\ \text{atm L mol}^{-1}\ \text{K}^{-1})(873\ \text{K})} = 1.40\text{x}10^{-2}\ \text{mol}$$

$$\text{and the average molar mass} = \left(\frac{2.65\ \text{g}}{1.40\text{x}10^{-2}\ \text{mol}}\right) = 189\ \text{g mol}^{-1}$$

Let the fraction of the mixture that is $AlCl_3$ be x, then the fraction of $Al_2Cl_6$ is (1-x), and we can write:

$$x(133.3) + (1-x)(266.6) = 189 \; ; \; 133.3x = 77.6 \; ; \; x = 0.58$$

Thus, mol $AlCl_3$ = $0.58(1.40\text{x}10^{-2}$ mol) = $\underline{0.81\text{x}10^{-2}\ \text{mol}}$

mol $Al_2Cl_6$ = $0.42(1.40\text{x}10^{-2}$ mol) = $\underline{0.59\text{x}10^{-2}\ \text{mol}}$

and we can now calculate the partial pressures, which are proportional to the mole fractions:

$$P_{AlCl_3} = 0.58(1\ \text{atm}) = \underline{0.58\ \text{atm}} \; ; \; P_{Al_2Cl_6} = 0.42(1\ \text{atm}) = \underline{0.42\ \text{atm}}$$

(d)

$$K_p = \frac{(P_{AlCl_3})^2}{(P_{Al_2Cl_6})} = \frac{(0.58\ \text{atm})^2}{(0.42\ \text{atm})} = \underline{0.80\ \text{atm}}$$

$$K_p = K_c(RT) \; ; \; K_c = \frac{0.80\ \text{atm}}{(0.0821\ \text{atm L mol}^{-1}\ \text{K}^{-1})(873\ \text{K})} = \underline{1.1 \times 10^{-2}\ \text{mol L}^{-1}} \text{ at } 600^{\circ}\text{C}$$

Alternatively, we could have calculated $K_c$ directly from the molar concentrations:

$$[AlCl_3] = 0.81\text{x}10^{-2}\ \text{mol L}^{-1}$$

$$[Al_2Cl_6] = 0.59\text{x}10^{-2}\ \text{mol L}^{-1}$$

$$K_c = \frac{[AlCl_3]^2}{[Al_2Cl_6]} = \frac{(0.81\text{x}10^{-2}\ \text{mol L}^{-1})^2}{(0.59\text{x}10^{-2}\ \text{mol L}^{-1})} = \underline{1.1\text{x}10^{-2}\ \text{mol L}^{-1}}$$

## CHAPTER 14

1. (a) to (d) are aqueous solutions of strong acids; (e) and (f) are aqueous solutions of strong bases.
   (a) $10^{-5}$ M $H_3O^+$(aq); $10^{-5}$ M $NO_3^-$(aq)
   (b) 0.0023 M $H_3O^+$(aq); 0.0023 M $Cl^-$(aq)
   (c) 0.113 M $H_3O^+$(aq); 0.113 M $ClO_4^-$(aq)
   (d) 0.034 M $H_3O^+$(aq); 0.034 M $Br^-$(aq)
   (e) $10^{-3}$ M $Na^+$(aq); $10^{-3}$ M $OH^-$(aq)
   (f) 0.145 M $Ba^{2+}$(aq); 0.290 M $OH^-$(aq)

2. All of the solutions are aqueous solutions of simple strong acids and bases, except for (f), where the reaction is
   $$Na_2O(aq) + H_2O(l) \longrightarrow 2Na^+(aq) + 2OH^-(aq)$$
   (a) 0.0234 M $H_3O^+$(aq); 0.0234 M $I^-$(aq)
   (b) $10^{-5}$ M $Ca^{2+}$(aq); $2x10^{-5}$ M $OH^-$(aq)
   (c) 0.204 M $Li^+$(aq); 0.204 M $OH^-$(aq)
   (d) 0.342 M $H_3O^+$(aq); 0.342 M $Br^-$(aq)
   (e) 0.0023 M $Sr^{2+}$(aq); 0.0046 M $OH^-$(aq)
   (f) $2x10^{-4}$ M $Na^+$(aq); $2x10^{-4}$ M $OH^-$(aq)

3. The common strong acids are the hydrogen halides - hydrochloric acid, HCl(aq), hydrobromic acid, HBr(aq), and hydroiodic acid, HI(aq) - nitric acid, $HNO_3$(aq), perchloric acid, $HClO_4$(aq), and sulfuric acid, $H_2SO_4$(aq), (in as far as its first stage of ionization is concerned).
   Important weak acids include hydrofluoric acid, HF(aq), acetic (ethanoic) acid, $CH_3CO_2H$(aq), hydrocyanic acid, HCN(aq), and hypochlorous acid, HOCl(aq), all of which are monoprotic acids, carbonic acid, $H_2CO_3$(aq), and hydrogen sulfide, $H_2S$(aq), which are diprotic acids, and phosphoric acid, $H_3PO_4$(aq), which is a triprotic acid. Anions that behave as weak acids, such as hydrogen sulfate ion, $HSO_4^-$(aq), and strongly hydrated metal cations, such as $Al(H_2O)_6^{3+}$, could also be included in the list.

4. Alkali metal hydroxides and the soluble alkaline earth hydroxides all behave as strong bases, as do their oxides (since the $O^{2-}$ ion is quantitatively protonated by $H_2O$ to give $OH^-$). Less common strong bases include the amides, nitrides, and hydrides of the alkali and alkaline earth metals, such as $NaNH_2$, $Li_3N$, and $MgH_2$.
   The commonest weak base is ammonia, $NH_3$(aq). Other examples include derivatives of ammonia, such as methylamine, $CH_3NH_2$, dimethylamine, $(CH_3)_2NH$, trimethylamine, $(CH_3)_3N$, and aminobenzene (aniline), $C_6H_5NH_2$. The list could also include other hydrides, such as water, $H_2O$, and phosphine, $PH_3$, and salts of the anions of weak acids, such as those containing phosphate ion, $PO_4^{3-}$, carbonate, $CO_3^{2-}$, fluoride, $F^-$, and cyanide, $CN^-$.

5. In solving all problems of this kind, follow the steps given in the text.
   1) Write the balanced equation for the reaction in aqueous solution.

2. Write the expession for the equilibrium constant
3. Write expressions for the concentration of each species initially and when the system has achieved equilibrium.
4. Substitute these expressions in the equilibrium constant expression, and solve for any unknowns. Thus

| | $HCN(aq) + H_2O(l) \rightleftharpoons$ | $H_3O^+(aq)$ + | $CN^-(aq)$ | |
|---|---|---|---|---|
| initially | 0.010 | 0 | 0 | $mol\ L^{-1}$ |
| at eq'm | 0.010-x | x | x | $mol\ L^{-1}$ |

$$K_a = \frac{[H_3O^+][CN^-]}{[HCN]} = \frac{x^2}{0.010-x} = 4.9x10^{-10}\ mol\ L^{-1}$$

Assuming x << 0.010, $x^2 = 4.9x10^{-12}$ ; $x = 2.2x10^{-6}$

and the assumption is justified.

Thus: $[H_3O^+] = \underline{2.2 \times 10^{-6}\ mol\ L^{-1}}$

% dissociation of acid = $(\frac{2.2x10^{-6}\ mol\ L^{-1}}{0.010\ mol\ L^{-1}}) \times 100\% = \underline{0.022\%}$

6. This problem is similar to Problem 5.

| | $HF(aq) + H_2O(l) \rightleftharpoons$ | $H_3O^+(aq)$ + | $F^-(aq)$ | |
|---|---|---|---|---|
| initially | 0.060 | 0 | 0 | $mol\ L^{-1}$ |
| at eq'm | 0.060-x | x | x | $mol\ L^{-1}$ |

$$K_a = \frac{[H_3O^+][F^-]}{[HF]} = \frac{x^2}{0.060-x} = 3.5x10^{-4}\ mol\ L^{-1}$$

Assuming x << 0.060, is not justified in this case.

Solving the quadratic equation gives

$[H_3O^+] = \underline{4.4 \times 10^{-3}\ mol\ L^{-1}}$

% dissociation of acid = $(\frac{4.4x10^{-3}\ mol\ L^{-1}}{0.0\ 56}) \times 100\% = \underline{7.8\%}$

7. We need to calculate $[H_3O^+]$ for each of the solutions.

$HNO_3(aq)$ is a strong acid, thus $[H_3O^+] = 0.0010\ mol\ L^{-1} = \underline{1.0x10^{-3}\ mol\ L^{-1}}$

For the solution of the weak acid acetic acid, $CH_3CO_2H(aq)$:

| | $CH_3CO_2H(aq) + H_2O(l) \rightleftharpoons$ | $H_3O^+(aq)$ + | $CH_3CO_2^-(aq)$ | |
|---|---|---|---|---|
| initially | 0.200 | 0 | 0 | $mol\ L^{-1}$ |
| at eq'm | 0.200-x | x | x | $mol\ L^{-1}$ |

$$K_a = \frac{[H_3O^+][CH_3CO_2^-]}{[CH_3CO_2H]} = \frac{x^2}{0.200-x} = 1.8x10^{-5}\ mol\ L^{-1}$$

Assuming x << 0.200, $x^2 = 3.6 \times 10^{-6}$ ; $x = 1.9 \times 10^{-3}$

and the assumption is justified.

Thus, $[H_3O^+] = \underline{1.9 \times 10^{-3}\ mol\ L^{-1}}$ which is greater than the $[H_3O^+]$ in 0.0010 mol $L^{-1}$ $HNO_3(aq)$. 0.200 M acetic acid has the higher concentration of $H_3O^+$ ions.

8. For solutions of weak bases the calculations are similar to those for solutions of weak acids (see e.g., Problems 5, 6, and 7).

| | $NH_3(aq) + H_2O(l)$ | $NH_4^+(aq)$ | $+ OH^-(aq)$ | |
|---|---|---|---|---|
| initially | 0.040 | 0 | 0 | mol $L^{-1}$ |
| at eq'm | 0.040-x | x | x | mol $L^{-1}$ |

$$K_b = \frac{[NH_4^+][OH^-]}{[NH_3]} = \frac{x^2}{0.040-x} = 1.8\text{x}10^{-5}\ mol\ L^{-1}$$

Assuming x << 0.040, $x^2 = 7.2\text{x}10^{-7}$ ; $x = 8.5\text{x}10^{-4}$

and the assumption is justified.

Thus: $[OH^-] = \underline{8.5 \times 10^{-4}\ mol\ L^{-1}}$

$$\%\ \text{dissociation of base} = \left(\frac{8.5\text{x}10^{-4}\ mol\ L^{-1}}{0.040\ mol\ L^{-1}}\right) \times 100\% = \underline{2.1\%}$$

9. This problem is similar to Problem 8.

| | $C_6H_5NH_2(aq) + H_2O(l) \rightleftharpoons$ | $C_6H_5NH_3^+(aq)$ | $+ OH^-(aq)$ |
|---|---|---|---|
| initially | 0.080 | 0 | 0 mol $L^{-1}$ |
| at eq'm | 0.080-x | x | x mol $L^{-1}$ |

$$K_b = \frac{[C_6H_5NH_3^+][OH^-]}{[C_6H_5NH_2]} = \frac{x^2}{0.080-x} = 4.3\text{x}10^{-10}\ mol\ L^{-1}$$

Assuming x << 0.080 ; $x^2 = 3.44\text{x}10^{-11}$ ; $x = 5.9\text{x}10^{-6}$

and the assumption is justified.

Thus: $[OH^-] = \underline{5.9 \times 10^{-6}\ mol\ L^{-1}}$

$$\%\ \text{dissociation of base} = \left(\frac{5.9\text{x}10^{-6}\ mol\ L^{-1}}{0.080\ mol\ L^{-1}}\right) \times 100\% = \underline{0.007\%}$$

10. $pH = -\log_{10}[H_3O^+]$, where $[H_3O^+]$ is in mol $L^{-1}$ but pH has no units:

(a) $pH = -\log 10^{-5} = -(-5) = \underline{5.0}$ (b) $pH = -\log 0.0023 = \underline{2.64}$

(c) $pH = -\log 0.113 = \underline{0.95}$ (d) $pH = -\log 0.034 = \underline{1.47}$

(e) $[H_3O^+][OH^-] = 10^{-14}\ mol^2\ L^{-2}$; $[H_3O^+] = \frac{10^{-14}}{[OH^-]}$

and taking log's of both sides of this expression gives

$$\underline{pH = 14.00 + \log [OH^-]}$$

$pH = 14.00 - 3.00 = \underline{11.00}$

(f) $pH = 14.00 - 0.54 = \underline{13.46}$

11. (a) pH = -log 0.0234 = 1.63

(b) pH = 14.00 + log $[OH^-]$ = 14.00 - 4.70 = 9.30

(c) pH = 14.00 + log $[OH^-]$ = 14.00 - 0.69 = 13.31

(d) pH = -log 0.342 = 0.47

(e) pH = 14.00 + log $[OH^-]$ = 14.00 - 2.34 = 11.66

(f) pH = 14.00 + log $(2\text{x}10^{-4})$ = 14.00 - 3.70 = 10.30

12. We need first to calculate $[H_3O^+]$ in 0.0050 M HF(aq).

$$HF(aq) + H_2O(l) \rightleftharpoons H_3O^+(aq) + F^-(aq)$$

| | | | | |
|---|---|---|---|---|
| initially | 0.0050 | 0 | 0 | mol $L^{-1}$ |
| at eq'm | 0.0050-x | x | x | mol $L^{-1}$ |

$$K_a = \frac{[H_3O^+][F^-]}{[HF]} = \frac{x^2}{0.0050-x} = 3.5\text{x}10^{-4} \text{ mol L}^{-1}$$

Assuming x << 0.0050, $x^2 = 1.75\text{x}10^{-6}$ ; x = $1.32\text{x}10^{-3}$, but the assumption x << 0.0050 is not justified, so we have to solve the quadratic equation, or, alternatively, we can use the method of successive approximations.

To a first approximation x = $1.32\text{x}10^{-3}$ = 0.00132, and we can write

$$\frac{x^2}{0.0050-0.0013} = 3.5\text{x}10^{-4}; \; x^2 = 1.30\text{x}10^{-6} \; ; \; x = 1.14\text{x}10^{-3}$$

and repeating the procedure gives:

$$\frac{x^2}{0.0050-0.0011} = 3.5\text{x}10^{-4}; \; x^2 = 1.37\text{x}10^{-6} \; ; \; x = \underline{1.17\text{x}10^{-3}}$$

This value of x is substantially the same as that obtained at the second stage of approximation and is an acceptable value for x. (Note that if we solve the quadratic equation we obtain x = $1.16\text{x}10^{-3}$)

Thus: $[H_3O^+] = 1.17\text{x}10^{-3}$ mol $L^{-1}$ ; pH = -log $[H_3O^+]$ = 2.93

13. From the given pH, we can calculate the $[H_3O^+]$ in 0.100 M HOCl(aq).

pH = 4.2 ; $[H_3O^+] = 10^{-4.2}$ = $6.3\text{x}10^{-5}$ mol $L^{-1}$, and we can write:

$$HOCl(aq) + H_2O(l) \rightleftharpoons H_3O^+(aq) + OCl^-(aq)$$

| | | | | |
|---|---|---|---|---|
| initially | 0.100 | 0 | 0 | mol $L^{-1}$ |
| at eq'm | 0.100-x | x | x | mol $L^{-1}$ |

and x = $6.3\text{x}10^{-5}$ mol $L^{-1}$. Thus, 0.100-x = 0.0999 and we can write:

$$K_a = \frac{[H_3O^+][OCl^-]}{[HOCl]} = \frac{(6.3\text{x}10^{-5} \text{ mol L}^{-1})^2}{0.0999 \text{ mol L}^{-1}} = \underline{4.0 \times 10^{-8} \text{ mol L}^{-1}}$$

$$pK_a = -\log K_a = \underline{7.40}$$

14. $pK_a = 3.08$; $K_a = 10^{-3.08} = 8.3\times10^{-4}$ mol $L^{-1}$, and for lactic acid:

| | HA(aq) + $H_2O(l)$ ⇌ | $H_3O^+(aq)$ + | $A^-(aq)$ | |
|---|---|---|---|---|
| initially | $1.0\times10^{-3}$ | 0 | 0 | mol $L^{-1}$ |
| at eq'm | $(1.0\times10^{-3})-x$ | x | x | mol $L^{-1}$ |

$$K_a = \frac{[H_3O^+][A^-]}{[HA]} = \frac{x^2}{(1.0\times10^{-3})-x} = 8.3\times10^{-4} \text{ mol L}^{-1}$$

assuming $x << 1\times10^{-3}$, $x^2 = 8.3\times10^{-7}$; $x = 9.1\times10^{-4}$, which is not admissable, and we have to solve the quadratic equation,

$$x^2 + (8.3\times10^{-4})x - (8.3\times10^{-7}) = 0 \; ; \quad x = \underline{5.9\times10^{-4}}$$

$$[H_3O^+] = \underline{5.9 \times 10^{-4} \text{ mol L}^{-1}} \; ; \; pH = \underline{3.23}$$

15. In each case we first determine the stoichiometric composition of the solution after reaction, so we first calculate the moles of reactants and products initially and at equilibrium.

(a) initial mol KOH(aq) = (25.00 mL)(0.10 mol $L^{-1}$) = 2.50 mmol

initial mol $HNO_3$(aq) = (50.0 mL)(0.080 mol $L^{-1}$) = 4.00 mmol

| | KOH(aq) + | $HNO_3(aq)$ ⟶ | $KNO_3(aq)$ + $H_2O(l)$ | |
|---|---|---|---|---|
| initially | 2.50 | 4.00 | 0 | mmol |
| at eq'm | 0 | 1.50 | 2.50 | mmol |

At equilibrium the solution contains 1.50 mmol of the strong acid $HNO_3$(aq) in a total volume of (25.0 + 50.0) mL = 75.00 mL

$$[H_3O^+] = \left(\frac{1.50 \text{ mmol HNO}_3}{75.0 \text{ mL}}\right) = \underline{0.020 \text{ mol L}^{-1}} \; ; \; \underline{pH = 1.70}$$

(b) initial mol HCl(aq) = (25.0 mL)(0.13 mol $L^{-1}$) = 3.25 mmol

initial mol NaOH(aq) = (35.0 mL)(0.12 mol $L^{-1}$) = 4.20 mmol

| | HCl(aq) + | NaOH(aq) ⟶ | NaCl(aq) + $H_2O(l)$ | |
|---|---|---|---|---|
| initially | 3.25 | 4.20 | 0 | mmol |
| at eq'm | 0 | 0.95 | 3.25 | mmol |

At equilibrium the solution contains 0.95 mmol of the strong base NaOH(aq) in (25.0 + 35.0) mL = 60.0 mL

$$[OH^-] = \left(\frac{0.95 \text{ mmol NaOH}}{60.0 \text{ mL}}\right) = \underline{0.016 \text{ mol L}^{-1}}$$

$$pH = 14.00 + \log [OH^-] = 14.00 - 1.80 = \underline{12.20}$$

(c) initial mol $HNO_3$(aq) = (35.0 mL)(0.050 mol $L^{-1}$) = 1.75 mmol

initial mol NaOH(aq) = (70.0 mL)(0.025 mol $L^{-1}$) = 1.75 mmol

| | $HNO_3(aq) + NaOH(aq)$ | $\rightarrow$ | $NaNO_3(aq) + H_2O(l)$ | |
|---|---|---|---|---|
| initially | 1.75 | 1.75 | 0 | mmol |
| at eq'm | 0 | 0 | 1.75 | mmol |

In this case the final solution contains only 1.75 mmol of the neutral salt $NaNO_3$ and is therefore neutral:

$$\underline{pH = 7.00}$$

16. Since $pK_a = -\log K_a$, or $pK_a = 10^{-K_a}$, the greater the magnitude of $pK_a$ the smaller is the value of $K_a$, hence, the order of acid strengths is:

| | $CCl_3CO_2H$ | > | $CHCl_2CO_2H$ | > | $CH_2ClCO_2H$ | > | $CH_3CO_2H$ | |
|---|---|---|---|---|---|---|---|---|
| $pK_a$: | 0.7 | | 1.3 | | 2.9 | | 4.8 | |
| Ka: | 0.20 | | $5.0\times10^{-2}$ | | $1.3\times10^{-3}$ | | $1.6\times10^{-5}$ | $(mol\ L^{-1})$ |

For each acid, we have:

| | $HA(aq) + H_2O(l)$ | $\rightleftharpoons$ | $H_3O^+(aq)$ | $+ A^-(aq)$ | |
|---|---|---|---|---|---|
| initially | 0.10 | | 0 | 0 | $mol\ L^{-1}$ |
| at eq'm | 0.10-x | | x | x | $mol\ L^{-1}$ |

$$K_a = \frac{[H_3O^+][A^-]}{[HA]} = \frac{x^2}{0.10-x}$$

and we can solve the equation for each acid according to the usual methods (Note that only for the solution of $CH_3CO_2H$ is the approximation $x << 0.10$ valid; in the other cases we have to solve the quadratic equations)

| acid | $[H_3O^+]$ $mol\ L^{-1}$ | pH |
|---|---|---|
| $CCl_3CO_2H$ | 0.073 | 1.1 |
| $CHCl_2CO_2H$ | 0.050 | 1.3 |
| $CH_2ClCO_2H$ | 0.011 | 2.0 |
| $CH_3CO_2H$ | $1.26\times10^{-3}$ | 2.9 |

17. KF is fully dissociated in solution into $K^+(aq)$ and $F^-(aq)$ ions. The former has no acid or base properties in water, but $F^-$ is the conjugate base of the weak acid HF(aq) and is thus a weak base in water:

| | $F^-(aq) + H_2O(l)$ | $\rightleftharpoons$ | $HF(aq)$ | $+ OH^-(aq)$ | |
|---|---|---|---|---|---|
| initially | 0.0050 | | 0 | 0 | $mol\ L^{-1}$ |
| at eq'm | 0.0050-x | | x | x | $mol\ L^{-1}$ |

$$K_b(F^-) = \frac{x^2}{0.0050-x} = \frac{K_w}{K_a(HF)} = \frac{10^{-14}\ mol^2\ L^{-2}}{3.5 \times 10^{-4}\ mol\ L^{-1}} = \underline{2.9\times10^{-11}\ M}$$

Assuming $x << 0.0050$, $x^2 = 1.45\times10^{-13}$; $\underline{x = 3.81\times10^{-7}}$

$[OH^-] = 3.81\times10^{-7}\ mol\ L^{-1}$

$pH = 14.00 + \log[OH^-] = 14.00 - 6.42 = \underline{7.58}$

18. (a) NaCN is fully dissociated into $Na^+(aq)$ and $CN^-(aq)$; the former has no acid or base properties, but the latter is the conjugate base of the weak acid HCN(aq), and is therefore a weak base.

$$CN^-(aq) + H_2O(l) \rightleftharpoons HCN(aq) + OH^-(aq)$$

| | | | | |
|---|---|---|---|---|
| initially | 0.10 | 0 | 0 | $mol\ L^{-1}$ |
| at eq'm | 0.10-x | x | x | $mol\ L^{-1}$ |

$$K_b(CN^-) = \frac{x^2}{0.10-x} = \frac{K_w}{K_a(HCN)} = \frac{10^{-14}\ mol^2\ L^{-2}}{4.9\times10^{-10}\ mol\ L^{-1}} = \underline{2.0\times10^{-5}\ M}$$

$x << 0.10$, $x^2 = 2.0\times10^{-6}$ ; $x = \underline{1.41\times10^{-3}}$

$[OH^-] = 1.41\times10^{-3}\ mol\ L^{-1}$ ; $pH = 14.00 + \log [OH^-] = \underline{11.2}$

(b) KCl is the salt of a strong acid and a strong base and will give a neutral solution of $K^+(aq)$ and $Cl^-(aq)$; $\underline{pH = 7.00}$

(c) LiOH is a strong base in aqueous solution and the solution will contain 0.10 M $OH^-$. $pH = 14.00-\log[OH^-] = \underline{13.00}$

(d) HBr is a strong acid in aqueous solution and the solution will contain 0.10 M $H_3O^+$. $pH = -\log 0.10 = \underline{1.00}$

(e) $NH_3$ is a weak base in aqueous solution.

$$NH_3(aq) + H_2O(l) \rightleftharpoons NH_4^+(aq) + OH^-(aq)$$

| | | | | |
|---|---|---|---|---|
| initially | 0.10 | 0 | 0 | $mol\ L^{-1}$ |
| at eq'm | 0.10-x | x | x | $mol\ L^{-1}$ |

$$K_b = \frac{[NH_4^+][OH^-]}{[NH_3]} = \frac{x^2}{0.10-x} = 1.8\times10^{-5}\ mol\ L^{-1}$$

$x << 0.10$, $x^2 = 1.8\times10^{-6}$; $\underline{x = 1.34\times10^{-3}}$

$[OH^-] = 1.34\times10^{-3}\ mol\ L^{-1}$; $pH = 14.00 + \log [OH^-] = \underline{11.13}$

(f) $NH_4Cl$ is fully dissociated into $NH_4^+(aq)$ and $Cl^-(aq)$. The former is the conjugate acid of the weak base $NH_3$ and behaves as a weak acid. $Cl^-$ is the conjugate base of the strong acid HCl and has no base properties in water. Thus

$$NH_4(aq) + H_2O(l) \rightleftharpoons H_3O^+(aq) + NH_3(aq)$$

| | | | | |
|---|---|---|---|---|
| initially | 0.10 | 0 | 0 | $mol\ L^{-1}$ |
| at eq'm | 0.10-x | x | x | $mol\ L^{-1}$ |

$$K_a(NH_4^+) = \frac{x^2}{0.10-x} = \frac{K_w}{K_b(NH_3)} = \frac{10^{-14}\ mol^2\ L^{-2}}{1.8\times10^{-5}\ mol\ L^{-1}} = 5.6\times10^{-10}\ M$$

$x << 0.10$, $x^2 = 5.6\times10^{-11}$; $\underline{x = 7.5\times10^{-6}}$

$[H_3O^+] = 7.5\times10^{-6}\ mol\ L^{-1}$; $\underline{pH = 5.12}$

(g) $K_2O(s)$ reacts with water as a strong base, according to

$$K_2O(s) + H_2O(l) \rightarrow 2K^+(aq) + 2OH^-(aq)$$

Thus the solution contains 0.20 M $OH^-(aq)$, and has a pH given by

$pH = 14.00 + \log [OH^-] = \underline{13.30}$

Thus, for 0.10 M solutions we have the following pH's:

NaCN, 11.2; KCl, 7.00; LiOH, 13.00; HBr, 1.00; $NH_3$, 11.13; $NH_4Cl$, 5.12, and $K_2O$, 13.30, or in order of increasing pH:

$$HBr < NH_4Cl < KCl < NH_3 < NaCN < LiOH < K_2O$$

19. $NH_4Cl(aq)$ is fully dissociated into $NH_4^+(aq)$ and $Cl^-(aq)$ ions. $NH_4^+$ is the conjugate acid of the weak base $NH_3$ and behaves as a weak acid. $Cl^-$ is the conjugate base of the strong acid HCl and therefore has no basic properties in water.

| | $NH_4^+(aq) + H_2O(l)$ | $\rightleftharpoons$ $H_3O^+(aq)$ | $+ NH_3(aq)$ | |
|---|---|---|---|---|
| initially | 0.020 | 0 | 0 | mol $L^{-1}$ |
| at eq'm | 0.020-x | x | x | mol $L^{-1}$ |

$$K_a(NH_4^+) = \frac{[H_3O^+][NH_3]}{[NH_4^+]} = \frac{x^2}{0.020-x} = \frac{K_w}{K_b(NH_3)} = \frac{10^{-14}\ mol^2\ L^{-2}}{1.8x10^{-5}\ mol\ L^{-1}}$$

$$= 5.6x10^{-10}\ mol\ L^{-1}$$

assume x << 0.020, $x^2 = 1.12x10^{-11}$ ; $x = 3.35x10^{-6}$ mol $L^{-1}$

$[H_3O^+] = [NH_3] = \underline{3.4x10^{-6}\ mol\ L^{-1}}$;

$[NH_4^+] = [0.020-(3.4x10^{-6})] = \underline{0.020\ mol\ L^{-1}}$

$$[OH^-] = \frac{K_w}{[H_3O^+]} = \frac{10^{-14}\ mol^2\ L^{-2}}{3.4x10^{-6}\ mol\ L^{-1}} = \underline{2.9x10^{-9}\ mol\ L^{-1}}$$

20. (a)

| | $CH_3CO_2H(aq) + H_2O(l)$ | $\rightleftharpoons$ $H_3O^+(aq)$ | $+ CH_3CO_2^-(aq)$ | |
|---|---|---|---|---|
| initially | 0.010 | 0 | 0 | mol $L^{-1}$ |
| at eq'm | 0.010-x | x | x | mol $L^{-1}$ |

$$K_a = \frac{[H_3O^+][CH_3CO_2^-]}{[CH_3CO_2H]} = \frac{x^2}{0.010-x} = 1.8x10^{-5}\ mol\ L^{-1}$$

For x << 0.010, $x^2 = 1.8x10^{-7}$ ; $x = 4.24x10^{-4}$ (assumption justified)

$[H_3O^+] = 4,24x10^{-4}$ mol $L^{-1}$; pH = $\underline{3.37}$

(b)

| | $HF(aq) + H_2O(l)$ | $\rightleftharpoons$ $H_3O^+(aq)$ | $+ F^-(aq)$ | |
|---|---|---|---|---|
| initially | 0.10 | 0 | 0 | mol $L^{-1}$ |
| at eq'm | 0.10-x | x | x | mol $L^{-1}$ |

$$K_a = \frac{[H_3O^+][F^-]}{[HF]} = \frac{x^2}{0.10-x} = 3.5x10^{-4}\ mol\ L^{-1}$$

For x << 0.10, $x^2 = 3.5x10^{-5}$; $x = 5.92x10^{-3}$ (assumption justified)

$[H_3O^+] = 5.92x10^{-3}$ mol $L^{-1}$ ; $\underline{pH = 2.23}$

(c)

| | $NH_3(aq) + H_2O(l)$ | $NH_4^+(aq)$ | $+ OH^-(aq)$ | |
|---|---|---|---|---|
| initially | 0.0030 | 0 | 0 | mol $L^{-1}$ |
| at eq'm | 0.0030-x | x | x | mol $L^{-1}$ |

$$K_b = \frac{[NH_4^+][OH^-]}{[NH_3]} = \frac{x^2}{0.0030-x} = 1.8\text{x}10^{-5} \text{ mol L}^{-1}$$

$x << 0.0030$ is not justified. Solving the quadratic equation gives

$x = 2.24\text{x}10^{-4}$ mol $L^{-1}$; $[OH^-] = 2.24\text{x}10^{-4}$ mol $L^{-1}$

pH = 14.00 + log $[OH^-]$ = 14.00 - 3.65 = 10.35

(d) $CH_3CO_2Na$ is the salt of a weak acid and a strong base that is fully dissociated in aqueous solution into $Na^+(aq)$ and $CH_3CO_2^-(aq)$ ions. $Na^+(aq)$ has no acid or base properties but $CH_3CO_2^-(aq)$ is the conjugate base of the weak acid $CH_3CO_2H(aq)$ and is a weak base.

| | $CH_3CO_2^-(aq) + H_2O(l)$ | $\rightleftharpoons$ | $CH_3CO_2H(aq)$ | $+ OH^-(aq)$ | |
|---|---|---|---|---|---|
| initially | 0.10 | | 0 | 0 | mol $L^{-1}$ |
| at eq'm | 0.10-x | | x | x | mol $L^{-1}$ |

$$K_b(CH_3CO_2^-) = \frac{[CH_3CO_2H][OH^-]}{[CH_3CO_2^-]} = \frac{x^2}{0.10-x} = \frac{K_w}{K_a(CH_3CO_2H)}$$

$$= \frac{10^{-14} \text{ mol}^2 \text{ L}^{-2}}{1.8\text{x}10^{-5} \text{ mol L}^{-1}} = 5.6\text{x}10^{-10} \text{ mol L}^{-1}$$

For $x << 0.10$, $x^2 = 5.6\text{x}10^{-11}$; $x = 7.48\text{x}10^{-6}$ (assumption justified)

Thus, $[OH^-] = 7.48\text{x}10^{-6}$ mol $L^{-1}$

pH = 14.00 + log $[OH^-]$ = 14.00 - 5.13 = 8.87

(e) $NH_4Cl$ is the salt of a weak base and a strong acid that is fully dissociated in aqueous solution to $NH_4^+(aq)$ and $Cl^-(aq)$ ions. $Cl^-(aq)$ has no basic properties but $NH_4^+$ is the conjugate acid of the weak base $NH_3$ and behaves as a weak acid.

| | $NH_4^+(aq) + H_2O(l)$ | $\rightleftharpoons$ | $H_3O^+(aq)$ | $+ NH_3(aq)$ | |
|---|---|---|---|---|---|
| initially | 0.20 | | 0 | 0 | mol $L^{-1}$ |
| at eq'm | 0.20-x | | x | x | mol $L^{-1}$ |

$$K_a(NH_4^+) = \frac{[H_3O^+][NH_3]}{[NH_4^+]} = \frac{K_w}{K_b(NH_3)} = \frac{10^{-14} \text{ mol}^2 \text{ L}^{-2}}{1.8\text{x}10^{-5} \text{ mol L}^{-1}}$$

$$= 5.6\text{x}10^{-10} \text{ mol L}^{-1} = \frac{x^2}{0.20-x}$$

$x << 0.20$, $x^2 = 1.12\text{x}10^{-10}$; $x = 1.06\text{x}10^{-5}$ (assumption justified)

$[H_3O^+] = 1.06\text{x}10^{-5}$ mol $L^{-1}$ ; pH = 4.97

21. For each example we need to react the conjugate base of the cation of the salt with the conjugate acid of the anion, in the molar amounts determined by the appropriate balanced equation.

(a) Ammonium nitrate, $NH_4^+NO_3^-$

$$NH_3(aq) + HNO_3(aq) \longrightarrow NH_4NO_3(aq)$$

(b) Ammonium chloride, $NH_4^+Cl^-$

$$NH_3(aq) + HCl(aq) \longrightarrow NH_4Cl(aq)$$

(c) Calcium sulfate, $Ca^{2+}SO_4^{2-}$

$$Ca(OH)_2(aq) + H_2SO_4(aq) \longrightarrow CaSO_4(aq) + 2H_2O(l)$$

(d) Potassium acetate, $CH_3CO_2^-K^+$

$$KOH(aq) + CH_3CO_2H(aq) \longrightarrow CH_3CO_2K(aq) + H_2O(l)$$

(e) Aluminum chloride, $Al^{3+}(Cl^-)_3$

$$Al(OH)_3(s) + 3HCl(aq) \longrightarrow AlCl_3(aq) + 3H_2O(l)$$

(f) Lithium iodide, $Li^+I^-$

$$LiOH(aq) + HI(aq) \longrightarrow LiI(aq) + H_2O(l)$$

Both $NH_4NO_3$ and $NH_4Cl$ are ammonium salts formed from strong acids. Neither the $NO_3^-$ ion nor the $Cl^-$ ion have any basicity in water, but $NH_4^+$ is the conjugate acid of the weak base $NH_3$, and behaves as a weak acid in aqueous solution. Both $NH_4NO_3(aq)$ and $NH_4Cl(aq)$ are acidic, due to

$$NH_4^+(aq) + H_2O(l) \rightleftharpoons H_3O^+(aq) + NH_3(aq)$$

Aqueous solutions of (a) and (b) are acidic.

LiI is a salt formed from a strong base and a strong acid. Neither the cation nor the anion have acid or base properties in aqueous solution. A solution of LiI is neutral.

$CH_3CO_2K$ and $CaSO_4$ are salts that contain neutral cations and basic anions, because $CH_3CO_2^-$ is the conjugate base of the weak acid $CH_3CO_2H$, and $SO_4^{2-}$ is the conjugate base of the weak acid $HSO_4^-$. Each of the anions reacts with water to give a basic solution,

$$CH_3CO_2^-(aq) + H_2O(l) \rightleftharpoons CH_3CO_2H(aq) + OH^-(aq)$$

$$SO_4^{2-}(aq) + H_2O(l) \rightleftharpoons HSO_4^-(aq) + OH^-(aq)$$

although the latter solution is only very weakly basic. Solutions (c) and (d) are basic.

The small highly charged $Al^{3+}$ ion is strongly hydrated in aqueous solution and is present in solution as the $Al(H_2O)_6^{3+}(aq)$ ion, which behaves as a weak acid. $Cl^-$ is the conjugate base of the strong acid HCl and has no basic properties in water.

$$Al(H_2O)_6^{3+}(aq) + H_2O(l) \rightleftharpoons Al(H_2O)_5OH^{2+}(aq) + H_3O^+(aq)$$

Thus, solution (e) is acidic.

22. In considering the reactions of ionic compounds of metals, consider the reactions of the metal cation with water (if any) and the reaction of the anion with water (if any).

(a) $Na_2O$, sodium oxide, is $(Na^+)_2O^{2-}$. $Na^+(aq)$ has no acid or base properties in water but the oxide ion, $O^{2-}$, behaves as a strong base.

$$O^{2-}(aq) + H_2O(l) \longrightarrow 2OH^-(aq)$$

In solution, $Na^+(aq)$ is a spectator ion. The balanced equation is,

$$Na_2O(s) + H_2O(l) \longrightarrow 2Na^+(aq) + 2OH^-(aq)$$

(b) $K_2S$, potassium sulfide, is $(K^+)_2S^{2-}$, containing the neutral $K^+$ ion and the sulfide ion, $S2-$, which is the conjugate base of the weakly acidic $HS^-$ ion

$$S^{2-}(aq) + H_2O(l) \rightleftharpoons HS^-(aq) + OH^-(aq)$$

(c) $Na_2NH_2$, sodium amide, contains the $Na^+$ and $NH_2^-$ ions. $Na^+(aq)$ is neutral but $NH_2^-(aq)$, the conjugate base of ammonia, $NH_3(aq)$, is a strong base in water.

$$NH_2^-(aq) + H_2O(l) \longrightarrow NH_3(aq) + OH^-(aq)$$

All ionic amides behave similarly. Overall we can write:

$$NaNH_2(aq) + H_2O(l) \longrightarrow Na^+(aq) + NH_3(aq) + OH^-(aq)$$

23. (a) Carbon dioxide, $CO_2$, is an acidic nonmetal oxide, which reacts with water to give a solution of carbonic acid, $H_2CO_3(aq)$, a weak diprotic acid in water.

$$CO_2(g) + H_2O(l) \rightleftharpoons H_2CO_3(aq)$$
$$H_2CO_3(aq) + H_2O(l) \rightleftharpoons H_3O^+(aq) + HCO_3^-(aq)$$
$$HCO_3^-(aq) + H_2O(l) \rightleftharpoons H_3O^+(aq) + CO_3^{2-}(aq)$$

The solution is acidic

(b) Sulfur trioxide, $SO_3$, is an acidic nonmetal oxide, which reacts with water to give a solution of the strong diprotic acid sulfuric acid, $H_2SO_4(aq)$. The solution is acidic.

$$SO_3(g) + H_2O(l) \longrightarrow H_2SO_4(aq)$$
$$H_2SO_4(aq) + H_2O(l) \longrightarrow H_3O^+(aq) + HSO_4^-(aq)$$
$$HSO_4^-(aq) + H_2O(l) \rightleftharpoons H_3O^+(aq) + SO_4^{2-}(aq)$$

(c) Calcium oxide, $Ca^{2+}O^{2-}$, is a metal oxide which behaves as a strong base in water. The solution is basic.

$$CaO(s) + H_2O(l) \longrightarrow Ca^{2+}(aq) + 2OH^-(aq)$$

(d) Phosphorus trichloride, $PCl_3(l)$, is a nonmetal chloride that behaves as a Lewis acid in water in which it is hydrolyzed to phosphorous acid, $H_3PO_3(aq)$ and HCl(aq). The solution is acidic.

$$PCl_3(l) + 6H_2O(l) \longrightarrow H_3PO_3(aq) + 3H_3O^+(aq) + 3Cl^-(aq)$$

(e) Orthorhombic sulfur, $S_8(s)$, is insoluble in water.

(f) Lithium hydride, $Li^+H^-$, contains the neutral $Li^+$ ion and the hydride ion, $H^-$, which is a strong base in water:

$$H^-(aq) + H_2O(l) \longrightarrow H_2(g) + OH^-(aq)$$

All ionic hydrides behave similarly. The overall equation is

$$LiH(s) + H_2O(l) \longrightarrow Li^+(aq) + OH^-(aq) + H_2(g)$$

The solution is basic.

24. $[H_3O^+]$ in a solution of pH 2.00 is $10^{-2}$ mol $L^{-1}$; that in a solution of pH 3.00 is $10^{-3}$ mol $L^{-1}$. When equal volumes of the two solutions are mixed the resulting $[H_3O^+]$ is:

$$[H_3O^+] = \left(\frac{0.01 + 0.001}{2}\right) = 0.0055 \text{ mol L}^{-1} \quad ; \text{pH} = 2.26$$

25. For a solution of $HNO_2(aq)$ of pH 2.00, $[H_3O^+] = 0.01$ mol $L^{-1}$. Thus

$$HNO_2(aq) + H_2O(l) \rightleftharpoons H_3O^+(aq) + NO_2^-(aq)$$

| | | | | |
|---|---|---|---|---|
| initially | $C_a$ | 0 | 0 | mol $L^{-1}$ |
| at eq'm | $C_a$-0.01 | 0.01 | 0.01 | mol $L^{-1}$ |

and for $pK_a = 3.35$, $K_a(HNO_2) = 10^{-3.35} = 4.5\text{x}10^{-4}$ mol $L^{-1}$; thus

$$K_a = \frac{[H_3O^+][NO_2^-]}{[HNO_2]} = \frac{(0.01)^2}{C_a - 0.01} = 4.5\text{x}10^{-4} \text{ mol L}^{-1}$$

i.e., $1\text{x}10^{-4} = (4.5\text{x}10^{-4})C_a - (4.5\text{x}10^{-6})$; $C_a = (\frac{1.045\text{x}10^{-6}}{4.5\text{x}10^{-4}}) = \underline{0.23 \text{ M}}$

The nitrous acid concentration is <u>0.23 M</u>.

26. In each case $K_w = [H_3O^+][OH^-]$, and for pure water $[H_3O^+] = [OH^-]$

i.e., $[H_3O^+] = (K_w)^{1/2}$ ; pH = -log $[H_3O^+]$, thus:

| t ($^oC$) | $K_w$x$10^{14}$(mol$^2$ L$^{-2}$) | $[H_3O^+]$ (mol $L^{-1}$) | pH |
|---|---|---|---|
| 0 | 0.115 | 3.39x$10^{-8}$ | 7.47 |
| 10 | 0.293 | 5.41x$10^{-8}$ | 7.27 |
| 30 | 1.471 | 1.21x$10^{-7}$ | 6.92 |
| 50 | 5.476 | 2.34x$10^{-7}$ | 6.63 |
| 100 | 51.3 | 7.16x$10^{-7}$ | 6.15 |

27. $pK_a = 3.10$; $K_a = 7.94\text{x}10^{-4}$ mol $L^{-1}$

$$HA(aq) + H_2O(aq) \rightleftharpoons H_3O^+(aq) + A^-(aq)$$

| | | | | |
|---|---|---|---|---|
| initially | 0.10 | 0 | 0 | mol $L^{-1}$ |
| at eq'm | 0.10-x | x | x | mol $L^{-1}$ |

$$K_a = \frac{[H_3O^+][A^-]}{[HA]} = \frac{x^2}{0.10-x} = 7.94\text{x}10^{-4} \text{ mol L}^{-1}$$

For x << 0.10, $x^2 = 7.94\text{x}10^{-5}$ ; $x = 8.91\text{x}10^{-3}$, but the approximation does not give a sufficiently accurate value of x. Successive approximations, or solving the quadratic equation gives

$x = 8.52\text{x}10^{-3}$ ; $[H_3O^+] = 8.5\text{x}10^{-3}$ mol $L^{-1}$; pH = <u>2.07</u>

28. <u>0.20 M KCN</u> The solution contains initially 0.20 M $K^+(aq)$ and 0.20 M $CN^-(aq)$; $K^+(aq)$ is a neutral cation but $CN^-$ is the conjugate base of the weak acid HCN, and is a weak base.

$$CN^-(aq) + H_2O(l) \rightleftharpoons HCN(aq) + OH^-(aq)$$

| | | | | |
|---|---|---|---|---|
| initially | 0.20 | 0 | 0 | mol $L^{-1}$ |
| at eq'm | 0.20-x | x | x | mol $L^{-1}$ |

$$K_b(CN^-) = \frac{[HCN][OH^-]}{[CN^-]} = \frac{x^2}{0.20-x} = \frac{K_w}{K_a(HCN)} = \frac{10^{-14}\ mol^2\ L^{-2}}{4.9x10^{-10}\ mol\ L^{-1}}$$

$$= \underline{2.04x10^{-5}\ mol\ L^{-1}}$$

$x << 0.20$, $x^2 = 4.1x10^{-6}$; $x = 2.02x10^{-3}$, (assumption justified).

$[OH^-] = 2.02x10^{-3}\ mol\ L^{-1}$; $pH = 14.00 + \log [OH^-] = \underline{11.31}$

(b) 0.50 M $NH_3$ (weak base); $[OH^-] = 3{,}0x10^{-3}\ mol\ L^{-1}$; pH = 11.48

(c) 0.10 M HBr (strong acid); $[H_3O^+] = 0.10\ mol\ L^{-1}$; pH = 1.00

(d) 0.10 M $NH_4ClO_4$ (salt of a weak base and a strong acid). $ClO_4^-$ has no acid or base properties because its conjugate acid is the strong acid $HClO_4$. $NH_4^+$ is a weak acid ($K_a = 5.6x10^{-10}\ mol\ L^{-1}$). $[H_3O^+] = 7.48x10^{-6}\ mol\ L^{-1}$; pH = 5.12

(e) 0.30 M $NaNH_2$ $Na^+(aq)$ has no acid-base properties; $NH_2^-$, amide ion, is a strong base in water; $NH_2^- + H_2O \longrightarrow NH_3 + OH^-$. $[OH^-] = 0.30$ M, pH = 13.48

(f) 0.33 M LiI (salt of a strong base and a strong acid). The solution is neutral, pH = 7.00

29. First calculate the moles of $H_3PO_4$.

$$Mol\ H_3PO_4 = (30.0\ g\ H_3PO_4)(\frac{1\ mol\ H_3PO_4}{97.99\ g\ H_3PO_4}) = \underline{0.306\ mol}$$

and the balanced equation is,

$$H_3PO_4(aq) + 3NaOH(aq) \longrightarrow Na_3PO_4(aq) + 3H_2O(l)$$

Thus, $$mol\ NaOH = (0.306\ mol\ H_3PO_4)(\frac{3\ mol\ NaOH}{1\ mol\ H_3PO_4}) = \underline{0.918\ mol}$$

$$Volume\ of\ 0.200\ M\ NaOH(aq) = (0.918\ mol\ NaOH)(\frac{1\ L}{0.200\ mol\ NaOH}) = \underline{4.59\ L}$$

(Note that the volume of NaOH is independent of the original volume of $H_3PO_4$). The final solution contains $Na_3PO_4(aq)$, the salt of a strong base and a weak acid. $Na^+(aq)$ has no acid-base properties. $PO_4^{3-}$ is the conjugate base of the weak acid $HPO_4^{2-}$ and behaves as a weak base, so the final solution is basic.

$$PO_4^{3-}(aq) + H_2O(l) \rightleftharpoons HPO_4^{2-}(aq) + OH^-(aq)$$

30. Aqueous sodium carbonate, $Na_2CO_3(aq)$, contains $Na^+(aq)$ and $CO_3^{2-}(aq)$ ions. $Na^+(aq)$ has no acid-base properties but $CO_3^{2-}$ is the conjugate base of the weak acid $HCO_3^-$, hydrogen carbonate ion, and thus behaves as a weak base.

$$CO_3^{2-}(aq) + H_2O(l) \rightleftharpoons HCO_3^-(aq) + OH^-(aq)$$

Adding $Na_2CO_3(s)$ to swimming pools will remove any $H_3O^+(aq)$ present and increase the pH.

31. For any acid HA and its conjugate base $A^-$, the relationship between $K_a(HA)$ and $K_b(A^-)$ is $K_a(HA) \times K_b(A^-) = K_w$, or in logarithmic form

$$pK_a + pK_b = pK_w = 14.00\ (25^oC)$$

(a) $pK_a(H_2CO_3) = 6.48$; $pK_b(HCO_3^-) = 7.52$; hydrogen carbonate ion.
(b) $pK_a(H_2S) = 7.04$; $pK_b(HS^-) = 6.96$; hydrogen sulfide ion.
(c) $pK_a(HNO_2) = 3.35$; $pK_b(NO_2^-) = 10.65$; nitrite ion.
(d) $pK_a(H_2SO_3) = 1.77$; $pK_b(HSO_3^-) = 12.23$; hydrogen sulfite ion.

32. Formic acid as the structure H-C(=O)OH in which only the H attached to O is acidic. The equilibria involved are,

(a) $$HCO_2H(aq) + H_2O(l) \rightleftharpoons H_3O^+(aq) + HCO_2^-(aq)$$, and

(b) $$HCO_2^-(aq) + H_2O(l) \rightleftharpoons HCO_2H(aq) + OH^-(aq)$$

Calculating the pH of each solution by the usual methods, using (a) $K_a(HCO_2H) = 2.09\text{x}10^{-4}$ mol $L^{-1}$, and (b) $K_b(HCO_2^-) = K_w/K_a$ i.e., $K_b(HCO_2^-) = 4.79\text{x}10^{-11}$ mol $L^{-1}$, gives

(a) pH = 2.34, and (b) pH = 8.34

33. An aqueous solution of $AlCl_3$ contains the weakly acidic $Al(H_2O)_6^{3+}(aq)$ ion and the neutral $Cl^-(aq)$ ion, and is therefore acidic.

| | $Al(H_2O)_6^{3+}$ + $H_2O$ ⇌ | $Al(H_2O)_5OH^{2+}$ + | $H_3O^+$ | |
|---|---|---|---|---|
| initially | 0.100 | 0 | 0 | mol $L^{-1}$ |
| at eq'm | 0.100-x | x | x | mol $L^{-1}$ |

From Table 14.1, the $K_a$ is $7.2\text{x}10^{-6}$ mol $L^{-1}$, so that

$$K_a = \frac{x^2}{0.100-x} = 7.2\text{x}10^{-6} \text{ mol } L^{-1};\ x << 0.100,\ x^2 = 7.2\text{x}10^{-7}$$

$$x = 8.5\text{x}10^{-4};\ pH = 3.07$$

34. Sodium hydrogen sulfate, $NaHSO_4$, contains neutral $Na^+$ ions and $HSO_4^-$ ions. The latter is the conjugate base of the strong acid $H_2SO_4$ and thus has no basic properties in water, but it is also the conjugate acid of the weak base $SO_4^{2-}$ and behaves as a weak acid in solution.

$$HSO_4^-(aq) + H_2O(l) \rightleftharpoons H_3O^+(aq) + SO_4^{2-}(aq)$$

By the usual methods, with $K_a(HSO_4^-) = 1.2\text{x}10^{-2}$ mol $L^{-1}$,
$[H_3O^+] = 0.0292$ mol $L^{-1}$ ; pH = 1.53

35. Phosphorous acid $H_3PO_3$ is a dibasic acid (see Chapter 8).
(a) Since there is a large difference in magnitude between the $K_a$ values for the first and second stages of ionization, the second is expected to contribute negligibly to the pH of 0.10 M $H_3PO_3(aq)$. Using the usual methods, and solving the quadratic equation gives: $[H_3O^+] = 0.05$ mol $L^{-1}$ , and pH = 1.30, for the ionization

$$H_3PO_3(aq) + H_2O(l) \rightleftharpoons H_2PO_3^-(aq) + H_3O^+(aq)$$

Checking the assumption that the second stage of ionization

$$H_2PO_3^-(aq) + H_2O(l) \rightleftharpoons HPO_3^{2-}(aq) + H_3O^+(aq)$$

is negligible, we find that its contribution changes $[H_3O^+]$ to 0.0505 mol $L^{-1}$ and the pH remains at 1.30.
(b) For a solution of the salt $NaH_2PO_3(aq)$, the solution contains the neutral cation $Na^+(aq)$ and the anion $H_2PO_3^{2-}(aq)$, which is the

conjugate base of the acid $H_3PO_3$, but also the conjugate acid of the $HPO_3^{2-}$ anion. The solution will be basic or acidic depending on which of $K_b(H_2PO_3^-)$ and $K_a(H_2PO_3^-)$ is greater. From the data given $K_a(H_2PO_3^-)$ is $2x10^{-5}$ mol $L^{-1}$ ; $pK_a(H_2PO_3^-) = 4.70$, and

$$pK_a(H_3PO_3) + pK_b(H_2PO_3^-) = 1.30 + pK_b(H_2PO_3^-) = 14.00$$

$$\text{i.e., } pK_b(H_2PO_3^-) = 12.70 >> pK_a(H_2PO_3^-)$$

Thus, the $H_2PO_3^-$ ion is a much stronger acid than it is a base, so that in $NaH_2PO_3(aq)$ the equilibrium to be considered is

$$H_2PO_3^-(aq) + H_2O(l) \rightleftharpoons H_3O^+(aq) + HPO_3^{2-}(aq)$$

| | | | | |
|---|---|---|---|---|
| initially | 0.10 | 0 | 0 | mol $L^{-1}$ |
| at eq'm | 0.10-x | x | x | mol $L^{-1}$ |

$$K_a = \frac{[H_3O^+][HPO_3^{2-}]}{[H_2PO_3^-]} = \frac{x^2}{0.10-x} = 2x10^{-5} \text{ mol } L^{-1}$$

$x << 0.10$, $x^2 = 2x10^{-6}$ ; $x = 1.4x10^{-3}$; $\underline{pH = 2.85}$

36. (a) An indicator in solution behaves as an acid or a base; unless the smallest concentration possible is added to a solution under investigation, its reaction with the system will change the pH by more than a negligible amount.

(b) For an indicator, we have the equilibrium

$$HIn(aq) + H_2O(l) \rightleftharpoons H_3O^+(aq) + In^-(aq)$$

$K_a = [H_3O^+]\frac{[In^-]}{[HIn]}$ and taking -log of both sides gives:

$$pK_a = pH - \log\frac{[In^-]}{[HIn]} \quad \underline{\text{or}} \quad pH = pK_a + \log\frac{[In^-]}{[HIn]}$$

(i) $pH = 4.2 + \log\frac{5}{1} = 4.2 + 0.7 = \underline{4.9}$

(ii) $pH = 4.2 + \log\frac{1}{1} = 4.2 + 0 = \underline{4.2}$

(iii) $pH = 4.2 + \log\frac{1}{5} = 4.2 - 0.7 = \underline{3.5}$

37. Reference to Table 14.7 gives us the information that methyl red is yellow in its base form and is effective in the pH range 4.2 to 6.2 Thus its solution will be definitely yellow at the higher pH of 6.2 Bromothymol blue is yellow in its acid form and blue in its base form, and is effective in the pH range 6.0 to 7.8. Thus, with this indicator the solution will be definitely yellow at the lower pH of 6.0, from which we can conclude that the pH of the solution must be in the range

$$6.0 < pH < 6.2$$

38. See Problem 36 for the relevant equations.

The $pK_a$ for methyl red is 5.0, so for pH = 5.0, we have

$$pH = 5.0 = pK_a + \log\frac{[In^-]}{[HIn]} = 5.0 + \log\frac{[In^-]}{[HIn]}$$

Thus, for the solution, $\log \frac{[In^-]}{[HIn]} = 0, \quad \frac{[In^-]}{[HIn]} = 1$

The indicator is half in its $In^-$ (yellow) form, and half in its HIn (red) form; the solution will be orange.

39. In each case equal volumes of acid and base (since they are all 0.1 M solutions) are needed to give the equivalence point, at which point the solution contains only the salt of the acid and base, which must have a concentration of 0.05 M. Thus, in each case we consider the pH of an 0.05 M solution of the respective salt. In general the indicator changes color in the range $pK_a \pm 1$.

(a) At the equivalence point we have an 0.05 M solution of NaCl(aq), containing neutral $Na^+$(aq) and neutral $Cl^-$(aq) ions. Thus the pH of the solution is 7.0 and an indicator with a $pK_a$ close to 7.0 would be suitable, e.g., Bromothymol blue, $pK_a$ 7.1.

(b) At the equivalence point we have an 0.005 M solution of KF(aq), for which the pH is calculated by the normal methods (see e.g., Problem 17) to be 8.23. Thus, an indicator with $pK_a$ close to 8.2 would be suitable, for example, thymol blue, $pK_a$ 8.2.

(c) At the equivalence point we have an 0.05 M solution of $NH_4Cl$(aq), which contains the acidic $NH_4^+$ ion and the neutral $Cl^-$ ion

| | $NH_4^+(aq) + H_2O(l) \rightleftharpoons$ | $H_3O^+(aq)$ + | $NH_3(aq)$ | |
|---|---|---|---|---|
| initially | 0.05 | 0 | 0 | mol $L^{-1}$ |
| at eq'm | 0.05-x | x | x | mol $L^{-1}$ |

$$K_a(NH_4^+) = \frac{K_w}{K_b(NH_3)} = 5.6\text{x}10^{-10} \text{ mol}^2 \text{ L}^{-2} = \frac{x^2}{0.05-x}$$

$x << 0.05$, $x^2 = 2.8\text{x}10^{-11}$ ; $x = 5.3\text{x}10^{-6}$ ; pH = 5.3

An indicator with $pK_a$ close to 5.1 would be suitable, for example, methyl red, $pK_a$ = 5.0.

(d) At the equivalence point we have an 0.05 M solution of the salt $CH_3NH_3Cl$ containing the acidic $CH_3NH_3^+$ ion and the neutral $Cl^-$ ion

| | $CH_3NH_3^+(aq) + H_2O(l) \rightleftharpoons$ | $H_3O^+(aq)$ + | $CH_3NH_2(aq)$ | |
|---|---|---|---|---|
| initially | 0.05 | 0 | 0 | mol $L^{-1}$ |
| at eq'm | 0.05-x | x | x | mol $L^{-1}$ |

$$K_a(CH_3NH_3^+) = \frac{K_w}{K_b(CH_3NH_2)} = \frac{10^{-14} \text{ mol}^2 \text{ L}^{-2}}{3.9\text{x}10^{-4} \text{ mol L}^{-1}} = 2.6\text{x}10^{-11} \text{mol L}^{-1}$$

$$\frac{x^2}{0.05-x} = 2.6\text{x}10^{-11}; \; x << 0.05, \; x^2 = 1.3\text{x}10^{-12} \; ; \; x = 1.14\text{x}10^{-6}$$

Thus, pH = 5.9, and a suitable indicator would be methyl red, which changes color in the pH range 4.2 to 6.2.

40. Comparison with the data in Table 14.4 gives

| Indicator | Observed Color | Conclusion | |
|---|---|---|---|
| | | Indicator in HIn or $In^-$ form? | pH range |
| Methyl violet | violet | $In^-$ | > 3.0 |
| Methyl orange | yellow | $In^-$ | > 4.4 |
| Methyl red | orange | HIn & $In^-$ | 4.2 < pH < 6.2 |
| Bromothymol blue | yellow | HIn | < 6.0 |

Methyl red is red in its acid form and yellow in its base form; the observed orange color indicates that the indicator must be close to half in its acid form and half in its base form:

$$pK_a = pH - \log \frac{[In^-]}{[HIn]} \quad \text{(see Problem 36)}$$

Thus, for $[In^-]/[HIn] = 1$, $pH = pK_a = 5.0$ for methyl red

$\underline{pH = 5.0}$ or $[H_3O^+] = \underline{10^{-5} \text{ mol L}^{-1}}$

and for the 0.10 M solution of the acid HA,

$$HA(aq) + H_2O(l) \rightleftharpoons H_3O^+(aq) + A^-(aq)$$

| | HA(aq) | $H_3O^+$(aq) | $A^-$(aq) | |
|---|---|---|---|---|
| initially | 0.10 | 0 | 0 | mol $L^{-1}$ |
| at eq'm | $0.10-10^{-5}$ | $10^{-5}$ | $10^{-5}$ | mol $L^{-1}$ |

$$K_a = \frac{[H_3O^+][A^-]}{[HA]} = \frac{(10^{-5}\text{ mol L}^{-1})^2}{0.10 \text{ mol L}^{-1}} = \underline{10^{-9} \text{ mol L}^{-1}}$$

41. We first calculate the pH of 0.10 M KF(aq), which contains neutral $K^+$(aq) ions and weakly basic $F^-$(aq) ions (see e.g., Problem 17). This calculation gives pH = $\underline{8.23}$ . Thus we need to consider addition of a few drops of each indicator to a solution of pH $\underline{8.2}$

As we saw in Problem 36, the relationship between the $pK_a$ of the indicator and the pH of the solution is $pK_a = pH - \log [In^-]/[HIn]$, where $[In^-]$ and [HIn] are the concentrations of the base form and the acid form of the indicator, respectively. Thus for a given indicator and a given pH,

$$\log \frac{[In^-]}{[HIn]} = pH - pK_a$$

| Indicator | $pK_a$ | $\log[In^-]/[HIn]$ | $[In^-]/[HIn]$ | Color |
|---|---|---|---|---|
| (a) Thymol blue | 8.2 | 0 | 1.0 | green* |
| (b) Phenolphthalein | 9.5 | -1.3 | 0.05 | colorless |
| (c) Bromothymol blue | 7.1 | 1.1 | 13 | blue |

*50% blue/50% yellow = green

42. The pH of 0.10 M $NH_4Cl(aq)$ was calculated in Problem 18(f) as 5.12. This problem is similar to Problem 41.

| | Indicator | $pK_a$ | $\log[In^-]/[HIn]$ | $[In^-]/[HIn]$ | Color |
|---|---|---|---|---|---|
| (a) | Methyl red | 5.0 | 0.1 | 1.2 | orange* |
| (b) | Methyl orange | 4.2 | 0.9 | 8 | yellow |
| (c) | Bromothymol blue | 7.1 | -2.0 | 0.01 | yellow |

*yellow + red = orange

43. $pH = pK_a + \log [In^-]/[HIn]$ (see Problem 36). For $[In^-] = [HIn]$, the $pH = pK_a$ for the indicator, and the indicator is half in the blue base form and half in the yellow acid form. This must occur when the $[H_3O^+]$ is approximately half-way between that represented by pH 7.9 and pH 9.4.

At pH 7.9, $[H_3O^+] = 1.26\times10^{-8}$ mol $L^{-1}$, and

at pH 9.4, $[H_3O^+] = 3.98\times10^{-10}$ mol $L^{-1}$

and $[H_3O^+]$ at the half-way point is $6.50 \times 10^{-9}$ mol $L^{-1}$, at pH 8.2.

i.e., The $pK_a$ of the indicator is approximately 8.2.

44. Buffer solutions contain roughly equivalent amounts of a weak acid and a salt of the weak acid, or a weak base and a salt of the weak base. Rather than remembering equations such as the Henderson-Hasselbalch equation, it is preferable to start with the $K_a$ expression for the weak acid (or $K_b$ for the weak base) and remember that for a buffer solution the concentrations of the acid [HA] and the base $[A^-]$ are their stoichiometric amounts in the solution. For example, for this problem

$$K_b = \frac{[NH_4^+]}{[NH_3]}[OH^-] = \left(\frac{0.25 \text{ mol L}^{-1}}{0.30 \text{ mol L}^{-1}}\right)[OH^-] = 1.8\times10^{-5} \text{ mol L}^{-1}$$

$$[OH^-] = (1.8\times10^{-5} \text{ mol L}^{-1})\left(\frac{25}{30}\right) = 1.5\times10^{-5} \text{ mol L}^{-1}$$

$$pH = 14.00 + \log [OH^-] = 9.18$$

45. (See Problem 44 for the method)

$$K_a(HCN) = 4.9\times10^{-10} \text{ mol L}^{-1} = [H_3O^+]\frac{[CN^-]}{[HCN]} = [H_3O^+] \; ; \quad pH = 9.31$$

46. $pH = 9.0$, $[H_3O^+] = 10^{-9}$ mol $L^{-1}$, and here it is more convenient to use the $K_a(NH_4^+)$ expression (see Problem 44).

$$K_a(NH_4^+) = [H_3O^+]\frac{[NH_3]}{[NH_4^+]} = (10^{-9} \text{ mol L}^{-1})\left(\frac{\text{mol } NH_3}{\text{mol } NH_4Cl}\right) = 5.6\times10^{-10} \text{ mol L}^{-1}$$

$$\left(\frac{\text{mol } NH_3}{\text{mol } NH_4Cl}\right) = 0.56$$

Thus, $\left(\frac{\text{mass } NH_3}{\text{mass } NH_4Cl}\right) = \left(\frac{\text{mol } NH_3}{\text{mol } NH_4Cl}\right)\left(\frac{17.03 \text{ g } NH}{1 \text{ mol } NH_3}\right)\left(\frac{1 \text{ mol } NH_4Cl}{53.49 \text{ g } NH_4Cl}\right)$

$= 0.56\left(\frac{17.03}{53.49}\right) = \underline{0.178}$

47. See Problem 44 for the method.
First calculate the concentrations of weak acid and its conjugate base in the system. Initially there are

$(25.00 \text{ mL NaOH})(0.100 \text{ mol L}^{-1}) = 2.50 \text{ mmol NaOH}$, and

$(50.00 \text{ mL } CH_3CO_2H)(0.100 \text{ mol L}^{-1}) = 5.00 \text{ mmol } CH_3CO_2H$

| | $CH_3CO_2H(aq)$ + | $NaOH(aq) \longrightarrow$ | $CH_3CO_2Na(aq) + H_2O(l)$ | |
|---|---|---|---|---|
| initially | 5.00 | 2.50 | 0 | mmol |
| at eq'm | 2.50 | 0 | 2.50 | mmol |

The solution contains equal amounts of $CH_3CO_2H$ and $CH_3CO_2Na$, and is therefore a buffer solution, for which

$$K_a = [H_3O^+]\frac{[CH_3CO_2^-]}{[CH_3CO_2H]} = [H_3O^+]\left(\frac{2.50 \text{ mmol}}{2.50 \text{ mmol}}\right) = [H_3O^+] = 1.8\times10^{-5} \text{ mol L}^{-1}$$

$\underline{pH = 4.74}$

Note that the actual concentrations of weak acid and salt do not have to be calculated; the ratio of their concentrations is the same as the ratio of the number of moles of each.

48. First calculate mol $NH_3(aq)$ and mol $NH_4Cl(aq)$.

mol $NH_3$ = $(15.0 \text{ mL})(0.0100 \text{ mol L}^{-1}) = 0.150 \text{ mmol } NH_3$

mol $NH_4Cl$ = $(25.0 \text{ mL})(0.0100 \text{ mol L}^{-1}) = 0.250 \text{ mmol } NH_4Cl$

$$K_b(NH_3) = [OH^-]\left(\frac{\text{mol } NH_4^+}{\text{mol } NH_3}\right) = [OH^-]\left(\frac{0.250 \text{ mmol}}{0.150 \text{ mmol}}\right) = 1.8\times10^{-5} \text{ mol L}^{-1}$$

$[OH^-] = 1.08\times10^{-5} \text{ mol L}^{-1}$; $pH = 14.00 + \log [OH^-] = \underline{9.03}$

Note that since here the solutions have the same concentration, the volumes could simply have been divided to give $[NH_4^+]/[NH_3]$.

49. Adding base to this buffer solution will change the amounts of $NH_4^+$ & $NH_3$ in the system. Calculate initial mol $NH_3$ and $NH_4^+$, mol of added base, and hence mol $NH_3$ and mol $NH_4^+$ after reaction.

initial mol $NH_3$ = $(100 \text{ mL})(0.18 \text{ mol L}^{-1})\left(\frac{1 \text{ L}}{1000 \text{ mL}}\right) = \underline{0.018 \text{ mol}}$

initial mol $NH_4Cl$ = $(1000 \text{ mL})(0.10 \text{ mol L}^{-1})\left(\frac{1 \text{ L}}{1000 \text{ mL}}\right) = \underline{0.010 \text{ mol}}$

mol of added NaOH(aq) = $(1.00 \text{ mL})(1.00 \text{ mol L}^{-1})\left(\frac{1 \text{ L}}{1000 \text{ mL}}\right) = \underline{0.001 \text{ mol}}$

| | $NH_4^+(aq)$ + | $OH^-(aq) \rightleftharpoons$ | $NH_3(aq) + H_2O(l)$ | |
|---|---|---|---|---|
| initially | 0.010 | 0.001 | 0.018 | mol |
| at eq'm | 0.009 | 0 | 0.019 | mol |

Thus, there are two buffer solutions, and initial buffer solution containing 0.010 mol $NH_4^+$ and 0.018 mol $NH_3$ (1), and a final solution containing 0.009 mol $NH_4^+$ and 0.019 mol $NH_3$ (2).

(1) $K_b(NH_3) = [OH^-](\frac{0.010}{0.018}) = 1.85\times10^{-5}$ mol $L^{-1}$; $[OH^-]$ = $\underline{3.24\times10^{-5}\ M}$

pH = 14.00 + log $[OH^-]$ = 14.00 - 4.49 = $\underline{9.51}$

(2) $K_b(NH_3) = [OH^-](\frac{0.009}{0.019}) = 1.8\times10^{-5}$ mol $L^{-1}$; $[OH^-]$ = $\underline{3.80\times10^{-5}\ M}$

pH = 14.00 + log $[OH^-]$ = $\underline{9.58}$

The pH changes from $\underline{9.51 \text{ to } 9.58}$.

50. For a buffer solution containing a weak acid HA and its conjugate base $A^-$,

$K_a = [H_3O^+]\frac{[A^-]}{[HA]}$ where $[A^-]$ is the initial concentration of salt and [HA] is the initial concentration of acid.

Taking -log of both sides, gives

$pK_a = pH - \log\frac{[base]}{[acid]}$ which is the Henderson-Hasselbalch equation.

For acetic acid/acetate buffer ($pK_a$ = 4.74),

$4.74 = 4.50 - \log(\frac{\text{mol } CH_3CO_2^-}{\text{mol } CH_3CO_2H})$ and $\log(\frac{\text{mol } CH_3CO_2^-}{\text{mol } CH_3CO_2H}) = \underline{-0.24}$

Mole ratio $CH_3CO_2H/CH_3CO_2^-$ = $\underline{1.74}$

51. (a) pH = 5.00, $[H_3O^+] = 1.00\times10^{-5}$ mol $L^{-1}$

$K_a(CH_3CO_2H) = [H_3O^+](\frac{\text{mol base}}{\text{mol acid}}) = 1.8\times10^{-5}$ mol $L^{-1}$

Mole ratio base/acid = $K_a/[H_3O^+] = (1.8\times10^{-5} M)/(1.00\times10^{-5}\ M) = \underline{1.8}$

and mol acid = (1.00 L)(0.200 mol $L^{-1}$) = 0.200 mol $CH_3CO_2H$.

Thus, $\frac{\text{mol base (salt)}}{\text{mol acid}} = \frac{\text{mol salt}}{0.200 \text{ mol}} = 1.8$; mol salt = $\underline{0.360 \text{ mol}}$

Mass of $CH_3CO_2Na$ = (0.360 mol)($\frac{82.03 \text{ g salt}}{1 \text{ mol}}$) = $\underline{29.5 \text{ g}}$

$\underline{29.5 \text{ g } CH_3CO_2Na}$ dissolved in 1.00 L 0.200 M $CH_3CO_2H(aq)$ gives a $\underline{\text{buffer solution}}$ of $\underline{\text{pH } 5.00}$

(b) The added HCl(aq) will react with $CH_3CO_2^-(aq)$ ions, and initial mol $CH_3CO_2H(aq)$ = $\underline{0.200 \text{ mol}}$, and initial mol $CH_3CO_2Na(aq)$ = $\underline{0.360 \text{ mol}}$.

Mol HCl(aq) added = (1.00 mL)(12.0 mol $L^{-1}$)(1 L/1000 mL) = $\underline{0.012 \text{ mol}}$

| | $CH_3CO_2^-(aq)$ + | $H_3O^+(aq)$ ⇌ | $CH_3CO_2H(aq)$ + | $H_2O(l)$ |
|---|---|---|---|---|
| initially | 0.360 | 0.012 | 0.200 | mol |
| at eq'm | 0.348 | 0 | 0.212 | mol |

and $K_a = [H_3O^+](\frac{\text{mol salt}}{\text{mol acid}}) = [H_3O^+](\frac{0.348}{0.212}) = 1.8\times10^{-5}$ mol $L^{-1}$

$[H_3O^+] = 1.1\times10^{-5}$ mol $L^{-1}$; $\underline{pH = 4.96}$

52. The pH of a buffer solution depends only on the $K_a$ of the acid and the mole ratio acid/conjugate base in solution. Thus we need only to calculate moles of formic acid and moles of sodium formate.

$$\text{mol } HCO_2H = (1.00\text{ g})\left(\frac{1\text{ mol acid}}{46.03\text{ g acid}}\right) = 0.0217\text{ mol}$$

$$\text{mol } HCO_2Na = (1.00\text{ g})\left(\frac{1\text{ mol salt}}{68.01\text{ g salt}}\right) = 0.0147\text{ mol}$$

$$pK_a = 3.68,\ K_a = 2.09\times10^{-4}\text{ mol L}^{-1}$$

$$K_a = 2.09\times10^{-4}\text{ mol L}^{-1} = [H_3O^+]\frac{[\text{salt}]}{[\text{acid}]} = [H_3O^+]\left(\frac{0.0147\text{ mol}}{0.0217\text{ mol}}\right)$$

$$[H_3O^+] = 3.08\times10^{-4}\text{ mol L}^{-1}$$

$$pH = \underline{3.51}$$

The mol ratio of salt to acid is independent of the volume of the solution and will be unchanged after addition of 100 mL (or any other amount) of distilled water. The pH is unaffected and remains at 3.51.

53. Let us first calculate the mol ratio of HOCN and NaOCN to give a buffer solution of pH 3.7. In this case pH = $pK_a$, so the mol ration must be 1.0.

In general the two ways to make up a buffer solution (containing equal moles of HOCN and NaOCN) are

1) Dissolve an equal number of moles of HOCN and NaOCN in a suitable volume of distilled water (the actual volume of water is immaterial).

2) Starting with an aqueous solution of HOCN of known concentration, add to it sufficient of a sodium hydroxide solution to neutralize one-half of the acid.

| | HOCN(aq) + | NaOH(aq) → | NaOCN(aq) + $H_2O$(l) | |
|---|---|---|---|---|
| initially | n | n/2 | 0 | mol |
| at eq'm | n/2 | 0 | n/2 | |

Such a solution contains equal moles of HOCN and NaOCN and has a pH equal to the $pK_a$ of the acid.

54. Here, dihydrogen phosphate ion, $H_2PO_4^-$, is the acid, and hydrogen phosphate ion, $HPO_4^{2-}$, is the conjugate base.

$$HPO_4^{2-} + H^+ \rightleftharpoons H_2PO_4^-$$

Thus, we need to calculate the moles of $HPO_4^{2-}$ and the moles of $H_2PO_4^-$ in the solution.

$$\text{mol } Na_2HPO_4 = (3.55\text{ g})\left(\frac{1\text{ mol salt}}{142.0\text{ g salt}}\right) = \underline{0.0250\text{ mol}}$$

$$\text{mol } KH_2PO_4 = (3.40\text{ g})\left(\frac{1\text{ mol salt}}{136.1\text{ g salt}}\right) = \underline{0.0250\text{ mol}}$$

$$K_a(H_2PO_4^-) = 6.2\times10^{-8}\text{ mol L}^{-1} = [H_3O^+]\left(\frac{\text{mol base}}{\text{mol acid}}\right) = [H_3O^+]\left(\frac{0.0250\text{ mol}}{0.0250\text{ mol}}\right)$$

$[H_3O^+] = 6.2\times10^{-8}$ mol $L^{-1}$ ; pH = 7.21

Since the pH of the buffer solution depends only on the mol ratio base/acid, it is unaffected by dilution; pH = 7.21

55. The pH of this buffer solution depends on the $K_a$ of $C_6H_5NH_3^+$ and the mol ratio $C_6H_5NH_2/C_6H_5NH_3^+$

$$\text{mol } C_6H_5NH_2 = (25.00 \text{ mL})(0.020 \text{ mol L}^{-1}) = 0.500 \text{ mmol}$$

$$\text{mol } C_6H_5NH_3^+ = (10.0 \text{ mL})(0.030 \text{ mol L}^{-1}) = 0.300 \text{ mmol}$$

Thus:

$$K_a(C_6H_5NH_3^+) = \frac{K_w}{K_b(C_6H_5NH_2)} = \frac{10^{-14} \text{ mol}^2 \text{ L}^{-2}}{4.3\times10^{-10} \text{ mol L}^{-1}} = \underline{2.3\times10^{-5} \text{ mol L}^{-1}}$$

$$= [H_3O^+]\frac{[C_6H_5NH_2]}{[C_6H_5NH_3^+]} = [H_3O^+](\frac{0.500 \text{ mmol}}{0.300 \text{ mmol}})$$

$$[H_3O^+] = (\frac{0.300}{0.500})(2.3\times10^{-5} \text{ mol L}^{-1}) = 1.4\times10^{-5} \text{ mol L}^{-1}$$

$$\underline{pH = 4.85}$$

Addition of the strong acid $HNO_3$ to the buffer solution converts some of the aniline to anilinium salt.

Moles of added $HNO_3$ = (1.00 mL)(0.040 mol $L^{-1}$) = 0.040 mmol

| | $C_6H_5NH_2$ + | $HNO_3$ ⟶ | $C_6H_5NH_3^+$ + | $NO_3^-$ | |
|---|---|---|---|---|---|
| initially | 0.500 | 0.040 | 0.300 | 0 | mmol |
| at eq'm | 0.460 | 0 | 0.340 | 0.040 | mmol |

$$K_a = [H_3O^+](\frac{0.460}{0.340}) = 2.3\times10^{-5} \text{ mol L}^{-1} \text{ ; } [H_3O^+] = \underline{1.7\times10^{-5} \text{ mol L}^{-1}}$$

$$\underline{pH = 4.77}$$

Addition of the strong base KOH(aq) to the buffer solution converts some of the anilinium salt to aniline.

| | $C_6H_5NH_3^+$ | + KOH ⟶ | $C_6H_5NH_2$ | + $K^+$ + $H_2O$ | |
|---|---|---|---|---|---|
| initially | 0.300 | 0.060 | 0.500 | 0 | mmol |
| at eq'm | 0.240 | 0 | 0.560 | 0.060 | mmol |

$$K_a = [H_3O^+](\frac{0.560}{0.240}) = 2.3\times10^{-5} \text{ mol L}^{-1} \text{ ; } [H_3O^+] = \underline{9.9\times10^{-6} \text{ mol L}^{-1}}$$

$$\underline{pH = 5.01}$$

56. (a) The strength of an acid refers to its extent of ionization which is related to its $K_a$ and measured by the pH of a solution; the concentration is the amount of acid per unit volume, which is independent of $K_a$.

(b) The equivalence point in an acid-base titration refers to the point at which moles of acid = moles of base, i.e., it corresponds to a solution containing only the salt of the acid and base. The end-point in a titration is an experimental matter. It is the point in the titration where a given indicator changes color,

The end-point only coincides with the equivalence point if the indicator selected is one with a $pK_a$ close in value to the pH at the equivalent point of the titration

57. In each case the solution at the equivalence point contains a 0.050 M solution of the salt of the respective acid and base.

(a) The solution contains 0.05M NaCl(aq), completely dissociated in solution into $Na^+$(aq) and $Cl^-$(aq) ions. $Na^+$(aq) has no acid-base properties (since it is the cation of a group 1 metal), and $Cl^-$(aq) has no basic properties since it is the conjugate base of the strong acid HCl(aq). The solution is neutral with a pH of 7.00. A suitable indicator to detect the equivalence point should have a $pK_a$ close to 7. Bromothymol blue ($pK_a$ 7.1) would be ideal. However, since the pH changes very rapidly in the vicinity of the equivalence point, the choice of indicator is not as critical for strong acid-strong base titrations, such as this, as it is for other kinds of acid-base titrations.

(b) The solution at the equivalence point contains 0.05 M $CH_3CO_2Na$(aq) completely dissociated in solution into $Na^+$(aq) and $CH_3CO_2^-$(aq) ions. $Na^+$(aq) has no acid-base properties but acetate ion, $CH_3CO_2^-$, is the conjugate base of the weak acid $CH_3CO_2H$ and is a weak base in aqueous solution. Thus, the pH of the salt solution is > 7. The actual pH is calculated by the usual methods (see, for example, Problem 20(d)). This gives pH = 8.72 at the equivalence point. A suitable indicator should have a $pK_a$ close to 8.7. Thymol blue, $pK_a$ 8.2, would be suitable.

(c) The solution contains 0.05 M $NH_4Cl$(aq) completely dissociated into $NH_4^+$(aq) and $Cl^-$(aq) ions. $Cl^-$(aq), the conjugate base of the strong acid HCl(aq) has no basic properties, but $NH_4^+$(aq) is the conjugate acid of the weak base $NH_3$(aq), and is a weak acid. Thus, the pH of the salt solution is < 7. The pH of the solution is calculated by the usual methods (see for example, Problem 20(e)). This gives pH = 5.28 at the equivalence point. A suitable indicator should have a $pK_a$ close to 5.3. A suitable indicator is methyl red, $pK_a$ 5.0.

58. In each case, calculate the concentration of salt solution present at the equivalence point of the titration (where moles of base = moles of acid), and hence the pH at the equivalence point, and select an indicator with a $pK_a$ close to this pH. The problem is similar to Problem 57.

(a) The solution at the equivalence point contains 0.033 M $KNO_3$(aq) - the salt of a strong acid and a strong base, so the pH is 7.00. Bromothymol blue ($pK_a$ 7.1) is an ideal indicator for this titration, as it is for all titrations between a strong acid and a strong base. Many other indicators would also be suitable because of the rapid change of pH in the vicinity of the equivalence point in a strong acid-strong base titration.

(b) The solution contains 0.067 M KF(aq), the salt of a weak acid and a strong base. The solution will be basic, pH > 7, because of the behaviour of $F^-(aq)$ as a weak base (see Problem 17). The actual pH is <u>8.14</u>. A suitable indicator is thymol blue ($pK_a$ 8.2).

(c) The solution contains 0.050 M $CH_3NH_3Cl$. $Cl^-(aq)$ is a neutral anion and $CH_3NH_3^+(aq)$ is a weak acid, since it is the conjugate acid of the weak base $CH_3NH_2$, methylamine. We can calculate the pH of the solution by the usual methods.

$K_b(CH_3NH_2)$ is $3.9 \times 10^{-4}$ mol $L^{-1}$, which gives $K_a(CH_3NH_3^+)$ the value:

$$K_a = 2.6 \times 10^{-11} \text{ mol L}^{-1}$$

and for 0.050 M $CH_3NH_3^+(aq)$ we obtain <u>pH = 5.95</u>. Thus, a suitable indicator is methyl red ($pK_a$ 5.2).

59. In calculating the pH at various points in the titration, the nature of the solution at any given point falls into the following categories:
   1. Initially - a solution of a weak acid
   2. Between the initial point and the equivalence point - a series of buffer solutions (base reacts with acid to give a salt of the weak acid but weak acid remains in excess - thus these solutions contain a weak acid and a salt of the weak acid)
   3. At the equivalence point - a solution of a salt
   4. Beyond the equivalence point - a solution of a salt containing excess base

   Cases 1 to 3 have already been dealt with in many examples. We will deal with case 4 explicitly later.

(a) <u>Initially</u> 0.10 M acetic acid - a solution of a weak acid

| | $HA(aq)$ + $H_2O(l)$ ⇌ | $H_3O^+(aq)$ + | $A^-(aq)$ | |
|---|---|---|---|---|
| initially | 0.10 | 0 | 0 | mol $L^{-1}$ |
| at eq'm | 0.10-x | x | x | mol $L^{-1}$ |

$$K_a = \frac{[H_3O^+][A^-]}{[HA]} = \frac{x^2}{0.10-x} = 1.8 \times 10^{-5} \text{ mol L}^{-1}$$

$x << 0.10$, $x^2 = 1.8 \times 10^{-6}$; $x = 1.34 \times 10^{-3}$ (assumption justified)

$[H_3O^+] = 1.34 \times 10^{-3}$ mol $L^{-1}$ ; <u>pH = 2.87</u>

(b) <u>After addition of 15.63 mL of base</u>

initial moles of acid = (25.00 mL)(0.10 mol $L^{-1}$) = 2.50 mmol

moles of added base = (15.63 mL)(0.08 mol $L^{-1}$) = 1.25 mmol

| | $HA(aq)$ + | $KOH(aq)$ → | $KA(aq)$ + $H_2O(l)$ | |
|---|---|---|---|---|
| initially | 2.50 | 1.25 | 0 | mmol |
| at eq'm | 1.25 | 0 | 1.25 | mmol |

Thus the solution is a <u>buffer solution</u> containing 1.25 mmol of acid and 1.25 mmol of salt (conjugate base), for which

$$K_a = [H_3O^+]\frac{[A^-]}{[HA]} = [H_3O^+]\left(\frac{1.25 \text{ mmol}}{1.25 \text{ mmol}}\right) = 1.8 \times 10^{-5} \text{ mol L}^{-1}$$

$[H_3O^+] = 1.8 \times 10^{-5}$ mol $L^{-1}$ ; <u>pH = 4.74</u>

(c) At the equivalence point

The initial moles of HA were 2.50 mmol, so the equivalence point is reached when 2.50 mmol KOH have been added, i.e.,

$$(2.50 \text{ mmol KOH})\left(\frac{1 \text{ L}}{0.08 \text{ mol KOH}}\right) = 31.25 \text{ mL KOH}$$

and the solution contains 2.50 mmol of the salt KA in a total volume of (25.00 + 31.25) = 56.25 mL, giving a salt concentration of

$$\left(\frac{2.50 \text{ mmol}}{56.25 \text{ mL}}\right) = \underline{0.044 \text{ mol L}^{-1}}$$

Thus, we need to calculate the pH of a 0.044 mol $L^{-1}$ solution of potassium acetate, containing the neutral $K^+(aq)$ and the weakly basic $CH_3CO_2^-$ ion.

| | $A^-(aq) + H_2O(l)$ | $\rightleftharpoons$ $HA(aq)$ | $+ OH^-(aq)$ | |
|---|---|---|---|---|
| initially | 0.044 | 0 | 0 | mol $L^{-1}$ |
| at eq'm | 0.044-x | x | x | mol $L^{-1}$ |

$$K_b(A^-) = \frac{K_w}{K_a(HA)} = 5.6\text{x}10^{-10} \text{ mol L}^{-1} = \frac{x^2}{0.044-x}$$

x << 0.044, $x^2 = 2.46\text{x}10^{-11}$; $x = 4.96\text{x}10^{-6}$ (assumption justified)

$[OH^-] = 4.96\text{x}10^{-6}$ mol $L^{-1}$ ; pH = 14.00 + log $[OH^-]$ = $\underline{8.70}$.

(d) Beyond the equivalence point

At the initial point in the titration we had 25.00 mL acid containing 2.50 mmol HA. After the addition of 50.00 mL KOH(aq), the moles of base added is

$$(50.00 \text{ mL})(0.08 \text{ mol L}^{-1}) = 4.00 \text{ mmol KOH}$$

| | $HA(aq)$ | $+ KOH(aq)$ | $\longrightarrow KA(aq)$ | $+ H_2O(l)$ |
|---|---|---|---|---|
| initially | 2.50 | 4.00 | 0 | mmol |
| after reaction | 0 | 1.50 | 2.50 | mmol |

KOH is in excess, and this excess $OH^-(aq)$ will largely repress the basic reaction of the acetate ion.

| | $A^-(aq) + H_2O(l)$ | $\rightleftharpoons$ $HA(aq)$ | $+ OH^-(aq)$ | |
|---|---|---|---|---|
| initially | 0.033 | 0 | 0.020 | mol $L^{-1}$* |
| at eq'm | 0.033-x | x | 0.020+x | mol $L^{-1}$ |

*calculated from the mmol above and the total volume of the solution of 75.0 mL.

Thus, $$K_b(A^-) = \frac{[HA][OH^-]}{[A^-]} = \frac{x(0.020+x)}{0.033-x} = 5.6\text{x}10^{-10} \text{ mol L}^{-1}$$

x << 0.020, $x = 9.2\text{x}10^{-10}$ mol $L^{-1}$, which is negligible compared to $[OH^-]$ = 0.020 mol $L^{-1}$ from the excess KOH(aq).

$[OH^-]$ = 0.020 mol $L^{-1}$ ; pH = 14.00 + log $\underline{[OH^-] = 12.30}$

60. At the equivalence point in Problem 59, pH = 8.70 the value of $pK_a$ for a suitable indicator should be close to 8.70, in practice an indicator with $pK_a$ 8.7±1 would be suitable (for example, thymol blue or phenolphthalein).

61. This problem is similar to Problem 59, except that we start with a solution of a weak base and titrate it with a strong acid.

(a) initially 0.100 M $NH_3(aq)$ - a weak base for which the pH can be determined according to the method given in Problem 8. pH = 11.13

(b) After addition of 10.00 mL of acid

At the start of the titration we have (25.00 mL)(0.100 mol $L^{-1}$)

= 2.50 mmol of $NH_3(aq)$

Moles of added acid = (10.00 mL)(0.100 mol $L^{-1}$) = 1.00 mmol

| | $NH_3(aq)$ + | $HCl(aq)$ → | $NH_4Cl(aq)$ | |
|---|---|---|---|---|
| initially | 2.50 | 1.00 | 0 | mmol |
| at eq'm | 1.50 | 0 | 1.00 | mmol |

and the solution contains 1.50 mmol $NH_3(aq)$ and 1.00 mmol $NH_4Cl(aq)$ - a weak base and a salt of the weak base, and is thus a buffer solution, for which

$$K_a(NH_4^+) = \frac{K_w}{K_b(NH_3)} = 5.6x10^{-10}\ \text{mol L}^{-1} = [H_3O^+]\frac{[NH_3]}{[NH_4^+]}$$

$$= [H_3O^+](\frac{1.50\ \text{mmol}}{1.00\ \text{mmol}})\ ;\ [H_3O^+] = 3.7x10^{-10}\ \text{mol L}^{-1}$$

pH = 9.43

(c) After addition of 12.50 mL of acid

This is also a buffer solution, containing 1.25 mmol $NH_4Cl(aq)$ and 1.25 mmol $NH_3(aq)$, for which

$$K_a(NH_4^+) = 5.6x10^{-10}\ \text{mol L}^{-1} = [H_3O^+](\frac{1.25\ \text{mmol}}{1.25\ \text{mmol}})$$

$$[H_3O^+] = 5.6x10^{-10}\ \text{mol L}^{-1}\ ;\ pH = 9.25$$

(d) At the equivalence point

moles of added acid = initial moles of base = 2.50 mmol HCl(aq) and the solution contains 2.50 mmol of $NH_4Cl(aq)$

Volume of acid to achieve equivalence point = (2.50 mmol)$(\frac{1\ L}{0.100\ mol})$

= 25.00 mL

Thus the solution contains 2.50 mmol $NH_4Cl$ in a total volume of

(25.00 + 25.00) mL = 50.00 mL

Concentration of $NH_4Cl(aq)$ = $(\frac{2.50\ \text{mmol}}{50.00\ \text{mL}})$ = 0.050 mol $L^{-1}$

The solution contains only 0.050 M $NH_4Cl(aq)$ and the pH is calculated according to the method given in Problem 20(e); pH = 5.28

(e) After addition of 40.00 mL of acid

This point is beyond the equivalence point, so the solution contains $NH_4Cl(aq)$ plus excess of strong acid, which essentially completely represses the ionization of $NH_4^+$ (as excess base did for the solution of sodium acetate in Problem 59(d)). Thus, all that is required is to calculate the concentration of excess acid. Initially there are 25.00 mL of a solution containing 2.50 mmol $NH_3(aq)$, and 40.00 mL of 0.100 M HCl(aq), or 4.00 mmol HCl(aq), are added to it.

| | $NH_3(aq)$ | + HCl(aq) | $\longrightarrow NH_4Cl(aq)$ | |
|---|---|---|---|---|
| initially | 2.50 | 4.00 | 0 | mmol |
| at eq'm | 0 | 1.50 | 2.50 | mmol |

At equilibrium, the solution contains 1.50 mmol HCl(aq) in a total volume of (25.00+40.00) = 65.00 mL of solution. Thus,

$$[H_3O^+] = \left(\frac{1.50 \text{ mmol HCl}}{65.00 \text{ mL HCl}}\right) = 0.023 \text{ mol L}^{-1} \quad ; \quad \underline{pH = 1.64}$$

62. The pH at the equivalence point is 5.28 (Problem 61(d)), so the more suitable indicator should have a $pK_a$ close to 5.3. Of methyl red ($pK_a$ 5.0) and thymol blue ($pK_a$ 8.2), methyl red is the more suitable. Thymol blue would in fact change color well before the equivalence point.

63. The method is similar to that given in Problem 59. Since both the HOCl(aq) and the KOH(aq) have the concentration, (d) corresponds to the equivalence point in this titration. The calculated pH values using $K_a(HOCl) = 3.1 \times 10^{-8}$ mol $L^{-1}$ are as follows:

(a) 0.010 M HOCl(aq); pH = 4.75.
(b) Buffer solution containing equal moles of HOCl and KOCl, pH = 7.51.
(c) Buffer solution containing mol ratio HOCl:KOCl = 1:19, pH = 8.79.
(d) 0.005 M KOCl(aq), pH = 9.60.
(e) 0.0049 M KOCl(aq) + 0.00024 M KOH(aq), pH = 10.40.
(f) 0.0040 M KOCl(aq)+0.0020 M KOH(aq), pH = 11.30.

64. (a) The solution is a solution of $CH_3CO_2H(aq)$ of concentration $C_a$. pH = 2.72, and $[H_3O^+] = 1.91 \times 10^{-3}$ mol $L^{-1}$.

| | $HA(aq) + H_2O(l) \rightleftharpoons$ | $H_3O^+(aq)$ | $+ A^-(aq)$ | |
|---|---|---|---|---|
| initially | $C_a$ | 0 | 0 | mol $L^{-1}$ |
| at eq'm | $C_a$-0.00191 | 0.00191 | 0.00191 | mol $L^{-1}$ |

$$K_a(HA) = \frac{[H_3O^+][A^-]}{[HA]} = \frac{(0.00191 \text{mol L}^{-1})^2}{(C_a - 0.00191 \text{ mol L}^{-1})} = 1.8 \times 10^{-5} \text{ mol L}^{-1}$$

$$\underline{C_a = 0.2046 \text{ M}}$$

(b) The equivalence point pH = 8.92 corresponds to $[H_3O^+] = 1.20 \times 10^{-9}$ M and the solution contains only sodium acetate, for which the concentration $C_s$ can be calculated.

$$A^-(aq) + H_2O(l) \rightleftharpoons HA(aq) + OH^-(aq)$$

| | | | | |
|---|---|---|---|---|
| initially | $C_s$ | 0 | 0 | mol $L^{-1}$ |
| at eq'm | $C_s - x$ | x | x | mol $L^{-1}$ |

where $x = [OH^-] = K_w/[H_3O^+] = 8.32\times10^{-6}$ mol $L^{-1}$.

$$\text{Thus, } K_b(A^-) = \frac{x^2}{C_s - x} = \frac{K_w}{K_a(HA)} = 5.6\times10^{-10} \text{ mol L}^{-1}$$

and substituting for x gives $C_s$ = <u>0.124 M</u>

Since at the equivalence point, mol of salt = initial mol of acid

$$= (50.00 \text{ mL})(0.2046 \text{ mol L}^{-1}) = 10.23 \text{ mmol}$$

and if the volume of added base is V mL, then

$$C_s = \frac{10.23 \text{ mmol salt}}{(50.00 + V) \text{ mL}} = 0.124 \text{ mol L}^{-1}; \quad \underline{V = 32.5 \text{ mL}}$$

The volume of base required to reach the equivalence point is

<u>32.5 mL</u>

(c) The concentration of base is calculated from mol of base and the volume of base.

$$\text{Concentration of base} = \left(\frac{10.23 \text{ mmol base}}{32.5 \text{ mL of base}}\right) = \underline{0.315 \text{ M}}$$

65. At the half-equivalence point, the solution contains half the initial mol of acid and an equal number of mol of the salt of the weak acid. Thus, it is a buffer solution containing equal mol of weak acid and the salt of the weak acid, for which

$$K_a = [H_3O^+]\frac{[A^-]}{[HA]} = [H_3O^+]\left(\frac{n \text{ mol salt}}{n \text{ mol acid}}\right) = [H_3O^+]$$

i.e., at the <u>half-equivalence point</u>, <u>$pK_a(HA)$ = pH</u>.

66. Use the shorthand notation $H_3A$ for the triprotic phosphoric acid.

$H_3A + H_2O \rightleftharpoons H_3O^+ + H_2A^-$ ; $K_a(1) = 7.5\times10^{-3}$ mol $L^{-1}$

$H_2A^- + H_2O \rightleftharpoons H_3O^+ + HA^{2-}$ ; $K_a(2) = 6.2\times10^{-8}$ mol $L^{-1}$

$HA^{2-} + H_2O \rightleftharpoons H_3O^+ + A^{3-}$ ; $K_a(3) = 3.1\times10^{-13}$ mol $L^{-1}$

Since the three $K_a$ values are well separated in magnitude, the system at various points in the titration can be treated as if the acid $H_3A$ is replaced by the acid $H_2A^-$, which in turn is replaced by the acid $HA^{2-}$.

The initial mol of acid = (10.00 mL)(0.100 mol $L^{-1}$) = <u>1.00 mmol</u>

and the moles of base added is calculated for each point in the titration, which enables the initial concentrations of of $H_3A$, $H_2A^-$, $HA^{2-}$, and $A^{3-}$, to be calculated for each of these points, as is shown in the following table. (a) is a solution of $H_3A$, (c) is a solution of $H_2A^-$, and (e) is a solution of $HA^{2-}$, for which the pH's can be calculated by the normal methods for weak acids. Solutions (b), (d), and (f) are buffer solutions and can also be treated using the usual methods for buffer solutions.

| | mL[a] | mmol[a] | mL[b] | $[H_3A]^c$ | $[H_2A^-]^c$ | $[HA^{2-}]^c$ | $[A^{3-}]^c$ | $[H_3O^+]$ | pH |
|---|---|---|---|---|---|---|---|---|---|
| (a) | 0.00 | 0.00 | 10.00 | 0.100 | - | - | - | $8.3\times10^{-2}$ | 1.08 |
| (b) | 4.00 | 0.50 | 14.00 | 0.036 | 0.036 | - | - | $7.5\times10^{-3}$ | 2.12 |
| (c) | 8.00 | 1.00 | 18.00 | - | 0.056 | - | - | $5.9\times10^{-5}$ | 4.23 |
| (d) | 12.00 | 1.50 | 22.00 | - | 0.023 | 0.023 | - | $6.2\times10^{-8}$ | 7.21 |
| (e) | 16.00 | 2.00 | 26.00 | - | - | 0.038 | - | $1.3\times10^{-10}$ | 9.89 |
| (f) | 20.00 | 2.50 | 30.00 | - | - | 0.017 | 0.017 | $2.1\times10^{-13}$ | 12.68 |

[a]base added; [b]total volume of solution; [c]initial concentrations

67. The $K_b$ values for $CO_3^{2-}$ and $HCO_3^-$ are given in Table 14.1.
Consider the equilibria,

$$CO_3^{2-}(aq) + H_2O(l) \rightleftharpoons HCO_3^-(aq) + OH^-(aq);\ K_b(1) = 2.1\times10^{-4}\ \text{mol L}^{-1}$$
$$HCO_3^-(aq) + H_2O(l) \rightleftharpoons H_2CO_3(aq) + OH^-(aq);\ K_b(2) = 2.5\times10^{-8}\ \text{mol L}^{-1}.$$

Since the two $K_b$ values differ by a large factor, disregard the second as a first approximation.

| | $CO_3^{2-}(aq) + H_2O(l) \rightleftharpoons$ | $HCO_3^-(aq)$ + | $OH^-(aq)$ | |
|---|---|---|---|---|
| initially | 0.200 | 0 | 0 | mol L$^{-1}$ |
| at eq'm | 0.200-x | x | x | mol L$^{-1}$ |

$$K_b(1) = \frac{[HCO_3^-][OH^-]}{[CO_3^{2-}]} = \frac{x^2}{0.200-x} = 2.1\times10^{-4}\ \text{mol}^2\ \text{L}^{-2}$$

$x << 0.200$, $x^2 = 4.2\times10^{-5}$; $x = 6.5\times10^{-3}$ (approximation justified).

Then, for the second equilibrium,

| | $HCO_3^-(aq) + H_2O(l) \rightleftharpoons$ | $H_2CO_3(aq)$ + | $OH^-(aq)$ | |
|---|---|---|---|---|
| initially | 0.0065 | 0 | 0.0065 | mol L$^{-1}$ |
| at eq'm | 0.0065-x | x | 0.0065+x | mol L$^{-1}$ |

$$K_b(2) = \frac{[H_2CO_3][OH^-]}{[HCO_3^-]} = \frac{x(0.0065+x)}{(0.0065-x)} = 2.5\times10^{-8}\ \text{mol L}^{-1}$$

$x << 0.0065$, $x = 2.5\times10^{-8}$ (assumption justified).

Thus, the concentrations of species at equilibrium are

$[CO_3^{2-}]$ = <u>0.194 mol L$^{-1}$</u>; $[HCO_3^-]$ = <u>0.0065 mol L$^{-1}$</u>

$[OH^-]$ = <u>0.0065 mol L$^{-1}$</u>; $[H_3O^+]$ = <u>$1.6\times10^{-12}$ mol L$^{-1}$</u>

68. (a) KOH(aq) is a strong base; $[OH^-] = 0.010$ M; <u>pH = 12.00</u>

(b) KCl(aq) is a solution of a neutral salt; <u>pH = 7.00</u>

(c) $NH_3(aq)$ is a weak base ($K_b = 1.8\times10^{-5}$ mol L$^{-1}$); <u>pH = 10.63</u>

(d) HF(aq) is a weak acid ($K_a = 3.5\times10^{-4}$ mol L$^{-1}$); <u>pH = 2.73</u>

(e) KF(aq) is a salt with a neutral cation and a basic anion;

$F^-(aq) + H_2O(l) \rightleftharpoons HF(aq) + OH^-(aq)$; $K_b(F^-) = 2.9x10^{-11}$ mol $L^{-1}$.

$\underline{pH = 7.73}$

(f) $HNO_3(aq)$ is a solution of a strong acid; $[H_3O^+] = 0.01$ mol $L^{-1}$;

$\underline{pH = 2.00}$

69. (a) Solution contains a strong acid and the salt of a strong acid, which is not a buffer solution.

(b) Solution contains a weak acid and the salt of a weak acid in equal concentration, which constitutes a buffer solution.

(c) Calculation of the composition of this solution shows that it contains 0.025 M KOH(aq) and 0.050 M $CH_3CO_2K(aq)$, potassium acetate. It is a solution of a strong base and the salt of a weak acid, which is not a buffer solution.

(d) Calculation of the composition of this solution shows that it contains 0.025 M $CH_3CO_2K(aq)$ and 0.025 M $CH_3CO_2H(aq)$. It is a solution containing a weak acid and the salt of a weak acid in equal concentrations, and is thus a buffer solution.

70. $pK_a$(phenol) = 9.80; $K_a = 1.58x10^{-10}$ mol $L^{-1}$.

| | $C_6H_5OH(aq)$ + $H_2O(l)$ $\rightleftharpoons$ | $H_3O^+(aq)$ + | $C_6H_5O^-(aq)$ | |
|---|---|---|---|---|
| initially | $C_a$ | 0 | 0 | mol $L^{-1}$ |
| at eq'm | $C_a$-x | x | x | mol $L^{-1}$ |

where $x = [H_3O^+] = 1.26x10^{-5}$ mol $L^{-1}$ (pH 4.90).

Thus, $K_a = \frac{[H_3O^+][C_6H_5O^-]}{[C_6H_5OH]} = \frac{x^2}{C_a - x} = 1.58x10^{-10}$ mol $L^{-1}$.

and for $x \ll C_a$, $C_a = \frac{(1.26 \times 10^{-5} \text{ mol } L^{-1})^2}{(1.58 \times 10^{-10} \text{ mol } L^{-1})} = \underline{1.00 \text{ mol } L^{-1}}$

71. At the equivalence point, mol of added base = initial mol acid, which occurs after the addition of 10.00 mL of strong base. Call this n mol of strong base. Thus, after addition of 5.00 mL of base, the amount of base added must be n/2 mol.

| | HA(aq) + | NaOH(aq) $\rightarrow$ | NaA(aq) + $H_2O(l)$ | |
|---|---|---|---|---|
| initially | n | n/2 | 0 | mol |
| at eq'm | n/2 | 0 | n/2 | mol |

and the solution is a buffer solution containing an equal number of moles of the weak acid and the salt of the weak acid, for which

$pH = pK_a = \underline{5.00}$

72. The moles of added NaOH are proportional to the volume of NaOH added, and at the equivalence point, where moles of added base = initial moles of acid, moles of added base are proportional to the 12.00 mL of base. Suppose the initial moles of acid = n mol, then the concentration of base is (n mol)/12 mL.

After addition of only 5.00 mL of base,

| | $HA(aq)$ + | $NaOH(aq)$ ⟶ | $NaA(aq) + H_2O(l)$ | |
|---|---|---|---|---|
| initially | n | $5mL(\frac{n}{12\ mL})$ | 0 | mol |
| at eq'm | $\frac{7n}{12}$ | 0 | $\frac{5n}{12}$ | mol |

Thus, this is a buffer solution, for which

$K_a(HA) = [H_3O^+]\frac{[A^-]}{[HA]} = [H_3O^+](\frac{5}{7})$ ; and $[H_3O^+] = 10^{-6}$ mol $L^{-1}$.

Thus, $K_a$ = $\underline{7.1x10^{-7}\ mol\ L^{-1}}$

# CHAPTER 15

1. In their reactions the alkali metals are oxidized to their $M^+$ cations. The reactivity increases as the ease of oxidation increases, as measured, for example, by their ionization energies, which decrease with increasing atomic number from Li to Cs. In their reactions with water at room temperature, lithium reacts relatively slowly, sodium is more reactive and the metal melts, the reaction of potassium is sufficiently vigorous to ignite the hydrogen that is produced, and rubidium and cesium react so vigorously that their reactions with water are dangerously explosive. In each case , the metal is oxidized to $M^+(aq)$ and water is reduced to $H_2(g)$; the other product is a $OH^-(aq)$:

$$2M(s) + 2H_2O(l) \rightarrow 2M^+(aq) + 2OH^-(aq) + H_2(g)$$

2. All these reactions give ionic products; the valence of Li (group 1) is 1, and the valence of Ca (group 2) is 2:

(a) $2Li(s) + Br_2(l) \longrightarrow 2LiBr(s)$ - lithium bromide; $Li^+ \; :\ddot{\underset{..}{Br}}:^-$

$Ca(s) + Br_2(l) \longrightarrow CaBr_2(s)$ - calcium bromide; $Ca^{2+} \; [:\ddot{\underset{..}{Br}}:^-]_2$

(b) $2Li(s) + S(s) \xrightarrow{heat} Li_2S(s)$ - lithium sulfide; $[Li^+]_2 :\ddot{\underset{..}{S}}:^{2-}$

$Ca(s) + S(s) \xrightarrow{heat} CaS(s)$ - calcium sulfide; $Ca^{2+} \; :\ddot{\underset{..}{S}}:^{2-}$

(c) $6Li(s) + N_2(g) \xrightarrow{heat} 2Li_3N(s)$ - lithium nitride; $[Li^+]_3 :\ddot{\underset{..}{N}}:^{3-}$

$3Ca(s) + N_2(g) \xrightarrow{heat} Ca_3N_2(s)$ - calcium nitride; $[Ca^{2+}]_3 [:\ddot{\underset{..}{N}}:^{3-}]_2$

3. In each of these reactions the alkali metal behaves as a reducing agent and is oxidized to $M^+$, and the other reactant is reduced to an anion.

(a) $2K(s) + Br_2(l) \longrightarrow 2KBr(s)$ potassium bromide

(b) $4Li(s) + O_2(g) \longrightarrow 2Li_2O(s)$ lithium oxide

(c) $2Na(s) + H_2(g) \longrightarrow 2NaH(s)$ sodium hydride

(d) $6Li(s) + N_2(g) \longrightarrow 2Li_3N_2(s)$ lithium nitride

(g) $2K(s) + 2H_2O(l) \longrightarrow 2KOH(aq) + H_2(g)$ potassium hydroxide, hydrogen

In each of these reactions the alkali metal cation remains unchanged and the anion, which is a strong base in water, is protonated to give its conjugate acid.

(e) $LiH(s) + H_2O(l) \longrightarrow LiOH(aq) + H_2(g)$ lithium hydroxide, hydrogen

(f) $Li_3N(s) + 3H_2O(l) \longrightarrow 3LiOH(aq) + NH_3(g)$ lithium hydroxide, ammonia.

4. Sodium hydrogen carbonate, $Na^+HCO_3^-$, contains the hydrogen carbonate ion, $HCO_3^-$ , the conjugate base of carbonic acid, $H_2CO_3$. But since carbonic acid is a diprotic acid, $HCO_3^-$ is also the conjugate acid of the $CO_3^{2-}$, carbonate ion. A buffer solution must contain a weak acid to react with any base that is added, and a weak base to react with any acid that is added, and clearly a solution of sodium hydrogen carbonate can behave in this way:

$$HCO_3^-(aq) + OH^-(aq) \longrightarrow CO_3^{2-}(aq) + H_2O(l)$$

$$HCO_3^-(aq) + H_3O^+(aq) \longrightarrow H_2CO_3(aq) + H_2O(l)$$

($H_2CO_3(aq)$ is unstable and decomposes to $CO_2(g)$ and water). The common name for sodium hydrogen carbonate is "sodium bicarbonate", or "bicarbonate of soda".

5. Lime is calcium oxide, CaO(s), containing $Ca^{2+}$ and $O^{2-}$ (oxide) ions. In water, $O^{2-}$ acts as a strong base and is fully protonated to give $OH^-$, thus calcium hydroxide is formed:

$$CaO(s) + H_2O(l) \longrightarrow Ca(OH)_2(s)$$

Lime is a relatively inexpensive base that is readily obtained simply by heating calcium carbonate (limestone):

$$CaCO_3(s) \longrightarrow CaO(s) + CO_2(g)$$

It is used industrially for treating acid soil, in making glass and cement, and to remove acidic impurities, such as silica, $SiO_2$, in metallurgical processes

6. In each of these reactions, magnesium is oxidized to $Mg^{2+}$ cations and the nonmetal is reduced to a monatomic anion.

(a) $2Mg(s) + O_2(g) \longrightarrow 2MgO(s)$ $Mg^{2+}O^{2-}$ - magnesium oxide

(b) $Mg(s) + S(s) \longrightarrow MgS(s)$ $Mg^{2+}S^{2-}$ - magnesium sulfide

(c) $3Mg(s) + N_2(g) \longrightarrow Mg_3N_2$ $[Mg^{2+}]_3[N^{3-}]_2$ - magnesium nitride

7. Calcium hydride is the ionic compound $Ca^{2+}(H^-)_2$, in which the hydride ion, $H^-$, is a strong base in water, so the reaction is

$$CaH_2(s) + 2H_2O(l) \longrightarrow Ca(OH)_2(s) + 2H_2(g)$$

First we calculate the number of moles of $H_2(g)$ formed by reaction of 2.00 g of CaH2(s) and then convert this to the volume of hydrogen at STP by using the ideal gas equation.

$$\text{moles of } H_2(g) = (2.00 \text{ g } CaH_2)\left(\frac{1 \text{ mol } CaH_2}{42.10 \text{ g } CaH_2}\right)\left(\frac{2 \text{ mol H}}{1 \text{ mol } CaH_2}\right) = 9.50\times10^{-2} \text{ mol}$$

$$PV = nRT; \quad V = \frac{nRT}{P} = \frac{(9.50\times10^{-2} \text{ mol})(0.0821 \text{ atm L mol}^{-1} \text{ K}^{-1})(273 \text{ K})}{1 \text{ atm}}$$

$$= \underline{2.13 \text{ L}}$$

8. Both sodium and magnesium are in period 3 with their valence electrons in the n = 3 shell. The core charge of sodium is +1 (group 1) and that of magnesium is +2 (group 2); thus, it is easier to remove an electron from a neutral Na atom than it is to remove an electron from a neutral Mg atom. The second electron is removed either from a $Na^+(g)$ ion or from a $Mg^+(g)$ ion. For magnesium this second electron is from the n = 3 shell, but for sodium it is from the inner n = 2 shell, which is much closer to the nucleus than the n = 3 shell; thus the second electron is much more difficult to remove from sodium than it is from magnesium.

First ionization energy: Mg > Na; second ionization energy: Na > Mg

9. All the compounds contain $Na^+$ ions, which have no acid or base properties in water. Thus, the compounds that react are those with negative ions that behave as bases in water.

(a) Sodium chloride: $Cl^-$ is the conjugate base of the strong acid HCl(aq) and has no basic properties in water - no reaction

(b) Sodium hydride: $H^-$ is the conjugate base of $H_2$ and behaves as a strong base in water

$$NaH(s) + H_2O(l) \longrightarrow Na^+(aq) + OH^-(aq) + H_2(g)$$

(c) Sodium hydroxide: $OH^-$ is the strongest base that can exist in water - no reaction

(d) Sodium oxide: $O^{2-}$ ion is the conjugate base of $OH^-(aq)$ which has no acidic properties in water; thus $O^{2-}$ is a strong base

$$Na_2O(s) + H_2O(l) \longrightarrow 2Na^+(aq) + 2OH^-(aq)$$

(e) Sodium sulfate: $SO_4^{2-}$ is the conjugate base of the $HSO_4^-$ ion, a moderately strong acid, and thus is protonated to a small extent in water

$$Na_2SO_4(s) \longrightarrow 2Na^+(aq) + SO_4^{2-}(aq)$$

$$SO_4^{2-}(aq) + H_2O(l) \rightleftharpoons HSO_4^-(aq) + OH^-(aq)$$

(f) Sodium carbonate: $CO_3^{2-}$ is the conjugate base of the weak acid $HCO_3^-$ and behaves as a weak base in water

$$Na_2CO_3(s) \longrightarrow 2Na^+(aq) + CO_3^{2-}(aq)$$

$$CO_3^{2-}(aq) + H_2O(l) \rightleftharpoons HCO_3^-(aq) + OH^-(aq)$$

10. In each case the metal behaves as a reducing agent and is oxidized to $M^+$ (alkali metals) or $M^{2+}$ (alkaline earth metals), and the nonmetal or water is reduced.

(a) $Mg(s) + Cl_2(g) \longrightarrow MgCl_2(s)$

(b) $2Ca(s) + O_2(g) \longrightarrow 2CaO(s)$

(c) $Sr(s) + H_2(g) \longrightarrow SrH_2(s)$

(d) $Mg(s) + H_2(g) \longrightarrow MgH_2(s)$

(e) $Ca(s) + 2H_2O(l) \longrightarrow Ca(OH)_2(s) + H_2(g)$

11. $Be(OH)_2$ and $Mg(OH)_2$ are insoluble and the solubility of the remaining hydroxides increases from Ca to Ba to Sr. All the cations have two positive charges, $M^{2+}$, and the size of the $M^{2+}$ ion increases from $Be^{2+}$ to $Ba^{2+}$. The $OH^-$ ion has a relatively small ionic radius. Thus the ionic bonds between $OH^-$ ions and the smaller alkaline earth metal cations are very strong, both by virtue of their +2 charges and size and $Be(OH)_2$ and $Mg(OH)_2$ are insoluble, but for the larger cations the ionic bonds become progressively weaker as the size of the cation increases; $Ca(OH)_2$ is slightly soluble and $Sr(OH)_2$ and $Ba(OH)_2$ more soluble in that order.

12. Although calcium carbonate, $CaCO_3(s)$ is very insoluble, calcium hydrogen carbonate, $Ca(HCO_3)_2(s)$ is relatively soluble. Carbonate ion, $CO_3^{2-}$, is

the conjugate base of the hydrogen carbonate ion, $HCO_3^-$, and is readily converted to $HCO_3^-$ by rain water, which contains dissolved $CO_2(g)$, and is a dilute solution of carbonic acid.

$$CaCO_3(s) + CO_2(aq) + H_2O(l) \rightleftharpoons Ca^{2+}(aq) + 2HCO_3^-(aq)$$

Thus the $CaCO_3(s)$ gradually dissolves as the $Ca(HCO_3)_2$ is washed away by running water and a cave is gradually formed. By the same process, water seeping through the roof of a cave is saturated with $Ca(HCO_3)_2$ and also contains dissolved $CO_2(g)$. As a drop of this water evaporates, it loses $CO_2(g)$ and the above equilibrium shifts to the left (Le Châtelier's principle) and insoluble $CaCO_3$ is deposited, so that gradually a stalactite hanging from the roof of the cave is formed. The stalagmites that form on the floor below the stalactites are due to the same process, as a consequence of the evaporation of water drops that fall to the floor

13. (a) $CaCO_3(s) \longrightarrow CaO(s) + CO_2(g)$; calcium oxide (lime)
(b) $Ca(OH)_2(s) \longrightarrow CaO(s) + H_2O(l)$; calcium oxide (lime)
(c) $2NaHCO_3(s) \longrightarrow Na_2CO_3(s) + H_2O(g) + CO_2(g)$; sodium carbonate
(d) $MgCl_2.6H_2O(s) \longrightarrow MgO(s) + 2HCl(g) + 5H_2O(g)$; magnesium oxide

14. Both $CaCO_3(s)$ and $Ca(HCO_3)_2(s)$ decompose to $CaO(s)$ on heating:

$$CaCO_3(s) \longrightarrow CaO(s) + CO_2(g) \quad .... (1)$$

$$Ca(HCO_3)_2(s) \longrightarrow CaO(s) + 2CO_2(g) + H_2O(g) \quad .... (2)$$

Since water is produced only in the second reaction,

$$\text{mol } Ca(HCO_3)_2 = (0.200 \text{ g } H_2O)\left(\frac{1 \text{ mol } H_2O}{18.02 \text{ g } H_2O}\right)\left(\frac{1 \text{ mol } Ca(HCO_3)_2}{1 \text{ mol } H_2O}\right)$$

$$= \underline{0.0111 \text{ mol}}$$

and the moles of $CO_2(g)$ produced is twice the number of mol of $Ca(HCO_3)_2$ plus the number of mol of $CaCO_3$.

$$\text{mol } CO_2 = (1.500 \text{ g } CO_2)\left(\frac{1 \text{ mol } CO_2}{44.01 \text{ g } CO_2}\right) = \underline{0.0341 \text{ mol}}$$

Hence: $\text{mol } CaCO_3 = [0.0341 - 2(0.0111)] = \underline{0.0119 \text{ mol}}$

Now we can convert mol of $CaCO_3$, and mol of $Ca(HCO_3)_2$, to grams, and find the mass of $CaO(s)$ in the original mixture by difference,

$$\text{mass of } CaCO_3(s) = (0.0119 \text{ mol})\left(\frac{100.1 \text{ g } CaCO_3}{1 \text{ mol } CaCO_3}\right) = \underline{1.19 \text{ g}}$$

$$\text{mass of } Ca(HCO_3)_2 = (0.0111 \text{ mol})\left(\frac{162.1 \text{ g } Ca(HCO_3)_2}{1 \text{ mol } Ca(HCO_3)_2}\right) = \underline{1.80 \text{ g}}$$

$$\text{mass \% } CaCO_3 = (1.19 \text{ g})\left(\frac{100\%}{10.00 \text{ g}}\right) = \underline{11.9\%}$$

$$\text{mass \% } Ca(HCO_3)_2 = (1.80 \text{ g})\left(\frac{100\%}{10.00 \text{ g}}\right) = \underline{18.0\%}$$

By difference: mass % CaO = (100-11.9-18.0) = $\underline{70.1\%}$

15. We first calculate the mol of Mg in one metric ton, and multiply by the unit conversion factor 1 L/0.05 mol Mg , and 100%/70%, to take care of the efficiency factor.

$$\text{L of seawater} = (10^3\ \text{kg Mg})\left(\frac{1000\ \text{g}}{1\ \text{kg}}\right)\left(\frac{1\ \text{mol Mg}}{24.31\ \text{g Mg}}\right)\left(\frac{1\ \text{L}}{0.05\ \text{mol Mg}}\right)\left(\frac{100}{70}\right)$$

$$= \underline{1.2 \times 10^6\ \text{L}}$$

16. To obtain washing soda, $Na_2CO_3.10H_2O(s)$, trona ore, $Na_5(CO_3)_2(HCO_3).2H_2O$, is heated to give anhydrous $Na_2CO_3$:

$$2Na_5(CO_3)_2(HCO_3).2H_2O(s) \longrightarrow 5Na_2CO_3(s) + CO_2(g) + 5H_2O(g)$$

which is the recrystallized from water to give $Na_2CO_3.10H_2O$. Thus.

Maximum mass of washing soda, $Na_2CO_3.10H_2O(s)$ =

$$(10^3\ \text{kg trona})\left(\frac{1000\ \text{g}}{1\ \text{kg}}\right)\left(\frac{1\ \text{mol trona}}{332.0\ \text{g trona}}\right)\left(\frac{5\ \text{mol soda}}{2\ \text{mol trona}}\right)\left(\frac{286.2\ \text{g soda}}{1\ \text{mol soda}}\right)$$

$$= (2\text{x}10^6\ \text{g soda})\left(\frac{1\ \text{kg}}{1000\ \text{g}}\right)\left(\frac{1\ \text{ton}}{1000\ \text{kg}}\right) = \underline{2\ \text{metric ton}}$$

17. $O^{2-}$, $H^-$, and $N^{3-}$, are all strong bases in water, and $CO_3^{\ 2-}$ is the conjugate base of the weak acid $HCO_3^{\ -}(aq)$ and behaves as a weak base:

(a) $O^{2-}(aq) + H_2O(l) \longrightarrow 2OH^-(aq)$

(b) $H^-(aq) + H_2O(l) \longrightarrow H_2(g) + OH^-(aq)$

(c) $N^{3-}(aq) + 3H_2O(l) \longrightarrow NH_3(aq) + 3OH^-(aq)$

(d) $CO_3^{\ 2-}(aq) + H_2O(l) \rightleftharpoons HCO_3^{\ -}(aq) + OH^-(aq)$

18. (a) Francium is the heaviest alkali metal, the hydride is FrH(s) with the Lewis structure $Fr^+\ :H^-$

(b) Cesium has a valence of 1, and oxygen has a normal valence of 2; $Cs_2O$, with the Lewis structure $[Cs^+]_2\ :\ddot{\underset{..}{O}}:^{2-}$

(c) Barium has a valence of 2, and oxygen has a normal valence of 2; BaO(s) with the Lewis structure $Ba^{2+}\ :\ddot{\underset{..}{O}}:^{2-}$

(d) Radium is the heaviest alkaline earth metal, the sulfate is $RaSO_4$ with the Lewis structure:

$$Ra^{2+}\quad \begin{array}{c} :\ddot{O}:^- \\ | \\ :\ddot{O}{=}S{=}\ddot{O}: \\ | \\ :\underset{..}{O}:^- \end{array}$$

(e) Sodium and lithium have valences of 1, sulfur has a valence of 2; $Na_2S(s)$ and $Li_2S(s)$ with the Lewis structures $[Na^+]_2\ :\ddot{\underset{..}{S}}:^{2-}$ and $[Li^+]_2\ :\ddot{\underset{..}{S}}:^{2-}$

19. $Mg^{2+}$ is a much smaller ion than $Ba^{2+}$ and is bonded to the six $H_2O$ molecules and exists as $Mg(H_2O)_6^{\ 2+}$ in $MgCl_2.6H_2O$. Because of the high charge/radius ratio of the $Mg^{2+}$ ion the electron pairs of six water molecules are strongly attracted to the $Mg^{2+}$ ion, which increases the electronegativity of the O atoms and makes the O-H bonds

more polar than in free $H_2O$, and thus more acidic. When $MgCl_2.6H_2O(s)$ is heated, rather than simple dehydration occurring, two of the acid protons from the water molecules are transferred to the $Cl^-$ ions, which gives volatile HCl(g) and the equilibrium

$$Mg(H_2O)_6^{2+} + 2Cl^- \rightleftharpoons Mg(OH)_2(H_2O)_4(s) + 2HCl(g)$$

shifts to the right until the reaction is complete. Eventually the hydrated $Mg(OH)_2$ loses water to give MgO(s) and water.

$$Mg(OH)_2(H_2O)_4(s) \longrightarrow MgO(s) + 5H_2O(g)$$

In contrast, the water molecules in $BaCl_2.2H_2O(s)$ are not very strongly bonded to the large $Ba^{2+}$ ion, and $Ba(H_2O)_2^{2+}$ does not behave as a weak acid. Thus, when it is heated the vapor pressure of water in contact with the hydrate increases, and as the water vapor escapes the dehydration of $BaCl_2.2H_2O(s)$ is eventually complete.

$$BaCl_2.2H_2O \rightleftharpoons BaCl_2(s) + 2H_2O(g)$$

20. The green flame strongly suggests that the solid E contains the $Ba^{2+}$ ion, which is confirmed by the formation of the white precipitate F when dilute $H_2SO_4(aq)$ is added to an aqueous solution of E.

$$Ba^{2+}(aq) + SO_4^{2-}(aq) \longrightarrow BaSO_4(s)$$

i.e., the white precipitate F is $BaSO_4(s)$. The gas obtained by heating A must be $CO_2(g)$, because it gives a white precipitate when bubbled through $Ca(OH)_2(s)$, which is the limewater test for $CO_2(g)$,

$$Ca(OH)_2(aq) + CO_2(g) \longrightarrow CaCO_3(s) + H_2O(l)$$

which suggests that A is a compound containing barium which when heated gives $CO_2(g)$, leading to the conclusion that A is very probably $BaCO_3(s)$.

$$BaCO_3(s) \longrightarrow BaO(s) + CO_2(g)$$

On this assumption, 1.00 g of $BaCO_3(s)$ is $(1.00\ g)(\frac{1\ mol}{197.3\ g})$ of $BaCO_3(s)$, $= 5.07\times10^{-3}$ mol $BaCO_3(s)$, which should give the same number of moles of $CO_2(g)$ when strongly heated. This is confirmed by calculating moles of $CO_2(g)$ from the gas data.

$$n = \frac{PV}{RT} = \frac{(750\ mm\ Hg)(\frac{1\ atm}{760\ mm\ Hg})(125\ mL)(\frac{1\ L}{1000\ mL})}{(0.0821\ atm\ L\ mol^{-1}\ K^{-1})(298\ K)} = \underline{5.04\times10^{-3}\ mol}$$

Thus, B, must be BaO(s), which dissolves in water to give a basic solution,

$$BaO(s) + H_2O(l) \longrightarrow Ba(OH)_2(aq)$$

which turns red litmus blue. Addition of excess HCl(aq) neutralizes the $Ba(OH)_2(aq)$ and the solution then turns blue litmus red.

$$Ba(OH)_2(aq) + 2HCl(aq) \longrightarrow BaCl_2(aq) + 2H_2O(l)$$

On evaporation to dryness the salt $BaCl_2.2H_2O(s)$ is obtained, which must be E. Finally, an aqueous solution of E gives a white precipitate of $BaSO_4(s)$, F, with dilute $H_2SO_4(aq)$,

$$BaCl_2(aq) + H_2SO_4(aq) \longrightarrow BaSO_4(s) + 2HCl(aq)$$

In summary: A is $BaCO_3(s)$; B is BaO(s); C is $CO_2(g)$; D is $CaCO_3(s)$;
E is $BaCl_2.2H_2O$ and F is $BaSO_4(s)$

21. The pH of the lake before addition of limestone, $CaCO_3(s)$, is 3.9, and the required pH is 6.3. Thus, sufficient limestone must be added to change the $[H_3O^+]$ of the lake from,

$$10^{-3.9}\ \text{mol L}^{-1} = 1.26\times10^{-4}\ \text{M} \ \underline{\text{to}}\ 10^{-6.3} = 5.01\times10^{-7}\ \text{M}$$

Carbonate ion, $CO_3^{2-}$, from the limestone reacts with $H_3O^+(aq)$ in the lake to give $HCO_3^-(aq)$ and, finally, $H_2CO_3(aq)$. If no $CO_2(g)$ is lost, the lake will finally constitute a buffer solution, containing $H_2CO_3(aq)$ and its conjugate base $HCO_3^-(aq)$, for which

$$K_a(H_2CO_3) = [H_3O^+]\left(\frac{[HCO_3^-]}{[H_2CO_3]}\right) = [H_3O^+]\left(\frac{\text{mol } HCO_3^-}{\text{mol } H_2CO_3}\right) = 4.3\times10^{-7}\ \text{mol L}^{-1}$$

i.e., $$\left(\frac{\text{mol } HCO_3^-}{\text{mol } H_2CO_3}\right) = \left(\frac{4.3\times10^{-7}\ \text{mol L}^{-1}}{5.0\times10^{-7}\ \text{mol L}^{-1}}\right) = \underline{0.86}$$

If we consider the reaction of $CO_3^{2-}$ with $H_3O^+$ to give $HCO_3^-$ and its reaction with $H_3O^+$ to give $H_2CO_3$, as separate reactions, the stoichiometry is represented by,

(1) $$CO_3^{2-}(aq) + H_3O^+(aq) \longrightarrow HCO_3^-(aq) + H_2O(l)$$

| | $CO_3^{2-}$ | $H_3O^+$ | $HCO_3^-$ | |
|---|---|---|---|---|
| initially | x | x | 0 | mol $L^{-1}$ |
| finally | 0 | 0 | x | mol $L^{-1}$ |

and

(2) $$CO_3^{2-}(aq) + 2H_3O^+(aq) \longrightarrow H_2CO_3(aq) + 2H_2O(l)$$

| | $CO_3^{2-}$ | $H_3O^+$ | $H_2CO_3$ | |
|---|---|---|---|---|
| initially | y | 2y | 0 | mol $L^{-1}$ |
| finally | 0 | 0 | 2y | mol $L^{-1}$ |

where $x + 2y = 1.26\times10^{-4}\ \text{mol L}^{-1}$ ..... (a)

and $\dfrac{x}{2y} = 0.86$ ..... (b)

From equation (b), $x = 1.72y$, and substituting for x in equation (a) gives,

$$3.72y = 1.26\times10^{-4}\ \text{mol L}^{-1};\ y = 3.39\times10^{-5}\ \text{mol L}^{-1}$$

Hence, initial $[CO_3^{2-}] = x+y = 2.72y = \underline{9.21\times10^{-5}\ \text{mol L}^{-1}}$

The amount of $CaCO_3(s)$ that has to be added to give the required buffer solution is $9.21\times10^{-5}$ mol per liter of lake water. To obtain the total number of moles, multiply this by the volume of the cylindrical lake in liters.

Volume of lake = (circular area)(average depth)

$$= \pi\left[(2.1\ \text{km})\left(\frac{1000\ \text{m}}{1\ \text{km}}\right)\right]^2(8.2\ \text{m}) = 1.14\times10^{8}\ \text{m}^3$$

$$= (1.14\times10^{8}\ \text{m}^3)\left(\frac{10\ \text{dm}}{1\ \text{m}}\right)^3\left(\frac{1\ \text{L}}{1\ \text{dm}^3}\right) = \underline{1.14\times10^{11}\ \text{L}}$$

Hence, the mass of $CaCO_3(s)$ to be added is given by

$$\left(\frac{9.21\times10^{-5}\text{ mol } CaCO_3}{1\text{ L}}\right)(1.14\times10^{11}\text{ L})\left(\frac{100.1\text{ g } CaCO_3}{1\text{ mol } CaCO_3}\right)\left(\frac{1\text{ kg}}{1000\text{ g}}\right)$$

$$= \underline{1.05\times10^{6}\text{ kg}}\ CaCO_3(s)$$

22. (a) The most important naturally occurring compound of calcium is calcium carbonate, $CaCO_3(s)$, as limestone, marble, chalk, etc., Others of importance are phosphate rock, $Ca_3(PO_4)_2(s)$, calcium fluoride, $CaF_2(s)$, (fluorite), and calcium sulfate, $CaSO_4.2H_2O$, (gypsum).

(b) (i) Calcium reacts with hydrogen when hydrogen gas is passed over heated calcium.

(ii) Calcium reacts rather slowly with cold water, and more vigorously on heating.

(iii) Calcium reacts with nitrogen on heating strongly to give calcium nitride.

$3Ca(s) + N_2(g) \longrightarrow Ca_3N_2$

(c) (i) $Ca(s) + H_2(g) \longrightarrow CaH_2$ (calcium hydride)

(ii) $Ca(s) + 2H_2O(l) \longrightarrow Ca(OH)_2(aq) + H_2(g)$

calcium hydroxide ... hydrogen

(iii) $3Ca(s) + N_2(g) \longrightarrow Ca_3N_2(s)$

calcium nitride

<u>Lewis structures:</u>

| $Ca^{2+}\ [:H^-]_2$ | $Ca^{2+}\ [H\text{-}\ddot{\underset{..}{O}}:^-]_2$ | H-H | $[Ca^{2+}]_3[:\ddot{\underset{..}{N}}:^{3-}]_2$ |
|---|---|---|---|
| calcium hydride | calcium hydroxide | hydrogen | calcium nitride |

23. (a) The alkali metals readily lose their single ($ns^1$) valence shell electron to form $M^+$ cations; they are good reducing agents. The increase in reactivity from Li to Cs in descending the group is correlated with the decrease in the first ionization energy in descending the group.

(b) Alkali metals have very low first ionization energies and therefore readily form $M^+$ ions, transferring their single valence shell electron to more electronegative atoms (nonmetals) to give ionic compounds.

(c) The simple oxides have the empirical formula $M_2O$, with the Lewis structure $(M^+)_2:\ddot{\underset{..}{O}}:^{2-}$. They are all readily soluble in water, in which the $O^{2-}$ ion behaves as a strong base,

$$M_2O(s) + H_2O(l) \longrightarrow 2M^+(aq) + 2OH^-(aq)$$

(d) Because of their small charge/radius ratios, none of the alkali metal cations are strongly hydrated in aqueous solution; thus, they have no acidic properties and, because they have no unshared electron pairs, they cannot behave as bases. Thus, they have no acid-base properties in aqueous solution. None of the anions derived from strong acids (conjugate bases of strong acids) have basic properties in water because their conjugate acids are fully dissociated. Thus alkali metal salts of strong acids give neutral aqueous solutions (pH 7.00).

24. (a) They readily lose their two $ns^2$ valence shell electrons to form $M^{2+}$ ions and thus behave as strong reducing agents. Since the core charge of +2 is the same for all of the alkaline earth metals, the first and second ionization energies decrease in descending the group and the reducing power (reactivity) increases in descending the group.

(b) Because of its very small size, beryllium has a somewhat greater electronegativity than the other alkali earth metals and has less of a tendency to form $M^{2+}$ ions. The remainder readily transfer their $ns^2$ valence electrons to more electronegative atoms (nonmetals) to ionic compounds.

(c) $Be^{2+}$ and $Mg^{2+}$ are small ions with relatively large charge/radius ratios. Thus, their ionic bonds with $OH^-$ are very strong and difficult to break. The charge/radius ratio decreases down the group as the ions become progressively larger, so that the ionic bonds with $OH^-$ also become weaker in going from $Ca(OH)_2(s)$ to $Sr(OH)_2(s)$ to $Ba(OH)_2(s)$, and their solubilities increase in the same order.

(d) In aqueous solution, only the $Be^{2+}$ and $Mg^{2+}$ ions have sufficiently large charge/radius ratios to be strongly hydrated. The strong attraction of these ions for lone pairs of water molecules increases the polarity of their O-H bonds, so that these hydrated ions, $Be(H_2O)_4{}^{2+}$ and $Mg(H_2O)_6{}^{2+}$ behave as weak acids in aqueous solution. The cations formed by the remainder of the alkaline earth metals are less strongly hydrated and they have no acidic or basic properties. The anions of strong acids have no basic properties, since the conjugate acids are strong. Thus, these salts give neutral solutions (pH 7).

25. As a consequence of its large charge/radius ratio, the very small $Be^{2+}$ ion attracts electron pairs strongly, to a maximum of four pairs that fill its valence shell with an octet of electrons. Thus the bonds in compounds such as $BeCl_2$ are better described as polar covalent rather than ionic, and the Be atom in $BeCl_2$ attracts two additional electron pairs to complete its valence shell, to form, for example, the complex $BeCl_4{}^{2-}$ ion. $BeCl_2.4H_2O(s)$ is composed of $Be(H_2O)_4{}^{2+}$ and $Cl^-$ ions, with $Be^{2+}$ covalently bonded in a tetrahedral arrangement to the four $H_2O$ molecules, each of which donates an electron pair to the $Be^{2+}$ ion. $Be(H_2O)_4{}^{2+}$ behaves as a weak acid, because the polarity of the OH bonds

is increased when water donates an electron pair to Be, so Be salts such as $BeCl_2.4H_2O$ give weakly acidic solutions (pH < 7). $Be(OH)_2$ dissolves both in acid and base and is amphoteric, because with acid it can form the $Be(H_2O)_4^{2+}$ ion, and with base it forms the $Be(OH)_4^{2-}$ ion.

$$Be(OH)_2(s) + 2H_3O^+(aq) \longrightarrow Be(H_2O)_4^{2+}(aq)$$

$$Be(OH)_2(s) + 2OH^-(aq) \longrightarrow Be(OH)_4^{2-}(aq)$$

In each of these ions the valence shell of Be is filled with an octet of electrons

26. Magnesium metal in air rapidly becomes coated with a thin layer of insoluble MgO(s) while calcium becomes coated with a thin layer of the more soluble CaO(s). Thus, in water, the surface protective layer of magnesium makes it difficult for water molecules to penetrate to the metal surface, while the surface coating on calcium dissolves and makes the metal surface available for reaction. In dilute acid, both MgO and CaO react readily with the $H_3O^+(aq)$ from the acid, so that both metal surfaces are soon exposed to reaction. Moreover, since $H_3O^+(aq)$ is a stronger oxidizing agent than $H_2O$, the reaction of both metals with acid is vigorous.

27. $MgCO_3(s)$ consists of $Mg^{2+}$ and $CO_3^{2-}$ ions in an ionic lattice, which strongly attract each other, so that $MgCO_3(s)$ is very insoluble. When $CO_2(g)$ is bubbled through a suspension of $MgCO_3(s)$ in water, the reaction

$$MgCO_3(s) + CO_2(g) + H_2O(l) \longrightarrow Mg(HCO_3)_2(aq)$$

occurs to give a solution of soluble magnesium hydrogen carbonate. A solution of $CO_2(g)$ in water gives the weak acid $H_2CO_3(aq)$, which reacts with $CO_3^{2-}$ to give the soluble $HCO_3^-$ ion: $CO_3^{2-} + H_2CO_3 \longrightarrow 2HCO_3^-$

or $H_2CO_3 + H_2O \rightarrow H_3O^+ + HCO_3^-$ ; $H_3O^+ + CO_3^{2-} \longrightarrow HCO_3^- + H_2O$

28. In each case Ca metal behaves as an oxidizing agent and is oxidized to the $Ca^{2+}$ ion. The other reactant is reduced.

(a) $Ca(s) + 2H_2O(l) \longrightarrow Ca(OH)_2(s) + H_2(g)$ slow in the cold, more rapid on heating

(b) $Ca(s) + Cl_2(g) \longrightarrow CaCl_2(s)$ slow in the cold, more rapid on heating

(c) $Ca(s) + H_2(g) \xrightarrow{heat} CaH_2(s)$

(d) $2Ca(s) + O_2(g) \longrightarrow 2CaO(s)$ on burning Ca in oxygen (or air)

(e) $Ca(s) + 2HBr(g) \longrightarrow CaBr_2(s) + H_2(g)$ rapid on heating

29. In each case we write the balanced equation for the formation of the aqueous solution and the $K_{sp}$ expression is given by the ionic product of the concentrations of the products of the reaction.

(a) $Fe(OH)_3(s) \rightleftharpoons Fe^{3+}(aq) + 3OH^-(aq)$; $K_{sp} = [Fe^{3+}][OH^-]^3$ ($mol^4\ L^{-4}$)

(b) $Ca_3(PO_4)_2(s) \rightleftharpoons 3Ca^{2+}(aq) + 2PO_4^{3-}(aq)$

$$K_{sp} = [Ca^{2+}]^3[PO_4^{3-}]^2 \quad (mol^5\ L^{-5})$$

30. (see Problem 29)

(a) $AgCl(s) \rightleftharpoons Ag^{+}(aq) + Cl^{-}(aq)$; $K_{sp} = [Ag^{+}][Cl^{-}]$ $(mol^{2}\ L^{-2})$

(b) $BaF_2(s) \rightleftharpoons Ba^{2+}(aq) + 2F^{-}(aq)$; $K_{sp} = [Ba^{2+}][F^{-}]^{2}$ $(mol^{3}\ L^{-3})$

(c) $Cr(OH)_3 \rightleftharpoons Cr^{3+}(aq) + 3OH^{-}(aq)$; $K_{sp} = [Cr^{3+}][OH^{-}]^{3}$ $(mol^{4}\ L^{-4})$

(d) $Bi_2S_3(s) \rightleftharpoons 2Bi^{3+}(aq) + 3S^{2-}(aq)$; $K_{sp} = [Bi^{3+}]^{2}[S^{2-}]^{3}$ $(mol^{5}\ L^{-5})$

(e) $Cu(OH)_2(s) \rightleftharpoons Cu^{2+}(aq) + 2OH^{-}(aq)$; $K_{sp} = [Cu^{2+}][OH^{-}]^{2}$ $(mol^{3}\ L^{-3})$

31. (see Problem 29) and we have to first write the formulas of the compounds.

(a) $AgF(s) \rightleftharpoons Ag^{+}(aq) + F^{-}(aq)$; $K_{sp} = [Ag^{+}][F^{-}]$

$CaF_2(s) \rightleftharpoons Ca^{2+}(aq) + 2F^{-}(aq)$ $K_{sp} = [Ca^{2+}][F^{-}]^{2}$

$PbF_2(s) \rightleftharpoons Pb^{2+}(aq) + 2F^{-}(aq)$ $K_{sp} = [Pb^{2+}][F^{-}]^{2}$

(b) $AgOH(s) \rightleftharpoons Ag^{+}(aq) + OH^{-}(aq)$ $K_{sp} = [Ag^{+}][OH^{-}]$

$Mg(OH)_2(s) \rightleftharpoons Mg^{2+}(aq) + 2OH^{-}(aq)$ $K_{sp} = [Mg^{2+}][OH^{-}]^{2}$

$Al(OH)_3(s) \rightleftharpoons Al^{3+}(aq) + 3OH^{-}(aq)$ $K_{sp} = [Al^{3+}][OH^{-}]^{3}$

(c) $Ag_2SO_4(s) \rightleftharpoons 2Ag^{+}(aq) + SO_4^{2-}(aq)$ $K_{sp} = [Ag^{+}]^{2}[SO_4^{2-}]$

$SrSO_4(s) \rightleftharpoons Sr^{2+}(aq) + SO_4^{2-}(aq)$ $K_{sp} = [Sr^{2+}][SO_4^{2-}]$

32. We have first to convert the solubility of $PbSO_4(s)$ in g $L^{-1}$ to mol $L^{-1}$.

$$0.030\ g\ L^{-1} = \left(\frac{0.060\ g\ PbSO_4}{2\ L}\right)\left(\frac{1\ mol\ PbSO4}{303.3\ g\ PbSO_4}\right) = 9.9\times10^{-5}\ mol\ L^{-1}$$

| | $PbSO_4(s) \rightleftharpoons Pb^{2+}(aq)$ | + $SO_4^{2-}(aq)$ | |
|---|---|---|---|
| initially | 0 | 0 | mol $L^{-1}$ |
| at eq'm | $9.9\times10^{-5}$ | $9.9\ \times10^{-5}$ | mol $L^{-1}$ |

$$K_{sp}(PbSO_4) = [Pb^{2+}][SO_4^{2-}] = (9.9\ \times10^{-5}\ mol\ L^{-1})^{2}$$
$$= \underline{9.8\times10^{-9}\ mol^{2}\ L^{-2}}$$

33.

| | $MgF_2(s) \rightleftharpoons Mg^{2+}(aq)$ | + $2F^{-}(aq)$ | |
|---|---|---|---|
| initially | 0 | 0 | mol $L^{-1}$ |
| at eq'm | 0.0012 | 0.0024 | mol $L^{-1}$ |

$$K_{sp}(MgF_2) = [Mg^{2+}][F^{-}]^{2} = (0.0012)(0.0024)^{2}\ mol^{3}\ L^{-3}$$
$$= \underline{6.9\times10^{-9}\ mol^{3}\ L^{-3}}$$

34. We have first to convert the concentrations to mol $L^{-1}$.

$$4.41\ pg\ L^{-1} = (4.41\ pg)\left(\frac{1\ g}{10^{12}\ pg}\right)\left(\frac{1\ mol\ PbS}{239.3\ g\ PbS}\right)\ L^{-1} = 1.84\times10^{-14}\ mol\ L^{-1}$$

$$16.8\ mg\ L^{-1} = (16.8\ mg)\left(\frac{1\ g}{10^{3}\ mg}\right)\left(\frac{1\ mol\ CaF_2}{78.08\ g\ CaF_2}\right)\ L^{-1} = 2.15\times10^{-4}\ mol\ L^{-1}$$

$$5.62\ \mu g\ L^{-1} = (5.62\ \mu g)\left(\frac{1\ g}{10^{6}\ \mu g}\right)\left(\frac{1\ mol\ Cr(OH)_3}{103.0\ g\ Vr(OH)_3}\right)\ L^{-1} = 5.46\times10^{-8}\ mol\ L^{-1}$$

$$PbS(s) \rightleftharpoons Pb^{2+}(aq) + S^{2-}(aq)$$

| | | | |
|---|---|---|---|
| initially | 0 | 0 | mol $L^{-1}$ |
| at eq'm | $1.84\times10^{-14}$ | $1.84\times10^{-14}$ | mol $L^{-1}$ |

$$K_{sp}(PbS) = [Pb^{2+}][S^{2-}] = (1.84\times10^{-14}\ \text{mol L}^{-1})^2 = \underline{3.39\times10^{-28}\ \text{mol}^2\ \text{L}^{-2}}$$

$$CaF_2(s) \rightleftharpoons Ca^{2+}(aq) + 2F^-(aq)$$

| | | | |
|---|---|---|---|
| initially | 0 | 0 | mol $L^{-1}$ |
| at eq'm | $2.15\times10^{-4}$ | $4.30\times10^{-4}$ | mol $L^{-1}$ |

$$K_{sp}(CaF_2) = [Ca^{2+}][F^-]^2 = (2.15\times10^{-4})(4.30\times10^{-4})^2$$
$$= \underline{3.98\times10^{-11}\ \text{mol}^3\ \text{L}^{-3}}$$

$$Cr(OH)_3(s) \rightleftharpoons Cr^{3+}(aq) + 3OH^-(aq)$$

| | | | |
|---|---|---|---|
| initially | 0 | 0 | mol $L^{-1}$ |
| at eq'm | $5.46\times10^{-8}$ | $1.64\times10^{-7}$ | mol $L^{-1}$ |

$$K_{sp}(Cr(OH)_3) = [Cr^{3+}][OH^-]^3 = (5.46\times10^{-8})(1.64\times10^{-7})^3$$
$$= \underline{2.39\times10^{-28}\ \text{mol}^4\ \text{L}^{-4}}$$

35. (a) $\underline{MgCO_3}$ $\quad MgCO_3(s) \rightleftharpoons Mg^{2+}(aq) + CO_3^{2-}(aq)$

| | | | |
|---|---|---|---|
| initially | 0 | 0 | mol $L^{-1}$ |
| at eq'm | x | x | mol $L^{-1}$ |

$$K_{sp}(MgCO_3) = [Mg^{2+}][CO_3^{2-}] = x^2 = 1.0\times10^{-5}\ \text{mol}^2\ \text{L}^{-2}$$
$$\text{solubility} = x = \underline{3.2\times10^{-3}\ \text{mol L}^{-1}}$$

(b) $\underline{AgCl}$ $\quad AgCl(s) \rightleftharpoons Ag^+(aq) + Cl^-(aq)$

| | | | |
|---|---|---|---|
| initially | 0 | 0 | mol $L^{-1}$ |
| at eq'm | x | x | mol $L^{-1}$ |

$$K_{sp}(AgCl) = [Ag^+][Cl^-] = x^2 = 1.8\times10^{-10}\ \text{mol}^2\ \text{L}^{-2}$$
$$\text{solubility} = x = \underline{1.3\times10^{-5}\ \text{mol L}^{-1}}$$

(c) $\underline{Al(OH)_3}$ $\quad Al(OH)_3(s) \rightleftharpoons Al^{3+}(aq) + 3OH^-(aq)$

| | | | |
|---|---|---|---|
| initially | 0 | 0 | mol $L^{-1}$ |
| at eq'm | x | 3x | mol $L^{-1}$ |

$$K_{sp}(Al(OH)_3) = [Al^{3+}][OH^-] = x(3x)^3 = 27x^4 = 3\times10^{-34}\ \text{mol}^4\ \text{L}^{-4}$$
$$\text{solubility} = x = \underline{2\times10^{-9}\ \text{mol L}^{-1}}$$

(d) $\underline{PbI_2}$ $\quad PbI_2(s) \rightleftharpoons Pb^{2+}(aq) + 2I^-(aq)$

| | | | |
|---|---|---|---|
| initially | 0 | 0 | mol $L^{-1}$ |
| at eq'm | x | 2x | mol $L^{-1}$ |

$$K_{sp} = [Pb^{2+}][I^-]^2 = x(2x)^2 = 4x^3 = 7.9\times10^{-9}\ \text{mol}^3\ \text{L}^{-3}$$
$$\text{solubility} = x = \underline{1.3\times10^{-3}\ \text{mol L}^{-1}}$$

36. If the solubility of silver chloride is $\underline{x}$ mol $L^{-1}$, which is the concentration of AgCl(aq) in a saturated solution, then

$$AgCl(s) \rightleftharpoons Ag^{+}(aq) + Cl^{-}(aq)$$

| | | | |
|---|---|---|---|
| initially | 0 | 0 | mol $L^{-1}$ |
| at eq'm | x | x | mol $L^{-1}$ |

and $K_{sp}(AgCl) = [Ag^{+}][Cl^{-}] = x^2 = 1.8x10^{-10}\ mol^2\ L^{-2}$; $x = \underline{1.3x10^{-5}\ M}$

$$\text{Mass of AgCl in 250 mL} = \left(\frac{1.3x10^{-5}\ mol}{1\ L}\right)\left(\frac{143.4\ g\ AgCl}{1\ mol\ AgCl}\right)\left(\frac{1\ L}{1000\ mL}\right)(250\ mL)$$

$$= \underline{4.6 \times 10^{-4}\ g}$$

37. This problem is similar to Problem 36.

$$Pb(OH)_2(s) \rightleftharpoons Pb^{2+}(aq) + 2OH^{-}(aq)$$

| | | | |
|---|---|---|---|
| initially | 0 | 0 | mol $L^{-1}$ |
| at eq'm | x | 2x | mol $L^{-1}$ |

$$K_{sp}(Pb(OH)_2) = [Pb^{2+}][OH^{-}]^2 = x(2x)^2 = 4x^3 = 6x10^{-16}\ mol^3\ L^{-3}$$

$$x = [Pb^{2+}] = \underline{5.3x10^{-6}\ mol\ L^{-1}}$$

38. In this type of problem, calculate the reaction quotient, Q, which has the same form as the $K_{sp}$, and compare its value to that of $K_{sp}$.

$Q > K_{sp}$ .... a precipitate will form

$Q < K_{sp}$ .... no precipitate will form

In the solution,

$$[Pb^{2+}] = (0.10\ mol\ L^{-1})\left(\frac{50\ mL}{150\ mL}\right) = 0.033\ mol\ L^{-1}$$

$$[Cl^{-}] = (0.05\ mol\ L^{-1})\left(\frac{100\ mL}{150\ mL}\right) = 0.033\ mol\ L^{-1}$$

$$\text{Thus, } Q = [Pb^{2+}][Cl^{-}]^2 = (0.033)(0.033)^2 = \underline{3.6x10^{-5}\ mol^3\ L^{-3}}$$

$$Q > K_{sp}(PbCl_2) = 1.7x10^{-5}\ mol^3\ L^{-3}$$

and a <u>precipitate is expected to form.</u>

39. This problem is similar to Problem 38.

(a) $[Mg^{2+}] = (0.10\ mol\ L^{-1})\left(\frac{10\ mL}{20\ mL}\right) = 0.050\ mol\ L^{-1}$, and

$$[NH_3]_{sol'n} = (2.0\ mol\ L^{-1})\left(\frac{10\ mL}{20\ mL}\right) = 1.0\ mol\ L^{-1}$$

and calculate $[OH^{-}]$ in 1.0 mol $L^{-1}$ $NH_3(aq)$.

$$NH_3(aq) + H_2O(l) \rightleftharpoons NH_4^{+}(aq) + OH^{-}(aq)$$

| | | | | |
|---|---|---|---|---|
| initially | 1.0 | 0 | 0 | mol $L^{-1}$ |
| at eq'm | 1.0-x | x | x | mol $L^{-1}$ |

$$K_b(NH_3) = \frac{[NH_4^+][OH^-]}{[NH_3]} = \frac{x^2}{1.00-x} = 1.8\times10^{-5}\ mol\ L^{-1}$$

$x << 0.10$, $x^2 = 1.8\times10^{-5}$; $x = 4.2\times10^{-3}$ (assumption justified)

$[OH^-] = 4.2\times10^{-3}\ mol\ L^{-1}$

$$Mg(OH)_2(s) \rightleftharpoons Mg^{2+}(aq) + 2OH^-(aq)$$

$Q = [Mg^{2+}][OH^-]^2 = (0.050)(4.2\times10^{-3})^2 = 8.8\times10^{-7}\ mol^3\ L^{-3}$

$>> K_{sp}(Mg(OH)_2) = 7.1\times10^{-12}\ mol^3\ L^{-3}$

A precipitate will form

(b) $[Sr^{2+}] = (0.10\ mol\ L^{-1})((\frac{10.0\ mL}{20.0\ mL}) = 0.050\ mol\ L^{-1}$

$[NH_3] = (2.0\ mol\ L^{-1})(\frac{10\ mL}{20\ mL}) = 1.0\ mol\ L^{-1}$

and for 1.0 M $NH_3(aq)$, $[OH^-] = 4.2\times10^{-3}\ mol\ L^{-1}$ (see part (a))

$$Sr(OH)_2(s) \rightleftharpoons Sr^{2+}(aq) + 2OH^-(aq)$$

$Q = [Sr^{2+}][OH^-]^2 = (0.050)(4.2\times10^{-3})^2 = 8.8\times10^{-7}\ mol^3\ L^{-3}$

$<< K_{sp}(Sr(OH)_2) = 3.2\times10^{-4}\ mol^3\ L^{-3}$

No precipitate of $Sr(OH)_2(s)$ will form in this solution

(c) Firstly we must calculate $[OH^-]$ in the buffer solution:

$$K_b(NH_3) = 1.8\times10^{-5}\ mol\ L^{-1} = [OH^-]\frac{[NH_4^+]}{[NH_3]} = [OH^-](\frac{0.10\ mol\ L^{-1}}{0.10\ mol\ L^{-1}})$$

$[OH^-] = 1.8\times10^{-5}\ mol\ L^{-1}$

In each case, if x mol $L^{-1}$ is the solubility of $M(OH)_2(s)$, we have

| | $M(OH)_2(s) \rightleftharpoons$ | $M^{2+}(aq)$ | $+ 2OH^-(aq)$ | |
|---|---|---|---|---|
| initially | | 0 | y | mol $L^{-1}$ |
| at eq'm | | x | y | mol $L^{-1}$ |

where $y = 1.8\times10^{-5}\ mol\ L^{-1}$, which will be essentially unchanged when $M(OH)_2(s)$ dissolves, since the solution is a buffer solution.

Thus, $K_{sp}(M(OH)_2) = [M^{2+}][OH^-]^2 = x(1.8\times10^{-5})^2 = 3.2\times10^{-10}\ x$

i.e., $K_{sp}(Mg(OH)_2) = 7.1\times10^{-12}\ mol^3\ L^{-3} = [(3.2\times10^{-10})mol^2\ L^{-2}]x$

$x = 2.2\times10^{-2}\ mol\ L^{-1}$

$K_{sp}(Sr(OH)_2) = 3.2\times10^{-4}\ mol^3\ L^{-3} = [(3.2\times10^{-10})\ mol^2L^{-2}]x$

$x = 1\times10^6\ mol\ L^{-1}$

40. We have

| | $Fe(OH)_2(s) \rightleftharpoons$ | $Fe^{2+}(aq)$ | $+ 2OH^-(aq)$ | |
|---|---|---|---|---|
| initially | | 0 | 0 | mol $L^{-1}$ |
| at eq'm | | x | 2x | mol $L^{-1}$ |

$K_{sp}(Fe(OH)_2(s)) = [Fe^{2+}][OH^-]^2 = x(2x)^2 = 4x^3 = 7.9\times10^{-15}\ mol^3\ L^{-3}$

solubility = $x = 1.3\times10^{-5}\ mol\ L^{-1}$

$$Fe(OH)_3(s) \rightleftharpoons Fe^{3+}(aq) + 3OH^-(aq)$$

| | | | | |
|---|---|---|---|---|
| initially | | 0 | 0 | mol $L^{-1}$ |
| at eq'm | | x | 3x | mol $L^{-1}$ |

$K_{sp}(Fe(OH)_3) = [Fe^{3+}][OH^-]^3 = x(3x)^3 = 27x^4 = 1.6\text{x}10^{-39}$ mol$^4$ L$^{-4}$

solubility = x = $\underline{8.8\text{x}10^{-11}\text{ mol L}^{-1}}$

Iron(III) hydroxide is less soluble than iron(II) hydroxide.

41. As in Problem 38, we first calculate the concentrations of ions in solution and then compare the reaction quotient Q with the $K_{sp}$.

(a) $[Ba^{2+}] = 0.00050$ mol $L^{-1}$ ; $[OH^-] = 0.500$ mol $L^{-1}$

$$Ba(OH)_2(s) \rightleftharpoons Ba^{2+}(aq) + 2OH^-(aq)$$

$Q = [Ba^{2+}][OH^-]^2 = (0.0005)(0.500)^2 = 1.3\text{x}10^{-4}$ mol$^3$ L$^{-3}$

$<< K_{sp}(Ba(OH)_2(s)) = 3\text{x}10^{-4}$ mol$^3$ L$^{-3}$

No precipitate will form in this solution.

(b) $[Ba^{2+}] = 0.500$ mol $L^{-1}$ ; $[OH^-] = 0.500$ mol $L^{-1}$

$Q = [Ba^{2+}][OH^-]^2 = (0.500)(0.500)^2 = 0.125$ mol$^3$ L$^{-3}$

$> K_{sp}(Ba(OH)_2(s)) = 3\text{x}10^{-4}$ mol$^3$ L$^{-3}$

A precipitate will form in this solution.

42. The solubility of AgCl(s) is very small and we can assume to a first approximation that all of the $Ag^+(aq)$ ion is precipitated as AgCl(s), since $Cl^-(aq)$ is in excess.

$$Ag^+(aq) + Cl^-(aq) \rightleftharpoons AgCl(s)$$

| | | | | |
|---|---|---|---|---|
| initially | 0.010 | 1.00 | 0 | mol $L^{-1}$ |
| at eq'm | 0.010-x | 1.00-x | | mol $L^{-1}$ |

$K_{sp}(AgCl) = 1.8\text{x}10^{-10}$ mol$^2$ L$^{-2}$ = (0.010-x)(1.00-x), and solving the quadratic equation gives x = 1.00 $\underline{or}$ 0.010

The first solution is inappropriate; taking the second, since we have 100 mL solution

mass of AgCl(s) = $(0.010\text{ mol L}^{-1})(100\text{ mL})(\frac{1\text{ L}}{1000\text{ mL}})(\frac{143.4\text{ g AgCl}}{1\text{ mol AgCl}})$

= $\underline{0.143\text{ g AgCl}}$

The final concentration of $Cl^-(aq)$ is 1.00-x = 0.990 mol $L^{-1}$. Thus:

$K_{sp}(AgCl) = [Ag^+][Cl^-] = x(0.99) = 1.8\text{x}10^{-10}$ mol $L^{-1}$

x = $\underline{1.9\text{x}10^{-10}\text{ mol L}^{-1}}$ = $[Ag^+]$

43. (a) The initial concentrations are.

$[Pb^{2+}] = (2.00\text{ mol L}^{-1})(\frac{50\text{ mL}}{100\text{ mL}}) = 1.00\text{ mol L}^{-1}$

$[I^-] = (4.00\text{x}10^{-3}\text{ mol L}^{-1})(\frac{50\text{ mL}}{100\text{ mL}}) = 2.00\text{x}10^{-3}\text{ mol L}^{-1}$

$$PbI_2(s) \rightleftharpoons Pb^{2+}(aq) + 2I^-(aq)$$

$Q = [Pb^{2+}][I^-]^2 = (1.00)(2.00\times10^{-3}) = \underline{4.00\times10^{-6}\ mol^3\ L^{-3}}$

$> K_{sp}(PbI_2) = 7.9\times10^{-9}\ mol^3\ L^{-3}$

$Q > K_{sp}$, so a precipitate of $PbI_2(s)$ will form

(b) The precipitate is $PbI_2(s)$ and the only other possibility is $NaNO_3(s)$ which is a soluble salt. To calculate the mass of $PbI_2(s)$ formed, we write

| | $Pb^{2+}(aq)$ + | $2I^-(aq)$ $\rightleftharpoons$ | $PbI_2(s)$ | |
|---|---|---|---|---|
| initially | 1.00 | 0.00200 | - | $mol\ L^{-1}$ |
| at eq'm | 1.00-x | (0.00200-2x) | - | $mol\ L^{-1}$ |

$K_{sp} = [Pb^{2+}][I^-]^2 = (1.00-x)(0.00200-2x)^2 = 7.9\times10^{-9}\ mol^3\ L^{-3}$

$x << 1.00$, $(0.00200-2x)^2 = 7.9\times10^{-9}$; $0.00200-2x = 8.9\times10^{-5}$,

whence: x = 0.00096, (assumption justified)

Thus the amount of $PbI_2(s)$ that is precipitated is

$(0.00096\ mol\ L^{-1})(100\ mL)\left(\frac{1\ L}{1000\ mL}\right)\left(\frac{461.0\ g\ PbI_2}{1\ mol\ PbI_2}\right) = \underline{0.0443\ g}$

(c) $[Pb^{2+}] = 0.100-x = \underline{0.999\ mol\ L^{-1}}$

$[I^-] = 0.00200-2x = \underline{0.00008\ mol\ L^{-1}}$

$[Na^+] = \underline{0.00200\ mol\ L^{-1}}$

$[NO_3^-] = \underline{2.00\ mol\ L^{-1}}$

44. (a)

| | $CaF_2(s)$ $\rightleftharpoons$ | $Ca^{2+}(aq)$ + | $2F^-(aq)$ | |
|---|---|---|---|---|
| initially | x | 0 | 0 | $mol\ L^{-1}$ |
| at eq'm | 0 | x | 2x | $mol\ L^{-1}$ |

$K_{sp}(CaF_2) = [Ca^{2+}][F^-]^2 = x(2x)^2 = 4x^3 = 3.9\times10^{-11}\ mol^3\ L^{-3}$

$\underline{x = 2.1\times10^{-4}\ mol\ L^{-1}}$

(b)

| | $CaF_2(s)$ $\rightleftharpoons$ | $Ca^{2+}(aq)$ + | $2F^-(aq)$ | |
|---|---|---|---|---|
| initially | | 0.010 | 0 | $mol\ L^{-1}$ |
| at eq'm | | 0.010+x | 2x | $mol\ L^{-1}$ |

x must be less than $2.1\times10^{-4}$, so $x << 0.010$

$K_{sp} = (0.01+x)(2x)^2 = 0.01(2x)^2 = 3.9\times10^{-11}$ ; $\underline{x = 3.1\times10^{-5}\ mol\ L^{-1}}$

(c)

| | $CaF_2(s)$ $\rightleftharpoons$ | $Ca^{2+}(aq)$ + | $2F^-(aq)$ | |
|---|---|---|---|---|
| initially | | 0 | 0.100 | $mol\ L^{-1}$ |
| at eq'm | | x | 0.100+2x | $mol\ L^{-1}$ |

$x << 0.100$, $K_{sp} = x(0.100)^2 = 3.9\times10^{-11}$; $\underline{x = 3.9\times10^{-9}\ mol\ L^{-1}}$

The solubilities of $CaF_2(s)$ are

$\underline{2.1\times10^{-4}\ mol\ L^{-1}}$ in pure water; $\underline{3.1\times10^{-5}\ mol\ L^{-1}}$ in 0.01 M $CaCl_2$, and $\underline{3.9\times10^{-9}\ mol\ L^{-1}}$ in 0.100 M NaF

45. If the solubility of $BaF_2$ is x mol $L^{-1}$, we have

$$BaF_2(s) \rightleftharpoons Ba^{2+}(aq) + 2F^-(aq)$$

| | | | |
|---|---|---|---|
| initially | 0.20 | 0 | mol $L^{-1}$ |
| at eq'm | 0.20+x | 2x | mol $L^{-1}$ |

$K_{sp}(BaF_2) = 1.7\text{x}10^{-6}\ \text{mol}^3\ L^{-3} = [Ba^{2+}][F^-]^2 = (0.20+x)(2x)^2$

assume x << 0.20, $0.80x^2 = 1.7\text{x}10^{-6}$; $x = \underline{1.5\text{x}10^{-3}\ \text{mol}\ L^{-1}}$

Taking this as the first approximation, and substituting 0.20+x = 0.2015 gives $0.81x^2 = 1.7\text{x}10^{-6}$; $x = \underline{1.5\text{x}10^{-3}\ \text{mol}\ L^{-1}}$

The solubility of $BaF_2(s)$ in a solution already 0.20 M in $Ba^{2+}(aq)$ is $\underline{1.5\text{x}10^{-3}\ \text{mol}\ L^{-1}}$ or $\underline{0.26\ g\ L^{-1}}$.

46. $MgCO_3(s)$ dissolves to give $Mg^{2+}(aq)$ and $CO_3^{2-}(aq)$, so that the initial concentrations of both of these ions are $5.0\text{x}10^{-9}$ mol $L^{-1}$.

$$Mg^{2+}(aq) + CO_3^{2-}(aq) \rightleftharpoons MgCO_3(s)$$

| | | | | |
|---|---|---|---|---|
| initially | $5\text{x}10^{-9}$ | $5\text{x}10^{-9}$ | - | mol $L^{-1}$ |
| finally | $(5\text{x}10^{-9})+x$ | $5\text{x}10^{-9}$ | - | mol $L^{-1}$ |

and for the $MgCO_3$ to remain in solution Q cannot exceed $K_{sp}(MgCO_3)$, so that the limiting value of x is given by

$[Mg^{2+}][CO_3^{2-}] = [(5\text{x}10^{-9})+x](5\text{x}10^{-9}) = K_{sp} = 3.5\text{x}10^{-8}$ mol $L^{-1}$

$\underline{x = 7.0\ \text{mol}\ L^{-1}}$

Concentration of added $MgCl_2$ = 7.0 mol $L^{-1}$

Amount that can be added to 2.00 L of solution before $MgCO_3(s)$ precipitates is 14.0 mol.

47. We consider the affect of the added substances on the equilibrium:

$$Fe(OH)_3(s) \rightleftharpoons Fe^{3+}(aq) + 3OH^-(aq)$$

(a) Addition of KOH(aq) increases the $[OH^-]$; according to Le Châtelier's principle the amount of $Fe(OH)_3(s)$ has to increase in response to this change.

(b) Addition of HCl(aq) gives $H_3O^+(aq)$ that reacts with the $OH^-(aq)$ to give water, which decreases $[OH^-]$. According to Le Châtelier's principle, $Fe(OH)_3(s)$ will decrease as more of it dissolves to replace the $OH^-$ that is removed.

48. (a)

$$Ca_3(PO_4)_2(s) \rightleftharpoons 3Ca^{2+}(aq) + 2PO_4^{3-}(aq)$$

| | | | |
|---|---|---|---|
| initially | 3x | 2x | mol $L^{-1}$ |
| at eq'm | (0.1+3x) | 2x | mol $L^{-1}$ |

$K_{sp} = [Ca^{2+}]^3[PO_4^{3-}]^2 = (0.1+3x)^3(2x)^2 = 1.3\text{x}10^{-32}\ \text{mol}^5\ L^{-5}$

x << 0.1, $(0.1)^3(2x)^2 = (4\text{x}10^{-3})x^2 = 1.3\text{x}10^{-32}$

$\underline{x = 1.8\text{x}10^{-15}\ \text{mol}\ L^{-1}}$

(b) $$Ca_3(PO_4)_2 \rightleftharpoons 3Ca^{2+}(aq) + 2PO_4^{3-}(aq)$$

| | | | |
|---|---|---|---|
| initially | 3x | 3x | mol $L^{-1}$ |
| at eq'm | 3x | (0.1+2x) | mol $L^{-1}$ |

$$K_{sp} = [Ca^{2+}]^3[PO_4^{3-}]^2 = (3x)^3(0.1+2x)^2 = 1.3x10^{-32}\ mol^5\ L^{-5}$$

$$x << 0.1, \quad (3x)^3(0.1)^2 = 0.27x^3 = 1.3x10^{-32}$$

$$\underline{x = 3.6x10^{-11}\ mol\ L^{-1}}$$

(c) $$K_{sp} = [Ca^{2+}]^3[PO_4^{3-}]^2 = (3x)^3(2x)^2 = 108x^5 = 1.3x10^{-32}\ mol^5\ L^{-5}$$

$$\underline{x = 1.6x10^{-7}\ mol\ L^{-1}}$$

49. A salt precipitates from solution when Q is just greater than its $K_{sp}$. In this problem we calculate the $[Mg^{2+}]$ just needed to satisfy $Q = K_{sp}$.

$$Q = K_{sp}(MgF_2) = 6.6x10^{-9}\ mol^3L^{-3} = [Mg^{2+}][F^-]^2 = [Mg^{2+}](0.0010\ M)^2$$

$$[Mg^{2+}] = \mathbf{6.6x10^{-3}}\ mol\ L^{-1}$$

$$Q = K_{sp}(MgCO_3) = [Mg^{2+}][CO_3^{2-}] = 1.0x10^{-5}\ mol^2\ L^{-2} = [Mg^{2+}](0.010\ M)$$

$$[Mg^{2+}] = \underline{1.0x10^{-3}\ mol\ L^{-1}}$$

Magnesium carbonate precipitates first. (Note that we have assumed that the concentrated solution of $MgCl_2(aq)$ does not significantly change the volume of the original solution).

50. This problem is similar to Problem 49:

$$Q = K_{sp}(PbSO_4) = 6.3x10^{-7}\ mol^2\ L^{-2} = [Pb^{2+}][SO_4^{2-}] = [Pb^{2+}](0.10\ M)$$

$$[Pb^{2+}] = \underline{6.3x10^{-6}\ mol\ L^{-1}}$$

$$Q = K_{sp}(PbF_2) = 3.6x10^{-8}\ mol^3\ L^{-3} = [Pb^{2+}][F^-]^2 = [Pb^{2+}](0.10\ M)^2$$

$$[Pb^{2+}] = \underline{3.6x10^{-6}\ mol\ L^{-1}}$$

Thus, $\underline{PbF_2(s)\ precipitates\ first}$. When $PbSO_4(s)$ starts to precipitate, $[Pb^{2+}] = 6.3x10^{-6}\ mol\ L^{-1}$

$$K_{sp}(PbF_2) = [Pb^{2+}][F^-]^2 = (6.3x10^{-6}\ mol\ L^{-1})[F^-]^2 = 3.6x10^{-8}\ mol^3\ L^{-3}$$

$$[F^-]^2 = 5.7x10^{-3}\ mol^2\ L^{-2}\ ;\ [F^-] = 0.075\ mol\ L^{-1}$$

$$\%\ F^-\ ion\ remaining\ in\ solution = (\frac{0.075\ M}{0.100\ M}) \times 100\% = \underline{75\%}$$

51. In qualitative analysis the aim is to separate cations as efficiently as possible. Although $PbCl_2(s)$ is among the least soluble of the metal chlorides it is fairly soluble in water at 25°C. 3 M HCl(aq) contains 3 M $Cl^-(aq)$, which because of the common ion effect reduces the solubility of $PbCl_2(s)$, so that more of the $Pb^{2+}(aq)$ is separated under these conditions than is the case in pure water.

$$PbCl_2(s) \rightleftharpoons Pb^{2+}(aq) + 2Cl^-(aq)$$

| | | | |
|---|---|---|---|
| initially | 0 | 3 | mol $L^{-1}$ |
| at eq'm | x | 3+2x | mol $L^{-1}$ |

$$K_{sp} = [Pb^{2+}][Cl^-]^2 = x(3+2x)^2 \quad .... (1)$$

and the value of $K_{sp}(PbCl_2)$ can be obtained from the solubility of $PbCl_2(s)$ in water.

| | $PbCl_2(s) \rightleftharpoons$ | $Pb^{2+}(aq)$ | + $2Cl^-(aq)$ | |
|---|---|---|---|---|
| initially | | 0 | 0 | mol $L^{-1}$ |
| at eq'm | | 0.016 | 0.032 | mol $L^{-1}$ |

$$\text{and } K_{sp}(PbCl_2) = [Pb^{2+}][Cl^-]^2$$
$$= (0.016)(0.032)^2 = \underline{1.64 \times 10^{-5} \text{ mol}^3 \text{ L}^{-3}}$$

and substituting this value in equation (1) above gives,

$$x(3+2x)^2 = 1.64 \times 10^{-5} \text{ mol}^3 \text{ L}^{-3}$$

$$x << 3, \quad 9x = 1.64 \times 10^{-5} \; ; \quad \underline{x = 1.8 \times 10^{-6} \text{ mol L}^{-1}}$$

In conclusion, the solubility of $PbCl_2(s)$ is reduced from 0.016 M in pure water to $1.8 \times 10^{-6}$ M in 3 M HCl(aq), which achieves a much more efficient separation of the lead from solution, than would be the case in pure water.

52. The concentration of $Fe^{2+}(aq)$ is 0.005 mol $L^{-1}$. Thus,

| | $Fe^{2+}(aq)$ | + $2OH^-(aq) \rightleftharpoons$ | $Fe(OH)_2(s)$ | |
|---|---|---|---|---|
| at eq'm | 0.005 | x | - | mol $L^{-1}$ |

$$\text{and } K_{sp}(Fe(OH)_2) = [Fe^{2+}][OH^-]^2 = (0.005 \text{ mol L}^{-1})x^2$$
$$= 7.9 \times 10^{-15} \text{ mol}^3 \text{ L}^{-3}$$

$$x = [OH^-] = 1.26 \times 10^{-6} \text{ mol L}^{-1}$$

$$pH = 14.00 + \log [OH^-] = \underline{8.10}$$

53. Magnesia is a saturated solution of $Mg(OH)_2(aq)$ in equilibrium with $Mg(OH)_2(s)$. The concentration of this solution is calculated from $K_{sp}(Mg(OH)_2) = 7.1 \times 10^{-12}$ mol$^3$ L$^{-3}$.

| | $Mg(OH)_2(s) \rightleftharpoons$ | $Mg^{2+}(aq)$ | + $2OH^-(aq)$ | |
|---|---|---|---|---|
| initially | | 0 | 0 | mol $L^{-1}$ |
| at eq'm | | x | 2x | mol $L^{-1}$ |

and substituting in the $K_{sp}$ expression,

$$K_{sp}(Mg(OH)_2) = [Mg^{2+}][OH^-]^2 = x(2x)^2 = 4x^3 = 7.1\times10^{-12}\ mol^3\ L^{-3}$$

$$x = 1.21\times10^{-4}\ mol\ L^{-1};\ [OH^-] = 2x = 2.4\times10^{-4}\ mol\ L^{-1}$$

$$pH = 14.00 + \log\ [OH^-] = \underline{10.38}$$

54. Decreasing the pH increases $[H_3O^+]$ which will affect the concentration of the anion of a salt in solution if it is a weak base.

$$A^-(aq) + H_3O^+(aq) \rightleftharpoons HA(aq) + H_2O(l)$$

(a) Copper(II) sulfide, $Cu_2S(s)$, will increase in solubility as the pH is decreased, due to the formation of $HS^-$ and $H_2S$ from the $S^{2-}$ ion, which decreases $[S^{2-}]$; in acidic solution it will dissolve completely and $H_2S(g)$ will be evolved from the solution.

(b) Silver iodide No change, assuming constant volume, because $I^-$ is the conjugate base of a strong acid and its concentration will remain unchanged.

(c) Magnesium carbonate, $MgCO_3(s)$, will increase in solubility as $CO_3^{2-}(aq)$ forms $HCO_3^-(aq)$ and ultimately $H_2CO_3(aq)$ $(= CO_2(g) + H_2O(l))$ and $CO_2(g)$ is lost from the solution; at low pH it dissolves completely.

(d) Copper(II) hydroxide, $Cu(OH)_2(s)$, increases its solubility as $[OH^-]$ in the solution decreases with decrease in pH; it dissolves in acid.

(e) Calcium fluoride, $CaF_2(s)$, increases in solubility as $[F^-]$ decreases due to the reaction $F^-(aq) + H_3O^+(aq) \rightleftharpoons HF(aq) + H_2O(l)$.

(f) Lead(II) sulfate, $PbSO_4(s)$, increases it solubility (provided the pH is not decreased by adding $H_2SO_4(aq)$), due to a small decrease in $[SO_4^{2-}]$ because of its conversion to $HSO_4^-(aq)$.

55.

$$Pb(OH)_2(s) \rightleftharpoons Pb^{2+}(aq) + 2OH^-(aq)$$

| | $Pb^{2+}$ | $OH^-$ | |
|---|---|---|---|
| initially | 0 | $10^{-6}$ | $mol\ L^{-1}$ |
| at eq'm | x | $10^{-6}$ | $mol\ L^{-1}$ |

Although some $OH^-$ is produced from the $Pb(OH)_2$, the pH remains the same in the solution since it is a buffer solution. Thus,

$$K_{sp}(Pb(OH)_2) = 6\times10^{-16}\ mol^3\ L^{-3} = [Pb^{2+}][OH^-]^2 = x(10^{-6}\ mol\ L^{-1})^2$$

$$\underline{x = 6\times10^{-4}\ mol\ L^{-1}}$$

56.

$$Ni(OH)_2(s) \rightleftharpoons Ni^{2+}(aq) + 2OH^-(aq)$$

| | $Ni^{2+}$ | $OH^-$ | |
|---|---|---|---|
| initially | 0 | $10^{-6}$ | $mol\ L^{-1}$ |
| at eq'm | 0.0020 | $10^{-6}$ | $mol\ L^{-1}$ |

$$K_{sp}(Ni(OH)_2 = [Ni^{2+}][OH^-]^2 = (0.0020\ mol\ L^{-1})(10^{-6}\ mol\ L^{-1})^2$$

$$= \underline{2.0\times10^{-15}\ mol^3\ L^{-3}}$$

At pH = 7, $[H_3O^+] = 10^{-7}\ mol\ L^{-1}$; $[OH^-] = 10^{-7}\ mol\ L^{-1}$

$$Ni(OH)_2(s) \rightleftharpoons Ni^{2+}(aq) + 2OH^-(aq)$$

| | $Ni^{2+}$ | $OH^-$ | |
|---|---|---|---|
| at eq'm | x | $10^{-7}$ | $mol\ L^{-1}$ |

$$K_{sp} = x(10^{-7})^2 = 2.0\times10^{-15}\ mol^3\ L^{-3};\ x = \underline{0.20\ M}$$

57. $[H_3O^+] = 10^{-5.5}$ mol L$^{-1}$; $[OH^-] = 10^{-8.5}$ mol L$^{-1}$

$$M(OH)_2(s) \rightleftharpoons M^{2+}(aq) + 2OH^-(aq)$$

| | $M^{2+}$ | $OH^-$ |
|---|---|---|
| initially | 0 | $10^{-8.5}$ mol L$^{-1}$ |
| at eq'm | x | $10^{-8.5}$ mol L$^{-1}$ |

$$K_{sp}(M(OH)_2) = [M^{2+}][OH^-]^2 = x(10^{-8.5})^2 = \underline{10^{-17}x \text{ mol}^3 \text{ L}^{-3}}$$

$\underline{Cu(OH)_2(s)}$ $10^{-17}$ x = $4.8\times10^{-20}$ mol$^3$ L$^{-3}$ ; $\underline{x = 4.8\times10^{-3} \text{ mol L}^{-1}}$

$\underline{Pb(OH)_2(s)}$ $10^{-17}$x = $6\times10^{-16}$ mol$^3$ L$^{-3}$ ; $\underline{x = 60 \text{ mol L}^{-1}}$

Because $Pb(OH)_2$ is very soluble at pH 5.5, while $Cu(OH)_2$ is insoluble addition of acid to a mixture of the hydroxides will dissolve $Pb(OH)_2$ completely before any appreciable amount of $Cu(OH)_2$ dissolves.

58. In pure water, $Zn(OH)_2(s) \rightleftharpoons Zn^{2+}(aq) + 2OH^-(aq)$

| | $Zn^{2+}$ | $OH^-$ | |
|---|---|---|---|
| initially | 0 | 0 | mol L$^{-1}$ |
| at eq'm | x | 2x | mol L$^{-1}$ |

$$K_{sp}(Zn(OH)_2) = 2.0\times10^{-12} \text{ mol}^2 \text{ L}^{-2} = x(2x)^2 \text{ ; } \underline{x = 7.9\times10^{-5} \text{ mol L}^{-1}}$$

In 0.100 M NaOH(aq), $Zn(OH)_2(s) \rightleftharpoons Zn^{2+}(aq) + 2OH^-(aq)$

| | $Zn^{2+}$ | $OH^-$ | |
|---|---|---|---|
| initially | 0 | 0.10 | mol L$^{-1}$ |
| at eq'm | x | 0.10+2x | mol L$^{-1}$ |

$$K_{sp}(Zn(OH)_2) = 2.0\times10^{-12} \text{ mol}^2 \text{ L}^2 = x(0.10+2x)^2$$

x << 0.10, $\underline{x = 2.0\times10^{-10} \text{ mol L}^{-1}}$

59. For $H_2S(aq)$, we have

$$H_2S(aq) + 2H_2O(l) \rightleftharpoons 2H_3O^+(aq) + S^{2-}(aq)$$

| | $H_2S$ | $H_3O^+$ | $S^{2-}$ | |
|---|---|---|---|---|
| initially | 0.10 | 0 | 0 | mol L$^{-1}$ |
| at eq'm | 0.10-x | $10^{-4}$ | x | mol L$^{-1}$ |

$$K_a(H_2S) = \frac{[H_3O^+]^2[S^{2-}]}{[H_2S]} = \frac{(10^{-4})^2 x}{0.10-x} = \frac{10^{-8}x}{0.10-x} = 1.0\times10^{-19} \text{ mol}^2 \text{ L}^{-2}$$

x << 0.1, $\underline{x = 10^{-12} \text{ mol L}^{-1}}$; i.e., $\underline{[S^{2-}] = 10^{-12} \text{ mol L}^{-1}}$

For each insoluble sulfide we can calculate the concentration of metal ion when $[S^{2-}] = 10^{-12}$ mol L$^{-1}$:

(a) $\underline{CuS(s)}$ $K_{sp} = [Cu^{2+}][S^{2-}] = [Cu^{2+}](10^{-12} \text{ M}) = 6\times10^{-36} \text{ M}^2$

$[Cu^{2+}]$ = solubility of CuS(s) = $\underline{6\times10^{-24} \text{ mol L}^{-1}}$

(b) $\underline{FeS(s)}$ $K_{sp} = [Fe^{2+}][S^{2-}] = [Fe^{2+}](10^{-12} \text{ M}) = 8\times10^{-19} \text{ M}^2$

$[Fe^{2+}]$ = solubility of FeS(s) = $\underline{8\times10^{-7} \text{ mol L}^{-1}}$

(c) $\underline{PbS(s)}$ $K_{sp} = [Pb^{2+}][S^{2-}] = [Pb^{2+}][10^{-12} \text{ M}] = 3\times10^{-28} \text{ M}^2$

$[Pb^{2+}]$ = solubility of PbS(s) = $\underline{3\times10^{-16} \text{ M}}$

(d) $\underline{Ag_2S(s)}$ $K_{sp} = [Ag^+]^2[S^{2-}] = [Ag^+]^2(10^{-12} \text{ M}) = 8\times10^{-51} \text{ M}^3$

$[Ag^+] = \underline{9\times10^{-20} \text{ mol L}^{-1}}$

In each case if this concentration of metal ion is exceeded, the sulfide will precipitate from solution.

60. In each case we calculate $[OH^-]$ in a saturated solution:

(a) $Al(OH)_3(s)$ $\quad Al(OH)_3(s) \rightleftharpoons Al^{3+}(aq) + 3OH^-(aq)$

| | | | |
|---|---|---|---|
| initially | 0 | 0 | $mol\ L^{-1}$ |
| at eq'm | x | 3x | $mol\ L^{-1}$ |

$K_{sp}(Al(OH)_3) = [Al^{3+}][OH^-]^3 = x(3x)^3 = 27x^4 = 3x10^{-34}\ mol^4\ L^{-4}$

$x = 1.8x10^{-9}\ mol\ L^{-1}$, however this cannot be the $[OH^-]$ in this case because $[OH^-]$ in this solution cannot be less than that in pure water ($10^{-7}\ mol\ L^{-1}$); i.e., $[OH^-]$ from $Al(OH)_3(s)$ must be << $[OH^-]$ from water.

$K_{sp}(Al(OH)_3) = x(10^{-7})^3 = 3x10^{-34}\ mol^4\ L^{-4}$; $x = 3x10^{-13}\ mol\ L^{-1}$; pH = 7.00

(b) $Ca(OH)_2(s)$ $\quad Ca(OH)_2(s) \rightleftharpoons Ca^{2+}(aq) + 2OH^-(aq)$

| | | | |
|---|---|---|---|
| initially | 0 | 0 | $mol\ L^{-1}$ |
| at eq'm | x | 2x | $mol\ L^{-1}$ |

$K_{sp}(Ca(OH)_2) = [Ca^{2+}][OH^-]^2 = x(2x)^2 = 4x^3 = 6.5x10^{-6}\ mol^3\ L^{-3}$

$x = 0.012\ mol\ L^{-1}$ ; pH = 12.08

(c) $Sr(OH)_2(s)$

$K_{sp}(Sr(OH)_2) = [Sr^{2+}][OH^-]^2 = x(2x)^2 = 4x^3 = 3.2x10^{-4}\ mol^3\ L^{-3}$

$x = 0.043\ mol\ L^{-1}$ ; pH = 12.63

(d) $Mg(OH)_2(s)$

$K_{sp} = [Mg^{2+}][OH^-]^2 = x(2x)^2 = 4x^3 = 7.1x10^{-12}\ mol^3\ L^{-3}$

$x = 1.2x10^{-4}\ mol\ L^{-1}$ ; pH = 10.08

(e) $Fe(OH)_2(s)$

$K_{sp} = [Fe^{2+}][OH^-]^2 = x(2x)^2 = 4x^3 = 7.9x10^{-15}\ mol^3\ L^{-3}$

$x = 1.25x10^{-5}\ mol\ L^{-1}$ ; pH = 9.10

61. The reactions are

$$MgCl_2.xH_2O(aq) + 2AgNO_3(aq) \longrightarrow 2AgCl(s) + Mg(NO_3)_2(aq) + xH_2O$$

and

$$MgCl_2.xH_2O(s) \longrightarrow MgCl_2(s) + xH_2O(g)$$

(a) Mol of AgCl(s) = $(20.0\ mL\ AgNO_3)(0.10\ mol\ L^{-1})$ = 2.00 mmol AgCl

mass of $MgCl_2$ in sample = $(2.00\ mmol\ AgCl)\left(\frac{1\ mol\ MgCl_2}{2\ mol\ AgCl}\right)\left(\frac{95.21\ g\ MgCl_2}{1\ mol\ MgCl_2}\right)$

= 0.0952 g $\quad$ (= $1.000x10^{-3}$ mol)

By difference, mass of $H_2O$ in 0.203 g $MgCl_2.xH_2O$

= (0.203 - 0.095) = 0.108 g = $(0.108\ g\ H_2O)\left(\frac{1\ mol\ H_2O}{18.02\ g\ H_2O}\right)$

= $5.99x10^{-3}\ mol\ H_2O$

Ratio mol $H_2O$ : mol $MgCl_2$ = $\left(\frac{5.99x10^{-3}\ mol}{1.00x10^{-3}\ mol}\right)$ = 6.0

Mol ratio = ratio of molecules = 6 , i.e., x = 6

Empirical formula is $MgCl_2.6H_2O$

(b) From the empirical formula $MgCl_2.6H_2O$, calculate the mass % water in the sample.

$$\text{mass \% } H_2O = (6\text{ mol } H_2O)\left(\frac{18.02\text{ g } H_2O}{1\text{ mol } H_2O}\right)\left(\frac{1\text{ mol } MgCl_2.6H_2O}{203.3\text{ g } MgCl_2.6H_2O}\right) \times 100\%$$
$$= \underline{53.2\%}$$

i.e., the mass loss on heating in a stream of dry HCl(g) must be due to the complete dehydration of $MgCl_2.6H_2O(s)$ to $MgCl_2(s)$.

(c) In air, HCl(g) and water are lost.

$$MgCl_2.6H_2O(s) \rightleftharpoons MgO(s) + 2HCl(g) + 5H_2O(g)$$

(d) $MgCl_2.6H_2O(s)$ contains the weakly acidic $Mg(H_2O)_6^{2+}$ ion. When heated in air, protons are transferred from two of the $H_2O$ molecules to two $Cl^-$ ions to give HCl(g). As the HCl(g) so formed escapes into the atmosphere, the equilibrium shifts to the right, as predicted by Le Châtelier's principle, and eventually the reaction goes to completion. In a stream of HCl(g), the equilibrium shifts to the left, but the vapor pressure of $H_2O(g)$ in equilibrium with the salt is unaffected. As this water vapor is swept away in the stream of HCl(g), the equilibrium

$$MgCl_2.6H_2O(s) \rightleftharpoons MgCl_2(s) + 6H_2O(g)$$

shifts to the right and eventually the dehydration to $MgCl_2(s)$ goes to completion.

62. Both $CaCO_3(s)$ and $MgCO_3(s)$ decompose to give the respective oxide and $CO_2(g)$.

$$MCO_3(s) \longrightarrow MO(s) + CO_2(g)$$

Suppose there are initially x mol $CaCO_3$ and y mol $MgCO_3$, then the mass of the limestone is given by

$$\text{Mass} = (x\text{ mol } CaCO_3)\left(\frac{100.1\text{ g } CaCO_3}{1\text{ mol } CaCO_3}\right) + (y\text{ mol } MgCO_3)\left(\frac{84.32\text{ g } MgCO_3}{1\text{ mol } MgCO_3}\right)$$
$$= \underline{2.634}\text{ g}$$

<u>or</u>, $100.1x + 84.32y = 2.634$ ..... (1)

and after heating to constant mass, the mass is given by

$$\text{Mass} = (x\text{ mol CaO})\left(\frac{56.08\text{ g CaO}}{1\text{ mol CaO}}\right) + (y\text{ mol MgO})\left(\frac{40.31\text{ g MgO}}{1\text{ mol MgO}}\right) = \underline{1.288\text{ g}}$$

<u>or</u>, $56.08x + 40.31y = 1.288$ ..... (2)

Solving equations (1) and (2) for x gives x = <u>0.00349 mol</u>

$$\text{Mass of } CaCO_3(s) \text{ in sample} = (0.00349\text{ mol } CaCO_3)\left(\frac{100.1\text{ g } CaCO_3}{1\text{ mol } CaCO_3}\right)$$
$$= \underline{0.349\text{ g}}$$

$$\text{Mass \% } CaCO_3(s) = (0.349\text{ g})\left(\frac{100\%}{2.634\text{ g}}\right) = \underline{13.2\text{ mass \%}}$$

63. (a) $MgSO_4.7H_2O(s)$, epsom salt, is very soluble in water. An appropriate reaction for its preparation would be to dissolve the $MgCO_3(s)$ in the requisite amount of dilute sulfuric acid.

$$MgCO_3(s) + H_2SO_4(aq) \longrightarrow MgSO_4(aq) + H_2O(l) + CO_2(g)$$

$MgSO_4.7H_2O(s)$ could then be obtained from the ensuing solution by evaporation of the water and recrystallization from a concentrated aqueous solution.

(b) To prepare $BaSO_4(s)$, use can be made of its very low solubility. Treatment of $BaCO_3(s)$ with HCl(aq) gives a solution of $BaCl_2(aq)$

$$BaCO_3(s) + 2HCl(aq) \longrightarrow BaCl_2(aq) + H_2O(l) + CO_2(g)$$

$BaSO_4(aq)$ is precipitated on addition of excess $H_2SO_4(s)$ and can be filtered off, washed, and dried.

$$BaCl_2(aq) + H_2SO_4(aq) \longrightarrow BaSO_4(s) + 2HCl(aq)$$

64. Sodium (i) Carbon in an insufficiently strong oxidizing agent to reduce $Na_2O$ to sodium, and in (ii) the electrolysis of NaCl(aq), water is reduced, rather than $Na^+(aq)$. Electrolysis of molten NaCl(l) (iii) is the commercial method for obtaining sodium; $Na^+$ ions are reduced by electrons and $Cl^-$ ions give up electrons to form $Cl_2(g)$. (Sodium was first prepared by Davy in 1807 by the electrolysis of fused NaOH). Strongly heating $Na_2O$ in air (iv) gives no reaction. The only metals that are more reactive than Na are the alkali metals below it in group 1. Thus Na could be produced by reducing NaCl with K, Rb, or Cs.
Calcium (i) Heating CaO with C gives calcium carbide, $CaC_2$, but does not result in the reduction of CaO to Ca metal. In (ii), the electrolysis of $CaCl_2(aq)$, water is reduced to $H_2(g)$ rather than $Ca^{2+}(aq)$. Electrolysis of molten $CaCl_2$ (iii) produces calcium metal. (iv) Strongly heating the oxide, CaO(s), results in no reaction. $CaCl_2$ may be reduced by a more reactive metal than Ca (v); any of the alkali metals are suitable.

65. Alkali metals ions are singly charged, $M^+$, and alkaline earth metal ions are doubly charged, $M^{2+}$, and in both series the sizes of the ions increase in going from the top to the bottom of the groups. In terms of the strength of their attraction for polar water molecules, the strength is expected to be greatest for the smallest ions of highest charge; a good guide to the strength of interaction is the charge to radius ratio of the cation. For the alkali metals this is greatest for $Li^+$ and $Na^+$, so that salts containing these ions often separate from solution as hydrated salts, but salts containing the larger $K^+$, $Rb^+$, and $Cs^+$ ions are usually anhydrous, unless the anions are hydrated. Because of the +2 charge on the alkaline earth metal cations, and their relatively smaller ionic radii relative to the alkali metal cation in the same period, due to their core charges of +2 compared to +1 for the group 1 cations, the charge:radius ratios are much greater for the group 2 cations than they are for the group 1 cations. Thus their salts often contain water of hydration when they separate from aqueous solution.

66. The $Be^{2+}$ ion is a highly charged ion of small radius with a large charge: radius ratio, which thus attracts polar water molecules strongly. $Be^{2+}$ is in period 2 and has an empty valence shell which is capable of being filled with a maximum of eight electrons (octet rule). In water it strongly attracts an unshared pair from each of four water molecules and forms the $Be(H_2O)_4^{2+}$ ion, with $AX_4$ tetrahedral geometry. Thus, the sulfate

separates on crystallization from aqueous solution as $BeSO_4.4H_2O$ containing $Be(H_2O)_4^{2+}$ ions and $SO_4^{2-}$ ions, and this salt is very soluble because it contains two relatively large ions between which the ionic bonds are relatively weak. Because of the strong attraction of the Be atom in $Be(H_2O)_4^{2+}$ for the electron pairs of the Be-O bonds, the O-H bonds of the $H_2O$ ligands are more polar than the O-H bonds of free water molecules, and $Be(H_2O)_4^{2+}$ behaves as a weak acid in water. Thus, an aqueous solution of $BeSO_4.4H_2O$ contains weakly acidic $Be(H_2O)_4^{2+}$ ions and neutral $SO_4^{2-}$ ions, and gives an acidic aqueous solution.

67. Among the given substances, (g), copper(II) nitrate, is unique in that it has a blue color, and barium carbonate, (b), is the only insoluble salt. Another simple physical property is the color that certain of the salts impart to a bunsen burner flame.
   (b) barium carbonate - green
   (c), (d), and (h) - all sodium salts - yellow
   (f) calcium nitrate - brick red

The sodium salts can be distinguished from the pH of their aqueous solutions: (c) $NaHSO_4$ contains the acidic $HSO_4^-$ ion and gives an acidic solution; (d) $Na_2CO_3$ contains the basic $CO_3^{2-}$ ion and gives a basic solution, while (h) sodium chloride contains the neutral $Cl^-$ ion and gives a neutral solution. Thus we have identified, and labeled, (b), (c), (d), (f), (g), and (h), leaving (a), ammonium carbonate, (e) ammonium chloride, and (i) silver nitrate to be identified. On heating ammonium salts evolve $NH_3(g)$, which can be identified by its odor and its basic reaction on moistened indicator paper. Thus (i), silver nitrate can be identified, and an aqueous solution of $NH_4Cl$, (e), distinguished from an aqueous solution of (a), ammonium carbonate, by adding an aqueous solution of $AgNO_3(aq)$, which precipitates white $AgCl(s)$ from the chloride solution, and the carbonate solution gives a brown precipitate of $Ag_2O(s)$.

68. (a) Limestone, $CaCO_3(s)$, dissolves in rain water (a dilute solution of carbonic acid, $H_2CO_3(aq)$), due to the formation of soluble $Ca(HCO_3)_2(aq)$.

$$CaCO_3(s) + [H_2O(l) + CO_2(aq)] \rightleftharpoons Ca(HCO_3)_2(aq)$$

(b) When the solution is boiled, insoluble $CaCO_3(s)$ is reprecipitated due to the reversal of the above equilibrium as $H_2CO_3(aq)$ decomposes to $H_2O(l)$ and $CO_2(g)$:

$$Ca(HCO_3)_2(aq) \rightleftharpoons CaCO_3(s) + CO_2(g) + H_2O(l)$$

(c) Hard water contains appreciable amounts of $Ca^{2+}(aq)$ and $Mg^{2+}(aq)$ together with $Cl^-(aq)$, $SO_4^{2-}(aq)$, and $HCO_3^-(aq)$ ions.

(d) Hard water can be "softened" by removing the $Ca^{2+}(aq)$ and $Mg^{2+}(aq)$ ions as their insoluble carbonates:

(i) By adding $Na_2CO_3(aq)$

(ii) If the hardness is due to $Ca(HCO_3)_2(aq)$, by boiling the water to precipitate $CaCO_3(s)$, or adding sufficient lime, $Ca(OH)_2$, to convert all of the hydrogen carbonate to carbonate,

$$Ca^{2+}(aq) + 2HCO_3^-(aq) + Ca(OH)_2(aq) \longrightarrow 2CaCO_3(s) + 2H_2O(l)$$

69. (a) The initial concentrations of $Ba^{2+}(aq)$, and $Sr^{2+}(aq)$, can be used to calculate the minimum $[SO_4^{2-}]$ required to precipitate each of the sulfates.

$$M^{2}(aq) + SO_4^{2-}(aq) \rightleftharpoons MSO_4(s)$$

$K_{sp}(BaSO_4) = [Ba^{2+}][SO_4^{2-}] = 3.2x10^{-7}\ mol^2\ L^{-2}$

$= (1.00\ mol\ L^{-1})[SO_4^{2-}]$; $[SO_4^{2-}] = \underline{3.2x10^{-7}\ mol\ L^{-1}}$

and

$K_{sp}(SrSO_4) = [Sr^{2+}][SO_4^{2-}] = 2\ x10^{-5}\ mol^2\ L^{-2}$

$= (0.001\ mol\ L^{-1})[SO_4^{2-}]$; $[SO_4^{2-}] = \underline{2x10^{-2}\ mol\ L^{-1}}$

$BaSO_4(s)$ will precipitate first. (Allowance has not specifically been made for the dilution of the solutions by the added $Na_2SO_4(aq)$, but this will not change the qualitative validity of the conclusion).

(b) The calculations are similar to those in part (a).

$K_{sp}(BaSO_4) = [Ba^{2+}][SO_4^{2-}] = (0.001\ mol\ L^{-1})[SO_4^{2-}] = 3.2x10^{-7}\ M^2$

$[SO_4^{2-}] = \underline{3.2x10^{-4}\ mol\ L^{-1}}$

$K_{sp}(SrSO_4) = [Sr^{2+}][SO_4^{2-}] = (1.00\ mol\ L^{-1})[SO_4^{2-}] = 2\ x10^{-5}\ M^2$

$[SO_4^{2-}] = \underline{2\ x10^{-5}\ mol\ L^{-1}}$

$SrSO_4(s)$ will precipitate first. (Note change in $K_{sp}$ in 2nd printing of text).

70.

| Element | Color | Element | Color |
|---|---|---|---|
| Li | red | Ca | orange-red |
| Na | yellow | Sr | deep red |
| K | lilac | Ba | pale green |
| Rb | purple | | |
| Cs | blue | | |

71. Quantitatively, since the molar mass of NaOH is 40.00 g $mol^{-1}$, the molar mass of the white opaque solid must be,

$$\left(\frac{155\%}{100\%}\right)(40.00\ g\ mol^{-1}) = \underline{62.00\ g\ mol^{-1}}$$

or some multiple thereof. NaOH(s) is a hygroscopic solid that readily absorbs water from the atmosphere (deliquesces) to give a concentrated aqueous solution. On long exposure to air the solution absorbs $CO_2(g)$ and the solution gradually evaporates and eventually colorless transparent crystals of $Na_2CO_3.10H_2O(s)$ crystallize. On standing, the $Na_2CO_3.10H_2O(s)$ crystals crumble to a white powder of composition $Na_2CO_3.H_2O$, which is the white opaque solid.

$$NaOH(s) \xrightarrow{water} NaOH(aq)$$

$$2NaOH(aq) + CO_2(g) \longrightarrow Na_2CO_3(aq) + H_2O(l)$$

$$Na_2CO_3(aq) + 10H_2O(l) \longrightarrow Na_2CO_3.10H_2O(s)$$

$$Na_2CO_3.10H_2O(s) \longrightarrow Na_2CO_3.H_2O(s) + 9H_2O(g)$$

Thus, the overall reaction is

$$2NaOH(s) + CO_2(g) \longrightarrow Na_2CO_3.H_2O(s)$$

The molar mass of $Na_2CO_3.H_2O(s)$ is 124.0 g $mol^{-1}$. Thus, the expected increase in mass is given by

$$\left(\frac{1 \text{ mol } Na_2CO_3.H_2O}{2 \text{ mol NaOH}}\right)\left(\frac{2 \text{ mol NaOH}}{80.00 \text{ g NaOH}}\right)\left(\frac{124.0 \text{ g } Na_2CO_3.H_2O}{1 \text{ mol } Na_2CO_3.H_2O}\right)$$

= <u>1.550</u>, or an increase in mass of 55%, as observed.

72. (a) (i)

$$PbCl_2(s) \rightleftharpoons Pb^{2+}(aq) + 2Cl^-(aq)$$

at eq'm — $x$ $2x$ mol $L^{-1}$

$K_{sp}(PbCl_2) = [Pb^{2+}][Cl^-]^2 = x(2x)^2 = 4x^3 = 1.7\text{x}10^{-5}$ mol$^3$ L$^{-3}$

$x$ = <u>0.016 mol L$^{-1}$</u> = solubility of $PbCl_2(s)$

(ii)

$$PbCl_2(s) \rightleftharpoons Pb^{2+}(aq) + 2Cl^-(aq)$$

at eq'm — $x$ $0.10+2x$ mol $L^{-1}$

$K_{sp}(PbCl_2) = x(0.10+2x)^2 = 1.7\text{x}10^{-5}$ mol$^3$ L$^{-3}$; $x$ = <u>1.7x10$^{-3}$ mol L$^{-1}$</u>

(iii)

$$PbCl_2(s) \rightleftharpoons Pb^{2+}(aq) + 2Cl^-(aq)$$

at eq'm — $0.20+x$ $2x$ mol $L^{-1}$

$K_{sp}(PbCl_2) = (0.20+x)(2x)^2 = 1.7\text{x}10^{-5}$ mol$^3$ L$^{-3}$ ;

$x$ = <u>4.6x10$^{-3}$ mol L$^{-1}$</u>

Thus, $PbCl_2(s)$ is <u>least soluble</u> in 0.10 M NaCl(aq) at 25°C.

(b) The common ion effect is the decrease in the solubility of a salt in the presence of an ion that is common to one of the ions of the salt.

(c) According to Le Chatelier's principle, addition of a common ion to a system in equilibrium causes the position of the equilibrium to shift, in such a way that the effect of an increase in the concentration of the common ion is minimized, which is achieved by some of the common ion being precipitated in the form of an insoluble salt.

73. First calculate the solubility of AgCl(s) at each temperature.

$$AgCl(s) \rightleftharpoons Ag^+(aq) + Cl^-(aq)$$

at eq'm — $x$ $x$ mol $L^{-1}$

$$K_{sp}(AgCl) = [Ag^+][Cl^-] = x^2$$

At 5°C, $x^2 = 2.10\text{x}10^{-11}$ mol$^2$ L$^{-2}$; <u>$x = 4.58\text{x}10^{-6}$ mol L$^{-1}$</u>

At 100°C, $x^2 = 2.15\text{x}10^{-10}$ mol$^2$ L$^{-2}$; <u>$x = 1.47\text{x}10^{-5}$ mol L$^{-1}$</u>

The difference in the solubilities ($x$) at the two temperatures is <u>1.01 x 10$^{-5}$ mol L$^{-1}$</u>. i.e., 1.01x10$^{-5}$ mol AgCl(s) will crystallize from the solution.

Mass of AgCl(s) = $(1.01\text{x}10^{-5} \text{ mol AgCl})\left(\frac{143.4 \text{ g AgCl}}{1 \text{ mol AgCl}}\right)$ = <u>1.45x10$^{-3}$ g</u>

74. (a) The reactions of group 1 and group 2 metals are related to their behaviour as reducing agents; they have low ionization energies and readily lose electrons to give $M^+$ and $M^{2+}$ ions, respectively. The ease with shich electrons are lost is related to core charge and the atomic sizes. For the elements of a particular group, the core charge is the same but the distances of the valence electrons from the nucleus increase in descending each group. Thus, electrons are more easily lost in going from the top of each group to the bottom, which parallels their reactivities.

(b) In forming metallic bonds in the solid metals, there is only a single $ns^1$ electron available per alkali metal atom and the metallic bonding is relatively weak. To melt a solid, sufficient energy has to be supplied to lengthen and weaken the bonds so that the atoms can slide past each other, which is small for these metals. In contrast, to vaporize the metals the metallic bonds all have to be broken, which takes a much greater amount of energy than that needed to melt the metal. Thus, the boiling points are much higher than the melting points.

(c) The alkali metals with one valence electron per atom for metallic bonding form relatively weak metallic bonds; the atoms are not very strongly attracted to each other and the number of atoms per unit volume (the density) is relatively low. Metals such as the alkaline earths and Al and Cu have more electrons per atom for metallic bonding and the atoms are held together considerably more strongly in the metals; there are more atoms per unit volume and consequently these metals have higher densities than the alkali metals.

(d) Alkali metal salts are generally quite soluble because they contain singly charged $M^+$ ions with a single positive charge so that the ionic bonds in the solids are not very difficult to break. Li is the smallest of the alkali metal ions and $F^-$ is the smallest of the halide ions, so that the ionic bonding is stronger in LiF(s) than in almost any other alkali metal salt, which accounts for the relatively low solubility.

75. Calcium oxide, CaO(s), in a basic metal oxide consisting of $Ca^{2+}$ and $O^{2-}$ ions. Each of the reactions involves the reaction of the basic $O^{2-}$ ion with an acidic nonmetal oxide to give a salt.

(a) $CaO(s) + H_2O(l) \longrightarrow Ca(OH)_2(s)$ calcium hydroxide

(b) $CaO(s) + CO_2(g) \longrightarrow CaCO_3(s)$ calcium carbonate

(c) $CaO(s) + SO_2(g) \longrightarrow CaSO_3(s)$ calcium sulfite

(d) $CaO(s) + SiO_2(s) \longrightarrow CaSiO_3(s)$ calcium silicate

(e) $6CaO(s) + P_4O_{10}(s) \longrightarrow 2Ca_3(PO_4)_2(s)$ calcium phosphate

76. Magnesium oxide, MgO(s) consists of $Mg^{2+}$ and $O_2^-$ ions. $Mg^{2+}$ is a very small ion with a large charge/radius ratio, so the ionic bonds with $O^{2-}$ are very strong and MgO(s) has a negligible solubility in water, as does the $Mg(OH)_2(s)$ produced.

$$MgO(s) + H_2O(l) \rightleftharpoons Mg(OH)_2(s)$$

$$Mg(OH)_2(s) \rightleftharpoons Mg^{2+}(aq) + 2OH^-(aq)$$

Solutions of HCl(aq) and $HNO_3$(aq) contain $H_3O^+$(aq) ions, and the above equilibria are shifted to the right when $H_3O^+$(aq) reacts with $OH^-$(aq) in a reaction that goes to completion.

$$H_3O^+(aq) + OH^-(aq) \rightarrow 2H_2O(l)$$

Thus, insoluble MgO(s) is soluble in these dilute acids.

$$MgO(s) + 2HCl(aq) \longrightarrow Mg^{2+}(aq) + 2Cl^-(aq) + H_2O(l)$$

$$MgO(s) + 2HNO_3(aq) \longrightarrow Mg^{2+}(aq) + 2NO_3^-(aq) + H_2O(l)$$

77. (a) The solubilities of the chlorides in (i) water, and (ii) 2 M HCl(aq) are calculated by the usual methods (see for example Problems 44 and 48).

| Chloride | Solubility (mol $L^{-1}$) in water | in 2 M HCl(aq) |
|---|---|---|
| $PbCl_2$(s) | $1.6 \times 10^{-2}$ | $4.3 \times 10^{-6}$ |
| AgCl(s) | $1.3 \times 10^{-5}$ | $9.0 \times 10^{-11}$ |
| $Hg_2Cl_2$(s) | $6.7 \times 10^{-7}$ | $3.0 \times 10^{-19}$ |

(b) Due to the common-ion effect, the solubilities of these chlorides are reduced to very small values in 2M HCl(aq), which ensures the efficient removal of their metal ions from solution.

# CHAPTER 16

1. Greater entropy is associated with greater disorder or randomness.

(a) The water molecules in liquid water are relatively free to move around each other and have a less ordered arrangement than the regular array of water molecules in ice; water has the greater entropy.
(b) A pack of cards randomly shuffled has a greater entropy than the more orderly arrangement of the cards in suits.
(c) A collection of jigsaw pieces is more disordered, and has a greater entropy, than the completed puzzle.
(d) Solid $NH_4Cl$ is a crystalline solid with a structure consisting of a regular repeating pattern of $NH_4^+$ and $Cl^-$ ions, while in an aqueous solution, these ions move about randomly. The solution has the greater entropy.

2. (a) Ethanol vapor has a more random arrangement than the ethanol molecules in the liquid, and has the greater entropy; $\Delta S > 0$.
(b) Mg(s) and $O_2$(g) are replaced by MgO(s); the overall degree of disorder decreases, because a gas and a solid are replaced by a solid; $\Delta S < 0$.
(c) A solid reactant is replaced by gases, resulting in increased disorder; $\Delta S > 0$.
(d) Four moles of gaseous reactants are replaced by two moles of gaseous products, resulting in an overall decrease in entropy; $\Delta S < 0$.
(e) A solid is replaced by a solid and a gas, resulting in an increased degree of disorder and an increase in entropy; $\Delta S > 0$.

3. (a) A solid is replaced by a solid and a gas, resulting in an increased disorder and an increase in entropy; $\Delta S > 0$.
(b) Two moles of gas are replaced by one mole of solid, resulting in a large decrease in randomness, or increased order; $\Delta S < 0$.
(c) One mole of solid and 1 mole of gas are replaced by 1 mole of solid, resulting in a decrease in disorder; $\Delta S < 0$.

4. The hydration of $Al^{3+}$ in water gives $Al(H_2O)_6^{3+}$(aq). When water molecules are coordinated to an aluminum ion, the disorder in the system decreases and so the entropy of the system decreases. Thus, an aqueous solution of $Al^{3+}$ has a lower entropy after the hydration of the ions. The ions are hydrated because the hydration of the $Al^{3+}$ ions involves the formation of polar Al-O bonds and is an exothermic process. Thus, there is an increase in the entropy of the surroundings which outweighs the entropy decrease in the system, so that there is an overall increase in the entropy of the system plus surroundings (universe).

5. (a) When a system moves to a lower state of energy, it gives out heat to the surroundings in an exothermic process; thus, the entropy of the surroundings increases. For the process to be spontaneous, the entropy of the system plus surroundings (universe) must increase, so that the entropy of the system must also increase, or if it decreases, the decrease must be less than $\Delta S_{surroundings}$.

(b) For a spontaneous process, the entropy of the system and its surroundings must increase. For an exothermic process and an increase in the entropy of the system the reaction is always spontaneous, because the entropy of the system plus that of the surroundings always increases. If the reaction is endothermic, the decrease in the entropy of the surroundings must be less than the increase in the entropy of the system for the process to be spontaneous.

6. (a) 3 moles of gas gives 2 moles of gas; the disorder in the system decreases; $\Delta S < 0$.
(b) 1 mole of solid and 1 mole of gas is replaced by 1 mole of solid; the disorder of the system decreases; $\Delta S < 0$.
(c) 9 moles of gas is replaced by 10 moles of gas; the disorder in the system increases; $\Delta S > 0$.
(d) 3 moles of gas is replaced by 1 mole of gas and 2 moles of liquid; the disorder in the system decreases; $\Delta S < 0$.
(e) 1 mole of gas is replaced by 2 mole of gas; the disorder in the system increases; $\Delta S > 0$.

7. (a) 1 mole of gas and 1 mole of liquid is replaced by 2 moles of gas; the disorder in the system increases; $\Delta S > 0$.
(b) 1 mole of solid and 1 mole of gas is replaced by 1 mole of solid and 1 mole of liquid; the disorder in the system decreases; $\Delta S < 0$.
(c) 3 moles of gas is replaced by 2 moles of liquid; the disorder in the system decreases; $\Delta S < 0$.
(d) 9 moles of gas is replaced by 4 moles of gas and 6 moles of liquid; the disorder in the system decreases; $\Delta S < 0$.
(e) 2 moles of gas is replaced by 5 moles of gas; the disorder in the system increases; $\Delta S > 0$.

8. (a) He(g) expands at 298 K and thus the disorder increases; $\Delta S > 0$.
(b) $I_2(g)$ is converted to $I_2(s)$ with a decrease in the disorder of the system; $\Delta S < 0$.
(c) 1 mole of solid $H_2O$ and 10 moles of liquid $H_2O$ are replaced by 11 moles of liquid $H_2O$, increasing the disorder of the system; $\Delta S > 0$.

9. To calculate $\Delta S^o$ for a reaction, use the equation

$$\Delta S^o = \sum [S^o(\text{products})] - \sum [S^o(\text{reactants})]$$

and $S^o$ values from Table 16.1 and Appendix B.

(a) $\Delta S^o = [S^o(CO_2,g)] - [S^o(C,s,\text{graphite}) + S^o(O_2,g)]$
$= [213.7] - [(5.8) + (205.0)] = \underline{2.9 \text{ J K}^{-1} \text{ mol}^{-1}}$

(b) $\Delta S^o = [2S^o(CO_2,g) + 3S^o(H_2O,l)] - [S^o(C_2H_5OH,l) + 3S^o(O_2,g)]$
$= [2(213.7)+3(70.0)] - [(160.7)+3(205.0)] = \underline{-138.3 \text{ J K}^{-1} \text{ mol}^{-1}}$

(c) $\Delta S^o = [6S^o(CO_2,g)+6S^o(H_2O,l)] - [S^o(C_6H_{12}O_6,s) + 6S^o(O_2,g)]$
$= [6(213.7)+6(70.0)] - [(182.4)+6(205.0)] = \underline{289.8 \text{ J K}^{-1} \text{ mol}^{-1}}$

(d) $\Delta S^o = [2S^o(HI,g)] - [S^o(H_2,g) + S^o(I_2,s)]$
$= [2(206.5)] - [(130.6)+(116.1)] = \underline{166.3 \text{ J K}^{-1} \text{ mol}^{-1}}$

10. The calculation of $\Delta S^o$ uses the method given in Problem 9:

(a) $\Delta S^o = [S^o(CaO,s) + S^o(CO_2,g)] - [S^o(CaCO_3,s)]$

$= [(38.1) + (213.7)] - [(92.9)] = \underline{158.9 \text{ J K}^{-1} \text{ mol}^{-1}}$

$\Delta S^o > 0$ because 1 mol solid gives 1 mol solid plus 1 mol gas, resulting in increased disorder in the system

(b) $\Delta S^o = [2S^o(BrF_3,g)] - [S^o(Br_2,l) + 3S^o(F_2,g)]$

$= [2(292.4)] - [(152.2) + 3(202.7)] = \underline{-175.5 \text{ J K}^{-1} \text{ mol}^{-1}}$

$\Delta S^o < 0$ because 1 mol liquid plus 3 mol gas gives 2 mol gas, resulting in a decrease in the disorder in the system

(c) $\Delta S^o = [2S^o(CO_2,g)] - [2S^o(CO,g) + S^o(O_2,g)]$

$= [2(213.7)] - [2(197.6) + (205.0)] = \underline{-172.8 \text{ J K}^{-1} \text{ mol}^{-1}}$

$\Delta S^o < 0$ because 3 mol of gas is replaced by 2 mol gas, resulting in a decrease in the entropy of the system

(d) $\Delta S^o = [S^o(CO,g) + S^o(H_2,g)] - [S^o(C,s,\text{graphite}) + S^o(H_2O,l)]$

$= [(197.6) + (130.6)] - [(5.8) + (70.0)] = \underline{252.4 \text{ J K}^{-1} \text{ mol}^{-1}}$

$\Delta S^o > 0$ because 1 mol solid and 1 mol liquid are replaced by 2 mol gas, resulting in an increased disorder in the system

(e) $\Delta S^o = [2S^o(NaCl,s)] - [2S^o(Na,s) + S^o(Cl_2,g)]$

$= [2(72.5)] - [2(51.3) + (223.0)] = \underline{-180.6 \text{ J K}^{-1} \text{ mol}^{-1}}$

$\Delta S^o < 0$ because 2 mol solid and 1 mol of gas is replaced by 2 mol solid, resulting in a decrease in the disorder in the system

11. See Problem 9 for the method:

(a) $\Delta S^o = [S^o(SO_2,g)] - [S^o(S,s,\text{ rhombic}) + S^o(O_2,g)]$

$= [(248.1)] - [(32.0) + (205.0)] = \underline{11.1 \text{ J K}^{-1} \text{ mol}^{-1}}$

(b) $\Delta S^o = [2S^o(NO,g)] - [S^o(N_2,g) + S^o(O_2,g)]$

$= [2(210.6)] - [(191.5) + (205.0)] = \underline{24.7 \text{ J K}^{-1} \text{ mol}^{-1}}$

(c) $\Delta S^o = [S^o(P_4O_{10},s)] - [S^o(P_4,s) + 5S^o(O_2,g)]$

$= [(231)] - [(41.1) + 5(205.0)] = \underline{-835.1 \text{ J K}^{-1} \text{ mol}^{-1}}$

(d) $\Delta S^o = [2S^o(Fe_2O_3,s)] - [4S^o(Fe,s) + 3S^o(O_2,g)]$

$= 2(87.4) - [4(27.3) + 3(205.0)] = \underline{-549.4 \text{ J K}^{-1} \text{ mol}^{-1}}$

(e) $\Delta S^o = [2S^o(H_2O,l)] - [2S^o(H_2,g) + S^o(O_2,g)]$

$= [2(70.0)] - [2(130.6) + (205.0)] = \underline{-326.2 \text{ J K}^{-1} \text{ mol}^{-1}}$

12. See Problem 9 for the method:

(a) $\Delta S^o = [S^o(PCl_5,g)] - [S^o(PCl_3,g) + S^o(Cl_2,g)]$

$= [(364.5)] - [(311.7) + (223.0)] = \underline{-170.2 \text{ J K}^{-1} \text{ mol}^{-1}}$

(b) $\Delta S^o = [S^o(H_2SO_4,l)] - [S^o(H_2O,l) + S^o(SO_3,g)]$

$= [(145.9)] - [(70.0) + (256.6)] = \underline{-180.7 \text{ J K}^{-1} \text{ mol}^{-1}}$

(c) $\Delta S^\circ = [S^\circ(C_2H_6,g)] - [S^\circ(C_2H_2,g) + 2S^\circ(H_2,g)]$

$= [(229.5)] - [(200.8)+2(130.6)] = \underline{-232.5\ J\ K^{-1}\ mol^{-1}}$

(d) $\Delta S^\circ = [2S^\circ(CO,g)] - [2S^\circ(C,s,graphite) + S^\circ(O_2,g)]$

$= [2(197.6)] - [2(5.8)+(205.0)] = \underline{178.6\ J\ K^{-1}\ mol^{-1}}$

(e) $\Delta S^\circ = [2S^\circ(HBr,g)] - [S^\circ(H_2,g) + S^\circ(Br_2,l)]$

$= [2(198.6)] - [(130.6)+(152.2)] = \underline{114.4\ K\ J^{-1}\ mol^{-1}}$

13. $\Delta H^\circ$, $\Delta S^\circ$, and $\Delta G^\circ$ for a reaction are all calculated from the standard values in the same way:

$$\Delta = \Sigma[(\text{products})] - \Sigma[(\text{reactants})]$$

$\Delta H^\circ = [2\Delta H^\circ_f(Fe,s) + 3\Delta H^\circ_f(CO,g)] - [\Delta H^\circ_f(Fe_2O_3,s) + 3\Delta H^\circ_f(C,s,gr)]$

$= [2(0) + 3(-110.5)] - [(-824) + 2(0)] = \underline{493\ kJ}$

$\Delta G^\circ = [2\Delta G^\circ_f(Fe,s) + 3\Delta G^\circ_f(CO,g)] - [\Delta G^\circ_f(Fe_2O_3,s) + 3\Delta G^\circ_f(C,s,gr)]$

$= [2(0) + 3(-137.2)] - [(-742.2) + 3(0)] = \underline{330.6\ kJ}$

$S^\circ = [2S^\circ(Fe,s) + 3S^\circ(CO,g)] - [S^\circ(Fe_2O_3,s) + 3S^\circ(C,s,gr)]$

$= [2(27.3) + 3(197.6)] - [(87.4) + 3(5.8)] = \underline{542.6\ J\ K^{-1}}$

Using $\Delta H^\circ$ and $\Delta S^\circ$ to calculate a value for $\Delta G^\circ$ gives,

$\Delta G^\circ = \Delta H^\circ - T\Delta S^\circ = (493\ kJ) - (298\ K)(542.6\ J\ K^{-1})(\frac{1\ kJ}{1000\ J})$

$= (493-161.7) = \underline{331\ kJ}$

which is the same as the value of $\Delta G^\circ$ by direct calculation.
For the reaction at 25°C, $\Delta G^\circ > 0$, and thus the reaction is <u>not</u> spontaneous, but rather it is the reverse reaction that is spontaneous.

$\Delta S^\circ > 0$, so the entropy change favors the spontaneity of the reaction, but $\Delta H^\circ$ is also > 0 (the reaction is endothermic) and this dominates and works against the spontaneity of the reaction.

14. $\Delta G^\circ = [2(\Delta G^\circ_f(HCl,g)] - [\Delta G^\circ_f(H_2,g) + \Delta G^\circ_f(Cl_2,g)] = \underline{-190.6\ kJ}$

$\Delta G^\circ < 0$, so the reaction as written is <u>spontaneous</u>. Calculating $\Delta S^\circ$,

$\Delta S^\circ = [2S^\circ(HCl,g)] - [S^\circ(H_2,g) + S^\circ(Cl_2,g)]$

$= [2(186.8)] - [(130.6) + (223.0)] = \underline{20.0\ J\ K^{-1}}$

and $\Delta H^\circ$ may be calculated either directly from $\Delta H^\circ_f$ values, or from

$\Delta G^\circ = \Delta H^\circ - T\Delta S^\circ$, <u>or</u> $\Delta H^\circ = \Delta G^\circ + T\Delta S^\circ$

i.e., $\Delta H^\circ = -190.6\ kJ + (298\ J)(20.0\ J\ K^{-1})(\frac{1\ kJ}{1000\ J}) = \underline{-184.6\ kJ}$

Both entropy and enthalpy changes contribute to the spontaneity of the reaction but the $\Delta H$ dominates because $T.\Delta S$ is only 6.0 kJ.

15. A reaction is spontaneous as written if the $\Delta G^\circ$ is negative.

(a) $\Delta G^\circ = [3\Delta G^\circ_f(CO_2,g)+4\Delta G^\circ_f(H_2O,g)] - [\Delta G^\circ_f(C_3H_8,g)+5\Delta G^\circ_f(O_2,g)]$

$= [3(-137.2)+4(-228.6)] - [(-23.4)+5(0)] = \underline{-1302.6\ kJ}$

$\Delta G^\circ$ is negative; the reaction is <u>spontaneous</u>.

(b) $\Delta G^\circ = [2\Delta G_f^\circ(NO_2,g)] - [\Delta G_f^\circ(N_2O_4,g)]$
$= [2(51.3)] - [(97.8)] = \underline{4.8\ kJ}$

$\Delta G^\circ$ is positive; the reaction is not spontaneous.

(c) $\Delta G^\circ = [2\Delta G_f^\circ(CH_2Cl_2,l)] - [\Delta G_f^\circ(CH_4,g) + \Delta G_f^\circ(CCl_4,l)]$
$= [2(-67.3)] - [(-50.8) + (-65.3)] = \underline{-18.5\ kJ}$

$\Delta G^\circ$ is negative; the reaction is spontaneous.

16. For $SO_2(g)$, for example, the reaction for its decomposition is

$$SO_2(g) \longrightarrow S(s) + O_2(g)$$

which is the reverse of the reaction which defines $\Delta G_f^\circ(SO_2,g)$. Thus the $\Delta G^\circ$ for the decomposition has the value of $-\Delta G_f^\circ(SO_2,g)$. Similarly for the other decompositions the $\Delta G^\circ$ values are minus the $\Delta G_f^\circ$ values given. For a spontaneous decomposition, $\Delta G^\circ$ must be negative. Among these compounds, only $NO_2(g)$ will have a spontaneous tendency to decompose to its elements at 298 K.

17. Calculation of $\Delta H^\circ$ and $\Delta S^\circ$ by the usual methods, using data from Appendix B gives:

$$\underline{\Delta H^\circ = -938.2\ kJ} \quad \text{and} \quad \underline{\Delta S^\circ = 163\ J\ K^{-1}}$$

and assuming that these do not change significantly at temperatures other than standard temperature (25°C), $\Delta G = \Delta H - T\Delta S$, gives a negative value for $\Delta G^\circ$ at all temperatures, so the reaction is spontaneous at all temperatures. Since the reaction is exothermic, increased temperature favors the reverse reaction (Le Châtelier's principle), so that the high temperature (and the catalyst) must serve to increase the rate of the reaction.

18. From $\Delta G = \Delta H - T\Delta S$,

(a) $\Delta H > 0$, $-T\Delta S < 0$; $\Delta G$ is either positive or negative depending on the values of $\Delta H$ and $T\Delta S$. At low temperature T is small, so $\Delta G$ will most often be positive at low temperature.
(b) $\Delta H < 0$, $-T\Delta S < 0$, and $\Delta G$ is negative under all circumstances
(c) $\Delta H < 0$, $-T\Delta S > 0$, and $\Delta G$ is either negative or positive depending on the values of $\Delta H$ and $T\Delta S$. At low temperature, T is small, so $\Delta G$ will be most often negative at low temperature.
(d) $\Delta H > 0$ and $-T\Delta S > 0$, and $\Delta G$ is positive under all circumstances.

19. (See Problem 18). Provided T is sufficiently large, the $-T\Delta S$ term will dominate and determine the sign of $\Delta G$.
(a) $-T\Delta S < 0$; $\Delta G$ is negative.
(b) $-T\Delta S < 0$; $\Delta G$ is always negative.
(c) $-T\Delta S > 0$; $\Delta G$ is positive.
(d) $-T\Delta S > 0$; $\Delta G$ is always positive.

20. (a) $\Delta G^\circ = [2\Delta G_f^\circ(NaCl,s) + \Delta G_f^\circ(F_2,g)] - [2\Delta G_f^\circ(NaF,s) + \Delta G_f^\circ(Cl_2,g)]$
$= [2(-384.3) + (0)] - [2(-546.3) + (0)] = \underline{324\ kJ}$

(b) $\Delta G^\circ = [2\Delta G_f^\circ(NaCl,s) + \Delta G_f^\circ(BR_2,l)] - [2\Delta G_f^\circ(NaBr,s) + \Delta G_f^\circ(Cl_2,g)]$
$= [2(-384.3) + (0)] - [2(-349.1) + (0)] = \underline{-70.4\ kJ}$

(c) $\Delta G^\circ = [\Delta G_f^\circ(Pb,s) + 2\Delta G_f^\circ(ZnO,s)] - [\Delta G_f^\circ(PbO_2,s) + 2\Delta G_f^\circ(Zn,s)]$

$= [(0) + 2(-320.5)] - [(-217.4) + 2(0)] = \underline{-423.6\ kJ}$

(d) $\Delta G^\circ = [2\Delta G_f^\circ(Al,s) + \Delta G_f^\circ(Fe_2O_3,s)] - [\Delta G_f^\circ(Al_2O_3,s) + 2\Delta G_f^\circ(Fe,s)]$

$= [2(0) + (-742.2)] - [(-1582) + 2(0)] = \underline{839.8\ kJ}$

Reaction (a) is not spontaneous as written, but the reverse reaction is spontaneous, i.e. $F_2(g)$ is a stronger oxidizing agent than $Cl_2(g)$, and from reaction (b), which is spontaneous as written, $Cl_2(g)$ is a stronger oxidizing agent than $Br_2(l)$, confirming the expected order of oxidizing power: $F_2 > Cl_2 > Br_2$. Reaction (c) is spontaneous as written; Zn is a stronger reducing agent than lead. Reaction (d) is not spontaneous as written, but the reverse reaction is spontaneous; Al is a stronger reducing agent than Fe.

21. (a) The balanced equation is:

$$C(s,\ diamond) + O_2(g) \longrightarrow CO_2(g)$$

under standard conditions, for which

$\Delta G^\circ = [\Delta G_f^\circ(CO_2,\ g)] - [\Delta G_f^\circ(C,s,diamond) + \Delta G_f^\circ(O_2,g)]$

$= [(-394.4)] - [(2.9) + (0)] = \underline{-397.3\ kJ\ mol^{-1}}$

(b) The negative sign of $\Delta G^\circ$ indicates that the reaction is spontaneous as written.

(c) Although the reaction occurs spontaneously it is in fact a very slow reaction. Thermodynamics tells us that this process is spontaneous but nothing about the rate at which the reaction proceeds to equilibrium.

22. For the calculation of $K_p$ we can use the equation

$$\underline{\Delta G = -RT\ \ln K_p}$$

where $\Delta G$ is the Gibb's free energy change for the reaction, which can be rearranged to give:

$$\underline{K_p = e^{-\Delta G/RT}} \qquad (R = 8.314\ J\ K^{-1}\ mol^{-1})$$

Thus, we first calculate $\Delta G^\circ$ for the reaction as written:

$\Delta G^\circ = (2\Delta G_f^\circ(SO_3,g))] - [2\Delta G_f^\circ(SO_2,g) + \Delta G_f^\circ(O_2,g)]$

$= [2(-371.1)] - [2(-300.1) + (0)] = \underline{-142\ kJ}$

$$-\frac{\Delta G^\circ}{RT} = \frac{-(-142\ kJ\ mol^{-1})(\frac{1000\ J}{1\ kJ})}{(8.314\ J\ K^{-1}\ mol^{-1})(298\ K)} = \underline{57.3}$$

$$K_p = e^{+57.3} = \underline{7.7\times10^{24}\ atm^{-1}}$$

The reaction as written is spontaneous, as indicated by the negative sign of $\Delta G^\circ$ (and the very large value of $K_p$).

23. $\Delta S^\circ = [2S^\circ(CO,g)] - [S^\circ(C,s,gr) + S^\circ(CO_2,g)]$

$= [2(197.6)] - [(5.8) + (213.7)] = \underline{175.7\ J\ K^{-1}}$

and $\Delta H^o = [2\Delta H_f^o(CO,g)] - [\Delta H_f^o(C,s,gr) + \Delta H_f^o(CO_2,g)]$

$= [2(-110.5)] - [(0) + (-392.5)] = \underline{171.5 \text{ kJ}}$

At 700°C (973 K) we can write:

$$\Delta G = \Delta H^o - T\Delta S^o = (171.5 \text{ kJ}) - (973 \text{ K})(175.7 \text{ J K}^{-1})(\frac{1 \text{ kJ}}{1000 \text{ J}})$$
$$= \underline{0.5 \text{ kJ}}$$

$$-\frac{\Delta G}{RT} = \frac{-(0.5 \text{ kJ})(\frac{1000 \text{ J}}{1 \text{ kJ}})}{(8.314 \text{ J K}^{-1} \text{ mol}^{-1})(973 \text{ K})} = -0.062$$

$$K_p = e^{-0.062} = \underline{0.94 \text{ atm}}$$

24. $\Delta G^o = [\Delta G_f^o(PCl_3,g) + \Delta G_f^o(Cl_2,g)] - [\Delta G_f^o(PCl_5,g)]$

$= [(-267.8) + (0)] - [(-305.0)] = \underline{37.2 \text{ kJ}}$

$$-\frac{\Delta G^o}{RT} = \frac{-(37.2 \text{ kJ})(\frac{1000 \text{ J}}{1 \text{ kJ}})}{(8.314 \text{ J K}^{-1})(298 \text{ K})} = -15.0$$

$$K_p = e^{-15.0} = \underline{3.1 \times 10^{-7} \text{ atm}}$$

25. $\Delta G^o = [2\Delta G_f^o(NO_2,g)] - [\Delta G_f^o(N_2O_4,g)]$

$= [2(51.3)] - [(97.8)] = \underline{4.8 \text{ kJ}}$

$$-\frac{\Delta G^o}{RT} = \frac{-(4.8 \text{ kJ})(\frac{1000 \text{ J}}{1 \text{ kJ}})}{(8.314 \text{ J K}^{-1})(298 \text{ K})} = -1.94 \; ; \quad K_p = \underline{0.14 \text{ atm}}$$

26. $\Delta G^o = [\Delta G_f^o(CO_2,g) + 2\Delta G_f^o(H_2O,g)] - [\Delta G_f^o(CH_4,g) + 2\Delta G_f^o(O_2,g)]$

$= [(-394.4) + 2(-228.6)] - [(-50.8) + 2(0)] = \underline{-800.8 \text{ kJ}}$

(a) $\Delta G$ is negative, so the reaction as written is spontaneous.

(b) $$-\frac{\Delta G}{RT} = \frac{-(-800.8 \text{ kJ})(\frac{1000 \text{ J}}{1 \text{ kJ}})}{(8.314 \text{ J K}^{-1})(298 \text{ K})} = 323 \; ; \; K_p = e^{323} = \underline{10^{140}}$$

(c) The reaction is very slow at room temperature and proceeds to equilibrium very slowly, unless initiated by a spark, when methane ignites and burns readily at high temperature.

27. $\Delta G^o = [2\Delta G_f^o(CO_2,g) + 4\Delta G_f^o(H_2O,l)] - [2\Delta G_f^o(CH_3OH,l) + 3\Delta G_f^o(O_2,g)]$

$= [2(-394.4) + 4(-237.2)] - [2(-166.4)+3(0)] = \underline{-1404.8 \text{ kJ}}$

(a) $\Delta G$ is negative, so the reaction as written is spontaneous

(b) $$-\frac{\Delta G}{RT} = \frac{-(-1404.8 \text{ kJ})(\frac{1000 \text{ J}}{1 \text{ kJ}})}{(8.314 \text{ J K}^{-1})(298 \text{ K})} = 567 \; ; \; K_p = e^{567} = \underline{1.8\text{x}10^{247} \text{ atm}^{-1}}$$

(c) Products are favored since $K_p >> 1$.

(d) Since in going from reactants to products the number of moles of gas decreases, increased pressure favors the forward reaction (Le Châtelier's principle).

(e) The reaction is highly exothermic, so increasing the temperature favors the reactants, (le Châtelier's principle); the value of $K_p$ will decrease with increasing temperature.

28\. (a) Since $\Delta G > 0$, the equilibrium constant will be very small and only a very small amount of $C_2H_2(g)$ will be present in the equilibrium mixture at room temperature; this is not a practical route for the synthesis of ethyne at room temperature.

(b) The reaction will be spontaneous at high temperature if $\Delta G$ becomes negative at high temperature, which we can determine by calculating $\Delta H^o$ and $\Delta S^o$ for the reaction, and then examining what happens to $\Delta G = \Delta H^o - T\Delta S^o$ at high temperature, assuming that both $\Delta H^o$ and $\Delta S^o$ are temperature independent.

$$\Delta H^o = [\Delta H_f^o(C_2H_2,g)] - [2\,\Delta H_f^o(C,s,gr) + \Delta H_f^o(H_2,g)]$$
$$= [(228.0)] - [2(0) + (0)] = \underline{228.0 \text{ kJ}}$$

$$\Delta S^o = [S^o(C_2H_2,g)] - [2S^o(C,s,gr) + S^o(H_2,g)]$$
$$= [(200.8)] - [2(5.8) + (130.6)] = \underline{58.6 \text{ J K}^{-1}}$$

Thus: $\Delta G = 228.0 \text{ kJ} - T(0.0586 \text{ kJ K}^{-1})$

and $\Delta G$ becomes negative above T = 3891 K.

(c) At 1200 K, $\Delta G = [228.0 \text{ kJ} - (1200 \text{ K})(0.0586 \text{ kJ K}^{-1}) = \underline{157.7 \text{ kJ}}$

$$-\frac{\Delta G}{RT} = \frac{-(157.7 \text{ kJ})(\frac{1000 \text{ J}}{1 \text{ kJ}})}{(8.341 \text{ J K}^{-1})(1200 \text{ K})} = -15.8;\ K_p = e^{-15.8} = \underline{1.4\text{x}10^{-7}}$$

29\.
$$N_2O_4(g) \rightleftharpoons 2NO_2(g)$$

$$K_p = \frac{(P_{NO_2})^2}{P_{N_2O_4}} = \frac{(0.020 \text{ atm})^2}{0.040 \text{ atm}} = \underline{0.01 \text{ atm}} \ll 0.14 \text{ atm}$$

The reaction is not at equilibrium and more $N_2O_4(g)$ has to dissociate to $NO_2(g)$ for equilibrium to be achieved.

30\. (a) $$\Delta G^o = [2\,\Delta G_f^o(Fe(CO)_5,g) + 3\,\Delta G_f^o(CO_2,g)] - [\Delta G_f^o(Fe_2O_3,s) + 13\,\Delta G_f^o(CO.g)]$$
$$= [2(-697.3) + 3(-394.4)] - [(-742.2) + 13(-137.2)] = \underline{-52.0 \text{ kJ}}$$

$$-\frac{\Delta G}{RT} = \frac{-(-52.0 \text{ kJ})(\frac{1000 \text{ J}}{1 \text{ kJ}})}{(8.314 \text{ J K}^{-1})(298 \text{ K})} = 21.0\ ;\ K_p = e^{21.0} = \underline{1.3\text{x}10^9 \text{ atm}^{-8}}$$

The reaction is spontaneous at 298 K.

(b) $$\Delta G^o = [\Delta G_f^o(HCN,g) + \Delta G_f^o(NH_3,g)] - [\Delta G_f^o(N_2,g) + \Delta G_f^o(CH_4,g)]$$
$$= [(124.7) + (-16.4)] - [(0) + (-50.8)] = \underline{159.1 \text{ kJ}}$$

$$-\frac{\Delta G}{RT} = \frac{-(159.1 \text{ kJ})(\frac{1000 \text{ J}}{1 \text{ kJ}})}{(8.314 \text{ J K}^{-1})(298 \text{ K})} = -64.2\ ;\ K_p = e^{-64.2} = \underline{1.6\text{x}10^{-28}}$$

The reaction is not spontaneous at 298 K.

31\. We can first calculate $\Delta G^o$ for the reaction in the usual way:

$$\Delta G^o = [2\,\Delta G_f^o(CO_2,g)] - [2\,\Delta G_f^o(CO,g) + \Delta G_f^o(O_2,g)]$$
$$= [2(-394.4)] + [2(-137.2) + (0)] = \underline{-514.4 \text{ kJ}}$$

$$-\frac{\Delta G^o}{RT} = -\frac{(-514.4\ kJ)(\frac{1000\ J}{1\ kJ})}{(8.314\ J\ K^{-1})(298\ K)} = 208$$

$$K_p = e^{208} = \underline{10^{90}\ atm^{-1}}$$

$\Delta G$ for the particular reaction conditions is calculated from

$$\Delta G = RT\ \ln(\frac{Q}{K_p})$$

$$\text{where } Q = \frac{(P_{CO_2})^2}{(P_{CO})^2(P_{O_2})} = \frac{(0.020\ atm)^2}{(0.020\ atm)^2(0.020\ atm)} = 50\ atm^{-1}$$

Thus,

$$\Delta G = (8.31\ J\ K^{-1})(298\ K)(\ln \frac{50}{10^{90}}) = \underline{-503.5\ kJ} < \Delta G^o$$

Since $\Delta G$ is negative, the reaction will proceed spontaneously from left to right until equilibrium is achieved.

32. Calculate the $\Delta G^o$ value for each reaction, then for the coupled reaction, the $\Delta G^o$ value is the sum of these $\Delta G^o$ values.

(a) $\Delta G^o = [2\ \Delta G_f^o(CH_3OH,l) + 3\Delta G_f^o(O_2,g)] - [2\ \Delta G_f^o(CO_2,g) + 4\ \Delta G_f^o(H_2O,l)$
$= [2(-166.4) + 3(0)] - [2(-394.4) + 4(-237.2] = \underline{1404.8\ kJ}$

(b) $\Delta G^o = [2\Delta G_f^o(CO_2,g)] - [2\ \Delta G_f^o(C,s,gr) + 2\ \Delta G_f^o(O_2,g)]$
$= [2(-394.4)] - [2(0) + (0)] = \underline{-788.8\ kJ}$

For the coupled reaction,

$$2C(s) + 4H_2O(l) \longrightarrow 2CH_3OH(l) + O_2(g)$$

$\Delta G^o = +1404.8 + (-788.8) = \underline{616.0\ kJ}$, which is a positive value.

Thus, the coupled reaction would <u>not</u> be spontaneous.

33. In an endothermic reaction heat is transferred from the surroundings to the system, so that the entropy of the surroundings decreases. For a spontaneous reaction, the total entropy of the system plus surroundings must increase, so the decrease in the entropy of the surroundings must be smaller than the increase in the entropy of the system. At higher temperatures, there is more randomness in the surroundings than at low temperature. Thus, the transfer of a given amount of heat from the surroundings to the system causes a smaller entropy decrease than at lower temperatures. Therefore, the total entropy change is more likely to be positive at higher temperatures and will increase in magnitude with increasing temperature, so that the reaction has an increasing tendency to proceed at higher temperatures. Endothermic reactions are favored by an increase in temperature.

34. For a spontaneous reaction

$$\Delta S_{total} = \Delta S_{system} + \Delta S_{surroundings} > 0$$

and substituting $\Delta S_{surroundings} = -\Delta H_{system}/T$ gives

$$\Delta S_{total} = \Delta S_{system} - \frac{\Delta H_{system}}{T} > 0$$

Multiplying through by T, gives

$$T.\Delta S_{total} = T.\Delta S_{system} - \Delta H_{system} > 0$$

<u>or</u>

$$-T.\Delta S_{total} = -T.\Delta S_{system} + \Delta H_{system}$$

$$= \Delta H_{system} - T.\Delta S_{system} < 0$$

Thus, an alternative criterion for a spontaneous reaction is that

$$\Delta H_{system} - T.\Delta S_{system} < 0$$

and since $\Delta G = \Delta H - T.\Delta S$

$$\underline{\Delta G_{system} < 0}$$

The second law of thermodynamics may be stated in the form:

"the entropy of the universe, $\Delta S_{total}$, increases in any spontaneous process", <u>or</u> for a spontaneous process,

$$\underline{\Delta G_{system} < 0}$$

35. $\Delta G$ values must be calculated for each of the reactions in which the metal oxides are reduced by carbon to metal and carbon is oxidized, <u>either</u> to $CO_2(g)$, <u>or</u> to $CO(g)$, which is done by adding each of the given equations, and <u>either</u> one of the following equations:

$$2C(s) + O_2(g) \longrightarrow 2CO(g) \quad ; \quad \Delta G^o_f(CO,g) = 2(-250) = -500 \text{ kJ}$$

and

$$C(s) + O_2(g) \longrightarrow CO_2(g) \quad ; \quad \Delta G^o_f(CO_2,g) = -380 \text{ kJ}$$

(a) $2Al_2O_3(s) \longrightarrow 4Al(s) + 3O_2(g)$ $\quad \Delta G = -(-2250) = 2250$ kJ

$3[2C(s) + O_2(g) \longrightarrow 2CO(g)]$ $\quad \Delta G = 3(-500) = -1500$ kJ

$2Al_2O_3(s) + 6C(s) \longrightarrow 4Al(s) + 6CO(g)$ $\quad \Delta G = 750$ kJ

$2Al_2O_3(s) \longrightarrow 4Al(s) + 3O_2(g)$ $\quad \Delta G = -(-2250) = 2250$ kJ

$3(C(s) + O_2(g) \longrightarrow CO_2(g)]$ $\quad \Delta G = 3(-380) = -1140$ kJ

$2Al_2O_3(s) + 3C(s) \longrightarrow 4Al(s) + 3CO_2(g)$ $\quad \Delta G = 1110$ kJ

Since $\Delta G$ is positive in both cases, neither reaction is spontaneous

(b) $2FeO(s) \longrightarrow 2Fe(s) + O_2(g)$ $\quad \Delta G = -(-250) = 250$ kJ

$2C(s) + O_2(g) \longrightarrow 2CO(g)$ $\quad \Delta G = -500$ kJ

$2FeO(s) + 2C(s) \longrightarrow 2Fe(s) + 2CO(g)$ $\quad \Delta G = -250$ kJ

$2FeO(s) \longrightarrow 2Fe(s) + O_2(g)$ $\quad \Delta G = -(-250) = 250$ kJ

$C(s) + O_2(g) \longrightarrow CO_2(g)$ $\quad = -380$ kJ

$2FeO(s) + C(s) \longrightarrow 2Fe(s) + CO_2(g)$ $\quad \Delta G = -130$ kJ

Both reactions are spontaneous at 1500°C.

(c) $2PbO(s) \longrightarrow 2Pb(s) + O_2(g)$ $\quad \Delta G = -(-120) = 120$ kJ

$2C(s) + O_2(g) \longrightarrow 2CO(g)$ $\quad \Delta G = -500$ kJ

$2PbO(s) + 2C(s) \longrightarrow 2Pb(s) + 2CO(g)$ $\quad \Delta G = -380$ kJ

$2PbO(s) \longrightarrow 2Pb(s) + O_2(g)$ $\quad \Delta G = -(-120) = 120$ kJ

$C(s) + O_2(g) \longrightarrow CO_2(g)$ $\quad \Delta G = -380$ kJ

$2PbO(s) + C(s) \longrightarrow 2Pb(s) + CO_2(g)$ $\quad \Delta G = -260$ kJ

$\Delta G$ is negative for both reactions, so each reaction is spontaneous

(d) $2CuO(s) \longrightarrow 2Cu(s) + O_2(g)$ $\quad \Delta G = -(0) = 0$ kJ

$2C(s) + O_2(g) \longrightarrow 2CO(g)$ $\quad \Delta G = -500$ kJ

$2CuO(s) + 2C(s) \longrightarrow 2Cu(s) + 2CO(g)$ $\quad \Delta G = -500$ kJ

$2CuO(s) \longrightarrow 2Cu(s) + O_2(g)$ $\quad \Delta G = 0$ kJ

$C(s) + O_2(g) \longrightarrow CO_2(g)$ $\quad \Delta G = -380$ kJ

$2CuO(s) + C(s) \longrightarrow 2Cu(s) + CO_2(g)$ $\quad \Delta G = -380$ kJ

$\Delta G$ is negative for both reactions, so each reaction is spontaneous

$Al_2O_3$ cannot be reduced to metal by carbon; FeO(s), PbO(s), and CuO(s) are reduced by carbon under all circumstances.

36. For each of the reactions, calculate $\Delta G^o$, which is the change in free energy at 25°C.

$$2C(s,gr) + H_2(g) \longrightarrow C_2H_2(g) \quad .... (1)$$

$$4CH_4(g) + 3O_2(g) \longrightarrow 2C_2H_2(g) + 6H_2O(g) \quad .... (2)$$

(a) Reaction (1) is the reaction for the formation of $C_2H_2(g)$ from the elements in their standard states.

Thus, $\Delta G = G^o_f(C_2H_2,g) = \underline{209.2 \text{ kJ mol}^{-1}}$

and for reaction (2),

$$\Delta G^o = [2\Delta G^o_f(C_2H_2,g) + 6\Delta G^o_f(H_2O,g)] - [4\Delta G^o_f(CH_4,g) + 3\Delta G^o_f(O_2,g)]$$

$$= [2(209.2) + 6(-228.6)] - [4(-50.8) + 3(0)] = \underline{-749 \text{ kJ}}$$

$\Delta G^o$ for reaction (1) is positive so that the reaction is not spontaneous at room temperature. For reaction (2), $\Delta G^o$ is negative and the reaction will occur spontaneously, albeit very slowly.

(b) Addition of suitable catalysts cannot affect the spontaneity of the reactions or their equilibrium positions, but will speed up the rate at which equilibrium is achieved. In other words, a catalyst does not affect the thermodynamics of the reactions but would be beneficial in reaction (2) in producing acetylene more rapidly.

37. The balanced equation for the reaction is,

$$2PbO(s) + 2Zn(s) \longrightarrow 2Pb(s) + 2ZnO(s)$$

(a) From the given data, and data from Appendix B, calculate the $\Delta G^o$ for the reaction by first calculating the $\Delta H^o$ and $\Delta S^o$ values and then using $\Delta G^o = \Delta H^o - T\Delta S^o$.

$$\Delta H^o = [2\Delta H^o_f(Pb,s) + 2\Delta H^o_f(ZnO,s)] - [2\Delta H^o_f(PbO,s) + 2\Delta H^o_f(Zn,s)]$$

$$= [2(0) + 2(-350.5)] - [2(-217.9) + 2(0)] = \underline{-265.2 \text{ kJ}}$$

$$\Delta S^o = [2S^o(Pb,s) + 2S^o(ZnO,s)] - [2S^o(PbO,s) + 2S^o(Zn,s)]$$

$$= [2(64.9) + 2(43.6)] - [2(69.5) + 2(41.6)] = \underline{-5.2 \text{ J K}^{-1}}$$

Thus, at 298 K,

$$\Delta G^o = \Delta H^o - T\Delta S^o = -265.2 \text{ kJ} - (298 \text{ K})(-5.2 \text{ J K}^{-1})\left(\frac{1 \text{ kJ}}{1000 \text{ J}}\right)$$

$$= \underline{-263.7 \text{ kJ}}$$

(b) Since $\Delta G^o$ is negative, the reduction of PbO(s) to Pb(s) by Zn(s) is a spontaneous reaction.

38. (a) The bond energy data required is to be found in Table 6.3, and recall that

$$\Delta H^o = \Sigma(\text{B.E. reactants}) - \Sigma(\text{B.E. products})$$

$$N_2(g) + 2H_2(g) \longrightarrow N_2H_4(g) \qquad \ldots\ldots (1)$$

$$\Delta H^o = [BE(N{\equiv}N) + 2BE(H\text{-}H)] - [BE(N\text{-}N) + 4BE(N\text{-}H)]$$

$$= [(941) + 2(436)] - [(159) + 4(389)] = \underline{98\text{ kJ}}$$

$$C_2H_2(g) + 2H_2(g) \longrightarrow C_2H_6(g) \qquad \ldots\ldots (2)$$

$$\Delta H^o = [BE(C{\equiv}C) + 2BE(C\text{-}H) + 2BE(H\text{-}H)] - [BE(C\text{-}C) + 6BE(C\text{-}H)]$$

$$= BE(C{\equiv}C) + 2BE(H\text{-}H) - BE(C\text{-}C) - 4BE(C\text{-}H)$$

$$= 812 + 2(436) - 348 - 4(413) = \underline{-316\text{ kJ}}$$

(b) In each case 3 mol gas is replaced by 1 mol gas, so that in each case $\Delta S^o$ should have a negative sign, and from $\Delta G^o = \Delta H^o - T\Delta S^o$ we have:

Reaction (1): $\Delta G^o$ positive

Reaction (2): $\Delta G^o$ negative except at very high temperature

(c) Thus, reaction (1) is not spontaneous under any conditions, while reaction (2) should be spontaneous under nearly all conditions, which is consistent with the inertness of $N_2(g)$ and the ease of reaction of ethyne.

39. (a) We first calculate $\Delta G^o$ for the reaction, and then $K_p$:

$$\Delta G^o = [2\Delta G_f^o(SO_3,g)] - [2\Delta G_f^o(SO_2,g) + \Delta G_f^o(O_2,g)]$$

$$= [2(-371.1)] - [2(-300.1) + 0] = \underline{-142.0\text{ kJ}}$$

$$-\frac{\Delta G^o}{RT} = \frac{-(-142.0\text{ kJ})(\frac{1000\text{ J}}{1\text{ kJ}})}{(8.314\text{ J K}^{-1})(298\text{ K})} = 57.3;\quad K_p = e^{57.3} = \underline{7.7\text{x}10^{24}\text{ atm}^{-1}}$$

(b) $$\Delta H^o = [2\Delta H_f^o(SO_3,g)] - [2\Delta H_f^o(SO_2,g) + \Delta H_f^o(O_2,g)]$$

$$= [2(-395.7)] - [2(-296.8) + (0)] = \underline{-197.8\text{ kJ}}$$

$$\Delta S^o = [2S^o(SO_3,g)] - [2S^o(SO_2,g) + S^o(O_2(g)]$$

$$= [2(256.6)] - [2(248.1) + (205.0)] = \underline{-188\text{J K}^{-1}}$$

At 500 K, $\Delta G = \Delta H^o - T\Delta S^o = -197.8\text{ kJ} - (500\text{ K})(-188\text{J K}^{-1})(\frac{1\text{ kJ}}{1000\text{ J}})$

$$= \underline{-103.8\text{ kJ}}$$

$$-\frac{\Delta G}{RT} = \frac{-(-103.8\text{ kJ})(\frac{1000\text{ J}}{1\text{ kJ}})}{(8.314\text{ J K}^{-1})(500\text{ K})} = 25.0;\quad K_p = e^{25.0} = \underline{7.2\text{x}10^{10}\text{atm}^{-1}}$$

(c) $$K_c = \frac{[SO_3]^2}{[SO_2]^2[O_2]}\text{ mol}^{-1}\text{ L} = K_p(RT)$$

$$K_c^{298} = (7.7\times10^{24}\ \text{atm}^{-1})(0.0821\ \text{atm L mol}^{-1}\ \text{K}^{-1})(298\ \text{K})$$
$$= \underline{1.9\times10^{26}\ \text{mol}^{-1}\ \text{L}}$$

$$K_c^{500} = (7.2\times10^{10}\ \text{atm}^{-1})(0.0821\ \text{atm L mol}^{-1}\ \text{K}^{-1})(500\ \text{K})$$
$$= \underline{3.0\times10^{12}\ \text{mol}^{-1}\ \text{L}}$$

Note that as expected for an exothermic reaction, the value of the equilibrium constant decreases with increase in temperature.

40. (a) $\Delta G^o = [\Delta G_f^o(CH_3OH,l)] - [\Delta G_f^o(CO,g) + 2\Delta G_f^o(H_2,g)]$
$= [(-166.4)] - [(-137.2) + 2(0)] = \underline{-29.2\ \text{kJ}}$

$$-\frac{\Delta G}{RT} = \frac{-(-29.2\ \text{kJ})(\frac{1000\ \text{J}}{1\ \text{kJ}})}{(8.314\ \text{J K}^{-1})(298\ \text{K})} = 11.8\ ;\ K_p = e^{11.8} = \underline{1.3\times10^{5}\ \text{atm}^{-3}}$$

(b) $\Delta H^o = [\Delta H_f^o(CH_3OH,l)] - [\Delta H_f^o(CO,g) + 2\Delta H_f^o(H_2,g)]$
$= [(-239.1)] - [(-110.5) + 2(0)] = \underline{-128.6\ \text{kJ}}$

$\Delta S^o = [S^o(CH_3OH,l)] - [S^o(CO,g) + 2S^o(H_2,g)]$
$= [(126.8)] - [(197.6) + 2(130.6)] = \underline{-332\ \text{J K}^{-1}}$

$$\Delta G^o = \Delta H^o - T\Delta S^o = -128.6\ \text{kJ} - (298\ \text{K})(-332\ \text{J K}^{-1})(\frac{1\ \text{kJ}}{1000\ \text{J}})$$
$$= \underline{-29.7\ \text{kJ}}\ ;\ K_p = e^{12.0} = \underline{1.6\times10^{5}\ \text{atm}^{-3}}$$

(c) Both methods give essentially the same value for $\Delta G^o$, and $K_p$.

41. (a) $\Delta H^o = [3\Delta H_f^o(O_2,g)] - [2\Delta H_f^o(O_3,g)]$
$= [3(0)] - 2[(142.7)] = \underline{-285.4\ \text{kJ}}$

(b) $\Delta G^o = [3\Delta G_f^o(O_2,g)] - [2\Delta G_f^o(O_3,g)]$
$= [3(0)] - [2(163.2)] = \underline{-326.4\ \text{kJ}}$

42. (a) This is a heterogeneous equilibrium, for which

$$\underline{K_p = P_{H_2O(g)}}$$

(b) $\Delta G^o = [\Delta G_f^o(H_2O,g)] - [\Delta G_f^o(H_2O,l)] = [(-228.6)]-[(-237.2)]$
$= \underline{8.6\ \text{kJ}}$

$$-\frac{\Delta G^o}{RT} = \frac{-(8.6\ \text{kJ})(\frac{1000\ \text{J}}{1\ \text{kJ}})}{(8.314\ \text{J K}^{-1})(298\ \text{K})} = -3.47;\ K_p = e^{-3.47} = \underline{0.0311\ \text{atm}}$$

$$P_{H_2O(g)}^{25^oC} = K_p = (0.0311\ \text{atm})(\frac{760\ \text{mm Hg}}{1\ \text{atm}}) = \underline{23.6\ \text{mm Hg}}$$

which is to be compared with the value of 23.8 mm Hg given for the vapor pressure of water at 25°C in Table 12.4.

(c) At 25°C, $\Delta S^o = [S^o(H_2O,g)] - [S^o(H_2O,l)] = [(188.7)]-[(70)]$
$= \underline{118.7\ \text{J K}^{-1}}$

and $\Delta H^\circ = \Delta G^\circ - T\Delta S^\circ = 8.6\ kJ + (118.7\ J\ K^{-1})(298\ K)(\frac{1\ kJ}{1000\ J})$

$= \underline{44.0\ kJ}$ (which is the molar enthalpy of vaporization of 1 mol of $H_2O(l)$ )

Thus, at 30°C, (303 K),

$\Delta G = \Delta H^\circ - T\Delta S^\circ = 44.0\ kJ - (303\ K)(118.7\ J\ K^{-1})(\frac{1\ kJ}{1000\ J})$

$= \underline{8.0\ kJ}$

$$-\frac{\Delta G}{RT} = \frac{-(8.0\ kJ)(\frac{1000\ J}{1\ kJ})}{(8.314\ J\ K^{-1})(303\ K)} = \underline{-3.18};\quad K_p = e^{-3.18} = \underline{0.0416\ atm}$$

$$P^{30^\circ C}_{H_2O(g)} = K_p = (0.0416\ atm)(\frac{760\ mm\ Hg}{1\ atm}) = \underline{31.6\ mm\ Hg}$$

(which is to be compared to the value of 31.8 mm Hg given in Table 12.4).

43. (a) $\Delta G^\circ = [2\Delta G_f^\circ(HCl,g)] - [\Delta G_f^\circ(H_2,g) + \Delta G_f^\circ(Cl_2,g)]$

$= [2(-95.3)] - [(0) + (0)] = \underline{-190.6\ kJ}$

$$-\frac{\Delta G^\circ}{RT} = \frac{-(-190.6\ kJ)(\frac{1000\ J}{1\ kJ})}{(8.314\ J\ K^{-1})(298\ K)} = 76.9\ ;\ K_p = e^{76.9} = \underline{2.6\times10^{23}}$$

(b)

$$Q = \frac{(P_{HCl})^2}{(P_{H_2})(P_{Cl_2})} = \frac{(1\ atm)^2}{(1\ atm)(1\ atm)} = 1\ atm << K_p$$

Thus, for equilibrium to be established, $P_{HCl}$ has to increase and $P_{H_2}$ and $P_{Cl_2}$ have to decrease, and the direction of spontaneous change is from left to right.

(c) Experimentally, the reaction is spontaneous from left to right as predicted above, but the rate of establishment of equilibrium is very slow unless the reaction is initiated by a spark or bright light, when it goes rapidly to completion, consistent with the very large value of $K_p$.

44. (a) For the first reaction, $\Delta G^\circ$ is $\Delta G_f^\circ$ for $SO_2(g)$ = $\underline{-300.1\ kJ\ mol^{-1}}$

and for the second reaction,

$\Delta G^\circ = [\Delta G_f^\circ(SO_3,g)] - [\Delta G_f^\circ(SO_2,g) + \frac{1}{2}\Delta G_f^\circ(O_2,g)]$

$= [(-371.1)] - [(-300.1) + (0)] = \underline{-70\ kJ}$

| | |
|---|---|
| $S(s) + O_2(g) \longrightarrow SO_2(g)$ | $\Delta G^\circ = -300.1\ kJ$ |
| $SO_2(g) + \frac{1}{2}O_2(g) \longrightarrow SO_3(g)$ | $\Delta G^\circ = -70.0\ kJ$ |
| $S(s) + \frac{3}{2}O_2(g) \longrightarrow SO_3(g)$ | $\Delta G^\circ = -371.1\ kJ$ |

(b) Each of the reactions is spontaneous ($\Delta G^\circ$ negative).

(c) The coupled reaction is spontaneous ($\Delta G^\circ$ negative).

(d) $SO_3$ should predominate, because $\Delta G^\circ$ for its formation is much more negative than it is for the formation of $SO_2(g)$.

(e) In the production of acid rain, the most important reaction is $SO_2 + 1/2\ O_2 \longrightarrow SO_3$, followed by $SO_3 + H_2O \longrightarrow H_2SO_4$

# CHAPTER 17

1. (a) Electrolysis is a process by which electrical energy is used to produce a chemical change.
An electrolyte is a substance which when melted or dissolved in a solvent conducts an electric current by ionic conductance.
A nonelectrolyte is a substance which when melted or dissolved in a solvent is a nonconductor of electricity.

(b) In ionic conductance the electric current is carried by ions, and in electronic conductance the electric current is carried by electrons.

2. (i) The anode is the electrode at which oxidation occurs.

(ii) The cathode is the electrode at which reduction occurs.

(iii) An electrolyte is a substance that forms ions in its molten state or when dissolved in a suitable solvent and therefore conducts an electric current by ionic conduction.

(iv) The Faraday constant is the charge associated with 1 mol of electrons. 1 F = 96 485 C $mol^{-1}$.

3. (a) Molten $AlCl_3$ contains $Al^{3+}(l)$ and $Cl^-(l)$ ions. On electrolysis, $Al^{3+}$ is reduced at the cathode and $Cl^-$ is oxidized at the anode:

$2[Al^{3+} + 3e^- \longrightarrow Al(l)]$ cathode

$3(2Cl^- \longrightarrow Cl_2(g) + 2e^-]$ anode

$2AlCl_3 \longrightarrow 2Al(l) + 3Cl_2(g)$ overall

(b) In moderately concentrated solution of $AlCl_3(aq)$, containing $Al^{3+}(aq)$ and $Cl^-(aq)$ ions, water is more readily reduced than $Al^{3+}(aq)$ ions, and $Cl^-(aq)$ ions are more readily oxidized than water; water is reduced at the cathode and $Cl^-$ is oxidized at the anode:

$2H_2O(l) + 2e^- \longrightarrow H_2(g) + 2OH^-(aq)$ cathode

$2Cl^-(aq) \longrightarrow Cl_2(g) + 2e^-$ anode

$2H_2O(l) + 2Cl^-(aq) \longrightarrow H_2(g) + Cl_2(g) + 2OH^-(aq)$ overall

Note that in dilute aqueous solution, water is more readily oxidized than $Cl^-(aq)$, and the product at the anode is then $O_2(g)$.

4. (a) $MgBr_2(l)$ contains $Mg^{2+}$ and $Br^-$ ions. $Mg^{2+}$ is reduced at the cathode to $Mg(l)$ and $Br^-$ is oxidized to $Br_2(g)$ at the anode.

$Mg^{2+} + 2e^- \longrightarrow Mg(l)$ cathode

$2Br^- \longrightarrow Br_2(g) + 2e^-$ anode

$MgBr_2(l) \longrightarrow Mg(l) + Br_2(g)$ overall

(b) $Cu(NO_3)_2(aq)$ contains $Cu^{2+}(aq)$ and $NO_3^-(aq)$ ions. At the cathode, $Cu^{2+}(aq)$ is reduced to Cu(s), but $H_2O$ is more readily oxidized than $NO_3^-(aq)$ at the anode.

$2[Cu^{2+}(aq) + 2e^- \longrightarrow Cu(s)]$ cathode

$2H_2O(l) \longrightarrow O_2(g) + 4e^- + 4H^+(aq)$ anode

$2Cu^{2+}(aq) + 2H_2O(l) \longrightarrow 2Cu(s) + O_2(g) + 4H^+(aq)$ overall

(c) HI(aq) contains $H_3O^+(aq)$ and $I^-(aq)$ ions; $H_3O^+(aq)$ is reduced at the cathode and $I^-(aq)$ is oxidized at the anode.

| | |
|---|---|
| $2H_3O^+(aq) + 2e^- \longrightarrow H_2(g) + 2H_2O(l)$ | cathode |
| $2I^-(aq) \longrightarrow I_2(s) + 2e^-$ | anode |
| $2HI(aq) \longrightarrow H_2(g) + I_2(s)$ | overall |

In the presence of excess $I^-(aq)$, the $I_2(s)$ dissolves to form the brown $I_3^-(aq)$ ion.

$I_2(s) + I^-(aq) \longrightarrow I_3^-(aq)$

(d) $FeCl_3(l)$ contains $Fe^{3+}(l)$ and $Cl^-(l)$ ions; the former is reduced at the cathode and the latter is oxidized at the anode.

| | |
|---|---|
| $2[Fe^{3+}(l) + 3e^- \longrightarrow Fe(s)]$ | cathode |
| $3[2Cl^-(l) \longrightarrow Cl_2(g) + 2e^-]$ | anode |
| $2FeCl_3(l) \longrightarrow 2Fe(l) + 3Cl_2(g)$ | overall |

Note: In dilute solution, water is oxidized rather than $Cl^-(aq)$.

5. The electrodes are inert; $H_3O^+(aq)$ ions are reduced at the cathode and $Br^-(aq)$ ions are oxidized at the anode.

| | |
|---|---|
| $2H_3O^+(aq) + 2e^- \longrightarrow H_2(g) + 2H_2O(l)$ | cathode |
| $2Br^-(aq) \longrightarrow Br_2(l) + 2e^-$ | anode |
| $2HBr(aq) \longrightarrow H_2(g) + Br_2(l)$ | overall |

In the external circuit, electrons move from the anode to the cathode; in the solution, $H_3O^+(aq)$ ions move to the cathode where they are reduced and $Br^-(aq)$ ions move to the anode, where they are oxidized

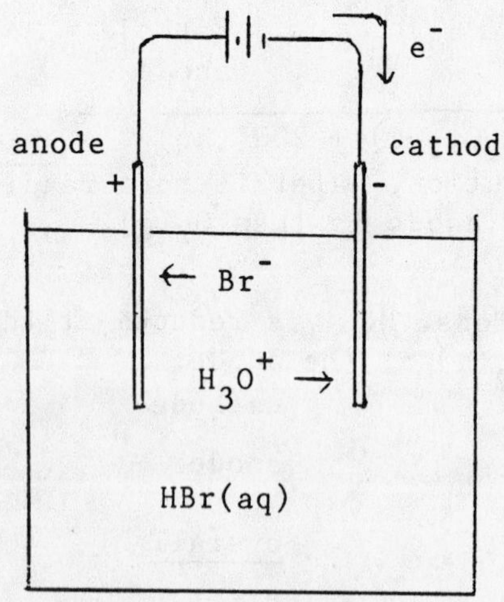


electrolytic cell

6. $CuSO_4(aq)$ contain $Cu^{2+}(aq)$ and $SO_4^{2-}(aq)$ ions. $Cu^{2+}(aq)$ ions move to the cathode where they are reduced to Cu(s) which is deposited on the Cu(s) electrode. $SO_4^{2-}(aq)$ ions carry the current to the anode where the Cu(s) electrode is oxidized rather than $SO_4^{2-}(aq)$ ions or water and Cu(s) dissolves to give $Cu^{2+}(aq)$ in solution.

$$Cu^{2+}(aq) + 2e^- \longrightarrow Cu(s) \qquad \text{reduction}$$

$$Cu(s) \longrightarrow Cu^{2+}(aq) + 2e^- \qquad \text{oxidation}$$

$$Cu^{2+}(aq) + Cu(s) \longrightarrow Cu(s) + Cu^{2+}(aq) \qquad \underline{\text{overall}}$$

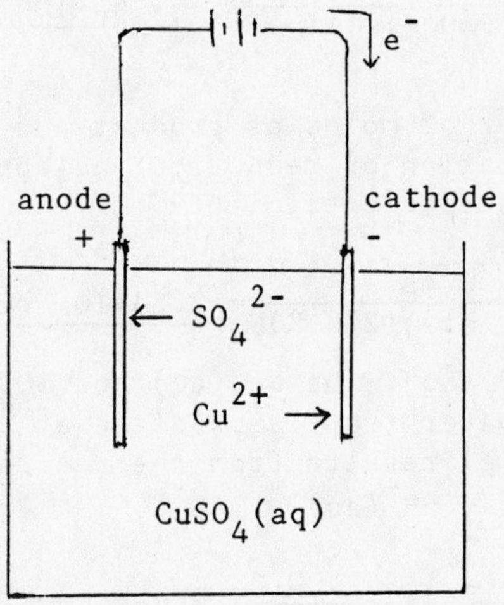


7. In each case, write the half-equation for the reduction and deduce from it the number of mol of electrons involved.

(a) $Cu^{2+}(aq) + 2e^- \longrightarrow Cu(s)$

$$\text{Coulombs} = (1 \text{ mol } Cu^{2+})\left(\frac{2 \text{ mol } e^-}{1 \text{ mol } Cu^{2+}}\right)\left(\frac{96500 \text{ C}}{1 \text{ mol } e^-}\right) = \underline{1.93\times10^5 \text{ C}}$$

(b) $Fe^{3+}(aq) + e^- \longrightarrow Fe^{2+}(aq)$

$$\text{Coulombs} = (1 \text{ mol } Fe^{3+})\left(\frac{1 \text{ mol } e^-}{1 \text{ mol } Fe^{3+}}\right)\left(\frac{96500 \text{ C}}{1 \text{ mol } e^-}\right) = \underline{9.65\times10^4 \text{C}}$$

(c) $MnO_4^-(aq) + 5e^- + 8H^+(aq) \longrightarrow Mn^{2+}(aq) + 4H_2O(l)$

$$\text{Coulombs} = (1 \text{ mol } MnO_4^-)\left(\frac{5 \text{ mol } e^-}{1 \text{ mol } MnO_4^-}\right)\left(\frac{96500 \text{ C}}{1 \text{ mol } e^-}\right) = \underline{4.83\times10^5 \text{ C}}$$

(d) $ClO_3^-(aq) + 6e^- + 6H^+(aq) \longrightarrow Cl^-(aq) + 3H_2O(l)$

$$\text{Coulombs} = (1 \text{ mol } ClO_3^-)\left(\frac{6 \text{ mol } e^-}{1 \text{ mol } ClO_3^-}\right)\left(\frac{96500 \text{ C}}{1 \text{ mol } e^-}\right) = \underline{5.79\times10^5 \text{ C}}$$

8. This problem is similar to Problem 7. First write the half-equation for the oxidation.

(a) $2H_2O(l) \longrightarrow O_2(g) + 4e^- + 4H^+(aq)$

$$\text{Coulombs} = (1 \text{ mol } H_2O)\left(\frac{4 \text{ mol } e^-}{2 \text{ mol } H_2O}\right)\left(\frac{96500 \text{ C}}{1 \text{ mol } e^-}\right) = \underline{1.93\times10^5 \text{ C}}$$

(b) $Cl_2 + 6H_2O \longrightarrow 2ClO_3^- + 10e^- + 12H^+$

$$\text{Coulombs} = (1\text{ mol } Cl_2)\left(\frac{10\text{ mol } e^-}{1\text{ mol } Cl_2}\right)\left(\frac{96\,500\text{ C}}{1\text{ mol } e^-}\right) = \underline{9.65\text{x}10^5\text{ C}}$$

(c) $Pb + 2H_2O \longrightarrow PbO_2 + 4e^- + 4H^+$

$$\text{Coulombs} = (1\text{ mol Pb})\left(\frac{4\text{ mol } e^-}{1\text{ mol Pb}}\right)\left(\frac{96\,500\text{ C}}{1\text{ mol } e^-}\right) = \underline{3.86\text{x}10^5\text{ C}}$$

(d) $2FeO + H_2O \longrightarrow Fe_2O_3 + 2e^- + 2H^+$

$$\text{Coulombs} = (1\text{ mol FeO})\left(\frac{2\text{ mol } e^-}{2\text{ mol FeO}}\right)\left(\frac{96\,500\text{ C}}{1\text{ mol } e^-}\right) = \underline{9.65\text{x}10^4\text{ C}}$$

9. In each case we calculate the number of moles of product and use the balanced half-reaction for oxidation or reduction, as appropriate, to calculate the number of coulombs of electricity:

(a)

$$n_{O_2} = \frac{PV}{RT} = \frac{(1\text{ atm})(50.0\text{ mL})\left(\frac{1\text{ L}}{1000\text{ mL}}\right)}{(0.0821\text{ atm L mol}^{-1}\text{ K}^{-1})(273\text{ K})} = \underline{2.23\text{x}10^{-3}\text{ mol}}$$

Although in a solution of $Na_2SO_4(aq)$, the $Na^+(aq)$ and $SO_4^{2-}(aq)$ ions carry the current, it is the water that is oxidized at the anode and reduced at the cathode; the $O_2(g)$ results from the reaction:

$$2H_2O(l) \longrightarrow O_2(g) + 4e^- + 4H^+(aq)$$

$$\text{Current} = (2.23\text{x}10^{-3}\text{ mol } O_2)\left(\frac{4\text{ mol } e^-}{1\text{ mol } O_2}\right)\left(\frac{96\,500\text{ C}}{1\text{ mol } e^-}\right) = \underline{861\text{ C}}$$

(b) The $Al^{3+}$ ions in $Al_2O_3$ are reduced to Al:

$$Al^{3+} + 3e^- \longrightarrow Al$$

$$\text{Current} = (50.0\text{ kg Al})\left(\frac{1000\text{ g}}{1\text{ kg}}\right)\left(\frac{1\text{ mol Al}}{26.98\text{ g Al}}\right)\left(\frac{3\text{ mol } e^-}{1\text{ mol Al}}\right)\left(\frac{96\,500\text{ C}}{1\text{ mol } e^-}\right)$$

$$= \underline{5.37\text{x}10^8\text{ C}}$$

(c)

$$Ca^{2+} + 2e^- \longrightarrow Ca$$

$$\text{Current} = (20.0\text{ g Ca})\left(\frac{1\text{ mol Ca}}{40.08\text{ g Ca}}\right)\left(\frac{2\text{ mol } e^-}{1\text{ mol Ca}}\right)\left(\frac{96\,500\text{ C}}{1\text{ mol } e^-}\right) = \underline{9.63\text{x}10^4\text{ C}}$$

(d) $AgNO_3(aq)$ is dissociated into $Ag^+(aq)$ and $NO_3^-(aq)$ ions:

$$Ag^+ + e^- \longrightarrow Ag$$

$$\text{Current} = (5.00\text{ g Ag})\left(\frac{1\text{ mol Ag}}{107.9\text{ g Ag}}\right)\left(\frac{1\text{ mol } e^-}{1\text{ mol Ag}}\right)\left(\frac{96\,500\text{ C}}{1\text{ mol } e^-}\right) = \underline{4.47\text{x}10^3\text{ C}}$$

10. $Ni^{2+}(aq)$ ions are reduced to Ni(s):

$$Ni^{2+} + 2e^- \longrightarrow Ni(s)$$

$$\text{Coulombs} = (1.00\text{ kg Ni})\left(\frac{1000\text{ g}}{1\text{ kg}}\right)\left(\frac{1\text{ mol Ni}}{58.70\text{ g Ni}}\right)\left(\frac{2\text{ mol } e^-}{1\text{ mol Ni}}\right)\left(\frac{96\,500\text{ C}}{1\text{ mol } e^-}\right)$$

$$= 3.288\text{x}10^6\text{C}$$

$$\text{Time} = (3.288\text{x}10^6\text{ C})\left(\frac{1\text{ As}}{1\text{ C}}\right)\left(\frac{1}{2.00\text{ A}}\right)\left(\frac{1\text{ hr}}{3600\text{ s}}\right)\left(\frac{1\text{ day}}{24\text{ hr}}\right) = \underline{19.0\text{ days}}$$

11. Water is more easily reduced than $Na^+(aq)$ and for a concentrated solution, $Cl^-(aq)$ is oxidized at the anode.

$$2H_2O + 2e^- \longrightarrow 2OH^- + H_2 \qquad \text{cathode}$$

$$2Cl^- \longrightarrow Cl_2 + 2e^- \qquad \text{anode}$$

$$\text{Mol of electrons} = (3.00\ \text{h})\left(\frac{3600\ \text{s}}{1\ \text{h}}\right)(0.200\ \text{A})\left(\frac{1\ \text{C}}{1\ \text{As}}\right)\left(\frac{1\ \text{mol e}^-}{96\ 500\ \text{C}}\right)$$

$$= \underline{0.0224\ \text{mol e}^-}$$

$$\text{Mass of NaOH} = (0.0224\ \text{mol e}^-)\left(\frac{2\ \text{mol NaOH}}{2\ \text{mol e}^-}\right)\left(\frac{40.00\ \text{g NaOH}}{1\ \text{mol NaOH}}\right) = \underline{0.896\ \text{g}}$$

$$\text{Mass of } Cl_2 = (0.0224\ \text{mol e}^-)\left(\frac{1\ \text{mol } Cl_2}{2\ \text{mol e}^-}\right)\left(\frac{70.90\ \text{g } Cl_2}{1\ \text{mol } Cl_2}\right) = \underline{0.794\ \text{g}}$$

12. The $Na^+(aq)$ and $SO_4^{2-}(aq)$ ions from the electrolyte $Na_2SO_4(aq)$ serve only to carry the current through the solution; water is reduced at the cathode and is oxidized at the anode.

$$2H_2O + 2e^- \longrightarrow 2OH^- + H_2 \qquad \text{cathode}$$

$$2H_2O \longrightarrow O_2 + 4e^- + 4H^+ \qquad \text{anode}$$

$$\text{Mol of electrons} = (4.00\ \text{A})(30\ \text{min})\left(\frac{60\ \text{s}}{1\ \text{min}}\right)\left(\frac{1\ \text{C}}{1\ \text{As}}\right)\left(\frac{1\ \text{mol e}^-}{96\ 500\ \text{C}}\right) - \underline{0.0746\ \text{mol}}$$

The volume of gas at 27°C and a pressure of (740-26.7) = 713 torr (allowing for the vapor pressure of water) is calculated by calculating the mol of gas produced from the mol of electrons used.

$$V_{H_2} = \frac{nRT}{P} = \frac{(0.0746\ \text{mol e}^-)\left(\frac{1\ \text{mol H}}{2\ \text{mol e}^-}\right)(0.0821\ \text{atm L mol}^{-1}\ \text{K}^{-1})(300\ \text{K})}{(713\ \text{torr})\left(\frac{1\ \text{atm}}{760\ \text{torr}}\right)}$$

$$= \underline{0.979\ \text{L}}\ H_2(g)$$

$$\text{and } V_{O_2} = (0.979\ \text{L } H_2)\left(\frac{1\ \text{mol } O_2}{2\ \text{mol } H_2}\right) = \underline{0.490\ \text{L}}\ O_2(g)$$

13. $Ag^+(aq)$ is reduced to $Ag(s)$ according to $Ag^+(aq) + e^- \longrightarrow Ag(s)$

The mass of silver is converted to moles of silver, and then the amount of current required and the time required, can be found as follows:

$$\text{Current required} = (16.0\ \text{g Ag})\left(\frac{1\ \text{mol Ag}}{107.9\ \text{g Ag}}\right)\left(\frac{1\ \text{mol e}^-}{1\ \text{mol Ag}}\right)\left(\frac{96500\ \text{C}}{1\ \text{mol e}^-}\right)$$

$$= \underline{1.431\text{x}10^4\ \text{C}}$$

$$\text{Time required} = (1.431\text{x}10^4\ \text{C})\left(\frac{1\ \text{As}}{1\ \text{C}}\right)\left(\frac{1}{6.00\ \text{A}}\right)\left(\frac{1\ \text{min}}{60\ \text{s}}\right) = \underline{39.7\ \text{min}}$$

14. $$Ag(aq) + e^- \longrightarrow Ag(s)$$

Calculate the mass of silver deposited from the volume of silver and its density, and hence the moles of Ag.

$$\text{mol Ag} = 2(24\ \text{cm})(12\ \text{cm})(0.020\ \text{mm})\left(\frac{1\ \text{cm}}{10\ \text{mm}}\right)\left(\frac{10.54\ \text{g}}{1\ \text{cm}^3}\right)\left(\frac{1\ \text{mol Ag}}{107.9\ \text{g Ag}}\right)$$

$= \underline{0.113 \text{ mol}}$ (remember the tray has two sides)

$$\text{Time} = (0.113 \text{ mol Ag})\left(\frac{1 \text{ mol } e^-}{1 \text{ mol Ag}}\right)\left(\frac{96\ 500 \text{ C}}{1 \text{ mol } e^-}\right)\left(\frac{1 \text{ As}}{1 \text{ C}}\right)\left(\frac{1}{7.65 \text{ A}}\right)\left(\frac{1 \text{ min}}{60 \text{ s}}\right)$$

$$= \underline{24 \text{ min}}$$

15. This problem is similar to Problem 14. $Cr^{3+}$ in $Cr_2(SO_4)_3(aq)$ is reduced to chromium metal

$$Cr^{3+} + 3e^- \longrightarrow Cr(s)$$

$$\text{mol Cr} = (0.10 \text{ m}^2)\left(\frac{100 \text{ cm}}{1 \text{ m}}\right)^2 (0.10 \text{ mm})\left(\frac{1 \text{ cm}}{10 \text{ mm}}\right)\left(\frac{7.1 \text{ g}}{1 \text{ cm}^3}\right)\left(\frac{1 \text{ mol Cr}}{52.00 \text{ g Cr}}\right)$$

$$= \underline{1.365 \text{ mol}}$$

$$\text{coulombs} = (1.365 \text{ mol Cr})\left(\frac{3 \text{ mol } e^-}{1 \text{ mol Cr}}\right)\left(\frac{96\ 500 \text{ C}}{1 \text{ mol } e^-}\right) = 3.952\text{x}10^5 \text{ C}$$

$$\text{Time} = \left(\frac{3.952\text{x}10^5 \text{ C}}{1.50 \text{ A}}\right)\left(\frac{1 \text{ As}}{1 \text{ C}}\right)\left(\frac{1 \text{ hr}}{3600 \text{ s}}\right)\left(\frac{1 \text{ day}}{24 \text{ hr}}\right) = \underline{3.05 \text{ days}}$$

16. The reductions are: $Cu^{2+} + 2e^- \longrightarrow Cu$, and $Ni^{2+} + 2e^- \longrightarrow Ni$

$$\text{Mass of Ni} = (10.0 \text{ g Cu})\left(\frac{1 \text{ mol Cu}}{63.55 \text{ g Cu}}\right)\left(\frac{2 \text{ mol } e^-}{1 \text{ mol Cu}}\right)\left(\frac{1 \text{ mol Ni}}{2 \text{ mol } e^-}\right)\left(\frac{58.70 \text{ g Ni}}{1 \text{ mol Ni}}\right)$$

$$= \underline{9.24 \text{ g Ni}}$$

17. In this reaction, $I^-(aq)$ is oxidized to $I_2(s)$ according to:

$$2I^-(aq) \longrightarrow I_2(s) + 2e^-$$

$$\text{Moles of electrons required} = (7.2428 \text{ g } I_2)\left(\frac{1 \text{ mol } I_2}{253.8 \text{ g } I_2}\right)\left(\frac{2 \text{ mol } e^-}{1 \text{ mol } I_2}\right)$$

$$= \underline{0.05707 \text{ mol } e^-} = \underline{0.05707 \text{ F}}$$

i.e., $0.05707 \text{ F} = (1.487 \text{ A})(3703.0 \text{ s})\left(\frac{1 \text{ C}}{1 \text{ As}}\right)$ ; $1 \text{ F} = \underline{9.648 \text{ x } 10^4 \text{ C}}$

The accepted value is 9.648 46 C $\text{mol}^{-1}$

18. $M^{2+}(aq) + 2e^- \longrightarrow M(s)$

$$\text{Mol of M} = (0.500 \text{ A})(2 \text{ hr})\left(\frac{3600 \text{ s}}{1 \text{ hr}}\right)\left(\frac{1 \text{ C}}{1 \text{ As}}\right)\left(\frac{1 \text{ mol M}}{2\text{mol } e^-}\right)\left(\frac{1 \text{ mol } e^-}{96\ 500 \text{ C}}\right)$$

$$= 0.01865 \text{ mol}$$

$$\text{Molar mass of M} = \left(\frac{1.98 \text{ g}}{0.01865 \text{ mol}}\right) = \underline{106 \text{ g mol}^{-1}}$$

Reference to the periodic table identifies the metal as palladium (which like Ni in the first transition metal series should form an $M^{2+}$ ion)

19. $Al^{3+} + 3e^- \longrightarrow Al(s)$

$$\text{Mass of Al} = (1.300\text{x}10^5 \text{ A})(60 \text{ s})\left(\frac{1 \text{ C}}{1 \text{ As}}\right)\left(\frac{1 \text{ mol } e^-}{96\ 500 \text{ C}}\right)\left(\frac{1 \text{ mol Al}}{2 \text{ mol } e^-}\right)\left(\frac{26.98 \text{ g Al}}{1 \text{ mol Al}}\right)$$

$$= \underline{726.9 \text{ g}}$$

For 100 such cells, per day, the mass of Al is given by:

$$(726.9\ g)(\frac{1\ kg}{1000\ g})(100)(\frac{1\ day}{1\ min})(\frac{24\ hr}{1\ day})(\frac{60\ min}{1\ hr}) = \underline{1.047x10^5\ kg\ Al}$$

For the oxidation of the carbon electrodes to $CO_2(g)$, we have:

$$C + 2O^{2-} \longrightarrow CO_2 + 4e^-$$

$$\text{Mass of C consumed} = (1.047x10^5\ kg\ Al)(\frac{1\ mol\ Al}{26.98\ g\ Al})(\frac{12.01\ g\ C}{1\ mol\ C})(\frac{3\ F}{4\ F})$$
$$= \underline{3.495x10^4\ kg\ C}$$

20. The electrode reactions are.

$$AgCl(s) + e^- \longrightarrow Ag(s) + Cl^-(aq) \quad \text{cathode}$$
$$Cu(s) \longrightarrow Cu^{2+}(aq) + 2e^- \quad \text{anode}$$

$$\text{Moles of electrons} = (0.700\ A)(1.600\ h)(\frac{3600\ s}{1\ h})(\frac{1\ mol\ e^-}{96\ 500\ C}) = \underline{0.0418\ mol}$$

$$g\ Ag = (0.0418\ mol\ e^-)(\frac{1\ mol\ Ag}{1\ F})(\frac{107.9\ g\ Ag}{1\ mol\ Ag}) = \underline{4.51\ g}$$

$$g\ Cu = (0.0418\ mol\ e^-)(\frac{1\ mol\ Cu}{2\ F})(\frac{63.55\ g\ Cu}{1\ mol\ Cu}) = \underline{1.33\ g}$$

21.
$$Fe^{3+}(aq) + 3e^- \longrightarrow Fe(s)$$
$$Cu^{2+}(aq) + 2e^- \longrightarrow Cu(s)$$

We do not need to know the current because from the mass of Fe deposited we can calculate moles Fe, and hence moles of electrons used:

$$\text{moles electrons} = (1.030\ g\ Fe)(\frac{1\ mol\ Fe}{55.85\ g\ Fe})(\frac{3\ mol\ e^-}{1\ mol\ Fe}) = \underline{0.05533\ mol}$$

$$g\ Cu = (0.05533\ mol\ e^-)(\frac{1\ mol\ Cu}{2\ mol\ e^-})(\frac{63.55\ g\ Cu}{1\ mol\ Cu}) = \underline{1.758\ g\ Cu}$$

22.
$$Zn^{2+}(aq) + 2e^- \longrightarrow Zn(s)$$

$$\text{moles electrons} = (1.000\ g\ Zn)(\frac{1\ mol\ Zn}{65.38\ g\ Zn})(\frac{2\ mol\ e^-}{1\ mol\ Zn}) = \underline{0.03059\ mol}$$

$$t = (0.3059\ mol\ e^-)(\frac{96\ 500\ C}{1\ F})(\frac{1\ As}{1\ C})(\frac{1}{8.000\ A})(\frac{1\ min}{60\ s}) = \underline{6.150\ min}$$

23. We need to calculate the moles of $OH^-(aq)$ produced at the cathode as the result of the reaction.

$$2H_2O(l) + 2e^- \longrightarrow H_2(g) + 2OH^-(aq)$$

$$mol\ OH^- = (2.500\ A)(1.500\ h)(\frac{3600\ s}{1\ h})(\frac{1\ mol\ e^-}{96\ 500\ C})(\frac{2\ mol\ OH^-}{2\ mol\ e^-}) = 0.1400\ mol$$

$$[OH^-] = 0.1400\ mol\ L^{-1}\ ;\ pH = 14.00 + \log\ [OH^-] = \underline{13.15}$$

24. In the electrolysis of KI(aq), $I^-(aq)$ ions are oxidized to iodine at the anode and water is reduced to hydrogen at the cathode:

$$2I^-(aq) \longrightarrow I_2(s) + 2e^- \quad \text{anode}$$
$$2H_2O(l) + 2e^- \longrightarrow H_2(g) + 2OH^-(aq) \quad \text{cathode}$$

The brown color occurs at the anode and is due to the reaction of the $I_2(s)$ produced reacting with excess $I^-(aq)$ to give the brown $I_3^-(aq)$ ion.

$$I_2(s) + I^-(aq) \longrightarrow \underset{\text{brown}}{I_3^-(aq)}$$

25. (a) At the cathode of the first cell. $Cu^{2+}(aq) + 2e^- \longrightarrow Cu(s)$

$$\text{moles of electrons} = (0.106 \text{ g Cu})\left(\frac{1 \text{ mol Cu}}{63.55 \text{ g Cu}}\right)\left(\frac{2 \text{ mol } e^-}{1 \text{ mol Cu}}\right)$$
$$= \underline{3.336\text{x}10^{-3} \text{ mol}}$$

$$\text{Current} = (3.336\text{x}10^{-3} \text{ mol } e^-)\left(\frac{96\ 500 \text{ C}}{1 \text{ mol } e^-}\right)\left(\frac{1 \text{ As}}{1 \text{ C}}\right)\left(\frac{1}{1.00 \text{ h}}\right)\left(\frac{1 \text{ h}}{3600 \text{ s}}\right)\left(\frac{1000 \text{ mA}}{1 \text{ A}}\right)$$
$$= \underline{89.4 \text{ mA}}$$

(b) At the cathode of the second cell. $Ag^+(aq) + e^- \longrightarrow Ag(s)$

$$\text{Mass Ag} = (3.336\text{x}10^{-3} \text{ mol } e^-)\left(\frac{1 \text{ mol Ag}}{1 \text{ mol } e^-}\right)\left(\frac{107.9 \text{ g Ag}}{1 \text{ mol Ag}}\right) = \underline{0.360 \text{ g}}$$

(c) In the third cell, sulfuric acid is electrolyzed. $H_3O^+(aq)$ from the solution is reduced at the cathode, and water rather than $SO_4^{2-}(aq)$ is oxidized at the anode:

$$2H_3O^+(aq) + 2e^- \longrightarrow H_2(g) + 2H_2O(l)$$
$$2H_2O(l) \longrightarrow O_2(g) + 4e^- + 4H^+(aq)$$

$$\text{Moles of } H_2 = (3.336\text{x}10^{-3} \text{ mol } e^-)\left(\frac{1 \text{ mol } H_2}{2 \text{ mol } e^-}\right) = \underline{1.668\text{x}10^{-3} \text{ mol}}$$

$$\text{Moles of } O_2 = (3.336\text{x}10^{-3} \text{ mol } e^-)\left(\frac{1 \text{ mol } O_2}{4 \text{ mol } e^-}\right) = \underline{0.834\text{x}10^{-3} \text{ mol}}$$

Thus, total moles of products = $2.502\text{x}10^{-3}$ mol, and the temperature is 293 K, and the pressure is (750 - 17.5) = 732 mm Hg, so

$$V = \frac{nRT}{P} = \frac{(2.502\text{x}10^{-3} \text{ mol})(0.0821 \text{ atm L mol}^{-1} \text{ K}^{-1})(293 \text{ K})}{(732 \text{ mm Hg})\left(\frac{1 \text{ atm}}{760 \text{ mm Hg}}\right)}$$
$$= \underline{0.0625 \text{ L}} \text{ or } \underline{62.5 \text{ mL}}$$

26. In the electrolysis of an aqueous solution of an alkali metal chloride, water is reduced at the cathode rather than $M^+(aq)$ ion. In concentrated solution $Cl^-(aq)$ is oxidized to $Cl_2(g)$.

$$2Cl^-(aq) \longrightarrow Cl_2(g) + 2e^-$$

(a) We can calculate the moles of $Cl^-$, and hence the moles of MCl, from the data for $Cl_2(g)$; ($P_{Cl_2}$ is 762.1-23.8) = 738.3 mm Hg):

$$n_{Cl_2} = \frac{PV}{RT} = \frac{(738.3 \text{ mm Hg})\left(\frac{1 \text{ atm}}{760 \text{ mm Hg}}\right)(114.2 \text{ mL})\left(\frac{1 \text{ L}}{1000 \text{ mL}}\right)}{(0.0821 \text{ atm L mol}^{-1} \text{ K}^{-1})(298 \text{ K})} = 0.00453 \text{ mol}$$

$$\text{moles of MCl} = (0.00453 \text{ mol } Cl_2)\left(\frac{2 \text{ mol MCl}}{1 \text{ mol } Cl_2}\right) = 0.00906 \text{ mol}$$

$$\text{molar mass MCl} = \left(\frac{0.5305 \text{ g}}{0.00906 \text{ mol}}\right) = \underline{58.6 \text{ g mol}^{-1}}$$

$$\text{Thus, molar mass of M} = (58.6\text{-}35.45) = \underline{23.1 \text{ g mol}^{-1}}$$

which identifies M as sodium (molar mass 22.99 g $mol^{-1}$).

(b) Coulombs = $(0.00453 \text{ mol } Cl_2)(\frac{2 \text{ mol e}^-}{1 \text{ mol } Cl_2})(\frac{96\,500 \text{ C}}{1 \text{ mol e}^-})$ = $8.74 \times 10^2$ C

$$\text{Time} = \frac{(8.74\times10^2 \text{ C})(\frac{1 \text{ As}}{1 \text{ C}})(\frac{1 \text{ min}}{60 \text{ s}})}{0.200 \text{ A}} = 72.8 \text{ min}$$

(c) anode: $2Cl^-(aq) \longrightarrow Cl_2(g) + 2e^-$ (oxidation)

cathode: $2H_2O(l) + 2e^- \longrightarrow 2OH^-(aq) + H_2(g)$ (reduction)

(d) The standard oxidation potential for $Cl^-(aq)$ is -1.36 V, while that for water is -1.23 V, since

$$Cl_2(g) + 2e^- \longrightarrow 2Cl^-(aq) \qquad E^o_{red} = +1.36 \text{ V}$$

$$O_2(g) + 4H^+(aq) + 4e^- \longrightarrow 2H_2O(l) \qquad E^o_{red} = +1.23 \text{ V}$$

Thus, under standard conditions ($[Cl^-]$ = 1 M), water should be preferentially oxidized to give $O_2(g)$, and, moreover, in a dilute solution ($[Cl^-]$ < 1 M], the tendency for $Cl^-(aq)$ to be oxidized is decreased (Nernst equation). However, because of the overvoltage effect, water is less easily oxidized than predicted. Thus, overall, oxygen is liberated in dilute solutions of $Cl^-(aq)$, and $Cl_2$ is liberated in more concentrated $Cl^-(aq)$ solutions, but the yield of $Cl_2$ is never quantitative - some water is always oxidized to $O_2$. The actual yield of $Cl_2$ depends on the conditions under which the electrolysis is carried out.

27. For each case, we calculate $E^o_{cell}$ for the half-reaction in which $Cr_2O_7^{2-}$ is reduced to $Cr^{3+}$ combined with the half-reaction in which the species in question is oxidized in aqueous solution, (see Table 17.1).

(a)

$$Cr_2O_7^{2-} + 6e^- + 14H^+ \longrightarrow 2Cr^{3+} + 7H_2O \qquad E^o_{red} = +1.33 \text{ V}$$

$$3[2F^- \longrightarrow F_2(g) + 2e^-] \qquad E^o_{ox} = -2.87 \text{ V}$$

$$Cr_2O_7^{2-} + 14H^+ + 6F^- \longrightarrow 2Cr^{3+} + 3F_2 + 7H_2O \qquad E^o_{cell} = -1.54 \text{ V}$$

$E^o_{cell}$ is negative for the reaction as written, and we conclude that the reverse reaction is the spontaneous reaction, so that $F^-(aq)$ will not be oxidized to $F_2(g)$ by $Cr_2O_7^{2-}(aq)$ in aqueous solution.

In general, the result is more easily obtained by considering the positions of each of the half-reactions written as reductions, as in the table of standard reduction potentials. The reactions at the top of the table have a greater tendency to proceed to the left than those at the bottom - only the half-reactions with $E^o_{red}$ values above that for the reduction of $Cr_2O_7^{2-}(aq)$ to $Cr^{3+}(aq)$ will proceed.

(b) $Cl^-(aq)$ will not be oxidized by $Cr_2O_7^{2-}(aq)$.

Of the remainder of the examples, all will be oxidized by $Cr_2O_7^{2-}(aq)$, except for $Mn^{2+}(aq)$. c), d), e), g), and h) are oxidized.

28. See Problem 27 for the method.

(a) The half-reactions in the table of standard reduction potentials are,

$$O_2(g) + 2H^+ + 2e^- \longrightarrow H_2O_2(aq) \qquad E^o_{red} = +0.68 \text{ V}$$

$$Cu^{2+} + 2e^- \longrightarrow Cu(s) \qquad E^o_{red} = +0.34 \text{ V}$$

To obtain the balanced equation we have to reverse the first equation and add it to the second equation.

$$H_2O_2(aq) \longrightarrow O_2(g) + 2H^+ + 2e^- \qquad E^o_{ox} = -E^o_{red} = -0.68\ V$$

$$Cu^{2+} + 2e^- \longrightarrow Cu(s) \qquad E^o_{red} = +0.34\ V$$

$$H_2O_2 + Cu^{2+} \longrightarrow O_2 + 2H^+ + Cu \qquad E^o_{cell} = \underline{-0.34\ V}$$

Since $E^o_{cell}$ is negative, the reaction does not proceed as written.

(b)

$$Ag^+ + e^- \longrightarrow Ag \qquad E^o_{red} = +0.80\ V$$

$$Fe^{2+} \longrightarrow Fe^{3+} + e^- \qquad E^o_{ox} = -0.77\ V$$

$$Ag^+ + Fe^{2+} \longrightarrow Ag + Fe^{3+} \qquad E^o_{cell} = \underline{+0.03\ V}$$

The reaction should proceed as written.

(c)

$$3[2I^- \longrightarrow I_2 + 2e^-] \qquad E^o_{ox} = -0.54\ V$$

$$2[NO_3^- + 3e^- + 4H^+ \longrightarrow NO + 2H_2O] \qquad E^o_{red} = +0.97\ V$$

$$6I^- + 4NO_3^- + 8H^+ \longrightarrow 3I_2 + 2NO + 4H_2O \qquad E^o_{cell} = \underline{+0.43\ V}$$

and the reaction proceeds as written

29. The method is the same as in Problem 28.

$$2[\ Cr_2O_7^{2-} + 6e^- + 14H^+ \longrightarrow 2Cr^{3+} + 7H_2O] \qquad E^o_{red} = +1.33\ V$$

$$3[\ 2H_2O \longrightarrow O_2 + 4e^- + 4H^+\ ] \qquad E^o_{ox} = -1.23\ V$$

$$2Cr_2O_7^{2-} + 16H^+ \longrightarrow 4Cr^{3+} + 3O_2 + 8H_2O \qquad E^o_{cell} = \underline{+0.10\ V}$$

$E^o_{cell}$ is positive and the oxidation should occur. Because no reaction is observed, its rate must be very slow.

30. (a)

$$3[Mg \longrightarrow Mg^{2+} + 2e^-\ ] \qquad E^o_{ox} = +2.36\ V$$

$$2[Cr^{3+} + 3e^- \longrightarrow Cr\ ] \qquad E^o_{red} = -0.74\ V$$

$$3Mg + 2Cr^{3+} \longrightarrow 3Mg^{2+} + 2Cr \qquad E^o_{cell} = \underline{+1.62\ V}$$

The reaction is spontaneous.

(b)

$$I_2 + 2e^- \longrightarrow 2I^- \qquad E^o_{red} = +0.54\ V$$

$$H_2 \longrightarrow 2H^+ + 2e^- \qquad E^o_{ox} = 0\ V$$

$$H_2(g) + I_2(aq) \longrightarrow 2H^+(aq) + 2I^-(aq) \qquad E^o_{cell} = \underline{+0.54\ V}$$

The reaction is spontaneous.

(c)

$$Cu \longrightarrow Cu^{2+} + 2e^- \qquad E^o_{ox} = -0.34\ V$$

$$2NO_3^- + 2e^- + 4H^+ \longrightarrow N_2O_4^* + 2H_2O \qquad E^o_{red} = +0.80\ V$$

$$Cu + 2NO_3^- + 4H^+ \longrightarrow Cu^{2+} + N_2O_4 + 2H_2O \qquad E^o_{cell} = \underline{+0.46\ V}$$

*$N_2O_4$ is used here rather than $NO_2$, which does not appear in Table 17.1. The reaction is spontaneous.

31. A metal displaces another from solution if it is a sufficiently strong reducing agent to reduce the other's ions. Provided a half-reaction is below another in Table 17.1, the lower reaction occurs as written as a reduction, while the reaction above occurs in the opposite direction, as an oxidation. For example, Al displaces $Zn^{2+}$, $Fe^{2+}$, and $Cu^{2+}$ from solution, but it does not displace $Mg^{2+}$ from solution. Thus the order is: Mg > Al > Zn > Fe > Cu

32. When $Fe^{2+}(aq)$ behaves as a reducing agent, it is oxidized to $Fe^{3+}(aq)$, in the half-reaction

$$Fe^{2+}(aq) \longrightarrow Fe^{3+}(aq) + e^- \quad E^o_{ox} = -0.77\ V$$

Thus, it can combine with any reduction half-reaction below it in Table 17.1 to give a positive $E^o_{cell}$.

(a) $Ag^+(aq)$ ; (e) $Br_2(aq)$, and (f) $MnO_4^-(aq)$ will be reduced.

33. $MnO_2(s) + 4H^+(aq) + 2e^- \longrightarrow Mn^{2+} + 4H_2O$ ; $E^o_{red} = +1.61\ V$, and this half-reaction can be combined with any of the reactions above it in Table 17.1, written as oxidations, to give a positive $E^o_{cell}$. Thus, it will oxidize all of the reactants in question.

34. (a) The half-reaction for reduction of $Fe^{3+}$ to $Fe^{2+}$ lies below the half-reaction for the reduction of $I_2$ to $I^-$, in Table 17.1; thus, $Fe^{3+}(aq)$ oxidizes $I^-$ to $I_2$.

(b) The half-reaction $Ag^+ + e^- \longrightarrow Ag$, lies below $Cu^{2+} + 2e^- \longrightarrow Cu$, and $Cu^+ + e^- \longrightarrow Cu$; thus, $Ag^+$ oxidizes copper to $Cu^{2+}$.

(c) $Fe^{3+}$ has to behave as an oxidizing agent, so the only possibility is for it to oxidize $Br^-$ to $Br_2$, but this reaction will not occur because the half-reaction for reduction of $Br_2$ to $Br^-$ lies below the half-reaction for the reduction of $Fe^{3+}$ to $Fe^{2+}$ in Table 17.1.

(d) No reaction will occur.

(e) As we saw in part (c), the half-equation for the reduction of $Br_2$ to $Br^-$ lies below that for the reduction of $Fe^{3+}$ to $Fe^{2+}$, so $Br_2$ will oxidize $Fe^{2+}$ to $Fe^{3+}$.

35. (a) The half-reaction for the reduction of $MnO_4^-$ to $Mn^{2+}$ lies below that for the reduction of $Cr_2O_7^{2-}$ to $Cr^{3+}$; thus the reaction will not occur under standard conditions.

(b) Since the product from the reduction of $O_2$ is not specified it must be $H_2O$, so for the half-reaction $E^o_{red} = +\ 1.23\ V$, which is below $E^o_{red}$ for $Br_2 + 2e^- \longrightarrow 2Br^-$ ; thus, $O_2$ will oxidize $Br_2$ to $Br^-$.

$$O_2 + 4e^- + 4H^+ \longrightarrow 2H_2O \qquad E^o_{red} = +1.23\ V$$
$$2[2Br^- \longrightarrow Br_2 + 2e^-] \qquad E^o_{ox} = -1.09\ V$$
$$O_2 + 4Br^- + 4H^+ \longrightarrow 2H_2O + 2Br_2 \qquad E^o_{cell} = \underline{+0.14\ V}$$

(c) The reaction will not occur under standard conditions.

36. For the half-reaction $O_2 + 4e^- + 4H^+ \longrightarrow 2H_2O$, $E^o_{red} = +1.23\ V$, so that any metal with $E^o_{ox} > -1.23\ V$ will be oxidized by $O_2(g)$ under standard conditions: i.e., (a), (b), (d) and (e).

37. $E^o_{red}$ for $2H^+ + 2e^- \longrightarrow H_2(g)$ is 0 V, so that any of the metal ions with standard reduction potentials > 0, (below that for $H^+(aq)$), will be reduced by $H_2$ under standard conditions: i.e., (a) $Au^{3+}$ and (d) $Ag^+$.

38. (a)

$$2Br^- \longrightarrow Br_2 + 2e^- \qquad E^o_{ox} = -1.09\ V$$
$$Cl_2 + 2e^- \longrightarrow 2Cl^- \qquad E^o_{red} = +1.36\ V$$
$$2Br^- + Cl_2 \longrightarrow Br_2 + 2Cl^- \qquad E^o_{cell} = +0.27\ V$$

A spontaneous reaction.

(b)

$$3[MnO_4^- + 5e^- + 8H_3O^+ \longrightarrow Mn^{2+} + 12H_2O] \qquad E^o_{red} = +1.49\ V$$
$$5[Au \longrightarrow Au^{3+} + 3e^-] \qquad E^o_{ox} = -1.50\ V$$
$$3MnO_4^- + 5Au + 24H_3O^+ \longrightarrow 3Mn^{2+} + 5Au^{3+} + 36\ H_2O \qquad E^o_{cell} = -0.01\ V$$

Not a spontaneous reaction.

(c)

$$2[Cr^{3+} + e^- \longrightarrow Cr^{2+}] \qquad E^o_{red} = -0.41\ V$$
$$H_2 \longrightarrow 2H^+ + 2e^- \qquad E^o_{ox} = 0.00\ V$$
$$2Cr^{3+} + H_2 \longrightarrow 2Cr^{2+} + 2H^+ \qquad E^o_{cell} = -0.41\ V$$

Not a spontaneous reaction.

(d)

$$2[AgBr + e^- \longrightarrow Ag + Br^-] \qquad E^o_{red} = +0.10\ V$$
$$Sn \longrightarrow Sn^{2+} + 2e^- \qquad E^o_{ox} = +0.16\ V$$
$$2AgBr + Sn \longrightarrow 2Ag + Sn^{2+} + 2Br^- \qquad E^o_{cell} = +0.26\ V$$

A spontaneous reaction.

(e)

$$2[Cu^+ \longrightarrow Cu^{2+} + e^-] \qquad E^o_{ox} = -0.15\ V$$
$$Cu^{2+} + 2e^- \longrightarrow Cu \qquad E^o_{red} = +0.34\ V$$
$$2Cu^+ \longrightarrow Cu^{2+} + Cu \qquad E^o_{cell} = +0.19\ V$$

A spontaneous reaction.

39. (a) no reaction ($E^o_{cell} = -0.93$ V).

(b)

$$Mg \longrightarrow Mg^{2+} + 2e^- \qquad E^o_{ox} = +2.36\ V$$
$$Fe^{2+} + 2e^- \longrightarrow Fe \qquad E^o_{red} = -0.44\ V$$
$$Mg + 2Fe^{2+} \longrightarrow Mg^{2+} + Fe \qquad E^o_{cell} = +1.92\ V$$

(c) no reaction ($E^o_{cell} = -0.34$ V).

(d)

$$2[Fe^{3+} + e^- \longrightarrow Fe^{2+}] \qquad E^o_{red} = +0.77\ V$$
$$Fe — Fe^{2+} + 2e^- \qquad E^o_{ox} = +0.44\ V$$
$$2Fe^{3+} + Fe \longrightarrow 3Fe^{2+} \qquad E^o_{cell} = +1.21\ V$$

(e) no reaction ($E^o_{cell} = -0.12$ V).

40. (a) The standard silver chloride electrode is assigned a potential of 0.00 V, rather than +0.22 V relative to 0.00 V for the standard hydrogen electrode, so that on the new scale any half-cell potential is obtained by subtracting 0.22 V from the value relative to the standard hydrogen electrode.

(i) $Cl_2(g) + 2e^- \longrightarrow 2Cl^-(aq)$ $E^o_{red} = (+1.36 - 0.22) = +1.14$ V

(ii) $2H^+(aq) + 2e^- \longrightarrow H_2(g)$ $E^o_{red} = (0.00 - 0.22) = -0.22$ V

(b)

$$Cl_2 + 2e^- \longrightarrow 2Cl^- \qquad E^o_{red} = +1.14 \text{ V}$$
$$H_2 \longrightarrow 2H^+ + 2e^- \qquad E^o_{ox} = +0.22 \text{ V}$$
$$Cl_2 + H_2 \longrightarrow 2H^+ + 2Cl^- \qquad E_{cell} = \underline{+1.36 \text{ V}}$$

(c) Calculation of $E^o_{cell}$ using $E^o_{red}$ values relative to 0.00 V for the standard hydrogen electrode gives $E^o_{cell} = +1.36\text{V} + 0.00 \text{ V} = \underline{1.36 \text{ V}}$

Changing the half-cell that is assigned a voltage of zero has no effect on $E^o_{cell}$.

41. (a) $E^o_{red}$ must be below $E^o_{red}$ for $Sn^{2+} + 2e^- \longrightarrow Sn(s)$, but above $E^o_{red}$ for $2H^+ + 2e^- \longrightarrow H_2(g)$; the only suitable reagent is $\underline{Pb^{2+}(aq)}$

(b) $E^o_{red}$ must be above $E^o_{red}$ for $Cr^{3+} + e^- \longrightarrow Cr^{2+}$ but below $E^o_{red}$ for $Cr^{3+} + 3e^- \longrightarrow Cr$; the only suitable reagent is $\underline{Fe}(s)$

42. Using the data in Table 17.1, the possible combinations that give spontaneous reactions are those for which $E^o_{cell}$ is positive.

$$H_2(g) + 2OH^- \longrightarrow 2H_2O + 2e^- \qquad E^o_{ox} = +0.83 \text{ V}$$

can be combined with all of the other reduction half-equations:

$$H_2(g) + 2OH^- \longrightarrow 2H_2O + 2e^- \qquad E^o_{ox} = +0.83 \text{ V}$$
$$2[Fe(OH)_3(s) + e^- \longrightarrow Fe(OH)_2(s) + OH^-] \qquad E^o_{red} = -0.56 \text{ V}$$
$$H_2(g) + 2Fe(OH)_3 \longrightarrow 2Fe(OH)_2 + 2H_2O \qquad E^o_{cell} = \underline{+0.27 \text{ V}}$$

$$H_2(g) + 2OH^- \longrightarrow 2H_2O + 2e^- \qquad E^o_{ox} = +0.83 \text{ V}$$
$$2[O_2(g) + e^- \longrightarrow O_2^-] \qquad E^o_{red} = -0.56 \text{ V}$$
$$H_2(g) + 2O_2(g) + 2OH^- \longrightarrow 2O_2^- + 2H_2O \qquad E^o_{cell} = \underline{+0.27 \text{ V}}$$

$$2[H_2(g) + 2OH^- \longrightarrow 2H_2O + 2e^-] \qquad E^o_{ox} = +0.83 \text{ V}$$
$$O_2(g) + 2H_2O + 4e^- \longrightarrow 4OH^- \qquad E^o_{red} = +0.40 \text{ V}$$
$$2H_2(g) + O_2(g) \longrightarrow 2H_2O \qquad E^o_{cell} = \underline{+1.23 \text{ V}}$$

Also.

$$4[Fe(OH)_2(s) + OH^- \longrightarrow Fe(OH)_3(s) + e^-] \qquad E^o_{ox} = +0.56 \text{ V}$$
$$O_2(g) + 2H_2O + 4e^- \longrightarrow 4OH^- \qquad E^o_{red} = +0.40 \text{ V}$$
$$4Fe(OH)_2 + O_2 + 2H_2O \longrightarrow 4Fe(OH)_3 \qquad E^o_{cell} = \underline{+0.96 \text{ V}}$$

and

$$4[O_2^- \longrightarrow O_2(g) + e^-] \qquad E^o_{ox} = +0.56\text{V}$$
$$O_2(g) + 2H_2O + 4e^- \longrightarrow 4OH^- \qquad E^o_{red} = +0.40 \text{ V}$$
$$4O_2^- + 2H_2O \longrightarrow 3O_2(g) + 4OH^- \qquad E^o_{cell} = \underline{+0.96 \text{ V}}$$

43. In terms of their ease of oxidation, Table 17.1 places these metals in the order Ca > Mg > Zn > Cu, which can be confirmed by a number of experiments, for example:
1) Only Ca metal reacts with water at an observable rate at room temperature, while Ca and Mg are oxidized by steam.
2) Ca, Mg, and Zn are all oxidized by dilute HCl(aq), ($H_3O^+$).
3) All of the metals are oxidized by hot concentrated sulfuric acid.
4) $Cu^{2+}$(aq) is displaced from solution as Cu by all of the other metals.
5) $Zn^{2+}$(aq) is displaced from solution as Zn by Ca and Mg.

44(a) The bronze-like finish is due to the deposition of Cu(s)

$$Cu^{2+}(aq) + 2e^- \longrightarrow Cu(s) \qquad E^o_{red} = +0.34 \text{ V}$$
$$Pb(s) \longrightarrow Pb^{2+}(aq) + 2e^- \qquad E^o_{ox} = +0.13 \text{ V}$$
$$Pb(s) + Cu^{2+}(aq) \longrightarrow Cu(s) + Pb^{2+}(aq) \qquad E^o_{cell} = \underline{+0.47 \text{ V}}$$

(b) The possibility is that the Cu(s) finish is replaced by Ag(s)

$$2[Ag^+(aq) + e^- \longrightarrow Ag(s)\ ] \qquad E^o_{red} = +0.80 \text{ V}$$
$$Cu(s) \longrightarrow Cu^{2+}(aq) + 2e^- \qquad E^o_{ox} = -0.34 \text{ V}$$
$$2Ag^+(aq) + Cu(s) \longrightarrow 2Ag(s) + Cu^{2+}(aq) \qquad E^o_{cell} = \underline{+0.46 \text{ V}}$$

$E^o_{cell}$ is positive, so the reaction is spontaneous and the bronze finish will be replaced by a silver finish.

45. (a) An electrochemical cell is set up in which the more easily oxidized metal, zinc, becomes the anode and the iron in the steel becomes the cathode. The anode reaction is $Zn \longrightarrow Zn^{2+} + 2e^-$, and water is reduced at the cathode (the steel propellor). Thus, the iron does not dissolve and the propellor is protected from corrosion.

(b) A galvanized garbage can is made of iron coated with zinc. Although zinc is more easily oxidized than iron it forms a tough layer of $Zn(OH)_2 \cdot xZnCO_3$(s) which protects the iron; even when the surface layer is broken it continues to protect the iron since the zinc becomes the anode of a cell in which the iron is the cathode. Iron is protected in tin cans by a coating of tin protected by a tough layer of tin oxide. Iron is more easily oxidized than tin and when a can is opened the iron is exposed to the atmosphere and becomes the anode of a cell in which tin is the cathode, which leads to the rapid formation of rust.

(c) Although the reaction occurs in basic solution ($CO_3^{2-}(aq) + H_2O \longrightarrow HCO_3^- + OH^-$) we can assume that the standared reduction potentials of $Ag^+$(aq) and $Al^{3+}$(aq) are in the same relative order that they are in acidic solution.

$Al^{3+} + 3e^- \longrightarrow Al(s)$; $E^o = -1.66$ V: $Ag^+ + e^- \longrightarrow Ag(s)$; $E^o = +0.80$ V

Thus, the reaction of $(Ag^+)_2S^{2-}$(s) with Al is expected to be spontaneous:

$$3Ag_2S(s) + 2Al(s) \longrightarrow 3Ag(s) + Al_2S_3(aq)$$

$E^o_{cell} \simeq (0.80 + 1.66) = \underline{2.46 \text{ V}}$

46. The cell is $Zn(s)|Zn^{2+}(aq) || Ag^{+}(aq)|Ag(s)$ in which zinc is the anode and silver the cathode. Your cell diagram should show electrons moving externally from Zn to Ag, and $Ag^{+}$ ions moving to the cathode, and anions (unspecified) moving to the anode.

| | | | |
|---|---|---|---|
| anode | $Zn \longrightarrow Zn^{2+} + 2e^-$ | $E^o_{ox}$ | $= +0.76$ V |
| cathode | $2[Ag^{+} + e^- \longrightarrow Ag\quad]$ | $E^o_{red}$ | $= 0.80$ V |
| | $Zn + 2Ag^{+} \longrightarrow Zn^{2+} + 2Ag$ | $E^o_{cell}$ | $= +1.56$ V |

47. (a) The electrons move externally from the Zn anode to the Ni cathode.

(b)

| | | | |
|---|---|---|---|
| anode | $Zn \longrightarrow Zn^{2+} + 2e^-$ | $E^o_{ox}$ | $= +0.76$ V |
| cathode | $Ni^{2+} + 2e^- \longrightarrow Ni$ | $E^o_{red}$ | $= +0.25$ V |
| | $Zn + Ni^{2+} \longrightarrow Zn^{2+} + Ni$ | $E^o_{cell}$ | $= +1.01$ V |

48. (a)

| | | |
|---|---|---|
| $2[Al \longrightarrow Al^{3+} + 3e^-\quad]$ | $E^o_{ox}$ | $= +1.66$ V |
| $3[Cu^{2+} + 2e^- \longrightarrow Cu\quad]$ | $E^o_{red}$ | $= +0.34$ V |
| $2Al + 3Cu^{2+} \longrightarrow 2Al^{3+} + 3Cu$ | $E^o_{cell}$ | $= +2.00$ V |

(b)

| | | |
|---|---|---|
| $Pb \longrightarrow Pb^{2+} + 2e^-$ | $E^o_{ox}$ | $= +0.13$ V |
| $2[Ag^{+} + e^- \longrightarrow Ag\quad]$ | $E^o_{red}$ | $= +0.80$ V |
| $Pb + 2Ag^{+} \longrightarrow Pb^{2+} + 2Ag$ | $E^o_{cell}$ | $= +0.93$ V |

(c)

| | | |
|---|---|---|
| $2[Ag(s) \longrightarrow Ag^{+} + e^-\quad]$ | $E^o_{ox}$ | $= -0.80$ V |
| $Cl_2 + 2e^- \longrightarrow 2Cl^-$ | $E^o_{red}$ | $= +1.36$ V |
| $2Ag + Cl_2 \longrightarrow 2Ag^{+} + 2Cl^-$ | $E^o_{cell}$ | $= +0.56$ V |

49. (a) $Fe(s)|Fe^{2+}(aq)|| H^{+}(aq)|H_2(g), Pt(s)$

(b) electrons.

(c) Reduction occurs at the Pt electrode, which is the cathode.

(d) $2H^{+}(aq) + 2e^- \longrightarrow H_2(g)$

(e) The electrons originally come from the Fe(s) anode when it dissolves to form $Fe^{2+}(aq)$; they flow through the external circuit from the Fe(s) anode to the Pt(s) cathode, where they combine with $H^{+}(aq)$ ions to give $H_2(g)$.

50. (a) $Zn(s)|Zn^{2+}(aq)||Br^{-}(aq)|Br_2(g),Pt(s)$

(b) $Pb(s)|Pb^{2+}(aq)||Ag^{+}(aq)|Ag(s)$

(c) $Pt(s)|Cu^{+}(aq), Cu^{2+}(aq)||Fe^{3+}(aq),Fe^{2+}(aq)|Pt(s)$

In each case, electrons flow from the anode to the cathode in the external circuit.

51. In each case an oxidation reaction occurs at the anode and a reduction reaction occurs at the cathode, and $E^o_{cell}$ has to have a positive value;

the species that is oxidized will be higher in Table 17.1 than the species that is reduced; electrons move from anode to cathode.

(a) cathode $Cu^{2+} + 2e^- \longrightarrow Cu$ $E^o_{red} = +0.34$ V

anode $Pb \longrightarrow Pb^{2+} + 2e^-$ $E^o_{ox} = +0.13$ V

$Cu^{2+} + Pb \longrightarrow Cu + Pb^{2+}$ $E^o_{cell} = +0.47$ V

anode $Pb(s)|Pb^{2+}(1\ M, aq)||Cu^{2+}(1\ M, aq)|Cu(s)$ cathode

(b) cathode $Cl_2 + 2e^- \longrightarrow 2Cl^-$ $E^o_{red} = +1.36$ V

anode $2[Ag + Cl^- \longrightarrow AgCl + e^-]$ $E^o_{ox} = -0.22$ V

$2Ag + Cl_2 \longrightarrow 2AgCl$ $E^o_{cell} = +1.14$ V

anode $AgCl(s)|NaCl(1\ M, aq)||\ NaCl(1\ M, aq)|Cl_2(g), Pt(s)$ cathode

(c) cathode $2H^+ + 2e^- \longrightarrow H_2(g)$ $E^o_{red} = 0.00$ V

anode $2I^- \longrightarrow I_2 + 2e^-$ $E^o_{ox} = -0.54$ V

$2H^+ + 2I^- \longrightarrow H_2 + I_2$ $E^o_{cell} = -0.54$ V

The reaction will not proceed as written and since no $I_2$ is present the reverse reaction cannot occur

anode $Pt(s)|HI(1\ M, aq)||HI(1\ M, aq)|H_2(g), Pt$ cathode

52. (a) $MnO_4^- + 5e^- + 8H^+ \longrightarrow Mn^{2+} + 4H_2O$ cathode (reduction)

$Fe^{2+} \longrightarrow Fe^{3+} + e^-$ anode (oxidation)

The platinum electrode dipping into the $KMnO_4(aq)$ solution is the anode; the platinum electrode dipping into the $FeSO_4(aq)$ solution is the cathode.

(b) Electrons move in the external circuit from the anode to the cathode.

(c) $Fe^{2+}(aq)$ ions move towards the cathode; $MnO_4^-$ ions move towards the anode.

(d) $E^o_{cell} = E^o_{red}(MnO_4^-(aq)) + E^o_{ox}(Fe^{2+}(aq)) = (+1.49) + (-0.77) = +0.72$ V

53. In this cell we have the equilibria.

$Zn(s) \rightleftharpoons Zn^{2+}(aq) + 2e^-$ $E^o_{ox} = +0.76$ V, and

$Cu^{2+}(aq) + 2e^- \rightleftharpoons Cu(s)$ $E^o_{red} = +0.34$ V

(a) Adding 2 M $Zn^{2+}(aq)$ to the $Zn(s)|Zn^{2+}(aq, 1\ M)$ half-cell increases $[Zn^{2+}(aq)]$, which shifts the equilibrium to the left and decreases the potential (voltage) of the cell.

(b) Adding Zn(s) has no effect.

(c) Addition of a few drops of NaOH(aq) to the $Zn(s)|Zn^{2+}(aq, 1\ M)$ half-cell will produce s precipitate of $Zn(OH)_2(s)$ and hence decrease $[Zn^{2+}]$

$$Zn^{2+}(aq) + 2OH^-(aq) \rightleftharpoons Zn(OH)_2(s)$$

which results in a greater tendency for the oxidation of Zn(s) to proceed and increases the voltage.

(d) Adding a large amount of $NH_3(aq)$ to the $Cu(s)|Cu^{2+}(aq)$ half-cell results in a decrease in $[Cu^{2+}(aq)]$ because of the formation of the complex ion $[Cu(NH_3)_4{}^{2+}](aq)$

$$Cu^{2+}(aq) + 4NH_3(aq) \rightleftharpoons Cu(NH_3)_4{}^{2+}(aq)$$

which decreases $[Cu^{2+}(aq)]$, resulting in less tendency for this reduction to occur, which decreases the voltage of the cell.

54. (a)

$$\begin{array}{llcl} 2[Ag^+(aq) + e^- \longrightarrow Ag(s)] & E^o_{red} & = & +0.80\ V \\ Cu(s) \longrightarrow Cu^{2+}(aq) + 2e^- & E^o_{ox} & = & -0.34\ V \\ \hline 2Ag^+(aq) + Cu(s) \longrightarrow 2Ag(s) + Cu^{2+}(aq) & E^o_{cell} & = & \underline{+0.46\ V} \end{array}$$

(b) We write the Nernst equation in the form

$$E = E^o - \frac{RT}{nF} \ln Q = E^o - \frac{0.0257}{2} \ln \frac{[Cu^{2+}]}{[Ag^+]^2}$$

$$= 0.46\ V - \frac{0.0257}{2} \ln \frac{(2.0)}{(0.05)^2} = (0.46-0.09)\ V = \underline{0.37\ V}$$

55. $I^-(aq)$ is oxidized in the reaction and an increase in $[I^-(aq)]$ will increase its tendency to be oxidized, i.e., the contribution of $E_{ox}$ to $E_{cell}$ increases, i.e., $E_{cell}$ increases.

56. For the reaction in question $Cl_2(aq)$ is oxidized to $HOCl(aq)$ and at the same time reduced to $Cl^-(aq)$:

$$\begin{array}{llcl} Cl_2(aq) + 2H_2O \longrightarrow 2HOCl(aq) + 2H^+(aq) + 2e^- & E^o_{ox} & = & -1.63\ V \\ Cl_2(aq) + 2e^- \longrightarrow 2Cl^-(aq) & E^o_{red} & = & +1.36\ V \\ \hline 2Cl_2(aq) + 2H_2O \longrightarrow 2HOCl(aq) + 2H^+(aq) + 2Cl^-(aq) & E^o_{cell} & = & \underline{-0.27\ V} \\ \underline{\text{or}}\ Cl_2(aq) + H_2O \longrightarrow HOCl(aq) + H^+(aq) + Cl^-(aq) & E^o_{cell} & = & \underline{-0.27\ V} \end{array}$$

For equilibrium, we have from the Nernst equation

$$E_{cell} = E^o_{cell} - \frac{RT}{nF} \ln Q = 0, \text{ and } Q = K; \quad \ln K = \frac{nFE^o}{RT} = \frac{nE^o}{0.0257}$$

For the required equilibrium, n = 1, since although we used $2e^-$ to balance the half-reactions, we eventually divided by 2, hence;

$$\ln K = \frac{E^o}{0.0257} = \frac{-0.27}{0.0257} = -10.5; \ K = e^{-10.5} \approx \underline{2.75 \times 10^{-5}\ mol\ L^{-1}}$$

57. We need first to calculate the standard cell potential.

$$\begin{array}{llcl} Sn^{2+} + 2e^- \longrightarrow Sn(s) & E^o_{red} & = & -0.16\ V \\ Pb(s) \longrightarrow Pb^{2+} + 2e^- & E^o_{ox} & = & +0.13\ V \\ \hline Sn^{2+} + Pb(s) \longrightarrow Sn(s) + Pb^{2+} & E^o_{cell} & = & \underline{-0.03\ V} \end{array}$$

and under non-standard conditions, with $[Sn^{2+}]$ = 1.00 M, $E_{cell}$ = 0.22 V, so we can use the Nernst equation to write

$$E = 0.22\ V = E^o - \frac{RT}{nF} \ln Q = -0.03 - \frac{0.0257}{2} \ln \frac{[Pb^{2+}]}{[Sn^{2+}]}$$

$$0.0129 \ln \frac{[Pb^{2+}]}{1.00} = -0.25 \ ; \ \ln [Pb^{2+}] = -19.4;$$

Hence, $[Pb^{2+}] = e^{-19.4} = \underline{3.8\times10^{-9}\ M}$

Since the concentration of $SO_4^{2-}$(aq) in the anode compartment is 1.00 M,

$$K_{sp}(PbSO_4) = [Pb^{2+}][SO_4^{2-}] = \underline{3.8\times10^{-9}\ mol^2\ L^{-2}}$$

58. Since $Cu^{2+}$ lies below $Fe^{2+}$ in Table 17.1, it will be reduced, and Fe(s) will be oxidized:

$$Cu^{2+} + 2e^- \longrightarrow Cu(s) \qquad E^o_{red} = +0.34\ V$$

$$Fe(s) \longrightarrow Fe^{2+} + 2e^- \qquad E^o_{ox} = +0.44\ V$$

$$Cu^{2+} + Fe(s) \longrightarrow Fe^{2+} + Cu(s) \qquad E^o_{cell} = \underline{+0.78\ V}$$

The reaction will proceed spontaneously as written

(a) Fe(s) will dissolve and Cu(s) will increase in mass
(b) The initial voltage is +0.78 V
(c) The $Fe^{2+}$(aq) solution will increase in concentration
(d) As the reaction proceeds towards equilibrium the voltage will decrease, eventually becoming zero at equilibrium

59. We calculated the standard cell potential for this cell to be -0.03 V in Problem 57, and we can use the Nernst equation to write

$$E = 0 = -0.03 - \frac{0.0257}{2}\ln\frac{[Pb^{2+}]}{[Sn^{2+}]}\ ; \qquad \ln\frac{0.001}{x} = -2.33$$

$$\frac{x}{0.001} = e^{2.33} = 10\ ; \quad x = \underline{0.010\ mol\ L^{-1}}$$

60. (a) The voltage will be $E^o_{cell}$ since these are standard conditions

$$Cu^{2+} + 2e^- \longrightarrow Cu(s) \qquad E^o_{red} = +0.34\ V$$

$$Pb(s) \longrightarrow Pb^{2+} + 2e^- \qquad E^o_{ox} = +0.13\ V$$

$$Cu^{2+} + Pb(s) \longrightarrow Cu(s) + Pb^{2+} \qquad E^o_{cell} = \underline{+0.47\ V}$$

(b) For this and the subsequent parts, we use the Nernst equation in the form

$$E_{cell} = E^o_{cell} - \frac{0.0257}{2}\ln\frac{[Pb^{2+}]}{[Cu^{2+}]}\ ; \quad E_{cell} = \underline{0.32\ V}$$

(c) $\underline{0.50\ V}$ (d) $\underline{0.54\ V}$

61. (a) $Ag^+$(aq) comes above $Br_2$ in Table 17.1, so Ag(s) will be oxidized:

$$Ag(s) \longrightarrow Ag^+ + e^- \qquad E^o_{ox} = -0.80\ V$$

$$Br_2 + 2e^- \longrightarrow 2Br^- \qquad E^o_{red} = +1.09\ V$$

$$2Ag(s) + Br_2 \longrightarrow 2Ag^+ + 2Br^- \qquad E^o_{cell} = \underline{+0.29\ V}$$

Since Ag is a solid and $Br_2$ is liquid, this is a heterogeneous equilibrium for which

$$Q = [Ag^+]^2[Br^-]^2$$

and we can write

$$E_{cell} = E^o_{cell} - \frac{0.0257}{2} \ln Q = E^o_{cell} - \frac{0.0257}{2} \ln\left[(0.01)^2(0.50)^2\right]$$

$$= +0.29 - 0.14 = \underline{0.15\ V}$$

(b) $K_c = [Ag^+]^2[Br^-]^2$

(c) At equilibrium $E_{cell} = 0$, and we can write

$$0 = +0.29 - \frac{0.0257}{2} \ln K_c \ ; \ K_c = e^{22.6} = \underline{6.5\text{x}10^9\ mol^4\ L^{-4}}$$

62. (a)

| | |
|---|---|
| $2[Fe^{3+} + e^- \longrightarrow Fe^{2+}]$ | $E^o_{red} = +0.77$ V |
| $Cu(s) \longrightarrow Cu^{2+} + 2e^-$ | $E^o_{ox} = -0.34$ V |
| $2Fe^{3+} + Cu(s) \longrightarrow 2Fe^{2+} + Cu^{2+}$ | $E^o_{cell} = \underline{+0.43\ V}$ |

$$Q = \frac{[Fe^{2+}]^2[Cu^{2+}]}{[Fe^{3+}]^2}$$

$$E_{cell} = E^o_{cell} - \frac{0.0257}{2} \ln Q = +0.43 - \frac{0.0257}{2} \ln\frac{(0.1)^2(0.5)}{(0.01)^2}$$

$$= +0.43 - 0.05 = \underline{0.38\ V}$$

(b) $$K_c = \frac{[Fe^{2+}]^2[Cu^{2+}]}{[Fe^{3+}]^2}$$

(c) $$E_{cell} = 0 = +0.43 - \frac{0.0257}{2} \ln K_c \ ; \ \ln K_c = 33.5$$

$$K_c = e^{33.5} = \underline{3.5\text{x}10^{14}}$$

63.

| | |
|---|---|
| $2[Ag^+ + e^- \longrightarrow Ag(s)]$ | $E^o_{red} = 0.80$ V |
| $H_2(g) \longrightarrow 2H^+ + 2e^-$ | $E^o_{ox} = 0.00$ V |
| $2Ag^+ + H_2(g) \longrightarrow 2H^+ + 2Ag(s)$ | $E^o_{cell} = 0.80$ V |

$$E_{cell} = 0.859\ V = 0.80\ V - \frac{0.0257}{2} \ln\frac{[H^+]^2}{[Ag^+]^2} = 0.80\ V - \frac{0.0257}{2} \ln\frac{[H^+]^2}{(0.10)^2}$$

$$\ln\frac{[H^+]^2}{(0.10)^2} = -4.59 \ ; \ \frac{[H^+]^2}{(0.10)^2} = 0.010 \ ; \ [H^+] = \underline{0.010\ mol\ L^{-1}}$$

$$\underline{pH = 2.00}$$

64.

| | |
|---|---|
| $Hg_2Cl_2(s) + 2e^- \longrightarrow 2Hg(l) + 2Cl^-(aq)$ | $E^o_{red} = 0.285$ V |
| $H_2(g) \longrightarrow 2H^+(aq) + 2e^-$ | $E^o_{ox} = 0.000$ V |
| $Hg_2Cl_2(s) + H_2(g) \longrightarrow 2Hg(l) + 2H^+(aq) + 2Cl^-(aq)$ | $E^o_{cell} = \underline{0.285\ V}$ |

$$E_{cell} = E^o_{cell} - \frac{0.0257}{2} \ln [H^+]^2[Cl^-]^2 \quad \text{and } [Cl^-] = 1.00\ M$$

$$0.823\ V = 0.285\ V - \frac{0.0257}{2} \ln [H^+]^2 \ ; \ \ln [H^+]^2 = -41.9$$

i.e., $[H^+]^2 = e^{-41.9} = 6.35\text{x}10^{-19}$ ;

$[H^+] = 7.97\text{x}10^{-10}$ mol $L^{-1}$ ; $\underline{pH = 9.10}$

At the equivalence point in the titration of $CH_3CO_2H(aq)$ with NaOH(aq) we have a solution of $CH_3CO_2Na(aq)$ of concentration $C_s$ mol $L^{-1}$.

$$CH_3CO_2^-(aq) + H_2O \rightleftharpoons CH_3CO_2H(aq) + OH^-(aq)$$

| | | | | |
|---|---|---|---|---|
| initially | $C_s$ | 0 | 0 | mol $L^{-1}$ |
| at eq'm | $C_s$-x | x | x | mol $L^{-1}$ |

$$K_b(CH_3CO_2^-) = \frac{K_w}{K_a(CH_3CO_2H)} = \frac{x^2}{C_s - x}$$

From the pH of 9.10, x = $[OH^-]$ = $1.26x10^{-5}$ mol $L^{-1}$ and $C_s$ may be calculated from the initial volume of $CH_3CO_2H(aq)$ in the titration and its concentration, and the volume of NaOH(aq) added to achieve the equivalence point and its concentration. Hence $K_a(CH_3CO_2H)$ may be obtained

65. The required equilibrium is AgI (s) $\longrightarrow$ $Ag^+(aq) + I^-(aq)$, for which we require the cell potential, which can be achieved from adding

$$Ag(s) \longrightarrow Ag^+ + e^- \qquad E^o_{ox} = -0.80\ V$$
$$AgI(s) + e^- \longrightarrow Ag(s) + I^- \qquad E^o_{red} = -0.15\ V$$
$$AgI(s) \longrightarrow Ag^+ + I^- \qquad E^o_{cell} = -0.95\ V$$

(a)
$$\ln K_{sp} = \frac{nE^o}{0.0257} = \frac{1(-0.95)}{0.0257} = -37.0\ ; \quad K_{sp} = e^{-37}$$
$$K_{sp}(AgI(s)) = [Ag^+][I^-] = \underline{8.5x10^{-17}\ mol^2\ L^{-2}}$$

(b) For the formation of 1 mol of AgI(s) from its elements in their standard states, the reaction is

$$Ag(s) + \frac{1}{2} I_2(s) \longrightarrow AgI(s)$$

which may be obtained by adding

$$Ag(s) \longrightarrow Ag^+ + e^- \qquad E^o_{ox} = -0.80\ V$$
$$\frac{1}{2} I_2(s) + e^- \longrightarrow I^- \qquad E^o_{red} = +0.54\ V$$
$$Ag(s) + \frac{1}{2} I_2(s) \longrightarrow AgI(s) \qquad E^o_{cell} = \underline{-0.26\ V}$$

66. $I^-(aq)$ is oxidized to $I_2(s)$, and $Fe^{3+}(aq)$ is reduced to $Fe^{2+}(aq)$, so we have:

$$2I^- \longrightarrow I_2 + 2e^- \qquad E^o_{ox} = -0.54\ V$$
$$2[Fe^{3+} + e^- \longrightarrow Fe^{2+}] \qquad E^o_{red} = +0.77\ V$$
$$2I^- + 2Fe^{3+} \longrightarrow I_2 + 2Fe^{2+} \qquad E^o_{cell} = \underline{+0.23\ V}$$

$$\ln K = \frac{nE^o}{0.0257} = \frac{2(+0.23)}{0.0257} = 17.9$$
$$K = e^{17.9} = \underline{5.9x10^7\ mol^{-2}\ L^2}$$

67. First we calculate the standard cell potential. $Mg^{2+}$ comes above $Ni^{2+}$ in Table 17.1, so $Ni^{2+}$ must be reduced and Mg(s) oxidized:

$$Ni^{2+} + 2e^- \longrightarrow Ni(s) \qquad E^o_{red} = -0.25\ V$$
$$Mg(s) \longrightarrow Mg^{2+} + 2e^- \qquad E^o_{ox} = +2.36\ V$$
$$Ni^{2+} + Mg(s) \longrightarrow Ni(s) + Mg^{2+} \qquad E^o_{cell} = +2.11\ V$$

$$E_{cell} = E^o_{cell} - \frac{0.0257}{2} \ln \frac{[Mg^{2+}]}{[Ni^{2+}]} = 2.11 - \frac{0.0257}{2} \ln(\frac{0.0500}{1.50})$$

$$= 2.11 + 0.04 = \underline{2.15\ V}$$

The overall cell reaction is $Ni^{2+} + Mg(s) \longrightarrow Ni(s) + Mg^{2+}$

cathode $Ni(s)|Ni^{2+}(aq, 1.50\ M||Mg^{2+}(0.0500\ M, aq|Mg(s)$ anode

68. In this cell, $Fe^{2+}(aq)$ is oxidized and $Ag^{+}(aq)$ is reduced:

$$Fe^{2+} \longrightarrow Fe^{3+} + e^- \qquad E^o_{ox} = -0.77\ V$$

$$Ag^+ + e^- \longrightarrow Ag(s) \qquad E^o_{red} = +0.80\ V$$

$$Ag^+ + Fe^{2+} \longrightarrow Ag(s) + Fe^{3+} \qquad E^o_{cell} = \underline{+0.03\ V}$$

(a) Increase in $[Ag^+]$ increases the tendency for $Ag^+$ to be reduced which increases $E_{cell}$.

(b) Increase in $[Fe^{3+}]$ decreases the tendency for $Ag^+$ to be reduced, which decreases $E_{cell}$.

(c) No change, because the ratio $[Fe^{3+}]/[Fe^{2+}]$ does not change.

(d) No change, because the amount of Ag(s) does not affect $E^o_{red}(Ag^+)$.

(e) Decrease in $[Fe^{2+}]$ decreases the tendency for $Ag^+$ to be reduced, which decreases $E_{cell}$.

(f) Addition of NaCl(aq) will precipitate $Ag^+$ as insoluble AgCl(s), thus decreasing $[Ag^+]$; $E_{cell}$ will decrease.

69. (a)

$$E_{cell} = E^o_{cell} - \frac{0.0257}{n} \ln \frac{[Cd^{2+}]}{[Pb^{2+}]} = E^o_{cell} - \frac{0.0257}{2} \ln (\frac{0.0250}{0.150})$$

$$E^o_{cell} = 0.293\ V + (-0.023\ V) = \underline{0.27\ V}$$

(b)

$$Cd(s) \longrightarrow Cd^{2+} + 2e^- \qquad E^o_{ox} = x\ \ V$$

$$Pb^{2+} + 2e^- \longrightarrow Pb(s) \qquad E^o_{red} = -0.13\ V$$

$$Cd(s) + Pb^{2+} \longrightarrow Cd^{2+} + Pb(s) \qquad E^o_{cell} = \underline{0.27\ V}$$

Hence, $x - 0.13\ V = 0.27\ V$; $x = \underline{0.40\ V}$

(c)

$$\ln K = \frac{nE^o}{0.0257} = \frac{2(0.27)}{0.0257} = 21.0\ ;\ K = e^{21.0} = \underline{1.3 \times 10^9}$$

70.

$$\underline{\Delta G = -nFE}$$

where n is the number of moles of electrons transferred, and F is the Faraday constant. Thus, in each case we calculate first the standard cell potential and then $E_{cell}$ for the given concentrations:

(a)

$$E_{cell} = E^o_{cell} - \frac{0.0257}{2} \ln \frac{[Fe^{2+}]}{[Cu^{2+}]} = E^o_{cell} - 0.03\ V$$

$$= 0.78\ V - 0.03\ V = \underline{0.75\ V}$$

$$\Delta G = -2(96\ 500\ C)(0.75\ V)(\frac{1\ J}{1\ VC})(\frac{1\ kJ}{1000\ J}) = \underline{-145\ kJ}$$

(b) $E_{cell} = E^o_{cell} - \frac{0.0257}{2} \ln \frac{[Zn^{2+}]}{[H^+]^2} = E^o_{cell} - 0.020 \text{ V}$

$E_{cell} = (0.76 - 0.020) \text{ V} = \underline{0.74 \text{ V}}$

$\Delta G = -2(96\ 500 \text{ C})(0.74 \text{ V})(\frac{1 \text{ J}}{1 \text{ V C}})(\frac{1 \text{ kJ}}{1000 \text{ J}}) = \underline{-143 \text{ kJ}}$

(c) $E_{cell} = E^o_{cell} - \frac{0.0257}{6} \ln \frac{[Cr^{3+}]^2}{[Ni^{2+}]^3} = = E^o_{cell} - 0.0 \text{ V}$

$E_{cell} = E^o_{cell} = \underline{0.49 \text{ V}}$

$\Delta G = -6(96\ 500 \text{ C})(0.49 \text{ V})(\frac{1 \text{ J}}{1 \text{ V C}})(\frac{1 \text{ kJ}}{1000 \text{ J}}) = \underline{-284 \text{ kJ}}$

(d) $E_{cell} = E^o_{cell} - \frac{0.0257}{1} \ln \frac{[V^{3+}]}{[V^{2+}][Ag^+]} = E^o_{cell} - 0.24 \text{ V}$

$E_{cell} = 1.06 \text{ V} - 0.24 \text{ V} = \underline{0.82 \text{ V}}$

$\Delta G = -1(96\ 500 \text{ C})(0.82 \text{ V})(\frac{1 \text{ J}}{1 \text{ V C}})(\frac{1 \text{ kJ}}{1000 \text{ J}}) = \underline{-79.1 \text{ kJ}}$

71. Calculate the solubility product constants of AgI(s) and $AgIO_3$(s) from the standard cell potentials for each of the cell reactions,

(1) $AgI(s) \longrightarrow Ag^+ + I^-$

(2) $AgIO_3(s) \longrightarrow Ag^+ + IO_3^-$

(1) For AgI(s)

$$Ag(s) \longrightarrow Ag^+ + e^- \qquad E^o_{ox} = -0.80 \text{ V}$$
$$AgI(s) + e^- \longrightarrow Ag(s) + I^- \qquad E^o_{red} = -0.15 \text{ V}$$
$$AgI(s) \longrightarrow Ag^+ + I^- \qquad E^o_{cell} = \underline{-0.95 \text{ V}}$$

(2) For $AgIO_3$(s)

$$Ag(s) \longrightarrow Ag^+ + e^- \qquad E^o_{ox} = -0.80 \text{ V}$$
$$AgIO_3(s) + e^- \longrightarrow Ag(s) + IO_3^- \qquad E^o_{red} = +0.35 \text{ V}$$
$$AgIO_3(s) \longrightarrow Ag^+ + IO_3^- \qquad E_{cell} = \underline{-0.45 \text{ V}}$$

(1) $\ln K_{sp}(AgI) = \frac{nE^o}{0.0257} = \frac{1(-0.95)}{0.0257} = -37.0$ ; $K_{sp} = e^{-37}$

$K_{sp}(AgI(s)) = [Ag^+][I^-] = \underline{8.5\text{x}10^{-17} \text{ mol}^2 \text{ L}^{-2}}$

(2) $\ln K_{sp}(AgIO_3) = \frac{nE^o}{0.0257} = \frac{1(-0.45)}{0.0257} = -17.5$ ; $K_{sp} = e^{-18}$

$K_{sp}(AgIO_3(s)) = [Ag^+][IO_3^-] = \underline{1.5\text{x}10^{-8} \text{ mol}^2 \text{ L}^{-2}}$

In each case, a solid will be precipitated if $Q > K_{sp}$.

(1) AgI(s) $\quad Q = [Ag^+][I^-] = 10^{-12} \text{ mol}^2 \text{ L}^{-2} > K_{sp}$; a precipitate forms

(2) $AgIO_3$(s) $\quad Q = [Ag^+][IO_3^-] = 10^{-12} \text{ mol}^2 \text{ L}^{-2} < K_{sp}$; no precipitate

72. (a) From the data in Table 17.1,

$$Zn(s) \longrightarrow Zn^{2+}(aq) + 2e^- \qquad E^o_{ox} = +0.76 \text{ V } \underline{\text{anode}}$$

$$O_2(g) + 4H^+(aq) + 4e^- \longrightarrow 2H_2O(l) \qquad E^o_{red} = +1.23 \text{ V } \underline{\text{cathode}}$$

(b) $E^o_{cell} = (+0.76 + 1.23) = \underline{1.99 \text{ V}}$ (close to 2 V)

(c) Calculate the electric charge needed to oxidize 5.0 g Zn(s) to $Zn^{2+}(aq)$.

$$\text{Coulombs} = (5.0 \text{ g Zn})(\frac{1 \text{ mol Zn}}{65.38 \text{ g Zn}})(\frac{2 \text{ mol } e^-}{1 \text{ mol Zn}})(\frac{96\ 500 \text{ C}}{1 \text{ mol } e^-}) = \underline{1.48\text{x}10^4 \text{ C}}$$

$$\text{Time} = (\frac{1.48\text{x}10^4 \text{ C}}{40 \text{ uA}})(\frac{1 \text{ As}}{1 \text{ C}})(\frac{10^6 \text{ uA}}{1 \text{ A}})(\frac{1 \text{ h}}{3600 \text{ s}})(\frac{1 \text{ day}}{24 \text{ h}})(\frac{1 \text{ year}}{365 \text{ day}}) = \underline{12 \text{ year}}$$

The zinc electrode would have to be replaced about once every 12 years.

73. (a) Carbon, oxidation number -3 in $C_2H_6$, is oxidized to carbon oxidation number +4 in $CO_2$, and the half-reaction in acid solution is,

$$C_2H_6 + 4H_2O \longrightarrow 2CO_2 + 14H^+ + 14e^-$$

$O_2$ is reduced to $H_2O$, and the half-reaction is

$$O_2 + 4H^+ + 4e^- \longrightarrow 2H_2O$$

Adding twice the first equation to seven times the second gives,

$$2C_2H_6 + 7O_2 \longrightarrow 4CO_2 + 6H_2O$$

in a process that involves $28e^-$ (14 $e^-$ per mole of $C_2H_6$ used).

(b) First calculate the moles of electrons required, and the moles of $C_2H_6$ consumed.

$$\text{mol } C_2H_6 = (6.00 \text{ h})(\frac{3600 \text{ s}}{1 \text{ h}})(0.50 \text{ A})(\frac{1 \text{ C}}{1 \text{ As}})(\frac{1 \text{ mol } e^-}{96\ 500 \text{ C}})(\frac{1 \text{ mol } C_2H_6}{14 \text{ mol } e^-})$$

$$= \underline{8.0\text{x}10^{-3} \text{ mol}}$$

$$V_{C_2H_6} = \frac{nRT}{P} = \frac{(8.0\text{x}10^{-3} \text{ mol})(0.0821 \text{ atm L mol}^{-1} \text{ K}^{-1})(273 \text{ K})}{1 \text{ atm}}$$

$$= \underline{0.18 \text{ L}}$$

$$V_{O_2} = (0.18 \text{ L } C_2H_6)(\frac{7 \text{ mol } O_2}{2 \text{ mol } C_2H_6}) = \underline{0.63 \text{ L}}$$

The amount of $O_2(g)$ at STP consumed at the cathode = $\underline{0.63 \text{ L}}$

# CHAPTER 18

1. (a) Nitric acid, $HNO_3(l)$ (b) Nitrous acid, $HNO_2(aq)$
   (c) Potassium nitrite, $KNO_2(s)$ (d) Nitrogen monoxide, $NO(g)$
   (e) Dinitrogen pentaoxide, $N_2O_5(s)$ (f) Hydrazine, $N_2H_4(l)$
   (g) Sodium azide, $NaN_3(s)$

2. (a) Ozone, $O_3(g)$ (b) Sodium peroxide, $Na_2O_2(s)$
   (c) Barium peroxide, $BaO_2(s)$ (d) Potassium superoxide, $KO_2(s)$
   (e) Hydrogen peroxide, $H_2O_2(l)$

3. (a) $NaNO_3$, sodium nitrate (b) $KNO_2$, potassium nitrite
   (c) $N_2O_4$, dinitrogen tetraoxide (d) $HN_3$, hydrazoic acid
   (e) $BaO_2$, barium peroxide (f) potassium superoxide, $KO_2$ (g) $N_2H_4$, hydrazine
   (h) $LiN_3$, lithium azide (i) lithium nitride, $Li_3N$

4. In each case the formula of the anhydride of the oxoacid is obtained by removing all the H from the oxoacid as $H_2O$.
   (a) $2HNO_3 \longrightarrow H_2O + N_2O_5$, dinitrogen pentaoxide.
   (b) $2HNO_2 \longrightarrow H_2O + N_2O_3$, dinitrogen trioxide.
   (c) $2H_3PO_3 \longrightarrow 3H_2O + P_2O_3$ gives the correct empirical formula; the anhydride has the molecular formula $P_4O_6$, tetraphosphorus hexaoxide.
   (d) $H_2SO_4 \longrightarrow H_2O + SO_3$, sulfur trioxide
   (e) $H_2CO_3 \longrightarrow H_2O + CO_2$, carbon dioxide

5. (a) Ammonia, $NH_3$, N(-3). (b) Hydrazine, $N_2H_4$, N(-2).
   (c) Nitrogen, $N_2$, N(0). (d) Nitric acid, $HNO_3$, N(+5).
   (e) Dinitrogen monoxide, $N_2O$, N(+1). (f) Nitrogen monoxide, NO, N(+2).
   (g) Nitrous acid, $HNO_2$, N(+3). (h) Dinitrogen tetraoxide, $N_2O_4$, N(+4).

6. (a) $2NH_3(g) + 3CuO(s) \longrightarrow N_2(g) + 3H_2O(g) + 3Cu(s)$
   N(-3) in $NH_3$ is oxidized to N(0) in $N_2$; Cu(+2) in CuO is reduced to Cu(0) in Cu(s). $NH_3$ is oxidized and CuO is reduced.
   (b) $NaNO_2(aq) + NH_4Cl(aq) \longrightarrow N_2(g) + NaCl(aq) + 2H_2O(l)$
   N(+3) in $NO_2^-$ is reduced to N(0) in $N_2$; N(-3) in $NH_4^+$ is oxidized to N(0) in $N_2$. $NO_2^-$ is reduced and $NH_4^+$ is oxidized.
   (c) $6NO(g) + 4NH_3(g) \longrightarrow 5N_2(g) + 6H_2O(g)$
   N(+2) in NO is reduced to N(0) in $N_2$; N(-3) in $NH_3$ is oxidized to N(0) in $N_2$. NO is reduced and $NH_3$ is oxidized.
   (d) $(NH_4)_2Cr_2O_7(s) \longrightarrow Cr_2O_3(s) + N_2(g) + 4H_2O(g)$
   N(-3) in $NH_4^+$ is oxidized to N(0) in $N_2$; Cr(+6) in $Cr_2O_7^{2-}$ is reduced to Cr(+3) in $Cr_2O_3$. $NH_4^+$ is oxidized and $Cr_2O_7^{2-}$ is reduced.

7. (a) $2KNO_3(s) \longrightarrow 2KNO_2(s) + O_2(g)$; N(+3).

(b) $NH_4NO_3(s) \longrightarrow N_2O(g) + 2H_2O(g)$; N(+1).

(c) $2Pb(NO_3)_2(s) \longrightarrow 2PbO(s) + 4NO_2(g) + O_2(g)$; N(+4)

8. (a)

$$HNO_2 + H_2O \longrightarrow HNO_3 + 2e^- + 2H^+ \qquad \text{oxidation}$$
$$2[HNO_2 + e^- + H^+ \longrightarrow NO + H_2O] \qquad \text{reduction}$$
$$\overline{3HNO_2 \longrightarrow HNO_3 + 2NO + H_2O}$$

(b)

$$4[HNO_3 + e^- + H^+ \longrightarrow NO_2 + H_2O] \qquad \text{reduction}$$
$$2H_2O \longrightarrow O_2 + 4e^- + 4H^+ \qquad \text{oxidation}$$
$$\overline{4HNO_3 \longrightarrow 4NO_2 + O_2 + 2H_2O}$$

(c) $C_6H_6 + HNO_3 + H_2SO_4 \longrightarrow C_6H_5NO_2 + H_3O^+ + HSO_4^-$

9. (a) $3Cu(s) + 8HNO_3(aq) \longrightarrow 3Cu(NO_3)_2(aq) + 2NO(g) + 4H_2O(l)$

$HNO_3(aq) + NH_3(g) \rightarrow NH_4^+(aq) + NO_3^-(aq)$

$Mg(OH)_2(s) + 2HNO_3(aq) \longrightarrow Mg^{2+}(aq) + 2NO_3^-(aq) + H_2O(l)$

(b) $Cu(s) + 4HNO_3(aq,conc) \longrightarrow Cu(NO_3)_2(aq) + 2NO_2(g) + 2H_2O(l)$

$Mg(OH)_2(s) + 2HNO_3(aq,conc) \longrightarrow Mg(NO_3)_2(aq) + 2H_2O(l)$

(c) $P_4O_{10}(s) + 4HNO_3(l) \longrightarrow 4HPO_3(s) + 2N_2O_5(g)$

$2H_2SO_4(l) + HNO_3(l) \longrightarrow NO_2^+ + H_3O^+ + 2HSO_4^-$

10. Ammonia, $NH_3(g)$ is oxidized with oxygen from air to give nitrogen monoxide, $NO(g)$,

$$4NH_3(g) + 5O_2(g) \xrightarrow{\text{catalyst}} 4NO(g) + 6H_2O(g)$$

and the NO(g) is oxidized with $O_2(g)$ from air to give $NO_2(g)$,

$$2NO(g) + O_2(g) \longrightarrow 2NO_2(g)$$

and, finally the $NO_2(g)$ is bubbled into water,

$$3NO_2(g) + H_2O(l) \longrightarrow 2HNO_3(aq) + NO(g)$$

This process is called the Ostwald process.

11. $SO_2(g)$ in aqueous solution behaves as an aqueous solution of sulfurous acid, $H_2SO_3(aq)$, but the white precipitate cannot be barium sulfite, $BaSO_3$ (s), because all sulfites are soluble in water. The only likely possibility is that the white precipitate is $BaSO_4(s)$, formed from $Ba^{2+}(aq)$ ions and $SO_4^{2-}(aq)$ ions resulting from the oxidation of $SO_2$ by $NO_3^-$ (aq).

$$3[SO_2 + 2H_2O \longrightarrow SO_4^{2-} + 2e^- + 4H^+] \qquad \text{oxidation}$$
$$2[NO_3^- + 3e^- + 4H^+ \longrightarrow NO + 2H_2O] \qquad \text{reduction}$$
$$\overline{3SO_2 + 2NO_3^- + 2H_2O \longrightarrow 3SO_4^{2-} + 2NO + 4H^+}$$

Which is followed by, $Ba^{2+}(aq) + SO_4^{2-}(aq) \longrightarrow BaSO_4(s)$

12. (a) This is the method for preparing pure nitric acid. $NO_3^-$ is a base in sulfuric acid and $NaNO_3$ reacts according to the equation

$$\underset{\text{base}}{NaNO_3} + \underset{\text{acid}}{H_2SO_4} \longrightarrow NaHSO_4 + HNO_3\uparrow \qquad \underline{\text{acid-base reaction}}$$

(b) Concentrated $HNO_3$ decomposes in bright light to give $NO_2(g)$, which colors the acid yellow-brown, and $O_2(g)$ in a disproportionation reaction in which N(+5) is reduced to N(+4) and O(-2) is oxidized to O(0).

$$4(HNO_3 + e^- \longrightarrow NO_2 + O^{2-} + H^+) \qquad \text{reduction}$$
$$2O^{2-} \longrightarrow O_2 + 4e^- \qquad \text{oxidation}$$
$$4HNO_3 \longrightarrow 4NO_2 + O_2 + 2H_2O \qquad \underline{\text{oxidation-reduction}}$$

(c) $Pb(NO_3)_2$ decomposes on heating to PbO, $NO_2$, and $O_2$ the oxidation state of Pb is unchanged (+2), but $NO_3^-$ disproportionates in a reaction in which N(+5) is reduced to N(+4) and O(-2) is oxidized to O(0).

$$4(NO_3^- + e^- \longrightarrow NO_2 + O^{2-}) \qquad \text{reduction}$$
$$2O^{2-} \longrightarrow O_2 + 4e^- \qquad \text{oxidation}$$
$$4NO_3^- \longrightarrow 4NO_2 + O_2 + 2O^{2-} \qquad \underline{\text{oxidation-reduction}}$$

and the final equation is obtained by adding $2Pb^{2+}$ to each side of this equation.

$$2Pb(NO_3)_2 \longrightarrow 2PbO + 4NO_2 + O_2$$

(d) Nitric acid is a strong base in sulfuric acid and the $H_2NO_3^+$ ion formed is unstable with respect to the nitronium ion, $NO_2^+$ and $H_2O$ into which it decomposes (c.f. isoelectronic $H_2CO_3$, carbonic acid). and since water is also a strong base in sulfuric acid, we can write

$$HNO_3 + H_2SO_4 \longrightarrow H_2NO_3^+ + HSO_4^- \qquad \underline{\text{acid-base}}$$
and $$H_2NO_3^+ + H_2SO_4 \longrightarrow NO_2^+ + H_3O^+ + HSO_4^- \qquad \underline{\text{acid-base}}$$

Adding these equations together gives the overall reaction:

$$HNO_3 + 2H_2SO_4 \longrightarrow NO_2^+ + H_3O^+ + 2HSO_4^-$$

(e) Nitrous acid is a weak acid in aqueous solution:

$$\underset{\text{acid}}{HNO_2} + \underset{\text{base}}{H_2O} \rightleftharpoons H_3O^+ + NO_2^- \qquad \underline{\text{acid-base reaction}}$$

(f) This is the reaction by which hydrazine is prepared in aqueous solution and also gives as biproducts the chloramines $H_2NCl$ and $HNCl_2$.

$$2NH_3 \longrightarrow N_2H_4 + 2e^- + 2H^+ \qquad \text{oxidation}$$
$$OCl^- + 2e^- + 2H^+ \longrightarrow Cl^- + H_2O \qquad \text{reduction}$$
$$2NH_3 + OCl^- \longrightarrow N_2H_4 + H_2O + Cl^- \qquad \underline{\text{oxidation-reduction}}$$

N(-3) in $NH_3$ is oxidized to N(-2) in $N_2H_4$ and Cl(+1) in $OCl^-$ is reduced to Cl(-1) in $Cl^-$. $NH_3$ is oxidized and $OCl^-$ is reduced.

(g) Hydrogen peroxide decomposes to $H_2O(l)$ and $O_2(g)$

$$2H_2O_2(l) \rightarrow 2H_2O(l) + O_2(g)$$

in a <u>disproportionation</u> reaction in which O(-1) in $H_2O_2$ is reduced to O(-2) in $H_2O$ and oxidized to O(0) in $O_2$. $H_2O_2$ is both reduced and oxidized, in this <u>oxidation-reduction</u> reaction.

(h) $BaO_2(s) + H_2SO_4(aq) \longrightarrow BaSO_4(s) + H_2O_2(aq)$

The peroxide ion, $O_2^{2-}$ is protonated twice to give $H_2O_2(aq)$ and $BaSO_4(s)$ is precipitated. This is an acid-base reaction in which the $O_2^{2-}$ ion is the base and $H_3O^+(aq)$ is the acid.

13. (a) Heating Cu(s) with dilute $HNO_3(aq)$ gives mainly NO(g):

$$3Cu(s) + 8HNO_3(aq) \longrightarrow 3Cu(NO_3)_2(aq) + 2NO(g) + 4H_2O(l)$$

(b) $NH_3(g)$ is the main product when $HNO_3(aq)$ is reduced with a reactive metal such as zinc. When dilute $HNO_3(aq)$ is heated with zinc -

$$4Zn(s) + 9HNO_3(aq) \longrightarrow 4Zn(NO_3)_2(aq) + NH_3(g) + 3H_2O(l)$$

(c) Heating Cu(s) with concentrated $HNO_3(aq)$ gives mainly $NO_2(g)$:

$$Cu(s) + 4HNO_3(aq) \longrightarrow Cu(NO_3)_2(aq) + 2NO_2(g) + 2H_2O(l).$$

14. (a) Ozone is produced in about 10% yield when an electric discharge is passed through $O_2(g)$.

$$3O_2(g) \longrightarrow 2O_3(g)$$

(b) Barium peroxide, $BaO_2(s)$, is prepared by passing $O_2(g)$ over BaO(s),

$$2BaO(s) + O_2(g) \longrightarrow 2BaO_2(s)$$

and hydrogen peroxide results when $BaO_2(s)$ is stirred with a cold aqueous solution of sulfuric acid, when $BaSO_4(s)$ is precipitated and can be filtered off to give a pure aqueous solution of $H_2O_2$.

15. (a) Dinitrogen oxide, $N_2O(g)$, results when solid ammonium nitrate is gently heated. The ammonium nitrate melts and then decomposes.

$$NH_4NO_3(l) \longrightarrow N_2O(g) + 2H_2O(g)$$

(b) Nitrogen monoxide is the main product when a metal such as copper is heated with dilute nitric acid.

$$3Cu(s) + 8HNO_3(aq) \longrightarrow 3Cu(NO_3)_2(aq) + 2NO(g) + 4H_2O(l)$$

(c) In the laboratory, a convenient method for the preparation of $N_2O_4(l)$ is by heating solid lead nitrate.

$$2Pb(NO_3)_2(s) \longrightarrow 2PbO(s) + 4NO_2(g) + O_2(g)$$

$N_2O_4(s)$ results when the $NO_2(g)$ is condensed by surrounding a collection flask with a freezing mixture, such as dry-ice/alcohol mixture. The white solid melts at -11°C to give mainly liquid $N_2O_4$, but at higher temperatures the equilibrium

$$N_2O_4(g) \rightleftharpoons 2NO_2(g)$$

shifts more and more in favour of $NO_2(g)$.

16. Of these nitrogen oxides, $N_2O$ and $N_2O_4$ have an even number of electrons while NO is an odd electron species. For each, several resonance structures can be written.

(a) $N_2O$ is isoelectronic with carbon dioxide, $CO_2$, with eight pairs of valence shell electrons, and $AX_2$ geometry.

$$:\ddot{O}=\overset{+}{N}=\ddot{N}:^- \quad \longleftrightarrow \quad ^-:\ddot{\underset{..}{O}}-\overset{+}{N}\equiv N:$$

(b) NO has 11 valence shell electrons. For an octet of electrons around each atom the Lewis structure is,

$$^-:\overset{..}{N}=\overset{.}{O}:^+ \longleftrightarrow :\overset{.}{N}=\overset{..}{O}: \qquad \underline{or} \qquad ^{\frac{1}{2}-}:\overset{.}{N}=\overset{.}{O}:^{\frac{1}{2}+}$$

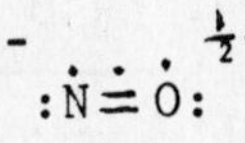

(c) $N_2O_4$ has 34 valence shell electrons, or 17 pairs. There are four resonance structures of the type,

$$^-:\overset{..}{O}: \quad + \quad + \quad \overset{..}{O}:$$
$$\overset{..}{O} = N - N$$
$$:\overset{..}{O} \qquad\qquad :\overset{..}{O}:^-$$

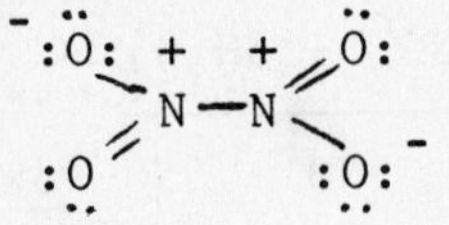

17. $N_2O_4$ is expected to react in the same way as $NO_2$, to give nitric acid, $HNO_3$(aq) and NO(g), as happens in the last stage of the preparation of nitric acid.

$$3N_2O_4(l) + 2H_2O(l) \longrightarrow 4HNO_3(aq) + 2NO(g)$$

This is a disproportionation (self oxidation-reduction) reaction. When $N_2O_4$(l) is dissolved in cold water, the reaction

$$N_2O_4(l) + H_2O(l) \longrightarrow HNO_2(aq) + HNO_3(aq)$$

occurs, but on raising the temperature the nitrous acid, $HNO_2$(aq), decomposes to $HNO_3$(aq) and NO(g).

$$3HNO_2(aq) \longrightarrow HNO_3(aq) + 2NO(g) + H_2O(l)$$

The sum of three times the second equation above and the last equation gives the first equation above.

18. (a) $PbS(s) + 4O_3(g) \longrightarrow PbSO_4(s) + 4O_2(g)$

(b)
$$2I^- \longrightarrow I_2 + 2e^- \qquad\qquad \text{oxidation}$$
$$O_3 + 2e^- + H_2O \longrightarrow O_2 + 2OH^- \qquad \text{reduction}$$
$$\overline{2I^- + O_3 + H_2O \longrightarrow I_2 + O_2 + 2OH^-}$$

(c)
$$2[Fe^{2+} \longrightarrow Fe^{3+} + e^-] \qquad\qquad \text{oxidation}$$
$$O_3 + 2e^- + 2H^+ \longrightarrow O_2 + H_2O \qquad \text{reduction}$$
$$\overline{2Fe^{2+} + O_3 + 2H^+ \longrightarrow 2Fe^{3+} + O_2 + H_2O}$$

19. (a) $2HgO(s) \longrightarrow 2Hg(l) + O_2(g)$

(b) $2PbO_2(s) \longrightarrow 2PbO(s) + O_2(g)$

(c) $2BaO_2(s) \longrightarrow 2BaO(s) + O_2(g)$

(d) $2KNO_3(s) \longrightarrow 2KNO_2(s) + O_2(g)$

20. (a) $4HNO_3 \longrightarrow 4NO_2 + 2H_2O + O_2$    <u>oxidation-reduction</u>

(b) The equation as written is balanced    <u>acid-base</u>

(c) $HNO_3 + 2H_2SO_4 \longrightarrow NO_2^+ + H_3O^+ + 2HSO_4^-$   <u>acid-base</u>

(d)
$$S + 4H_2O \longrightarrow H_2SO_4 + 6e^- + 6H^+ \qquad\qquad \text{oxidation}$$
$$2[HNO_3 + 3e^- + 3H^+ \longrightarrow NO + 2H_2O] \qquad \text{reduction}$$
$$\overline{S + 2HNO_3 \longrightarrow H_2SO_4 + 2NO} \qquad\qquad\qquad \text{oxidation-reduction}$$

(e) $3[I_2 + 6H_2O \longrightarrow 2HIO_3 + 10e^- + 10H^+]$ oxidation

$10[HNO_3 + 3e^- + 3H^+ \longrightarrow NO + 2H_2O]$ reduction

$3I_2 + 10\ HNO_3 \longrightarrow 6HIO_3 + 10\ NO + 2H_2O$ oxidation-reduction

(f) $P_4 + 16H_2O \longrightarrow 4H_3PO_4 + 20e^- + 20H^+$ oxidation

$5[2HNO_3 + 4e^- + 4H^+ \longrightarrow NO + NO_2 + 3H_2O]$ reduction

$P_4 + 10HNO_3 + H_2O \longrightarrow 4H_3PO_4 + 5NO + 5NO_2$ oxidation-reduction

(g) $4(Zn \longrightarrow Zn^{2+} + 2e^-]$ oxidation

$2HNO_3 + 8e^- + 8H^+ \longrightarrow N_2O + 5H_2O$ reduction

$4Zn + 2HNO_3 + 8H^+ \longrightarrow 4Zn^{2+} + N_2O + 5H_2O$

and adding $8NO_3^-$ to each side of the equation gives,

$4Zn + 10HNO_3 \longrightarrow 4Zn(NO_3)_2 + N_2O + 5H_2O$ oxidation-reduction

(h) $3[Fe^{2+} \longrightarrow Fe^{3+} + e^-]$ oxidation

$HNO_3 + 3e^- + 3H^+ \longrightarrow NO + 2H_2O$ reduction

$3Fe^{2+} + HNO_3 + 3H^+ \longrightarrow 3Fe^{3+} + NO + 2H_2O$ oxidation-reduction

21. (a) $8H_2S \longrightarrow S_8 + 16e^- + 16H^+$ oxidation

$8[NO_2 + 2e^- + 2H^+ \longrightarrow NO + H_2O]$ reduction

$8H_2S + 8NO_2 \longrightarrow S_8 + 8NO + 8H_2O$; $NO_2$ is an oxidizing agent.

(b) $3I^- \longrightarrow I_3^- + 2e^-$ oxidation

$NO_2 + 2e^- + H_2O \longrightarrow NO + 2OH^-$ reduction

$NO_2 + 3I^- + H_2O \longrightarrow I_3^- + NO + 2OH^-$; $NO_2$ is an oxidizing agent.

(c) $MnO_4^- + 5e^- + 8H^+ \longrightarrow Mn^{2+} + 4H_2O$ reduction

$5[NO_2 + H_2O \longrightarrow NO_3^- + e^- + 2H^+]$ oxidation

$MnO_4^- + 5NO_2 + H_2O \longrightarrow Mn^{2+} + 5NO_3^- + 2H^+$

or, adding $2H_2O$ to each side of the equation,

$MnO_4^- + 5NO_2 + 3H_2O \longrightarrow Mn^{2+} + 5NO_3^- + 2H_3O^+$ ; $NO_2$ is a reducing agent.

(d) $NO_2 + H_2O \longrightarrow HNO_3 + e^- + H^+$ reduction

$NO_2 + e^- + H^+ \longrightarrow HNO_2$ oxidation

$2NO_2 + H_2O \longrightarrow HNO_3 + HNO_2$, in which $NO_2$ behaves both as an oxidizing agent and as a reducing agent.

22. $2NaNO_3(l) \longrightarrow 2NaNO_2(s) + O_2(g)$; $2KNO_3(l) \longrightarrow 2KNO_2(s) + O_2(g)$

$NaNO_3(s) + Cu(s) \longrightarrow NaNO_2(s) + CuO(s)$

$KNO_3(s) + Pb(s) \longrightarrow KNO_2(s) + PbO(s)$

23.

$$\text{Mol NO} = \frac{PV}{RT} = \frac{(1\ \text{atm})(3.26\ \text{L})}{(0.0821\ \text{atm L mol}^{-1}\ \text{K}^{-1})(273\ \text{K})} = \underline{0.1454\ \text{mol}}$$

$$\text{Mol } NO_3^- = (0.1454\ \text{mol NO})\left(\frac{2\ \text{mol } NO_3^-}{2\ \text{mol NO}}\right) = \underline{0.1454\ \text{mol}}$$

$$[NO_3^-] = \left(\frac{0.1454\ \text{mol}}{2.00\ \text{L}}\right) = \underline{0.0727\ \text{mol L}^{-1}}$$

24. (a) 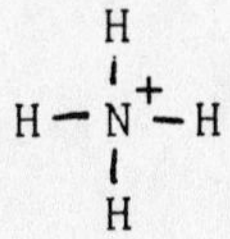
(b) 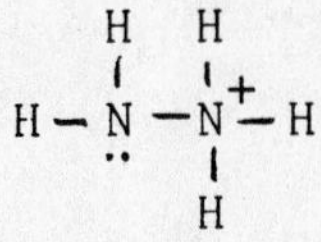
(c) 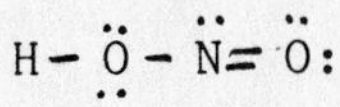

(d) 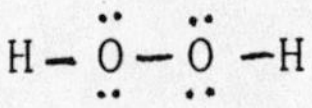
(e) 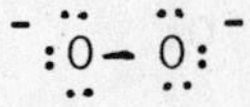
(f) 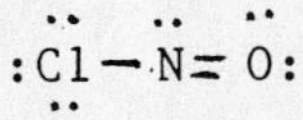

25. (a) 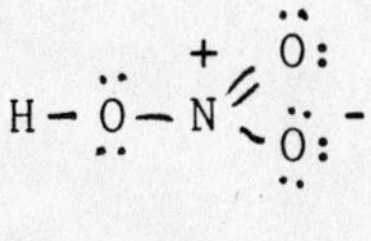

two resonance structures

(b) 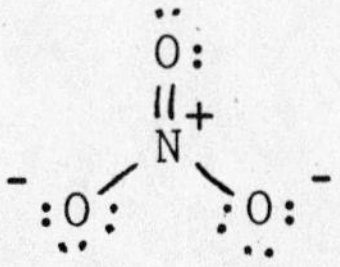

three resonance structures

(c) 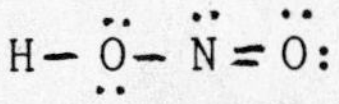

one structure

(d) 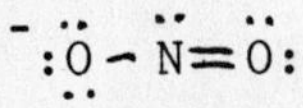

two resonance structures

(e) 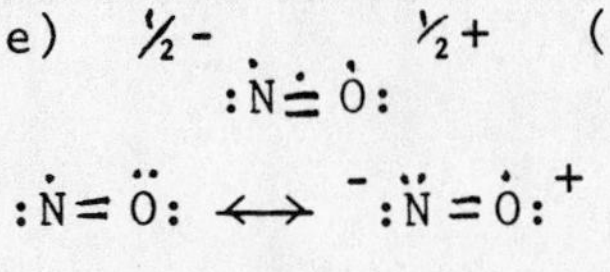

two resonance structures

(f) 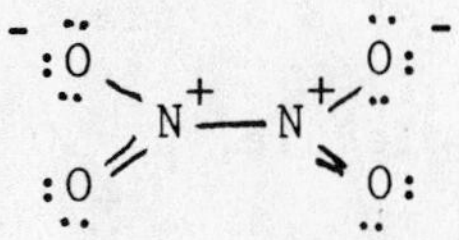

four resonance structures

(g) 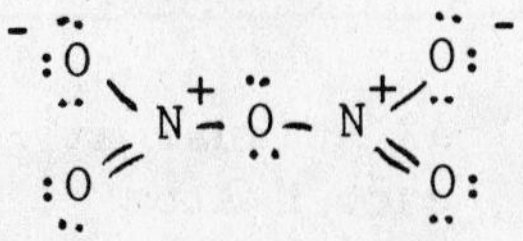

four resonance structures

26. $\underline{NO_2}$ 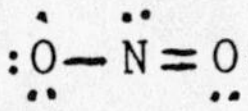 $AX_2E$, angular

$\underline{NO_2}^+$ 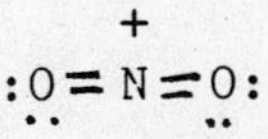 $AX_2$ linear

$\underline{NO_3}^-$ 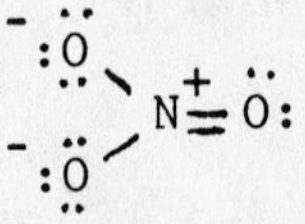 $AX_3$, triangular planar

$\underline{HNO_3}$ 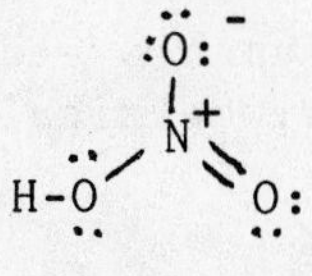 $AX_3$. triangular planar

$\underline{N_2O}$ 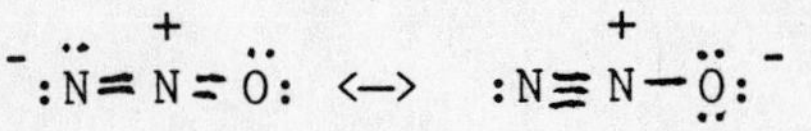 $AX_2$, linear

$\underline{ClNO}$ 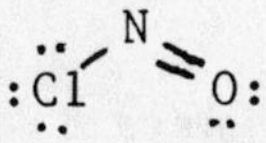 $AX_2E$, angular

$\underline{NF_3}$ 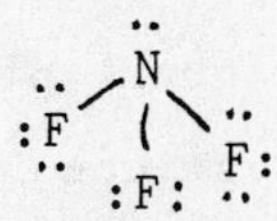 $AX_3E$, triangular pyramidal

HONO

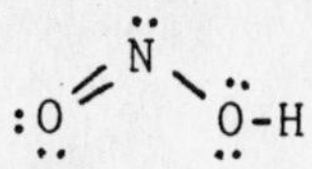

$:O=\overset{\cdot\cdot}{N}-\overset{\cdot\cdot}{O}-H$    AX$_2$E, angular

H$_3$CNO$_2$

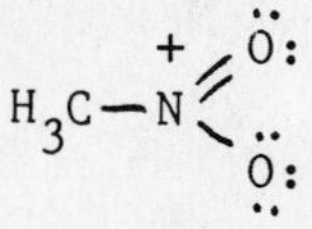

$H_3C-\overset{+}{N}\overset{\diagup\overset{\cdot\cdot}{O}:}{\underset{\diagdown\overset{\cdot\cdot}{O}:^-}{}}$    AX$_3$
triangular
planar at N.

27. (a) $2HNO_3 \longrightarrow N_2O_5 + H_2O$    (b) $2HNO_2 \longrightarrow N_2O_3 + H_2O$

(structure for N$_2$O$_5$)      (structure for N$_2$O$_3$)

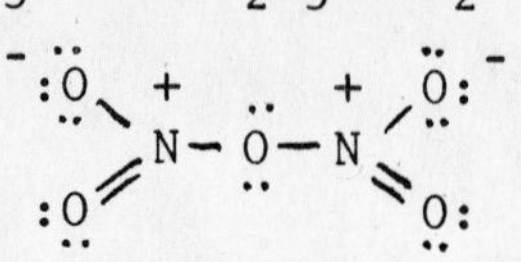

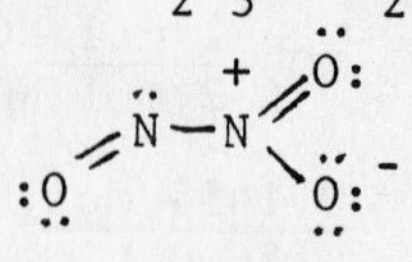

(c) $4H_3PO_4 \longrightarrow P_4O_{10} + 6H_2O$    (d) $H_2SO_4 \longrightarrow SO_3 + H_2O$

(structure for P$_4$O$_{10}$)      (structure for SO$_3$)

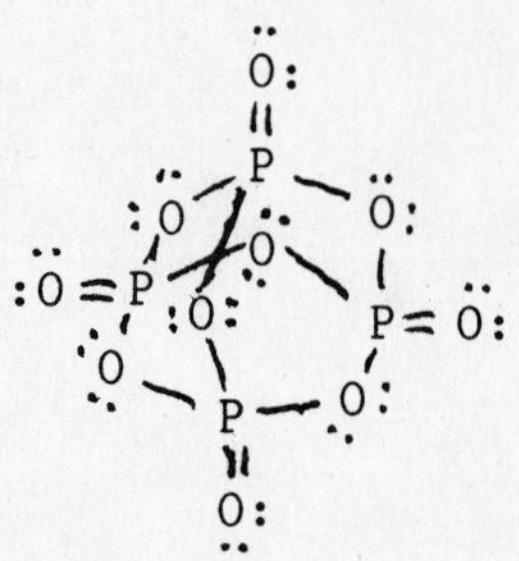

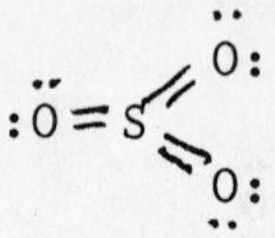

28. (a) NO, polar molecule, $\mu \neq 0$*

(b)   $:\overset{\cdot\cdot}{O}-\overset{\cdot\cdot}{N}=O: \longleftrightarrow :O=\overset{\cdot\cdot}{N}-\overset{\cdot\cdot}{O}:$

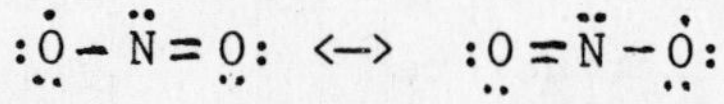

In the two resonance structures the NO bonds are polar and
the molecule is AX$_2$E angular; $\mu \neq 0$.*

(c) N$_2$, nonpolar molecule, $\mu = 0$

(d)   $:\overset{\cdot\cdot}{O}=\overset{+}{O}-\overset{\cdot\cdot}{O}:^- \longleftrightarrow ^-:\overset{\cdot\cdot}{O}-\overset{+}{O}=\overset{\cdot\cdot}{O}:$    AX$_2$ angular, $\mu \neq 0$.*

(e)   $^-:N=\overset{+}{N}=\overset{\cdot\cdot}{O}: \longleftrightarrow :N\equiv\overset{+}{N}-\overset{\cdot\cdot}{O}:^-$    AX$_2$ linear, $\mu \neq 0$.*

(f)   (structure)   four resonance   AX$_3$ planar at    $\mu = 0$.
                 structures      each N atom

(g)   (structure H-O-O-H)   AX$_2$E$_2$ angular at each O, $\mu \neq 0$*, because the
                   molecule is not trans planar.

(h)   (structure)   AX$_3$E at each N atom, $\mu \neq 0$; the molecule does not
                   have a symmetrical conformation.

(i)   (structure)   AX$_3$ at N, $\mu \neq 0$.*

(j)   $:\overset{\cdot\cdot}{Cl}-\overset{+}{N}\overset{\diagup\overset{\cdot\cdot}{O}:}{\underset{\diagdown\overset{\cdot\cdot}{O}:^-}{}}$   AX$_3$ trigonal planar at N, $\mu \neq 0$.*

*center of positive charge does not coincide with center of negative
charge, $\mu \neq 0$.

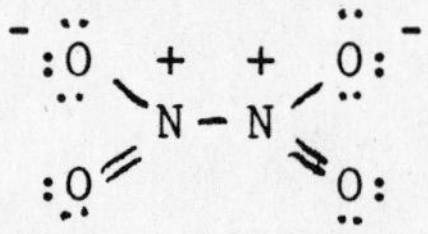

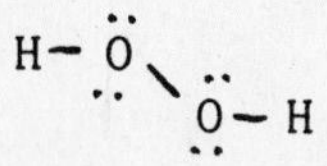

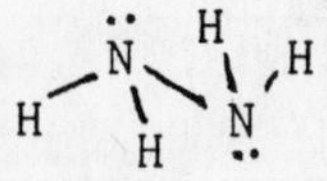

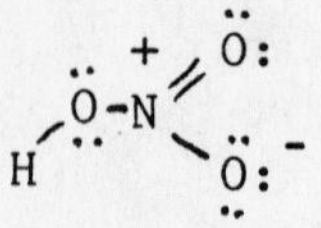

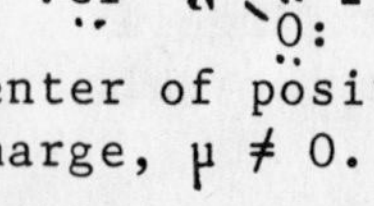

29. Each of the species has 16 valence electrons (8 pairs)

$N_2O \quad NO_2^+ \quad N_3^-$

30. $O_3$ has 18 (nine pairs) of valence electrons and the two possible resonance structures:

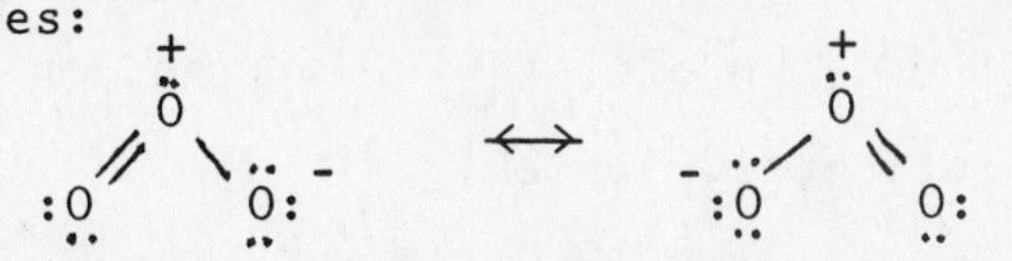

$O_3$ is not given the cyclic structure :O—O: because experiment shows an OOO bond angle of 120° and the distance between two of the O atoms is too large for them to be bonded together. Thus, the arrangement of the atoms is angular, as shown above. The two OO bonds are equal in length, and the OO bond distance is intermediate between that expected for a single bond and a double bond, which is consistent with the two resonance structures, which predict a bond order of 1.5.

31. $$2NO_2(g) \rightleftharpoons N_2O_4(g)$$

This question is an example of the application of Le Châtelier's principle.

(a) This would have no effect, because the partial pressures of $NO_2$ and $N_2O_4$ at equilibrium would not be affected.

(b) Increasing the volume at constant temperature is the same as reducing the partial pressures of $NO_2$ and $N_2O_4$ in the system. The system would respond by establishing a new position of equilibrium in which there would be an increase in the total moles of gases, to counteract the initial decrease in pressure. More $N_2O_4$ would dissociate into $NO_2$. The concentration of $N_2O_4$ would decrease.

(c) From the data in Appendix B.

$$\Delta H^o = [\Delta H_f^o(N_2O_4,g)] - [2\Delta H_f^o(NO_2,g)] = [(9.3)] - [2(33.20]$$
$$= \underline{-57.1 \text{ kJ}} \text{ (exothermic).}$$

Decreasing the temperature at constant volume would favor the forward exothermic reaction. The concentration of $N_2O_4$ would increase.

32. (a) $$HNO_2(aq) + H_2O(l) \rightleftharpoons H_3O^+(aq) + NO_2^-(aq)$$

| | $HNO_2$ | $H_3O^+$ | $NO_2^-$ | |
|---|---|---|---|---|
| initially | 0.020 | 0 | 0 | mol $L^{-1}$ |
| at eq'm | 0.020-x | x | x | mol $L^{-1}$ |

$$pK_a(HNO_2) = 3.14;\ K_a = 7.2\times10^{-4} \text{ mol L}^{-1} = \frac{[H_3O^+][NO_2^-]}{[HNO_2]}$$
$$= \frac{x^2}{0.020-x}$$

Solving the quadratic equation gives $x = [H_3O^+] = \underline{3.45\times10^{-3} \text{ mol L}^{-1}}$

$\underline{pH = 2.46.}$

(b) $NaNO_2(aq)$ contains the neutral $Na^+(aq)$ ion and the weakly basic $NO_2^-(aq)$ ion.

$$NO_2^-(aq) + H_2O(aq) \rightleftharpoons HNO_2(aq) + OH^-(aq)$$

| | $NO_2^-$ | $HNO_2$ | $OH^-$ | |
|---|---|---|---|---|
| at eq'm | 0.020-x | x | x | mol $L^{-1}$ |

$pK_b(NO_2^-) = 14.00 - 3.14 = 10.85$; $K_b(NO_2^-) = 1.4\times10^{-11}$ mol $L^{-1}$

$$K_b(NO_2^-) = \frac{[HNO_2][OH^-]}{[NO_2^-]} = \frac{x^2}{0.020-x} = 1.4\times10^{-11}$$

$x << 0.020$, $x^2 = 2.8\times10^{-13}$; $x = 5.3\times10^{-7}$ (assumption justified).

$[OH^-] = 5.3\times10^{-7}$ mol $L^{-1}$; pH = 6.28; pH = 14.00-pOH = <u>7.72</u>

( ) This is a buffer solution, for which,

$$K_a(HNO_2) = [H_3O^+]\frac{[NO_2^-]}{[HNO_2]} = [H_3O^+](\frac{0.010\ M}{0.010\ M}) = 7.2\times10^{-4}\ \text{mol L}^{-1}$$

$$\underline{pH = 3.14}$$

33. $4NH_3(g) + 5O_2(g) \longrightarrow 4NO(g) + 6H_2O(g)$, for which

$$\Delta H^o = [4\Delta H_f^o(NO,g) + 6\Delta H_f^o(H_2O,g)] - [4\Delta H_f^o(NH_3,g) + 5\Delta H_f^o(O_2,g)]$$
$$= [4(90.3) + 6(-241.8)] - [4(-46.2) + 5(0)] = \underline{-904.8\ kJ}$$

The reaction is <u>exothermic</u> and will be favored by low temperature. Since 9 mol of gases react to give 10 mol of gases, low pressure will favor the forward reaction.

34. Use the relationship, $\Delta H = \sum(\text{BE reactants}) - \sum(\text{BE products})$, applied to the reaction,

$$3O_2(g) \longrightarrow 2O_3(g) \quad \Delta H^o = 2\Delta H_f^o(O_3,g) = \underline{284\ kJ\ mol^{-1}}$$

3 :Ö=Ö:   2 :Ö=O$^+$—Ö:$^-$

$$\Delta H^o = 284\ kJ\ mol^{-1} = 3BE(O=O) - 4BE(OO) = 3(498) - 4BE(OO)$$
$$\underline{BE(OO) = 303\ kJ\ mol^{-1}}$$

35. $CO(g) + NO_2(g) \longrightarrow CO_2(g) + NO(g)$

$$\Delta H^o = [\Delta H_f^o(CO_2,g) + \Delta H_f^o(NO,g)] - [\Delta H_f^o(CO,g) + \Delta H_f^o(NO_2,g)]$$
$$= [(-393.5) + (90.3)] - [(-110.5) + (33.2)] = \underline{-225.9\ kJ}$$

36. The first step in the reaction is $NO(g) + O_3(g) \longrightarrow NO_2(g) + O_2(g)$,

$$\Delta H^o = [\Delta H_f^o(NO_2,g) + \Delta H_f^o(O_2,g)] - [\Delta H_f^o(NO,g) + \Delta H_f^o(O_3,g)]$$
$$= [(33.2) + (0)] - [(90.3) + (142.7)] = \underline{-199.8\ kJ}$$

The second step involves the reaction, $NO_2(g) + O(g) \longrightarrow O_2(g) + NO(g)$.

$$\Delta H^o = [\Delta H_f^o(O_2,g) + \Delta H_f^o(NO,g)] - [\Delta H_f^o(NO_2,g) + \Delta H_f^o(O,g)]$$
$$= [(0) + (90.3)] - [(33.2) + (249)] = \underline{-192\ kJ}$$

Note that $\Delta H_f^o(O,g)$ is one-half of the bond dissociation energy of $O_2(g)$.

37. (a) $N_2O_3(g) \longrightarrow NO(g) + NO_2(g)$

$$\Delta H^o = [\Delta H_f^o(NO,g) + \Delta H_f^o(NO_2,g)] - [\Delta H_f^o(CO,g) + \Delta H_f^o(NO_2,g)]$$
$$= [(90.3) + (33.2)] - [(83.7)] = \underline{39.8\ kJ}$$

$$\Delta S^\circ = [S^\circ(NO,g) + S^\circ(NO_2,g)] - [S^\circ(N_2O_3,g)]$$

$$= [(210.7) + (239.9)] - [(312.2)] = \underline{138.4 \text{ J K}^{-1}}$$

$$\Delta G^\circ = \Delta H^\circ - T\Delta S^\circ = 39.8 \text{ kJ} - (298 \text{ K})(138.4 \text{ J K}^{-1})(\frac{1 \text{ kJ}}{1000 \text{ J}})$$

$$= \underline{-1.4 \text{ kJ}}$$

(b) The reaction is <u>spontaneous</u>.

(c) $\Delta G^\circ = -RT \ln K_p$.

$$\ln K_p = -\frac{\Delta G^\circ}{RT} = -\frac{(-1.4 \text{ kJ})(\frac{1000 \text{ kJ}}{1 \text{ kJ}})}{(8.314 \text{ J K}^{-1})(298 \text{ K})} = 0.57$$

$$K_p = e^{0.57} = \underline{1.8 \text{ atm}}$$

(d) The decomposition will not occur when $\Delta G > 0$, so calculate the temperature for which $\Delta G = 0$, assuming that $\Delta H$ and $\Delta S$ are independent of temperature.

$$\Delta G = \Delta H - T\Delta S = (39.8 \text{ kJ}) - T(138.4 \text{ J K}^{-1})(\frac{1 \text{ kJ}}{1000 \text{ J}}) = 0$$

$$T = 288 \text{ K} = \underline{15^\circ\text{C}}$$

Thus, $N_2O_3$ is stable below $15^\circ$C

(e) <u>Lewis structures</u>

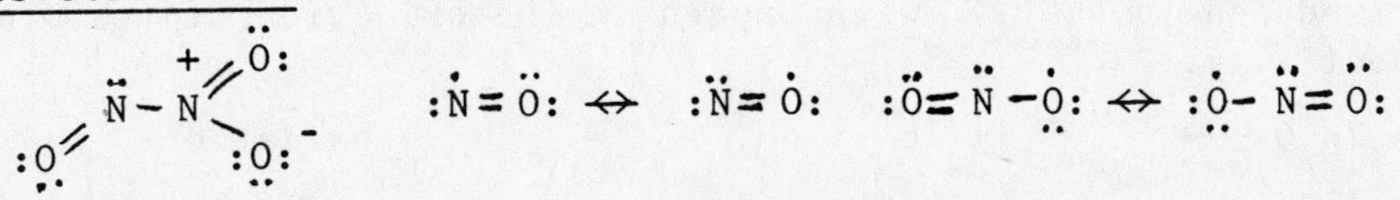

38. (a) The largest entropy change will be associated with the reaction that gives the larger number of moles of gases per mol of $N_2H_4(l)$ that is decomposed.

$N_2H_4(l) \longrightarrow N_2(g) + 2H_2(g)$ ; 1 mol $\longrightarrow$ 3 mol gases

$3N_2H_4(l) \longrightarrow 4NH_3(g) + N_2(g)$; 1 mol $\longrightarrow \frac{5}{3}$ mol gases

The first reaction will have the larger entropy change.

(b) $$\Delta H^\circ = [\Delta H_f^\circ(N_2,g) + 2\Delta H_f^\circ(H_2,g)] - [\Delta H_f^\circ(N_2H_4,l)]$$

$$= [(0) + 2(0)] - [(50.5 \text{ kJ})] = \underline{-50.5 \text{ kJ}}$$

$$S^\circ = [S^\circ(N_2,g) + 2S^\circ(H_2,g)] - [S^\circ(N_2H_4,l)]$$

$$= [(191.5) + 2(130.6)] - [(121.2)] = \underline{331.5 \text{ J K}^{-1}}$$

$$\Delta G^\circ = \Delta H^\circ - T\Delta S^\circ = -50.5 \text{ kJ} - (298 \text{ K})(331.5 \text{ J K}^{-1})(\frac{1 \text{ kJ}}{1000 \text{ J}})$$

$$= \underline{-103.8 \text{ kJ}}$$

(c) Yes, $\Delta G^\circ$ is negative.

39. (a) $NH_4^+$, N(-3) (b) $N_2H_4$, N(-2) (c) $KNO_2$, N(+3) (d) $NO_2$, N(+4)
(e) $NH_2OH$, N(-1) (f) $N_2O$ N(+1) (g) $NH_4NO_3$ N(+1)* (h) $LiN_3$ N($-\frac{1}{3}$)
(i) $Li_3N$ N(-3)

*or N(-3) & N(+5)

40. (a) $3(H_2S \longrightarrow S + 2e^- + 2H^+(aq))$ oxidation

$2[NO_3^- + 3e^- + 4H^+ \longrightarrow NO + 2H_2O]$ reduction

$3H_2S(aq) + 2NO_3^-(aq) + 2H_3O^+(aq) \longrightarrow 3S(s) + NO(g) + 4H_2O(l)$

(b) $4[Zn + 4OH^- \longrightarrow Zn(OH)_4^{2-} + 2e^-]$ oxidation

$NO_3^- + 8e^- + 6H_2O \longrightarrow NH_3 + 9OH^-$ reduction

$4Zn(s) + 7OH^-(aq) + NO_3^-(aq) + 6H_2O(l) \longrightarrow 4Zn(OH)_4^{2-}(aq) + NH_3(aq)$

(c) $3[P_4 + 16H_2O \longrightarrow 4H_3PO_4 + 20\ e^- + 20\ H^+]$ oxidation

$20[NO_3^- + 3e^- + 4H^+ \longrightarrow NO + 2H_2O]$ reduction

$3P_4(s) + 20NO_3^-(aq) + 8H_2O(l) + 20\ H_3O^+(aq) \longrightarrow 12H_3PO_4(aq) +$

$20\ NO(g) + 20H_2O(l)$

41. (a) $NO_3^- + e^- + 2H^+ \longrightarrow NO_2 + H_2O$

(b) $NO_3^- + 3e^- + 4H^+ \longrightarrow NO + 2H_2O$

(c) $2NO_3^- + 10e^- + 12H^+ \longrightarrow N_2 + 6H_2O$

(d) $NO_3^- + 8e^- + 10H^+ \longrightarrow NH_4^+ + 3H_2O$

42. Oxygen in $H_2O_2$ is in the -1 oxidation state and can be oxidized to $O_2(g)$ or reduced to $H_2O(l)$, with oxygen in the oxidation states 0 and -2, respectively.

(a) $SO_2 + 2H_2O \longrightarrow SO_4^{2-} + 2e^- + 4H^+$ oxidation

$H_2O_2 + 2e^- + 2H^+ \longrightarrow 2H_2O$ reduction

$SO_2 + H_2O_2 \longrightarrow SO_4^{2-} + 2H^+$

or $SO_2(g) + H_2O_2(l) + 2H_2O(l) \longrightarrow SO_4^{2-}(aq) + 2H_3O^+(aq)$

(b) $O_3 + 6e^- + 6H^+ \longrightarrow 3H_2O$ reduction

$3[H_2O_2 \longrightarrow O_2 + 2e^- + 2H^+]$ oxidation

$O_3(g) + 3H_2O_2(aq) \longrightarrow 3H_2O(l) + 3O_2(g)$

(c) $2I^- \longrightarrow I_2 + 2e^-$ oxidation

$H_2O_2 + 2e^- + 2H^+ \longrightarrow 2H_2O$ reduction

$2I^- + H_2O_2 + 2H^+ \longrightarrow I_2 + 2H_2O$

or $2I^-(aq) + H_2O_2(aq) + 2H_3O^+(aq) \longrightarrow I_2(s) + 4H_2O(l)$

(d) $2[Cr(OH)_3 + 5OH^- \longrightarrow CrO_4^{2-} + 3e^- + 4H_2O]$ oxidation

$3[H_2O_2 + 2e^- \longrightarrow 2OH^-]$ reduction

$2Cr(OH)_3(s) + 3H_2O_2(aq) + 4OH^-(aq) \longrightarrow 2CrO_4^{2-}(aq) + 8H_2O(l)$

(e) $NO_2^- + H_2O \longrightarrow NO_3^- + 2e^- + 2H^+$ oxidation

$H_2O + 2e^- + 2H^+ \longrightarrow 2H_2O$ reduction

$NO_2^-(aq) + H_2O_2(aq) \longrightarrow NO_3^-(aq) + H_2O(l)$

43. Nitrogen in $HNO_2$ has oxidation number +3. Since there are common oxidation states up to +5 (in e.g. $HNO_3$), and down to -3 (in e.g. $NH_3$), nitrous acid can be both oxidized and reduced. In other words it can behave either as a reducing agent or as an oxidizing agent.

Both oxidation and reduction occur in the disproportionation of nitrous acid in concentrated aqueous solution,

$$2HNO_2(aq) \longrightarrow NO(g) + NO_2(g) + H_2O(l)$$

In many reactions, $NO_2^-(aq)$ is oxidized to $NO_3^-(aq)$. Examples include reduction of $MnO_4^-(aq)$ to $Mn^{2+}(aq)$ in acid solution. In its behaviour as an oxidizing agent, nitrous acid is most commonly reduced to NO(g). Examples of reactions in which $HNO_2(aq)$ behaves as an oxidizing agent include reactions such as oxidation of $SO_2$ to $H_2SO_4$, oxidation of $H_2S$ to sulfur, and oxidation of $Fe^{2+}(aq)$ to $Fe^{3+}(aq)$, all in acid solution.

44. $O_3(g) + 2H^+ + 2e^- \longrightarrow O_2(g) + H_2O$ $E^o_{red} = 2.07$ V

and for oxygen in aqueous solution from Table 17.1,

$O_2(g) + 4H^+ + 4e^- \longrightarrow 2H_2O$ $E^o_{red} = 1.23$ V

To decide if a particular oxidation will occur under standard conditions each of the above reduction potentials is combined with the standard oxidation potential for a particular oxidation, $E^o_{ox} = -E^o_{red}$. If the result is a positive voltage the reaction is spontaneous under standard conditions.

(a) $2H_2O \longrightarrow H_2O_2(aq) + 2H^+ + 2e^-$, $E^o_{ox} = -1.78$ V

For oxidation with $O_3(g)$, $E^o_{cell} = +0.29$ V - spontaneous

For oxidation with $O_2(g)$, $E^o_{cell} = -0.55$ V - not spontaneous

(b) $Mn^{2+} + 4H_2O \longrightarrow MnO_4^- + 8H^+ + 5e^-$, $E^o_{ox} = -1.49$ V

For oxidation with $O_3(g)$, $E^o_{cell} = +0.58$ V - spontaneous

For oxidation with $O_2(g)$, $E^o_{cell} = -0.26$ V - not spontaneous

(c) $2Cr^{3+} + 7H_2O \longrightarrow Cr_2O_7^{2-} + 14H^+ + 6e^-$, $E^o_{ox} = -1.33$ V

For oxidation with $O_3(g)$, $E^o_{cell} = +0.74$ V - spontaneous

For oxidation with $O_2(g)$, $E^o_{call} = -0.10$ V - not spontaneous

(d) $Co^{2+} \longrightarrow Co^{3+} + e^-$, $E^o_{ox} = -1.81$ V

For oxidation with $O_3(g)$, $E^o_{cell} = +0.26$ V - spontaneous

For oxidation with $O_2(g)$, $E^o_{cell} = -0.58$ V - not spontaneous

In all of these cases, $O_3(g)$ is effective and $O_2(g)$ is ineffective.

45. An oxide that supports combustion and is prepared by gently heating ammonium nitrate is dinitrogen monoxide, $N_2O(g)$. Assuming that X is $N_2O(g)$, the reaction with white phosphorus is <u>either</u>

$$P_4(s) + 6N_2O(g) \longrightarrow P_4O_6(s) + 6N_2(g) \quad ..... \quad (1)$$

<u>or</u>

$$P_4(s) + 10N_2O(g) \longrightarrow P_4O_{10}(s) + 10N_2(g) \quad ..... \quad (2)$$

Both reactions are consistent with the observations that the volume of the product gas is equal to the initial volume of $N_2O(g)$ at the same temperature and pressure, and that the product gas does not support combustion, so that Y must be $N_2(g)$.

From the initial mass of $P_4$ of 0.1020 g, the anticipated mass of each oxide may be calculated:

$$(1)\ \text{mass } P_4O_6 = (0.1020\ \text{g } P_4)\left(\frac{1\ \text{mol } P_4}{123.9\ \text{g } P_4}\right)\left(\frac{219.9\ \text{g } P_4O_6}{1\ \text{mol } P_4O_6}\right) = \underline{0.1810\ \text{g}}$$

$$(2)\ \text{mass } P_4O_{10} = (0.1020\ \text{g } P_4)\left(\frac{1\ \text{mol } P_4}{123.9\ \text{g } P_4}\right)\left(\frac{283.9\ \text{g } P_4O_{10}}{1\ \text{mol } P_4O_{10}}\right) = \underline{0.2337\ \text{g}}$$

The calculation in (2) justifies the correctness of the assumptions, is consistent with $P_4O_{10}(s)$ as the oxide of phosphorus formed, and confirms that Y is $N_2(g)$. Since equal volumes of gases at the same temperature and pressure contain equal numbers of moles of gases, the expected ratio of densities is the ratio of the molar masses of the gases. i.e.,

$$\frac{\text{density } N_2O}{\text{density } N_2} = \left(\frac{44.02\ \text{g mol}^{-1}}{28.02\ \text{g mol}^{-1}}\right) = \underline{1.571}, \quad \underline{\text{as observed}}$$

The balanced equation for the decomposition of $NH_4NO_3(s)$ is

$$NH_4NO_3(s) \longrightarrow N_2O(g) + 2H_2O(g)$$

a reaction in which N, oxidation number -3, in $NH_4^+$, is oxidized to N, oxidation number 0, in $N_2$, and N, oxidation number +5, in $NO_3^-$ is reduced to N(0) in $N_2$. This is a self-disproportionation reaction in which $NO_3^-$ oxidizes $NH_4^+$ and $NH_4^+$ reduces $NO_3^-$.

<u>Lewis structures</u>

<u>X $N_2O$</u> $\quad ^{-}:\ddot{N}{=}\overset{+}{N}{=}\ddot{O}: \ \longleftrightarrow\ :N{\equiv}\overset{+}{N}{-}\ddot{\underset{..}{O}}:^{-}$ $\qquad$ <u>Y $N_2$</u> $\quad :N{\equiv}N:$

46. (a) $H_2O_2(aq)$ is conveniently prepared in the laboratory by the acid-base reaction between barium peroxide, $BaO_2(s)$, prepared from heating BaO(s) to red heat in a stream of $O_2(g)$, and dilute $H_2SO_4(aq)$.

$$BaO_2(s) + H_2SO_4(aq) \longrightarrow BaSO_4(s) + H_2O_2(aq)$$

The other product, $BaSO_4(s)$, is an insoluble salt and is removed from the mixture by filtration.

(b) (i) $PbS(s) + 4H_2O(l) \longrightarrow PbSO_4(s) + 8H^+(aq) + 8e^-$ $\quad$ oxidation

$4[H_2O_2(aq) + 2e^- + 2H^+(aq) \longrightarrow 2H_2O(l)$ $\quad$ reduction

$$PbS(s) + 4H_2O_2(aq) \longrightarrow PbSO_4(s) + 4H_2O(l)$$

The $H_2O_2(aq)$ behaves as an <u>oxidizing agent</u>.

(ii) $[Fe^{2+}(aq) \longrightarrow Fe^{3+}(aq) + e^-]$ oxidation

$H_2O_2(aq) + 2e^- + 2H_3O^+ \longrightarrow 4H_2O(l)$ reduction

---

$2Fe^{2+}(aq) + H_2O_2(aq) + 2H_3O^+(aq) \longrightarrow 2Fe^{3+}(aq) + 4H_2O(l)$

The $H_2O_2(aq)$ behaves as an <u>oxidizing agent</u>.

(iii) $3I^-(aq) \longrightarrow I_3^-(aq) + 2e^-$ oxidation

$H_2O_2(aq) + 2e^- + 2H_3O^+(aq) \longrightarrow 4H_2O(l)$ reduction

---

$3I^-(aq) + H_2O_2(aq) + 2H_3O^+(aq) \longrightarrow I_3^-(aq) + 4H_2O(l)$

The $H_2O_2(aq)$ behaves as an <u>oxidizing agent</u>.

(iv) $HOCl(aq) + 2e^- + 2H_3O^+(aq) \longrightarrow HCl(aq) + 3H_2O(l)$ reduction

$H_2O_2(aq) + 2H_2O(l) \longrightarrow O_2(g) + 2e^- + 2H_3O^+(aq)$ oxidation

---

$HOCl(aq) + H_2O_2(aq) \longrightarrow HCl(aq) + O_2(g) + H_2O(l)$

The $H_2O_2(aq)$ behaves as a <u>reducing agent</u>.

(v) $2[MnO_4^- + 5e^- + 8H_3O^+ \longrightarrow Mn^{2+}(aq) + 12H_2O(l)$ reduction

$5[H_2O_2(aq) + 2H_2O(l) \longrightarrow O_2(g) + 2e^- + 2H_3O^+(aq)$ oxidation

---

$2MnO_4^-(aq) + 5H_2O_2(aq) + 6H_3O^+(aq) \longrightarrow 2Mn^{2+}(aq) + 5O_2(g) + 14\,H_2O(l)$

The $H_2O_2(aq)$ behaves as a <u>reducing agent</u>.

47. (a) $N_2H_4(l) \longrightarrow N_2(g) + 2H_2(g)$ ..... (1)

$3N_2H_4(l) \longrightarrow N_2(g) + 4NH_3(g)$ ..... (2)

(b) The favored reaction will be that with the most negative $\Delta G^o$.

(1) $\Delta G^o$ is $- \Delta G^o_f\ (N_2H_4, l)$ = <u>$-149.2$ kJ mol$^{-1}$</u>

(2) $\Delta G^o = [\Delta G^o_f(N_2,g) + 4\,\Delta G^o_f(NH_3,g)] - [3\,\Delta G^o_f(N_2H_4,l)]$

$= [(0) + 4(-16.4)] - [3(149.2)] = -513.2$ kJ

<u>or $-171.1$ kJ mol$^{-1}$</u>

i.e., the <u>second</u> reaction; it has the more negative $\Delta G^o$.

48. The reaction described is the catalyzed oxidation of $NH_3(g)$ to $NO(g)$, which then reacts with oxygen in the air to gives brown $NO_2(g)$.

$4NH_3(g) + 5O_2(g) \longrightarrow 4NO(g) + 6H_2O(g)$

$2NO(g) + O_2(g) \longrightarrow 2NO_2(g)$ (brown fumes)

49. (a) Phosphorus is in group 5 and period 3, while nitrogen is in group 5 and period 2. Second period elements have only 2s and 2p orbitals available while third period elements have 3s, 3p, and 3d orbitals. Thus, nitrogen is limited to the valence state

2s [↑↓] 2p [↑|↑|↑]

and a valence of 3, while phosphorus has a valence of 3 in its ground state and a valence of 5 in its excited state.

2s [↑↓] 2p [↑|↑|↑] and 2s [↑] 2p [↑|↑|↑] 3d [↑| | | | ]

Thus, nitrogen is limited to a valence shell of eight electrons and the formation of compounds such $NF_3$, while phosphorus can expand its valence shell beyond the octet and with highly electronegative elements form compounds such as $PCl_5$ and $PF_5$, in addition to forming $PCl_3$ and $PF_3$.

(b) Phosphorus is considerably larger than nitrogen so that its lone pair is larger and less localized and therefore attracts a proton less strongly than does the nitrogen lone pair. Thus $PH_3$ is a weaker base than $NH_3$. Also the P-H bond is weaker than the N-H bond because of the larger size and smaller electronegativity of phosphorus compared to nitrogen. Thus the P-H bond is more easily broken than the N-H bond, so that $PH_4^+$ is a stronger acid than $NH_4^+$; in other words, $PH_3$ is a weaker base than $NH_3$.

(c) Nitrogen, $N_2(g)$, contains a very strong triple bond while phosphorus, $P_4$, is a tetrahedral molecules with single P-P bonds of much lower bond energy. Thus, it is very much easier to break a P-P bond in $P_4$ than it is to break the :N≡N: bond in $N_2$. Another important difference is that the N atoms in $N_2$ have valence shells filled with four pairs of electrons and obey the octet rule, while phosphorus in $P_4$ also has a valence shell with four pairs but can expand its valence shell to more than four pairs; it is therefore vulnerable to attack by Lewis bases while $N_2$ is not.

(d) White phosphorus is a molecular solid consisting of $P_4$ molecules, while red phosphorus consists of giant two-dimensional sheets of P atoms covalently bonded in a network structure. Each has 3 bonds to each P atom, but those in $P_4$ are bent and much weaker than those in red phosphorus.

50.

| Species | Valence electrons | Lewis Structure | Bond Order | Bond Length (pm) |
|---|---|---|---|---|
| $O_2$ | 12 | :O=O: | 2.00 | 121 |
| $O_3$ | 18 | :O=O⁺–O:⁻ | 1.50 | 128 |
| $O_2^-$ | 13 | ½- :O⋯O: ½- | 1.50 | 128 |
| $O_2^{2-}$ | 14 | ⁻:O–O:⁻ | 1.00 | 149 |

As anticipated from their bond orders, the bond lengths in $O_3$ (two resonance structures) correspond to bond orders of 1.50, and the bond length is the same as that in the $O_2^-$ ion, also with a bond order of 1.50. In general, bond length decreases with increased bond order, which is the relationship observed for all covalent bonds.

51.

| Species | Valence electrons | Lewis Structure | Bond Angle |
|---|---|---|---|
| $NO_2^+$ | 16 | :O=N⁺=O: | 180° |
| $NO_2$ | 17 | :O=N–O: ↔ :O–N=O: | 135° |
| $NO_2^-$ | 18 | :O=N–O:⁻ | 115° |

$NO_2^+$ is an $AX_2$ linear species and thus the ONO bond angle is 180°. Both $NO_2$ and $NO_2^-$ are represented by two resonance structures of the type shown and have $AX_2E$ angular geometry. In the case of $NO_2$ the E represents a single unpaired electron on N, while in $NO_2^-$ it represents a lone pair of electrons, and the respective NO bond orders are 1.75 and 1.50. A single unpaired electron takes up considerably less space in the valence shell of nitrogen than does a lone pair, which accounts for an ONO bond angle of 135° in $NO_2$, and a considerably smaller bond angle of 115° in $NO_2^-$.

52. $$NH_4NO_2(s) \longrightarrow N_2(g) + 2H_2O(g)$$

Mol of $N_2(g) = (10.0\ g\ NH_4NO_2)\left(\frac{1\ mol\ NH_4NO_2}{64.04\ g\ NH_4NO_2}\right)\left(\frac{1\ mol\ N_2}{1\ mol\ NH_4NO_2}\right) = \underline{0.156\ mol}$

$$V = \frac{nRT}{P} = \frac{(0.156\ mol)(0.0821\ atm\ L\ mol^{-1}\ K^{-1})(333\ K)}{(740\ torr)\left(\frac{1\ atm}{760\ torr}\right)} = \underline{4.38\ L}$$

53. $$2NaNO_3(s) \longrightarrow 2NaNO_2(s) + O_2(g)$$

$$\text{Mol } O_2 = \frac{PV}{RT} = \frac{(739\ torr)\left(\frac{1\ atm}{760\ torr}\right)(131.8\ mL)\left(\frac{1\ L}{1000\ mL}\right)}{(0.0821\ atm\ L\ mol^{-1}\ K^{-1})(298\ K)} = \underline{5.24\times10^{-3}\ mol}$$

Theoretical yield of $O_2(g) = (1.354\ g\ NaNO_2)\left(\frac{1\ mol\ NaNO_2}{69.00\ g\ NaNO_2}\right)\left(\frac{1\ mol\ O_2}{2\ mol\ NaNO_2}\right)$

$= \underline{9.812\times10^{-3}\ mol}$

Purity of sample $= \left(\frac{100\%}{9.812\times10^{-3}\ mol}\right)(5.24\times10^{-3}\ mol) = \underline{53.4\%}$

## CHAPTER 19

1. Since HCl(aq) is totally dissociated to $H_3O^+$(aq) and $Cl^-$(aq), the rate of the reaction can be measured by the rate at which $H_3O^+$ disappears from solution,

$$\text{Rate} = -\frac{\Delta[H_3O^+]}{\Delta t}$$

| t (min) | $[H_3O^+]$ (mol $L^{-1}$) | $\Delta t$ (min) | $\Delta[H_3O^+]$ (mol $L^{-1}$) | Average Rate (mol $L^{-1}$ $min^{-1}$) |
|---|---|---|---|---|
| 0 | 1.85 | | | |
| 80 | 1.66 | 80 | -0.19 | $2.4 \times 10^{-3}$ |
| 159 | 1.53 | 79 | -0.13 | $1.6 \times 10^{-3}$ |
| 314 | 1.31 | 155 | -0.22 | $1.4 \times 10^{-3}$ |
| 628 | 1.02 | 314 | -0.29 | $0.92 \times 10^{-3}$ |

Note that as time passes and $[H_3O^+]$ decreases, the average rate also decreases

2. Measure the rate of this reaction by the rate at which $SO_2Cl_2$ disappears,

$$\text{Rate} = -\frac{\Delta[SO_2Cl_2]}{\Delta t}$$

| t (min) | $[SO_2Cl_2]$ (mol $L^{-1}$) | $\Delta t$ (min) | $\Delta[SO_2Cl_2]$ (mol $L^{-1}$) | Average Rate (mol $L^{-1}$ $min^{-1}$) |
|---|---|---|---|---|
| 0 | 0.010 00 | | | |
| 20 | 0.009 70 | 20 | -0.000 30 | $1.5 \times 10^{-5}$ |
| 50 | 0.009 28 | 30 | -0.000 42 | $1.4 \times 10^{-5}$ |
| 100 | 0.008 61 | 50 | -0.000 67 | $1.3 \times 10^{-5}$ |
| 200 | 0.007 41 | 100 | -0.001 20 | $1.2 \times 10^{-5}$ |
| 400 | 0.005 49 | 200 | -0.001 92 | $0.96 \times 10^{-5}$ |
| 700 | 0.003 50 | 300 | -0.001 99 | $0.66 \times 10^{-5}$ |
| 1000 | 0.002 23 | 300 | -0.001 27 | $0.42 \times 10^{-5}$ |

3. (a) $$\text{Rate} = -\frac{\Delta[HI]}{\Delta t} = \frac{2\Delta[H_2]}{\Delta t} = \frac{2\Delta[I_2]}{\Delta t}$$

(b) $$\text{Rate} = -\frac{\Delta[NOCl]}{\Delta t} = \frac{\Delta[NO]}{\Delta t} = \frac{2\Delta[Cl_2]}{\Delta t}$$

(c) $$\text{Rate} = -\frac{\Delta[N_2O_5]}{\Delta t} = \frac{1}{2}\frac{\Delta[NO_2]}{\Delta t} = \frac{2\Delta[O_2]}{\Delta t}$$

4. (a) $$\text{Rate} = -\frac{\Delta[C_4H_{10}]}{\Delta t} = -\frac{2}{13}\frac{\Delta[O_2]}{\Delta t} = \frac{1}{4}\frac{\Delta[CO_2]}{\Delta t} = \frac{1}{5}\frac{\Delta[H_2O]}{\Delta t}$$

(b) Rate $= -\frac{\Delta[C_2H_6]}{\Delta t} = -\frac{1}{2}\frac{\Delta[H_2O]}{\Delta t} = \frac{1}{2}\frac{\Delta[CO]}{\Delta t} = \frac{1}{5}\frac{\Delta[H_2]}{\Delta t}$

(c) Rate $= -\frac{\Delta[Br^-]}{\Delta t} = -5\frac{\Delta[BrO_3^-]}{\Delta t} = -\frac{5}{6}\frac{\Delta[H_3O^+]}{\Delta t} = \frac{5}{3}\frac{\Delta[Br_2]}{\Delta t}$

5. Rate $= \frac{\Delta[N_2]}{\Delta t} = \frac{2}{6}\frac{\Delta[H_2O]}{\Delta t} = -\frac{2}{4}\frac{\Delta[NH_3]}{\Delta t} = -\frac{2}{3}\frac{\Delta[O_2]}{\Delta t} = \underline{0.27 \text{ mol L}^{-1} \text{ s}^{-1}}$

(a) Rate of formation of $H_2O = \frac{6}{2}\frac{\Delta[N_2]}{\Delta t} = \underline{0.81 \text{ mol L}^{-1} \text{ s}^{-1}}$

(b) Rate of consumption of $NH_3 = \frac{4}{2}\frac{\Delta[N_2]}{\Delta t} = \underline{0.54 \text{ mol L}^{-1} \text{ s}^{-1}}$

(c) Rate of consumption of $O_2 = \frac{3}{2}\frac{\Delta[N_2]}{\Delta t} = \underline{0.41 \text{ mol L}^{-1} \text{ s}^{-1}}$

6. (a) Rate $= -\frac{\Delta[O_2]}{\Delta t} = -\frac{\Delta[H_2]}{\Delta t} = \frac{\Delta[H_2O_2]}{\Delta t}$

(b) Rate $= -\frac{\Delta[O_2]}{\Delta t} = -\frac{\Delta[H_2]}{\Delta t} = \frac{\Delta[H_2O_2]}{\Delta t}$

(c) Rate $= -\frac{\Delta[O_2]}{\Delta t} = -\frac{1}{2}\frac{\Delta[NO]}{\Delta t} = \frac{1}{2}\frac{\Delta[NO_2]}{\Delta t}$

(d) Rate $= -\frac{\Delta[O_2]}{\Delta t} = -\frac{8}{4}\frac{\Delta[PH_3]}{\Delta t} = \frac{8}{6}\frac{\Delta[H_2O]}{\Delta t}$

7. Rate $= \frac{\Delta(p_{NO})}{\Delta t} = 1095 \text{ mm Hg s}^{-1}$

Using the ideal gas equation to convert pressure to mol $L^{-1}$,

$$\frac{n}{V} = \frac{P}{RT}$$

gives, $\frac{\Delta[NO]}{\Delta t} = \frac{1}{RT}\left(\frac{dp_{NO}}{dt}\right) = \frac{(1095 \text{ mm Hg})\left(\frac{1 \text{ atm}}{760 \text{ mm Hg}}\right) \text{ s}^{-1}}{(0.082 \text{ atm L mol}^{-1} \text{ K}^{-1})(500 \text{ K})}$

$= \underline{3.51 \times 10^{-2} \text{ mol L}^{-1} \text{ s}^{-1}}$

and $\frac{\Delta[NO]}{\Delta t} = -\frac{\Delta[NH_3]}{\Delta t} = -\frac{6}{5}\frac{\Delta[O_3]}{\Delta t}$

Rate of disappearance of $O_3 = -\frac{\Delta[O_3]}{\Delta t} = \frac{5}{6}\frac{\Delta[NO]}{\Delta dt} = \underline{2.92\text{x}10^{-2} \text{ mol L}^{-1} \text{ s}^{-1}}$

Rate of disappearance of $NH_3 = -\frac{\Delta[NH_3]}{\Delta t} = \frac{\Delta[NO]}{\Delta t} = \underline{3.51\text{x}10^{-2} \text{ mol L}^{-1} \text{ s}^{-1}}$

8. (a) The rate of a reaction is the rate at which reactants are consumed, or products are formed, measured by the change in the concentration of a reactant or a product with respect to time.

(b) The order of a reaction with respect to a particular reactant is the exponent of the concentration of that reactant as it appears in the experimentally determined rate law. For a reaction which is first order in X, and second order in Y, the rate law is

$$\text{Rate} = k[X][Y]^2$$

where k is the rate constant (constant for a particular temperature)

(c) $k = \dfrac{\text{rate}}{[X][Y]^2}$ (d) $k = \dfrac{\text{mol L}^{-1}\ t^{-1}}{(\text{mol L}^{-1})(\text{mol L}^{-1})^2} = \text{mol}^{-2}\ \text{L}^2\ t^{-1}$

9. For a reaction that is first order in A, first order in B, and second order in C, the rate equation is

$$\text{Rate} = k[A][B][C]^2$$

$$k = \frac{\text{Rate}}{[A][B][C]^2}\ \left(\frac{\text{mol L}^{-1}\ t^{-1}}{(\text{mol L}^{-1})^4}\right)$$ ; Units of k are $\text{mol}^{-3}\ \text{L}^3\ t^{-1}$

10. For a first order reaction, rate = k[A]

(a) $\text{Rate} = k[A] = (3.7\text{x}10^{-2}\ \text{s}^{-1})(0.040\ \text{mol L}^{-1}) = 1.5\text{x}10^{-3}\ \text{mol L}^{-1}\ \text{s}^{-1}$

(b) $\text{Rate} = (1.5 \text{ x } 10^{-3}\ \text{mol L}^{-1}\ \text{s}^{-1})\left(\dfrac{3600\ \text{s}}{1\ \text{h}}\right) = 54\ \text{mol L}^{-1}\ \text{h}^{-1}$

11. (a) $\text{Rate} = k[B]^2[C] = (4.0\text{x}10^{-3}\ \text{mol}^{-2}\ \text{L}^2\ \text{s}^{-1})(0.010\ \text{mol L}^{-1})^3$
$= 4.0\text{x}10^{-9}\ \text{mol L}^{-1}\ \text{s}^{-1}$

(b) $\text{Rate} = k[B]^2[C] = (4.0\text{x}10^{-3}\ \text{mol}^{-2}\ \text{L}^2\ \text{s}^{-1})(0.050\ \text{mol L}^{-1})^3$
$= 5.0\text{x}10^{-7}\ \text{mol L}^{-1}\ \text{s}^{-1}$

12. In general, rate = $k[\text{NOBr}]^x$, where x is the order of the reaction.

x = 0, rate = k, which the data shows to be incorrect, and for

x = 1, a plot of rate versus [NOBr] should be a straight line, and for

x = 2, a plot of rate versus $[\text{NOBr}]^2$ should be a straight line.

Plotting the data shows that x = 2; the reaction is second order in NOBr, and the rate law is

$$\text{Rate} = k[\text{NOBr}]^2$$

Alternatively,

For x = 1 $k = \dfrac{\text{Rate}}{[\text{NOBr}]}$ and for x = 2, $k = \dfrac{\text{Rate}}{[\text{NOBr}]^2}$ and calculation of each expression from the data shows that only for the second is a constant value obtained for k, so the reaction is second order.

13. Assume that the average rate calculated for a given time interval corresponds to the rate for the average concentration in that interval. Then plot rate versus $[SO_2Cl_2]$ and rate versus $[SO_2Cl_2]^2$. Only the former is a straight line, showing that the reaction is first order.

$$\text{Rate} = k[SO_2Cl_2]$$

14. Since doubling the $[Cl_2]$ increases the rate by a factor pf 2, the reaction must be first order in $[Cl_2]$ and since doubling both $[Cl_2]$ and [NO] increases the rate by a factor of 8, the effect of just doubling [NO] would be to increase the rate of the reaction by a factor of 4. Thus, the reaction is second order in NO, and third order overall, with the rate law

$$\text{Rate} = k[Cl_2][NO]^2$$

15. $$CH_3CHO(g) \longrightarrow CH_4(g) + CO(g)$$

Rate = $k[CH_3CHO]^x$, and when $[CH_3CHO]$ is increased by a factor of 2.00, the rate increases by a factor of 2.83.

Thus, $(2.00)^x = 2.83$, or $x = 1.50$.

The reaction is of order 1.5 in ethanal, with the rate law

$$\text{Rate} = k[CH_3CHO]^{1.5}$$

16. Rate = $k[NO]^x[H_2]^y$, and for the given data:

$$0.500 = k[0.150]^x[0.800]^y \quad \ldots\ldots (1)$$

$$0.125 = k[0.075]^x[0.800]^y \quad \ldots\ldots (2)$$

$$0.250 = k[0.150]^x[0.400]^y \quad \ldots\ldots (3)$$

From (1) and (2), doubling [NO] increases the rate by a factor of 4. Thus, $x = 2$. From (2) and (3), doubling $[H_2]$ increases the rate by a factor of 2. Thus, $y = 1$, and the rate law is

$$\text{Rate} = k[NO]^2[H_2]$$

The value of the rate constant, k, is obtained from the data for any of the experiments. For example, for Exp. 1.,

$$k = \frac{\text{rate}}{[NO]^2[H_2]} = \frac{0.500 \text{ mol L}^{-1} \text{ min}^{-1}}{(0.150 \text{ mol L}^{-1})^2(0.800 \text{ mol L}^{-1})}$$

$$= 27.8 \text{ mol}^{-2} \text{ L}^2 \text{ min}^{-1}$$

17. $$H_2(g) + Br_2(g) \longrightarrow 2HBr(g)$$

Rate = $k[H_2]^x[Br_2]^y$, and when $[H_2]$ is increased by a factor of 2, the rate increases by a factor of 2, so $x = 1$. When $[Br_2]$ is increased by a factor of 3.00, rate increases by a factor of 1.75. Thus,

$$1.75 = 3.00^y \; ; \; \underline{y = 0.50}$$

The reaction is first order in $H_2$ and of order 0.5 in $Br_2$, and the rate law is

$$\underline{\text{Rate} = k[H_2][Br_2]^{0.5}}$$

The reaction is not an elementary process, for which the orders with respect to the reactants would correspond to their stoichiometric coefficients in the balanced equation, i.e., for which the reaction would be first order in both $H_2$ and $Br_2$

18. $$2H_2O_2 \longrightarrow 2H_2O + O_2$$

(a) $t_{\frac{1}{2}} = \underline{17.0 \text{ min}}$

(b) (i) 51.0 min corresponds to three half-lives. The fraction of reactant remaining is

$$(\tfrac{1}{2})^3 = \tfrac{1}{8}$$

(ii) After 10 half-lives, the fraction of reactant remaining is

$$(\tfrac{1}{2})^{10} = \frac{1}{1024}$$

19. When the fraction of reactant remaining is $\frac{1}{4} = (\frac{1}{2})^2$, 2 half-lives must have passed. Therefore $t_{\frac{1}{2}} = \frac{8.0 \text{ min}}{2} = \underline{4.0 \text{ min}}$

When the % of reactant remaining is 3.125%, the fraction of reactant is $\frac{3.125}{100} = \frac{1}{32}$ or $(\frac{1}{2})^5$. Thus it takes 5 half-lives, or 20 min, to go from the initial concentration to 3.125% of the initial concentration.

20. For a first order reaction, a plot of $\ln[SO_2Cl_2]$ versus the time, t, should be a straight line. For a second order reaction a plot of $\frac{1}{[SO_2Cl_2]}$ versus time t should be linear. The plots show this to be a first order reaction. The slope of the straight line is

$$-k = -1.5\text{x}10^{-3} \text{ min}^{-1}\text{; } \underline{k = 1.5 \times 10^{-3} \text{ min}^{-1}}$$

21. The volume of $KMnO_4$(aq) used in the titration is directly proportional to the $[H_2O_2]$ remaining in solution at time t. Thus a plot of ln(volume $KMnO_4$) versus t should be a straight line (just as a plot of $\ln(H_2O_2]$ versus t would be be for a first order reaction.
The slope of the line is $-0.050 \text{ min}^{-1} = -k$. Thus $k = 0.050 \text{ min}^{-1}$, and since $t_{1/2} = \frac{\ln 2}{k}$, $t_{1/2} = \frac{0.693}{0.050 \text{ min}^{-1}} = \underline{13.9 \text{ min}}$

22. (a) The plot of $[N_2O_5]$ versus initial rate is linear, showing that the reaction is first order in $N_2O_5$, rate = $k[N_2O_5]$

(b) The rate constant may be calculated from any one of the four sets of data. Using the first set,

$$k = \frac{rate}{[N_2O_5]} = \left(\frac{1.26 \times 10^{-3} \text{ mol L}^{-1} \text{ min}^{-1}}{2.00 \text{ mol L}^{-1}}\right) = \underline{6.30\text{x}10^{-4} \text{ min}^{-1}}$$

(c) The half-life of the reaction is $t_{1/2} = \frac{\ln 2}{k} = 1.10\text{x}10^3$ min.

For the concentration to decrease from 2.00 mol $L^{-1}$ to 0.50 mol $L^{-1}$ (one-quarter of the initial concentration) two half--lives must have passed. Therefore, $\underline{t = 2.20\text{x}10^3 \text{ min}}$.

(d) Most conveniently, since $O_2(g)$ is evolved from the solution and $N_2O_5$ and $NO_2$ remain in solution, the rate of increase in $[O_2]$ could be followed by following the change in the pressure of the $O_2(g)$ evolved from the solution.

23. Fraction of DDT remaining = $\left(\frac{1 \text{ g DDT}}{1000 \text{ kg DDT}}\right)\left(\frac{1 \text{ kg}}{10^3 \text{ g}}\right) = 10^{-6}$

So that if the number of half=lives for the decrease to 1 g DDT to occur is x, then

$$\left(\frac{1}{2}\right)^x = 10^{-6} \; ; \; x = 19.9 \text{ half-lives}$$

or 19.9(10 years) = $\underline{200 \text{ years}}$

24. Use the integrated rate law, $\ln \frac{[C_o]}{[C_t]} = kt$, $k = 2.5\text{x}10^{-4} \text{ s}^{-1}$, and

$$t = (30 \text{ min})\left(\frac{60 \text{ s}}{1 \text{ min}}\right) = 1800 \text{ s}.$$

$$\ln \frac{200 \text{ torr}}{p \text{ torr}} = (2.5\text{x}10^{-4} \text{ s}^{-1})(1800 \text{ s}) = 0.45$$

$$\frac{200 \text{ torr}}{p \text{ torr}} = e^{0.45} = 1.57, \text{ and } \underline{p = 127 \text{ torr}}$$

i.e., after 30 min, $p_{azomethane} = \underline{127 \text{ torr}}$

| | $H_3CNNCH_3(g) \longrightarrow$ | $C_2H_6(g)$ + | $N_2(g)$ | |
|---|---|---|---|---|
| initially | 200 | 0 | 0 | torr |
| after 30 min | 127 | 73 | 73 | torr |

Thus, the final pressure is (127+73+73) torr = $\underline{273 \text{ torr}}$

25. From experiments 1 and 2, $[O_2]$ decreases by a factor of 2 and the rate also decreases by a factor of 2. Thus, the reaction is first order in $O_2$. From experiments 1 and 4, [NO] increases by a factor of 4 and the rate increases by a factor of 16. Thus the reaction is second order in NO.

$$\underline{\text{Rate} = k[NO]^2[O_2]}$$

$$k = \frac{\text{rate}}{[NO]^2[O_2]} = \frac{0.014 \text{ mol L}^{-1} \text{ s}^{-1}}{(0.010 \text{ mol L}^{-1})^2(0.020 \text{ mol L}^{-1})} = \underline{7.0\times10^3 \text{ mol}^{-2}\text{L}^2 \text{ s}^{-1}}$$

26. (a) $BrO_3^- + 5Br^- + 6H_3O^+ \longrightarrow 3Br_2 + 9H_2O$

(b) Rate = $k[BrO_3^-][Br^-][H_3O^+]^2$

27. (a) $t_{1/2} = \frac{\ln 2}{k}$ ; $k = \frac{0.693}{35.0 \text{ s}} = \underline{0.0198 \text{ s}^{-1}}$

(b) When 90% of the phosphine has decomposed, the amount remaining is 10%.

Thus, if the number of half-lives that have elapsed is x,

$(\frac{1}{2})^x = 0.10$ ; $x = \underline{3.32 \text{ half-lives}}$

Time elapsed = 3.32(35.0 s) = <u>116 s</u>

<u>or</u> <u>1 min 56 s</u>

28. Consider the Arrhenius equation, $k = Ae^{-E_a/RT}$, then

$$\ln k = \ln A - \frac{E_a}{RT}$$

and a plot of ln k versus $\frac{1}{T}$ should be a straight line with slope $-\frac{E_a}{R}$, <u>or</u> $E_a$ = -R(slope).

Here the slope = $-2.24\times10^4$ K,

and $E_a = (8.314 \text{ J K}^{-1} \text{ mol}^{-1})(\frac{1 \text{ kJ}}{10^3 \text{ J}})(2.24\times10^4 \text{ K}) = \underline{186 \text{ kJ mol}^{-1}}$

From the Arrhenius equation, for temperatures $T_1$ and $T_2$,

$$k_{T_1} = k_{T_2} e^{E_a/R(1/T_2 - 1/T_1)}$$

In this case, let $T_2$ = 302°C (575 K), and $T_2$ = 400°C (673 K), then

$$\frac{E_a}{R}(\frac{1}{T_2} - \frac{1}{T_2}) = \frac{(186 \text{ kJ mol}^{-1})(\frac{10^3 \text{ J}}{1 \text{ kJ}})}{8.314 \text{ J K}^{-1} \text{ mol}^{-1}} (\frac{1}{575} - \frac{1}{673}) \text{ K}^{-1} = 5.67$$

$$k_{T_1} = (1.18\times10^{-6} \text{ mol}^{-1} \text{ L s}^{-1})e^{5.67} = \underline{3.42\times10^{-4} \text{ mol}^{-1} \text{ L s}^{-1}}$$

29. $2HI(g) \rightarrow H_2(g) + I_2(g)$

$\Delta H^\circ = \Delta H_f^\circ(\text{products}) - \Delta H_f^\circ(\text{reactants})$

$= [\Delta H_f^\circ(H_2,g) + \Delta H_f^\circ(I_2,g)] - [2\,\Delta H_f^\circ(HI,g)]$

$= (0 + 62.4) - 2(26.4) = \underline{+9.6\ kJ}$

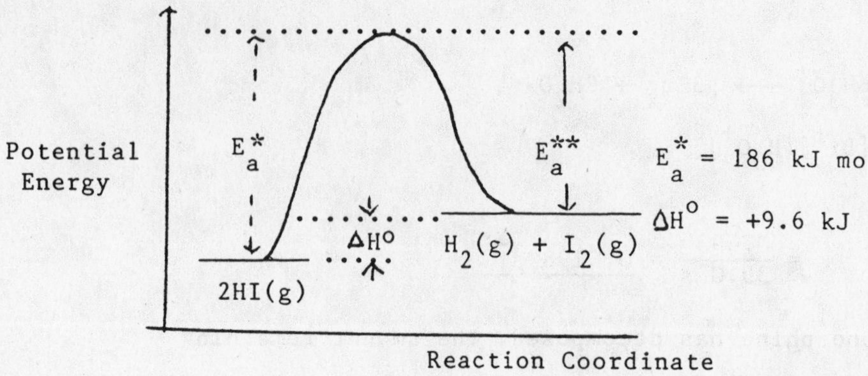


For the reverse reaction, $H_2(g) + I_2(g) \longrightarrow 2HI(g)$,

$$E_a^{**} = (186 - 9.6) = 176\ kJ\ mol^{-1}$$

30. Consider the bimolecular reaction $A + B \longrightarrow$ products.

(a) The greater the frequency of collisions between A and B, the greater is the chance that they will react, and therefore the greater is the rate. The rate is proportional to the number of collisions.

(b) The colliding molecules A and B must have a certain minimum kinetic energy before they will react. In collisions in which the molecules have less than this minimum kinetic energy, no reaction occurs. Since the number of collisions in a given time in which the molecules have sufficient kinetic energy increases with increasing temperature, the rate increases with increasing average kinetic energy of the collisions.

(c) Only collisions in which the colliding molecules have a certain orientation lead to reaction. For large molecules, only a very few out of all many possible orientations of the colliding molecules will lead to reaction. Thus, the less frequently the colliding molecules have the correct orientation, the slower the reaction.

31. Use the Arrhenius equation in the form $\ln(k_{T_1}/k_{T_2}) = \frac{E_a}{R}(1/T_2 - 1/T_2)$.

with $\frac{k_{T_1}}{k_{T_2}} = 10$, and $T_2 = 0^oC$ (273 K).

$$\ln \frac{k_{T_1}}{k_{T_2}} = 2.303 = \frac{(80\ kJ\ mol^{-1})(\frac{10^3\ J}{1\ kJ})}{8.314\ J\ mol^{-1}\ K^{-1}}\left(\frac{1}{273} - \frac{1}{T_1}\right) K^{-1}$$

$$T_1 = 292\ K,\ \underline{or}\ \underline{19^oC}$$

32. The problem is solved in the same way as Problem 31.

Here $\frac{k_{T_1}}{k_{T_2}} = 2$, and $T_2 = 27^oC$ (300 K), and $T_1 = 305$ K, $\underline{or}$ $\underline{32^oC}$

33. Use the equation in Problem 31, and solve for $E_a$,

$$E_a = R\left(\ln \frac{k_{T_1}}{k_{T_2}}\right)\left(\frac{T_1T_2}{T_1 - T_2}\right) \quad \text{with} \quad \frac{k_{T_1}}{k_{T_2}} = 3,\ T_1 = 323\ \text{K and } T_2 = 298\ \text{K}$$

$$E_a = (\ln 3)(8.314\ \text{J K}^{-1}\ \text{mol}^{-1})\left(\frac{1\ \text{kJ}}{10^3\ \text{J}}\right)\left(\frac{323 \times 298}{323 - 298}\right)\ \text{K} = \underline{35.2\ \text{kJ mol}^{-1}}$$

34. The rate of the reaction is proportional to the rate constant, k.

$$\frac{\text{rate catalyzed reaction}}{\text{rate uncatalyzed reaction}} = \frac{k_{cat}}{k} = \frac{Ae^{-100/RT}}{Ae^{-200/RT}} = e^{(-100+200)/RT}$$

$$= e^{100/RT} = e^{(100\ \text{kJ mol}^{-1})\left(\frac{10^3\ \text{J}}{1\ \text{kJ}}\right)/(8.314\ \text{J K}^{-1}\ \text{mol}^{-1})(373\ \text{K})}$$

$$= e^{32.2} = \underline{9.6 \times 10^{13}}$$

35. From the Arrhenius equation, $k = Ae^{-E_a/RT}$, $\ln k = \ln A - E_a/RT$, so that a plot of ln k versus 1/T is a straight line with a slope of $-E_a/RT$. Plotting the data gives, slope $= -1.40 \times 10^4$ K.

$$E_a = -R(\text{slope}) = -(8.314\ \text{J K}^{-1}\ \text{mol}^{-1})\left(\frac{1\ \text{kJ}}{10^3\ \text{J}}\right)(1.40 \times 10^4\ \text{K})$$

$$= \underline{116\ \text{kJ mol}^{-1}}$$

36. First find $E_a$ using the equation

$$E_a = R.\ln \frac{k_{T_1}}{k_{T_2}}\left(\frac{T_1T_2}{T_1 - T_2}\right)$$

$$= (8.314\ \text{J K}^{-1}\ \text{mol}^{-1})\left(\frac{1\ \text{kJ}}{10^3\ \text{J}}\right)\left[\ln\left(\frac{3.8 \times 10^{-3}\ \text{mol}^{-1}\ \text{L s}^{-1}}{1.1 \times 10^{-3}\ \text{mol}^{-1}\ \text{L s}^{-1}}\right)\right]$$

$$\times \left(\frac{1001 \times 838}{1001 - 838}\right)\ \text{K} = \underline{53.0\ \text{kJ mol}^{-1}}$$

Then use this value of $E_a$ to find k at 780°C (1053 K) using the equation derived in Problem 28.

$$\ln \frac{k_{T_2}}{k_{T_1}} = \frac{E_a}{R}\left(\frac{T_2 - T_1}{T_1T_2}\right) = \frac{(53.0\ \text{kJ mol}^{-1})\left(\frac{10^3\ \text{J}}{1\ \text{kJ}}\right)}{(8.314\ \text{J K}^{-1}\ \text{mol}^{-1})}\left(\frac{1053 - 838}{1053 \times 838}\right)\ \text{K}^{-1} = 1.55$$

$$\frac{k_{T_2}}{k_{T_1}} = 4.71,\quad k_{T_2} = (1.1 \times 10^{-3}\ \text{mol}^{-1}\ \text{L s}^{-1}) \times 4.71$$

$$= \underline{5.2 \times 10^{-3}\ \text{mol}^{-1}\ \text{L s}^{-1}}$$

37. At higher elevation, where the pressure is lower, water boils at a lower temperature. Thus the cooking process occurs at a lower temperature and proceeds at a lower rate, requiring a longer time for completion.

38. This problem is solved in the same way as Problem 36.

$$E_a = R.\ln \frac{k_{T_1}}{k_{T_2}} \left(\frac{T_1T_2}{T_1 - T_2}\right)$$

$$= (8.314\ J\ K^{-1}\ mol^{-1})\left(\frac{1\ kJ}{10^3 J}\right)\left[\ln\left(\frac{23\ mol^{-1}\ L\ s^{-1}}{1.3\ mol^{-1}\ L\ s^{-1}}\right)\left(\frac{798x698}{100}\right)K\right]$$

$$= \underline{133\ kJ\ mol^{-1}}$$

$$\ln \frac{k_{571}}{k_{698}} = \frac{E_a}{R}\left(\frac{T_2 - T_1}{T_1T_2}\right) = \frac{(133\ kJ\ mol^{-1})\left(\frac{10^3\ J}{1\ kJ}\right)}{8.314\ J\ K^{-1}\ mol^{-1}}\left(\frac{571-698}{698x571}\right) K^{-1} = \underline{-5.10}$$

$$\frac{k_{571}}{k_{698}} = 6.10x10^{-3} = \frac{k_{571}}{1.3\ mol^{-1}\ L\ s^{-1}}\ ;\quad k_{571} = \underline{7.9x10^{-3}\ mol^{-1}\ L\ s^{-1}}$$

39. A second order reaction is one that obeys a second order rate law, rate = k[A][B], or rate = $k[A]^2$. A bimolecular reaction is one in which the rate determining step involves collision between two molecular species (which may be reaction intermediates). Only if the reaction is an elementary single step process will a bimolecular reaction necessarily obey a second order rate law.

40.(a) Elementary process is the term used to describe each of the steps in a reaction mechanism. (b) A reaction mechanism is the series of successive reaction steps (elementary processes) by which reactants are converted in a reaction to products. (c) A rate limiting reaction is the slowest step in the series of reaction steps that constitute the mechanism of the reaction. It is the step that determines the rate of the overall reaction. (d) A reaction intermediate is a species formed in one step of the reaction mechanism that is used up in a subsequent step, so that it does not appear in the final products or in the balanced equation for the reaction.

41. The rate of a reaction is determined by the rate of the slowest step in the mechanism, the rate determining step. Thus, the expected rate law is

$$\text{rate} = k_3[COCl][Cl_2]$$

However, to experimentally test this rate law, it must be expressed in terms of reactant concentrations, rather than in terms of any intermediates, such as COCl or Cl.

From step 2, $K_2 = \frac{[COCl]}{[Cl][CO]}$ or $[COCl] = K_2[Cl][CO]$

and, from step 1, $K_1 = \frac{[Cl]^2}{[Cl_2]}$ or $[Cl] = K_1^{1/2}[Cl_2]^{1/2}$

Substituting into the above rate law gives,

$$\text{rate} = k_3[CO][Cl_2] = k_3K_2K_1^{1/2}[Cl_2]^{1/2}[CO][Cl_2] = \underline{k[Cl_2]^{3/2}[CO]}$$

The expected rate law is $\underline{rate = k[Cl_2]^{3/2}[CO]}$

A reaction intermediate is a species formed in one step of the reaction mechanism and used up in a following step, so that it does not appear in the final products, or the balanced equation for the reaction. CO and Cl are reaction intermediates.

42. The expected rate law is that for the slowest step in the mechanism,

$$\underline{rate = k_1[NO_2]^2}$$

If the reaction were an elementary process it would occur in a single step according to the balanced equation for the reaction, for which in this case,

$$\underline{rate = k[NO_2][CO]}$$

43. The expected rate law is that given by the slowest step in the mechanism (step 2), for which

$$rate = k_2[O_3][O]$$

From step 1,

$$K_1 = \frac{[O_2][O]}{[O_3]}, \text{ or } [O] = K_1\frac{[O_3]}{[O_2]}$$

Thus, $\underline{rate} = k_2K_1[O_3]^2[O_2]^{-1} = \underline{k[O_3]^2[O_2]^{-1}}$

The rate is inversely proportional to $[O_2]$ and since $O_2$ is a product in the first equilibrium step, any increase in $[O_2]$ will shift this equilibrium to the left, thereby decreasing the concentration of O atoms, the reaction intermediate. Since the rate of the rate determining step is proportional to [O], a decrease in its concentration will decrease the rate.

44. (a) The simultaneous collision of three molecules (with the correct steric orientation) is relatively improbable.

(b) The following mechanisms are both consistent with the observed rate law:

(i) $NO + NO \rightleftharpoons N_2O_2$ fast equilibrium

$N_2O_2 + O_2 \longrightarrow 2NO_2$ slow

$\underline{rate} = k_2[N_2O_2][O_2] = k_2K_1[NO]^2[O_2] = \underline{k[NO]^2[O_2]}$

(ii) $NO + O_2 \rightleftharpoons NO_3$ fast equilibrium

$NO_3 + NO \longrightarrow 2NO_2$ slow

$\underline{rate} = k_2[NO_3][NO] = k_2K_1[NO]^2[O_2] = \underline{k[NO]^2[O_2]}$

45. From the bond energy in $H_2(g) = 436\text{ kJ mol}^{-1}$, we can calculate the maximum wavelength of light needed to dissociate a $H_2(g)$ molecule into H(g) atoms.

Bond energy per molecule $= (436\text{ kJ mol}^{-1})\left(\frac{1\text{ mol}}{6.022\times10^{23}\text{ molecules mol}^{-1}}\right)$

Dissociation energy per $H_2$ molecule = $\underline{7.24 \times 10^{-22}\ kJ}$

and the corresponding wavelength is given by $\lambda = \frac{hc}{E}$

i.e., $\lambda = \dfrac{(6.63\times10^{-34}\ J\ s)(3.00\times10^{8}\ m\ s^{-1})}{(7.24\times10^{-22}\ kJ)(\frac{10^3\ J}{1\ kJ})} = \underline{2.75\times10^{-7}\ m}$

or $\lambda$ = 275 nm, which is outside the range of visible light (i.e., $\lambda$ = 400 nm to 700 nm). No light in the visible range supplies sufficient energy to dissociate $H_2(g)$ into 2H(g), so that the mechanism of the reaction cannot be explained by such a step in the mechanism.

46. (a) The molecularity of a reaction is the number of reactants in the reaction. In this case, steps 1,2, and 4 are unimolecular and step 3 is bimolecular.
(b) From the first (slow) step, rate = $k[N_2O_5]$, which is the experimental rate law.

47. (a) The expected rate law from the mechanism is

$$\text{Rate} = k_1[H_2O_2][I^-]$$

and the reaction is first order in $H_2O_2$, first order in $I^-$, and zero order in $H_3O^+$.
(b) Since the rate is independent of the $[H_3O^+]$, and $I^-(aq)$ is the anion of the strong acid HI(aq), changing the pH should have no effect on the rate.

48. A catalyst
(a) increases the rate of a reaction,
(b) lowers the activation energy of a reaction
(c) does not change the enthalpy change of a reaction
(d) allows a reaction to proceed at a given rate at a lower temperature, and
(e) does not affect the position of equilibrium

49. (a) A homogeneous catalyst is in the same phase as the reactants, and (b) a heterogeneous catalyst is in a different phase from the reactants.

50. A catalyst changes the mechanism of a reaction and does not appear in the balanced equation for the reaction, but it does appear in the rate law for the reaction. For example, the decomposition of $H_2O_2(aq)$ is catalyzed by a small amount of $I^-(aq)$, and the reaction occurs in two steps,

$$H_2O_2(aq) + I^-(aq) \longrightarrow IO^-(aq) + H_2O(l) \quad \text{Slow}$$

$$H_2O_2(aq) + IO^-(aq) \longrightarrow H_2O(l) + O_2(g) + I^-(aq) \quad \text{Fast}$$

Overall reaction $\quad 2H_2O_2(aq) \longrightarrow 2H_2O(l) + O_2(g)$

and the rate law is $\quad \text{Rate} = k[H_2O_2][I^-]$

51. From the Arrhenius equation, $k = pZe^{-E_a/RT}$, the rate of a reaction may be increased by lowering the activation energy, $E_a$, so that the exponential term $e^{-E_a/RT}$ increases in value. In other words, so that a greater proportion of the colliding molecules have sufficient energy to react, which is the main factor in homogeneous catalysis. Particularly in heterogeneous catalysis, a catalyst may also increase the value of the steric factor, p, by bringing a greater proportion of the molecules into suitable steric orientations for reaction to occur.

52. The base B does not appear in the balanced equation for the reaction and must therefore be a catalyst, and for the mechanism to be consistent with the observed rate law the rate limiting step has to involve reaction between a $(CH_3)_2CO$ molecule and a molecule of the base, B, which suggests the following mechanism.

$$B: + \; H-\underset{\underset{H}{|}}{\overset{\overset{H}{|}}{C}}-\underset{O}{\overset{||}{C}}-CH_3 \longrightarrow BH^+ + \; {}^-:\underset{\underset{H}{|}}{\overset{\overset{H}{|}}{C}}-\underset{O}{\overset{||}{C}}-CH_3 \qquad \text{slow}$$

$$I_2 + \; {}^-:CH_2-\underset{O}{\overset{||}{C}}-CH_3 \longrightarrow ICH_2-\underset{O}{\overset{||}{C}}-CH_3 + I^- \qquad \text{fast}$$

$$BH^+ \; \rightleftharpoons \; H^+ + B \qquad \text{fast equilibrium}$$

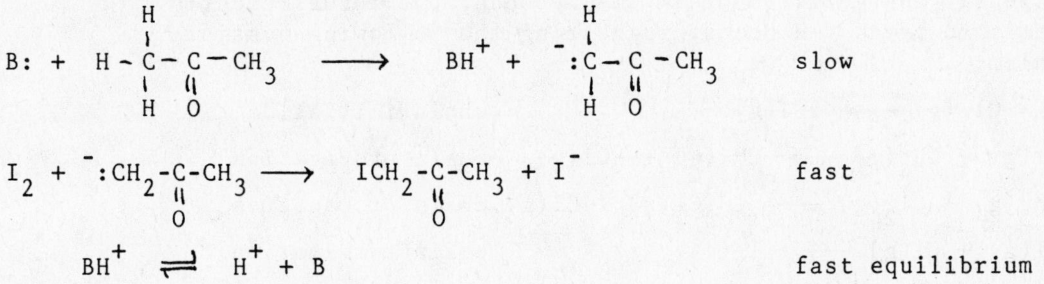

53. The catalyzed step in the mechanism should be spontaneous and have $E^o$ positive. From the given data, this is true only when $Fe^{3+}(aq)$ ions catalyze the reaction,

$$2Fe^{3+}(aq) + 2I^-(aq) \longrightarrow 2Fe^{2+}(aq) + I_2(s) \qquad E^o_{cell} = +0.23 \text{ V}$$

followed by

$$2Fe^{2+}(aq) + S_2O_8^{2-}(aq) \longrightarrow 2Fe^{3+}(aq) + 2SO_4^{2-}(aq) \quad E^o_{cell} = +1.24 \text{ V}.$$

For $Cr^{3+}(aq)$ ions, the first step of the mechanism would be

$$2Cr^{3+}(aq) + 2I^-(aq) \longrightarrow 2Cr^{2+}(aq) + I_2(s) \qquad E^o_{cell} = -0.95 \text{ V},$$

which is not a spontaneous reaction. Thus $Fe^{2+}(aq)$ ions catalyze the reaction but $Cr^{3+}(aq)$ ions do not catalyze the reaction.

54. Step 1 is a <u>chain initiation</u> which produces free radical intermediates. Steps 2 and 3 are <u>chain propagation</u> steps in which the number of free radicals generates remains constant. Step 4 is a <u>chain termination</u> step, which reduces the number of free radicals.

From Table 6.3, the Br-Br bond energy in $Br_2(g)$ is 190 kJ $mol^{-1}$, from which the minimum energy required to dissociate a $Br_2(g)$ molecule can be calculated, and, using the relationship $E = hc/\lambda$, the maximum wavelength of light to dissociate $Br_2$ molecules is found.

$$\lambda = \frac{hc}{E} = \frac{(6.63\times10^{-34} \text{ Js})(3.00\times10^8 \text{ m s}^{-1})(1 \text{ kJ}/1000 \text{ J})}{(190 \text{ kJ mol}^{-1})(\frac{1 \text{ mol } Br_2 \text{ molecules}}{6.022\times10^{23} \ Br_2 \text{ molecules}})} = \underline{6.30\times10^{-7}} \text{ m}$$

i.e., <u>630 nm</u>

55. (a) This is a chain reaction that could be initiated either by the photons of light interacting with $Cl_2(g)$ molecules to form Cl(g) atoms, or with $CH_4(g)$ molecules to form $CH_3$ radicals.

$$Cl_2(g) \longrightarrow 2Cl(g) \quad \underline{\text{or}} \quad CH_4(g) \longrightarrow CH_3(g) + H(g)$$

The photons must have sufficient energy to break the specific bond. Therefore, calculate the energy of one mole of photons from the given wavelength.

$$E = \frac{Nhc}{\lambda} = \frac{(6.022\text{x}10^{23}\ \text{photon mol}^{-1})(8.63\text{x}10^{-34}\ \text{J})(3.00\text{x}10^{8}\ \text{m s}^{-1})}{480\ \text{x}\ 10^{-9}\ \text{m}}$$

$$= 2.49\text{x}10^{5}\ \text{J mol}^{-1}\ \text{photons, } \underline{\text{or}}\ \ \underline{249\ \text{kJ mol}^{-1}\ \text{photons}}$$

The average C-H bond energy in $CH_4$ is 416 kJ $mol^{-1}$, while the bond energy of $Cl_2$ is 242 kJ $mol^{-1}$. Thus light of wavelength 480 nm is sufficiently energetic to break Cl-Cl bonds, but insufficiently energetic to break C-H bonds, suggesting the following possible mechanism:

| | | |
|---|---|---|
| (b) | $Cl_2(g) \longrightarrow 2Cl(g)$ | chain initiation |
| | $Cl(g) + CH_4(g) \longrightarrow CH_3(g) + HCl(g)$ | chain propagation |
| | $CH_3(g) + Cl_2(g) \longrightarrow CH_3Cl(g) + Cl(g)$ | chain propagation |
| | $Cl(g) + Cl(g) \longrightarrow Cl_2(g)$ | chain termination |
| | $CH_3(g) + CH_3(g) \longrightarrow C_2H_6(g)$ | chain termination |

56. (a) The order of a reaction is the sum of the exponents of the concentration terms in the rate law for the reaction. Molecularity refers to the number of reactant molecules in any step in the reaction mechanism for the reaction.

(b) From the integrated rate expression for a first order reaction,

$$\ln [A]_t = \ln [A]_o - k_1 t$$

where $[A]_o$ is the initial concentration at time t = 0, $[A]_t$ is the concentration at time t, and $k_1$ is the rate constant for the reaction, a plot of ln $[A]_t$ versus t gives a straight line of slope $-k_1$. From the integrated rate expression for a second order reaction,

$$\frac{1}{[A]_t} = \frac{1}{[A]_o} + k_2 t$$

where $[A]_o$ is the initial concentration at time t = 0, $[A]_t$ is the concentration at time t, and $k_2$ is the rate constant for the reaction, in the special case where the initial concentrations of the reactants that appear in the rate equation are arranged experimentally to be initially equal, a plot of $1/[A]_t$ versus t is a straight line with a slope of $k_2$. By constructing both plots from the rate data for a particular reaction, the one that is linear establishes the experimental order and gives the rate constant at a particular temperature.

(c) Order and molecularity are the same only for a single step reaction, or for any one step of a multistep reaction.

57. (a) Not all collisions between reactant molecules lead to the formation of products because (i) not all collisions occur with the reactant molecules suitably oriented in space for the appropriate new bonds to be formed, and (ii) not all collisions will be between species with sufficient energy to provide the activation energy for the reaction.

(b) An elementary process is unimolecular and follows a first order rate law if one reactant molecule is involved; it is bimolecular and follows a second order rate law if two reactant molecules are involved.

(c) Increasing the temperature increases the average kinetic energy of the molecules and therefore increases the proportion of the molecules that have sufficient energy to overcome the activation energy barrier to reaction. Addition of a catalyst does not affect the energy of the molecules, but nevertheless also increases the proportion of the molecules that have enough energy to react. A catalyst does this by providing an alternative mechanism for the reaction with a lower activation energy than that for the uncatalyzed reaction.

58. (a) $2Br^-(aq) + H_2O_2(aq) + 2H_3O^+(aq) \longrightarrow Br_2(aq) + 4H_2O(l)$

(b) Since doubling the concentration of $H_2O_2$ doubled the rate, and doubling the concentration of $Br^-$ also doubled the rate, the reaction must be first order in each of $H_2O_2$ and $Br^-$. Changing the pH from 1.00 to 0.400 changed the $[H_3O^+]$ from 0.100 mol $L^{-1}$ to 0.398 mol $L^{-1}$, or by a factor of 4. Since this change also increased the rate by a factor of 4, the reaction is also first order in $[H_3O^+]$, and the rate equation is

$$\underline{\text{Rate} = k[H_2O_2][Br^-][H_3O^+]}$$

(c) $$-\frac{d[Br^-]}{dt} = -\frac{2d[H_2O_2]}{dt} = +\frac{2d[Br_2]}{dt}$$

Thus, if the rate of disappearance of $Br^-$ is $7.2\times10^{-3}$ mol $L^{-1}$ $s^{-1}$, the rate of disappearance of $H_2O_2$ and the rate of appearance of $Br_2$ are both $\underline{3.6\times10^{-3} \text{ mol L}^{-1} \text{ s}^{-1}}$

(d) As the pH increases, the $[H_3O^+]$ decreases and the rate decreases.

(e) If the initial volume is doubled, all of the concentrations are halved and the initial rate decreases by a factor of 8.

(f)

| | |
|---|---|
| $H_2O_2 + H_3O^+ \rightleftharpoons H_3O_2^+ + H_2O$ | fast |
| $H_3O_2^+ + Br^- \longrightarrow HOBr + H_2O$ | slow |
| $2HOBr \longrightarrow Br_2 + H_2O_2$ | fast |

$\underline{\text{Rate}} = k_2[H_3O_2^+][Br^-] = k_2K[H_2O_2][H_3O^+][Br^-] = \underline{k[H_3O^+][H_2O_2][Br^-]}$

59. If step 2 is the rate determining step, the expected rate law would be

$$\text{Rate} = k_2[Cl][CHCl_3] = k_2K_1^{1/2}[Cl_2]^{1/2}[CHCl_3]$$
$$= k[Cl_2]^{1/2}[CHCl_3]$$

which is consistent with the observed rate law.

If step 3 is the slow rate determining step, the expected rate law would be

$$\text{Rate} = k_3[CCl_3][Cl] = k_3K_2[Cl]^2[CHCl_3][HCl]^{-1}$$
$$= k_3K_2K_1[Cl_2][CHCl_3][HCl]^{-1}$$
$$= k[Cl_2][CHCl_3][HCl]^{-1}$$

which is inconsistent with the experimental rate law.

60. (a) Reaction rate is the speed at which a reaction occurs, which is measured by the decrease in the concentration of one of the reactants, or the increase in one of the products, in a given time interval

(b) Reaction order is the power to which a concentration term is raised in the experimental rate law for the reaction

(c) The rate constant is the constant that appears in a rate law; it is the reaction rate when all of the concentrations of the reactants in the rate law are exactly 1 mol $L^{-1}$

(d) The half-life of a reaction is the time required for the concentration of a reactant to decrease to half of an initial value

(e) The molecularity of any step in a reaction mechanism is the number of reactant molecules. For a one-step reaction, the molecularity is the same as the order.

## CHAPTER 20

1. Electrons in the outer n = 5 shell must be arranged so that the number of unpaired electrons equals the valence state.

(a) Iodine

I(+1) [Kr]$4d^{10}$

I(+3) [Kr]$4d^{10}$

I(+5) [Kr]$4d^{10}$

I(+7) [Kr]$4d^{10}$

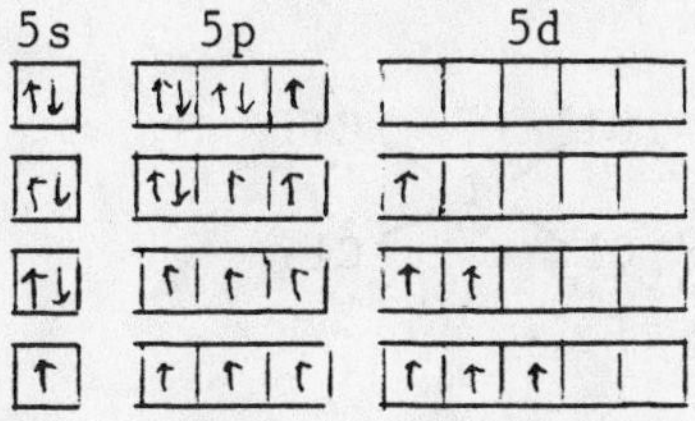


(b) Xenon

Xe(+2) [Kr]$4d^{10}$

Xe(+4) [Kr]$4d^{10}$

Xe(+6) [Kr]$4d^{10}$

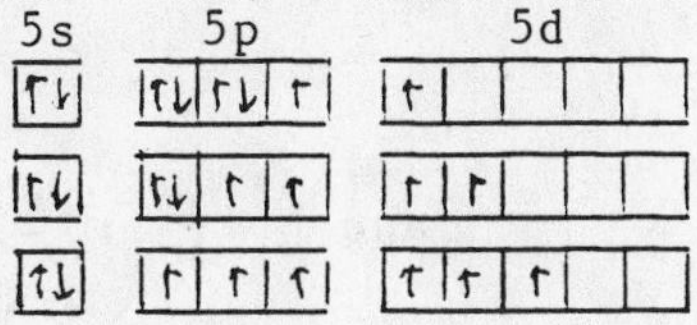


2. (a) $AX_5$ molecules are trigonal bipyramidal.
$AX_6$ molecules are octahedral.

(b) Examples of $AX_5$ molecules include $PCl_5$ and $PF_5$.
Examples of $AX_6$ molecules include $SF_6$ and $PCl_6^-$.

(c) $SeF_4$ | $BrF_3$ | $XeF_4$ | $ICl_2^-$

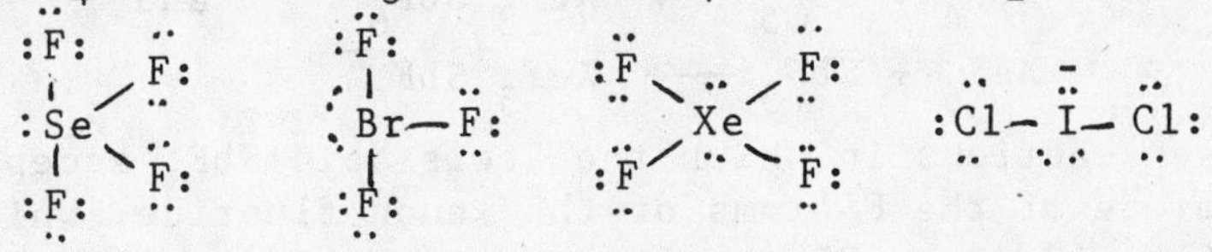


$AX_4E$ | $AX_3E_2$ | $AX_4E_2$ | $AX_2E_3$

disphenoidal | T-shaped | square planar | linear

3. In the gas phase $PCl_5(g)$ exists as discrete covalent $PCl_5$ molecules with $AX_5$ trigonal bipyramidal geometry. In the solid state, $PCl_5(s)$ contains $PCl_4^+$, $AX_4$, tetrahedral ions, and $PCl_6^-$, $AX_6$, octahedral ions.

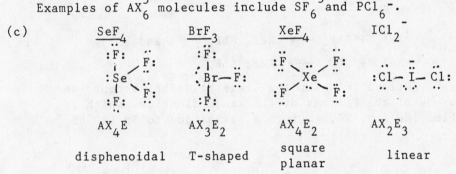

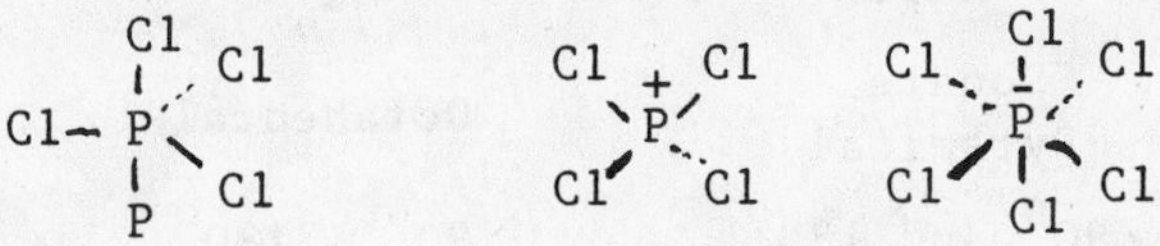


4. (a) Elements of the third an subsequent periods have low lying d orbitals in their valence shells in addition to s and p orbitals. Using the ns and three np orbitals an octet of electrons can be accommodated but use also of the nd orbitals permits expansion of of the valence shells to more than eight electrons.

(b) Phosphorus in group 5 forms $PF_3(g)$ and $PF_5(g)$. $PF_3$ is an $AX_3E$ triangular pyramidal molecules, while $PF_5$ is an $AX_5$ trigonal bipyramidal molecule. Sulfur in group 6 forms $SF_2$, $SF_4$, and $SF_6$. $SF_2$ is an $AX_2E_2$ angular molecule; $SF_4$ is an $AX_4E$ disphenoidal shape, and $SF_6$ is an $AX_6$ octahedral molecule.

5. $ICl_3$ has the $AX_3E_2$ structure. In $I_2Cl_6$, two $ICl_3$ fragments form a dimer containing Cl-I-Cl bridge bonds, and each I atom becomes $AX_4E_2$, square planar; all the atoms in $I_2Cl_6$ lie in the same plane with <u>cis</u> ClICl bond angles of approximately 90° and <u>trans</u> ClICl angles of approximately 180°.

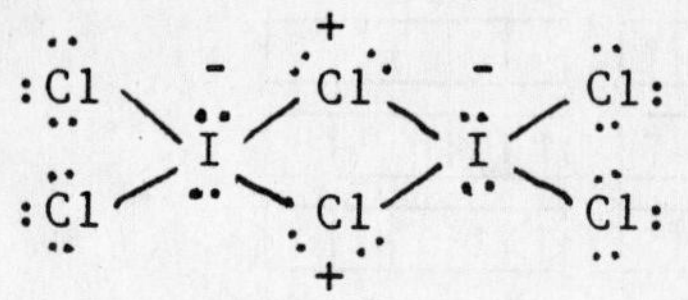

6. The Lewis structures are,

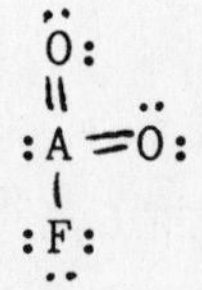 and 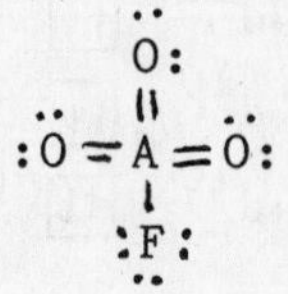

| | | |
|---|---|---|
| <u>Type</u>: | $AX_3E$ | $AX_4$ |
| <u>Shape</u>: | trigonal pyramidal | tetrahedral |

7. The reactions

$$XeF_4 + SbF_5 \longrightarrow XeF_3{}^+SbF_6{}^- \quad \text{and}$$
$$XeF_6 + SbF_5 \longrightarrow XeF_5{}^+SbF_6{}^-$$

are Lewis acid-base reactions in which the Lewis acid $SbF_5$ accepts an electron pair from one of the F atoms of the xenon fluoride, which is transferred as a fluoride ion, $F^-$, to form a sixth bond to Sb in $SbF_6{}^-$.

Lewis structures:

| | <u>$XeF_3{}^+$</u> | <u>$XeF_5{}^+$</u> | <u>$SbF_6{}^-$</u> |
|---|---|---|---|
| | 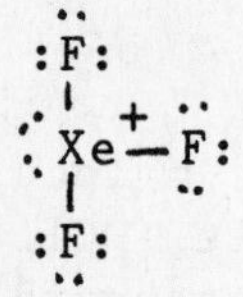 | 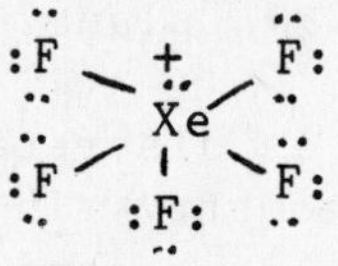 | 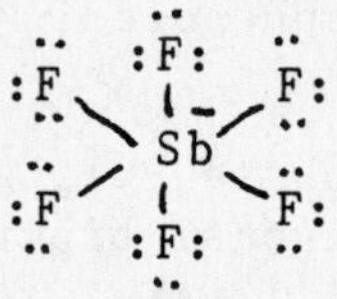 |
| Type: | $AX_3E_2$ | $AX_5E$ | $AX_6$ |
| Shape: | T-shaped | Square Pyramidal | Octahedral |
| Bond Angles: | ~ 90° & 180° | ~ 90° and 180° | 90° & 180° |

Note that in $XeF_3{}^+$ and $XeF_5{}^+$ the bonds adjacent to the lone pairs are distorted away slightly from the lone pairs, so that the bond angles are slightly smaller than the ideal angles, while $SbF_6{}^-$ is a regular octahedral species.

8. A molecule of the $AX_nE_m$ type has a non-zero dipole moment if the sum of the A-X bond dipole vectors is not equal to zero (so that the center of positive charge in the molecule does not coincide with the center of negative charge). To asses this, the expected molecular shape must be determined.

(a) $PF_5$, $AX_5$ trigonal bipyramidal, $\mu = 0$.

(b) $SeF_4$, $AX_4E$ disphenoidal shape, $\mu \neq 0$.

(c) $ClF_3$, $AX_3E_2$ T-shaped, $\mu \neq 0$.

(d) $BrF_5$, $AX_5E$ square based pyramidal shaped, $\mu \neq 0$.

(e) $XeF_2$, $AX_2E_3$ linear shaped, $\mu = 0$.

(f) $XeF_4$, $AX_4E_2$ square planar shaped, $\mu = 0$.

$SeF_4$, $ClF_3$, and $BrF_5$ are expected to have dipole moments.

9. (a) $SiF_6^{2-}$, $AX_6$ octahedral (b) $SiF_5^-$, $AX_5$ trigonal bipyramidal

(c) $PF_6^-$, $AX_6$ octahedral (d) $SeF_5^-$, $AX_5E$ square based pyramid

(e) $BrF_4^-$, $AX_4E_2$, square planar (f) $IF_4^-$, $AX_4E_2$ square planar

10. (a) $XeOF_4$, $AX_5E$ square based pyramidal, Oxidation state, Xe(+6).

(b) $XeO_2F_2$, $AX_4E$ disphenoidal shaped, oxidation state, Xe(+6).

(c) $XeO_3F_2$, $AX_5$ trigonal bipyramidal, oxidation state, Xe(+8).

11. The known <u>interhalogen</u> compounds are the following:

(a) ClF, BrF, $AXE_3$ linear diatomic molecules.

(b) $ClF_3$, $BrF_3$, and $IF_3$, $AX_3E_2$ T-shaped molecules.

(c) $ClF_5$, $BrF_5$, and $IF_5$, $AX_5E$ square based pyramidal molecules.

(d) $IF_7$ is an $AX_7$ molecule with the shape of a pentagonal bipyramid.

All of these molecules, except $IF_7$, have a dipole moment, $\mu \neq 0$.

12. (a) Each <u>axial</u> bond in an $AX_5$ trigonal bipyramidal molecule is $90^\circ$ away from three <u>equatorial</u> bonds, while each <u>equatorial</u> bond is $90^\circ$ away from two <u>axial</u> bonds and makes angles of $120^\circ$ with two others. To relieve the resulting inter electron pair repulsions, which are greatest in the axial directions, the axial bonds are lengthened slightly relative to the lengths of the equatorial bonds. In other words, there is less room in the axial positions of a trigonal bipyramidal molecule to accommodate electron pairs than there is in the equatorial positions, so that the axial bonds are always longer than the equatorial bonds.

(b) In $AX_4E$, $AX_3E_2$, and $AX_2E_3$ molecules, each based upon the trigonal bipyramidal arrangement of five electron pairs in the valence shell of A, the greater size of the lone pairs relative to that of the bond pairs is best accommodated by always placing them in equatorial positions to minimize the repulsions with other electron pairs, of which those at $90^\circ$ are the most important.

(c) In $AX_4E$ and $AX_3E_2$ molecules, the lone pairs repel the bond pairs causing the XAX bond angles to decrease to less than the ideal angles. In linear $AX_2E_3$ type molecules the lone pairs do not affect the XAX bond angle but cause the AX bond lengths to be unusually long.

13. First classify each oxoacid according to the $X(OH)_mO_n$ nomenclature. The acid strengths of oxoacids increases as the value of n (the number of O atoms forming double bonds to X) , with increasing electronegativity and increasing oxidation number of the central atom X.

| Acid | $X(OH)_mO_n$ | m | Electronegativity of X | Oxidation State |
| --- | --- | --- | --- | --- |
| $HClO_4$ | $HOClO_3$ | 3 | 2.8 | +7 |
| $H_2SO_4$ | $(HO)_2SO_2$ | 2 | 2.4 | +6 |
| $H_3PO_4$ | $(HO)_3PO$ | 1 | 2.1 | +5 |
| HClO | HOCl | 0 | 2.8 | +1 |

Thus, the order of acid strengths is

$$HClO_4 > H_2SO_4 > H_3PO_4 > HClO$$

Note that both $HClO_4$ and $H_2SO_4$ behave as strong acids in aqueous solution, $H_3PO_4(aq)$ is a weak acid and HClO(aq) is a very weak acid.

14. For the method of balancing oxidation-reduction equations see Chapter 10, page 500 of text.

(a) $4Na^+ + Cr_2(SO_4)_3 + 10\ OH^- + 3NaOCl \longrightarrow 2Na_2CrO_4 + 3SO_4^{2-} + 3NaCl + 5H_2O$

(b) $3H_2O + 6KI + NaClO_3 \longrightarrow 3I_2 + NaCl + 6KOH$

(c) $5I^- + IO_3^- + 6H_3O^+ \longrightarrow 3I_2 + 9H_2O$

15. $XeF_2 + BrO_3^- + 2OH^- \longrightarrow BrO_4^- + Xe + 2F^- + H_2O$

Perbromate ion, $BrO_4^-$, is an $AX_4$ type species with tetrahedral geometry.

16. (a) Cold NaOH(aq) $2NaOH + Cl_2 \longrightarrow 2Na^+ + Cl^- + ClO^- + H_2O$

(b) Hot NaOH(aq) $6NaOH + 3Cl_2 \longrightarrow 6Na^+ + 5Cl^- + ClO_3^- + 3H_2O$

17. $4KClO_3(s) \xrightarrow{heat} 3KClO_4(s) + KCl(s)$

$2KClO_3(s) \xrightarrow[MnO_2(s)\ catalyst]{heat} 2KCl(s) + 3O_2(g)$

18. Each of these reactions is a Lewis acid-base hydrolysis reaction and results in a solution of an oxoacid (or acidic oxide) and hydrogen halide.

(a) $PCl_5(s) + 4H_2O(l) \longrightarrow H_3PO_4(aq) + 5HCl(aq)$

(b) $IF_5(g) + 3H_2O(l) \longrightarrow HIO_3(aq) + 5HF(aq)$

(c) $SF_4(g) + 2H_2O(l) \longrightarrow SO_2(aq) + 4HF(aq)$

(d) $Br_2(l) + H_2O(l) \longrightarrow HOBr(aq) + HBr(aq)$

19. HOCl(aq) can be prepared by the reaction of $Cl_2(g)$ with water,

$$Cl_2(g) + 2H_2O(l) \rightleftharpoons HOCl(aq) + H_3O^+(aq) + Cl^-(aq)$$

The reaction can be shifted to the right by removing the $Cl^-(aq)$ from the solution as AgCl(s),

$$Ag_2O(s) + 2Cl^-(aq) + 2H_3O^+(aq) \longrightarrow 2AgCl(s) + 3H_2O(l)$$

Attempts to concentrate the solution by evaporation cause the HOCl(aq) to decompose.

$$2HOCl(aq) + 2H_2O(l) \longrightarrow 2H_3O^+(aq) + 2Cl^-(aq) + O_2(g)$$

20. (a) $3Br_2(l) + 6NaOH(\text{hot, conc}) \longrightarrow 5Br^-(aq) + BrO_3^-(aq) + 6Na^+(aq) + 3H_2O(l)$

(b) $Br_2(l) + H_2O_2(aq) \longrightarrow O_2(g) + 2HBr(aq)$

(c) $KBr(s) + H_3PO_4(\text{conc, aq}) \longrightarrow KH_2PO_4(aq) + HBr(g)$

(d) $5SO_2(g) + 2KIO_3(aq) + 12H_2O(l) \longrightarrow I_2(s) + 8H_3O^+(aq) + 5SO_4^{2-}(aq) + 2K^+(aq)$

21, and 22.

| | Substance | Name | Oxidation State of Halogen |
|---|---|---|---|
| (a) | HOBr | Hypobromous acid | +1 |
| (b) | $Ca(OCl)_2$ | Calcium hypochlorite | +1 |
| (c) | $KBrO_3$ | Potassium bromate | +5 |
| (d) | $KClO_2$ | Potassium chlorite | +3 |
| (e) | $Mg(ClO_4)_2$ | Magnesium perchlorate | +7 |
| (f) | $HIO_3$ | iodic acid | +5 |
| (g) | $HBrO_4$ | Perbromic acid | +7 |
| (h) | $H_5IO_6$ | Paraperiodic acid | +7 |

23. (a) $ClO_3^- + 6H_3O^+ + 6e^- \longrightarrow Cl^- + 9H_2O$

(b) $ClO_2^- + 4H_3O^+ + 4e^- \longrightarrow Cl^- + 6H_2O$

(c) $ClO^- + 2H_3O^+ + 2e^- \longrightarrow Cl^- + 3H_2O$

24. $2HIO_3(s) \xrightarrow{\text{heat}} I_2O_5(s) + H_2O(g)$

$2I_2O_5(s) \xrightarrow{\text{heat}} 2I_2(s) + 5O_2(g)$

25. Sulfur in period 3 has low lying 3d orbitals which can be utilized for bonding with highly electronegative ligands; it is large enough to accommodate more than eight electrons in its valence shell, as it does, for example, when it reacts with $F_2(g)$ to give $SF_4$ or $SF_6$. Oxygen in period 2 has no low lying d orbitals and is restricted to compounds in which the valence shell of oxygen is restricted to an octet of electrons. Thus, it cannot form $OF_4$ or $OF_6$, where the valence shell of oxygen would have to accommodate 10 and 12 electrons, respectively.

26. $$xI_2(s) + yCl_2(l) \longrightarrow 2I_xCl_y(s)$$

100 g of compound contains 54.5 g I and 45.5 g Cl

$$= (54.5 \text{ g I})\left(\frac{1 \text{ mol I}}{126.9 \text{ g I}}\right) : (45.5 \text{ g Cl})\left(\frac{1 \text{ mol Cl}}{35.45 \text{ g Cl}}\right)$$

$$= 0.429 \text{ mol I} : 1.28 \text{ mol Cl}$$

Ratio of moles (atoms) = 1.00 : 2.98

Thus, the <u>empirical formula</u> is $ICl_3$ (formula mass 233.3 u)

The molecular mass of 467 g $mol^{-1}$ is consistent with the molecular formula $I_2Cl_6$. (See Problem 5 for the Lewis structure)

27.

| | Species | Oxidation Number* | | Species | Oxidation Number* |
|---|---|---|---|---|---|
| (a) | $\underline{Cl}O_3^-$ | +5 | (b) | $\underline{Br}F_3$ | +3 |
| (c) | $H\underline{Cl}O_4$ | +7 | (d) | $\underline{Cl}O_2^-$ | +3 |
| (e) | $\underline{Cl}O_2$ | +4 | (f) | $H_5\underline{I}O_6$ | +7 |

*underlined atom.

28.

$$2ClO_3^- + 12H_3O^+ + 10e^- \longrightarrow Cl_2 + 18H_2O \qquad \underline{\text{reduction}}$$
$$10[Fe^{2+} \longrightarrow Fe^{3+} + e^-] \qquad \underline{\text{oxidation}}$$
$$10Fe^{2+} + 2ClO_3^- + 12H_3O^+ \longrightarrow 10Fe^{3+} + Cl_2 + 18H_2O$$

29. In a <u>disproportionation reaction</u>, one species acts as both the oxidizing agent and the reducing agent. Examples include:

$$Cl_2 + H_2O \longrightarrow HOCl + H_3O^+ + Cl^-$$
$$4KClO_3 \longrightarrow 3KClO_4 + KCl$$
$$6XeF_4 + 12H_2O \longrightarrow 2XeO_3 + 4Xe + 3O_2 + 24HF$$

30. Balance the equation for the reaction, omitting the spectator ion $Na^+$.

$$2IO_3^- + 5HSO_3^- + 2H_2O \longrightarrow I_2 + 5SO_4^{2-} + 3H_3O^+$$

$HSO_3^-$ reduces $IO_3^-$ with I in the +5 oxidation state to $I_2$ with I in the Oxidation state 0. It is important not to add excess $HSO_3^-$ since this can reduce $I_2$ further to $I^-$ (iodine oxidation state -1).

Mass of $NaHSO_3$(s) required is given by

$$(50\text{ kg NaIO}_3)\left(\frac{10^3\text{ g}}{1\text{ kg}}\right)\left(\frac{1\text{ mol NaIO}_3}{197.9\text{ g NaIO}_3}\right)\left(\frac{5\text{ mol NaHSO}_3}{2\text{ mol NaIO}_3}\right)\left(\frac{104.1\text{ g NaHSO}_3}{1\text{ mol NaHSO}_3}\right)$$

$= 6.57 \times 10^4$ g $NaHSO_3$(s), <u>or</u> <u>65.7 kg</u>

and the maximum possible yield of $I_2$(s) is

$$(50\text{ kg NaIO}_3)\left(\frac{10^3\text{ g}}{1\text{ kg}}\right)\left(\frac{1\text{ mol NaIO}_3}{197.9\text{ g NaIO}_3}\right)\left(\frac{1\text{ mol I}_2}{2\text{ mol NaIO}_3}\right)\left(\frac{253.8\text{ g I}_2}{1\text{ mol I}_2}\right)$$

$= 3.21 \times 10^4$ g $I_2$(s), <u>or</u> <u>32.1 kg</u>

31. (a) $Pb(s) + 2HCl(aq, conc) \xrightarrow{heat} PbCl_2(aq) + H_2(g)$

$PbCl_2$(s) is relatively soluble in hot solution but rather insoluble in the cold, especially in the presence of excess $Cl^-$(aq) ion, and crystallizes on cooling the solution.

(b) $2Fe(s) + 3Cl_2(g) \longrightarrow 2FeCl_3(s)$

Dry $Cl_2$(g) is passed over iron heated to red heat; $FeCl_3$(g) sublimes and is collected on a cool part of the apparatus as $FeCl_3$(s).

(c) Excess of red phosphorus is heated with a limited supply of $Cl_2(g)$ in the absence of oxygen or water; the $PCl_3(l)$ distils off and is collected.

$$2P(s) + 3Cl_2(g) \longrightarrow 2PCl_3(l)$$

(d) $CCl_4(l)$ results from the high temperature reaction of chlorine with methane, as the final product of a series of chain reactions.

$$CH_4(g) + Cl_2(g) \longrightarrow CH_3Cl(g) + HCl(g)$$
$$CH_3Cl(g) + Cl_2(g) \longrightarrow CH_2Cl_2(g) + HCl(g)$$
$$CH_2Cl_2(g) + Cl_2(g) \longrightarrow CHCl_3(g) + HCl(g)$$
$$CHCl_3(g) + Cl_2(g) \longrightarrow CCl_4(g) + HCl(g)$$

The HCl(g) can be dissolved in water and the mixture of chloromethanes separated by fractional distillation.

(e) Aluminum metal is dissolves in HCl(aq) to give a solution of $AlCl_3(aq)$. On evaporation of the solution and cooling, crystals of $AlCl_3.6H_2O$ are obtained.

$$2Al(s) + 6HCl(aq) \xrightarrow{\text{heat}} 2AlCl_3(aq) + 3H_2(g)$$
$$\text{evaporate}$$
$$AlCl_3.6H_2O(s)$$

32.

Initial moles $KIO_3 = (0.3574 \text{ g } KIO_3)(\frac{1 \text{ mol } KIO_3}{214.0 \text{ g } KIO_3}) = \underline{1.670\text{x}10^{-3} \text{ mol}}$

and the reaction is: $KIO_3(s) \longrightarrow \text{white solid} + O_2(g)$

$$\text{Mol } O_2(g) \text{ formed} = \frac{PV}{RT} = \frac{(755 \text{ mm Hg})(\frac{1 \text{ atm}}{760 \text{ mm Hg}})(61.5 \text{ mL})(\frac{1 \text{ L}}{1000 \text{ mL}})}{(0.0821 \text{ L atm mol}^{-1} \text{ K}^{-1})(298 \text{ K})}$$

$$= \underline{2.50\text{x}10^{-3} \text{ mol}}$$

$$\frac{\text{moles } KIO_3 \text{ reacted}}{\text{moles } O_2 \text{ formed}} = \frac{1.67\text{x}10^{-3} \text{ mol}}{2.50\text{x}10^{-3} \text{ mol}} = 0.668, \underline{\text{or}} \ \frac{2}{3}$$

i.e., in the balanced equation, $2KIO_3(s) \longrightarrow X(s) + 3O_2(g)$

<u>or</u> $2KIO_3(s) \longrightarrow 2KI(s) + 3O_2(g)$ <u>balanced</u>

33. (a) For the oxidation of $Cl^-(aq)$ to $ClO^-(aq)$,

$$Cl^-(aq) + H_2O(l) \longrightarrow ClO^-(aq) + 2e^- + 2H^+(aq)$$

$$\text{Mass of NaOCl} = (4.50 \text{ A})(60 \text{ min})(\frac{60 \text{ s}}{1 \text{ min}})(\frac{1 \text{ C}}{1 \text{ A s}})(\frac{1 \text{ mol } e^-}{96\ 500 \text{ C}})(\frac{1 \text{ mol NaOCl}}{2 \text{ mol } e^-})$$

$$\times (\frac{74.44 \text{ g NaOCl}}{1 \text{ mol NaOCl}}) = \underline{6.249 \text{ g NaOCl}}$$

(b) For the oxidation of $Cl^-(aq)$ to $ClO_3^-(aq)$,

$$Cl^-(aq) + 3H_2O(l) \longrightarrow ClO_3^-(aq) + 6e^- + 6H^+(aq)$$

$$\text{Mass of } KClO_3 = (3.05\ A)(145\ min)\left(\frac{60\ s}{1\ min}\right)\left(\frac{1\ C}{1\ A\ s}\right)\left(\frac{1\ mol\ e^-}{96\ 500\ C}\right)\left(\frac{1\ mol\ KClO_3}{6\ mol\ e^-}\right)$$

$$\times \left(\frac{122.5\ g\ KClO}{1\ mol\ KClO_3}\right) = \underline{5.614\ g\ KClO_3(s)}$$

34. (a) Sodium bromide is oxidized by sulfuric acid to brown nonpolar $Br_2$ which is extracted into the organic solvent tetrachloroethene on shaking.

$$2NaBr + 5H_2SO_4 \longrightarrow Br_2 + SO_2 + 2H_3O^+ + 2Na^+ + 4HSO_4^-$$

(b) The reactions are

$$KBrO_3(aq) + 3H_2S(g) \longrightarrow 3S(s) + Br^-(aq) + K^+(aq) + 3H_2O(l)$$

$$Br^-(aq) + Ag^+(aq) \longrightarrow AgBr(s) \quad \text{(pale yellow precipitate)}$$

$$AgBr(s) + 2NH_3(aq) \longrightarrow Ag(NH_3)_2^+(aq) + Br^-(aq)$$

Bromate ion oxidizes $H_2S$ to sulfur. $Br^-(aq)$ in the filtrate reacts with $Ag^+(aq)$ from $AgNO_3(aq)$ to give pale yellow insoluble $AgBr(s)$. The yellow $AgBr(s)$ dissolves in $NH_3(aq)$ due to the formation of the soluble $Ag(NH_3)_2^+(aq)$ complex ion.

(c) The gas must be $O_2(g)$ formed by the reaction

$$2KBrO_3(s) \longrightarrow 2KBr(s) + 3O_2(g)$$

(d) The reaction is

$$KBrO_3(a) + 5KBr(aq) + 6H_3O^+(aq) \longrightarrow 6Br_2(aq) + 6K^+(aq) + 9H_2O(l)$$

in which bromate ion oxidizes bromide ion to bromine. The nonpolar $Br_2$ that is formed is extracted into the organic layer.

35. The equilibrium constant is related to the standard cell potential by the equation

$$E^o_{cell} = \frac{RT}{nF} \ln K$$

Solving for the equilibrium constant K gives, $K = e^{nFE^o/RT} = e^{nE^o/0.0257}$

(a) $Cl_2 + 2Br^- \longrightarrow Br_2 + 2Cl^-$ $\qquad E^o_{cell} = \underline{+0.27\ V}$

$$K = e^{2E^o/0.0257} = e^{21} = \underline{1.3 \times 10^9}$$

(b) $Cl_2 + 2I^- \longrightarrow I_2 + 2Cl^-$ $\qquad E^o_{cell} = \underline{+0.82\ V}$

$$K = e^{2E^o/0.0257} = e^{64} = \underline{6.2 \times 10^{27}}$$

(c) $Br_2 + 2I^- \longrightarrow I_2 + 2Br^-$ $\qquad E^o_{cell} = \underline{+0.55\ V}$

$$K = e^{2E^o/0.0257} = e^{43} = \underline{4.7 \times 10^{18}}$$

36. (a) $NaIO_3(aq) + Cl_2(g) + 4NaOH(aq) \longrightarrow Na_3H_2IO_6(s) + 2NaCl(aq) + H_2O(l)$

(b) $Na_3H_2IO_6(s) \xrightarrow{\text{excess } HNO_3\text{(conc)}}$ white crystals

| | Na | I | O | |
|---|---|---|---|---|
| 100 g of white solid contains | 10.8 | 59.3 | 29.9 | g |
| = | $\frac{10.8}{22.99}$ | $\frac{59.3}{126.9}$ | $\frac{29.9}{16.00}$ | mol |
| Ratio of mol (atoms) = | 0.469 : | 0.466 : | 1.869 | |
| = | $\frac{0.469}{0.466}$ : | $\frac{0.466}{0.466}$ : | $\frac{1.869}{0.466}$ | |
| = | 1.01 : | 1.00 : | 4.01 | |

Thus, the empirical formula is $NaIO_4$
and the balanced equation for the reaction is

$$Na_3H_2IO_6(s) + 2HNO_3(aq) \longrightarrow NaIO_4(s) + 2NaNO_3(aq) + 2H_2O(g)$$

(c) $Na_3H_2IO_6$

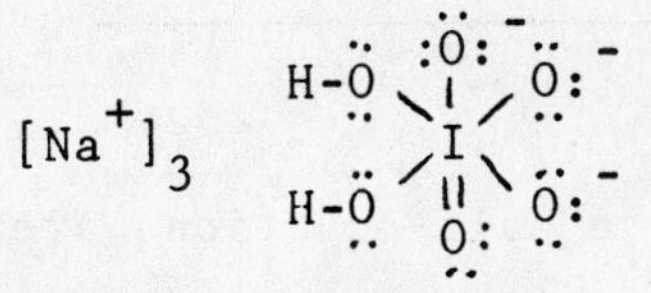

AX$_6$
octahedral

$NaIO_4$

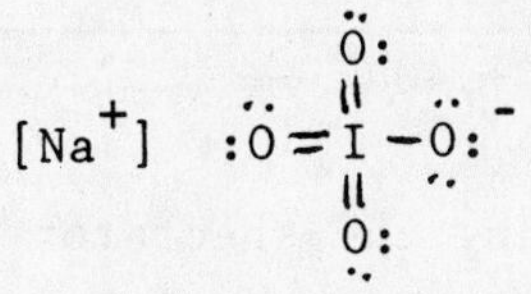

AX$_4$
tetrahedral

In each case, the Lewis structure of the anion is just one of the possible resonance structures.

37. (a) As $pK_a$ increases the acid strength decreases, i.e., in the order

$$ClOH > BrOH > IOH$$

The decreasing acid strength is related to the electronegativity of the halogen atom. The more electronegative is X in X-OH, the more it withdraws electrons from the oxygen atom and the more the polarity of the O-H bond, $^{\delta-}O\text{-}H^{\delta+}$, is increased, thus increasing the acidity. Since the polarity of the O-H bond is expected to decrease in the order ClOH > BrOH > IOH, the acid strength decreases in the same order.

(b) For the method see Chapter 14, Problems 5, 6, and 7).

0.010 M HOCl(aq), pH = 4.75

0.010 M HOBr(aq), pH = 5.31

0.010 M HOI(aq), pH = 6.32

38. (a) Determination of the empirical formula of each compound from its mass % composition by the usual methods (see e.g., Problem 36) gives,

$HIO_3$ and $I_2O_5$

(b) $HIO_3$

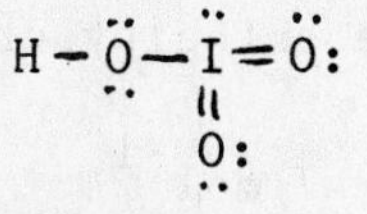

$AX_3E$ trigonal pyramidal

$I_2O_5$

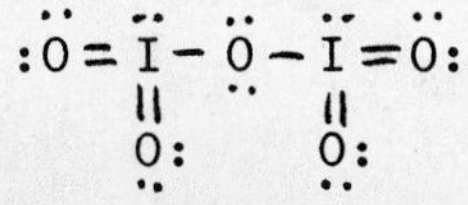

Each I atom is in an $AX_3E$ environment, with trigonal pyramidal geometry.

(c) $I_2(s) + 10HNO_3(\text{fuming}) \longrightarrow 2HIO_3(s) + 10NO_2(g) + 4H_2O(l)$

$2HIO_3(s) \longrightarrow I_2O_5(s) + O_2(g)$

(d) The reactions of $Cl_2(g)$ with hot and cold NaOH(aq), respectively, give $NaClO_3$ and NaOCl, in addition to NaCl. In each reaction, the Cl(0) in $Cl_2$ is oxidized and reduced in the same reaction. Such reactions where one reagent behaves both as an oxidizing agent and as a reducing agent are known as <u>disproportionation reactions</u>.

$$5[Cl_2 + 2e^- \longrightarrow 2Cl^-] \qquad \text{reduction}$$
$$Cl_2 + 12OH^- \longrightarrow 2ClO_3^- + 10e^- + 6H_2O \qquad \text{oxidation}$$
$$6Cl_2 + 12OH^- \longrightarrow 10Cl^- + 2ClO_3^- + 6H_2O \qquad \text{hot NaOH(aq)}$$

$$Cl_2 + 2e^- \longrightarrow 2Cl^- \qquad \text{reduction}$$
$$Cl_2 + 4OH^- \longrightarrow 2OCl^- + 2e^- \qquad \text{oxidation}$$
$$2Cl_2 + 4OH^- \longrightarrow 2Cl^- + 2OCl^- \qquad \text{cold NaOH(aq)}$$

39. (a) Omitting the spectator ion $K^+$, the half-equations are,

$$2[ClO_3^- + e^- + 2H^+ \longrightarrow ClO_2 + H_2O] \qquad \text{reduction}$$
$$ClO_3^- + H_2O \longrightarrow ClO_4^- + 2e^- + 2H^+ \qquad \text{oxidation}$$
$$3ClO_3^- + 2H^+ \longrightarrow 2ClO_2 + ClO_4^- + H_2O$$

and allowing for the behaviour of the water as a strong base in concentrated $H_2SO_4$, the final equation is,

$$3ClO_3^- + 3H_2SO_4 \longrightarrow 2ClO_2 + ClO_4^- + H_3O^+ + 3HSO_4^-$$

(b)
$$2[ClO_3^- + e^- + 2H^+ \longrightarrow ClO_2 + H_2O] \qquad \text{reduction}$$
$$H_2C_2O_4 \longrightarrow 2CO_2 + 2e^- + 2H^+ \qquad \text{oxidation}$$
$$2ClO_3^- + H_2C_2O_4 + 2H^+ \longrightarrow 2ClO_2 + 2CO_2 + 2H_2O$$

and adding $C_2O_4^{2-}$ to each side of the equation gives,

$$2ClO_3^- + 2H_2C_2O_4 \longrightarrow 2ClO_2 + 2CO_2 + 2H_2O + C_2O_4^{2-}$$

(c) $2AgClO_3 + Cl_2 \longrightarrow 2AgCl + 2ClO_2 + O_2$

40.
$$F_2(g) + H_2O(s) \xrightarrow{-40^\circ C} HOF(s) + HF(l)$$
$$2HOF(l) \xrightarrow{\text{below room T}} 2HF(l) + O_2(g)$$
$$HOF(l) + 2H_2O(l) \longrightarrow H_2O_2(aq) + H_3O^+(aq) + F^-(aq)$$
$$2HOF(l) + 2OH^-(aq) \longrightarrow O_2(g) + 2F^-(aq) + 2H_2O(l)$$
$$HOF(l) + HF(g) \longrightarrow F_2(g) + H_2O(l)$$

41. First write out all of the reactions for which the enthalpy changes are given:

(a) $\frac{1}{2}Cl_2(g) + \frac{1}{2}F_2(g) \longrightarrow ClF(g) \qquad \Delta H_f^o = -54 \text{ kJ mol}^{-1}$

(b) $\frac{1}{2}Br_2(l) + \frac{1}{2}F_2(g) \longrightarrow BrF(g) \qquad \Delta H_f^o = -58 \text{ kJ mol}^{-1}$

(c) $\frac{1}{2}Br_2(l) + \frac{5}{2}F_2(g) \longrightarrow BrF_5(g)$ $\qquad \Delta H_f^o = -429$ kJ mol$^{-1}$

(d) $\frac{1}{2}I_2(s) + \frac{7}{2}F_2(g) \longrightarrow IF_7(g)$ $\qquad \Delta H_f^o = -954$ kJ mol$^{-1}$

(e) $\frac{1}{2}Cl_2(g) + \frac{3}{2}F_2(g) \longrightarrow ClF_3(g)$ $\qquad \Delta H_f^o = -163$ kJ mol$^{-1}$

(f) $\frac{1}{2}Br_2(l) + \frac{3}{2}F_2(g) \longrightarrow BrF_3(g)$ $\qquad \Delta H_f^o = -256$ kJ mol$^{-1}$

(g) $\frac{1}{2}I_2(s) + \frac{5}{2}F_2(g) \longrightarrow IF_5(g)$ $\qquad \Delta H_f^o = -838$ kJ mol$^{-1}$

(h) $\frac{1}{2}F_2(g) \longrightarrow F(g)$ $\qquad \Delta H_f^o = 79.4$ kJ mol$^{-1}$

(i) $\frac{1}{2}Cl_2(g) \longrightarrow Cl(g)$ $\qquad \Delta H_f^o = 119.5$ kJ mol$^{-1}$

(j) $\frac{1}{2}Br_2(g) \longrightarrow Br(g)$ $\qquad \Delta H_f^o = 111.9$ kJ mol$^{-1}$

(k) $\frac{1}{2}I_2(g) \longrightarrow I(g)$ $\qquad \Delta H_f^o = 106.8$ kJ mol$^{-1}$

Then use these data to find the average bond energy of the F-halogen bond in each of the interhalogen compounds formed in reactions (a) through (g). To do this, write the reaction that represents breaking all of the bonds in the compound, and determine its $\Delta H^o$ from the data above. If the compound contains only one bond, the $\Delta H^o$ is the bond energy. If the compound contains more than one bond, divide the $\Delta H^o$ by the number of bonds to get the average bond energy.

$ClF(g) \longrightarrow Cl(g) + F(g)$ ; $\Delta H^o = \Delta H_h^o + \Delta H_i^o - \Delta H_a^o = \underline{252.9 \text{ kJ mol}^{-1}}$

$BrF(g) \longrightarrow Br(g) + F(g)$; $\Delta H^o = \Delta H_h^o + \Delta H_j^o - \Delta H_b^o = \underline{249.3 \text{ kJ mol}^{-1}}$

$BrF_5(g) \longrightarrow Br(g) + 5F(g)$

$\Delta H^o = 5\,\Delta H_h^o + \Delta H_j^o - \Delta H_c^o = 937.9$ kJ mol$^{-1}$

Average BrF bond energy = $\Delta H^o/5$ = $\underline{187.6 \text{ kJ mol}^{-1}}$

$IF_7(g) \longrightarrow I(g) + 7F(g)$

$\Delta H^o = 7\,\Delta H_h^o + \Delta H_k^o - \Delta H_d^o = 1616.6$ kJ mol$^{-1}$

Average IF bond energy = $\Delta H^o/7$ = $\underline{230.9 \text{ kJ mol}^{-1}}$

$ClF_3(g) \longrightarrow Cl(g) + 3F(g)$

$\Delta H^o = 3\,\Delta H_h^o + \Delta H_i^o - \Delta H_e^o = 520.7$ kJ mol$^{-1}$

Average ClF bond energy = $\Delta H^o/3$ = $\underline{173.6 \text{ kJ mol}^{-1}}$

$BrF_3(g) \longrightarrow Br(g) + 3F(g)$

$\Delta H^o = 3\,\Delta H_h^o + \Delta H_j^o - \Delta H_f^o = 606.1$ kJ mol$^{-1}$

Average BrF bond energy = $\underline{202.0 \text{ kJ mol}^{-1}}$

$IF_5(g) \longrightarrow I(g) + 5F(g)$

$\Delta H^o = 5\,\Delta H_h^o + \Delta H_k^o - \Delta H_g^o = 1314.8$ kJ mol$^{-1}$

Average IF bond energy = $\underline{268.4 \text{ kJ mol}^{-1}}$

42. This problem is solved by the same method as Problem 41.

| Compound | Bond Energy (kJ $mol^{-1}$) |
|---|---|
| ClF | 252.9 |
| $Cl_2$ | 239.0 |
| ClBr | 216.4 |
| ICl | 208.3 |

43. The dipole of a molecule is given by $u = Qr$. Here, u and r are given, so Q, the charge on each atom, may be calculated directly for these diatomic molecules. The charge in coulombs is then converted to units of the charge on the electron, by dividing Q by the charge on one electron ($1.6022 \times 10^{-19}$ C).

| Molecule | u (C m) | r (pm) | Q (C) | % of one electron charge |
|---|---|---|---|---|
| HF | $6.36 \times 10^{-30}$ | 91.7 | $6.93 \times 10^{-20}$ | 43.3 |
| HCl | $3.43 \times 10^{-30}$ | 127.4 | $2.69 \times 10^{-20}$ | 16.8 |
| HBr | $2.63 \times 10^{-30}$ | 141.4 | $1.86 \times 10^{-20}$ | 11.6 |
| HI | $1.27 \times 10^{-30}$ | 160.9 | $7.89 \times 10^{-20}$ | 4.9 |

In all cases the halogen is the negative end of the dipole. As the halogen becomes less electronegative, the charge separation decreases and the H-X bond becomes less polar.

44. (a) From the mass % elemental composition of bleaching powder, the empirical formula is $Ca_3O_4Cl_4H_2$, and the data suggest that it is an ionic compound containing $Ca^{2+}$, $Cl^-$, and $OCl^-$ ions, and $OH^-$.

$$1.000 \text{ g bleaching powder} = (1.000\text{ g})\left(\frac{1\text{ mol}}{327.9\text{ g}}\right) = \underline{3.050 \times 10^{-3}\text{ mol}}$$

and the reaction with $AgNO_3(aq)$ gives the amount of $Cl^-(aq)$ ion in the solution, since $Ag^+(aq) + Cl^-(aq) \longrightarrow AgCl(s)$.

$$\text{mol AgCl(s)} = (0.874\text{ g AgCl})\left(\frac{1\text{ mol AgCl}}{143.4\text{ g mol}^{-1}}\right) = \underline{6.095 \times 10^{-3}\text{ mol}}$$

Thus, in 1.000 g bleaching powder, X, $\frac{\text{mol Cl}^-\text{ ion}}{\text{mol X}} = \frac{6.095 \times 10^{-3}\text{ mol}}{3.050 \times 10^{-3}\text{ mol}}$

$$= \underline{2.0}$$

For the reaction of $OCl^-$ with $I^-(aq)$, the balanced equation is

$$OCl^-(aq) + 2I^-(aq) + 2H^+(aq) \longrightarrow Cl^-(aq) + I_2(s) + H_2O(l)$$

Thus, mol of $OCl^-$ in 1.000 g of bleaching powder is given by,

$$\text{mol OCl}^- = (121.9\text{ mL})\left(\frac{1\text{ L}}{1000\text{ mL}}\right)(0.100\text{ mol L}^{-1}\text{ I}^-)\left(\frac{1\text{ mol OCl}^-}{2\text{ mol I}^-}\right)$$
$$= \underline{6.10 \times 10^{-3}\text{ mol}}$$

and in 1.000 g of X, $\frac{\text{mol OCl}^-}{\text{mol X}} = \frac{6.10 \times 10^{-3}\text{ mol}}{3.050 \times 10^{-3}\text{ mol}} = \underline{2.0}$

Thus, we can rewrite the empirical formula $Ca_3O_4Cl_4H_2$ as

$$CaCl_2 \cdot Ca(OCl)_2 \cdot Ca(OH)_2 \quad \underline{\text{or}} \quad Ca_3Cl_2(OCl)_2(OH)_2$$

(b) For the formation of $OCl^-$ and $Cl^-$ in basic solution

$$Cl_2 + 4OH^- \longrightarrow 2OCl^- + 2e^- + 2H_2O \qquad \text{oxidation}$$

$$Cl_2 + 2e^- \longrightarrow 2Cl^- \qquad \text{reduction}$$

$$2Cl_2 + 4OH^- \longrightarrow 2OCl^- + 2Cl^- + 2H_2O$$

Thus, for the formation of bleaching powder from $Ca(OH)_2$ and $Cl_2$,

$$3Ca(OH)_2 + 2Cl_2 \longrightarrow CaCl_2.Ca(OCl)_2.Ca(OH)_2$$

45. (a) The enthalpies of vaporization increasing from He to Rn in descending the group, which implies that the intermolecular forces increase going down the group. This is what is expected in terms of the increasing polarizabilities of the atoms as they become progressively larger from He to Rn.

(b) London forces between the water molecules and the noble gas atoms increase progressively with the increasing polarizability of the noble gas atoms from He to Rn.

46. (a) $XeF_2$, Xe(+2) (b) $XeF_4$, Xe(+4) (c) $XeO_2F_2$, Xe(+6)
(d) $XeF_3^+$, Xe(+4) (e) $XeO_3$, Xe(+6) (f) $XeO_4$, Xe(+8)

47. The formation of compounds of Kr and Xe is related to their ability to form excited valence states utilizing low lying d orbitals that are unoccupied in the ground states. Neon is a second period element that has no low lying d orbitals, and its (n = 2) valence shell is restricted to the four pairs of electrons that it has in its ground state, because it has to obey the octet rule. Thus, neon forms no known compounds and none are expected.

48. Compound A contains only Xe and F, compound B contains Xe, F, and O, and compound C contains only Xe and O. Calculation of the empirical formulas from the mass % elemental compositions, by the usual method, gives:

(a) A $XeF_6$; B $XeOF_4$, and C $XeO_3$

(b) Compounds B and C are formed from A by the reactions,

$$XeF_6 + H_2O \longrightarrow XeOF_4 + 2HF, \quad \text{and}$$

$$XeOF_4 + 2H_2O \longrightarrow XeO_3 + 4HF$$

(c)

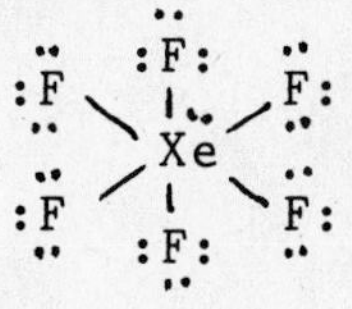

$AX_6E$
distorted octahedral

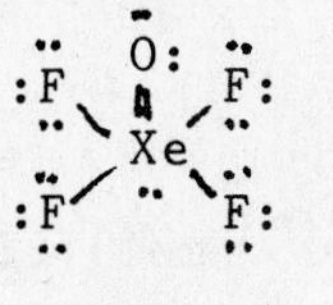

$AX_5E$
square bases pyramidal

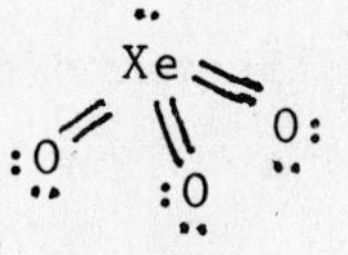

$AX_3E$
trigonal pyramidal

49.

$$XeO_3(s) \xrightarrow{\Delta H_1^o = -402 \text{ kJ}} Xe(g) + \frac{3}{2}O_2(g)$$

$$\Delta H_2^o = 80 \text{ kJ} \qquad XeO_3(s) \searrow XeO_3(g) \nearrow Xe(g) + \frac{3}{2}O_2(g) \qquad \Delta H_3^o$$

$\Delta H_1^o = \Delta H_2^o + \Delta H_3^o$; $\Delta H_3^o = (-402-80) = \underline{-482 \text{ kJ mol}^{-1}}$

and for the reaction, $XeO_3(g) \longrightarrow Xe(g) + \frac{3}{2} O_2(g)$

$$\Delta H_3^o = 3BE(Xe{=}O) - \frac{3}{2}BE(O{=}O) = 3BE(Xe{=}O) - \frac{3}{2}[2\Delta H_f^o(O,g)]$$
$$= 3BE(Xe{=}O) - 3(259.1 \text{ kJ}) = -482 \text{ kJ}$$

$3BE(Xe{=}O) = 265.2$ kJ; $\underline{BE(Xe{=}O) = 88.4 \text{ kJ mol}^{-1}}$

Compared to other bond energies in Table 6.3, Xe=O is one of the weakest covalent bonds (weaker than even F-F, O-O, and N-N single bonds). Thus, $XeO_3$ is expected to be very unstable with respect to its dissociation into Xe(g) and $O_2$(g).

50. $XeO_3$ oxidizes $Mn^{2+}$(aq) to $MnO_4^-$(aq), according to the equation

$$5XeO_3(aq) + 6Mn^{2+}(aq) + 27H_2O(l) \longrightarrow 5Xe(g) + 6MnO_4^-(aq) + 18H_3O^+$$

Thus the standard reduction potential of $XeO_3$(aq) to Xe(g) must be greater than the standard reduction potential of $MnO_4^-$(aq) to $Mn^{2+}$(aq).

51* $I_3^-$(aq) is the complex ion formed between $I^-$(aq) and $I_2$(s) in aqueous solution. Thus, the reactions are,

$$XeF_4(s) + 6NaI(aq) \longrightarrow Xe(g) + 4F^-(aq) + 6Na^+(aq) + 2I_3^-(aq)$$

and $$I_3^-(aq) + 2S_2O_3^{2-}(aq) \longrightarrow S_4O_6^{2-}(aq) + 3I^-(aq)$$

Moles of $S_2O_3^{2-}$ used in the titration is given by

$$(40.00 \text{ mL})(0.1000 \text{ mol L}^{-1})\left(\frac{1 \text{ L}}{1000 \text{ mL}}\right) = \underline{4.000\text{x}10^{-3} \text{ mol } S_2O_3^{2-}}$$

Thus,

$$\text{mol of } XeF_4 \text{ reacted} = (4.000\text{x}10^{-3} \text{ mol } S_2O_3^{2-})\left(\frac{1 \text{ mol } I_3^-}{2 \text{ mol } S_2O_3^{2-}}\right)\left(\frac{1 \text{ mol } XeF_4}{2 \text{ mol } I_3^-}\right)$$
$$= 1.000\text{x}10^{-3} \text{ mol } XeF_4$$

$$\text{mass of } XeF_4 \text{ reacted} = (1.000\text{x}10^{-3} \text{ mol})\left(\frac{207.3 \text{ g}}{1 \text{ mol}}\right) = \underline{0.2073 \text{ g}}$$

(a) Purity of $XeF_4 = \left(\frac{0.2073 \text{ g } XeF_4}{0.2100 \text{ g sample}}\right) \times 100\% = \underline{98.7\%}$

(b) Moles of Xe(g) liberated = moles of $XeF_4$ reacted.

Volume of Xe(g) at STP = $(1.00\text{x}10^{-3} \text{ mol})(22.4 \text{ L mol}^{-1})$

$= 2.24\text{x}10^{-2}$ L <u>or</u> <u>22.4 mL</u>

*<u>Note</u>: These solutions reflect changes made to the problem in the second printing of the text.

52. The group numbers for the families of main group elements are associated with the number of valence shell electrons. Accordingly, it is logical to designate 8 as the group number for the noble gases (except for He with only 2 valence electrons). Alternatively, if a filled shell of electrons is regarded as no longer constituting the valence shell, these elements could be logically designated as group 0. Although the higher members of the group (Kr and Xe) form some compounds utilizing the elements in excited valence states, the common valence is 0. For the higher groups in the periodic table, the common valence is given by 8 minus the group number, which again suggests group 8 as the designation , although the group 0 designation would also be logical if the group was placed to the left of group 1, because for the lower groups in the periodic table, the group number is equal to the common valence. However, the group number for a family of main group elements may also be associated with the maximum oxidation state that elements in that group exhibit in their polar covalent compounds. Although the lower members of the noble gases exist only in the 0 oxidation state, others, most notably Xe, form compounds with the noble gas atom with an oxidation state as high as +8. On balance, group 8 seems the better designation, especially in terms of their electron configurations, since, except for He they all have $ns^2\ np^6$ outer shells.

53. This reaction is reminiscent of the reaction between $H_2(g)$ and $Cl_2(g)$ to form HCl(g), which in bright sunlight is initiated by the dissociation of $Cl_2(g)$ into Cl(g) atoms. In bright sunlight $F_2(g)$ dissociates into F(g) atoms, which are much more reactive than $F_2$ molecules because they behave as free radicals.

The maximum wavelength of light that will initiate the reaction corresponds to that of a photon with just sufficient energy to break the F-F bond in a $F_2$ molecule, which is the F-F bond energy per molecule. From Table 6.3, BE(F-F) = 155 kJ $mol^{-1}$. Thus, for one $F_2$ molecule, the photon has to have a minimum energy $E_{photon}$ given by,

$$E_{photon} = \frac{(155\ \text{kJ mol}^{-1})(\frac{10^3\ \text{J}}{1\ \text{kJ}})}{6.022\times10^{23}\ \text{mol}^{-1}} = h\nu = \frac{hc}{\lambda} = 2.57\times10^{-19}\ \text{J}$$

$$\lambda = \frac{(6.63\times10^{-34}\ \text{J s})(3.00\times10^{8}\ \text{m s}^{-1})}{2.57\times10^{-19}\ \text{J}} = \underline{7.74\times10^{-7}\ \text{m}}$$

or 774 nm which corresponds to light at the red end of the visible spectrum.

## CHAPTER 21

1. $Cr^{3+}$ $[Ar]3d^3$ $Ni^{2+}$ $[Ar]3d^8$ $Cu^+$ $[Ar]3d^{10}$ $Co^{2+}$ $[Ar]3d^7$
   $Au^{3+}$ $[Xe]4f^{14}5d^8$ $Fe^{3+}$ $[Ar]3d^5$

2. Ti $[Ar]4s^23d^2$ Cr $[Ar]4s^13d^5$ Fe $[Ar]4s^23d^6$ Ni $[Ar]4s^23d^8$
   Zn $[Ar]4s^23d^{10}$ Au $[Xe]4f^{14}6s^15d^{10}$
   $Hg^{2+}$ $[Xe]4f^{14}5d^{10}$ $Cu^{2+}$ $[Ar]3d^9$ $Ag^+$ $[Kr]4d^{10}$ $Mn^{2+}$ $[Ar]3d^5$

3. $Cu^{2+}$, $d^9$ $Fe^{3+}$, $d^5$ $Mn^{2+}$, $d^5$ $Ag^+$, $d^{10}$ $V^{3+}$, $d^2$ $Ti^{2+}$, $d^2$

4. (a) $KMnO_4$, Mn(+7) (b) $K_2Cr_2O_7$, Cr(+6) (c) $Ag(CN)_2^-$, Ag(+1)
   (d) $[Co(NH_3)_4Cl_2]^+$, Co(+3) (e) $K_3[Cr(CN)_6]$, Cr(+3)
   (f) $Na_2CoCl_4$, Co(+2) (g) $K_2MnO_4$, Mn(+6) (h) MnO(OH), Mn(+3)
   (i) $VO_2Cl$, V(+5) (j) $TiO.SO_4$, Ti(+4)

5. $FeCl_4^-$, Fe(+3) (b) $Co(H_2O)_6^{3+}$, Co(+3) (c) $Al(OH)_4^-$, Al(+3)
   (d) $Ag(NH_3)_2^+$, Ag(+1) (e) $Fe(CN)_6^{4-}$, Fe(+2) (f) $Fe(CN)_6^{3-}$, Fe(+3)

6.

| Atomic Number | Element | Electron Configuration | Possible Ox. States |
|---|---|---|---|
| 22 | Ti | $[Ar]4s^23d^2$ | 0 to +4 |
| 25 | Mn | $[Ar]4s^23d^5$ | 0 to +7 |
| 27 | Co | $[Ar]4s^23d^7$ | 0 to +9 |

Commonly observed oxidation states are as follows:
Ti +2, +3, +4; Mn +2, +3, +4, +6, +7; Co +2, +3.

7. (a) $CrO_4^{2-}$, Cr(+6) (b) $Cr_2O_7^{2-}$, Cr(+6) (c) $MnO_4^-$, Mn(+7)
   (d) $VO_4^{3-}$, V(+5) (e) $VO^{2+}$, V(+4) (f) $FeO_4^{2-}$, Fe(+6)

8. Cr may have oxidation states from 0 to +6, of which +2, +3, and +6 are the most common.

   (a) $CrO_3$, +6 (b) $Cr_2O_3$, +3 (c) $CrF_2$, +2 (d) $PbCrO_4$, +6
   (e) $Cr_2(SO_4)_3$, +3 (f) $[Cr(H_2O)_5(NH_3)]Cl_2$, +2

9. (a) The ligands are $NH_3$ (neutral), $H_2O$ (neutral), and Cl (-1).
   (b) The oxidation state of Co is +2.
   (c) Replacement of two $H_2O$ ligands by two $Cl^-$ ligands would change the charge from +1 to -1.
   (d) The complex is of $AX_6$ type with octahedral geometry

10. (a) $[Zn(NH_3)_4]Cl_2$, coordination number of Zn = 4.

(b) $[Co(NH_3)_3Cl_3]$, coordination number of Co = 6.

(c) $[Co(NH_3)_5Cl]Cl_2$, coordination number of Co = 6.

(d) $K_2[FeCl_4]$, coordination number of Fe = 4.

11. The names of the compounds in Problem 10 are as follows:

(a) Tetramminezinc(II) chloride
(b) Trichlorotriamminecobalt(III)
(c) Chloropentamminecobalt(III) chloride
(d) Potassium tetrachloroferrate(II)

12. (a) trans-$[Cr(NH_3)_4Cl_2]^+$

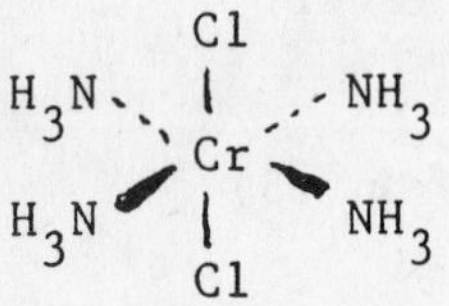


(b) $[Co(C_2O_4)_3]^{3-}$

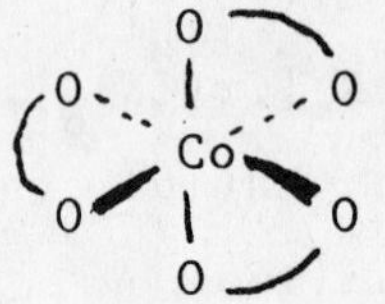


(c) $[Cr(C_2O_4)Br_4]^{3-}$

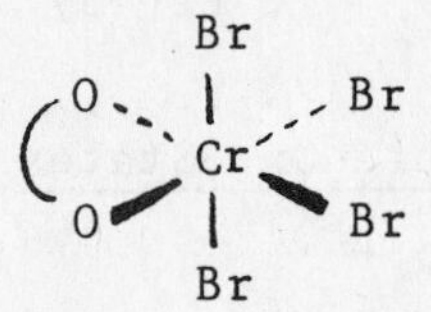


(d) cis-$[Pt(en)_2(CN)_2]^{2-}$

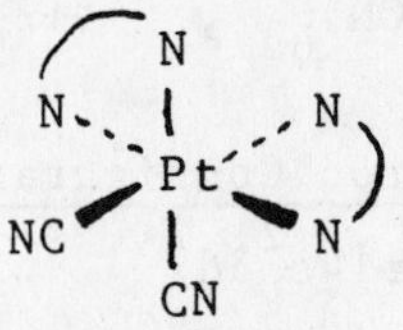


Note: O⌒O represents $\overset{O}{\underset{O}{\gtrless}}C-C\overset{O}{\underset{O}{\lessgtr}}$ ; N⌒N represents $H_2N-CH_2-CH_2-NH_2$

13. The names of the compounds in Problem 12 are as follows:

(a) trans- dichlorotetraamminechromium(III) cation
(b) trioxalatocobaltate(III) anion
(c) tetrabromooxalatochromate(III) anion
(d) cis- dicyanobisethylenediamineplatinate(0) anion

14. The chloride ions exist in solution as $Cl^-(aq)$ and react with $AgNO_3(aq)$,

$$Ag^+(aq) + Cl^-(aq) \longrightarrow AgCl(s)$$

while the remainder of the Cl must be complexed to the Pt and is not free to give AgCl(s) with $Ag^+(aq)$ from the $AgNO_3(aq)$.

| Empirical Formula | Mol AgCl(s) = Mol $Cl^-$ per Mol of Complex | Structural Formula |
|---|---|---|
| $PtCl_4 \cdot 6NH_3$ | 4 | $[Pt(NH_3)_6]Cl_4$ |
| $PtCl_4 \cdot 5NH_3$ | 3 | $[Pt(NH_3)_5Cl]Cl_3$ |
| $PtCl_4 \cdot 4NH_3$ | 2 | $[Pt(NH_3)_4Cl_2]Cl_2$ |
| $PtCl_4 \cdot 3NH_3$ | 1 | $[Pt(NH_3)_3Cl_2]Cl$ |
| $PtCl_4 \cdot 2NH_3$ | 0 | $[Pt(NH_3)_2Cl_4]$ |

15. Sulfate and bromide that are complexed to the metal are not free to react. Ionic $SO_4^{2-}$ and $Br^-$ react with $BaCl_2(aq)$ and $AgNO_3(aq)$, respectively, to give insoluble $BaSO_4(s)$ and insoluble $AgBr(s)$,

$$SO_4^{2-}(aq) + Ba^{2+}(aq) \longrightarrow BaSO_4(s)$$
$$Br^-(aq) + Ag^+(aq) \longrightarrow AgBr(s)$$

Since the <u>red</u> compound produces $AgBr(s)$ but not $BaSO_4(s)$, its structural formula must be $[Co(NH_3)_5SO_4]Br$, with the $SO_4^{2-}$ complexed to the metal and not free to react. The <u>violet</u> compound produces $BaSO_4(s)$ but not $AgBr(s)$; its structural formula is $[Co(NH_3)_5Br]SO_4$, with the $Br^-$ complexed to the metal and not free to react.

16. (a) $AgCl(s) + 2Na_2S_2O_3(aq) \longrightarrow Ag(S_2O_3)_2^{3-}(aq) + Cl^-(aq) + 4Na^+(aq)$

(b) $[Cr(en)_2Cl_2]Cl(aq) + 2H_2O(l) \longrightarrow [Cr(en)_2(H_2O)_2]Cl_3(aq)$

$[Cr(en)_2(H_2O)_2]Cl_3(aq) + 3AgNO_3(aq) \longrightarrow 3AgCl(s) + 3NO_3^-(aq) + [Cr(en)_2(H_2O)_2]^{3+}(aq)$

(c) $Ni(OH)_2(s) + 6NH_3(aq) \longrightarrow Ni(NH_3)_6^{2+}(aq) + 2OH^-(aq)$

(d) $CoSO_4(s) + 6H_2O(l) \longrightarrow Co(H_2O)_6^{2+}(aq) + SO_4^{2-}(aq)$  pale pink

$Co(H_2O)_6^{2+}(aq) + 4HCl(aq) \longrightarrow [CoCl_4(H_2O)_2]^{2-}(aq) + 2H_3O^+(aq)$
　pink　　　　　　　　　　　　　　　deep blue

17. $CaC_2O_4(s)$ dissolves in a solution of EDTA because of the formation of the very soluble and stable $[CaEDTA]^{2-}(aq)$ ion (with a very large formation constant). This is an example of the stability of a complex formed between a metal ion and a polydentate chelating agent.

$$CaC_2O_4(s) + EDTA^{4-}(aq) \rightleftharpoons [CaEDTA]^{2-}(aq) + C_2O_4^{2-}(aq)$$

18. $Zn(OH)_4^{2-}$ is formed when $Zn(OH)_2(s)$ dissolves in $NaOH(aq)$. Formally it can be thought of as formed from a $Zn^{2+}$ ion with a $d^{10}$ valence shell with four $OH^-$ ions coordinated to it. Since the d shell is filled, $Zn(OH)_4^{2-}$ is an $AX_4$ <u>tetrahedral</u> species. The formal charge on Zn is 2-.

$$
\begin{array}{c}
\overset{..}{\underset{..}{O}}-H \\
| \\
H-\overset{..}{\underset{..}{O}} - Zn^{2-}\, \overset{..}{\underset{..}{O}}-H \\
| \\
\overset{..}{O}-H
\end{array}
$$

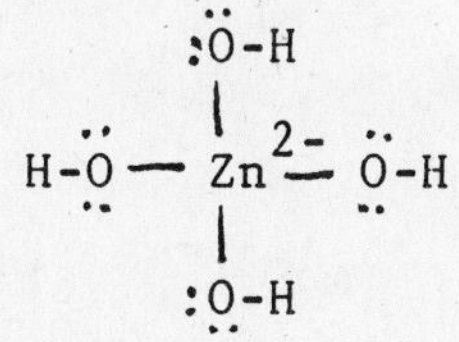

19. (a) From the analytical data, calculate the empirical formulas by the usual methods:

| | Element | g/100 g of compound | mol/100 g of compound | mol ratio | empirical formula |
|---|---|---|---|---|---|
| A | Co | 25.22 | 0.428 | 1.00 | |
| | Cl | 45.99 | 1.295 | 3.03 | $Co(NH_3)_4Cl_3$ |
| | $NH_3$ | 29.19 | 1.717 | 4.01 | |
| B | Co | 25.22 | 0.428 | 1.00 | |
| | Cl | 45.59 | 1.284 | 3.00 | $Co(NH_3)_4Cl_3$ |
| | $NH_3$ | 29.19 | 1.717 | 4.00 | |
| C | Co | 23.53 | 0.399 | 1.00 | |
| | Cl | 42.46 | 1.196 | 3.00 | $Co(NH_3)_5Cl_3$ |
| | $NH_3$ | 34.01 | 2.000 | 5.01 | |

| | | | | | |
|---|---|---|---|---|---|
| | Co | 22.03 | 0.374 | 1.00 | |
| D | Cl | 39.76 | 1.122 | 3.00 | $Co(NH_3)_6Cl_3$ |
| | $NH_3$ | 38.21 | 2.244 | 5.95 | |

(b) The mass of AgCl(s) precipitated enables moles of $Cl^-$ ion in each of the complexes to be calculated and the ratio of mol $Cl^-$ to mol of complex.

| | g complex | mol complex | g AgCl(s) | mol $Cl^-$ | mol ratio complex/$Cl^-$ |
|---|---|---|---|---|---|
| A | 0.232 | $9.94 \times 10^{-4}$ | 0.143 | $9.97 \times 10^{-4}$ | 1 : 1 |
| B | 0.255 | $1.09 \times 10^{-3}$ | 0.157 | $1.09 \times 10^{-3}$ | 1 : 1 |
| C | 0.226 | $9.02 \times 10^{-4}$ | 0.258 | $1.80 \times 10^{-3}$ | 1 : 2 |
| D | 0.348 | $1.30 \times 10^{-3}$ | 0.559 | $3.90 \times 10^{-3}$ | 1 : 3 |

Thus, we can rewrite the empirical formulas indicating the number of ionizable $Cl^-$ ions outside the formula of the complex Co ion.

A $[Co(NH_3)_4Cl_2]Cl$, cis or trans.

B The same as A but the other geometric isomer.

C $[Co(NH_3)_5Cl]Cl_2$.

D $[Co(NH_3)_6]Cl_3$.

Structures of complexes

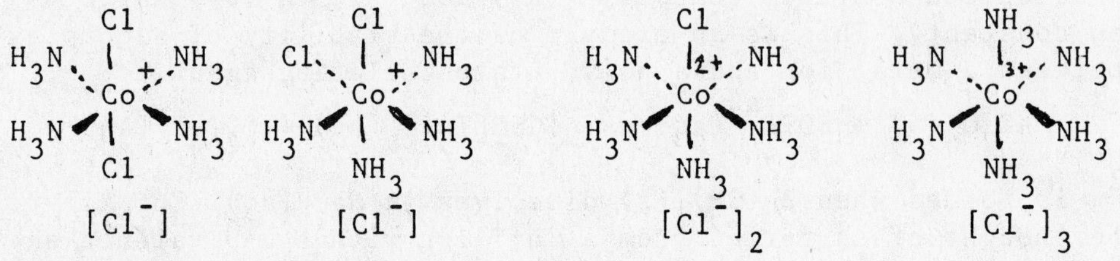

trans cis

A or B C D

20. $Fe^{3+}(aq) + SCN^-(aq) \rightleftharpoons Fe(SCN)^{2+}(aq)$

0.010 0.0003 x mol $L^{-1}$ at equilibrium

$$K_f = \frac{[Fe(SCN)^{2+}]}{[Fe^{3+}][SCN^-]} = 1 \times 10^3 \text{ mol L}^{-1}$$

$$x = [Fe(SCN)^{2+}] = K_f[Fe^{3+}][SCN^-]$$
$$= (1 \times 10^3 \text{ mol L}^{-1})(1 \times 10^{-2} \text{ mol L}^{-1})(3 \times 10^{-4} \text{ mol L}^{-1}]$$
$$= \underline{3 \times 10^{-3} \text{ mol L}^{-1}}$$

21. (a) $Na_2[Zn(OH)_4]$ (b) $Na[Al(OH)_4]$ (c) $[PtBr(H_2O)_3]Cl$
(d) $FeCl_4^{2-}$ (e) $[Ni(en)_3]Br_2$ (f) $K_2[Co(CN)_4]$

22. (a) potassium pentacyanomangante(IV)
(b) hexacyanoferrate(II)

(c) nitropentamminecobalt(III)
(d) tetramminecopper(II) sulfate
(e) hexamminecobalt(II) chloride
(f) tetracarbonylnickel(0)

23. $Zn(NH_3)_4Cl_2$ is <u>not</u> expected to be colored; it contains zinc with a $d^{10}$ configuration and therefore there is no possibility of transitions between d levels.

24. (a) A solution that appears yellow absorbs light of its complimentary color, which is violet. Violet light is in the wavelength region 380-450 nm.

(b) A solution of cobalt(III) in aqueous ammonia contains the yellow $Co(NH_3)_6^{3+}(aq)$ ion. When acid is added the $NH_3$ ligands are replaced by water molecules,

$$Co(NH_3)_6^{3+}(aq) + 6H_3O^+(aq) \longrightarrow Co(H_2O)_6^{3+}(aq) + 6NH_4^+(aq)$$

Since $H_2O$ is a weaker field ligand than $NH_3$, the energy of the electronic transitions are expected to decrease, so that the wavelength of the light absorbed should increase. The color of the light absorbed will shift from violet towards the red end of the visible spectrum. In fact $Co(H_2O)_6^{3+}(aq)$ is a blue complex which absorbs orange light of wavelength 600 nm.

25. See Problem 24. Note that this problem has no straightforward solution. An aqueous solution of manganese(II) nitrate, $Mn(NO_3)_2(aq)$, contains the pale pink $Mn(H_2O)_6^{2+}(aq)$ ion, while a aqueous solution of potassium hexacyanomanganate(II) contains the $Mn(CN)_6^{4-}(aq)$ ion, which is deep blue. Relative to $H_2O$, $CN^-$ is a strong-field ligand. which interacts much more strongly with the metal d orbitals; the light absorbed should be at a lower wavelength than that absorbed by the almost colorless $Mn(H_2O)_6^{2+}(aq)$ ion. The color of the light absorbed should shift towards the violet end of the spectrum; the transmitted light should be towards the red end of the spectrum, which is <u>contrary</u> to observation.

26. Write $[Co(NH_3)_4Cl_2]^+$ as $MX_4Y_2$. If such a complex ion was hexagonal planar, there should be <u>three</u> possible isomers:

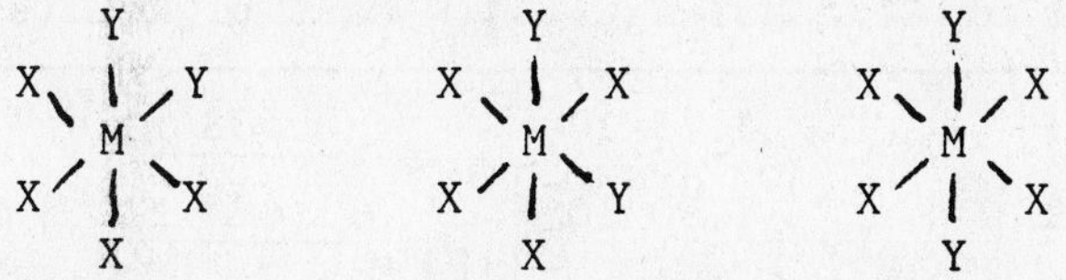


If such a complex ion was a <u>triangular prism</u>, there should also be <u>three</u> possible isomers:

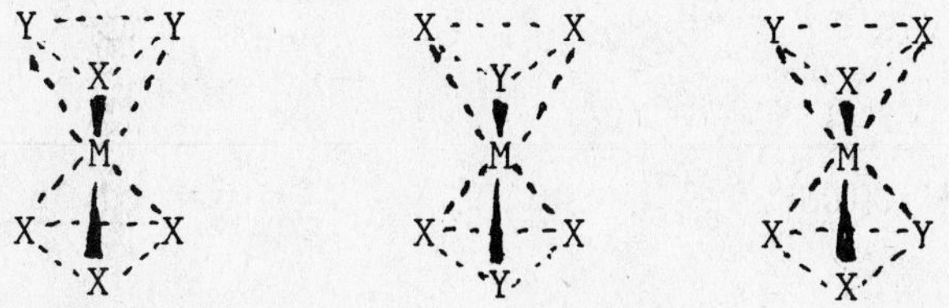


If such a complex ion was <u>octahedral</u>, there should be only <u>two</u> possible isomers, the <u>cis</u> and <u>trans</u> forms:

trans     cis

While it may seem that there are more possible isomers of the above structures, rotation of any other forms that may be drawn shows them to be superimposable with one of the above forms, and therefore identical to one of them.

27. (a) Compounds A, B, and C are structural isomers of formula $CrCl_3.6H_2O$ which differ in terms of the number of $H_2O$ and Cl ligands which are coordinated to the metal atom, and $Cl^-$ ions and $H_2O$ molecules of crystallization that are not part of the complex ion. Water of crystallization is lost on long exposure to a dehydrating agent, but strongly bonded $H_2O$ molecules are retained. Thus, from the mass of $H_2O$ lost on dehydration, the structures may be deduced:

| Compound | Mass of Compound | Moles of Compound | Mass of $H_2O$ lost | Moles of $H_2O$ lost | Mol $H_2O$ lost / Mol Compound |
|---|---|---|---|---|---|
| A | 10.00 g | 0.03752 | 1.35 g | $7.49\times10^{-2}$ | 2.00 |
| B | 10.00 g | 0.03752 | 0.68 g | $3.8\times10^{-2}$ | 1.01 |
| C | 10.00 g | 0.03752 | 0 | 0 | 0 |

Structural Formulas

A   $[Cr(H_2O)_4Cl_2]Cl.2H_2O$

B   $[Cr(H_2O)_5Cl]Cl_2.H_2O$

C   $[Cr(H_2O)_6]Cl_3$

(b) Only the $Cl^-$ anions will be precipitated when excess $AgNO_3(aq)$ is added to aqueous solutions of the compounds. In each case, 1.00 g of complex is $3.75\times10^{-3}$ mol.

| Compound | Number of $Cl^-$ | Mol AgCl(s) | Mass of AgCl(s) |
|---|---|---|---|
| A | 1 | $3.75\times10^{-3}$ | 0.538 g |
| B | 2 | $7.50\times10^{-3}$ | 1.08 g |
| C | 3 | $11.25\times10^{-3}$ | 1.61 g |

28.

$$Cu(H_2O)_4^{2+} + 2e^- \longrightarrow Cu(s) + 4H_2O \qquad E^o_{red} = +0.35\ V$$

$$Cu(s) + 4NH_3 \longrightarrow Cu(NH_3)_4^{2+} + 2e^- \qquad E^o_{ox} = +0.03\ V$$

$$Cu(H_2O)_4^{2+} + 4NH_3 \longrightarrow Cu(NH_3)_4^{2+} + 4H_2O \qquad E^o_{cell} = +0.38\ V$$

for which,

$$K_f = e^{nE^o/0.0257} = e^{2(0.38)/0.0257} = e^{29.6} = 7.2\times10^{12}$$

29. In an octahedral complex, four of the ligands are in the xy plane and the other two point along the z axis. Repulsions between the ligand electron pairs and metal electrons in the $d_{x^2-y^2}$ and $d_{z^2}$ orbitals of the metal, which are directed at the ligands, cause these two orbitals of the metal to be higher in energy than the $d_{xy}$, $d_{yz}$, and $d_{xz}$ orbitals that lie between the ligands, and the five d orbitals split into the two sets, $d_{x^2-y^2}$ and $d_{z^2}$, and $d_{xy}$, $d_{yz}$ and $d_{xz}$, with the latter three orbitals at a lower energy than the remaining two orbitals.

30. In the $Co(NH_3)_6^{3+}$ complex ions, Co has six d electrons and the splitting of the d orbital energy levels is large, so that all six electrons are accommodated in three pairs in the set of three d orbitals of lower energy. Thus, in this complex ion, there are no unpaired electrons and $Co(NH_3)_6^{3+}$ is diamagnetic. In the $Mn(NH_3)_6^{2+}$ complex ion, Mn has five electrons, and these are all unpaired according to the information given. Thus, the splitting of the d orbital energy levels must be small, favoring an arrangement in which each d orbital accommodates a single electron (Hund's rule). The five unpaired d electrons make the $Mn(NH_3)_6^{2+}$ complex ion paramagnetic.

31. (a) $NiCl_4^{2-}$ and $Ni(CN)_4^{2-}$ complex ions both contain Ni with a $d^8$ electron arrangement, but they differ in that $Cl^-$ is a very weak field ligand while $CN^-$ is a very strong field ligand. In $NiCl_4^{2-}$ the interaction between the ligands an the d shell of the metal is weak and the d electrons do not distort the shape predicted by the VSEPR model, so that this complex ion has the $AX_4$ tetrahedral shape. In $Ni(CN)_4^{2-}$, with strong field ligands, a square planar arrangement is preferred because all eight d electrons can be accommodated in the four lowest energy d orbitals (Figure 21.17). The interaction between the d shell and the four ligands is strong and the complex has a square planar rather than a tetrahedral shape.
(b) and (c) In $Ni(CN)_4^{2-}$ the eight d electrons are accommodated as four pairs in the four lowest energy d orbitals and the complex is diamagnetic. In $NiCl_4^{2-}$ three pairs of electrons are accommodated in the three lower energy orbitals but the remaining two d electrons occupy the two higher energy orbitals and are unpaired (Hund's rule). Thus $NiCl_4^{2-}$ has two unpaired electrons and is paramagnetic.

32.

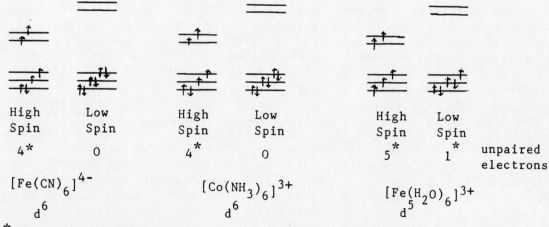

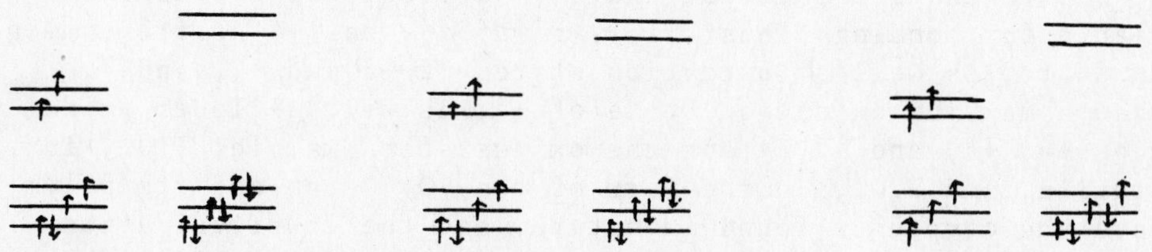

| | High Spin | Low Spin | High Spin | Low Spin | High Spin | Low Spin | |
|---|---|---|---|---|---|---|---|
| | 4* | 0 | 4* | 0 | 5* | 1* | unpaired electrons |
| | $[Fe(CN)_6]^{4-}$ $d^6$ | | $[Co(NH_3)_6]^{3+}$ $d^6$ | | $[Fe(H_2O)_6]^{3+}$ $d^5$ | | |

*paramagnetic species

33. With increasing oxidation number a transition metal takes on increasing nonmetal character as its electronegativity increases. Thus, the oxides become increasingly covalent and acidic. For example, chromium has common II, III, and VI oxidation states. Cr(II)O(s) is a basic oxide, Cr(III) oxide, $Cr_2O_3(s)$, is amphoteric, and $Cr(VI)O_3(s)$ is an acidic oxide.

34. As the oxidation state increases, an increasing number of valence shell electrons is involved in bonding and the electronegativity of the transition element increases. Thus, lower oxidation state compounds are typically metallic and ionic but in higher oxidation state compounds the trend is towards increasingly nonmetallic and covalent behaviour. Cr (see Problem 33), V, and Mn provide good illustrative examples. For example, vanadium forms compounds in the important +5, +4, +3, and +2 oxidation states. In its +5 oxidation state, vanadium behaves more like a nonmetal than a metal. $V_2O_5$ is an amphoteric oxide, dissolving in aqueous acid to give $VO_2^+(aq)$ and in aqueous base to give the vanadate, $VO_4^{3-}(aq)$, ion, which has a chemistry similar to that of the $PO_4^{3-}(aq)$, phosphate, ion. Reduction of a solution of $VO_2^+(aq)$ gives $VO(H_2O)_5^{2+}(aq)$ with vanadium in the +4 oxidation state, and further reduction gives $V(H_2O)_6^{3+}(aq)$ with vanadium in the +3 oxidation state, and finally $V(H_2O)_6^{2+}(aq)$, with vanadium in the +2 oxidation state.

35. If all of the outer electrons were available for metallic bonding, the melting points would be expected to continue to increase beyond chromium, with increase in the number of valence electrons and increasing core charge. However, this is apparently not the case, and the maximum at chromium and the decrease in melting point beyond is due to the decrease in size of the 3d orbitals with increasing core charge, so that they become progressively less available for metallic bonding (and for compound formation). The balance between the increasing number of outer shell electrons and their decreasing availability results in the strongest metallic bonding at Cr. The densities of the metals are also consistent with these same trends. There is a rapid increase in density from Sc to Cr and then only a slow increase between Cr and Cu and a drop from Cu to Zn, approximately paralleling the changes in the strength of metallic bonding and atomic mass.

36. For the early transition metals, oxidation states utilizing the 4s and all of the 3d electrons are common, but in going from left to right along the period, the core charges increase and the 3d electrons become less readily available for bonding. Thus, Ti, $[Ar]4s^23d^2$, has +4 as its commonest oxidation state, as well as oxidation states of +3 and +2, and V, $[Ar]4s^23d^3$, has a maximum oxidation state of +5, as well as lower oxidation states of +4, +3, and +2. Among the oxides, for example, $TiO_2$, in the form of rutile, and $V_2O_5$, in the form of the $VO_4^{3-}$ ion and its polymeric forms, are the compounds found in nature, and the lower oxidation state compounds result from their reduction but are readily oxidized back to the more stable Ti(IV) and V(V) atates. In contrast, Co, $[Ar]4s^23d^7$, and Ni, $[Ar]4s^23d^8$, have +2 as their commonest oxidation states, although Co is also found in the +3 oxidation state in many complexes. The +2 oxidation states utilize only the two 4s electrons for compound formation and contain the $Co^{2+}$ and $Ni^{2+}$ ions. The marked change in behaviour of the later members of the series compared to that of the early members is the consequence of the decreasing availability of the d shell electrons after manganese.

37. Sc, Ti, and V have the electron configurations $[Ar]4s^2 3d^1$, $[Ar]4s^2 3d^2$, and $[Ar]4s^2 3d^3$, respectively, and the $Sc^{3+}$, $Ti^{3+}$, and $V^{3+}$ ions have $d^0$, $d^1$, and $d^2$, configurations, respectively. Thus, d energy level transitions are possible for $Ti^{3+}$ and $V^{3+}$ but not for $Sc^{3+}$; $Ti^{3+}$ has a violet color, $V^{3+}$ has a green color, but $Sc^{3+}$ is colorless.

38. Cu has the electron configuration $[Ar]3d^{10} 4s^1$. Thus, $Cu^{2+}$ has a $d^9$ configuration, while $Cu^+$ has a $d^{10}$ configuration. Electron transitions between d energy levels are possible for $Cu^{2+}$ but not for $Cu^+$. Thus, the former is colored and the latter is colorless.

39. (a) When copper (II) sulfate, $CuSO_4.5H_2O(s)$ is dissolved in water it forms the blue $Cu(H_2O)_4^{2+}$ (aq) ion and colorless $SO_4^{2-}$(aq) ion.

(b) On addition of excess of concentrated HCl(aq) to the solution in part (a), the blue solution turns green due to the formation of the complex ion $CuCl_4^{2-}$(aq).

$$Cu(H_2O)_4^{2+}(aq) + 4Cl^-(aq) \rightleftharpoons CuCl_4^{2-}(aq) + 4H_2O$$

(c) On addition of excess $NH_3$(aq) to the solution in part (b), the green solution turns a deep blue, due to the formation of the complex ion $Cu(NH_3)_4^{2+}$(aq).

(d) On addition of excess KCN(aq) to the solution in part (a), the blue solution turns colorless due to the formation of the complex ion $Cu(CN)_4^{2-}$(aq). $CN^-$ binds more strongly than $H_2O$ to Cu(II).

40. (a) The ions with $d^{10}$ electron configurations give colorless aqueous solutions. i.e., $Zn^{2+}$, $[Ar]3d^{10}$ and $Cu^+$, $[Ar]3d^{10}$.

(b) The +2 oxidation state, corresponding to the ionization of the two 4s electrons is common to all of these metals; they all form $M^{2+}$ ions.

(c)

| blue | white | green |
|---|---|---|
| $Cu(OH)_2$ | $Mn(OH)_2$ | $Ni(OH)_2$ |
| | $Zn(OH)_2$ | $Cr(OH)_2$ |
| | $Fe(OH)_2$ | |

(d) $Fe^{2+}$(aq) is pale green. On addition of NaOH(aq) and $H_2O_2$(aq) to a solution containing $Fe^{2+}$(aq), $Fe^{2+}$(aq) is oxidized by the $H_2O_2$ to $Fe^{3+}$(aq), which gives a brown insoluble precipitate of $Fe(OH)_3$ with $OH^-$(aq).

(e) $Co(H_2O_6)^{2+}$(aq) is pale pink. On addition of excess of concentrated HCl(aq) the solution turns blue due to replacement of some of $H_2O$ ligands by $Cl^-$ ligands.

(f) Cr, Mn, and Fe form compounds in the +6 oxidation state, although such Fe compounds are rare. Cr(+6) occurs in compounds such as $CrO_3$, $Na_2CrO_4$, and $K_2Cr_2O_7$. Mn(+6) occurs in compounds such as $K_2MnO_4$, but unlike Cr(+6) this oxidation state of Mn is not very stable and $K_2MnO_4$ readily disproportionates to Mn(+4) and Mn(+7).

(g) $MnO_4^-$(aq) is a strong oxidizing agent. If a solution was a pale pink color the concentration of $MnO_4^-$(aq) could be only very dilute because this ion has an intense color. Addition of a reducing agent, such as $Fe^{2+}$(aq) in acid would rapidly turn the solution colorless. $Mn^{2+}$(aq) would not react with $Fe^{2+}$(aq).

41. Metal oxides are basic; nonmetal oxides are acidic, and semimetal oxides are amphoteric. For a given metal, as its oxidation state increases, the trend is for the character of the metal to become more like that of a nonmetal and for the oxides to go from basic or amphoteric to acidic.

(a) <u>MgO(s)</u> Metal oxide, <u>basic</u>.

$$MgO(s) + 2H_3O^+(aq) \longrightarrow Mg^{2+}(aq) + 3H_2O(l)$$

(b) <u>$SO_3(g)$</u> Nonmetal oxide, <u>acidic</u>

$$SO_3(g) + H_2O(l) \longrightarrow H_2SO_4(aq)$$

(c) <u>$CrO_3$</u> Oxide of high oxidation state Cr(VI), <u>acidic</u>.

$$CrO_3(s) + H_2O(l) \longrightarrow H_2CrO_4(aq)$$

$$2CrO_4^{2-}(aq) + 2H_3O^+(aq) \longrightarrow Cr_2O_7^{2-}(aq) + 3H_2O(l)$$

(d) <u>ZnO(s)</u> Oxide of low oxidation state Zn(II), <u>amphoteric</u>.

$$ZnO(s) + 2H_3O^+(aq) \longrightarrow Zn^{2+}(aq) + 3H_2O(l)$$

$$ZnO(s) + 2OH^-(aq) + H_2O(l) \longrightarrow Zn(OH)_4^{2-}(aq)$$

(e) <u>$Cr_2O_3(s)$</u> Oxide of intermediate oxidation state Cr(III), <u>amphoteric</u>.

$$Cr_2O_3(s) + 6H_3O^+(aq) \longrightarrow 2Cr^{3+}(aq) + 9H_2O(l)$$

$$Cr_2O_3(s) + 2OH^-(aq) + 3H_2O(l) \longrightarrow 2Cr(OH)_4^-(aq)$$

(f) <u>CuO(s)</u> Metal oxide, <u>basic</u>.

$$CuO(s) + 2H_3O^+(aq) \longrightarrow Cu^{2+}(aq) + 3H_2O(l)$$

(g) <u>$TiO_2(s)$</u> Oxide of high oxidation state Ti(IV), <u>acidic</u>.

$$TiO_2(s) + 2NaOH(molten) \longrightarrow Na_2TiO_3(s) + H_2O(g)$$

(h) <u>$V_2O_5(s)$</u> Oxide of high oxidation state V(V), <u>amphoteric</u>.

$$V_2O_5(s) + 2H_3O^+(aq) \longrightarrow 2VO_2^+(aq) + 3H_2O(l)$$

$$V_2O_5(s) + 6OH^-(aq) \longrightarrow 2VO_4^{3-}(aq) + 3H_2O(l)$$

42. (a) CuS(s) is much less soluble than ZnS(s), ($K_{sp}$ values $8.0 \times 10^{-37}$ and $1.6 \times 10^{-24}$ $mol^2\ L^{-2}$, respectively). In the presence of $H_3O^+(aq)$, the extent of dissociation of $H_2S(aq)$ to $S^{2-}(aq)$ ions is very small and CuS(s) precipitates but not ZnS(s).

(b) AgCl(s) is very insoluble but $Ag^+(aq)$ forms a very stable complex ion, $Ag(S_2O_3)_2^{3-}(aq)$, with $S_2O_3^{2-}(aq)$, and AgCl(s) dissolves because the reaction

$$AgCl(s) + 2S_2O_3^{2-}(aq) \rightleftharpoons Ag(S_2O_3)_2^{3-}(aq) + Cl^-(aq)$$

has a large equilibrium constant ($K = K_{sp}.K_f$).

(c) $FeCl_3(s)$ dissolves in water to give a solution containing $Fe(H_2O)_6^{3+}$ and $Cl^-$ ions. The former, because of the high formal charge on Fe, behaves as a weak acid, giving a solution with pH < 7.

$$Fe(H_2O)_6^{3+}(aq) + H_2O(l) \rightleftharpoons Fe(H_2O)_5OH^{2+}(aq) + H_3O^+(aq)$$

(d) Initially, the $Cu^{2+}(aq)$ ions in solution react with $OH^-(aq)$ to give pale blue $Cu(OH)_2(s)$, which dissolves in excess $NH_3(aq)$ due to the formation of the very stable deep blue $Cu(NH_3)_4^+(aq)$ ion.

$$Cu^{2+}(aq) + 2OH^-(aq) \longrightarrow Cu(OH)_2(s)$$
$$Cu(OH)_2(s) + 4NH_3(aq) \longrightarrow Cu(NH_3)_4^{2+}(aq) + 2OH^-(aq)$$

43. $Cr(H_2O)_6^{3+}(aq) + H_2O(l) \rightleftharpoons Cr(H_2O)_5OH^{2+}(aq) + H_3O^+(aq)$

| | | | |
|---|---|---|---|
| initially | 0.050 | 0 | 0 mol $L^{-1}$ |
| at eq'm | 0.050-x | x | x mol $L^{-1}$ |

$$pK_a = 3.70;\quad K_a = 2.0\times10^{-4}\ mol\ L^{-1} = \frac{[H_3O^+][Co(H_2O)_5OH^{2+}]}{[Co(H_2O)_6^{3+}]} = \frac{x^2}{0.050-x}$$

x << 0.050, is not justified, but solving the quadratic equation gives,
$x = [H_3O^+] = 0.00306\ mol\ L^{-1}$; pH = 2.51.

44. $Cr(H_2O)_6^{3+}(aq)$ ions react with $OH^-(aq)$ ions to give insoluble gray-green $Cr(OH)_3(s)$,

$$Cr^{3+}(aq) + 3OH^-(aq) \longrightarrow Cr(OH)_3(s)$$

$Cr(OH)_3(s)$ is an amphoteric hydroxide and dissolves in excess $OH^-(aq)$ to give a solution containing the green $Cr(OH)_4^-(aq)$ ion,

$$Cr(OH)_3(s) + OH^-(aq) \longrightarrow Cr(OH)_4^-(aq)$$

Boiling this solution with $Na_2O_2(s)$ oxidizes Cr(III) in $Cr(OH)_4^-(aq)$ to Cr(VI) in yellow $CrO_4^{2-}(aq)$,

$$2Cr(OH)_4^-(aq) + 3O_2^{2-}(aq) \longrightarrow 2CrO_4^{2-}(aq) + 4OH^-(aq) + 2H_2O$$

and on adding acid in excess (pH < 7), yellow $CrO_4^{2-}(aq)$ is converted to orange $Cr_2O_7^{2-}(aq)$ ions.

$$2CrO_4^{2-}(aq) + 2H_3O^+(aq) \longrightarrow Cr_2O_7^{2-}(aq) + 3H_2O(l)$$

45. (a) In each case, calculate the requisite standard cell potential.

(i) $Zn(s) \longrightarrow Zn^{2+}(aq) + 2e^-$ $\qquad E^o_{ox} = +0.76$ V
$Ni^{2+}(aq) + 2e^- \longrightarrow Ni(s)$ $\qquad E^o_{red} = -0.25$ V

$Zn(s) + Ni^{2+}(aq) \longrightarrow Zn^{2+}(aq) + Ni(s)$ $\qquad E^o_{cell} = +0.51$ V

The cell voltage is positive, so the reaction proceeds as written; Zn(s) displaces $Ni^{2+}(aq)$ from solution; Zn(s) will dissolve and Ni(s) is deposited on the surface of the zinc.

(ii) $Zn(s) \longrightarrow Zn^{2+}(aq) + 2e^-$ $\qquad E^o_{ox} = +0.76$ V
$2H_3O^+(aq) + 2e^- \longrightarrow H_2(g) + 2H_2O$ $\qquad E^o_{red} = 0.00$ V

$Zn(s) + 2H_3O^+(aq) \longrightarrow Zn^{2+}(aq) + 2H_2O + H_2(g)$, $\qquad E^o_{cell} = +0.76$ V

Zinc dissolves in 1.0 M HCl(aq) to give $Zn^{2+}(aq)$ in solution and $H_2(g)$ is evolved.

(b) As in part (a) calculate the standard cell potentials.

(i) $Ni(s) \longrightarrow Ni^{2+}(aq) + 2e^-$ $\qquad E^o_{ox} = +0.25$ V
$Zn^{2+}(aq) + 2e^- \longrightarrow Zn(s)$ $\qquad E^o_{red} = -0.76$ V

$Ni(s) + Zn^{2+}(aq) \longrightarrow Ni^{2+}(aq) + Zn(s)$ $\qquad E^o_{cell} = -0.51$ V

The negative voltage indicates that it is the reverse reaction that is spontaneous. In other words, there is <u>no reaction</u>.

(ii) $Zn(s) \longrightarrow Zn^{2+}(aq) + 2e^-$ $\qquad E^o_{ox} = +0.25\ V$

$2H_3O^+(aq) + 2e^- \longrightarrow H_2(g) + 2H_2O$ $\qquad E^o_{red} = 0.00\ V$

$Ni(s) + 2H_3O^+(aq) \longrightarrow Ni^{2+}(aq) + 2H_2O + H_2(g)$ $\qquad E^o_{cell} = \underline{+0.25\ V}$

Nickel dissolves in 1.0 M HCl(aq) to give $Ni^{2+}(aq)$ in solution and $H_2(g)$ is evolved.

46. (a) $2CrO_4^{2-}(aq) + 2H_3O^+(aq) \longrightarrow Cr_2O_7^{2-}(aq) + 3H_2O(l)$
yellow $\qquad$ orange

(b) $Cr_2O_7^{2-}(aq) + 2OH^-(aq) \longrightarrow 2CrO_4^{2-}(aq) + H_2O(l)$
orange $\qquad$ yellow

(c) $(NH_4)_2Cr_2O_7(s) \xrightarrow{heat} Cr_2O_3(s) + 4H_2O(g) + N_2(g)$ (chemical volcano)

(d) $SO_2(g)$ reduces orange $Cr_2O_7^{2-}(aq)$ to green $Cr^{3+}(aq)$,

$Cr_2O_7^{2-}(aq) + 3SO_2(g) + 2H_3O^+(aq) \longrightarrow 2Cr^{3+}(aq) + 3SO_4^{2-}(aq) + 3H_2O(l)$

(e) $H_2S(g)$ is oxidized to S(s), which forms a finely divided precipitate; orange $Cr_2O_7^{2-}(aq)$ is reduced to green $Cr^{3+}(aq)$.

$3H_2S(g) + Cr_2O_7^{2-}(aq) + 8H_3O^+(aq) \longrightarrow 3S(s) + 2Cr^{3+}(aq) + 15H_2O(l)$

47. Rust is hydrated $Fe_2O_3(s)$ which contains $Fe^{3+}$ ions which form a very stable chelated complex with oxalic acid in aqueous solution.

$Fe_2O_3(s) + 6H_3O^+(aq) + 6C_2O_4^{2-}(aq) \longrightarrow 2Fe(C_2O_4)_3^{3-}(aq) + 9H_2O(l)$

48. The $M^{3+}(aq)$ ion with the largest standard reduction potential for the reaction

$$M^{3+}(aq) + e^- \longrightarrow M^{2+}(aq)$$

is the most easily reduced, and is therefore the strongest oxidizing agent. Thus, the order of oxidizing strengths is

$Cr^{3+} < V^{3+} < Fe^{3+} < Mn^{3+} < Co^{3+} < Ti^{3+}$

Of these ions, only $Cr^{3+}(aq)$ and $V^{3+}(aq)$, whose reduction potentials are negative relative to 1 M $H_3O^+(aq)$ are stable in 1.00 M aqueous acid relative to their $M^{2+}(aq)$ ions.

49. $CrO_2Cl_2$ contains Cr(VI) and has the Lewis structure,

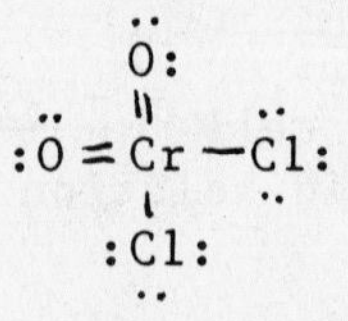

$AX_4$, tetrahedral

50. The gray-black compound contains 0.540 g Fe and 0.365 g S. i.e.,

$(0.540\ g\ Fe)\left(\frac{1\ mol\ Fe}{55.85\ g\ Fe}\right) = \underline{9.67\times10^{-3}\ mol\ Fe}$

$(0.365\ g\ S)\left(\frac{1\ mol\ S}{32.06\ g\ S}\right) = \underline{1.14\times10^{-3}\ mol\ S}$

Mol (atom) ratio Fe : S = 9.67 : 11.4 = 1.00 : 1.18, or 17 : 20.

The composition is not too far from the empirical formula FeS. That the analytical sulfur content is in excess of that required by the formation of pure FeS(s) is due to the presence of small amounts of $Fe^{3+}$, which requires additional $S^{2-}$ ions for charge balance. In other words, this is a nonstoichiometric compound containing both $Fe^{2+}$ and $Fe^{3+}$.

51. $$Ni(s) + 4CO(g) \longrightarrow Ni(CO)_4(g)$$

$$\text{Mol } Ni(CO)_4 = (2.50 \text{ g Ni})\left(\frac{1 \text{ mol Ni}}{58.70 \text{ g Ni}}\right)\left(\frac{1 \text{ mol } Ni(CO)_4}{1 \text{ mol Ni}}\right) = \underline{4.26\text{x}10^{-2} \text{ mol}}$$

$$\text{Volume of gas} = \frac{nRT}{P} = \frac{(4.26\text{x}10^{-2} \text{ mol})(0.0821 \text{ atm L mol}^{-1} \text{ K}^{-1})(353 \text{ K})}{2.00 \text{ atm}}$$

$$= \underline{0.617 \text{ L}}$$

$$\text{Mol } CO(g) = (\text{mol } Ni(CO)_4)\left(\frac{4 \text{ mol CO}}{1 \text{ mol } Ni(CO)_4}\right)$$

$$\text{Volume of } CO(g) = \frac{nRT}{P} = \frac{4(4.26\text{x}10^{-2} \text{ mol})(0.0821 \text{ atm L mol}^{-1} \text{ K}^{-1})(298 \text{ K})}{1 \text{ atm}}$$

$$= \underline{4.17 \text{ L}}$$

52. The zinc, but not the copper, in brass is oxidized to $Zn^{2+}(aq)$ with the evolution of $H_2(g)$.

$$Zn(s) + 2H_3O^+(aq) \longrightarrow Zn^{2+}(aq) + 2H_2O(l) + H_2(g)$$

$$\text{Mol } H_2 = \text{mol Zn} = \frac{PV}{RT} = \frac{(732.2 \text{ mm Hg})\left(\frac{1 \text{ atm}}{760 \text{ mm Hg}}\right)(102.8 \text{ mL})\left(\frac{1 \text{ L}}{1000 \text{ mL}}\right)}{(0.0821 \text{ atm L mol}^{-1} \text{ K}^{-1})(298 \text{ K})}$$

$= \underline{4.05\text{x}10^{-3} \text{ mol}}$ (Note that the pressure has been corrected for the partial pressure of $H_2O(g)$).

$$\text{Mass of Zn} = (4.05\text{x}10^{-3} \text{ mol Zn})\left(\frac{65.38 \text{ g Zn}}{1 \text{ mol Zn}}\right) = \underline{0.265 \text{ g}}$$

$$\text{Mass \% Zn} = (0.265 \text{ g Zn})\left(\frac{100 \%}{0.50 \text{ g}}\right) = \underline{53 \text{ mass \%}}$$

and, by difference, mass % Cu = 47 mass %

53. $$\text{Mass of NiS} = (10^3 \text{ kg steel})\left(\frac{10^3 \text{ g}}{1 \text{ kg}}\right)\left(\frac{18 \text{ g Ni}}{100 \text{ g steel}}\right)\left(\frac{1 \text{ mol Ni}}{58.70 \text{ g Ni}}\right)\left(\frac{1 \text{ mol NiS}}{1 \text{ mol Ni}}\right)$$

$$\text{x } \left(\frac{90.76 \text{ g NiS}}{1 \text{ mol NiS}}\right) = 278\text{x}10^3 \text{ g, } \underline{\text{or } 278 \text{ kg}}$$

54. $$\text{Mass \% Co} = \left(\frac{\text{mass Co}}{\text{mass Vit } B_{12}}\right) \text{ x } 100 = \left(\frac{58.93 \text{ g}}{1357 \text{ g}}\right) \text{ x } 100 = \underline{4.343} \%$$

$$\text{Mass of Co} = (1 \text{ µg Vit } B_{12})\left(\frac{1 \text{ g}}{10^6 \text{ µg}}\right)\left(\frac{4.343 \text{ g Co}}{100 \text{ g Vit } B_{12}}\right) = \underline{4.343\text{x}10^{-8} \text{ g}}$$

$$\underline{\text{or } 43.43 \text{ ng Co}}$$

55.

$$\text{Solubility} = \left(\frac{5.6\ \text{g Cr(OH)}_3}{1\ \text{L}}\right)\left(\frac{1\ \text{mol Cr(OH)}_3}{103.0\ \text{g Cr(OH)}_3}\right) = \underline{5.44\times10^{-2}\ \text{mol L}^{-1}}$$

$$Cr(OH)_3(s) \rightleftharpoons Cr^{3+}(aq) + 3OH^-(aq)$$

| | | | |
|---|---|---|---|
| in solution | 0.0544 | 0.1632 | mol $L^{-1}$ |

$$K_{sp}(Cr(OH)_3) = [Cr^{3+}][OH^-]^3 = (0.0544\ \text{mol L}^{-1})(0.1632\ \text{mol L}^{-1})^3$$

$$= \underline{2.36\times10^{-4}\ \text{mol}^4\ \text{L}^{-4}}$$

56. From Table 21.3, the density of Cr is 7.19 g $cm^{-3}$. Thus, the moles of Cr needed are given by:

$$\text{Mol of Cr} = (1.00\ \text{m}^2)\left(\frac{100\ \text{cm}}{1\ \text{m}}\right)^2(0.10\ \text{mm})\left(\frac{1\ \text{cm}}{10\ \text{mm}}\right)\left(\frac{7.19\ \text{g}}{1\ \text{cm}^3}\right)\left(\frac{1\ \text{mol Cr}}{52.00\ \text{g Cr}}\right)$$

$$= \underline{13.83\ \text{mol Cr}}$$

$$Cr^{3+} + 3e^- \longrightarrow Cr(s)$$

$$\text{Time needed} = \frac{(13.83\ \text{mol Cr})\left(\frac{3\ \text{mol e}^-}{1\ \text{mol Cr}}\right)\left(\frac{96\ 500\ \text{C}}{1\ \text{mol e}^-}\right)\left(\frac{1\ \text{A s}}{1\ \text{C}}\right)\left(\frac{1\ \text{h}}{3600\ \text{s}}\right)\left(\frac{1\ \text{day}}{24\ \text{h}}\right)}{1.5\ \text{A}}$$

$$= \underline{31\ \text{days}}$$

To plate out even a thickness of Cr of 0.1 mm, it takes a long time and is very expensive in terms of both raw materials and electrical energy.

57. (a)

$$Ag_3PO_4(s) \rightleftharpoons 3Ag^+(aq) + PO_4^{3-}(aq)$$

| | | | |
|---|---|---|---|
| In solution | 3x | x | mol $L^{-1}$ |

where x mol $L^{-1}$ is the solubility of $Ag_3PO_4(s)$, for which

$$K_{sp} = [Ag^+]^3[PO_4^{3-}] = (3x)^3x = 27x^4 = 1.8\times10^{-18}\ \text{mol}^4\ \text{L}^{-4}$$

$$x = \underline{1.6\times10^{-5}\ \text{mol L}^{-1}}$$

$$\text{Solubility} = \left(\frac{1.6\times10^{-5}\ \text{mol Ag}_3\text{PO}_4}{1\ \text{L}}\right)\left(\frac{418.7\ \text{g Ag}_3\text{PO}_4}{1\ \text{mol Ag}_3\text{PO}_4}\right) = \underline{6.7\times10^{-3}\ \text{g L}^{-1}}$$

(b)

$$Ag_3PO_4(s) \rightleftharpoons 3Ag^+(aq) + PO_4^{3-}(aq)$$

| | | | | |
|---|---|---|---|---|
| initially | - | 0.0010 | 0 | mol $L^{-1}$ |
| at eq'm | - | 0.0010+3x | x | mol $L^{-1}$ |

$$K_{sp} = [Ag^+]^3[PO_4^{3-}] = (0.0010+3x)^3x = 1.8\times10^{-18}\ \text{mol}^4\ \text{L}^{-4}$$

Assume x << 0.0010, $\underline{x = 1.8\times10^{-9}}$, and the assumption is justified.

$$\text{Solubility} = \left(\frac{1.8\times10^{-9}\ \text{mol L}^{-1}}{1\ \text{L}}\right)\left(\frac{418.7\ \text{g Ag}_3\text{PO}_4}{1\ \text{mol Ag}_3\text{PO}_4}\right) = \underline{7.5\times10^{-7}\ \text{g L}^{-1}}$$

58. Use the formula $\Delta T_f = iK_fm$

(a)

$$\text{mol HgCl}_2 = (1.084\ \text{g HgCl}_2)\left(\frac{1\ \text{mol HgCl}_2}{271.5\ \text{g HgCl}_2}\right) = \underline{3.993\times10^{-3}\ \text{mol}}$$

$$i = \frac{\Delta T_f}{K_f m} = \frac{0.075^\circ C}{(1.86^\circ C\ mol^{-1}\ kg)(\frac{3.993x10^{-3}\ mol}{100\ g\ H_2O})(\frac{1000\ g}{1\ kg})} = \underline{1.01}$$

(b) $$mol\ Hg(NO_3)_2 = (0.545\ g\ Hg(NO_3)_2)(\frac{1\ mol\ Hg(NO_3)_2}{324.6\ g\ Hg(NO_3)_2}) = \underline{1.68x10^{-3}}$$

$$i = \frac{\Delta T_f}{K_f m} = \frac{0.093^\circ C}{(1.86^\circ C\ mol^{-1}\ kg)(\frac{1.68x10^{-3}\ mol}{100\ g\ H_2O})(\frac{1000\ g}{1\ kg})} = \underline{3.0}$$

The results suggest that $HgCl_2$ is a nonelectrolyte in aqueous solution, while $Hg(NO_3)_2$ is an electrolyte and is fully dissociated into its ions.

$$Hg(NO_3)_2(s) \longrightarrow Hg^{2+}(aq) + 2NO_3^-(aq)$$

In other words, $HgCl_2$ is a covalently bonded compound, while $Hg(NO_3)_2$ is an ionic compound.

59. $$5Fe^{2+}(aq) + MnO_4^-(aq) + 8H_3O^+(aq) \longrightarrow 5Fe^{3+}(aq) + Mn^{2+}(aq) + 12H_2O(l)$$

$$Mol\ Fe^{2+} = (\frac{250\ mL\ solution}{25.00\ mL\ sol'n})(24.00\ mL\ KMnO_4)(\frac{1\ L}{1000\ mL})(\frac{0.020\ mol\ KMnO4}{1\ L})$$

$$x\ (\frac{5\ mol\ Fe^{2+}}{1\ mol\ MnO_4^-}) = \underline{2.4x10^{-2}\ mol\ Fe^{2+}}$$

and molar mass of $FeSO_4.xH_2O = (\frac{6.673\ g}{2.4x10^{-2}\ mol}) = \underline{278\ g\ mol^{-1}}$

Mass % $FeSO_4 = (151.9\ g\ FeSO_4)(\frac{100\%}{278\ g}) = \underline{54.6\%}$

The difference between the molar mass of $FeSO_4.xH_2O$ and $FeSO_4$ gives the mass of $H_2O$ in 1 mole of hydrate.

Mass of $H_2O$ = (278-151.9) = $\underline{126.1\ g}\ H_2O$

Mol of $H_2O$ in 1 mol $FeSO_4.xH_2O = (126.1\ g\ H_2O)(\frac{1\ mol\ H\ O}{18.02\ g\ H_2O}) = \underline{7.00} = \underline{x}$

60.

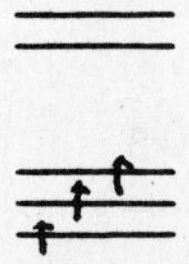

$Cr^{3+}$ [Ar]$3d^3$

Ruby

In ruby the crystal field splitting of the d energy levels is as shown, and the difference between the energy levels of the two sets of energy levels, $\Delta_o$, must be large, corresponding to the absorption of blue light, because the color is red. At high pressure the oxide ions (ligands) would be forced in closer to the $Cr^{3+}$ ion, thus increasing $\Delta_o$. As $\Delta_o$ increases, the wavelength of the light absorbed decreases. Thus, at high pressure, the wavelength of absorbed light is expected to be below 500 nm and the compound to transmit light in the orange, yellow, or colorless spectrum, depending on the magnitude of the pressure.

61. $$Ag_2CrO_4(s) \rightleftharpoons 2Ag^+(aq) + CrO_4^{2-}(aq)$$

| | | | | |
|---|---|---|---|---|
| initially | - | 0 | 0.10 | mol $L^{-1}$ |
| at eq'm | - | 2x | 0.10+x | mol $L^{-1}$ |

$K_{sp} = [Ag^+]^2[CrO_4^{2-}] = (2x)^2(0.10+x) = 1.7\times10^{-12}\ mol^3\ L^{-3}$

$x << 0.10,\quad 4x^2(0.10) = 1.7\times10^{-12};\ x = \underline{2.1\times10^{-6}}$

$$\text{Solubility} = 2.1\times10^{-6}\ mol\ L^{-1} = (2.1\times10^{-6}\ mol\ L^{-1})\left(\frac{331.8\ g}{1\ mol}\right)$$

$$= \underline{7.0\times10^{-4}\ g\ L^{-1}}$$

62. When equal volumes of 2.0 M solutions are mixed, the resulting solutions are 1.0 M, and are thus standard solutions.

(a) $2[Fe^{3+} + e^- \longrightarrow Fe^{2+}$ $\quad E^o_{red} = +0.77\ V$

$V^{3+} + 6H_2O \longrightarrow VO_2^+ + 2e^- + 4H_3O^+$ $\quad E^o_{ox} = -1.00\ V$

$2Fe^{3+} + V^{3+} + 6H_2O \longrightarrow 2Fe^{2+} + VO_2^+ + 4H_3O^+$ $\quad E^o_{cell} = \underline{-0.23\ V}$

$E^o_{cell}$ is negative; there will be no $\underline{\text{reaction.}}$

(b) $Fe^{2+} \longrightarrow Fe^{3+} + e^-$ $\quad E^o_{ox} = -0.77\ V$

$V^{3+} + e^- \longrightarrow V^{2+}$ $\quad E^o_{red} = +0.20\ V$

$Fe^{2+} + V^{3+} \longrightarrow Fe^{3+} + V^{2+}$ $\quad E^o_{cell} = \underline{-0.57\ V}$

$E^o_{cell}$ is negative; there will be no reaction.

(c) $Fe^{3+} + e^- \longrightarrow Fe^{2+}$ $\quad E^o_{red} = +0.77\ V$

$V^{2+} \longrightarrow V^{3+} + e^-$ $\quad E^o_{ox} = -0.20\ V$

$Fe^{3+} + V^{2+} \longrightarrow Fe^{2+} + V^{3+}$ $\quad E^o_{cell} = \underline{+0.57\ V}$

Yellow $Fe^{3+}(aq)$ is expected to oxidize violet $V^{2+}(aq)$ to pale green $Fe^{2+}(aq)$ and green $V^{3+}(aq)$.

63. Summarizing the reactions:

$$TiCl_4(l) \xrightarrow[NH_4Cl(aq)]{conc\ HCl(aq)} \text{yellow crystals} \xrightarrow[\text{aqueous solution}]{\text{heat}} TiO_2(s) + Cl^-$$

<u>0.1664 g yellow crystals contain:</u>

$$(0.0400\ g\ TiO_2)\left(\frac{1\ mol\ TiO_2}{79.90\ g\ TiO_2}\right)\left(\frac{1\ mol\ Ti}{1\ mol\ TiO_2}\right) = \underline{5.00\times10^{-4}\ mol\ Ti}$$

$$\left(\frac{250\ mL\ sol'n}{100\ mL\ sol'n}\right)(0.1708\ g\ AgCl)\left(\frac{1\ mol\ AgCl}{143.4\ g\ AgCl}\right)\left(\frac{1\ mol\ Cl}{1\ mol\ AgCl}\right) = \underline{2.98\times10^{-3}\ mol\ Cl}$$

$$(0.1664\ g\ cryst)\left(\frac{8.4\ g\ N}{100\ g\ cryst}\right)\left(\frac{1\ mol\ N}{14.01\ g\ N}\right) = \underline{1.0\times10^{-3}\ mol\ N}$$

Thus, mol (atom) ratio Ti : Cl : N = 5.00 : 29.8 : 10.00

= <u>1.0 : 6.0 : 2.0</u>

Since the crystals were obtained from $TiCl_4$ and $NH_4Cl$ in concentrated HCl(aq), the nitrogen must assuredly be present as ammonium ions, and the yellow crystals can be formulated as $(NH_4)_2TiCl_6.xH_2O$, where x remains to be determined. Since there is $5.00\times10^{-4}$ mol $(NH_4)_2TiCl_6$ in 0.1664 g of yellow crystals, mass of $(NH_4)_2TiCl_6$ in sample is given by

$$(5.00\times10^{-4}\ \text{mol}\ (NH_4)_2TiCl_6)\left(\frac{296.7\ \text{g}\ (NH_4)_2TiCl_6}{1\ \text{mol}\ (NH_4)_2TiCl_6}\right) = \underline{0.1484\ \text{g}}$$

Thus, mass of $H_2O$ in sample = (0.1664-0.1484) g = $\underline{0.0176\ \text{g}\ H_2O}$

$$\text{Mol}\ H_2O\ \text{in sample} = (0.0176\ \text{g}\ H_2O)\left(\frac{1\ \text{mol}\ H_2O}{18.02\ \text{g}\ H_2O}\right) = \underline{1.0\times10^{-3}\ \text{mol}}$$

and ratio of mol $H_2O$ : mol $(NH_4)_2TiCl_6$ = $1.0\times10^{-3}$ : $5.0\times10^{-4}$ = $\underline{2:1}$

Thus x =2, and the <u>empirical formula</u> of the yellow crystals is $\underline{(NH_4)_2TiCl_6.2H_2O}$

<u>Lewis structure:</u>

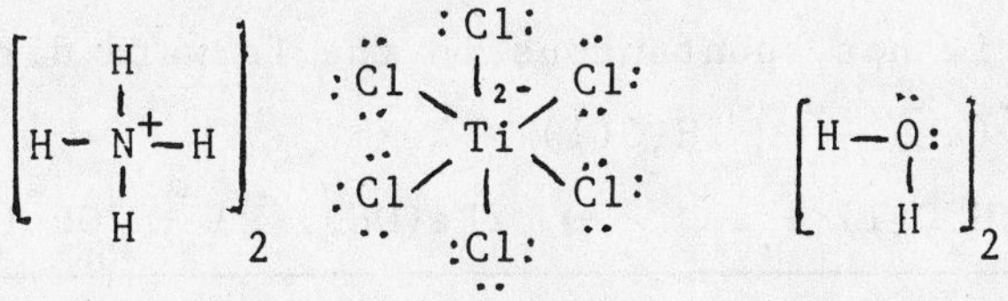

<u>containing the $AX_6$ octahedral $TiCl_6^{2-}$ ion.</u>

64. Calculate the energy for the transition of a single d electron and the corresponding wavelength.

$$\Delta_o = (182\ \text{kJ mol}^{-1})\left(\frac{1000\ \text{J}}{1\ \text{kJ}}\right)\left(\frac{1\ \text{mol photons}}{6.022\times10^{23}\ \text{photons}}\right) = h\nu = \frac{hc}{\lambda}$$

$$= \frac{(6.626\times10^{-34}\ \text{J s})(2.998\times10^{8}\ \text{m s}^{-1})(6.022\times10^{23}\ \text{photons})(1\ \text{kJ})}{(182\ \text{kJ})(1000\ \text{J})}$$

$$= 6.57\times10^{-7}\ \text{m},\ \underline{\text{or}}\ \ \underline{657\ \text{nm}}$$

65. Metals whose cations have positive reduction potentials for the reaction $M^{2+}(aq) + 2e^- \longrightarrow M(s)$, (above that for $2H^+(aq) + 2e^- \rightarrow H_2(g)$), will not be soluble in 1.00 M $H_3O(aq)$. Thus, the metals that will be soluble in 1.00 M acid, with the evolution of $H_2(g)$ are:

V, Cr, Mn, Fe, Co, Ni and Zn

66. (a) $Fe^{2+}(aq)$ is readily oxidized to $Fe^{3+}(aq)$ in air. On exposure to air an almost colorless (very pale green) solution of $FeSO_4(aq)$ turns yellow-brown as $Fe^{2+}(aq)$ is oxidized to yellow - brown $Fe^{3+}(aq)$.*

$4[Fe^{2+}(aq) \longrightarrow Fe^{3+}(aq) + e^-]$ $\quad E^o_{ox} = -0.77$ V

$O_2(g) + 4e^- + 4H^+(aq) \longrightarrow 2H_2O(\ell)$ $\quad E_{red} = +1.23$ V

$4Fe^{2+}(aq) + O_2(g) + H^+(aq) \longrightarrow 4Fe^{3+}(aq) + 2H_2O(\ell)$ $\quad E^o_{cell} = \underline{+0.46\ \text{V}}$

*Note that $Fe(H_2O)_6^{3+}$ is pale violet; the brown color is due to the formation of ions such as $Fe(H_2O)_5OH^{2+}$, since the hydrated ion is a weak acid.

Addition of $OH^-(aq)$ to fresh $FeSO_4(aq)$ precipitates insoluble white $Fe(OH)_2(s)$, which rapidly turns brown due to the oxidation of $Fe(OH)_2(s)$ to brown $Fe_2O_3(s)$ by air.

$$2[2Fe(OH)_2(s) + 2OH^-(aq) \longrightarrow Fe_2O_3(s) + 3H_2O + 2e^-] \quad E^o_{ox} = +0.56\ V$$
$$O_2(g) + 4e^- + 2H_2O(l) \longrightarrow 4OH^-(aq) \quad E^o_{red} = +0.40\ V$$
$$4Fe(OH)_2(s) + O_2(g) \longrightarrow 2Fe_2O_3(s) + 4H_2O(l) \quad E^o_{cell} = \underline{+0.96\ V}$$

(b) To decide if the reactions in part (a) can be reversed by bubbling $H_2(g)$ through the solutions, calculate $E^o_{cell}$ for each reaction:

(i) $$2(Fe^{3+}(aq) + e^- \longrightarrow Fe^{2+}(aq)) \quad E^o_{red} = -0.77\ V$$
$$H_2(g) \longrightarrow 2H^+(aq) + 2e^- \quad E^o_{ox} = 0.00\ V$$
$$2Fe^{3+}(aq) + H_2(g) \longrightarrow 2Fe^{2+}(aq) + 2H^+(aq) \quad E^o_{cell} = \underline{-0.77\ V}$$

The reaction is not spontaneous in the forward direction.

(ii) $$H_2(g) + 2OH^-(aq) \longrightarrow 2H_2O(l) + 2e^- \quad E^o_{ox} = +0.83\ V$$
$$Fe_2O_3(s) + 3H_2O(l) + 2e^- \longrightarrow 2Fe(OH)_2(s) + 2OH^-(aq) \quad E^o_{red} = -0.56\ V$$
$$Fe_2O_3(s) + H_2O(l) + H_2(g) \longrightarrow 2Fe(OH)_2(s) \quad E^o_{cell} = \underline{+0.27\ V}$$

The reaction is spontaneous under standard conditions and should occur.

67. (a) $\underline{CoCl_2.6H_2O}$, cobalt(II) chloride hexahydrate, containing octahedral $Co(H_2O)_6^{2+}$ ions and chloride ions.

(b) $\underline{Co(NH_3)_6Cl_3}$, hexamminecobalt(III) chloride, containing octahedral $Co(NH_3)_6^{3+}$ ions and chloride ions.

(c) $\underline{Co(NH_3)_5Cl_3}$, chloropentamminecobalt(III) chloride, containing octahedral $Co(NH_3)_5Cl^{2+}$ ions and chloride ions.

(d) $\underline{Co(NH_3)_4Cl_3}$, <u>cis</u>- or <u>trans</u>-dichlorotetraamminecobalt(III) chloride, containing octahedral $Co(NH_3)_4Cl_2^+$ ions, with the two Cl ligands <u>either</u> cis <u>or</u> trans, and chloride ions.

68. (a) $$Ag^+(aq) + 2CN^-(aq) \rightleftharpoons Ag(CN)_2^-(aq)$$

$$K_f = \frac{[Ag(CN)_2^-]}{[Ag^+]\ [CN^-]^2} = 1.0\times10^{21}\ mol^{-2}\ L^2$$

(b) $$Ag_2S(s) \rightleftharpoons 2Ag^+(aq) + S^{2-}(aq)$$

$$K_{sp} = [Ag^+]^2[S^{2-}] = 8\times10^{-51}\ mol^3\ L^{-3}$$

(i) In pure water, $[Ag^+] = 2x$, and $[S^{2-}] = x$, where x mol $L^{-1}$ is the solubility of $Ag_2S(s)$.

$$K_{sp} = (2x)^2x = 4x^3 = 8\times10^{-51}\ mol^3\ L^{-3}\ ;\ x = \underline{1.26\times10^{-17}\ mol\ L^{-1}}$$

$$Solubility = (1.26\times10^{-17}\ mol\ L^{-1})(247.9\ g\ mol^{-1})$$
$$= \underline{3\times10^{-15}\ g\ L^{-1}}$$

(ii) $Ag_2S(s) \rightleftharpoons 2Ag^+(aq) + S^{2-}(aq)$ $\qquad K_{sp}$

$\underline{2Ag^+(aq) + 4CN^-(aq) \rightleftharpoons 2Ag(CN)_2^-}$ $\qquad \underline{K_f^2}$

$Ag_2S(s) + 4CN^-(aq) \rightleftharpoons 2Ag(CN)_2^-(aq) + S^{2-}(aq)$

for which $K = K_{sp} \cdot K_f^2 = (8\times10^{51}\ mol^3\ L^{-3})(1.0\times10^{21}\ mol^{-2}\ L^2)^2$

$= \underline{8\times10^{-9}\ mol\ L^{-1}}$

If the solubility of $Ag_2S(s)$ is x mol $L^{-1}$, then $[Ag(CN)_2^-] = 2x$, and $[S^{2-}] = x$, with $[CN^-] = 3.0$ mol $L^{-1}$, and

$$K = \frac{[Ag(CN)_2^-]^2[S^{2-}]}{[CN^-]^4} = \frac{(2x)^2(x)}{(3.0)^4} = 8\times10^{-9}\ mol\ L^{-1}$$

whence, $4x^3 = 6.48\times10^{-7}$; $\underline{x = 5.5\times10^{-3}\ mol\ L^{-1}}$

Solubility of $Ag_2S(s) = (5.5\times10^{-5}\ mol\ L^{-1})(247.9\ g\ mol^{-1})$

$= \underline{1.4\times10^{-2}\ g\ L^{-1}}$

69. (a) 0.270 g A contains $(0.0396\ g\ Cr_2O_3)(\frac{1\ mol\ Cr_2O_3}{152.0\ g\ Cr_2O_3})(\frac{2\ mol\ Cr}{1\ mol\ Cr_2O_3})$

$= \underline{5.21\times10^{-4}\ mol\ Cr}$

and $(0.270\ g\ A)(\frac{73.53\ g\ I}{100.0\ g\ A})(\frac{1\ mol\ I}{126.9\ g\ I}) = \underline{1.564\times10^{-3}\ mol\ I}$

(b) $NH_3(aq) + HCl(aq) \longrightarrow NH_4Cl(aq)$

0.270 g A contains $(\frac{0.270\ g\ A}{0.229\ g\ A})(21.95\ mL)(\frac{0.100\ mol}{1\ L})(\frac{1\ L}{1000\ mL})$

$= \underline{2.59\times10^{-3}\ mol\ NH_3}$

Thus, mol (atom) ratio, $Cr:NH_3:I = 5.21 : 25.9 : 15.6$

$= 1.0 : 5.0 : 3.0$

which suggests the empirical formula $Cr(NH_3)_5I_3$ with Cr in the +3 oxidation state, which may be formulated as $[Cr(NH_3)_5I]I_2$, with the formula mass 517.9 u.

(c) The expected behaviour in aqueous solution is,

$[Cr(NH_3)_5I]I_2(s) \longrightarrow [Cr(NH_3)_5I]^{2+}(aq) + 2I^-(aq)$, $\quad \underline{i = 3}$

and the concentration of complex is $(1.004\ g)(\frac{1\ mol}{517.9\ g})(\frac{1}{10.00\ g})(\frac{1000\ g}{1\ kg})$

$= \underline{1.939\times10^{-3}\ mol\ kg^{-1}}$

Thus, the expected depression of f.p. is $\Delta T = iK_fm$

$= 3(1.86^\circ C\ kg^{-1})(1.939\times10^{-3}\ mol\ kg^{-1}) = \underline{1.08^\circ C}$, as observed,

which confirms the formula of the complex as $[Cr(NH_3)_5I]I_2$

# CHAPTER 22

1. (a) The structures of all silicates are based upon the $SiO_4^{4-}$ tetrahedron.
   (b) In aluminosilicates, some of the Si atoms of silicates are replaced by $Al^-$.

2. An example of a silicate with a layer structure is talc, with the empirical formula $Mg_3(Si_2O_5)_2(OH)_2$, containing the silicate ion of empirical formula $Si_2O_5^{2-}$, in which each $SiO_4$ unit shares three oxygen atoms with neighboring $SiO_4$ tetrahedra.

3. An amphibole contains two infinite chains of $SiO_4^{4-}$ tetrahedra joined by sharing oxygen atoms on alternate tetrahedra. Thus, the empirical formula is $Si_4O_{11}^{6-}$.

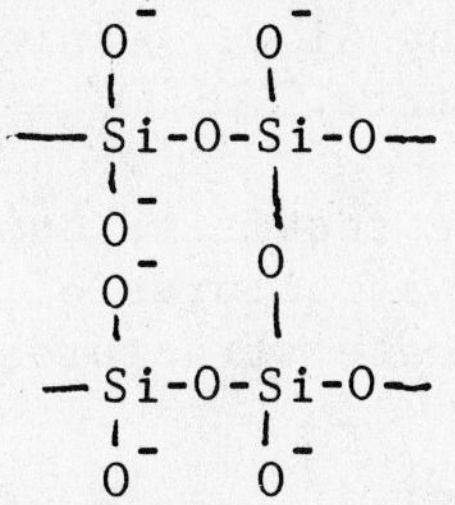

4. $Si_4O_{12}^{8-}$ has the Lewis structure

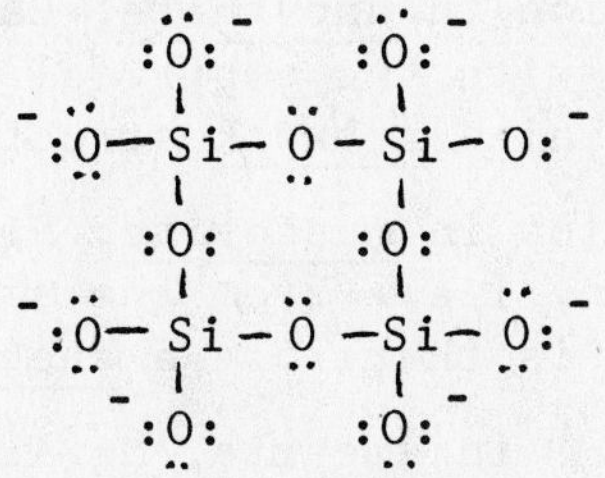

5. $Si_6O_{18}^{12-}$ is a cyclic anion with the Lewis structure

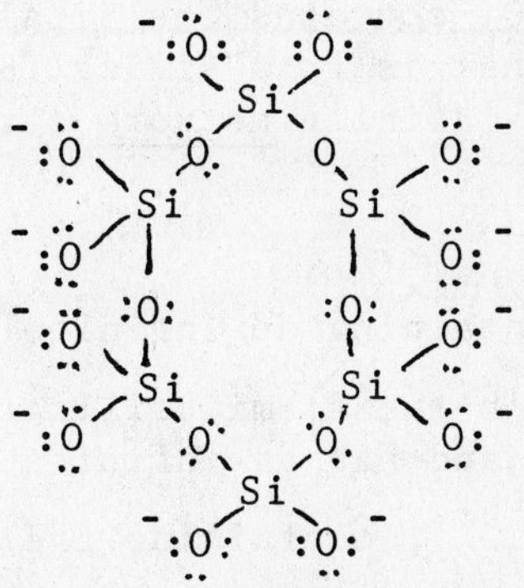

It is found in the mineral beryl, with the empirical formula

$$(Be^{2+})_3(Al^{3+})_2(Si_6O_{18}^{12-})$$

6. In each example, determine the empirical formula of the silicate anion, and hence how it is connected to similar units to form a chain, ring, double chain, sheet, etc.,).

For $SiO_4^{4-}$ joined to form an infinite chain, the empirical formula of the resulting anion is $(SiO_3^{2-})_n$. When similar units form a cyclic ring the empirical formula of the anion is $Si_nO_{3n}^{2n-}$. For $SiO_4^{4-}$ units joined to form a double chain, with the chains joined through alternate Si atoms the empirical formula is $Si_4O_{11}^{6-}$ (see Problem 4), and for $SiO_4^{4-}$ tetrahedra joined through three O atoms on each Si to form an infinite sheet, the empirical formula is $Si_2O_5^{2-}$. Thus,

(a) Gillespite, $BaFeSi_4O_{10}$ contains a silicate anion of empirical formula $Si_2O_5^{2-}$, which is a sheet anion.

(b) Dentitoite, $BaTiSi_3O_9$, contains the cyclic $Si_3O_9^{6-}$ anion.

(c) Chrysolite, $Mg_3Si_2O_5(OH)_4$, contains the $Si_2O_5^{2-}$ sheet anion, with four $OH^-$ ions per unit.

(d) Rhodonite, $CaMn_4Si_5O_{15}$, contains the cyclic $Si_5O_{15}^{10-}$ anion.

(e) Vermiculite, $Mg_3Si_4O_{10}(OH)_4 . xH_2O$, contains the $Si_2O_5^{2-}$ anion, with two $OH^-$ ions per unit, and has a sheet structure.

7. In aluminosilicates, Si atoms are replaced by the requisite number of $Al^-$ entities. The structures are thus related to the structures of the silicates, and may be deduced by considering the silicate structures on which they are based (see Problem 6).

(a) The aluminosilicate ion in anorthite, $Ca(Al_2Si_2O_8)$, is derived from the silicon compound with the empirical formula $SiO_2$, silica, and therefore has an infinite network structure.

(b) The aluminosilicate ion in muscovite, $Si_3AlO_{10}^{5-}$, is derived from the $Si_4O_{10}^{4-}$ silicate ion, i.e., from the silicate ion with the empirical formula $Si_2O_5^{2-}$, and is therefore a sheet anion.

(c) The aluminosilicate ion in amesite, $Mg_2Al(SiAlO_5)(OH)_4$, has the empirical formula $SiAlO_5^{3-}$, derived from the silicate anion $Si_2O_5^{2-}$, and is therefore a sheet anion.

(d) Thomsonite, $NaCa_2(Si_5Al_5O_{20}).6H_2O$, contains the $Si_5Al_5O_{20}^{5-}$ ion, which is derived from $Si_{10}O_{20}$, that is from silica, $SiO_2$. Thus, the aluminosilicate anion has an infinite network structure.

8. See Problems 6 and 7 for the method.

(a) Diopside, $Ca^{2+}$ $Mg^{2+}$ $(SiO_3^{2-})_2$ containing an infinite chain anion.

(b) Orthoclase, $K^+$ $(AlSi_3O_8^-)$ in which the aluminosilicate ion is derived from $Si_4O_8$, that is from silica, and has an infinite network structure.

(c) Hardystonite, $(Ca^{2+})_2$ $Zn^{2+}$ $(Si_2O_7^{6-})$, containing the disilicate ion, $Si_2O_7^{2-}$.

(d) Denitoite, $Ba^{2+}$ $Ti^{4+}$ $(Si_3O_9^{6-})$, containing the cyclic $Si_3O_9^{6-}$ anion.

9. The principal components of common glass are sodium oxide, $Na_2O$, calcium oxide, CaO, and silica, $SiO_2$.

10. Photochromic glass is glass that darkens on exposure to bright sunlight which becomes clear again in the shade. Silver chloride, AgCl(s), or silver bromide, AgBr(s), is incorporated into the glass, which is decomposed by sunlight to give dispersed microscopic black silver crystals and halogen atoms that remain trapped in the glass. In the absence of sunlight, adjacent silver atoms and halogen atoms recombine to reform silver halide.

11. (a) Like a crystalline solid, glass is a hard substance that does not flow to any appreciable extent at normal temperatures.
(b) Like a liquid, the atoms in the structure of glass are more randomly arranged than is the case in a crystalline solid, and the viscosity of glass increases as the temperature increases.

12. In silicon, each Si atom is surrounded by a tetrahedral arrangement of four other Si atoms. The unit cell is face-centered cubic with Si atoms at the lattice points and four additional Si atoms situated one-quarter of the way along each body diagonal from the four corners.

(a)

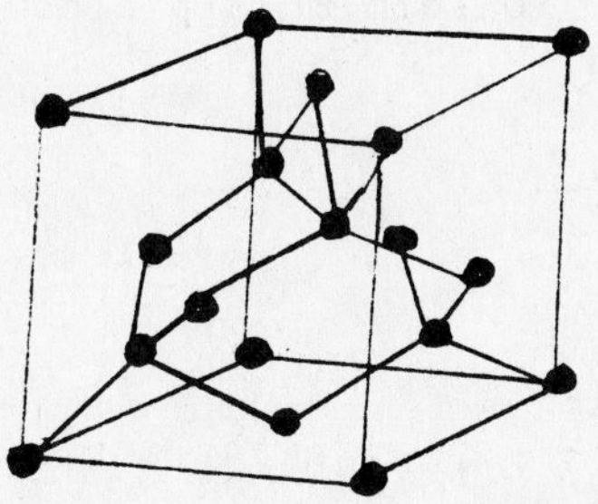

Unit cell of silicon

(b) Eight Si atoms at the corners of the cube are each shared with eight other units cells, for a total contribution to the unit cell of 8(1/8) = 1 Si atom.
Six Si atoms at the center of each of the six faces are shared between two unit cells, for a total contribution to the unit cell of 6(1/2) = 3 Si atoms, and 4 internal Si atoms belong entirely to the unit cell. Thus, the total number of Si atoms in the unit cell is 8.

(c) Each Si atom is covalently bonded to four Si nearest neighbors.

(d) (i)

$$\text{Cell volume} = (545\ \text{pm})^3\left(\frac{10^{-12}\ \text{m}}{1\ \text{pm}}\right)^3 = \underline{1.62\times10^{-28}\ \text{m}^3}$$

and the mass of the unit cell is the mass of 8 Si atoms. Thus,

$$\text{density} = \left(\frac{8\ \text{atoms}}{1\ \text{cell}}\right)\left(\frac{1\ \text{mol Si}}{6.022\times10^{23}\ \text{atoms}}\right)\left(\frac{28.09\ \text{g Si}}{1\ \text{mol Si}}\right)\left(\frac{1\ \text{cell}}{1.62\times10^{-28}\ \text{m}^3}\right)$$

$$= 2.30\times10^6\ \text{g m}^3,\ \underline{\text{or}}\ \ \underline{2.30\ \text{g cm}^{-3}}$$

(ii) The Si-Si bond distance is one-quarter of the length of the body diagonal of the cubic unit cell of edge 545 pm, and the length of the body diagonal is $\sqrt{3}$ a, where a is the length of the unit cell edge.

$$r_{Si\text{-}Si} = \frac{1}{4}[\sqrt{3}(545\ \text{pm})] = \underline{236\ \text{pm}}$$

(iii) The covalent radius of Si is one-half the Si-Si bond length, = 118 pm.

13. Zone-refining is a method for preparing ultrapure silicon for use in devices such as silicon chips. It depends on the greater solubility of impurities in molten silicon than in solid silicon. A short segment of a silicon rod is slowly moved through a heater so that the molten zone gradually traverses the entire length of the rod, concentrating the impurities in the molten zone as it moves. When the impure silicon reaches the end of the rod, it is allowed to solidify and then cut off. The process is repeated as many times as necessary.

14. Diamond and silicon both have the face-centered cubic structure described in Problem 12, with each atom bonded by single covalent bonds to four surrounding atoms in an $AX_4$ tetrahedral arrangement. In graphite, carbon has a layer structure in which each C atom is bonded to three C atoms in an $AX_3$ trigonal planar arrangement, and the planes of C atoms interact only through weak London forces. In the graphite structure each CC bond has a bond order of $1^1/_3$. Silicon has no allotropic form analogous to graphite, which is consistent with the fact that a third period element such as Si has much less tendency to form multiple bonds than does a second period element such as carbon.

15. Lewis structures:

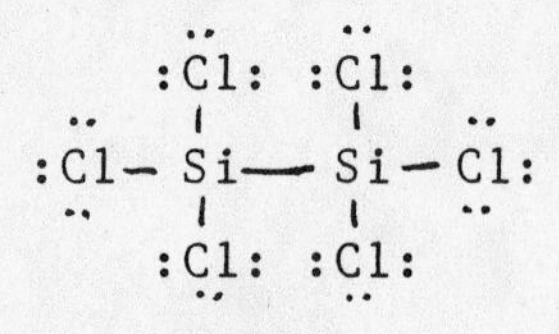

$Si_2Cl_6$

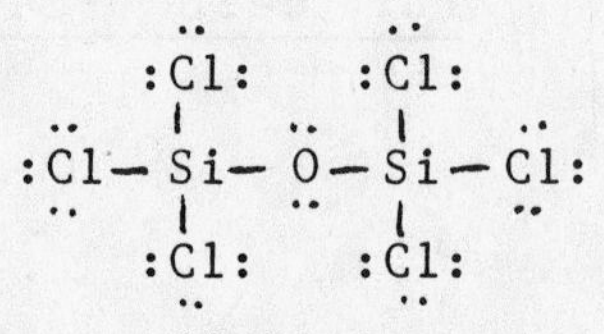

$Si_2Cl_6O$

$Si_2Cl_6$ has no dipole moment; its centers of negative and positive charge coincide at the center of the Si-Si bond, but $Si_2Cl_6O$ is $AX_2E_2$ angular at the O atom, so its centers of negative and positive charge do not coincide; it has a dipole moment.

16. Silicones are organosilicon synthetic polymers containing chains and/or rings of alternating Si and O atoms. The type of silicone depends on the nature of the parent organosilicic acid monomer that undergoes a condensation reaction (with the elimination of water) to form the silicone. $R_3SiOH$ monomers give dimers of the type $R_3Si\text{-}O\text{-}SiR_3$, while $R_2Si(OH)_2$ monomers give chains or rings of the type

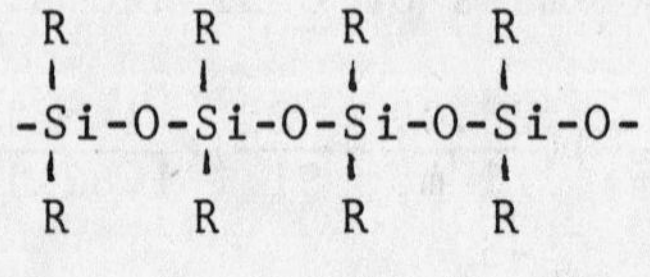

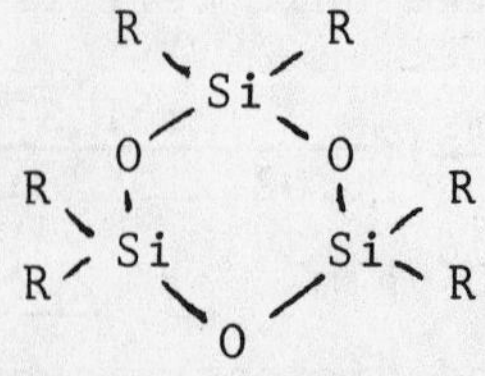

and larger rings. For the linear chains, the chain is terminated by introducing small amounts of $R_3SiOH$ -- and chains may be cross-linked by introducing small amounts of $RSi(OH)_3$. Normally, silicones are prepared by heating silicon with chloroalkanes, followed by hydrolysis of the resulting alkylsilicon chlorides with water. Among their important properties are their greater resistance to decomposition and chemical attack compared to their carbon analogues, their nontoxicity, and their properties as insulators. Liquid silicones have a large liquid range which makes them useful as lubricants at low temeperature and solid silicones retain their elasticity even at low temperatures. They are also useful water repellants.

17. The carbon analogues of silicones are the ketones, $RR'C=O$, which contain $C=O$ bonds and do not polymerise, while the silicone polymers are joined by $Si-O-Si$ bonds. The difference in structure is due to the greater tendency of carbon in the second period to form double bonds, compared to silicon in the third period.

18. (a) Hydrolysis of $(CH_3)_3SiCl$ gives $(CH_3)_3SiOH$, two molecules of which condense together to form the dimer $(CH_3)_2SiOSi(CH_3)_3$.

$$(CH_3)_3SiCl + H_2O \longrightarrow (CH_3)_3SiOH + HCl$$

$$(CH_3)_3SiOH + HOSi(CH_3)_3 \longrightarrow (CH_3)_3Si-O-Si(CH_3)_3 + H_2O$$

(b) Hydrolysis of $(CH_3)_2SiCl_2$ gives $(CH_3)_2Si(OH)_2$, n molecules of which condense together to form linear chain polymers, or cyclic polymers, with the molecular formula $[(CH_3)_2SiO]_n$.

$$(CH_3)_2SiCl_2 + 2H_2O \longrightarrow (CH_3)_2Si(OH)_2 + 2HCl$$

$$n(CH_3)_2Si(OH)_2 \longrightarrow \begin{matrix} CH_3 & CH_3 \\ | & | \\ \text{[-O-Si-O-Si-]}_{n/2} \\ | & | \\ CH_3 & CH_3 \end{matrix} + nH_2O$$

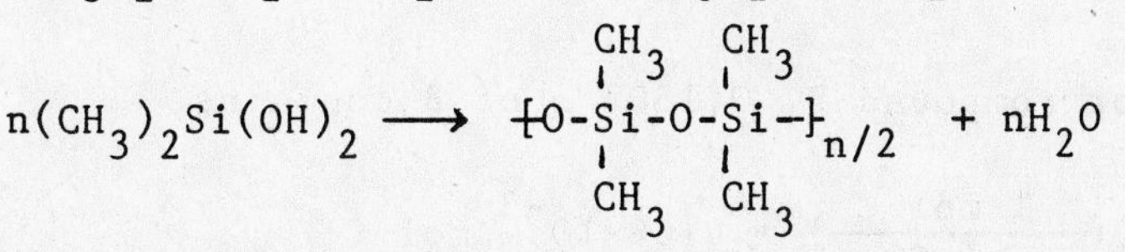

(c) Hydrolysis of $CH_3SiCl_3$ gives $CH_3Si(OH)_3$, n molecules of which condense together to give a polymer containing a double chain.

$$CH_3SiCl_3 + 3H_2O \longrightarrow CH_3Si(OH)_3 + 3HCl$$

$$nCH_3Si(OH)_3 \longrightarrow \begin{bmatrix} CH_3 & CH_3 \\ | & | \\ \text{-O-Si-O-Si-} \\ | & | \\ O & O \\ | & | \\ \text{-O-Si-O-Si-} \\ | & | \\ CH_3 & CH_3 \end{bmatrix}_{n/4} + nH_2O$$

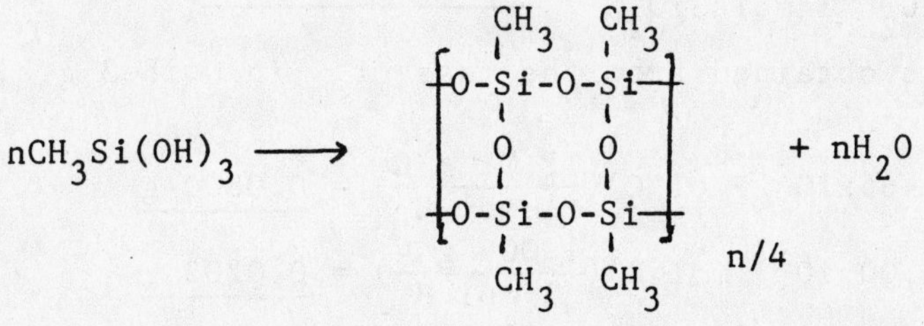

<u>Note</u> that a more complex 3D polymer is also possible

19. Silanes, with the general formula $Si_nH_{2n+2}$, are the silicon analogues of the alkanes. Compared to the carbon analogues, silanes are very much more reactive because Si, unlike C, has d-orbitals which can accept electron pairs from bases; for example, they react readily with $OH^-$ in aqueous solution to give $H_2(g)$ and $SiO(OH)_3^-$ ions.

$$SiH_4 + OH^-(aq) + 3H_2O(l) \longrightarrow SiO(OH)_3^- + 4H_2(g)$$

20. (a) $SiO_4^{4-}$

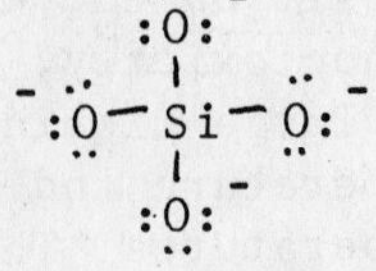

$AX_4$, tetrahedral.

(b) $(H_3Si)_2O$

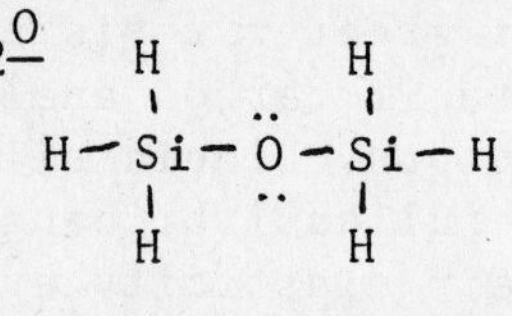

$AX_4$ tetrahedral at each Si atom and $AX_2E_2$, angular, at the O atom.

(c) $(CH_3)_2SiCl_2$

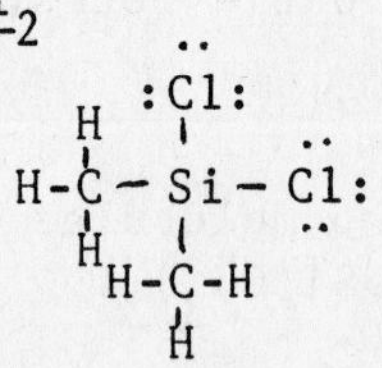

$AX_4$, tetrahedral

(d) $SiF_4$

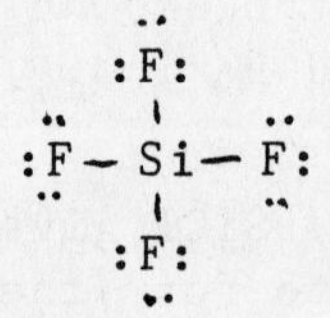

$AX_4$, tetrahedral

(e) $SiF_5^-$

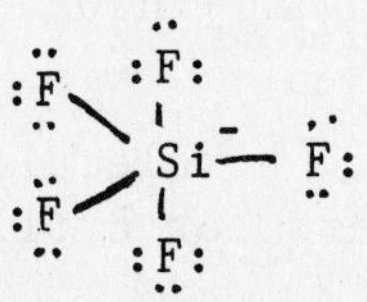

$AX_5$, trigonal bipyramid

(f) $SiF_6^{2-}$

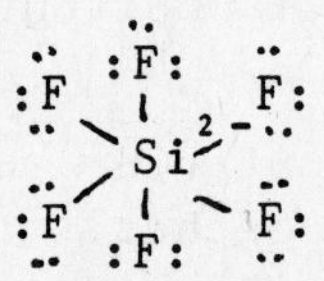

$AX_6$, octahedral

21. From the analytical data for compound B, 0.1803 g of B contains,

$$(0.2931\ g\ CO_2)\left(\frac{1\ mol\ CO_2}{44.01\ g\ CO_2}\right)\left(\frac{1\ mol\ C}{1\ mol\ CO_2}\right) = \underline{6.66\times10^{-3}\ mol\ C}$$

$$(0.1800\ g\ H_2O)\left(\frac{1\ mol\ H_2O}{18.02\ g\ H_2O}\right)\left(\frac{2\ mol\ H}{1\ mol\ H_2O}\right) = \underline{2.00\times10^{-2}\ mol\ H}$$

$$(0.1334\ g\ SiO_2)\left(\frac{1\ mol\ SiO_2}{60.09\ g\ SiO_2}\right)\left(\frac{1\ mol\ Si}{1\ mol\ SiO_2}\right) = \underline{2.22\times10^{-3}\ mol\ Si}$$

and the amount of oxygen is obtained from the mass of O in 0.1803 g B, by difference:

$$6.66\times10^{-3}\ mol\ C = (6.66\times10^{-3}\ mol\ C)\left(\frac{12.01\ g\ C}{1\ mol\ C}\right) = \underline{0.0800\ g}$$

$$2.00\times10^{-2}\ mol\ H = (2.00\times10^{-2}\ mol\ H)\left(\frac{1.008\ g\ H}{1\ mol\ H}\right) = \underline{0.0202\ g}$$

$$2.22\times10^{-3}\ mol\ Si = (2.22\times10^{-3}\ mol\ Si)\left(\frac{28.09\ g\ Si}{1\ mol\ Si}\right) = \underline{0.0624\ g}$$

Mass of C, H, and Si = (0.0800 + 0.0202 + 0.0624) g = <u>0.1626 g</u>

Mass of O = (0.1803 - 0.1626) g = <u>0.0177 g</u>

$$Mol\ of\ O = (0.0177\ g\ O)\left(\frac{1\ mol\ O}{16.00\ g\ O}\right) = \underline{1.11\times10^{-3}\ mol\ O}$$

Thus, for compound B,

Ratio of moles (atoms), C:H:Si:O = 6.66 : 20.0 : 2.22 : 1.11

Ratio of atoms = 6 : 18 : 2 : 1, so that the <u>empirical formula</u> of B is <u>$C_6H_{18}Si_2O$</u>, (formula mass 162.4 u).

Moles of B in 0.3345 g B = $\dfrac{PV}{RT}$

$$= \dfrac{(755 \text{ mm Hg})(\frac{1 \text{ atm}}{760 \text{ mm Hg}})(80.5 \text{ mL})(\frac{1 \text{ L}}{1000 \text{ mL}})}{(0.0821 \text{ atm L mol}^{-1} \text{ K}^{-1})(473 \text{ K})} = \underline{2.06 \times 10^{-3} \text{ mol}}$$

Molar mass of B = $(\dfrac{0.3345 \text{ g}}{2.06 \times 10^{-3} \text{ mol}}) = \underline{162 \text{ g mol}^{-1}}$

Thus, the <u>molecular formula</u> of B is also $C_6H_{18}Si_2O$, which may be formulated as $[(CH_3)_3Si]_2O$, with the <u>Lewis structure</u>,

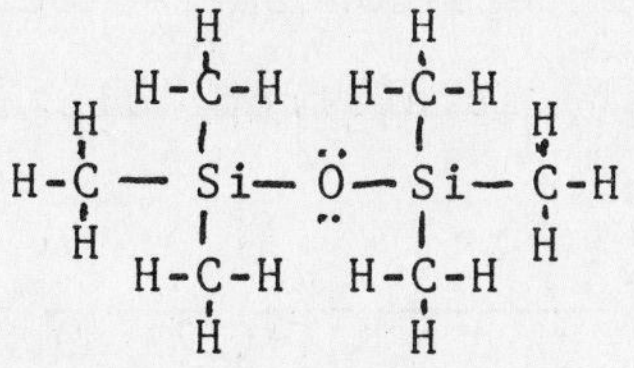

(hexamethyldisiloxane)

This siloxane results from the hydrolysis of $(CH_3)_3SiCl$, followed by the condensation of two $(CH_3)_3SiOH$ molecules. <u>A</u> is <u>trimethylchlorosilane</u>, $(CH_3)_3SiCl$, with the <u>Lewis structure</u>,

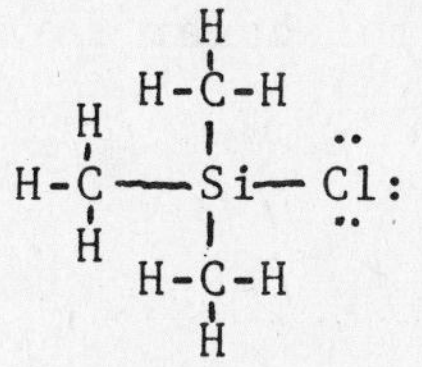

22. $(CH_3)_2SiCl_2 + 2H_2O \longrightarrow (CH_3)_2Si(OH)_2 + 2HCl$

$$n(CH_3)_2Si(OH)_2 \longrightarrow \begin{bmatrix} CH_3 & CH_3 \\ | & | \\ -Si-O-Si-O- \\ | & | \\ CH_3 & CH_3 \end{bmatrix}_{n/2} + nH_2O$$

23. (a) $3SiF_4(g) + 4H_2O(l) \longrightarrow Si(OH)_4(s) + 2H_2SiF_6(aq)$

$$\text{mol SiF}_4 = \dfrac{PV}{RT} = \dfrac{(756 \text{ mm Hg})(\frac{1 \text{ atm}}{760 \text{ mm Hg}})(50 \text{ mL})(\frac{1 \text{ L}}{1000 \text{ mL}})}{(0.0821 \text{ atm L mol}^{-1} \text{ K}^{-1})(283 \text{ K})}$$

$$= \underline{2.14 \times 10^{-3} \text{ mol SiF}_4}$$

$$\text{mol H}_2SiF_6 = (2.14 \times 10^{-3} \text{ mol SiF}_4)(\dfrac{2 \text{ mol H}_2SiF_6}{3 \text{ mol SiF}_4}) = \underline{1.43 \times 10^{-3} \text{ mol}}$$

Molarity of $H_2SiF_6 = (\dfrac{1.43 \times 10^{-3} \text{ mol}}{50 \text{ mL}})(\dfrac{1000 \text{ mL}}{1 \text{ L}}) = \underline{0.0286 \text{ M}}$

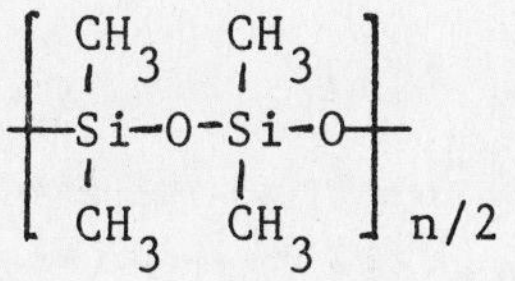

(b)    $BaCl_2(aq) + H_2SiF_6(aq) \longrightarrow BaSiF_6(s) + 2HCl(aq)$

$$mol\ BaSiF_6 = (mol\ BaCl_2)\left(\frac{1\ mol\ BaSiF_6}{1\ mol\ BaCl_2}\right)$$

$$= (20\ mL)\left(\frac{0.010\ mol}{1\ L}\right)\left(\frac{1\ L}{1000\ mL}\right) = \underline{2.0\times10^{-4}\ mol}$$

$$mass\ of\ BaSiF_6 = (2.0\times10^{-4}\ mol)\left(\frac{279.4\ g}{1\ mol}\right) = \underline{0.056\ g}$$

$$Remaining\ molarity\ of\ H_2SiF_6(aq) = \frac{[(1.43\times10^{-3})-(2.0\times10^{-4})]\ mol}{(70\ mL)\left(\frac{1\ L}{1000\ mL}\right)}$$

$$\underline{1.76\times10^{-2}\ M}$$

and for a solution of a <u>strong acid</u> of this concentration,

$[H_3O^+] = 3.52\times10^{-2}M$; pH = <u>1.45</u>  ($H_2SiF_6$ is a strong diprotic acid)

24.  Mol of gas in 1 L at STP $= \dfrac{PV}{RT} = \dfrac{(1\ atm)(1\ L)}{(0.0821\ atm\ L\ mol^{-1}\ K^{-1})(273\ K)}$

$$= \underline{4.46\times10^{-2}\ mol}$$

$$Molar\ mass = \left(\frac{1.23\ g}{4.46\times10^{-2}\ mol}\right) = \underline{27.6\ g\ mol^{-1}}$$

Since the compound contains only B and H, its molecular formula must be <u>$B_2H_6$</u> (molar mass = 27.67 g mol$^{-1}$).

25.        $2B(s) + 3H_2(g) \xrightarrow{\ \Delta H_1\ } B_2H_6(g)$

  $\Delta H_2 \downarrow \quad \Delta H_3 \downarrow \qquad \Delta H_4 \nearrow$

      $2B(g) + 6H(g)$

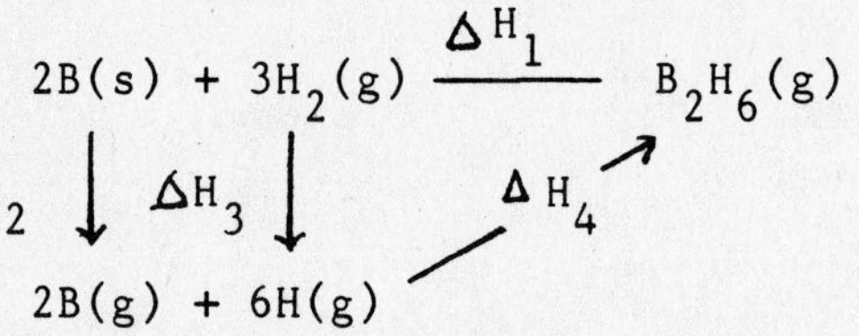

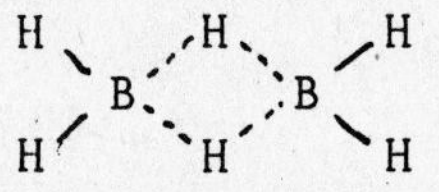

$$\Delta H_1 = \Delta H_2 + \Delta H_3 + \Delta H_4$$

i.e.,  $\Delta H_f^o(B_2H_6,g) = 2\,\Delta H_f^o(B,g) + 6\,\Delta H_f^o(H,g) - [4BE(B\text{-}H) + 4BE(B\cdots H)]$

and assume that BE(B-H) is the same as the average BE(B-H) of $BH_4^-$.

Thus, 36 kJ mol$^{-1}$ = [2(560)+6(218)] - [4(389)+4BE(B$\cdots$H)] kJ mol$^{-1}$

        BE(B$\cdots$H) = <u>209 kJ mol$^{-1}$</u>

The bond energy of each bridge bond is only approximately one-half that of a terminal B-H bond, consistent with each bridge bond having a formal bond order of 1/2 compared to 1 for each of the terminal B-H bonds.

26. $B(OH)_3$ behaves as a Lewis acid by reacting with $OH^-$ (from water or a base) to form the $B(OH)_4^-$ ion,

$$B(OH)_3 + H_2O \longrightarrow (OH)_3B.OH_2$$

$$(OH)_3B.OH_2 + H_2O \rightleftharpoons B(OH)_4^- + H_3O^+$$

and $Al(OH)_3$ behaves similarly, to give the $Al(OH)_4^-$ ion. However, since the electronegativity of Al is less than that of B, Al is more metallic than B, and it can also react with acid in aqueous solution to form the hydrated $Al(H_2O)_6^{3+}$ ion.

$$Al(OH)_3 + 3H_3O^+ \longrightarrow Al(H_2O)_6^{3+}$$

27. A <u>Lewis acid</u> is an electron pair acceptor, and a <u>Lewis base</u> is an electron pair donor.

Because boron in its trivalent compounds has an empty 2p orbital it can accept an electron pair from a base, and thus can behave as a Lewis acid. Examples include:

$$BF_3 + :NH_3 \longrightarrow F_3\overset{-}{B}\text{-}\overset{+}{N}H_3 \qquad BCl_3 + :\ddot{Cl}:^- \longrightarrow BCl_4^-$$

$$BF_3 + :\ddot{F}:^- \longrightarrow BF_4^- \qquad B(OH)_3 + {}^-:\ddot{O}\text{-}H \longrightarrow B(OH)_4^-$$

$$B_2H_6 + 2H:^- \longrightarrow 2BH_4^-$$

28.

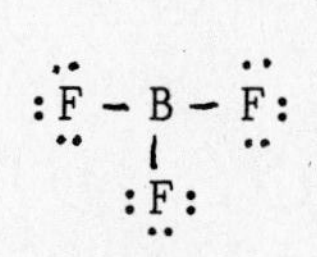

$AX_3$
trigonal
planar

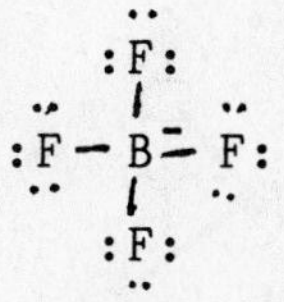

$AX_4$
tetrahedral

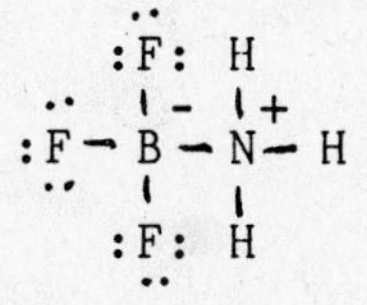

$AX_4$, tetrahedral, at both B and N

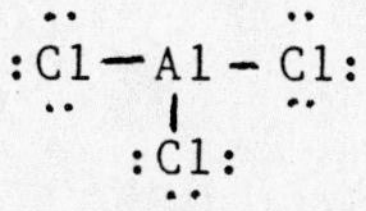

$AX_3$
trigonal
planar

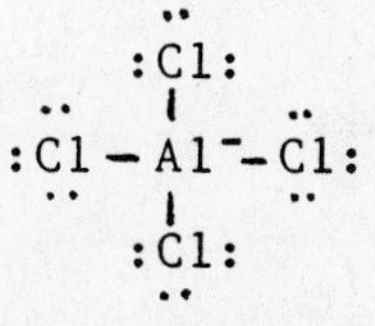

$AX_4$
tetrahedral

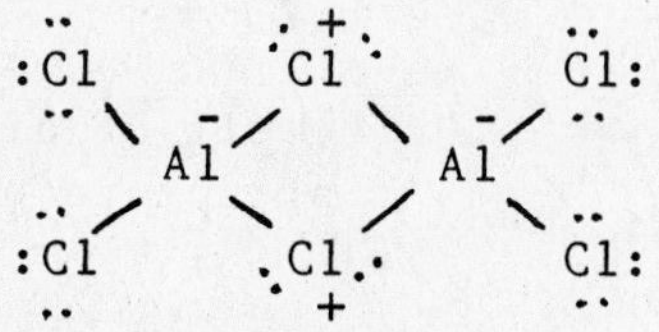

$AX_4$, tetrahedral, at both Al atoms

29.

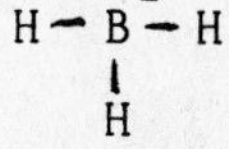

$AX_4$
tetrahedral

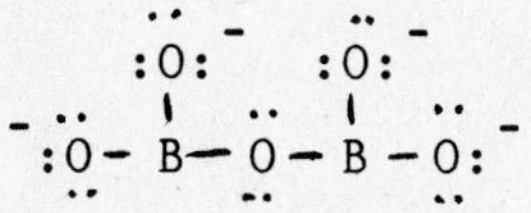

Each B is $AX_3$, trigonal planar, and the central O is $AX_2E_2$, angular

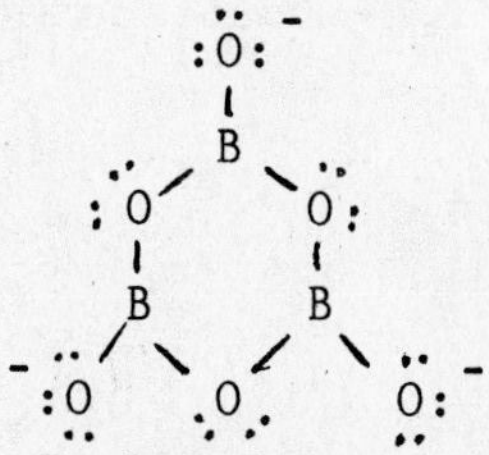

Each B is $AX_3$, trigonal planar, and each O is $AX_2E_2$, angular

30. (a) According to the $XO_m(OH)_n$ classification of oxoacids, boric acid, $B(OH)_3$, has m = 0, and is thus expected to be among the weakest of the oxoacids. In fact, it behaves as a Lewis acid and first accepts a lone pair from a water molecule to form $(HO)_3B\text{-}OH_2$, which then behaves as a very weak Brønsted acid because boron has a rather low electronegativity and the O-H bonds are not very polar.

$$(HO)_3B.OH_2 + H_2O \rightleftharpoons H_3O^+ + B(OH)_4^-$$

(b) $BCl_3$, containing B with an unfilled 2p orbital, behaves as a Lewis acid and accepts an electron pair from a water molecule (Lewis base) which leads eventually to its complete hydrolysis, and the resulting solution contains the strong acid HCl(aq).

$$BCl_3 + 6H_2O \longrightarrow B(OH)_3 + 3H_3O^+ + 3Cl^-$$

(c) Boric acid is a weak acid, so its anions, such as $BO_3^{3-}$, are expected to be weak bases.

$$BO_3^{3-} + H_2O \rightleftharpoons HBO_3^{2-} + OH^-$$

31. $(BN)_n$ is isoelectronic with $(C_2)_n$, and thus boron nitride can form structures analagous to the graphite and diamond allotropes of carbon. The graphite-like form contains six-membered $(BN)_3$ rings joined in infinite sheets and in the diamond-like form, the structure is an infinite three-dimensional network in which each B atom is at the center of a tetrahedral arrangement of four N atoms, and each N atom is at the center of a tetrahedral arrangement of four B atoms.

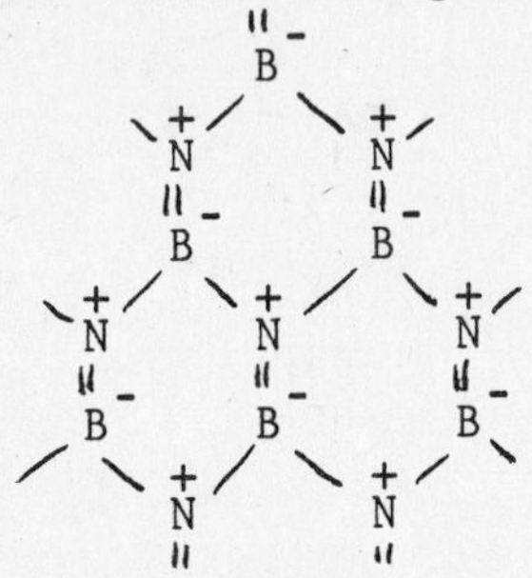

part of graphite-like (resonance) structure of BN

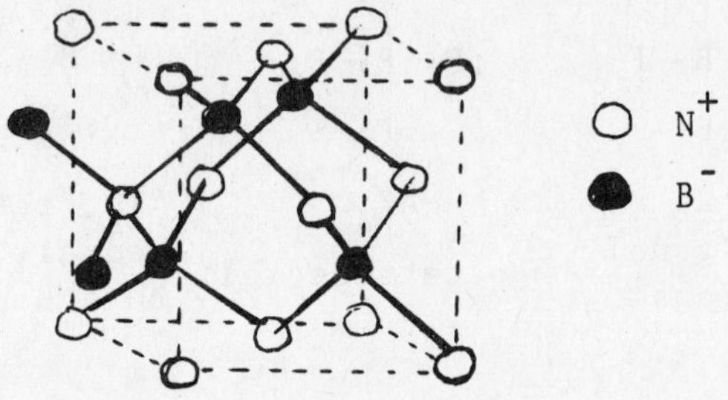


Unit cell of diamond-like structure of BN

32. $B_3N_3H_6$ is isoelectronic with benzene, $C_6H_6$, and may be depicted by resonance structures with planar six-membered $(BN)_3$ rings containing alternating B-N and B=N bonds. The expected isomers of $B_3N_3H_4Cl_2$ are analagous to those of dichlorobenzene, with the two Cl atoms in the 1,2-, 1,3-, or 1,4- positions of the ring.

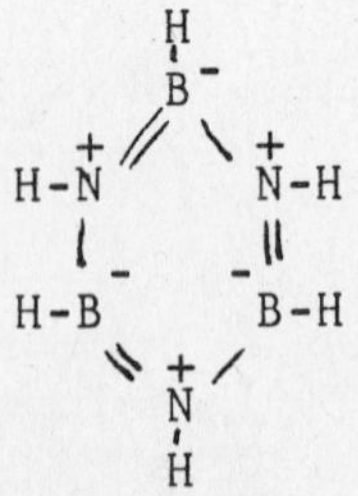

$B_3N_3H_3$

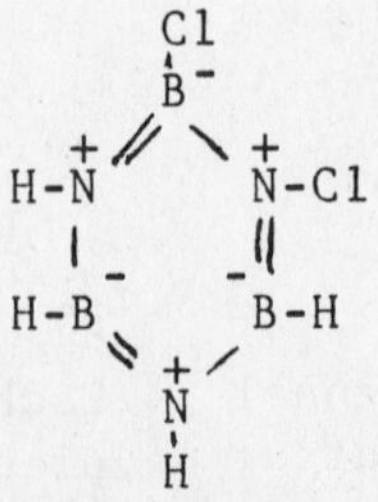

1,2-isomer

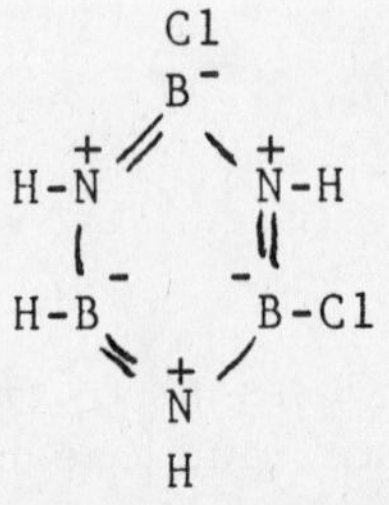

1,3-isomer

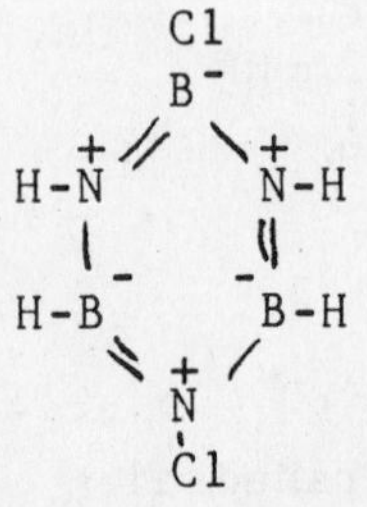

1,4-isomer

33. Calculate the empirical formula of X and its molecular formula.

| | B | O | C | H |
|---|---|---|---|---|
| Grams in 100 g of X = | 7.4 | 32.9 | 49.3 | 10.4 |

$$\text{Mol in 100 g of X} = \quad \frac{7.4}{10.81} \quad \frac{32.9}{16.00} \quad \frac{49.3}{12.01} \quad \frac{10.4}{1.008}$$

Ratio of mol (atoms) = $\quad 0.68 \; : \; 2.06 \; : \; 4.08 \; : \; 10.3$

$\qquad\qquad\qquad = \quad 1.0 \; : \; 3.0 \; : \; 6.0 \; : \; 15.0$

The <u>empirical formula</u> is $BO_3C_6H_{15}$ (formula mass = 146.0 u)

$$\text{Mol of X in 0.371 g} = \frac{PV}{RT} = \frac{(740 \text{ mm Hg})(\frac{1 \text{ atm}}{760 \text{ mm Hg}})(101.3 \text{ mL})(\frac{1 \text{ L}}{1000 \text{ mL}})}{(0.0821 \text{ atm L mol}^{-1} \text{ L}^{-1})(473 \text{ K})}$$

$$= 2.54 \times 10^{-3} \text{ mol}$$

$$\text{Molar mass of X} = (\frac{0.371 \text{ g}}{2.54 \times 10^{-3} \text{ mol}}) = \underline{146 \text{ g mol}^{-1}}$$

Thus, the <u>molecular formula</u> is also $BO_3C_6H_{15}$

X results from the reaction of an alcohol with an inorganic acid, which suggests that it is an <u>ester</u> of boric acid.

$$B(OH)_3 + 3ROH \longrightarrow B(OR)_3 + 3H_2O$$

and the function of the concentrated sulfuric acid is to react with the water formed, so that the reaction goes to completion. Thus the compound X is $B(OC_2H_5)_3$, with the Lewis structure,

```
        H H      ..        H H
        | |      ..        | |
    H-C-C-O-B-O-C-C-H
        | |   ..        | |
        H H  :O:        H H
              ..
             |
          H-C-H
             |
          H-C-H
             |
             H
```

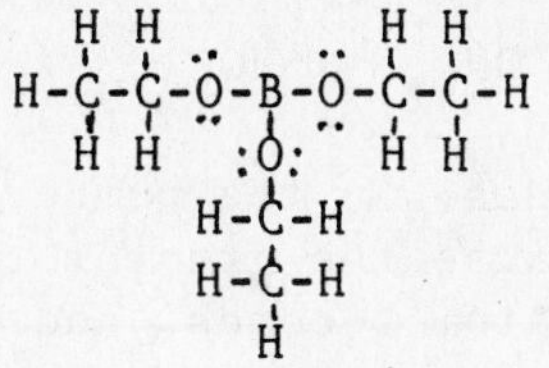

34. In descending group 4 of the periodic table, the electronegativity of the elements decreases; carbon is a nonmetal; silicon is a semimetal, and lead is a metal.

(a) The elements have the $ns^2np^2$ ground state electron configuration. Because of its relatively low electronegativity, lead can lose the two p electrons to form $Pb^{2+}$ but neither C nor Si form compounds where the element is in the ground state. Rather, they form covalent species utilizing their $ns^1np^3$ excited states, as can Pb with highly electronegative atoms, such as Cl, O, or F. Thus, the common valence states of C and Si are the C(IV) and Si(IV) valence states, while the common valence state of Pb is Pb(II), but lead can also form compounds of Pb(IV) which are not very stable and are readily reduced to Pb(II) species. The four-valent compounds are <u>covalent</u>, while Pb(II) compounds are <u>ionic</u>.

(b) The <u>oxides</u> are CO(g), $CO_2$(g), $SiO_2$(s), PbO(s), $PbO_2$(s), and $Pb_3O_4$(s). Because of its ability to form multiple bonds, both CO and $CO_2$ are small covalent molecules with the structures

$$^-:C\equiv O:^+ \qquad\qquad \text{and} \qquad\qquad :\overset{..}{O}=C=\overset{..}{O}:$$

CO(g) is readily oxidized to $CO_2$(g) and is a good reducing agent; it has no acid-base properties, while $CO_2$ behaves as a weak acid in water. Because of its inability to form multiple bonds, $SiO_2$(s) is an infinite covalent network solid in which each Si atom is bonded tetrahedrally to four O atoms. It is an acidic oxide and reacts slowly in basic aqueous solution to give a large number of silicate anions, all based on the $SiO_4^{4-}$ ion. PbO is amphoteric, consisting of $Pb^{2+}$ and $O^{2-}$ ions, while $PbO_2$(s) is insoluble in water. $Pb_3O_4$(s) behaves as $Pb(II)_2Pb(IV)O_4$.

(c) Carbon and silicon form the covalent chlorides $CCl_4$ and $SiCl_4$, both of which are liquids at room temperature. The former is rather unreactive, while the latter, because of the avaliability of unfilled d-orbitals on Si, is readily hydrolyzed to $Si(OH)_4$(s). Lead forms insoluble $PbCl_2$ which is an ionic chloride, $Pb^{2+}(Cl^-)_2$, and $PbCl_4$, which is a covalent yellow liquid which decomposes on warming to $PbCl_2$(s) and $Cl_2$(g).

35. (a) <u>Limestone</u>, $CaCO_3$(s)    (b) <u>Gypsum</u>, $CaSO_4 \cdot 2H_2O$(s)

(c) <u>Quartz</u>, $SiO_2$(s)        (d) <u>Bauxite</u>, $Al_2O_3 \cdot xH_2O$(s)

(e) <u>Talc</u>, $Mg_3(Si_2O_5)_2(OH)_2$(s)(e) <u>Beryl</u>, $Be_3Al_2Si_6O_{18}$(s)

36. Formulating these oxoacids in terms of the $XO_m(OH)_n$ nomenclature gives,

$Si(OH)_4$    m = 0;   $OP(OH)_3$   m = 1;  $O_2S(OH)_2$ m = 2; $HOClO_3$  m = 3
silicic acid      phosphoric acid    sulfuric acid    perchloric acid

The acid strength increases as the value of m increases. Thus, the expected order of increasing acidity is,

$$Si(OH)_4 < OP(OH)_3 < O_2S(OH)_2 < HOClO_3$$

Acid strength increases with the increase in the number of oxygen atoms doubly bonded to the central atom because the greater the value of m the more electronegative the central atom becomes, and this in turn makes the O-H bonds increasingly polar.

37. A <u>three-center</u> bond is one in which three atoms are bonded together by one pair of bonding electrons. The most familiar example is the B..H..B bridge bonds in diborane, $B_2H_6$, in which two boron atoms and a hydrogen atom are bonded by a single electron pair.

$$\begin{array}{ccccc} H & & H & & H \\ & B & \cdots & B & \\ H & & H & & H \end{array}$$

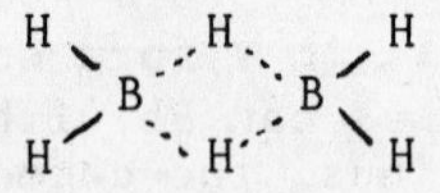

38. First write the reactions for which the $\Delta H_f^o$ values are given, and then use them to calculate the $\Delta H^o$ values for the requisite reactions in which the molecules are dissociated into atoms.

(i)

$$C(s) + 2H_2(g) \xrightarrow{\Delta H_1} CH_4(g)$$

$\Delta H_2 \downarrow \quad \Delta H_3 \downarrow \quad \Delta H_4$

$$C(g) + 4H(g)$$

$$H - \overset{\displaystyle H}{\underset{\displaystyle H}{C}} - H$$

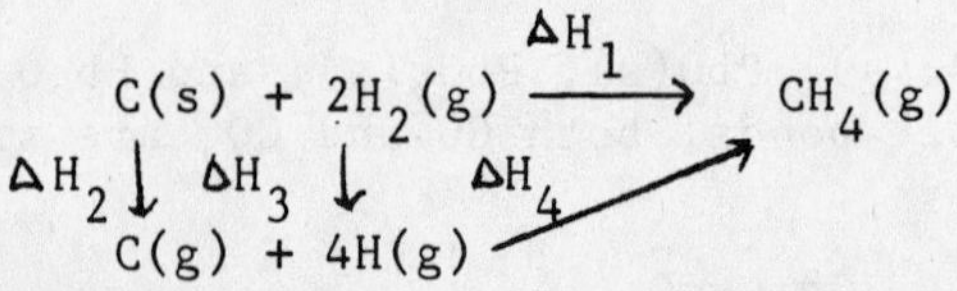

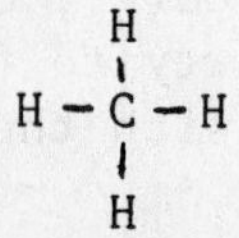

$$\Delta H_1 = \Delta H_2 + \Delta H_3 + \Delta H_4$$

$$\Delta H_f^o(CH_4,g) = \Delta H_f^o(C,g) + 4\,\Delta H_f^o(H,g) - [4BE(CH)]$$

$$-75 \text{ kJ mol}^{-1} = [(716) + 4(218)] - [4BE(CH)] \text{ kJ mol}^{-1}$$

$$4BE(CH) = 1663 \text{ kJ mol}^{-1}; \quad BE(C-H) = \underline{416 \text{ kJ mol}^{-1}}$$

(ii) Similarly, for $SiH_4$,

$$\Delta H_1 = \Delta H_2 + \Delta H_3 + \Delta H_4$$

$$\Delta H_f^o(SiH_4,g) = \Delta H_f^o(Si,g) + 4\,\Delta H_f^o(H,g) - [4BE(SiH)]$$

$$34 \text{ kJ mol}^{-1} = [(450) + 4(218)] - [4BE(SiH)] \text{ kJ mol}^{-1}$$

$$4BE(SiH) = 1288 \text{ kJ mol}^{-1}; \quad BE(SiH) = \underline{322 \text{ kJ mol}^{-1}}$$

(iii)

$$2C(s) + 3H_2(g) \xrightarrow{\Delta H_1} C_2H_6(g)$$

$$\Delta H_2 \downarrow \quad \Delta H_3 \downarrow \qquad \Delta H_4 \nearrow$$

$$2C(g) + 6H(g)$$

$$
\begin{array}{ccc}
& H & H \\
& | & | \\
H- & C- & C -H \\
& | & | \\
& H & H
\end{array}
$$

$$\Delta H_1 = \Delta H_2 + \Delta H_3 + \Delta H_4$$

$$\Delta H_f^o(C_2H_6,g) = 2\,\Delta H_f^o(C,g) + 6\,\Delta H_f^o(H,g) - [6BE(CH) + BE(CC)]$$

$$-84 \text{ kJ mol}^{-1} = [2(716) + 6(218)] - [6(416) + BE(CC)] \text{ kJ mol}^{-1}$$

$$BE(CC) = \underline{328 \text{ kJ mol}^{-1}}$$

(iv) Following the same method as in part (iii),

$$\Delta H_1 = \Delta H_2 + \Delta H_3 + \Delta H_4$$

$$\Delta H_f^o(Si_2H_6,g) = 2\,\Delta H_f^o(Si,g) + 6\,\Delta H_f^o(H,g) - [6BE(SiH) + BE(SiSi)]$$

$$80 \text{ kJ mol}^{-1} = [2(450) + 6(218)] - [6(322) + BE(SiSi] \text{ kJ mol}^{-1}$$

$$BE(SiSi) = \underline{196 \text{ kJ mol}^{-1}}$$

39.  $$B_2H_6(g) + 6H_2O(l) \longrightarrow 2B(OH)_3(aq) + 6H_2(g)$$

$$\text{Mol of } B_2H_6 = (1.50 \text{ g } B_2H_6)\left(\frac{1 \text{ mol } B_2H_6}{27.67 \text{ g } B_2H_6}\right) = \underline{5.42 \times 10^{-2} \text{ mol}}$$

$$\text{Mol of } B(OH)_3 \text{ in 1 L of solution} = (5.42 \times 10^{-2} \text{ mol } B_2H_6)\left(\frac{2 \text{ mol } B(OH)_3}{1 \text{ mol } B_2H_6}\right)$$

$$= \underline{0.108 \text{ mol}} \ B(OH)_3$$

$$B(OH)_3 + 2H_2O \rightleftharpoons H_3O^+ + B(OH)_4^-$$

| | | | | |
|---|---|---|---|---|
| initially | 0.108 | 0 | 0 | mol L$^{-1}$ |
| at eq'm | 0.108-x | x | x | mol L$^{-1}$ |

$$pK_a = 9.22; \quad K_a = 6.03 \times 10^{-10} \text{ mol L}^{-1} = \frac{[H_3O^+][B(OH)_4^-]}{[B(OH)_3]} = \frac{x^2}{0.108-x}$$

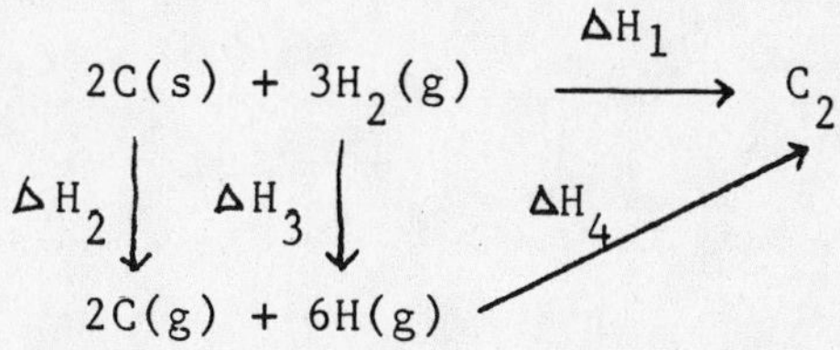

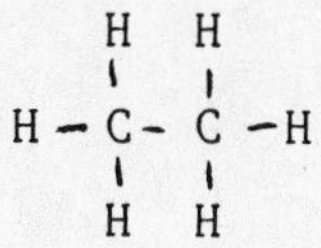

For $x \ll 0.108$, $x = [H_3O^+] = 8.1 \times 10^{-6}$ mol $L^{-1}$; pH = 5.09

40. Compared with the analogous silicon compounds, $GeO_2$ and $GeCl_4$ would be expected to have bonds of greater polarity because Ge is less electronegative than Si. Nevertheless, $GeO_2$ is expected to be, like, $SiO_2$, a covalent infinite network solid, because of the inability of Ge to form multiple bonds, and $GeCl_4$ is expected to be a covalent liquid containing tetrahedral $GeCl_4$ molecules. $GeO_2$ is expected to be an acidic oxide and $GeCl_4$ is expected to be readily hydrolyzed to germanic acid, $Ge(OH)_4$, because of the availability of unfilled d orbitals on Ge.

# CHAPTER 23

1. (a) Butane, 4 C atoms (b) 2-methylpropane, 3 C atoms

   (c) 2,2-dimethyloctane, 8 C atoms (d) 2,3-dimethylpentane, 5 C atoms

   (e) 2,2,5,5-tetramethylhexane, 6 C atoms

2. (a) The general formula is $C_nH_{2n+2}$ ; $C_8H_{18}$ for n = 8.

   (b) The general formula is $C_nH_{2n}$ ; $C_6H_{12}$ for n = 6.

   (c) The general formula is $C_nH_{2n-2}$ ; $C_5H_8$ for n = 5.

   (d) The general formula is $(CH_2)_n$ ; $C_6H_{12}$ for n = 6.

3.

| | Compound | Class | Lewis Structure | Molecular Formula |
|---|---|---|---|---|
| (a) | Methane | alkane | 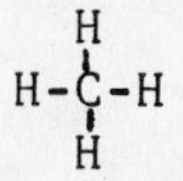 | $CH_4$ |
| (b) | Ethene | alkene | 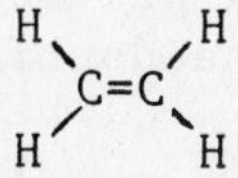 | $C_2H_4$ |
| (c) | Propyne | alkyne | 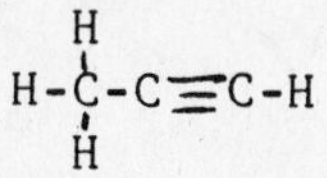 | $C_3H_4$ |
| (d) | Cyclobutane | cycloalkane | 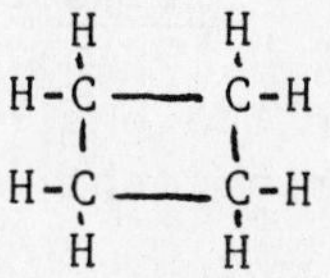 | $C_4H_8$ |
| (e) | Cyclopropene | cycloalkene | 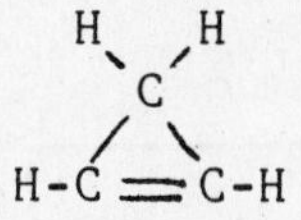 | $C_3H_4$ |

4. (a) 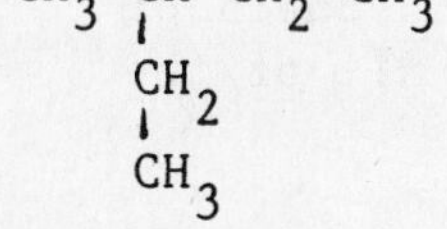

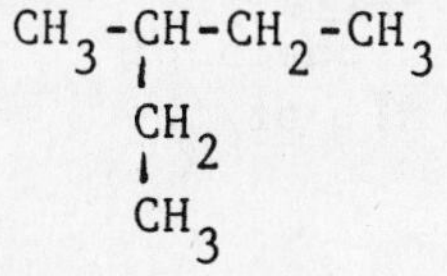

   "2-ethylbutane" has 5 C atoms in the longest continuous chain of C atoms and should be named

   3-methylpentane

   (b) 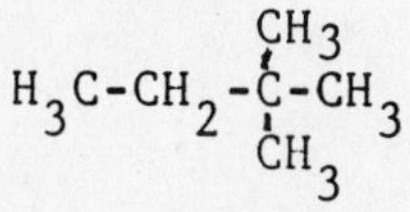

   "3-3-dimethyl butane" should be named giving the substituted carbon atom the lowest possible number in the continuous chain of 4 C atoms, and is correctly named

   2,2-dimethylbutane

   (c) 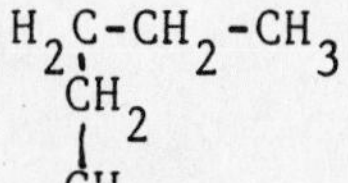

   "1-ethylpropane" has a continuous chain of 5 C atoms and is correctly named as

   pentane

(d)

CH$_3$
|
H$_3$C-C-CH$_3$
|
CH$_3$

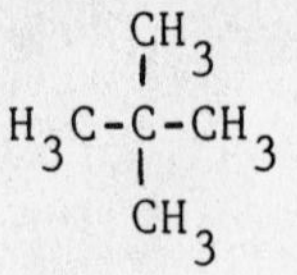

"2-2-dimethylpropane" is correct but here the 2,2- is superfluous because the 2 position is the only possible position for the two methyl substituents.

<u>dimethylpropane</u> would be unambiguous

(e)

H$_2$C-CH-CH$_3$
|   |
H$_3$C  CH$_3$

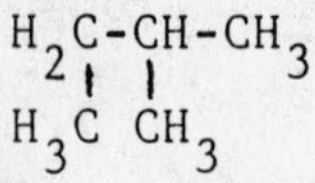

"1,2-dimethylpropane" has a continuous chain of <u>4</u> C atoms and should be named

<u>2-methylbutane</u>   or   <u>methylbutane</u>

5. The three isomers of C$_2$H$_2$Cl$_2$ (a substituted alkene) are

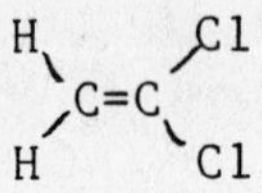

1,1-dichloroethene   <u>trans</u>-1,2-dichloro-   <u>cis</u>-1,2-dichloro-
ethene                          ethene

6. Alkenes such as 2-butene has a planar structure in which the two methyl groups can be on the same side of the double bond (cis), or on opposite sides of the double bond (trans).

H$_3$C        CH$_3$
 \        /
  C=C
 /        \
H          H

H$_3$C        H
 \        /
  C=C
 /        \
H          CH$_3$

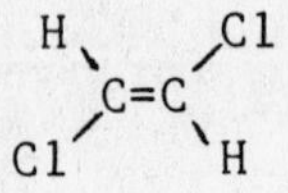

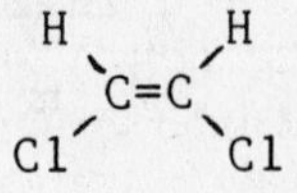

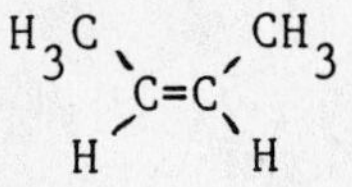

<u>cis</u>-2-butene              <u>trans</u>-2-butene

In an alkyne, such as 2-butyne the C atoms have a linear arrangement and there is only one possible arrangement of the methyl groups.

H$_3$C-C≡C-CH$_3$

2-butyne

7. Each of the isomers of heptane contains <u>7</u> C atoms. To deduce the structures, start with the isomer with a continuous chain of <u>7</u> C atoms; consider next the possible isomers with a continuous chain of <u>6</u> C atoms, and then those with <u>5</u> C atoms, and so on until all of the possible structures are deduced.

CH$_3$-CH$_2$-CH$_2$-CH$_2$-CH$_2$-CH$_2$-CH$_3$                         heptane

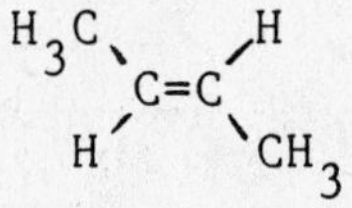

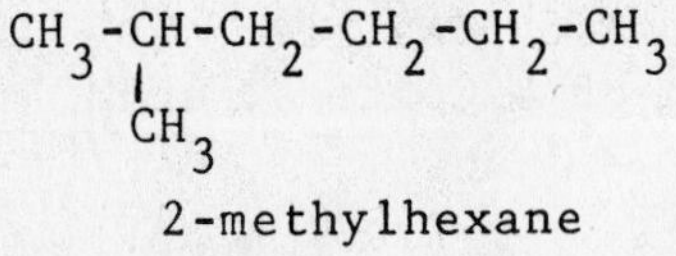

CH$_3$-CH-CH$_2$-CH$_2$-CH$_2$-CH$_3$        and      CH$_3$-CH$_2$-CH-CH$_2$-CH$_2$-CH$_3$
|                                                                          |
CH$_3$                                                                    CH$_3$

2-methylhexane                                  3-methylhexane

CH$_3$-CH-CH-CH$_2$-CH$_3$        CH$_3$-CH-CH$_2$-CH-CH$_3$              CH$_3$
|   |                              |              |                     |
CH$_3$CH$_3$                       CH$_3$          CH$_3$           CH$_3$-C-CH$_2$-CH$_2$-CH$_3$
                                                                        |
                                                                       CH$_3$

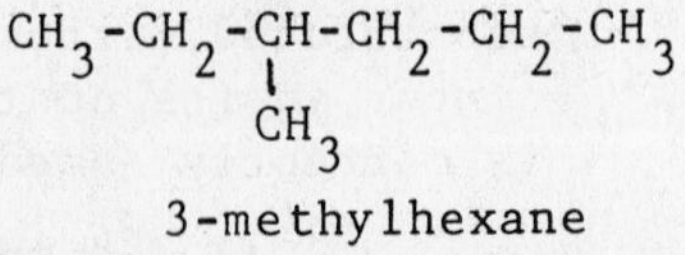

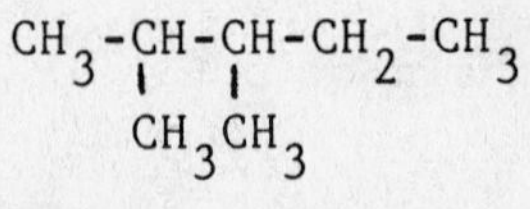

2,3-dimethylpentane   2,4-dimethylpentane   2,2-dimethylpentane

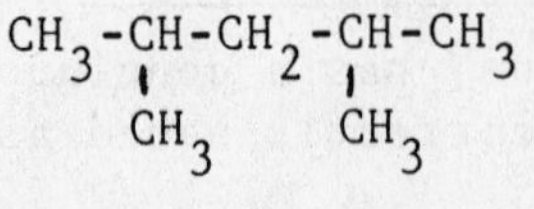

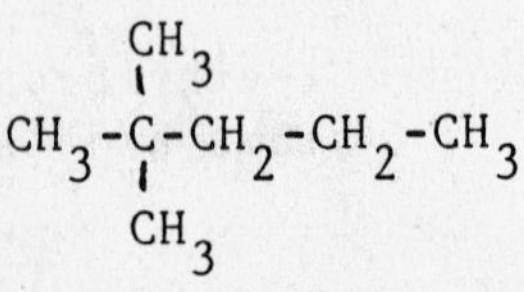

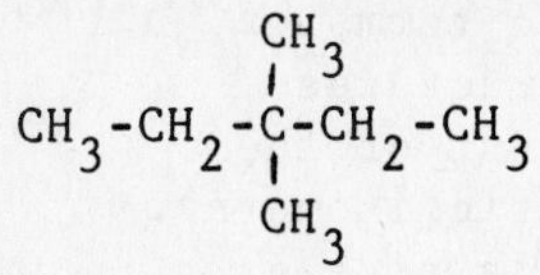
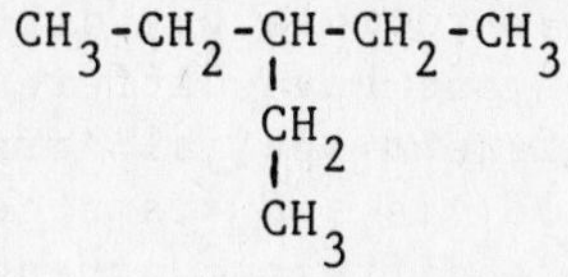
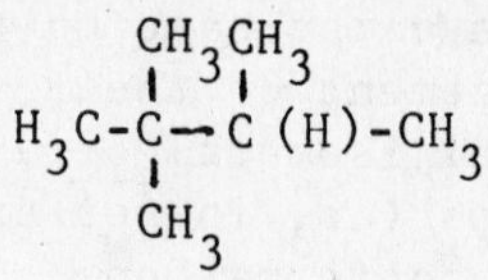

3,3-dimethylpentane    3-ethylpentane    2,2,3-trimethylbutane

8. $C_4H_8$ is the molecular formula of alkenes or cycloalkanes with <u>4</u> C atoms.

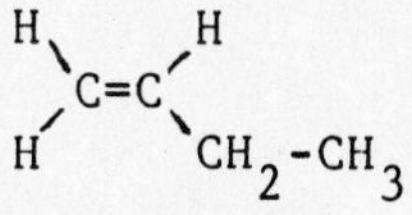
1-butene

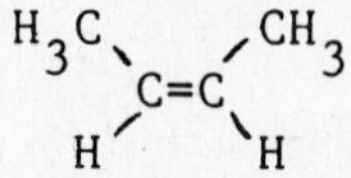
<u>cis</u>-2-butene

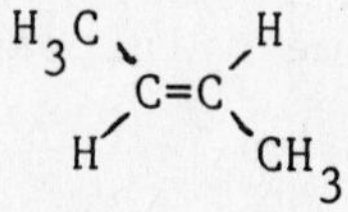
<u>trans</u>-2-butene

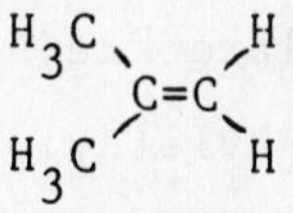
methylpropene

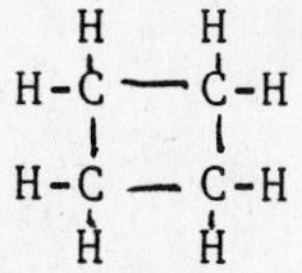
cyclobutane

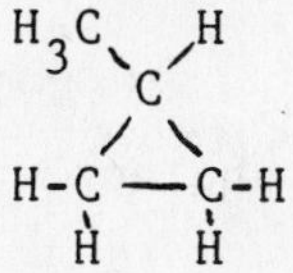
methylcyclopropane

9. (a) <u>1,3-cyclopentadiene</u>

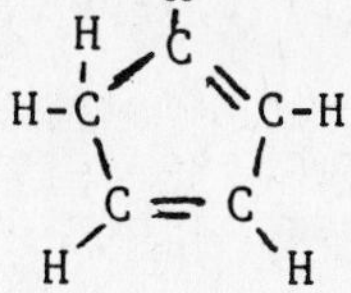

A cycloalkene with double bonds at positions 1 and 3.

(b) <u>1,3-cyclohexadiene</u>

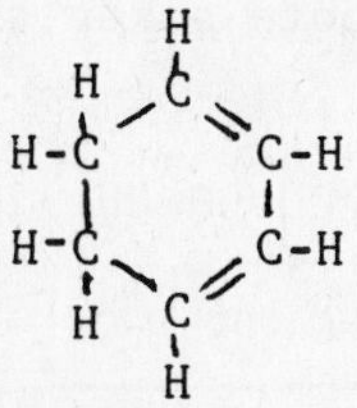

A cycloalkene with double bonds at positions 1 and 3.

(c) <u>3-chloro-2,4-hexadiene</u>

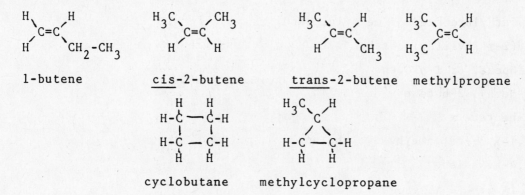

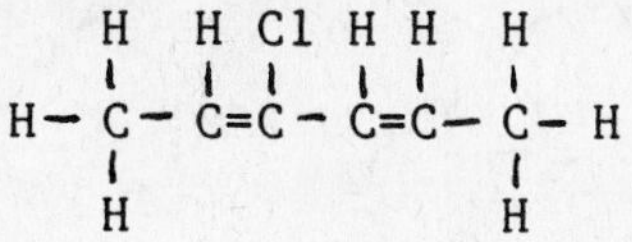

An alkene with double bonds at positions 2 and 4 and a Cl substituent at position 3.

10. (a) A saturated hydrocarbon has four single bonds to each C atom; an unsaturated hydrocarbon contains one or more double and/or triple carbon-carbon bonds.

(b) In a straight-chain alkane, all of the C atoms form one continuous chain; a branched chain alkane has shorter carbon chains branching off of the longest continuous chain of carbon atoms in the molecule.

(c) Aromatic hydrocarbons are organic compounds containing one or more benzene rings, some or all of which may be fused together.

(d) Isomers have the same molecular formula but differ in terms of the arrangement of their atoms and have different structures (structural isomers) <u>or</u> a different spatial arrangement of the same groups (as, in for example cis and trans geometric isomers). Molecular conformations are the different arrangements of the atoms of a molecule that can result from rotating their groups of atoms around single bonds. Isomers are different distinct molecular species, while conformers are easily interconverted simply by rotation about single bonds.

11. (a) 2-methylpropane, <u>or</u> methylpropane

(b) 4,(2-methylethyl)octane

(c) propene, <u>or</u> 1-propene

(d) 2-methyl-2-butene

(e) 2-heptene

(f) 1,1,2,3,-tetramethylcyclobutane

(g) 1,4-hexadiene

12. (a) 1-butene (b) 1-butyne (c) methyl-2-butene

(d) 1,5-hexadiene (e) <u>cis</u>-2-butene (f) 1,3-pentadiene

13. 2,3-dimethylbutane, <u>c</u>, contains only <u>six</u> carbon atoms and is therefore an isomer of hexane, rather than an isomer of heptane.

14. Isomers with the formula $C_5H_{10}$ include both alkenes and cycloalkanes, of which there is a total of <u>nine</u>.

$CH_3-CH_2-CH_2-CH=CH_2$ 1-pentene $CH_3-CH_2-CH=CH-CH_3$ 2-pentene (cis & trans)

$(CH_3)_2C=CH(CH_3)$ 2-methyl-2-butene $H_2C=C(CH_3)-CH_2-CH_3$ 2-methyl-1-butene

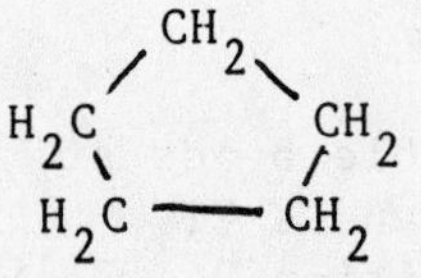
cyclopentane

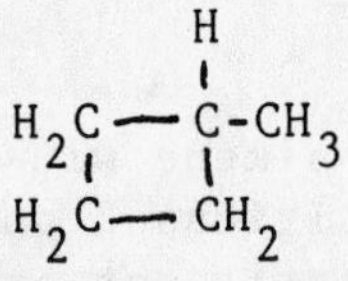
methylcyclobutane

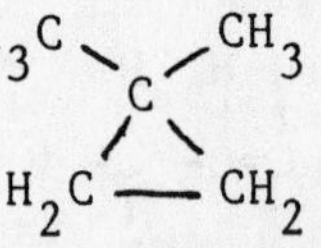
1,1-dimethyl-cyclopropane

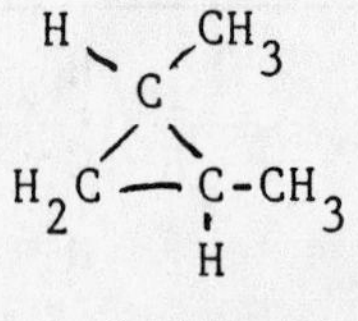
1,2-dimethyl-propane

15. (a) 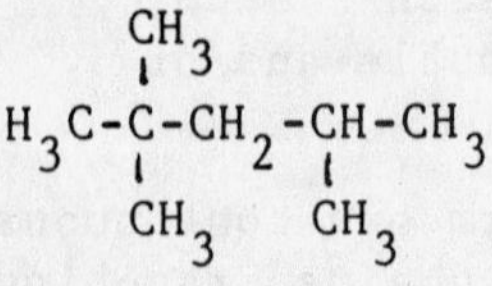 (b) 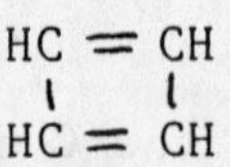 (c) 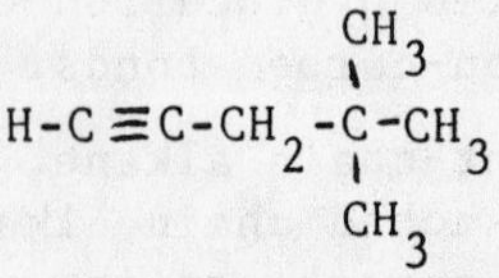

(d) 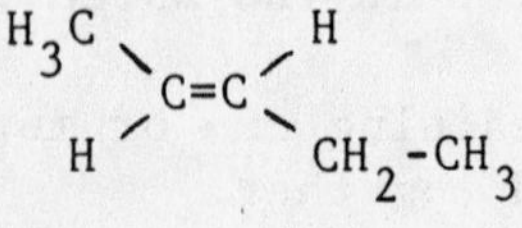 (e) $H_2C=CH-CH=CH_2$ (f) 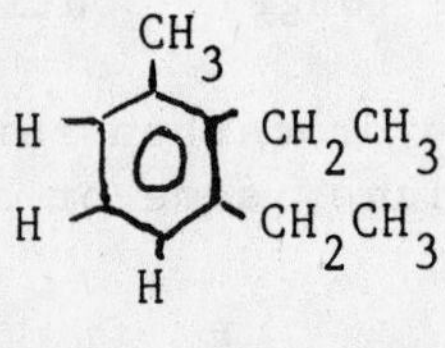

(g)

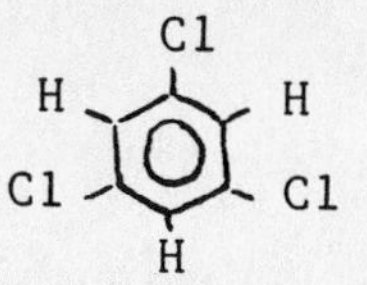

(h)

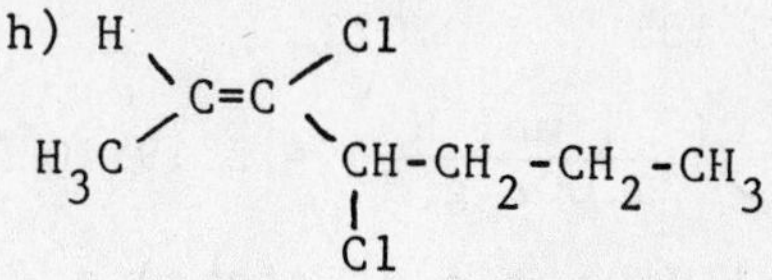

16. (a)

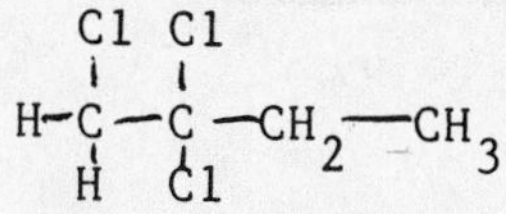

(b)

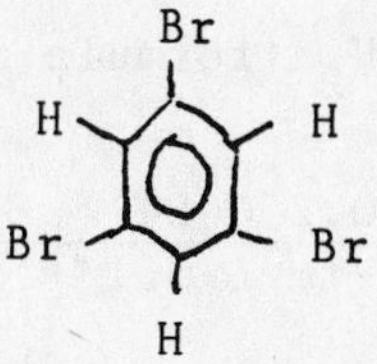

(c)

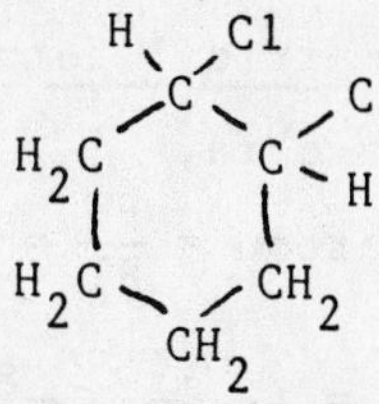

17. From the mass of $CO_2$ from the complete combustion of the sample, the moles of C in the hydrocarbon is calculated, and from the mass of $H_2O$ from the same sample, the moles of H in the hydrocarbon is calculated.

$$C_xH_y \xrightarrow{\text{excess } O_2} xCO_2 + \frac{y}{2}H_2O$$

$$\text{Mol of C} = (0.318 \text{ g } CO_2)\left(\frac{1 \text{ mol } CO_2}{44.01 \text{ g } CO_2}\right)\left(\frac{1 \text{ mol C}}{1 \text{ mol } CO_2}\right) = 7.23\text{x}10^{-3} \text{ mol}$$

$$\text{Mol of H} = (0.163 \text{ g } H_2O)\left(\frac{1 \text{ mol } H_2O}{18.02 \text{ g } H_2O}\right)\left(\frac{2 \text{ mol H}}{1 \text{ mol } H_2O}\right) = 1.81\text{x}10^{-2} \text{ mol}$$

Ratio of moles (atoms) = $7.23\text{x}10^{-3}$ mol C : $1.81\text{x}10^{-2}$ mol H
= 2.51 : 1.00, or 2.0 : 5.0

Thus, the empirical formula is $C_2H_5$ and from the gas data,

$$\text{mol hydrocarbon} = \frac{PV}{RT} = \frac{(1 \text{ atm})(0.250 \text{ L})}{(0.0821 \text{ atm L mol}^{-1} \text{ K}^{-1})(373 \text{ K})} = 8.16\text{x}10^{-3} \text{ mol}$$

$$\text{Molar mass} = \left(\frac{0.4743 \text{ g}}{8.16\text{x}10^{-3} \text{ mol}}\right) = 58.1 \text{ g mol}^{-1}$$

From the empirical formula mass of 17.05 u, and the molar mass, the molecular formula is $C_4H_{10}$

This is the molecular formula of an alkane, for which there are two isomers:

$H_3C\text{-}CH_2\text{-}CH_2\text{-}CH_3$ or $H_3C\text{-}CH(CH_3)\text{-}CH_3$

butane — 2-methylpropane

18. This problem is very similar to Problem 17. First calculate the mol of C and the mol of H in the sample.

$$\text{mol of C} = (0.4832 \text{ g } CO_2)\left(\frac{1 \text{ mol } CO_2}{44.01 \text{ g } CO_2}\right)\left(\frac{1 \text{ mol C}}{1 \text{ mol } CO_2}\right) = 1.098\text{x}10^{-2} \text{ mol}$$

and from the mass of the sample and the mass of its carbon content, the mass of H in the sample may be calculated.

$$\text{Mass of C} = (1.098\text{x}10^{-2} \text{ mol C})\left(\frac{12.01 \text{ g C}}{1 \text{ mol C}}\right) = 0.1319 \text{ g}$$

Mass of H = (0.1540-0.1319) = 0.0221 g

Mol of H = (0.0221 g H)$(\frac{1 \text{ mol H}}{1.008 \text{ g H}})$ = $2.19 \times 10^{-2}$ mol

Ratio of mol C : mol H (= ratio of atoms) = $1.098 \times 10^{-2}$ : $2.19 \times 10^{-2}$
$$= \quad 1.0 : 2.0$$

i.e., the <u>empirical formula</u> is $CH_2$ (formula mass 14.03 u)

From the gas data,

Mol hydrocarbon = $\frac{PV}{RT}$ = $\frac{(1 \text{ atm})(0.250 \text{ L})}{(0.0821 \text{ atm L mol}^{-1} \text{ K}^{-1})(373 \text{ K})}$ = $\underline{8.16 \times 10^{-3}}$ mol

Molar mass = $(\frac{0.4580 \text{ g}}{8.16 \times 10^{-3} \text{ mol}})$ = $\underline{56.1 \text{ g mol}^{-1}}$

From the empirical formula mass and the molecular mass, the <u>molecular formula</u> is $C_4H_8$, for which there is <u>six</u> possible isomers (for which the names and structures were given in Problem 8). They could be separated by fractional distillation, mass spectrometry, or gas chromatography.

19. A <u>functional group</u> is an atom or a group of atoms that replaces a hydrogen atom in a hydrocarbon.

(a) $-C\overset{\displaystyle \nearrow O}{\underset{\displaystyle \searrow H}{}}$  aldehyde  (b) $-C\overset{\displaystyle \nearrow O}{\underset{\displaystyle \searrow CH_3}{}}$ ketone  (c) $-C\overset{\displaystyle \nearrow O}{\underset{\displaystyle \searrow OH}{}}$ carboxylic acid

(d) $-C\overset{\displaystyle \nearrow O}{\underset{\displaystyle \searrow NH_2}{}}$ amide  (e) $-NH_2$ primary amino  (f) $-OH$  secondary alcohol

(g) $-C\overset{\displaystyle \nearrow O}{\underset{\displaystyle \searrow OH}{}}$ carboxylic acid, & $-NH_2$ amino ($\alpha$-amino acid)

(h) $-OH$  phenol

20. (a) ethanal (acetaldehyde)  (b) 2-butanone (methylethyl ketone)

(c) propanoic acid  (d) 2-methylpropanamide

(e) amino-2,2-dimethylpropane (tert-butylamine)  (f) 3-pentanol

(g) 2-aminopropanoic acid (glycine)

(h) benzenol (phenol, or hydroxybenzene)

21. (a) methanol (methyl alcohol)  (b) ethanol (ethyl alcohol)

(c) methanoic acid (formic acid)  (d) 2-butanol  (e) propanal

(f) 1,2-ethanediol  (g)  propanone (acetone)

(h) 2-methylpropanal  (i) 2-bromo,3-chloropropane

22. (a) $H_3C-\overset{\displaystyle OH}{\underset{\displaystyle CH_3}{C}}-CH_2-CH_3$  (b) $H_3C-\overset{}{\underset{\displaystyle OH}{CH}}-CH_2-CH_2-\overset{}{\underset{\displaystyle CH_3}{CH}}-CH_3$  (c) $HO-CH_2-\overset{}{\underset{\displaystyle Cl}{CH}}-CH_3$

(d) $\overset{\displaystyle HO}{\underset{\displaystyle H}{}}C\overset{\displaystyle CH_2-\overset{\displaystyle Cl}{CH}}{\underset{\displaystyle CH_2-CH_2}{\diagup \diagdown CH_2}}$  (e) $HO-CH_2-CH_2-CH_2-OH$

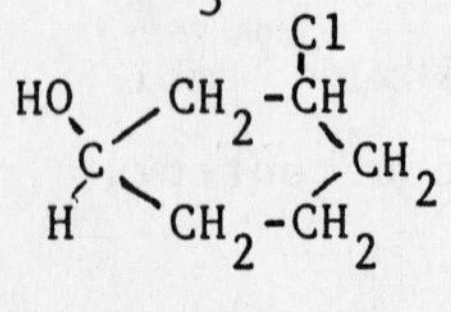

23. (a) $CH_3CH_2CH_2C(H)=O$ (b) $CH_3-C(=O)-CH_2CH_2CH_3$ (c) $CH_3-C(=O)-CH(CH_3)-CH_3$

(d) $CH_3CH_2CH_2-C(CH_3)_2-CH_2-C(H)=O$ (e) $CH_3-C(=O)-CH_3$ (f) $H-C(=O)-H$

24. (a) $CH_3CH_2CH_2OH$ 1-propanol (b) $CH_3-CH(OH)-CH_3$ 2-propanol

(c) $CH_3CH_2C(=O)-H$ propanal (d) $H_3C-C(=O)-CH_3$ propanone

(e) $CH_3CH_2C(OH)=O$ propanoic acid (f) $CH_3CH_2-O-CH_3$ methoxyethane

(g) $CH_3CH_2C(NH_2)=O$ propanamide (h) $CH_3CH_2CH_2NH_2$ propylamine

(i) $H_3C-\ddot{N}(H)-CH_2CH_3$ ethylmethylamine (j) $H_3C-C\equiv C-H$ propyne

25. Primary alcohols have a $-CH_2-OH$ group, secondary alcohols have a $C-C(H)(OH)-C$ group, and tertiary alcohols have a $C-C(C)(OH)-C$ group.

(a) 2-methyl-2-butanol, <u>tertiary</u> (b) 2-butanol, <u>secondary</u>

(c) 2,2-dimethylpropanol, <u>primary</u> (d) 2-propanol, <u>secondary</u>

(e) 3-pentanol, <u>secondary</u>

26. (a) methylmethanoate $CH_3-C(=O)-O-CH_3$

(b) ethylbutanoate $CH_3CH_2CH_2-C(=O)-O-CH_2CH_3$

(c) butylethanoate $CH_3-C(=O)-O-CH_2CH_2CH_2CH_3$

(d) 2-propylpropanoate $CH_3CH_2-C(=O)-O-C(H)(CH_3)_2$

(e) ethylphosphate $HO-P(=O)(OH)-O-CH_2CH_3$

27. (a) The products of combustion of an alkane are $CO_2(g)$ and $H_2O(l)$.

$$CH_4(g) + 2O_2(g) \longrightarrow CO_2(g) + 2H_2O(l)$$

(b) In an elimination reaction, one molecule decomposes (usually at high temperature) into two others; (a molecule is eliminated).

$$C_2H_6(g) \longrightarrow C_2H_4(g) + H_2(g) \quad \text{(a cracking reaction)}$$

(c) $H_2O$ is eliminated from a primary alcohol to give an alkene.

$$C_2H_5OH(l) \xrightarrow{H_2SO_4} C_2H_4(g) + H_2O$$

(d) In an addition reaction two molecules combine to give one molecule and a triple CC bond is converted to a double bond, or a double bond to a single bond. For example:

$$H_2C{=}CH_2(g) + Br_2(l) \longrightarrow H_2\overset{Br}{\overset{|}{C}}-\overset{Br}{\overset{|}{C}}H_2$$

28. (a) Cracking is a process whereby alkanes, when heated to a high temperature, are converted to shorter chain alkanes and alkenes (and/or hydrogen).

$$C_4H_{10}(g) \longrightarrow C_2H_4(g) + C_2H_6(g)$$

(b) A polymerization reaction is one in which monomer units are linked together to form a longer (polymer) chain. For example, propene can be polymerized to polypropene.

$$n\ CH_3\text{-}CH{=}CH_2 \longrightarrow \text{-}(CH_2\text{-}\underset{CH_3}{\underset{|}{CH}})_n\text{-}$$

(c) An addition reaction is one by which two molecules combine to give a single molecule; addition across a multiple bond is an addition reaction. For example,

$$H_2C{=}CH_2(g) + HCl(g) \longrightarrow H_2\underset{Cl}{\underset{|}{C}}\text{-}CH_3$$

(d) An elimination reaction is one in which a single molecule decomposes to give two others.

$$CH_3CH_2CH_2OH \longrightarrow CH_3CH{=}CH_2 + H_2O$$

(e) In a substitution reaction, one group in a molecule is replaced by another.

$$CH_3I + OH^- \longrightarrow CH_3OH + I^-$$

(f) A free radical mechanism is one that involves a reactant that is a species with an odd number of electrons (free radical). Many chain reactions proceed by such a mechanism.

$$Cl_2 \xrightarrow{light} 2Cl\cdot$$
$$\cdot Cl + CH_4 \longrightarrow HCl + \cdot CH_3$$
$$\cdot CH_3 + Cl_2 \longrightarrow CH_3Cl + \cdot Cl$$
$$\cdot Cl + \cdot Cl \longrightarrow Cl_2$$

29. (a) Ethanol is a weak acid; with $OH^-$ ion,

$$C_2H_5OH + OH^- \longrightarrow H_2O + C_2H_5O^- \quad \text{ethoxide ion}$$

(b) Ethanol is a weak base; with $H_3O^+$ ion,

$$C_2H_5OH + H_3O^+ \longrightarrow C_2H_5OH_2^+ + H_2O \quad \text{ethanoxonium ion}$$

(c) The reaction is analagous with that with water,

$$2C_2H_5OH + 2Na \longrightarrow 2C_2H_5O^-Na^+ + H_2(g) \quad \text{sodium ethoxide}$$

(d) Ethanol is dehydrated by excess concentrated $H_2SO_4$.

$$C_2H_5OH + H_2SO_4 \longrightarrow C_2H_4 + H_3O^+ + HSO_4^- \quad \text{ethene}$$

30. Alkanes are unreactive for the following reasons:

(a) Both the C-H and C-C bonds are strong (high bond energies) and not easily broken except at high temperature, so elimination reactions are rare.
(b) All of their atoms are saturated, i.e., have as many atoms bonded to them as possible, so addition reactions are impossible.
(c) They have no lone pairs of electrons, and so they do not act as Lewis bases.
(d) The duets of the hydrogens and the octets of the carbons are satisfied, so they do not act as Lewis acids.
(e) The C-H bonds are insufficiently polar for them to behave as Brønsted-Lowry acids.

By contrast, alkenes and alkynes are unsaturated compounds that readily undergo addition reactions. The presence of a multiple bond also increases the polarity of adjacent C-H bonds, so that alkynes, for example, behave as weak Brønsted-Lowry acids.

31. (a) $Br_2$ adds across the double bond,

$$C_6H_{10} + Br_2 \longrightarrow C_6H_{10}Br_2 \quad \text{1,2-dibromocyclohexane}$$

(b) This is a combustion reaction,

$$C_3H_8 + 5O_2 \longrightarrow 3CO_2 + 4H_2O \quad \text{carbon dioxide and water}$$

(c) This is a reaction in which hydrogen is eliminated.

$$C_2H_5OH \xrightarrow[\text{catalyst}]{\text{heat}} CH_3CHO + H_2 \quad \text{ethanal and hydrogen}$$

(d) -OH is replaced by $-NH_2$ and water is eliminated,

$$CH_3OH + NH_3 \longrightarrow CH_3NH_2 + H_2O \quad \text{methylamine and water}$$

32. (a) Ethyne decolorizes a solution of $Br_2$ by undergoing the addition reaction

$$HC\equiv CH + 2Br_2 \longrightarrow CHBr_2\text{-}CHBr_2$$

Ethane, $C_2H_6$, a saturated hydrocarbon will not add $Br_2$.

(b) Propane burns in oxygen while $CO_2$ is inert and does not react.

$$C_3H_8 + 5O_2 \longrightarrow 3CO_2 + 4H_2O$$

Alternatively, if the gases are bubbled into limewater, $Ca(OH)_2(aq)$, only $CO_2(g)$ gives a white precipitate of insoluble $CaCO_3(s)$.

$$Ca(OH)_2 + CO_2 \longrightarrow CaCO_3(s) + H_2O$$

(c) Ethene decolorizes a solution of $Br_2$ while saturated $C_3H_8$ does not.

$$H_2C{=}CH_2 + Br_2 \longrightarrow CH_2Br\text{-}CH_2Br$$

33. Aldehydes result from the oxidation of primary alcohols, and ketones from the oxidation of secondary alcohols, in reactions where two H atoms are removed from the alcohol - one from the -OH groups and the other from the substituted carbon atom.

(a) $CH_3CH_2OH \longrightarrow CH_3CHO + H_2O$    (b) $CH_3CH(OH)CH_3 \longrightarrow CH_3-\overset{O}{\overset{\|}{C}}-CH_3 + H_2O$

    <u>ethanol</u>        ethanal          <u>2-propanol</u>       propanone

(c) $(CH_3)_2CHCH_2OH \longrightarrow (CH_3)_2CH-CHO + H_2O$

    <u>2-methyl-1-propanol</u>   2-methylpropanal

(d) $CH_3CH_2CH_2CH(OH)CH_3 \longrightarrow CH_3CH_2CH_2-\overset{O}{\overset{\|}{C}}-CH_3$

    <u>2-pentanol</u>               2-pentanone

34. These are all reactions between an alcohol and a carboxylic acid, which result in the formation of an ester with the elimination of water.

(a) $CH_3-\overset{\|}{\underset{O}{C}}-OH + CH_3-\underset{OH}{\overset{\cdot}{C}H}-CH_3 \xrightarrow{-H_2O} CH_3-\underset{O}{\overset{\|}{C}}-O-\underset{H}{\overset{\cdot}{C}}(CH_3)_2$   2-propylethanoate

(b) $CH_3-\overset{\|}{\underset{O}{C}}-OH + CH_3CH_2CH_2CH_2OH \xrightarrow{-H_2O} CH_3-\underset{O}{\overset{\|}{C}}-O-CH_2CH_2CH_2CH_3$   butylethanoate

(c) $\bigcirc\!\!\!\!-\overset{\|}{\underset{O}{C}}-OH + CH_3OH \xrightarrow{-H_2O} \bigcirc\!\!\!\!-\overset{\|}{\underset{O}{C}}-O-CH_3$    methylbenzoate

(d) $H-\overset{\|}{\underset{O}{C}}-OH + CH_3CH_2OH \xrightarrow{-H_2O} H-\overset{\|}{\underset{O}{C}}-O-CH_2-CH_3$   ethylmethanoate

35. (a) $CH_4(g) + H_2O(g) \xrightarrow[\text{Ni catalyst}]{700-800^\circ C} CO(g) + 3H_2(g)$

(b) $CO(g) + 2H_2(g) \xrightarrow[\text{Al}_2O_3 \text{ catalyst}]{\substack{240-260^\circ C \\ 50-100 \text{ atm}}} CH_3OH(g)$

(c) $CH_3OH(g) \xrightarrow[\text{Cu catalyst}]{550-600^\circ C} CH_2O(g) + H_2(g)$

(d) $CH_3OH(g) + NH_3(g) \xrightarrow[\text{Al}_2O_3 \text{ catalyst}]{400^\circ C} CH_3NH_2(g) + H_2O(g)$

36. (a) $2C_2H_5OH(g) + O_2(g) \xrightarrow{\text{catalyst}} 2CH_3CHO(g) + 2H_2O(g)$

(b) $CH_3CH_2CH_2OH(l) + NaOH(s) \longrightarrow CH_3CH_2CH_2O^-Na^+(s) + H_2O(l)$

(c) $2CH_3OH(l) + 2Na(s) \longrightarrow 2CH_3O^-Na^+(s) + H_2(g)$

(d) $C_2H_5OH(l) + H_2SO_4(conc) \longrightarrow H_2C=CH_2(g) + H_3O^+ + HSO_4^-$

(e) $CH_3CH=CH_2(g) + H_2O(g) \xrightarrow[\text{Catalyst}]{\text{Heat, Pressure}} CH_3CH_2CH_2OH(g)^*$

       * and $CH_3CH(OH)CH_3(g)$

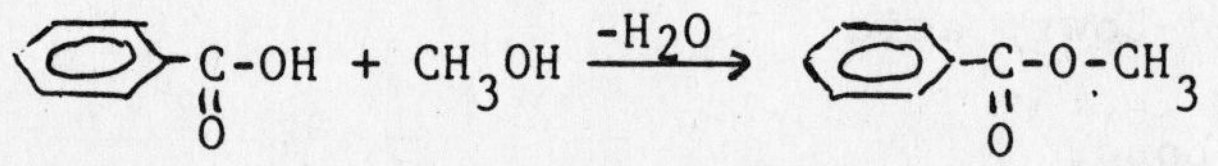

37. (a) $3CH_3CH_2CHO + K_2Cr_2O_7 + 8H_3O^+ \longrightarrow 3CH_3CH_2CO_2H + 2Cr^{3+} + 2K^+ + 12H_2O$

    (b) $CH_3CH_2CO_2H + (CH_3)_2CHOH \longrightarrow CH_3CH_2\overset{\text{O}}{\underset{||}{C}}-O-C(H)(CH_3)_2 + H_2O$

    (c) $CH_3OH + NH_3 \xrightarrow{\text{heat}} CH_3NH_2 + H_2O$

    (d) $CH_3NH_2 + H_2O \rightleftharpoons CH_3NH_3^+ + OH^-$   (weak base at room temperature)

    (e) $CH_3NH_2 + HCl \longrightarrow CH_3NH_3^+ + Cl^-$

    (f) $CH_3CO_2H + C_2H_5NH_2 \longrightarrow CH_3-\overset{\text{O H}}{\underset{|| \,\,|}{C-N}}-CH_2CH_3 + H_2O$

38. Synthesis gas is prepared from the reaction of $CH_4(g)$ with steam,

$$CH_4(g) + H_2O(g) \xrightarrow[\text{Ni catalyst}]{700-800^\circ C} CO(g) + 3H_2(g)$$

and $CH_3OH$ is prepared from the reaction,

$$CO(g) + 2H_2(g) \xrightarrow[\substack{50-100 \text{ atm} \\ Al_2O_3 \text{ catalyst}}]{240-260^\circ C} CH_3OH(g)$$

Thus, <u>overall</u>,  $CH_4(g) + H_2O(g) \longrightarrow CH_3OH(g) + H_2(g)$

39. Methylmethanoate is formed from methanol and methanoic acid,

$$CH_3OH + HCO_2H \longrightarrow CH_3O-\overset{\text{O}}{\underset{||}{C}}-H + H_2O$$

methyl methanoate

40. (a) $CO(g) + 2H_2(g) \xrightarrow{\text{ZnO catalyst}} H_3C-OH$   <u>methanol</u>

    (b) $\underset{H_3C}{\overset{H_3C}{>}}C=C\overset{CH_3}{\underset{H}{<}} + Br_2 \text{---} H_3C-\overset{H_3C \,\, H}{\underset{Br \,\, Br}{\underset{|\,\,\,|}{\overset{|\,\,\,|}{C-C}}}}-CH_3$   2,3-dibromo-2-methylbutane

    (c) $\underset{H_3C}{\overset{H_3C}{>}}C=C\overset{CH_3}{\underset{H}{<}} + H_2 \text{---} H_3C-\overset{H}{\underset{CH_3}{\underset{|}{\overset{|}{C}}}}-CH_2-CH_3$   2-methylbutane

    (d) $CH_3OH + C_2H_5OH \xrightarrow{H_2SO_4} H_3C-O-C_2H_5 + H_2O$  methoxyethane

                                               <u>or</u> methyl ethyl ether

    (e) $2CH_3CH_2CH_2OH + 2Na \longrightarrow 2CH_3CH_2CH_2O^-Na^+ + H_2(g)$

                                  sodium propoxide

41. Ethene is converted to vinyl chloride, $CH_2=CHCl$ by first reacting it with chlorine,

$$H_2C=CH_2 + Cl_2 \longrightarrow ClCH_2-CH_2Cl$$

and then heating the 1,2-dichloroethane to eliminate $HCl(g)$

$$ClCH_2-CH_2Cl \longrightarrow CH_2=CHCl + HCl$$

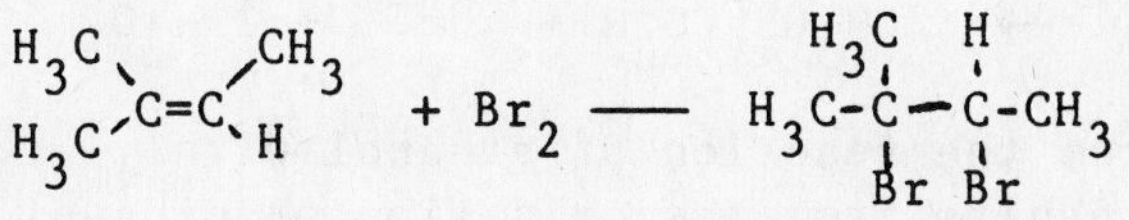

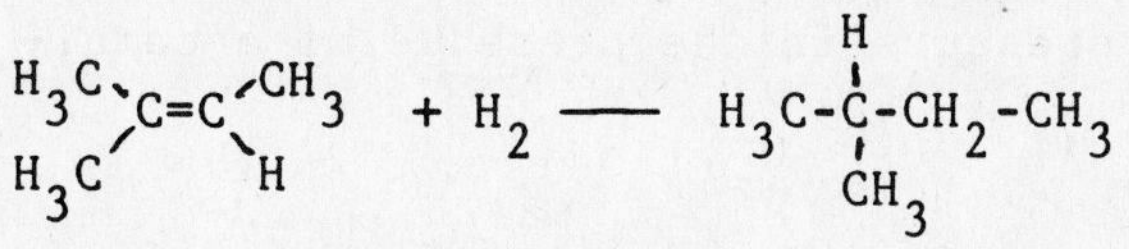

42. Addition of water to ethene at 300°C and 70 atm pressure in the presence of $H_3PO_4$ catalyst gives ethanol,

$$H_2C{=}CH_2 + H{-}OH \longrightarrow CH_3CH_2OH$$

which may be oxidized by acidified $K_2Cr_2O_7(aq)$ to ethanoic acid,

$$3C_2H_5OH + 2Cr_2O_7^{2-} + 16H_3O^+ \longrightarrow 3CH_3CO_2H + 4Cr^{3+} + 27H_2O$$

Reaction of the ethanol with ethanoic acid gives ethyl ethanoate.

$$CH_3CH_2OH + CH_3CO_2H \longrightarrow CH_3CH_2{-}O{-}\underset{\underset{O}{\parallel}}{C}{-}CH_3$$

43. (a) $CH_3{-}\underset{H}{C}{=}CH_2 + H_2O \xrightarrow[\text{catalyst}]{\text{high T}} CH_3{-}\overset{OH}{\underset{H}{C}}{-}CH_3$

propene    water

(b) $H_3C{-}C{\equiv}C{-}CH_3 + 2Br_2 \longrightarrow H_3C{-}\overset{Br}{\underset{Br}{C}}{-}\overset{Br}{\underset{Br}{C}}{-}CH_3$

2-butyne    bromine

(c) $H_3C{-}C{\equiv}C{-}CH_3 + H_2 \longrightarrow H_3C{-}\underset{H}{C}{=}\underset{H}{C}{-}CH_3$

2-butyne    hydrogen

(d) $H_3C{-}C{\equiv}C{-}H + HBr \longrightarrow H_3C{-}\underset{Br}{C}{=}CH_2$

propyne    hydrogen bromide

44. (a) Reaction of ethene with $Cl_2$ gives 1,2-dichloroethane,

$$H_2C{=}CH_2 + Cl_2 \longrightarrow H_2\underset{Cl}{C}{-}\underset{Cl}{C}H_2$$

(b) Propanoic acid is formed when 1-propanol is oxidized, with, for example, acidified $KMnO_4(aq)$,

$$5CH_3CH_2CH_2OH + 4MnO_4^- + 12H_3O^+ \longrightarrow 5CH_3CH_2CO_2H + 4Mn^{2+} + 23H_2O$$

(c) Ethyl ethanoate is prepared from the reaction of ethanol with ethanoic acid. Ethanoic acid results from the oxidation of ethanal with air at 60-80°C and 5 atm pressure in the presence of a catalyst

$$2CH_3CHO \longrightarrow 2CH_3CO_2H$$

and ethanol is formed when ethanal is reduced, with for example hydrogen in the presence of a nickel catalyst,

$$CH_3CHO + H_2 \longrightarrow CH_3CH_2OH$$

Finally, the ethanol and ethanoic acid are reacted to give the required ester.

$$CH_3CH_2OH + CH_3CO_2H \longrightarrow CH_3CH_2O{-}\overset{\overset{O}{\parallel}}{C}{-}CH_3$$

45. For each molecular species, first draw the Lewis structure and then count up the number of bonds (n) and the number of unshared pairs of electrons (m) on the designated atom:

(a) $^{*}CH_3OH$

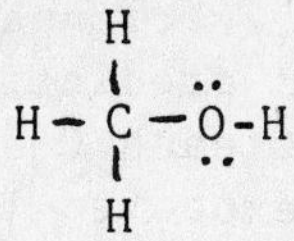

$AX_4$ tetrahedral

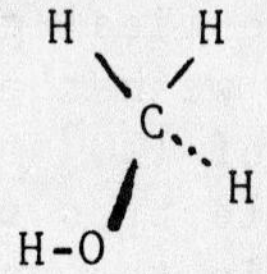

(b) $H_3C\text{-}\overset{*}{O}\text{-}CH_3$

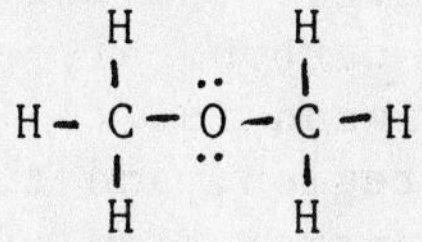

$AX_2E_2$ angular

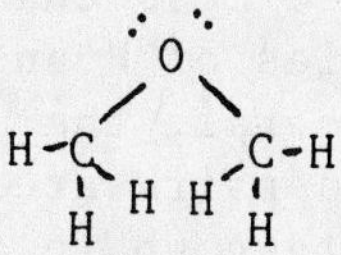

(c) $H\overset{*}{C}O_2H$

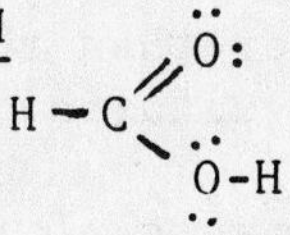

$AX_3$ trigonal planar

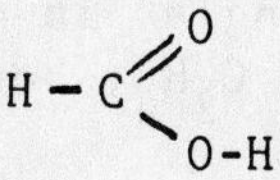

(d) $H\overset{*}{C}O_2^-$

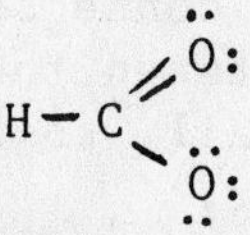

$AX_3$ trigonal planar

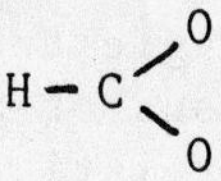

(two resonance structures)

(e) $CH_3\overset{*}{C}O_2H$

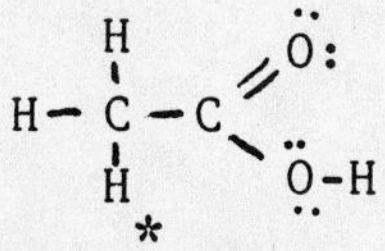

$AX_3$ trigonal planar

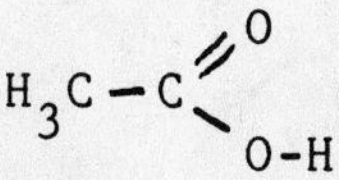

(f) $(CH_3)_2\overset{*}{C}O$

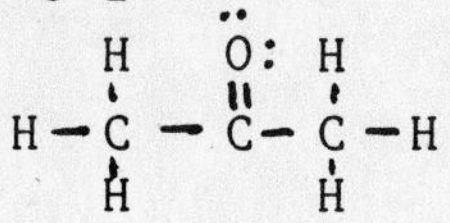

$AX_3$ trigonal planar

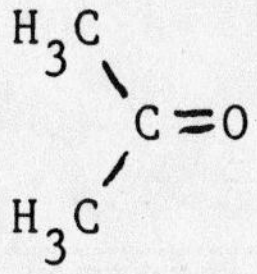

(g) $CH_3\overset{*}{N}H_2$

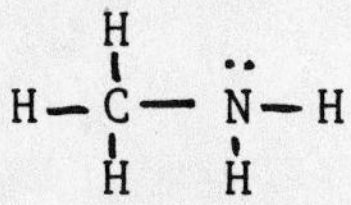

$AX_3E$ trigonal pyramidal

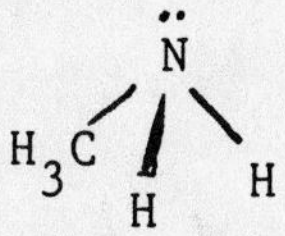

(h) $(CH_3)_4\overset{*}{N}^+$

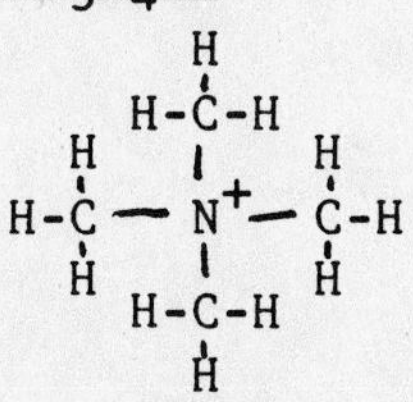

$AX_4$ tetrahedral

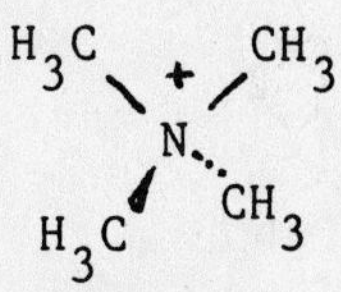

(i) $(CH_3)_3\overset{*}{N}$

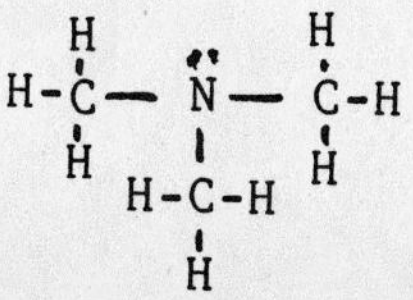

$AX_3E$ trigonal pyramidal

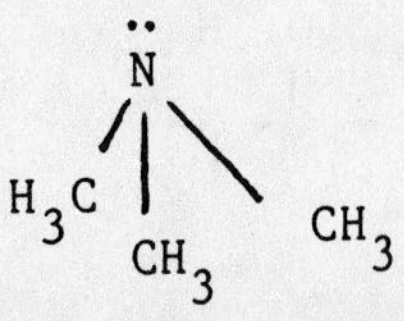

46. In both alkenes the ideal geometry at each C atom is $AX_3$, trigonal planar, consistent with the observed planar geometry and predicting ideal bond angles of 120° for each of the bond angles at each C atom. However, the two electron pairs of the double bond are expected to occupy more space in the valence shell of carbon than the single electron pairs of each of the single bonds, so that the CCX bond angles are expected to be greater than 120° and the XCX bond angles are expected to be less than 120°, as observed. However, the relative electronegativities of H and F, relative to that of carbon, also have an effect. In general, the more electronegative a ligand is, the more strongly its bond pairs are drawn into the bonding region, and the less space they take up in the valence shell of a central atom. Thus, C-F bond pairs are expected to take up less space than C-H bond pairs and the FCF bond angles in $C_2F_4$ are expected to be smaller than the HCH bond angles in $C_2H_4$, consistent with the observed structures.

47. A molecular conformation is one of many possible arrangements of the atoms and the groups in a molecule that are possible as a consequence of rotation about single bonds. The energy difference between the different conformations of a molecule are usually very small, so that one conformation is relatively easily converted to another.

(a) Two important different conformations of ethane are the staggered and eclipsed forms.

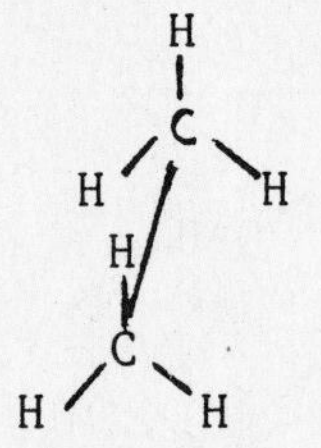

eclipsed

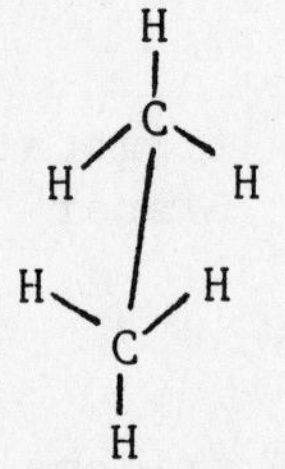

staggered

(b) Two important conformations of cyclohexane, $C_6H_{12}$, are the boat and chair forms.

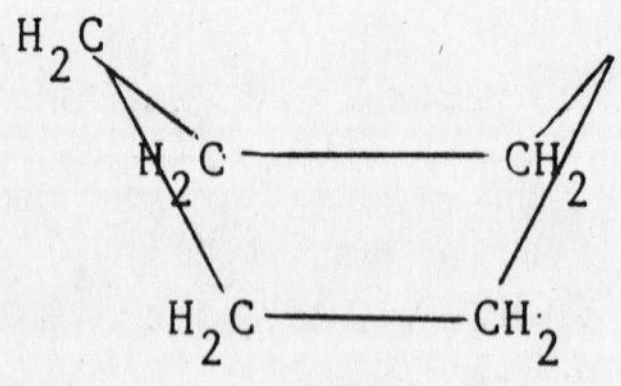

"boat"

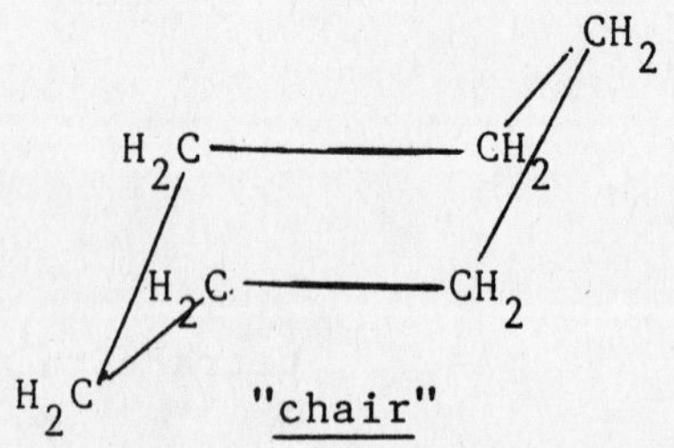

"chair"

48. (a) Two Lewis structures can be written for the amide group.

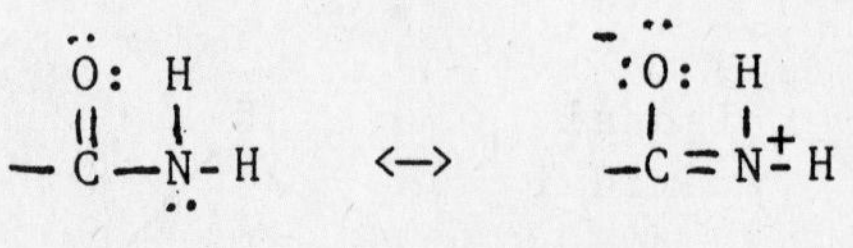

Thus, the actual structure has double bond character in both the CO and CN bonds, and the geometry at both C and N is $AX_3$ trigonal planar. The partial double bond character of both the CO and CN bonds restricts rotation out of the plane. In terms of hybrid orbitals, the planar O-C-$NH_2$ framework is rationalized by forming $\sigma$ bonds using $sp^2$ hybrid orbitals on C, N, and O. This leaves a 2p orbital on each of the atoms perpendicular to the plane of the group, which may be overlapped sideways to give a $\pi$-type orbital to accommodate the delocalized pair of electrons.

(b) The explanation is similar to that in part (a). Two resonance structures may be written for the carboxylate group;

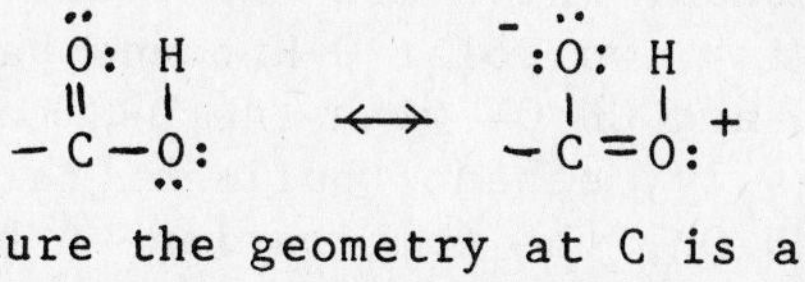

In each structure the geometry at C is always $AX_3$ planar. Alternatively, a $\sigma$-$\pi$ model may be used.

49. (a) <u>CO</u> $^-$:C≡O:$^+$ (b) <u>$CO_3^{2-}$</u>

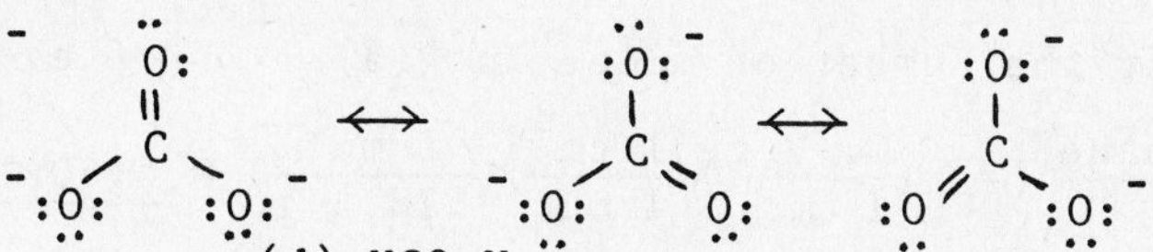

(c) <u>$HCO_2^-$</u>

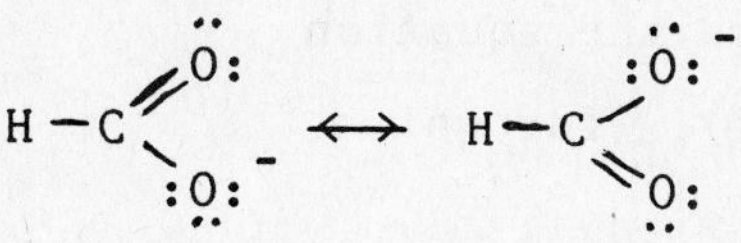

(d) <u>$HCO_2H$</u>

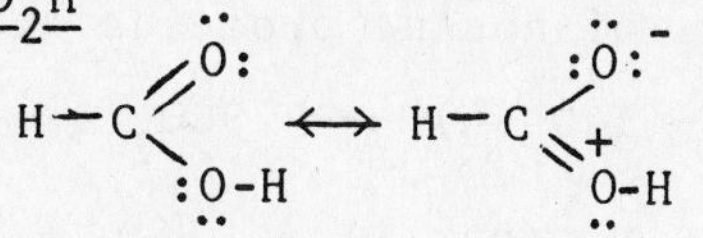

(e) <u>$HCONH_2$</u>

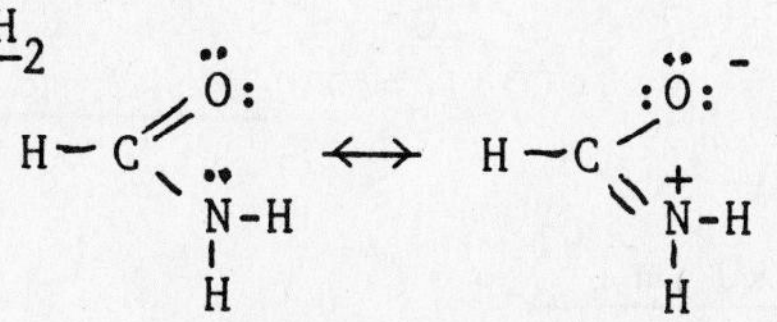

(f) <u>$C_6H_6$</u>

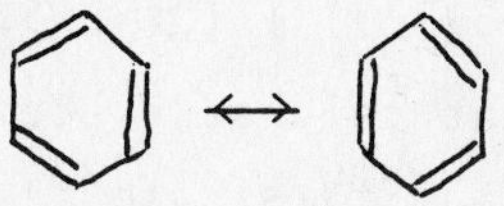

50. (a) <u>$CH_4$</u> The C atoms has $AX_4$ tetrahedral geometry, requiring a set of four $sp^3$ hybrid orbitals on C.

(b) <u>$H_2C{=}CH_2$</u> Each C atom has $AX_3$ trigonal planar geometry, requiring a set of three $sp^2$ hybrid orbitals on each C atom.

(c) <u>HC≡CH</u> Each C atom has $AX_2$ linear geometry, requiring a set of two sp hybrid orbitals on each C atom.

(d) <u>$H_3C$-O-$CH_3$</u> The O atom has $AX_2E_2$ angular geometry, requiring a set of four $sp^3$ hybrid orbitals on the O atom to accommodate the two lone pairs of electrons and to form two CO bonds.

51. The boiling point of ethanol, $C_2H_5OH$, is significantly higher than that of dimethyl ether, $(CH_3)_2O$, because only the former has a polar O-H bond that can form intermolecular hydrogen bonds. A simple chemical test that would distinguish ethanol from dimethyl ether is its reaction with sodium, towards which ether is inert.

$$2C_2H_5OH + 2Na \longrightarrow 2C_2H_5O^-Na^+ + H_2(g)$$

Another simple test is addition to an acidified solution of $K_2Cr_2O_7$(aq) which oxidizes $C_2H_5OH$ to $CH_3CO_2H$, so that orange $Cr_2O_7^{2-}$(aq) is replaced by green $Cr^{3+}$(aq), but dimethyl ether is again unreactive.

52. Boiling points increase as the strength of the intermolecular forces in the liquids increase. The weakest intermolecular forces are found in $CH_4$ because this is a nonpolar substances the molecules of which attract each other by induced dipole-induced dipole (London) forces. Ethane, $C_2H_6$ is similar but has a larger polarizability than $CH_4$ because of its larger number of atoms, and has a higher boiling point. Next in order of increase of intermolecular forces is dimethyl ether, $(CH_3)_2O$, which with one more atom than $C_2H_6$ is expected to have a greater polarizability, and because it has an angular geometry at the O atom it also has a small dipole moment and dipole-dipole forces of attraction. In order of increasing boiling points, methanol, $CH_3OH$, ethanol, $C_2H_5OH$, and water, $H_2O$, are all polar molecules but they also form intermolecular hydrogen bonds, which are expected to be the dominant intermolecular forces. Water with two polar O-H bonds has more hydrogen bonds per molecule than $C_2H_5OH$ and $CH_3OH$ with one O-H bond each, and thus the highest boiling point. Finally, ethanol boils higher than methanol because with more atoms than $CH_3OH$, the combination of hydrogen bonds and London forces is greater.

53. Consider the number of moles in (a) 1 mL of each compound:

$$\frac{\text{Mol nonane}}{1\text{ mL}} = \left(\frac{0.72\text{ g}}{1\text{ cm}^3}\right)\left(\frac{1\text{ cm}^3}{1\text{ mL}}\right)\left(\frac{1\text{ mol}}{128.4\text{ g}}\right) = \underline{5.60\times10^{-3}\text{ mol}}\ C_9H_{20}\text{ per mL}$$

Combustion of nonane proceeds according to the equation

$$C_9H_{20}(l) + 14O_2(g) \longrightarrow 9CO_2(g) + 10H_2O(g),\quad \text{for which}$$

$$\Delta H^\circ = [9\,\Delta H_f^\circ(CO_2,g) + 10\,\Delta H_f^\circ(H_2O,g)] - [\Delta H_f^\circ(C_9H_{20},l) + 14\,\Delta H_f^\circ(O_2,g)]$$
$$= [9(-393.5) + 10(-285.8)] - [(-275) + 14(0)] = \underline{-6124\text{ kJ mol}^{-1}}$$

Thus, the heat released per mL of nonane is

$$(-6124\text{ kJ mol}^{-1})(5.6\times10^{-3}\text{ ml}) = \underline{34.4\text{ kJ ml}^{-1}}$$

Similarly, for ethanol,

$$\frac{\text{Mol ethanol}}{1\text{ mL}} = \left(\frac{0.79\text{ g}}{1\text{ cm}^3}\right)\left(\frac{1\text{ cm}^3}{1\text{ mL}}\right)\left(\frac{1\text{ mol}}{46.07\text{ g}}\right) = \underline{1.71\times10^{-2}\text{ mol}}\ C_2H_5OH\text{ per mL}$$

Combustion of ethanol proceeds according to the equation

$$C_2H_5OH(l) + \tfrac{7}{2}O_2(g) \longrightarrow 2CO_2(g) + 3H_2O(l),\ \text{for which}$$

$$\Delta H^\circ = [2\,\Delta H_f^\circ(CO_2,g) + 3\,\Delta H_f^\circ(H_2O,l)] - [\Delta H_f^\circ(C_2H_5OH,l) + \tfrac{7}{2}\,\Delta H_f^\circ(O_2,g)]$$
$$= [2(-393.5) + 3(-285.8)] - [(-277) + \tfrac{7}{2}(0)] = \underline{-1367\text{ kJ mol}^{-1}}$$

Thus, the heat released per mL of ethanol is

$$(-1367\text{ kJ mol}^{-1})(1.71\times10^{-2}\text{ mol}) = \underline{23.4\text{ kJ mL}^{-1}}$$

(b) Consider 1 g of each compound, which corresponds to,

$$(1\text{ g nonane})\left(\frac{1\text{ mol}}{128.4\text{ g}}\right) = \underline{7.79\times10^{-3}\text{ mol}}\text{ nonane}$$

$$(1\text{ g ethanol})\left(\frac{1\text{ mol}}{46.07\text{ g}}\right) = \underline{2.17\times10^{-2}\text{ mol}}\text{ ethanol}$$

Heat released per g nonane = $(6124\text{ kJ mol}^{-1})(7.79\times10^{-3}\text{ mol}) = \underline{47.7\text{ kJ}}$

Heat released per g ethanol = $(1367\text{ kJ mol}^{-1})(2.17\times10^{-2}\text{ mol}) = \underline{29.7\text{ kJ}}$

Nonane is clearly the more economical of these fuels since it costs less per liter and its combustion releases more heat, on both a per mL and a per g basis.

54. (a) $pK_b(CH_3NH_2) = 3.34$; $K_b(CH_3NH_2) = 4.57\times10^{-4}$ mol $L^{-1}$

| | $CH_3NH_2$ + $H_2O$ ⇌ | $CH_3NH_3^+$ | + $OH^-$ | |
|---|---|---|---|---|
| initially | 0.10 | 0 | 0 | mol $L^{-1}$ |
| at eq'm | 0.10-x | x | x | mol $L^{-1}$ |

$$K_b = \frac{[CH_3NH_3^+][OH^-]}{[CH_3NH_2]} = \frac{x^2}{0.10-x} = 4.57\times10^{-4} \text{ mol L}^{-1}$$

$x << 0.10$, $x = [OH^-] = 6.76\times10^{-3}$ mol $L^{-1}$; pOH = 2.17; <u>pH = 11.83</u>

(b) For a conjugate acid-base pair, $K_a.K_b = K_w$, or $pK_a + pK_b = 14.00$

$$pK_a(CH_3NH_3^+) = 14.00-3.34 = \underline{10.66}$$

(c) First calculate the moles of base and acid, before and after reaction.

$$\text{mol } CH_3NH_2 = (24.5 \text{ mL})(0.090 \text{ mol L}^{-1})(\frac{1 \text{ L}}{1000 \text{ mL}}) = 2.20\times10^{-3} \text{ mol}$$

$$\text{mol HCl} = (10.0 \text{ mL})(0.100 \text{ mol L}^{-1})(\frac{1 \text{ L}}{1000 \text{ mL}}) = 1.00\times10^{-3} \text{ mol}$$

| | $CH_3NH_2(aq)$ + | $HCl(aq)$ ⟶ | $CH_3NH_3^+(aq)$ + | $Cl^-(aq)$ | |
|---|---|---|---|---|---|
| initially | $2.20\times10^{-3}$ | $1.00\times10^{-3}$ | 0 | 0 | mol |
| finally | $1.20\times10^{-3}$ | 0 | $1.00\times10^{-3}$ | $1.00\times10^{-3}$ | mol |

Thus, the final solution contains both $CH_3NH_2$ and $CH_3NH_3^+$ and is a buffer solution, for which

$$K_a(CH_3NH_3^+) = 2.19\times10^{-11} \text{ mol L}^{-1} = [H_3O^+]\frac{[CH_3NH_2]}{[CH_3NH_3^+]}$$

$$= [H_3O^+](\frac{\text{mol } CH_3NH_2}{\text{mol } CH_3NH_3^+}) = [H_3O^+](\frac{1.20}{1.00})$$

$[H_3O^+] = 1.83\times10^{-11}$ mol $L^{-1}$; pH = <u>10.74</u>

55. Mol C in 100 g compound = $(82.6 \text{ g})(\frac{1 \text{ mol C}}{12.01 \text{ g C}})$ = <u>6.88 mol</u>

Mol H in 100 g compound = $(17.4 \text{ g})(\frac{1 \text{ mol H}}{1.008 \text{ g H}})$ = 17.3 mol

Ratio of mol C : mol H = ratio of atoms = 6.88 : 17.3 = 1.00 : 2.51 <u>or 2 : 5</u>

Thus, the <u>empirical formula</u> is $C_2H_5$

From the gas data,

$$n = \frac{PV}{RT}\ ; \quad n = \frac{(750 \text{ mm Hg})(\frac{1 \text{ atm}}{760 \text{ mm Hg}})(200 \text{ mL})(\frac{1 \text{ L}}{1000 \text{ mL}})}{(0.0821 \text{ atm L mol}^{-1} \text{ K}^{-1})(298 \text{ K})} = \underline{8.07\times10^{-3} \text{ mol}}$$

Thus, molar mass = $(\frac{0.470 \text{ g}}{8.07\times10^{-3} \text{ mol}})$ = <u>58.2 g mol$^{-1}$</u>

Since the empirical formula mass is 29.06 u, the <u>molecular formula</u> is $C_4H_{10}$.

There is no unique structure for $C_4H_{10}$; both of the isomers

$CH_3$-$CH_2$-$CH_2$-$CH_3$ and $CH_3$-CH-$CH_3$ (with $CH_3$ on the central C)

butane    2-methylpropane

have the molecular formula $C_4H_{10}$.

56. 100 g of compound contains $(85.6\text{ g C})(\frac{1\text{ mol C}}{12.01\text{ g C}}) = 7.13\text{ mol C}$

100 g of compound contains $(14.4\text{ g H})(\frac{1\text{ mol H}}{1.008\text{ g H}}) = 14.3\text{ mol H}$

Ratio of mol C : mol H = ratio of atoms = 7.13 : 14.3 = 1.00 : 2.00

Thus, the empirical formula is $CH_2$ (formula mass = 14.03 u).

From the gas data,

$$n = \frac{PV}{RT} = \frac{(760\text{ mm Hg})(\frac{1\text{ atm}}{760\text{ mm Hg}})(250\text{ mL})(\frac{1\text{ L}}{1000\text{ mL}})}{(0.0821\text{ atm L mol}^{-1}\text{ K}^{-1})(373\text{ K})} = 8.16\text{x}10^{-3}\text{ mol}$$

$$\text{Molar mass} = (\frac{0.687\text{ g}}{8.16\text{x}10^{-3}\text{ mol}}) = 84.2\text{ g mol}^{-1}$$

Thus, the molecular formula is $C_6H_{12}$ (molar mass 84.16 g $mol^{-1}$)

There are many possible structures; $C_6H_{12}$ represents the structure of a number of alkenes and cycloalkanes with 6 carbon atoms, the most familiar of which are 1-hexene, 2-hexene, and 3-hexene, and cyclohexane.

57. This problem is solved in the same way as Problems 55 and 56. The empirical formula is $CH_2$ (formula mass 14.03 u), so that the molar mass, which is 56.1 g $mol^{-1}$ experimentally, gives the molecular formula $C_4H_8$. The structures of the six isomers with this molecular formula were given in Problem 8.

58. (a) Using the methods of Problems 55 and 56, the empirical formula of the compound is $C_2H_3$ (empirical formula mass = 27.04 u).

In the reaction with 1.00 g of the hydrocarbon, the mol $H_2$ consumed is given by,

$$n = \frac{(1\text{ atm})(0.906\text{ L})}{(0.0821\text{ atm L mol}^{-1}\text{ K}^{-1})(298\text{ K})} = 3.71\text{x}10^{-2}\text{ mol}$$

and from the given data, the molar mass of the unsaturated hydrocarbon is 54.1 g $mol^{-1}$, consistent with the molecular formula $C_4H_6$, with a molar mass of 54.08 g $mol^{-1}$.

Since 1.00 g of the compound (= $1.85\text{x}10^{-2}$ mol) reacts with $3.71\text{x}10^{-2}$ mol $H_2$(g), 1 mol of the compound reacts with 2 mol $H_2$(g). This suggests that the compound contains either one triple bond or two double bonds.

(b) The isomers with molecular formula $C_4H_6$ and two double bonds or a triple bond are:

$CH_2$=CH-CH=$CH_2$    HC≡C-$CH_2$-$CH_3$    and    $H_3$C-C≡C-$CH_3$

1,3-butadiene    1-butyne    2-butyne

(c) Reaction of each of the three isomers with $Br_2$ gives a different product.

$$CH_2=CH-CH=CH_2 + 2Br_2 \longrightarrow CH_2Br-CHBr-CHBr-CH_2Br$$

1,2,3,4-tetrabromobutane

$$HC \equiv C-CH_2-CH_3 + 2Br_2 \longrightarrow CHBr_2-CBr_2-CH_2-CH_3$$

1,1,2,2-tetrabromobutane

$$H_3C-C \equiv C-CH_3 + 2Br_2 \longrightarrow H_3C-CBr_2-CBr_2-CH_3$$

2,2,3,3-tetrabromobutane

59. From the gas data, calculate the total moles of gases in the mixture.

$$n = \frac{PV}{RT} = \frac{(744 \text{ mm Hg})(\frac{1 \text{ atm}}{760 \text{ mm Hg}})}{(0.0821 \text{ atm L mol}^{-1} \text{ K}^{-1})(298 \text{ K})} = \underline{0.0400 \text{ mol}}$$

And for the combustion reactions, write

$$CH_4 \xrightarrow{O_2} CO_2 + 2H_2O \quad \underline{and} \quad C_nH_{2n} \xrightarrow{O_2} nCO_2 + nH_2O$$

Now calculate the mol of C and the mol of H in the mixture:

$$\text{Mol C} = (2.641 \text{ g CO}_2)(\frac{1 \text{ mol CO}_2}{44.01 \text{ g CO}_2})(\frac{1 \text{ mol C}}{1 \text{ mol CO}_2}) = \underline{0.0600 \text{ mol C}}$$

$$\text{Mol H} = (1.442 \text{ g H}_2O)(\frac{1 \text{ mol H}_2O}{18.02 \text{ g H}_2O})(\frac{2 \text{ mol H}}{1 \text{ mol H}_2O}) = \underline{0.1600 \text{ mol H}}$$

Thus, for a mixture containing x mol $CH_4$ and y mol $C_nH_{2n}$, we have:

From the total moles of gas,     $x + y = 0.0400$  .... (1)

From the moles of C,             $x + ny = 0.0600$  .... (2)

From the moles of H,            $4x + 2ny = 0.1600$  .... (3)

and solving equations (1), (2), and (3), for x, y, and n, gives:

$x = \text{mol CH}_4 = \underline{0.0200 \text{ mol}}$ ; $\underline{y = 0.0200 \text{ mol}}$, and $y = \underline{2}$

i.e., the $\underline{\text{alkene}}$ is $C_2H_4$, $\underline{\text{ethene}}$

Use the given mass of the mixture to confirm these results:

Mass of gas mixture = $(0.0200 \text{ mol CH}_4)(\frac{16.04 \text{ g}}{1 \text{ mol}}) + (0.0200)(\frac{28.05 \text{ g}}{1 \text{ mol}})$

$= \underline{0.882 \text{ g}}$

Mass of $CH_4$ = $(0.0200 \text{ mol CH}_4)(\frac{16.04 \text{ g}}{1 \text{ mol}}) = \underline{0.321 \text{ g}}$

Mass % $CH_4$ = $(0.321 \text{ g})(\frac{100 \text{ \%}}{0.882 \text{ g}}) = \underline{36.4\%}$ ; mass : $C_2H_4 = \underline{63.6\%}$

60. Since $C_5H_{12}O$ gives on oxidation the ketone $C_5H_{10}O$, it must be a secondary alcohol, since primary alcohols give aldehydes and tertiary alcohols give neither aldehydes nor ketones on oxidation. Possible secondary alcohols with the molecular formula $C_5H_{10}O$ are:

```
H3C-CH-CH2-CH2-CH3     CH3-CH2-CH-CH2-CH3     H3C-CH-CH-CH3
     |                          |                    |  |
     OH                         OH                   OH CH3

  2-pentanol               3-pentanol          3-methyl-2-butanol
```

61. Calculate both the moles of carbon and the moles of hydrogen, and their masses in 256 mg of the compound, and then the amount of oxygen by difference.

$$\text{mol C} = (0.512 \text{ g } CO_2)\left(\frac{1 \text{ mol } CO_2}{44.01 \text{ g } CO_2}\right)\left(\frac{1 \text{ mol C}}{1 \text{ mol } CO_2}\right) = 1.16\text{x}10^{-2} \text{ mol C}$$

$$\text{mol H} = (0.209 \text{ g } H_2O)\left(\frac{1 \text{ mol } H_2O}{18.02 \text{ g } H_2O}\right)\left(\frac{2 \text{ mol H}}{1 \text{ mol } H_2O}\right) = 2.32\text{x}10^{-2} \text{ mol H}$$

$$\text{mass of C} = (1.16\text{x}10^{-2} \text{ mol})\left(\frac{12.01 \text{ g}}{1 \text{ mol}}\right) = \underline{0.139 \text{ g C}}$$

$$\text{mass of H} = (2.32\text{x}10^{-2} \text{ mol})\left(\frac{1.008 \text{ g}}{1 \text{ mol}}\right) = \underline{0.0234 \text{ g H}}$$

$$\text{Mass of O} = (0.256-0.139-0.023) = \underline{0.094 \text{ g}}$$

$$\text{mol O} = (0.094 \text{ g O})\left(\frac{1 \text{ mol O}}{16.00 \text{ g O}}\right) = 5.9\text{x}10^{-3} \text{ mol O.}$$

Mol ratio C : H : O = atom ratio = 1.16 : 2.32 : 0.59 = 2.0 : 4.0 : 1.0

Thus, the empirical formula is $C_2H_4O$ (formula mass 44.05 u).

From the gas data and the mass of 156 mg, the experimental molar mass is 44.1 g $mol^{-1}$, so that the molecular formula is also $C_2H_4O$.

The most likely structure is $CH_3CHO$, ethanal, an aldehyde, which may be confirmed by its oxidation by $Ag(NH_3)_2^+(aq)$ to ethanoic acid (silver mirror test), or its effect on $K_2Cr_2O_7(aq)$ or $KMnO_4(aq)$ in acid solution.

62. 100 g compound contains $(52.1 \text{ g C})\left(\frac{1 \text{ mol C}}{12.01 \text{ g C}}\right) = 4.35 \text{ mol C}$

100 g compound contains $(13.1 \text{ g H})\left(\frac{1 \text{ mol H}}{1.008 \text{ g H}}\right) = 13.0 \text{ mol H}$

100 g compound contains $(34.8 \text{ g O})\left(\frac{1 \text{ mol O}}{16.00 \text{ g O}}\right) = 2.17 \text{ mol O}$

Ratio of moles (atoms) C : H : O = 4.35 : 13.0 : 2.17 = 2.0 : 6.0 : 1.0

Thus, the empirical formula is $C_2H_6O$. Calculation of the molar mass from the gas data gives molar mass = 46.0 g $mol^{-1}$, so that the molecular formula is also $C_2H_6O$.

Dimethyl ether, $(CH_3)_2O$, and ethanol, $C_2H_5OH$, are isomers of $C_2H_6O$, but only the latter reacts with sodium,

$$2C_2H_5OH + 2Na \longrightarrow 2C_2H_5O^-Na^+ + H_2(g)$$

to give sodium ethoxide and hydrogen.

$$\text{Mol of } H_2 = (0.250 \text{ g ethanol})\left(\frac{1 \text{ mol ethanol}}{46.07 \text{ g ethanol}}\right)\left(\frac{1 \text{ mol } H_2}{2 \text{ mol ethanol}}\right)$$
$$= \underline{2.71\text{x}10^{-3} \text{ mol}}$$

$$\text{Volume of } H_2 = \frac{nRT}{P} = \frac{(2.71\text{x}10^{-3} \text{ mol})(0.0821 \text{ atm L mol}^{-1} \text{ K}^{-1})(298 \text{ K})}{(730 \text{ torr})\left(\frac{1 \text{ atm}}{760 \text{ torr}}\right)}$$
$$= 6.90 \text{ x } 10^{-2} \text{ L or } 69.0 \text{ mL}$$

63. Calculation of the empirical formula and the molar mass by the usual methods gives $C_3H_6Cl_2$ (empirical formula mass 113.0 u) and 111 g mol$^{-1}$, so that the <u>molecular formula</u> is also $C_3H_6Cl_2$.

The molecular formula corresponds to that of a substituted propane, for which there are three possible isomers.

$$\begin{matrix} Cl & & Cl & Cl & & Cl \\ | & & | & | & & | \\ Cl-C-CH_2-CH_3 & & H-C-C-CH_3 & & H_3C-C-CH_3 \\ | & & | & | & & | \\ H & & H & H & & Cl \end{matrix}$$

   1,1-dichloropropane   1,2-dichloro-   2,2-dichloropane
                                propane

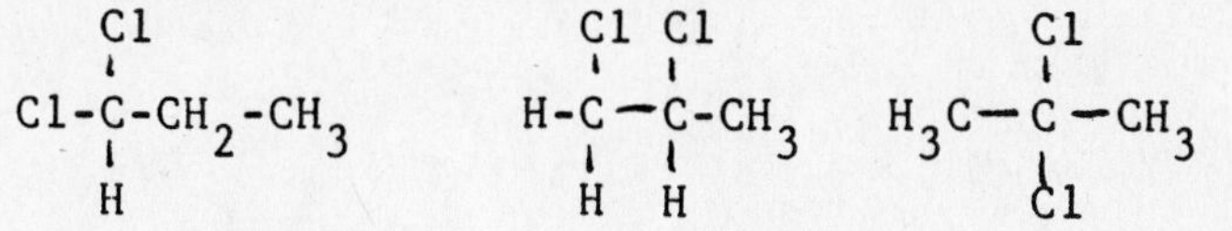

64. (a) $R-CH_2OH \xrightarrow{O} RCO_2H$ ; $RCO_2H + NaOH \longrightarrow RCO_2^-Na^+ + H_2O$

Mol $RCH_2OH$ = mol $RCO_2H$ = $(83.3 \text{ mL})(\frac{1 \text{ L}}{1000 \text{ mL}})(0.20 \text{ mol L}^{-1})$
$$= \underline{1.67 \times 10^{-2} \text{ mol}}$$

Molar mass of $RCH_2OH$ = $(\frac{1.00 \text{ g}}{1.67 \times 10^{-2} \text{ mol}})$ = $\underline{59.9 \text{ g mol}^{-1}}$

and for $C_nH_{2n+1}OH$, n = 3, and the <u>molecular formula</u> is $C_3H_7OH$.

(b) The alcohol is $H_3C-CH_2-CH_2-OH$, 1-propanol, which on oxidation gives firstly the aldehyde <u>propanal</u>, $CH_3CH_2CHO$, and then the carboxylic acid $CH_3CH_2CO_2H$, <u>propanoic acid</u>.

(c) The other alcohol with the molecular formula $C_3H_7O$ is 2-propanol, with the structure

$$\begin{matrix} H_3C-CH-CH_3 \\ | \\ OH \end{matrix}$$

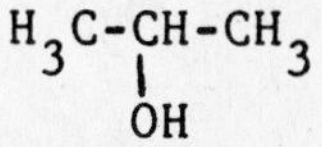

Oxidation of a secondary alcohol gives a ketone,

$$H_3C-CH(OH)-CH_3 \xrightarrow{O} H_3C-\underset{\underset{O}{\|}}{C}-CH_3$$

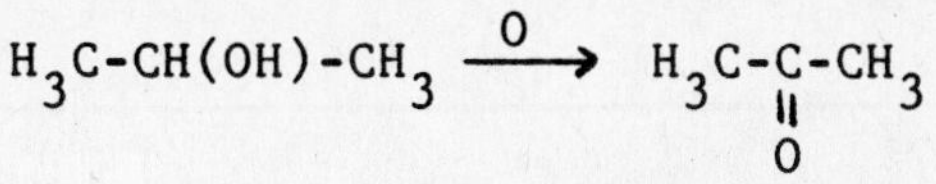

2-propanone (acetone)

Ketones cannot be oxidized farther, except to $CO_2$ and water upon combustion.

65. Primary alcohols are oxidized first to aldehydes and then to carboxylic acids. Secondary alcohols are oxidized to ketones.

(a) <u>Butanol</u>   $H_3C-CH_2-CH_2-CHO$, <u>butanal</u>, & $H_3C-CH_2-CH_2-CO_2H$, <u>butanoic acid</u>.

(b) <u>Propanal</u>   $H_3C-CH_2-CO_2H$, <u>propanoic acid</u>.

(c) <u>2-Methyl-1-butanol</u>   $\underset{\underset{CH_3}{|}}{CH_3-CH_2-CH}-CHO$  and  $\underset{\underset{CH_3}{|}}{CH_3-CH_2-CH}-CO_2H$

                          2-methyl butanal       2-methylbutanoic acid.

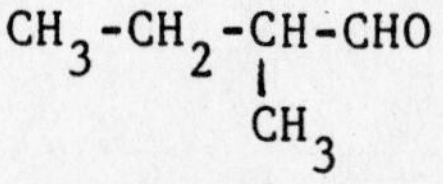

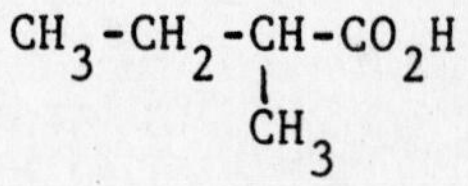

(d) <u>3-Methyl-2-butanol</u>   $\underset{\underset{CH_3}{|}}{H_3C-CH}-\underset{\underset{O}{\|}}{C}-CH_3$   3-methyl-2-butanone.

(e) <u>3,3-Dimethylbutanal</u>  $(CH_3)_3C-CH_2-CO_2H$  3,3-dimethylbutanoic acid.

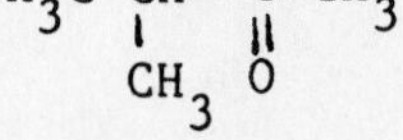

## CHAPTER 24

1. (a) An <u>addition polymer</u> is a molecule consisting of a number of repeating units (monomers), and which contains all of the atoms of the monomers from which it is composed. Examples include polyethylene (from ethylene monomers), polypropene (from propene monomers), and Teflon,(polytetrafluoroethene), (from tetrafluoroethene monomers).

   (b) A <u>condensation polymer</u> is a molecule formed from monomers by a condensation reaction, in which small molecules such as $H_2O$ are eliminated. Examples include Nylon, a polyamide formed by condensing a dicarboxylic acid and a diamine, and Dacron, a polyester formed by condensing 1,2-ethanediol with terephthalic acid (1,4-benzene dicarboxylic acid).

   (c) An <u>α-amino acid</u> is an organic compound containing an amino, $-NH_2$, functional group and a carboxylic acid, $-CO_2H$, functional group, attached to the <u>same</u> carbon atom. Examples include glycine, $H_2N-CH_2-CO_2H$, alanine, $H_2N-CH(CH_3)-CO_2H$, and all of the molecules given in Table 24.2.

   (d) A <u>sugar</u> is the common name for monomeric or dimeric carbohydrate molecules (monosaccharides, or disaccharides). Examples include the monomers glucose, with a six-membered ring of 5 C atoms and an O atom, and fructose, with a five-membered ring of 4 C atoms and an O atom; both with the formula $C_6H_{12}O_6$, and the dimer sucrose, formed by condensing a glucose molecule with a fructose molecule, with the formula $C_{12}H_{22}O_{11}$.

   (e) A <u>polypeptide</u> is a polyamide type condensation polymer formed from α-amino acid monomers. All proteins, such as insulin and hemoglobin, are examples.

2. (a) Polymerization of chloroethene, $CHCl=CH_2$, gives <u>polyvinyl chloride</u>, PVC, (polychloroethene).

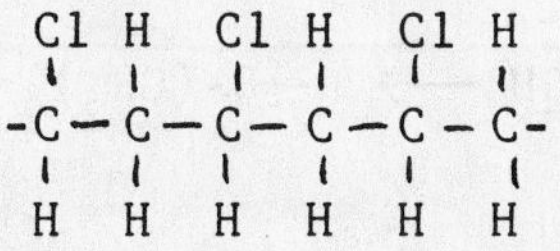

   (b) Polymerization of phenylethylene, $C_6H_5C(H)=CH_2$, gives <u>polystyrene</u>, (polyphenylethene).

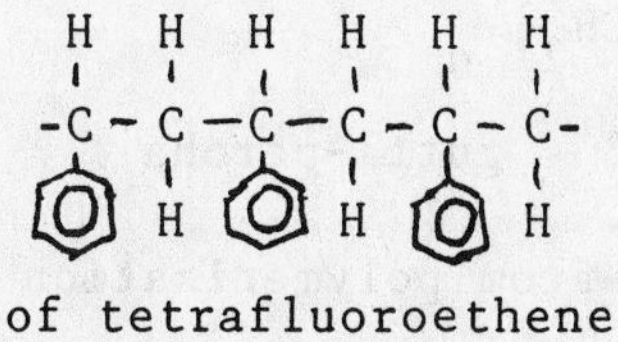

   (c) Polymerization of tetrafluoroethene, $F_2C=CF_2$, gives <u>Teflon</u>, (polytetrafluoroethene).

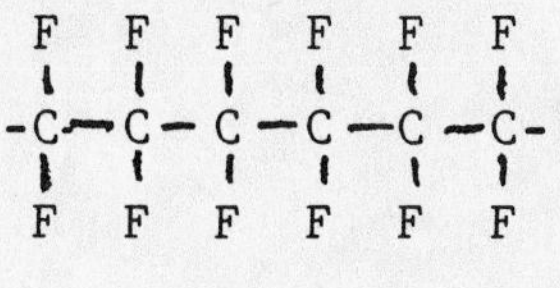

3. (a) <u>Teflon</u> is an addition polymer formed from tetrafluoroethene $F_2C=CF_2$, monomers.

   (b) <u>Saran</u> is an addition polymer formed from 1,1-dichloroethene, $Cl_2C=CH_2$, monomers.

   (c) <u>PVC</u>, polyvinyl chloride, is an addition polymer formed from chloroethene, $CHCl=CH_2$, (vinyl chloride), monomers.

   (d) <u>Nylon</u> is a condensation polymer formed from 1,6-hexanedioic acid $HOOC-(CH_2)_4COOH$, and 1,6-hexanediamine, $H_2N-(CH_2)_6-NH_2$, monomers.

   (e) <u>Dacron</u> is a condensation polymer formed from 1,2-ethanediol, $HOCH_2CH_2OH$, and 1,4-benzene dicarboxylic acid, $HOOC-C_6H_4-COOH$, monomers.

4. (a) Propene gives the addition polymer <u>polypropene</u>.

$$nCH_3-CH=CH_2 \longrightarrow -[CH_2-\overset{\overset{\displaystyle CH_3}{|}}{CH}]-_n$$

   (b) 1,3-Butadiene gives the addition polymer polybutadiene (synthetic rubber).

$$nCH_2=CH-CH=CH_2 \longrightarrow -[CH_2-CH=CH-CH_2-]_n$$

   (c) 1,6-Hexanediamine & 1,4-hexanedioic acid give the condensation polymer Nylon-66, a polyamide.

$$nH_2N-(CH_2)_6-NH_2 + nHOOC-(CH_2)_4-COOH \longrightarrow$$

$$-[\underset{\underset{\displaystyle H}{|}}{N}-\underset{\underset{\displaystyle O}{||}}{C}-(CH_2)_4-\underset{\underset{\displaystyle O}{||}}{C}-\underset{\underset{\displaystyle H}{|}}{N}-(CH_2)_6]-_n + nH_2O$$

   (d) 1,2-Ethanediol and terephthalic acid give the condensation polymer Dacron, a polyester.

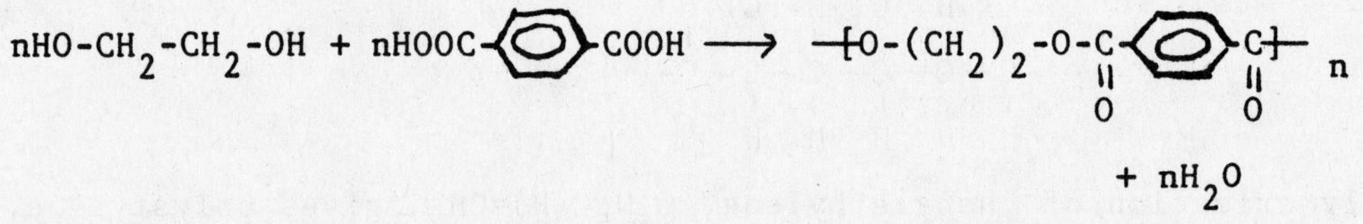

$$nHO-CH_2-CH_2-OH + nHOOC-\langle\bigcirc\rangle-COOH \longrightarrow -[O-(CH_2)_2-O-\underset{\underset{\displaystyle O}{||}}{C}-\langle\bigcirc\rangle-\underset{\underset{\displaystyle O}{||}}{C}]-_n$$

$$+ nH_2O$$

5. 

$$n\,\,\underset{\underset{\displaystyle H}{|}}{\overset{\overset{\displaystyle H}{|}}{C}}=\underset{\underset{\displaystyle CH=CH_2}{|}}{\overset{\overset{\displaystyle CH_3}{|}}{C}} \longrightarrow -[CH_2-\underset{\underset{\displaystyle CH_3}{|}}{C}=\overset{\overset{\displaystyle H}{|}}{C}-CH_2]-_n \quad \text{gutta-percha}$$

6. The addition polymer Orlon results from polymerization of acrylonitrile monomers, $H_2C=\underset{\underset{\displaystyle CN}{|}}{C}-H$.

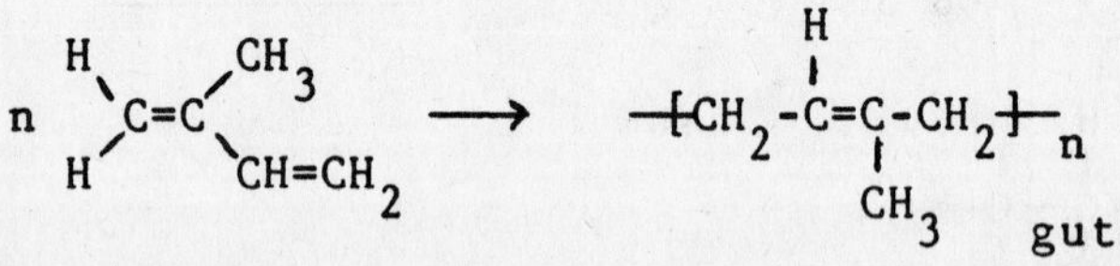

7. (a)

$$nHOCH_2CH_2OH + nH_3CO{-}\underset{\underset{O}{\|}}{C}{-}CH_2{-}\underset{\underset{O}{\|}}{C}{-}OCH_3 \longrightarrow$$

$$\left[CH_2{-}CH_2{-}O{-}\underset{\underset{O}{\|}}{C}{-}CH_2{-}\underset{\underset{O}{\|}}{C}{-}O\right]_n + nCH_3OH$$

(b)

$$nHOCH_2CH_2CH_2OH + nH_5C_2O{-}\underset{\underset{O}{\|}}{C}{-}CH_2{-}CH_2{-}\underset{\underset{O}{\|}}{C}{-}OC_2H_5 \longrightarrow$$

$$\left[CH_2{-}CH_2{-}CH_2{-}O{-}\underset{\underset{O}{\|}}{C}{-}CH_2{-}CH_2{-}\underset{\underset{O}{\|}}{C}{-}O\right]_n + n\ C_2H_5OH$$

8. The basic amino groups in the polypeptide chain of nylon are protonated by $H_3O^+(aq)$ and the C-N bonds are then broken by attack by water (hydrolysis), to give the monomer units.

$$-\underset{\underset{O}{\|}}{C}{-}\underset{\underset{H}{|}}{\ddot{N}}{-} + H_3O^+ \longrightarrow -\underset{\underset{O}{\|}}{C}{-}\underset{\underset{H}{|}}{\overset{\overset{H}{|}}{N}}{-}^{+} + H_2O \longrightarrow -\underset{\underset{O}{\|}}{C}{-}OH + H{-}\underset{\underset{H}{|}}{\overset{\overset{H}{|}}{N}}{-}^{+}$$

9. To undergo an addition polymerization, a monomer molecule must have a multiple bond. On polymerization, a triple bond is converted to a double bond, or a double bond is converted to a single bond. Strained rings may also undergo such reactions.

10. (a) Polymer chains in polyethylene, $(CH_2{-}CH_2)_n$, are attracted by London forces.

(b) Terylene, is a polyester containing polar C=O groups. The polymer chains are attracted by London forces and by dipole-dipole forces.

(c) Polyamides contain polar C=O bonds and N-H groups and the polymer chains can attract each other through hydrogen bonds, London forces and dipole-dipole interactions, the most important of which are N-H...O=C hydrogen bonds.

11. A <u>nucleotide</u> is a condensation polymer formed by the condensation of phosphoric acid, a sugar, and a nitrogen containing base.

12. The α and β forms of glucose differ in the orientation of the -OH group on the C-1 atom of the six-membered ring.

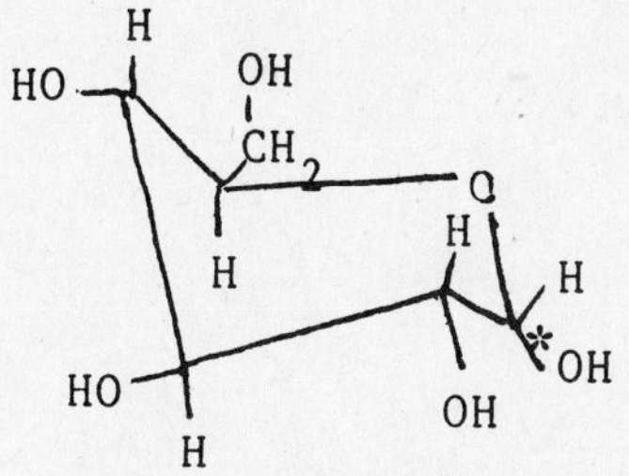


α-Glucose

$^*C_1$

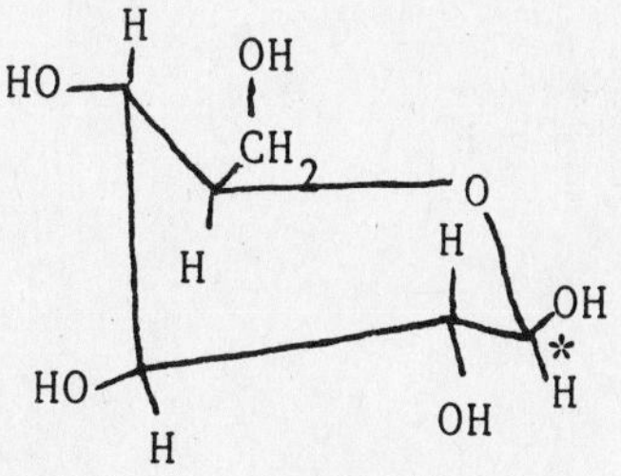


β-Glucose

13. (a) α-Aminocarboxylic acid is another name for an α-amino acid, which is an organic compound with an amino group, $-NH_2$, and a carboxylic acid group, $-CO_2H$, on the same carbon atom.

(b)

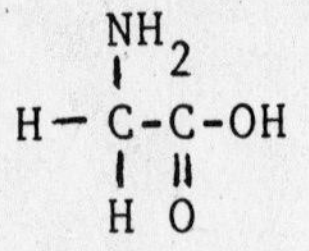

glycine

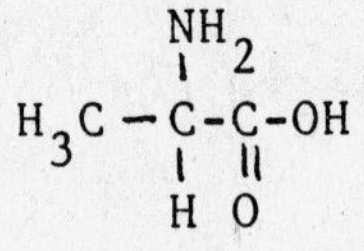

alanine

14. (a) Sugars have empirical formula of the type $C_xH_{2y}O_y$, which can be written as $C_x(H_2O)_y$, i.e., as a hydrate of carbon. They were originally thought to be carbon hydrates - hence the name carbohydrate. However, we now know the name to be inappropriate because they contain no $H_2O$ units in their molecular structures.

(b) (i) glucose, or fructose

(ii) sucrose

(iii) starch, cellulose

(c) Glucose and fructose are structural isomers of formula $C_6H_{12}O_6$. They differ in that glucose contains a six-membered ring, whereas fructose has a five membered ring. For the structures of α and β glucose, see Problem 12. Fructose has the structure

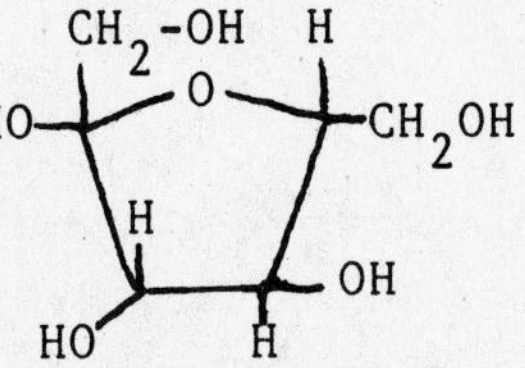

15. Starch is a mixture of straight-chained and branched polymers of α-glucose. Cellulose is a straight-chained polymer of β-glucose.

16. Starch is a condensation polymer formed from α-glucose; thus, the repeat unit is $(C_6H_{12}O_6 - H_2O)$, or $C_6H_{10}O$ , with a molar mass of 162.1 g $mol^{-1}$. Thus, the number of glucose residues in a starch molecule of molecular mass $2.57\times10^5$ u, or molar mass $2.57\times10^5$ g, is

$$\frac{2.57\times10^5 \text{ g}}{162.1 \text{ g residue}^{-1}} = \underline{1590 \text{ residues}}$$

17.

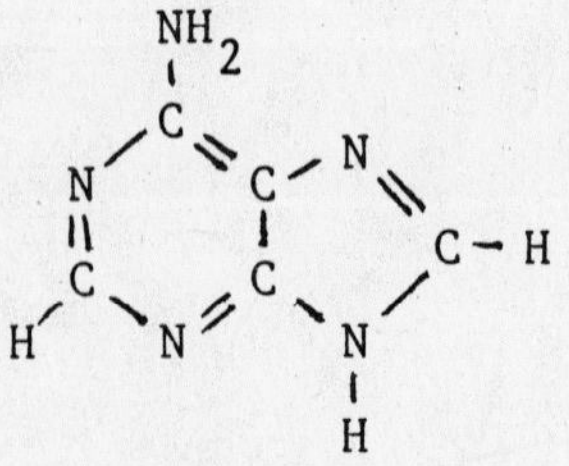

Adenine

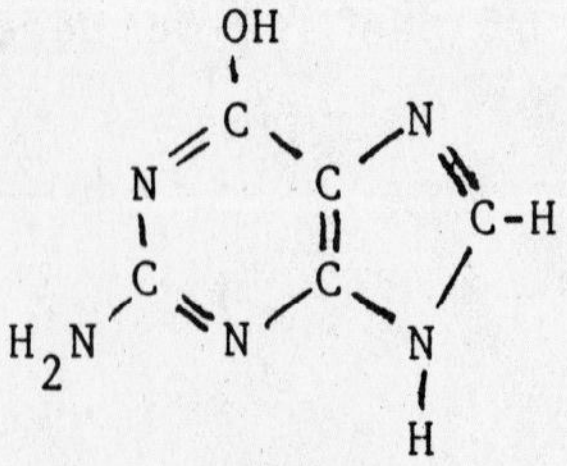

Guanine

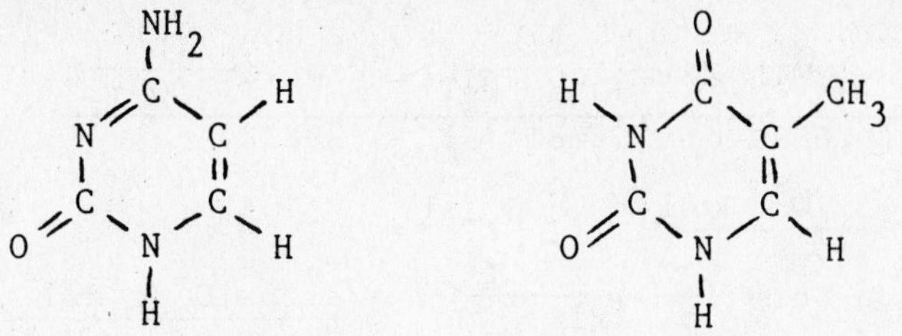

Cytosine Thymine

Adenine and guanine are purines, and cytosine and thymine are pyrimidines.

18. When linked by hydrogen bonds (see Figure 24.20), adenine and thymine are separated by a distance of 1.1 nm, and guanine and cytosine when linked by hydrogen bonds give exactly the same 1.1 nm separation. Thus, DNA strands, when so linked, have a constant separation. No other pairing of pairs of the four bases by hydrogen bonding would give the same separation and thus the same structural stability to the DNA molecule.

19. The primary structure of a protein is the sequence in which the $\alpha$-amino acids are linked together in the polypeptide chain. The secondary structure is the conformation of the polypeptide chains, helical or sheet, which is primarily determined by the hydrogen bonding interactions. The tertiary structure is the form resulting from the folding of the helical chains upon themselves.

20. An enzyme is a globular protein that acts as a catalyst in increasing the rate of a biochemical reaction. It does so by holding the reactant molecules close to each other and in the correct steric orientations for the specific reaction to occur.

21. This is an application of Raoult's law,

$$p = X.p^o$$

where p is the vapor pressure of the solvent in the solution, $p^o$ is the vapor pressure of the pure solvent, and X is the mole fraction of the solvent. Thus, the vapor pressure of the solution containing a nonvolatile solute (sucrose) is directly proportional to the mole fraction of the solvent in the solution.

$$X_{solvent} = 0.9 = \frac{\text{mol } H_2O}{\text{mol } H_2O + \text{mol sucrose}}$$

and mol $H_2O$ = $(100 \text{ g } H_2O)(\frac{1 \text{ mol } H_2O}{18.02 \text{ g } H_2O})$ = 5.55 mol

Thus, $\frac{5.55}{5.55 + \text{mol sucrose}} = 0.90$; mol sucrose = 0.62 mol

Mass of sucrose, $C_{12}H_{22}O_{11}$, $= (0.62 \text{ mol})(\frac{342.3 \text{ g sucrose}}{1 \text{ mol}}) = 2.1\times10^2$ g

22. $\Pi V = nRT$

$$n = \frac{\Pi V}{RT} = \frac{(10.1 \text{ mm Hg})\left(\frac{1 \text{ atm}}{760 \text{ mm Hg}}\right)(100 \text{ mL})\left(\frac{1 \text{ L}}{1000 \text{ mL}}\right)}{(0.0821 \text{ atm L mol}^{-1} \text{ K}^{-1})(298 \text{ K})}$$

$$= \underline{5.43\text{x}10^{-5} \text{ mol}}$$

$$\text{Average molar mass of nylon-66} = \frac{1.772 \text{ g}}{5.43\text{x}10^{-5} \text{ mol}} = \underline{3.26\text{x}10^{4} \text{ g mol}^{-1}}$$

23. A functional group is an atom or group of atoms that replaces a hydrogen atom of a hydrocarbon.

(a) Lucite, methyl ester, $-C\overset{\diagup OCH_3}{\underset{\diagdown\!\!\diagdown O}{}}$

(b) Nylon-66, amide, $-\underset{O}{\underset{\|}{C}}-\underset{H}{\underset{|}{N}}-$

(c) 1,2-Ethanediol, alcohol, -OH

(d) Terephthalic acid, carboxylic acid, $-\overset{O}{\overset{\|}{C}}-OH$

(e) Fructose, alcohol, -OH, and ether linkage, C-O-C, in the ring.

(f) Glycine, amino, $-NH_2$, and carboxylic acid, $-\underset{O}{\underset{\|}{C}}-OH$

24.

| ethylamine | ethanoic acid | glycine |
|---|---|---|
| $CH_3CH_2-NH_2$ | $CH_3-C\overset{\diagup\!\!\diagup O}{\underset{\diagdown OH}{}}$ | $H_2N-CH_2-C\overset{\diagup\!\!\diagup O}{\underset{\diagdown OH}{}}$ |
| m.p. $-81^{o}C$ | m.p. $17^{o}C$ | m.p. $233^{o}C$ |

As bond polarity increases (O-H > N-H), the strength of dipole-dipole interactions increases. Also, as the number of atoms capable of participating in hydrogen bonding increases, the strength of hydrogen bonding intermolecular attractions increases. Both of these factors support the increase in the observed melting points.

25. The double helix of DNA is held together by the two hydrogen bonds that form between an adenine base of one strand and a thymine base of the other strand, and by the three hydrogen bonds that form between a guanine base of one strand and a cytosine base of the other strand.

26. (a) The nitrogen bases found in DNA are the purines adenine and guanine, and the pyrimidines cytosine and thymine.

(b) In DNA, the base pairs found on complementary strands are A-T and G-C, and these pairs link the strands together by hydrogen bonds.

(c) The base pairs are found on the inside of the helix (as is shown in Figure 24.20).

27. (a) An enzyme is a protein molecule whose function is to catalyze a metabolic reaction.

(b) In the absence of enzymes, most metabolic reactions would proceed very slowly, and life as we know it would not exist.

(c) The mechanism by which a reaction proceeds in the presence of an enzyme has a lower activation energy barrier than the uncatalyzed reaction.

(d) The lock and key theory of enzyme function compares the active site of an enzyme to a lock and the reactant molecules to a key which specifically fits only that lock. The reaction is then facilitated by the precise orientation of reactive groups into steric positions that are favorable to reaction.

28* (a) Total HCl added = $(50.0\ mL)(\frac{1\ L}{1000\ mL})(0.0500\ mol\ L^{-1})$
$= \underline{2.50 \times 10^{-3}\ mol}$

Excess HCl = mol NaOH titrated

$= (30.57\ mL)(\frac{1\ L}{1000\ mL})(0.0600\ mol\ L^{-1}) = \underline{1.83 \times 10^{-3}\ mol}$

Mol HCl neutralized by $NH_3$ = $(2.50-1.83) \times 10^{-3}$ mol = $\underline{6.7 \times 10^{-4}\ mol}$

(b) Mol N in sample = mol $NH_3$ formed = mol HCl neutralized by $NH_3$
$= \underline{6.7 \times 10^{-4}\ mol}$

Mass of N in sample = $(6.7 \times 10^{-4}\ mol\ N)(\frac{14.01\ g\ N}{1\ mol\ N}) = \underline{9.4 \times 10^{-3}\ g}$

(c) Mass % N in sample = $(9.4 \times 10^{-3}\ g)(\frac{100\%}{0.0500\ g}) = \underline{18.8\ mass\%}$

Glycine, $H_2N\text{-}CH_2\text{-}CO_2H$, has molar mass 75.07 g $mol^{-1}$

Mass % N in glycine = $(14.01\ g\ mol^{-1})(\frac{100\%}{75.07\ g\ mol^{-1}}) = 18.66\%$

Thus, the observed mass% N in sample is very close to that expected for glycine.

*Problem 29 in first printing (Problem 28 in first printing has been deleted in second printing)

## CHAPTER 25

1. The subsript represents the atomic number, Z, the number of protons, while the superscript represents the mass number, the number of protons plus neutrons (nucleons).

| | | protons | neutrons |
|---|---|---|---|
| (a) | $^{6}_{3}Li$ | 3 | 3 |
| (b) | $^{13}_{6}C$ | 6 | 7 |
| (c) | $^{94}_{40}Zr$ | 40 | 54 |
| (d) | $^{137}_{56}Ba$ | 56 | 81 |

2. Locate each element in the periodic table, to deduce the atomic number, and then proceed as in Problem 1. Ne is the noble gas in period 2 and has the configuration 2.8, for Z = 10; Sr is the alkaline earth in period 5, with the configuration 2.8.8.18.2; W is the fourth transition metal in period 6, with the configuration $[Xe]6s^{2}4f^{14}5d^{4}$, for Z = 74, and Cm has Z = 96.

| | | protons | neutrons |
|---|---|---|---|
| (a) | $^{22}Ne$ | 10 | 12 |
| (b) | $^{88}Sr$ | 38 | 50 |
| (c) | $^{92}Sr$ | 38 | 54 |
| (d) | $^{180}W$ | 74 | 106 |
| (e) | $^{242}Cm$ | 96 | 146 |

3. (a) An $\alpha$-particle has a mass close to 4u and a charge of +2e.

   (b) A $\beta$-particle is an electron with a mass close to 0 u, and a charge of -e. (The actual mass is ~ $5.5\times10^{-4}$ u).

   (c) A positron has a mass close to 0 u and a charge of +e.

4. (a) An $\alpha$-particle is a helium nucleus, $^{4}_{2}He$. Thus, its emission gives a product nucleus with two protons less than the parent, which corresponds to a isotope with atomic number Z-2, where Z is the atomic number of the parent, i.e., an isotope two groups to the left in the periodic table. For example,

$$^{238}_{92}U \longrightarrow \, ^{234}_{90}Th + \, ^{4}_{2}He$$

   (b) A $\beta$-particle is an electron emitted from the nucleus of the parent isotope, which results in the formation of an additional proton in the nucleus of the product and an atomic number of Z+1 , where Z is the atomic number of the parent; i.e., the isotope formed is one group to the right in the periodic table. For example,

$$^{234}_{90}Th \longrightarrow \, ^{234}_{91}Pa + \, ^{o}_{-1}e$$

5. (a) Emission of a positron, ${}^{0}_{+1}e$, results from the conversion of a nuclear proton to a neutron, which decreases the nuclear charge of the product, and hence its atomic number, by 1. The product isotope will be located one group to the <u>left</u> in the periodic table.
   (b) Capture of an electron converts a proton in the nucleus to a neutron, which also decreases the nuclear charge of the product, and hence its atomic number, by 1. The product isotope will be located one group to the <u>left</u> in the periodic table.

6. In nuclear reactions, mass and charge are conserved.

(a) ${}^{80}_{35}Br \longrightarrow {}^{80}_{36}Kr + {}^{0}_{-1}e$ (b) ${}^{80}_{35}Br \longrightarrow {}^{80}_{34}Se + {}^{0}_{+1}e$

(c) ${}^{80}_{35}Br + {}^{0}_{-1}e \longrightarrow {}^{80}_{34}Se$

7. ${}^{233}_{90}Th \longrightarrow {}^{233}_{91}Pa + {}^{0}_{-1}e$ ; ${}^{233}_{91}Pa \longrightarrow {}^{233}_{92}U + {}^{0}_{-1}e$

The ${}^{233}U$ isotope is formed.

8. In each case, mass and charge are conserved.

(a) ${}^{32}_{15}P \longrightarrow {}^{32}_{16}S + {}^{0}_{-1}e$ (b) ${}^{15}_{8}O \longrightarrow {}^{15}_{7}N + {}^{0}_{+1}e$

(c) ${}^{52}_{26}Fe \longrightarrow {}^{52}_{25}Mn + {}^{0}_{+1}e$ (d) ${}^{218}_{87}Fr \longrightarrow {}^{214}_{85}At + {}^{4}_{2}He$

(e) ${}^{50}_{26}Fe \longrightarrow {}^{50}_{27}Co + {}^{0}_{-1}e$ (f) ${}^{122}_{53}I \longrightarrow {}^{122}_{54}Xe + {}^{0}_{-1}e$

9. (a) ${}^{207}_{84}Po + {}^{0}_{-1}e \longrightarrow {}^{207}_{83}Bi$ (b) ${}^{207}_{84}Po \longrightarrow {}^{0}_{+1}e + {}^{207}_{83}Bi$

(c) ${}^{207}_{84}Po \longrightarrow {}^{4}_{2}He + {}^{203}_{82}Pb$

10.

$$
{}^{231}_{90}Th \xrightarrow[{}^{0}_{-1}e]{} {}^{231}_{91}Pa \xrightarrow[{}^{4}_{2}He]{} {}^{227}_{89}Ac \xrightarrow[{}^{4}_{2}He]{} {}^{223}_{87}Fr \xrightarrow[{}^{0}_{-1}e]{} {}^{223}_{88}Ra \xrightarrow[{}^{4}_{2}He]{} {}^{219}_{86}Rn
$$

$$
\xrightarrow[{}^{4}_{2}He]{} {}^{215}_{84}Po \xrightarrow[{}^{4}_{2}He]{} {}^{211}_{82}Pb \xrightarrow[{}^{0}_{-1}e]{} {}^{211}_{83}Bi \xrightarrow[{}^{0}_{-1}e]{} {}^{211}_{84}Po \xrightarrow[{}^{4}_{2}He]{} {}^{207}_{82}Pb
$$

11. ${}^{235}_{92}U \longrightarrow {}^{4}_{2}He + {}^{231}_{90}Th$ ; ${}^{231}_{90}Th \longrightarrow {}^{0}_{-1}e + {}^{231}_{91}Pa$ ;

${}^{231}_{91}Pa \longrightarrow {}^{4}_{2}He + {}^{227}_{89}Ac$ ; ${}^{227}_{89}Ac \longrightarrow {}^{0}_{-1}e + {}^{227}_{90}Th$

12. (a) $^{35}_{17}Cl + ^{1}_{1}H \longrightarrow ^{32}_{16}S + ^{4}_{2}He$ (b) $^{15}_{7}N + ^{1}_{1}H \longrightarrow ^{12}_{6}C + ^{4}_{2}He$

(c) $^{12}_{6}C + ^{12}_{6}C \longrightarrow ^{1}_{1}H + ^{23}_{11}Na$ (d) $^{4}_{2}He + ^{3}_{2}He \longrightarrow ^{7}_{4}Be + \gamma$

13. (a) $^{32}_{16}S + ^{1}_{o}n \longrightarrow ^{1}_{1}H + ^{32}_{15}P$ (b) $^{7}_{4}Be + ^{o}_{-1}e \longrightarrow ^{7}_{3}Li$

(c) $^{81}_{37}Rb \longrightarrow ^{o}_{+1}e + ^{81}_{36}Kr$ (d) $^{1}_{1}H + ^{11}_{5}B \longrightarrow 3^{4}_{2}He$

(e) $^{235}_{92}U + ^{1}_{o}n \longrightarrow ^{135}_{54}Xe + 2^{1}_{o}n + ^{99}_{38}Sr$

14. (a) $^{179}_{75}Re + ^{o}_{-1}e \longrightarrow ^{179}_{74}W$ (b) $^{149}_{62}Sm \longrightarrow ^{4}_{2}He + ^{145}_{60}Nd$

(c) $^{228}_{88}Ra \longrightarrow ^{o}_{-1}e + ^{228}_{89}Ac$ (d) $^{25}_{12}Mg + ^{1}_{1}H \longrightarrow ^{26}_{13}Al + \gamma$

(e) $^{26}_{12}Mg + ^{1}_{1}H \longrightarrow ^{26}_{13}Al + ^{1}_{o}n$

15. (a) $^{66}_{32}Ge \longrightarrow ^{o}_{+1}e + ^{66}_{31}Ga$

(b) $t_{1/2}$ = 2.50 h; 12.50 h = $5t_{1/2}$, and a 50 mg sample is reduced to

$$(50.0 \text{ mg})(\tfrac{1}{2})^5 = \underline{1.56 \text{ mg}}$$

16. Radioactive decay occurs according to a first order process, for which

$$\text{Rate} = -\frac{d[R]}{dt} = k_1[R], \text{ and the integrated first order rate law is,}$$

$$\ln [R]_t = \ln [R]_o - k_1 t \quad \ldots (1)$$

Thus, at the half-life of the reaction, $t_{1/2}$, where $[R]_t = [R]_o/2$

$$\ln [R]_o - \ln [R]_o/2 = \ln 2 = k_1 t_{1/2} = 0.693$$

$$k_1 = \frac{0.693}{30.2 \text{ yr}} = \underline{2.29\times10^{-2} \text{ yr}^{-1}}$$

Rearranging equation (1) gives $\ln([R]_t/[R]_o) = -k_1 t$

and for 1.0% of activity remaining, $[R]_t/[R]_o = 0.01$, whence

$$t = \frac{-\ln 0.01}{2.29\times10^{-2} \text{ yr}^{-1}} = \underline{201 \text{ yr}}$$

17. $kt_{1/2} = \ln 2 = 0.693$; $k = \dfrac{0.693}{t_{1/2}}$

$\underline{^{131}I}$  $t_{1/2} = 8.0$ days,  $k = \frac{0.693}{8.0 \text{ day}} = \underline{8.66\text{x}10^{-2} \text{ day}^{-1}}$

$\underline{^{90}Sr}$  $t_{1/2} = 19.9$ yr,  $k = \frac{0.693}{19.9 \text{ yr}} = \underline{3.48\text{x}10^{-2} \text{ yr}^{-1}}$

(a) Using the integrated first order rate equation,

$$\ln \frac{[R]_t}{[R]_o} = kt \text{ , and } \frac{[R]_t}{[R]_o} = 10\% = 0.10 \text{ ; } \quad t = \frac{-\ln 0.10}{k}$$

For $^{131}I$, $t = \frac{-\ln 0.10}{8.66\text{x}10^{-2} \text{ day}^{-1}} = \underline{27 \text{ day}}$ and

for $^{90}Sr$, $t = \frac{-\ln 0.10}{3.48\text{x}10^{-2} \text{ yr}^{-1}} = \underline{66.1 \text{ yr}}$

(b) $t = \frac{-\ln 0.01}{k}$

For $^{131}I$, $t = \frac{-\ln 0.01}{8.66\text{x}10^{-2} \text{ day}^{-1}} = \underline{53 \text{ day}}$ and

for $^{90}Sr$, $t = \frac{-\ln 0.01}{3.48\text{x}10^{-2} \text{ yr}^{-1}} = \underline{132 \text{ yr}}$

$^{90}Sr$ will have the most serious long-term effect.

18. Plot counts $\text{min}^{-1}$ versus t in minutes, and determine $t_{1/2}$, which is the time at which the radioactivity has diminished to one-half of its initial activity (334 counts $\text{min}^{-1}$). From the graph, $t_{1/2} = \underline{12.4 \text{ min}}$.

19. From the integrated first order rate equation,

$$-\ln \frac{[R]_t}{[R]_o} = kt \text{ ; } -\ln \frac{[R]_t}{[R]_o} = -\ln \left(\frac{2.09 \text{ mg}}{3.40 \text{ mg}}\right) = 0.487 = k(10.0 \text{ day})$$

$\underline{k = 0.0487 \text{ day}^{-1}}$

$kt_{1/2} = \ln 2 = 0.693$;  $t_{1/2} = \frac{0.693}{0.0487 \text{ day}^{-1}} = \underline{14.2 \text{ day}}$

20. From the integrated first order rate equation,

$$-\ln \frac{[R]_t}{[R]_o} = kt \text{ ; } -\ln \frac{[R]_t}{[R]_o} = -\ln 0.070 = 2.66 = k(4.5\text{x}10^{9} \text{ yr})$$

$\underline{k = 5.9\text{x}10^{-10} \text{ yr}^{-1}}$

$kt_{1/2} = \ln 2 = 0.693$;  $t_{1/2} = \frac{0.693}{5.9\text{x}10^{-10} \text{ yr}^{-1}} = \underline{1.2\text{x}10^{9} \text{ yr}}$

21. Living carbon has 15.3 disintegrations $\text{min}^{-1}$, and $t_{1/2}$ for $^{14}C$ is 5730 yr.

$kt_{1/2} = \ln 2 = 0.693$;  $k = \frac{0.693}{t_{1/2}} = \frac{0.693}{5730 \text{ yr}} = \underline{1.21\text{x}10^{-4} \text{ yr}^{-1}}$

From the integrated first order rate equation,

$$-\ln \frac{[R]_t}{[R]_o} = kt \; ; \; -\ln \left(\frac{2.4}{15.3}\right) = -\ln (0.157) = 1.85 = kt$$

$$t = \frac{1.85}{1.21\times10^{-4}\ yr^{-1}} = \underline{1.53\times10^{4}\ yr}$$

The artists lived 15300 years ago, i.e., in about 13 000 B.C.

22. $kt_{1/2} = \ln 2 \; ; \; k = \frac{0.693}{t_{1/2}} = \frac{0.693}{12.4\ h} = \underline{5.59\times10^{-2}\ h^{-1}}$

(a) After 30 h,

$$-\ln \frac{[R]_t}{[R]_o} = kt; \quad -\ln \frac{[R]_t}{1.10\times10^{9}\ d\ min^{-1}} = (5.59\times10^{-2}\ h^{-1})(30.0\ h) = \underline{1.677}$$

$$\frac{[R]_t}{1.10\times10^{9}\ d\ min^{-1}} = 0.187 \; ; \; [R]_t = \underline{2.06\times10^{8}\ d\ min^{-1}}$$

(b)

$$-\ln \frac{[R]_t}{[R]_o} = -\ln \frac{1.0\times10^{5}\ d\ min^{-1}}{1.10\times10^{9}\ d\ min^{-1}} = -\ln (9.1\times10^{-5}) = \underline{9.30}$$

$$= kt = (5.59\times10^{-2}\ h^{-1})t \; ; \; t = \underline{166\ h}$$

23*. Living carbon has a disintegration rate of 15.3 disintegrations $min^{-1}$ per gram of carbon, or 382.5 d $min^{-1}$ per 25.0 g carbon, and $t_{1/2}$ = 5730 yr; k = $1.21\times10^{-4}$ $yr^{-1}$ (see Problem 21).

From the integrated first order rate equation,

$$-\ln \frac{[R]_t}{[R]_o} = -\ln \left(\frac{260\ d\ min^{-1}}{383\ d\ min^{-1}}\right) = 0.387 = kt = (1.21\times10^{-4}\ yr^{-1})t$$

$$t = \frac{0.387}{1.21\times10^{-4}\ yr^{-1}} = \underline{3.20\times10^{3}\ yr}$$

The age of the Chinese artifact is $\underline{3.20\times10^{3}\ years}$.

24*. This is similar to problem 23. Living carbon has 765 d $min^{-1}$ per 50.0 g of carbon, and $t_{1/2}$ = 5730 yr; k = $1.21\times10^{-4}$ $yr^{-1}$.

$$-\ln \left(\frac{650\ d\ min^{-1}}{765\ d\ min^{-1}}\right) = 0.163 = kt = (1.21\times10^{-4}\ yr^{-1})t$$

$$t = \frac{0.163}{1.21\times10^{-4}\ yr^{-1}} = \underline{1.35\times10^{3}\ yr}$$

The person buried in the casket died in approximately 540 A.D.

*These solutions reflect changes made to the problem in the second printing of the text.

25. Mass of proton = 1.007 28 u; mass of neutron = 1.008 66 u, and mass of electron = 0.000 548 58 u.

$^{35}Cl$ contains 17 protons, 18 neutrons, and 17 electrons. Thus, the sum of the masses of its elementary particles is,

17(1.007 28 u) + 18(1.008 66 u) + 17(0.000 548 58 u) = 35.289 0 u,

and the mass defect is (35.289 0 - 34.968 9) = 0.320 1 u.

Converting this to binding energy, using $E = mc^2$, gives,

$E = (0.320\ 1\ u)(2.998 \times 10^{8}\ m\ s^{-1})^2(1.66 \times 10^{-27}\ kg\ u^{-1}) = 4.78 \times 10^{-11}\ J$,

or since there are 35 nucleons, $1.36 \times 10^{-12}$ J nucleon$^{-1}$.

26. See Problem 25 for the method.

| Atom | protons | neutrons | electrons | mass of components* | atomic mass* | mass defect* | binding energy |
|---|---|---|---|---|---|---|---|
| $^{20}_{10}Ne$ | 10 | 10 | 10 | 20.164 90 | 19.992 44 | 0.17246 | 25.73 pJ |
| $^{64}_{30}Zn$ | 30 | 34 | 30 | 64.529 34 | 63.929 14 | 0.60020 | 89.55 pJ |
| $^{61}_{28}Ni$ | 28 | 33 | 28 | 61.505 02 | 60.930 06 | 0.57496 | 85.78 pJ |
| $^{226}_{88}Ra$ | 88 | 138 | 88 | 227.884 1 | 226.025 4 | 1.8587 | 277.3 pJ |

*atomic units, u

and the binding energies per nucleon are:

$^{20}Ne$, 1.287 pJ; $^{64}Zn$, 1.399 pJ; $^{61}Ni$, 1.406 pJ; $^{226}Ra$, 1.227 pJ

27. See Problem 25 for the method.

| Atom | protons | neutrons | electrons | mass of components* | atomic mass* | mass defect* | binding energy |
|---|---|---|---|---|---|---|---|
| $^{2}_{1}H$ | 1 | 1 | 1 | 2.016 49 | 2.014 10 | 0.00239 | 0.3566 pJ |
| $^{16}_{8}O$ | 8 | 8 | 8 | 16.131 92 | 15.994 91 | 0.13701 | 20.440 pJ |
| $^{56}_{26}Fe$ | 26 | 30 | 26 | 56.463 38 | 55.934 90 | 0.52848 | 78.850 pJ |
| $^{125}_{52}Te$ | 52 | 73 | 52 | 126.039 3 | 124.904 4 | 1.1349 | 169.33 pJ |
| $^{238}_{92}U$ | 92 | 146 | 92 | 239.984 7 | 238.050 8 | 1.9339 | 288.54 pJ |

*atomic units, u

and the binding energies per nucleon are:

$^{2}H$, 0.1783 pJ; $^{16}O$, 1.278 pJ; $^{56}Fe$, 1.408 pJ; $^{125}Te$, 1.355 pJ, and

$^{238}U$, 1.212 pJ.

28. Plot binding energy per nucleon versus mass number ; the graph clearly shows a maximum at $^{56}Fe$.

29. Use the method of Problem 25.

The binding energy of $^{14}N$ is 1.20 pJ nucleon$^{-1}$, and that of $^{15}N$ is 1.23 pJ nucleon$^{-1}$. $^{15}N$ is more stable; $^{14}N$ is more abundant.

30. To calculate the energy of the reaction $^{7}_{3}Li + ^{1}_{1}H \longrightarrow ^{4}_{2}He + ^{4}_{2}He$, first calculate the change in mass for the reaction, and then convert that mass to its energy equivalent, using $E = mc^2$.

$$m = (2\text{x}4.00260\ u) - (7.01600\ u) - (1.00783\ u) = \underline{-0.01863\ u}$$

$$E = mc^2 = (-0.01863\ u)(2.998\text{x}10^{8}\ m\ s^{-1})^2(1.66\text{x}10^{-27}\ kg\ u^{-1})$$
$$= \underline{-2.7796\text{x}10^{-12}\ J}$$

The energy is exothermic; energy is released.

31. The atomic mass of $^{2}_{1}H$ is 2.014 10 u, that of $^{3}_{2}He$ is 3.016 03 u, and the neutron mass is 1.008 66 u. Thus, the mass change in the reaction is,

$$2(2.014\ 10\ u) - (3.01603\ u) - (1.008\ 66\ u) = \underline{-0.003\ 51\ u}, \text{ or}$$

for 1 mol $D_2$, the mass change is

$$\left(\frac{0.00351\ u}{2\ \text{atoms D}}\right)\left(\frac{2\ \text{atoms D}}{1\ \text{molecule}\ D_2}\right)\left(\frac{6.022\text{x}10^{23}\ \text{molecules}\ D_2}{1\ \text{mol}\ D_2}\right) = \underline{2.11\text{x}10^{21}\ u\ mol^{-1}}$$

and the energy equivalent is,

$$E = mc^2 = (2.11\text{x}10^{21}\ u\ mol^{-1})(2.998\text{x}10^{8}\ m\ s^{-1})^2\left(\frac{1.66\text{x}10^{-27}\ kg}{1\ u}\right)$$
$$= \underline{3.15\text{x}10^{11}\ J\ mol^{-1}\ D_2}$$

For the reaction, $D_2(g) + \frac{1}{2}O_2(g) \longrightarrow D_2O(l)$, the enthalpy of combustion is equal to the enthalpy of formation of 1 mol of $D_2O(l)$, which is -285.8 kJ mol$^{-1}$ (assuming it to be the same as for $H_2O(l)$).

Thus, the ratio of the energy released per mol $D_2$ in the nuclear reaction to that released in the combustion reaction is,

$$\left(\frac{3.15\text{x}10^{11}\ J}{285.8\ kJ}\right)\left(\frac{1\ kJ}{10^3\ J}\right) = \underline{1.1\ \text{x}10^{6}}$$

The fusion reaction releases $\sim 10^6$ times more energy per mol of $D_2$ than does the combustion reaction.

32. (a) The critical mass is the minimum mass of fissionable material for which a branching chain nuclear reaction can occur.
(b) A moderator is used to slow down the neutrons in a nuclear reactor so that just one of the 2.4 neutrons produced in the fission of a $^{235}U$ nucleus is captured by another fissionable nucleus.
(c) The control rods are made of substances, such as graphite, that strongly absorb neutrons, and are used to adjust the rate of a fission reaction.

33. Common examples with the required characteristics among the lighter isotopes are, for example:

(a) ${}^{12}_{6}C$, 6 neutrons and 6 protons. (b) ${}^{11}_{5}B$, 6 neutrons and 5 protons.

(c) ${}^{13}_{6}C$, 7 neutrons and 6 protons. (d) ${}^{14}_{7}N$, 7 neutrons and 7 protons.

34. Use $E = mc^2$ to find the mass equivalent of the energy.

$$m = \frac{3.9\times10^{23}\ J\ s^{-1}}{(2.998\times10^{8}\ m\ s^{-1})^2}\left(\frac{1\ kg\ m^2\ s^{-2}}{1\ J}\right) = \underline{4.3\times10^{6}\ kg}$$

35. Although there is no change in the concentrations of $H_2$, $I_2$, and HI in equilibrium at a particular temperature, the forward reaction

$$H_2(g) + I_2(g) \longrightarrow 2HI(g)$$

and the reverse reaction

$$2HI(g) \longrightarrow H_2(g) + I_2(g)$$

are both occurring, but at the same rate. This is easily demonstrated by adding a trace of radioactive ${}^{131}I_2$ to the reaction mixture, because it is observed that this isotope becomes rapidly distributed between the $I_2$ and the HI in the equilibrium mixture. Thus, when the reaction mixture is treated with excess NaOH(aq), for example, and the iodide ion resulting from the reaction

$$HI(g) + NaOH(aq) \longrightarrow NaI(aq) + H_2O(l)$$

is precipitated as AgI(s), the latter is found to be radioactive. When a similar experiment is performed using radioactive $H^{131}I$, it is also found that the radioactive iodine is rapidly distributed between the HI(g) and $I_2(g)$ in the equilibrium mixture. Thus, the dynamic nature of the equilibrium is established.

36*. $${}^{232}_{90}Th + {}^{12}_{6}C \longrightarrow {}^{240}_{96}Cm + 4\,{}^{1}_{0}n$$

37* (a) The binding energy of a nucleus is the energy needed to overcome the forces that hold the nuclear particles together.

(b) A branching chain reaction is a fission reaction which is caused by a neutron which produces more than one neutron, so that the fission reaction can be maintained.

(c) Critical mass is the minimum mass of fissionable material for which a branching chain reaction can occur.

*Note that these are respectively Problems 37 and 38 in the first printing.

(d) A fission reaction is a nuclear reaction in which a radioactive heavy nucleus splits into two lighter nuclei with the release of a large amount of energy.

(e) A fusion reaction is a nuclear reaction in which two light nuclei combine to form a heavier nucleus with the release of a large amount of energy.

# Notes